Lecture Notes in Computer Science 7431

Commenced Publication in 1973
Founding and Former Series Editors:
Gerhard Goos, Juris Hartmanis, and Jan van Leeuwen

George Bebis   Richard Boyle
Bahram Parvin   Darko Koracin
Charless Fowlkes   Sen Wang
Min-Hyung Choi   Stephan Mantler
Jürgen Schulze   Daniel Acevedo
Klaus Mueller   Michael Papka (Eds.)

Advances in Visual Computing

8th International Symposium, ISVC 2012
Rethymnon, Crete, Greece, July 16-18, 2012
Revised Selected Papers, Part I

 Springer

Volume Editors

George Bebis, E-mail: bebis@cse.unr.edu

Richard Boyle, E-mail: richard.boyle@nasa.gov

Bahram Parvin, E-mail: parvin@hpcrd.lbl.gov

Darko Koracin, E-mail: darko@dri.edu

Charless Fowlkes, E-mail: fowlkes@ics.uci.edu

Sen Wang, E-mail: sen.wang@kodak.com

Min-Hyung Choi, E-mail: min.choi@ucdenver.edu

Stephan Mantler, E-mail: step@stephanmantler.com

Jürgen Schulze, E-mail: jschulze@ucsd.edu

Daniel Acevedo, E-mail: daniel.acevedo@kaust.edu.sa

Klaus Mueller, E-mail: mueller@cs.sunysb.edu

Michael Papka, E-mail: papka@anl.gov

ISSN 0302-9743   e-ISSN 1611-3349
ISBN 978-3-642-33178-7   e-ISBN 978-3-642-33179-4
DOI 10.1007/978-3-642-33179-4
Springer Heidelberg Dordrecht London New York

Library of Congress Control Number: 2012945624

CR Subject Classification (1998): I.3-5, H.5.2, I.2.10, J.3, F.2.2, I.3.5

LNCS Sublibrary: SL 6 – Image Processing, Computer Vision, Pattern Recognition, and Graphics

Typesetting: Camera-ready by author, data conversion by Scientific Publishing Services, Chennai, India

Printed on acid-free paper

Springer is part of Springer Science+Business Media (www.springer.com)

Preface

It is with great pleasure that we welcome you to the proceedings of the 8th International Symposium on Visual Computing (ISVC 2012) that was held in Rethymnon, Crete, Greece. ISVC provides a common umbrella for the four main areas of visual computing including vision, graphics, visualization, and virtual reality. The goal is to provide a forum for researchers, scientists, engineers, and practitioners throughout the world to present their latest research findings, ideas, developments, and applications in the broader area of visual computing.

This year, the program consisted of 11 oral sessions, one poster session, seven special tracks, and six keynote presentations. The response to the call for papers was very good; we received over 200 submissions for the main symposium from which we accepted 68 papers for oral presentation and 35 papers for poster presentation. Special track papers were solicited separately through the Organizing and Program Committees of each track. A total of 45 papers were accepted for oral presentation in the special tracks.

All papers were reviewed with an emphasis on potential to contribute to the state of the art in the field. Selection criteria included accuracy and originality of ideas, clarity and significance of results, and presentation quality. The review process was quite rigorous, involving two–three independent blind reviews followed by several days of discussion. During the discussion period we tried to correct anomalies and errors that might have existed in the initial reviews. Despite our efforts, we recognize that some papers worthy of inclusion may have not been included in the program. We offer our sincere apologies to authors whose contributions might have been overlooked.

We wish to thank everybody who submitted their work to ISVC 2012 for review. It was because of their contributions that we succeeded in having a technical program of high scientific quality. In particular, we would like to thank the ISVC 2012 Area Chairs, the organizing institutions (UNR, DRI, LBNL, and NASA Ames), the industrial sponsors (BAE Systems, Intel, Ford, Hewlett Packard, Mitsubishi Electric Research Labs, Toyota, General Electric), the international Program Committee, the special track organizers and their Program Committees, the keynote speakers, the reviewers, and especially the authors that contributed their work to the symposium. In particular, we would like to express our appreciation to BAE Systems and Riad Hammoud for their sponsorship of the "best" paper award this year.

July 2012

George Bebis
Richard Boyle
Bahram Parvin
Darko Koracin
Charless Fowlkes
Sen Wang
Min-Hyung Choi
Stephan Mantler
Jürgen Schulze
Daniel Acevedo
Klaus Mueller
Michael Papka

Organization

ISVC 2012 Steering Committee

Bebis George   University of Nevada, Reno, USA
Boyle Richard   NASA Ames Research Center, USA
Parvin Bahram   Lawrence Berkeley National Laboratory, USA
Koracin Darko   Desert Research Institute, USA

ISVC 2012 Area Chairs

Computer Vision

Fowlkes Charless   University of California at Irvine, USA
Wang Sen   Kodak Research Labs, USA

Computer Graphics

Choi Min-Hyung   University of Colorado Denver, USA
Mantler Stephan   VRVis Research Center, Austria

Virtual Reality

Schulze Jurgen   University of California at San Diego, USA
Acevedo Daniel   KAUST, Saudi Arabia

Visualization

Mueller Klaus   Stony Brook University, USA
Papka Michael   Argonne National Laboratory, USA

Publicity

Albu Branzan Alexandra   University of Victoria, Canada

Local Arrangements

Zaboulis, Xenophon   Institute of Computer Science, FORTH, Greece

Special Tracks

Porikli, Fatih   Mitsubishi Electric Research Labs, USA

ISVC 2012 Keynote Speakers

Faloutsos Petros   York University, Canada
Coquillart Sabine   INRIA, France
Schmid Cordelia   INRIA, France
Cremers Daniel   Technical University of Munich Germany
Asari Vijayan   University of Dayton, USA
Randy Goebel   University of Alberta, Canada

ISVC 2012 International Program Committee

(Area 1) Computer Vision

Abidi Besma   University of Tennessee at Knoxville, USA
Abou-Nasr Mahmoud   Ford Motor Company, USA
Agaian Sos   University of Texas at San Antonio, USA
Aggarwal J.K.   University of Texas, Austin, USA
Albu Branzan Alexandra   University of Victoria, Canada
Amayeh Gholamreza   Eyecom, USA
Agouris Peggy   George Mason University, USA
Argyros Antonis   University of Crete, Greece
Asari Vijayan   University of Dayton, USA
Athitsos Vassilis   University of Texas at Arlington, USA
Basu Anup   University of Alberta, Canada
Bekris Kostas   University of Nevada at Reno, USA
Bensrhair Abdelaziz   INSA-Rouen, France
Bhatia Sanjiv   University of Missouri-St. Louis, USA
Bimber Oliver   Johannes Kepler University Linz, Austria
Bioucas Jose   Instituto Superior Técnico, Lisbon, Portugal
Birchfield Stan   Clemson University, USA
Boufama Boubakeur   University of Windsor, Canada
Bourbakis Nikolaos   Wright State University, USA
Brimkov Valentin   State University of New York, USA
Campadelli Paola   Università degli Studi di Milano, Italy
Cavallaro Andrea   Queen Mary, University of London, UK
Charalampidis Dimitrios   University of New Orleans, USA
Chellappa Rama   University of Maryland, USA
Chen Yang   HRL Laboratories, USA
Cheng Hui   Sarnoff Corporation, USA
Cochran Steven Douglas   University of Pittsburgh, USA
Chung, Chi-Kit Ronald   The Chinese University of Hong Kong,
                              Hong Kong
Cremers Daniel   Technical University of Munich, Germany
Cui Jinshi   Peking University, China
Dagher Issam   University of Balamand, Lebanon

Darbon Jerome	CNRS-Ecole Normale Superieure de Cachan, France
Debrunner Christian	Colorado School of Mines, USA
Demirdjian David	Vecna Robotics, USA
Duan Ye	University of Missouri-Columbia, USA
Doulamis Anastasios	Technical University of Crete, Greece
Dowdall Jonathan	510 Systems, USA
El-Ansari Mohamed	Ibn Zohr University, Morocco
El-Gammal Ahmed	University of New Jersey, USA
Eng How Lung	Institute for Infocomm Research, Singapore
Erol Ali	Ocali Information Technology, Turkey
Fan Guoliang	Oklahoma State University, USA
Fan Jialue	Northwestern University, USA
Ferri Francesc	Universitat de València, Spain
Ferryman James	University of Reading, UK
Foresti GianLuca	University of Udine, Italy
Fukui Kazuhiro	The University of Tsukuba, Japan
Galata Aphrodite	The University of Manchester, UK
Georgescu Bogdan	Siemens, USA
Goh Wooi-Boon	Nanyang Technological University, Singapore
Guerra-Filho Gutemberg	University of Texas Arlington, USA
Guevara, Angel Miguel	University of Porto, Portugal
Gustafson David	Kansas State University, USA
Hammoud Riad	BAE Systems, USA
Harville Michael	Hewlett Packard Labs, USA
He Xiangjian	University of Technology, Sydney, Australia
Heikkilä Janne	University of Oulu, Filand
Hongbin Zha	Peking University, China
Hou Zujun	Institute for Infocomm Research, Singapore
Hua Gang	IBM T.J. Watson Research Center, USA
Imiya Atsushi	Chiba University, Japan
Jia Kevin	IGT, USA
Kamberov George	Stevens Institute of Technology, USA
Kampel Martin	Vienna University of Technology, Austria
Kamberova Gerda	Hofstra University, USA
Kakadiaris Ioannis	University of Houston, USA
Kettebekov Sanzhar	Keane Inc., USA
Kim Tae-Kyun	Imperial College London, UK
Kimia Benjamin	Brown University, USA
Kisacanin Branislav	Texas Instruments, USA
Klette Reinhard	Auckland University, New Zealand
Kokkinos Iasonas	Ecole Centrale Paris, France
Kollias Stefanos	National Technical University of Athens, Greece
Komodakis Nikos	Ecole Centrale de Paris, France

Kozintsev, Igor	Intel, USA
Kuno Yoshinori	Saitama University, Japan
Kim Kyungnam	HRL Laboratories, USA
Latecki Longin Jan	Temple University, USA
Lee D.J.	Brigham Young University, USA
Li Chunming	Vanderbilt University, USA
Li Xiaowei	Google Inc., USA
Lim Ser N.	GE Research, USA
Lin Zhe	Adobe, USA
Lisin Dima	VidoeIQ, USA
Lee Hwee Kuan	Bioinformatics Institute, A*STAR, Singapore
Lee Seong-Whan	Korea University, Korea
Leung Valerie	ONERA, France
Li Shuo	GE Healthecare, Canada
Li Wenjing	STI Medical Systems, USA
Loss Leandro	Lawrence Berkeley National Lab, USA
Luo Gang	Harvard University, USA
Ma Yunqian	Honyewell Labs, USA
Maeder Anthony	University of Western Sydney, Australia
Makrogiannis Sokratis	NIH, USA
Maltoni Davide	University of Bologna, Italy
Maybank Steve	Birkbeck College, UK
Medioni Gerard	University of Southern California, USA
Melenchón Javier	Universitat Oberta de Catalunya, Spain
Metaxas Dimitris	Rutgers University, USA
Miller Ron	Wright Patterson Air Force Base, USA
Ming Wei	Konica Minolta Laboratory, USA
Mirmehdi Majid	Bristol University, UK
Monekosso Dorothy	University of Ulster, UK
Morris Brendan	University of Nevada, Las Vegas, USA
Mulligan Jeff	NASA Ames Research Center, USA
Murray Don	Point Grey Research, Canada
Nait-Charif Hammadi	Bournemouth University, UK
Nefian Ara	NASA Ames Research Center, USA
Nicolescu Mircea	University of Nevada, Reno, USA
Nixon Mark	University of Southampton, UK
Nolle Lars	The Nottingham Trent University, UK
Ntalianis Klimis	National Technical University of Athens, Greece
Or Siu Hang	The Chinese University of Hong Kong, Hong Kong
Papadourakis George	Technological Education Institute, Greece
Papanikolopoulos Nikolaos	University of Minnesota, USA
Pati Peeta Basa	CoreLogic, India
Patras Ioannis	Queen Mary University, London, UK

Pavlidis Ioannis	University of Houston, USA
Petrakis Euripides	Technical University of Crete, Greece
Peyronnet Sylvain	LRI, University Paris-Sud, France
Pinhanez Claudio	IBM Research, Brazil
Piccardi Massimo	University of Technology, Australia
Pietikäinen Matti	LRDE/University of Oulu, Filand
Pitas Ioannis	Aristotle University of Thessaloniki, Greece
Porikli Fatih	Mitsubishi Electric Research Labs, USA
Prabhakar Salil	Digital Persona Inc., USA
Prati Andrea	University IUAV of Venice, Italy
Prokhorov Danil	Toyota Research Institute, USA
Pylvanainen Timo	Nokia Research Center, USA
Qi Hairong	University of Tennessee at Knoxville, USA
Qian Gang	Arizona State University, USA
Raftopoulos Kostas	National Technical University of Athens, Greece
Regazzoni Carlo	University of Genoa, Italy
Regentova Emma	University of Nevada, Las Vegas, USA
Remagnino Paolo	Kingston University, UK
Ribeiro Eraldo	Florida Institute of Technology, USA
Robles-Kelly Antonio	National ICT Australia (NICTA), Australia
Ross Arun	West Virginia University, USA
Samal Ashok	University of Nebraska, USA
Samir Tamer	Ingersoll Rand Security Technologies, USA
Sandberg Kristian	Computational Solutions, USA
Sarti Augusto	DEI Politecnico di Milano, Italy
Savakis Andreas	Rochester Institute of Technology, USA
Schaefer Gerald	Loughborough University, UK
Scalzo Fabien	University of California at Los Angeles, USA
Scharcanski Jacob	UFRGS, Brazil
Shah Mubarak	University of Central Florida, USA
Shi Pengcheng	Rochester Institute of Technology, USA
Shimada Nobutaka	Ritsumeikan University, Japan
Singh Rahul	San Francisco State University, USA
Skurikhin Alexei	Los Alamos National Laboratory, USA
Souvenir, Richard	University of North Carolina - Charlotte, USA
Su Chung-Yen	National Taiwan Normal University, Taiwan (R.O.C.)
Sugihara Kokichi	University of Tokyo, Japan
Sun Zehang	Apple, USA
Syeda-Mahmood Tanveer	IBM Almaden, USA
Tan Kar Han	Hewlett Packard, USA
Tan Tieniu	Chinese Academy of Sciences, China
Tavakkoli Alireza	University of Houston - Victoria, USA
Tavares, Joao	Universidade do Porto, Portugal

Teoh Eam Khwang	Nanyang Technological University, Singapore
Thiran Jean-Philippe	Swiss Federal Institute of Technology Lausanne (EPFL), Switzerland
Tistarelli Massimo	University of Sassari, Italy
Tong Yan	University of South Carolina, USA
Tsechpenakis Gabriel	University of Miami, USA
Tsui T.J.	Chinese University of Hong Kong, Hong Kong
Trucco Emanuele	University of Dundee, UK
Tubaro Stefano	DEI . Politecnico di Milano, Italy
Uhl Andreas	Salzburg University, Austria
Velastin Sergio	Kingston University London, UK
Veropoulos Kostantinos	GE Healthcare, Greece
Verri Alessandro	Università di Genova, Italy
Wang C.L. Charlie	The Chinese University of Hong Kong, Hong Kong
Wang Junxian	Microsoft, USA
Wang Song	University of South Carolina, USA
Wang Yunhong	Beihang University, China
Webster Michael	University of Nevada, Reno, USA
Wolff Larry	Equinox Corporation, USA
Wong Kenneth	The University of Hong Kong, Hong Kong
Xiang Tao	Queen Mary, University of London, UK
Xue Xinwei	Fair Isaac Corporation, USA
Xu Meihe	University of California at Los Angeles, USA
Yang Ming-Hsuan	University of California at Merced, USA
Yang Ruigang	University of Kentucky, USA
Yi Lijun	SUNY at Binghampton, USA
Yu Ting	GE Global Research, USA
Yu Zeyun	University of Wisconsin-Milwaukee, USA
Yuan Chunrong	University of Tübingen, Germany
Zabulis Xenophon	Foundation for Research and Technology - Hellas (FORTH), Greece
Zhang Yan	Delphi Corporation, USA
Cheng Shinko	HRL Labs, USA
Zhou Huiyu	Queen's University Belfast, UK

(Area 2) Computer Graphics

Abd Rahni Mt Piah	Universiti Sains Malaysia, Malaysia
Abram Greg	Texas Advanced Computing Center, USA
Adamo-Villani Nicoletta	Purdue University, USA
Agu Emmanuel	Worcester Polytechnic Institute, USA
Andres Eric	Laboratory XLIM-SIC, University of Poitiers, France
Artusi Alessandro	CaSToRC Cyprus Institute, Cyprus
Baciu George	Hong Kong PolyU, Hong Kong

Balcisoy Selim Saffet	Sabanci University, Turkey
Barneva Reneta	State University of New York, USA
Belyaev Alexander	Heriot-Watt University, UK
Benes Bedrich	Purdue University, USA
Berberich Eric	Max Planck Institute, Germany
Bilalis Nicholas	Technical University of Crete, Greece
Bimber Oliver	Johannes Kepler University Linz, Austria
Bohez Erik	Asian Institute of Technology, Thailand
Bouatouch Kadi	University of Rennes I, IRISA, France
Brimkov Valentin	State University of New York, USA
Brown Ross	Queensland University of Technology, Australia
Bruckner Stefan	Vienna University of Technology, Austria
Callahan Steven	University of Utah, USA
Capin Tolga	Bilkent University, Turkey
Chaudhuri Parag	Indian Institute of Technology Bombay, India
Chen Min	University of Oxford, UK
Cheng Irene	University of Alberta, Canada
Chiang Yi-Jen	Polytechnic Institute of New York University, USA
Comba Joao	Univ. Fed. do Rio Grande do Sul, Brazil
Crawfis Roger	Ohio State University, USA
Cremer Jim	University of Iowa, USA
Crossno Patricia	Sandia National Laboratories, USA
Culbertson Bruce	HP Labs, USA
Dana Kristin	Rutgers University, USA
Debattista Kurt	University of Warwick, UK
Deng Zhigang	University of Houston, USA
Dick Christian	Technical University of Munich, Germany
DiVerdi Stephen	Adobe, USA
Dingliana John	Trinity College, Ireland
El-Sana Jihad	Ben Gurion University of The Negev, Israel
Entezari Alireza	University of Florida, USA
Fabian Nathan	Sandia National Laboratories, USA
Fiorio Christophe	Université Montpellier 2, LIRMM, France
De Floriani Leila	University of Genova, Italy
Fuhrmann Anton	VRVis Research Center, Austria
Gaither Kelly	University of Texas at Austin, USA
Gao Chunyu	Epson Research and Development, USA
Geist Robert	Clemson University, USA
Gelb Dan	Hewlett Packard Labs, USA
Gotz David	IBM, USA
Gooch Amy	University of Victoria, Canada
Gu David	Stony Brook University, USA
Guerra-Filho Gutemberg	University of Texas Arlington, USA

Habib Zulfiqar	COMSATS Institute of Information Technology, Lahore, Pakistan
Hadwiger Markus	KAUST, Saudi Arabia
Haller Michael	Upper Austria University of Applied Sciences, Austria
Hamza-Lup Felix	Armstrong Atlantic State University, USA
Han JungHyun	Korea University, Korea
Hand Randall	Lockheed Martin Corporation, USA
Hao Xuejun	Columbia University and NYSPI, USA
Hernandez Jose Tiberio	Universidad de los Andes, Colombia
Huang Jian	University of Tennessee at Knoxville, USA
Huang Mao Lin	University of Technology, Australia
Huang Zhiyong	Institute for Infocomm Research, Singapore
Hussain Muhammad	King Saud University, Saudi Arabia
Jeschke Stefan	Vienna University of Technology, Austria
Joaquim Jorge	Instituto Superior Técnico, Portugal
Jones Michael	Brigham Young University, USA
Julier Simon J.	University College London, UK
Kakadiaris Ioannis	University of Houston, USA
Kamberov George	Stevens Institute of Technology, USA
Ko Hyeong-Seok	Seoul National University, Korea
Klosowski James	AT&T Labs, USA
Kobbelt Leif	RWTH Aachen, Germany
Kolingerova Ivana	University of West Bohemia, Czech Republic
Lai Shuhua	Virginia State University, USA
Lee Chang Ha	Chung-Ang University, Korea
Levine Martin	McGill University, Canada
Lewis R. Robert	Washington State University, USA
Li Frederick	University of Durham, UK
Lindstrom Peter	Lawrence Livermore National Laboratory, USA
Linsen Lars	Jacobs University, Germany
Loviscach Joern	Fachhochschule Bielefeld, University of Applied Sciences, Germany
Magnor Marcus	TU Braunschweig, Germany
Martin Ralph	Cardiff University, UK
Meenakshisundaram Gopi	University of California-Irvine, USA
Mendoza Cesar	Natural Motion Ltd., USA
Metaxas Dimitris	Rutgers University, USA
Mudur Sudhir	Concordia University, Canada
Myles Ashish	University of Florida, USA
Nait-Charif Hammadi	University of Dundee, UK
Nasri Ahmad	American University of Beirut, Lebanon
Noh Junyong	KAIST, Korea
Noma Tsukasa	Kyushu Institute of Technology, Japan
Okada Yoshihiro	Kyushu University, Japan

Olague Gustavo	CICESE Research Center, Mexico
Oliveira Manuel M.	Univ. Fed. do Rio Grande do Sul, Brazil
Owen Charles	Michigan State University, USA
Ostromoukhov Victor M.	University of Montreal, Canada
Pascucci Valerio	University of Utah, USA
Patchett John	Los Alamos National Lab, USA
Peters Jorg	University of Florida, USA
Pronost Nicolas	Utrecht University, The Netherlands
Qin Hong	Stony Brook University, USA
Rautek Peter	Vienna University of Technology, Austria
Razdan Anshuman	Arizona State University, USA
Renner Gabor	Computer and Automation Research Institute, Hungary
Rosen Paul	University of Utah, USA
Rosenbaum Rene	University of California at Davis, USA
Rudomin, Isaac	ITESM-CEM, Mexico
Rushmeier, Holly	Yale University, USA
Sander Pedro	The Hong Kong University of Science and Technology, Hong Kong
Sapidis Nickolas	University of Western Macedonia, Greece
Sarfraz Muhammad	Kuwait University, Kuwait
Scateni Riccardo	University of Cagliari, Italy
Schaefer Scott	Texas A&M University, USA
Sequin Carlo	University of California-Berkeley, USA
Shead Tinothy	Sandia National Laboratories, USA
Sourin Alexei	Nanyang Technological University, Singapore
Stamminger Marc	REVES/INRIA, France
Su Wen-Poh	Griffith University, Australia
Szumilas Lech	Research Institute for Automation and Measurements, Poland
Tan Kar Han	Hewlett Packard, USA
Tarini Marco	Università dell'Insubria (Varese), Italy
Teschner Matthias	University of Freiburg, Germany
Umlauf Georg	HTWG Constance, Germany
Vanegas Carlos	Purdue University, USA
Wald Ingo	University of Utah, USA
Walter Marcelo	UFRGS, Brazil
Wimmer Michael	Technical University of Vienna, Austria
Woodring Jon	Los Alamos National Laboratory, USA
Wylie Brian	Sandia National Laboratory, USA
Wyman Chris	University of Calgary, Canada
Wyvill Brian	University of Iowa, USA
Yang Qing-Xiong	University of Illinois at Urbana, Champaign, USA
Yang Ruigang	University of Kentucky, USA

Ye Duan	University of Missouri-Columbia, USA
Yi Beifang	Salem State University, USA
Yin Lijun	Binghamton University, USA
Yoo Terry	National Institutes of Health, USA
Yuan Xiaoru	Peking University, China
Zhang Jian Jun	Bournemouth University, UK
Zeng Jianmin	Nanyang Technological University, Singapore
Zara Jiri	Czech Technical University in Prague, Czech Republic

(Area 3) Virtual Reality

Alcañiz Mariano	Technical University of Valencia, Spain
Arns Laura	Purdue University, USA
Balcisoy Selim	Sabanci University, Turkey
Behringer Reinhold	Leeds Metropolitan University, UK
Benes Bedrich	Purdue University, USA
Bilalis Nicholas	Technical University of Crete, Greece
Blach Roland	Fraunhofer Institute for Industrial Engineering, Germany
Blom Kristopher	University of Barcelona, Spain
Bogdanovych Anton	University of Western Sydney, Australia
Borst Christoph	University of Louisiana at Lafayette, USA
Brady Rachael	Duke University, USA
Brega Jose Remo Ferreira	Universidade Estadual Paulista, Brazil
Brown Ross	Queensland University of Technology, Australia
Bues Matthias	Fraunhofer IAO in Stuttgart, Germany
Capin Tolga	Bilkent University, Turkey
Chen Jian	Brown University, USA
Cooper Matthew	University of Linköping, Sweden
Coquillart Sabine	INRIA, France
Craig Alan	NCSA University of Illinois at Urbana-Champaign, USA
Cremer Jim	University of Iowa, USA
Edmunds Timothy	University of British Columbia, Canada
Egges Arjan	Universiteit Utrecht, The Netherlands
Encarnao L. Miguel	ACT Inc., USA
Figueroa Pablo	Universidad de los Andes, Colombia
Fox Jesse	Stanford University, USA
Friedman Doron	IDC, Israel
Fuhrmann Anton	VRVis Research Center, Austria
Gobron Stephane	EPFL, Switzerland
Gregory Michelle	Pacific Northwest National Lab, USA
Gupta Satyandra K.	University of Maryland, USA
Haller Michael	FH Hagenberg, Austria
Hamza-Lup Felix	Armstrong Atlantic State University, USA

Herbelin Bruno	EPFL, Switzerland
Hinkenjann Andre	Bonn-Rhein-Sieg University of Applied Sciences, Germany
Hollerer Tobias	University of California at Santa Barbara, USA
Huang Jian	University of Tennessee at Knoxville, USA
Huang Zhiyong	Institute for Infocomm Research (I2R), Singapore
Julier Simon J.	University College London, UK
Kaufmann Hannes	Vienna University of Technology, Austria
Kiyokawa Kiyoshi	Osaka University, Japan
Klosowski James	AT&T Labs, USA
Kozintsev	Igor, Intel, USA
Kuhlen Torsten	RWTH Aachen University, Germany
Lee Cha	University of California, Santa Barbara, USA
Liere Robert van	CWI, The Netherlands
Livingston A. Mark	Naval Research Laboratory, USA
Malzbender Tom	Hewlett Packard Labs, USA
Molineros Jose	Teledyne Scientific and Imaging, USA
Muller Stefan	University of Koblenz, Germany
Olwal Alex	MIT, USA
Owen Charles	Michigan State University, USA
Paelke Volker	Institut de Geomàtica, Spain
Peli Eli	Harvard University, USA
Pettifer Steve	The University of Manchester, UK
Piekarski Wayne	Qualcomm Bay Area R & D, USA
Pronost Nicolas	Utrecht University, The Netherlands
Pugmire Dave	Los Alamos National Lab, USA
Qian Gang	Arizona State University, USA
Raffin Bruno	INRIA, France
Raij Andrew	University of South Florida, USA
Reitmayr Gerhard	Graz University of Technology, Austria
Richir Simon	Arts et Metiers ParisTech, France
Rodello Ildeberto	University of Sao Paulo, Brazil
Sandor Christian	University of South Australia, Australia
Santhanam Anand	University of California at Los Angeles, USA
Sapidis Nickolas	University of Western Macedonia, Greece
Sherman Bill	Indiana University, USA
Slavik Pavel	Czech Technical University in Prague, Czech Republic
Sourin Alexei	Nanyang Technological University, Singapore
Steinicke Frank	University of Münster, Germany
Suma Evan	University of Southern California, USA
Stamminger Marc	REVES/INRIA, France
Srikanth Manohar	Indian Institute of Science, India
Vercher Jean-Louis	Université de la Méditerranée, France

Wald Ingo   University of Utah, USA
Wither Jason   University of California, Santa Barbara, USA
Yu Ka Chun   Denver Museum of Nature and Science, USA
Yuan Chunrong   University of Tübingen, Germany
Zachmann Gabriel   Clausthal University, Germany
Zara Jiri   Czech Technical University in Prague,
                               Czech Republic
Zhang Hui   Indiana University, USA
Zhao Ye   Kent State University, USA

(Area 4) Visualization

Andrienko Gennady   Fraunhofer Institute IAIS, Germany
Avila Lisa   Kitware, USA
Apperley Mark   University of Waikato, New Zealand
Balázs Csébfalvi   Budapest University of Technology and
                               Economics, Hungary
Brady Rachael   Duke University, USA
Benes Bedrich   Purdue University, USA
Bilalis Nicholas   Technical University of Crete, Greece
Bonneau Georges-Pierre   Grenoble Université, France
Bruckner Stefan   Vienna University of Technology, Austria
Brown Ross   Queensland University of Technology, Australia
Bühler Katja   VRVis Research Center, Austria
Callahan Steven   University of Utah, USA
Chen Jian   Brown University, USA
Chen Min   University of Oxford, UK
Chiang Yi-Jen   Polytechnic Institute of New York University,
                               USA
Cooper Matthew   University of Linköping, Sweden
Chourasia Amit   University of California - San Diego, USA
Coming Daniel   Desert Research Institute, USA
Daniels Joel   University of Utah, USA
Dick Christian   Technical University of Munich, Germany
DiVerdi Stephen   Adobe, USA
Doleisch Helmut   SimVis GmbH, Austria
Duan Ye   University of Missouri-Columbia, USA
Dwyer Tim   Monash University, Australia
Entezari Alireza   University of Florida, USA
Ertl Thomas   University of Stuttgart, Germany
De Floriani Leila   University of Maryland, USA
Fujishiro Issei   Keio University, Japan
Geist Robert   Clemson University, USA
Gotz David   IBM, USA
Grinstein Georges   University of Massachusetts Lowell, USA
Goebel Randy   University of Alberta, Canada

Görg Carsten	University of Colorado at Denver, USA
Gregory Michelle	Pacific Northwest National Lab, USA
Hadwiger Helmut Markus	KAUST, Saudi Arabia
Hagen Hans	Technical University of Kaiserslautern, Germany
Hamza-Lup Felix	Armstrong Atlantic State University, USA
Healey Christopher	North Carolina State University at Raleigh, USA
Hege Hans-Christian	Zuse Institute Berlin, Germany
Hochheiser Harry	University of Pittsburgh, USA
Hollerer Tobias	University of California at Santa Barbara, USA
Hong Lichan	University of Sydney, Australia
Hong Seokhee	Palo Alto Research Center, USA
Hotz Ingrid	Zuse Institute Berlin, Germany
Huang Zhiyong	Institute for Infocomm Research (I2R), Singapore
Jiang Ming	Lawrence Livermore National Laboratory, USA
Joshi Alark	Yale University, USA
Julier Simon J.	University College London, UK
Kohlhammer Jörn	Fraunhofer Institut, Germany
Kosara Robert	University of North Carolina at Charlotte, USA
Laramee Robert	Swansea University, UK
Lee Chang Ha	Chung-Ang University, Korea
Lewis R. Robert	Washington State University, USA
Liere Robert van	CWI, The Netherlands
Lim Ik Soo	Bangor University, UK
Linsen Lars	Jacobs University, Germany
Liu Zhanping	University of Pennsylvania, USA
Ma Kwan-Liu	University of California at Davis, USA
Maeder Anthony	University of Western Sydney, Australia
Malpica Jose	Alcala University, Spain
Masutani Yoshitaka	The University of Tokyo Hospital, Japan
Matkovic Kresimir	VRVis Research Center, Austria
McCaffrey James	Microsoft Research / Volt VTE, USA
Melançon Guy	CNRS UMR 5800 LaBRI and INRIA Bordeaux Sud-Ouest, France
Miksch Silvia	Vienna University of Technology, Austria
Monroe Laura	Los Alamos National Labs, USA
Morie Jacki	University of Southern California, USA
Mudur Sudhir	Concordia University, Canada
Museth Ken	Linköping University, Sweden
Paelke Volker	Institut de Geomàtica, Spain
Peikert Ronald	Swiss Federal Institute of Technology Zurich, Switzerland
Pettifer Steve	The University of Manchester, UK

Pugmire Dave	Los Alamos National Lab, USA
Rabin Robert	University of Wisconsin at Madison, USA
Raffin Bruno	Inria, France
Razdan Anshuman	Arizona State University, USA
Rhyne Theresa-Marie	North Carolina State University, USA
Rosenbaum Rene	University of California at Davis, USA
Santhanam Anand	University of California at Los Angeles, USA
Scheuermann Gerik	University of Leipzig, Germany
Shead Tinothy	Sandia National Laboratories, USA
Shen Han-Wei	Ohio State University, USA
Sips Mike	Stanford University, USA
Slavik Pavel	Czech Technical University in Prague, Czech Republic
Sourin Alexei	Nanyang Technological University, Singapore
Thakur Sidharth	Renaissance Computing Institute (RENCI), USA
Theisel Holger	University of Magdeburg, Germany
Thiele Olaf	University of Mannheim, Germany
Toledo de Rodrigo	Petrobras PUC-RIO, Brazil
Tricoche Xavier	Purdue University, USA
Umlauf Georg	HTWG Constance, Germany
Viegas Fernanda	IBM, USA
Wald Ingo	University of Utah, USA
Wan Ming	Boeing Phantom Works, USA
Weinkauf Tino	Max-Planck-Institut für Informatik, Germany
Weiskopf Daniel	University of Stuttgart, Germany
Wischgoll Thomas	Wright State University, USA
Wylie Brian	Sandia National Laboratory, USA
Xu Wei	Stony Brook University, USA
Yeasin Mohammed	Memphis University, USA
Yuan Xiaoru	Peking University, China
Zachmann Gabriel	Clausthal University, Germany
Zhang Hui	Indiana University, USA
Zhao Ye	Kent State University, USA
Zheng Ziyi	Stony Brook University, USA
Zhukov Leonid	Caltech, USA

ISVC 2012 Special Tracks

1. 3D Mapping, Modeling and Surface Reconstruction

Organizers

Nefian Ara	Carnegie Mellon University/NASA Ames Research Center, USA
Edwards Laurence	NASA Ames Research Center, USA
Huertas Andres	NASA Jet Propulsion Lab, USA

2. Computational Bioimaging

Organizers

Tavares João Manuel R.S.	University of Porto, Portugal
Natal Jorge Renato	University of Porto, Portugal
Cunha Alexandre	Caltech, USA

3. Optimization for Vision, Graphics and Medical Imaging

Organizers

Komodakis Nikos	University of Crete, Greece
Kohli Pushmeet	Microsoft Research Cambridge, UK
Kumar Pawan	Ecole Centrale de Paris, France
Maeder Anthony	University of Western Sydney, Australia
Carsten Rother	Microsoft Research Cambridge, UK

4. Unconstrained Biometrics: Advances and Trends

Organizers

Proença Hugo	University of Beira Interior, Covilhã, Portugal
Du Yingzi	Indiana University-Purdue University Indianapolis, Indianapolis, USA
Scharcanski Jacob	Federal University of Rio Grande do Sul Porto Alegre, Brazil
Ross Arun	West Virginia University, USA

5. Intelligent Environments: Algorithms and Applications

Organizers

Bebis George	University of Nevada, Reno, USA
Nicolescu Mircea	University of Nevada, Reno, USA
Bourbakis Nikolaos	Wright State University, USA
Tavakkoli Alireza	University of Houston, Victoria, USA

6. Object Recognition

Organizers

Scalzo Fabien	University of California at Los Angeles, USA
Salgian Andrea	The College of New Jersey, USA

7. Face Processing and Recognition

Organizers

Hussain Muhammad	King Saud Univesity, Saudi Arabia
Muhammad Ghulam	King Saud Univesity, Saudi Arabia
Bebis George	University of Nevada, Reno, USA

Organizing Institutions and Sponsors

Table of Contents – Part I

ST: Computational Bioimaging I

Computer Graphics I

Calibration and 3D Vision

Object Recognition

Illumination, Modeling, and Segmentation

Visualization I

ST: 3D Mapping, Modeling and Surface Reconstruction

Motion and Tracking

Computer Graphics II

ST: Optimization for Vision, Graphics and Medical Imaging

HCI and Recognition

Visualization II

Table of Contents – Part II

ST: Intelligent Environments: Algorithms and Applications

Applications

Visualization III

Virtual Reality

ST: Face Processing and Recognition

Poster

Simulation of the Abdominal Wall and Its Arteries after Pneumoperitoneum for Guidance of Port Positioning in Laparoscopic Surgery

J. Bano[1,2], A. Hostettler[1], S.A. Nicolau[1], C. Doignon[2], H.S. Wu[3],
M.H. Huang[3], L. Soler[1], and J. Marescaux[1]

[1] IRCAD, Virtual-Surg, Place de l'Hopital 1, 67091 Strasbourg Cedex, France
[2] LSIIT (UMR 7005 CNRS), University of Strasbourg, Parc d'Innovation,
Boulevard S. Brant, BP 10412 67412 Illkirch Cedex, France
[3] IRCAD Taiwan, Medical Imaging Team, 1-6 Lugong Road, Lukang 505, Taiwan
jordan.bano@etu.unistra.fr

Abstract. During laparoscopic surgery, the trocar insertion can injure arteries of the abdominal wall. Although these arteries are visible in a preoperative computed tomography [CT] with contrast medium, it is difficult for the surgeon to estimate their true intraoperative positions since the pneumoperitoneum dramatically stretches the abdominal wall. A navigation system showing the artery position would thus be very helpful for the surgeon. We present in this paper a method to simulate the position of the abdominal wall and its arteries after pneumoperitoneum. Our method requires a segmented preoperative CT image and an intraoperative surface reconstruction of the skin. The intraoperative skin surface allows us to compute a displacement field of the abdominal wall's outer surface that we propagate to estimate the artery position.

Our simulation was evaluated using two sets of pig CT images, before and after pneumoperitoneum. Results show that our method provides an estimation of the abdominal wall and artery positions with an average error of respectively 2 mm and 6 mm which fits the clinical application constraint. In the near future, we will focus on viscera movement simulation after pneumoperitoneum using our abdominal wall shape prediction.

Keywords: Predictive simulation, pneumoperitoneum, skin tracking.

1 Introduction

1.1 Clinical Context

Laparoscopic surgery becomes more and more a common surgical technique. It consists in inserting an endoscopic camera and several tools through small incisions in the abdominal skin. Advantages of this minimally invasive technique for the patient are currently well-known. Unfortunately, the intervention is usually more difficult than in open surgery, mostly due to the loss of 3D depth and tactile perceptions. In order to help surgeons, numerous navigation systems have

G. Bebis et al. (Eds.): ISVC 2012, Part I, LNCS 7431, pp. 1–11, 2012.

been developed allowing the display of non-visible structures. However, most of these systems register a preoperative rigid model without taking deformation due to the gas injection (the so called pneumoperitoneum) into account. Indeed, pneumoperitoneum distorts the skin shape and the abdominal viscera move over several centimeters [1] (Fig. 1). Then, movement induces limitations for navigation systems, which are usually based on a preoperative static image acquired before gas injection [2–4].

The work presented in this paper focuses on the movement prediction of the abdominal wall and the epigastric arteries after pneumoperitoneum. We underline that the estimation of abdominal wall position is a crucial step since its arteries can be injured [5–7]. Our long term goal is to accurately estimate the abdominal wall and viscera positions after gas injection to guide surgeons during trocar positioning at the beginning of an intervention.

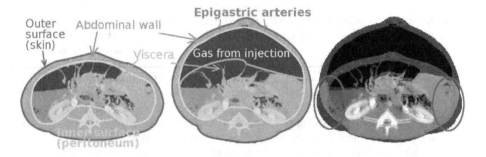

Fig. 1. Image before (*left*) and after pneumoperitoneum (*middle*), and the fusion of these two images (*right*). One can see the significant movement of the abdominal wall which contains arteries. On the right image, we highlight the shape deformation of the cavity containing the viscera. It is thus clear that one cannot apply a rigid registration to predict the deformation.

1.2   Previous Work

The position of the abdominal wall and viscera after gas injection can be obtained in two main ways: by simulating the deformation of a 3D virtual model of the patient using mechanical information from a preoperative image [8] or by acquiring intraoperative images during the intervention [9–12].

To our knowledge, only Kitasaka et al.[8] deal with pneumoperitoneum simulation. They simulate gas injection in the peritoneum by applying forces, whose direction seems to be antero-posterior, on the inner surface of only a part of the abdominal wall. Recently, they evaluated the accuracy of their results on eight patients by comparing thirteen skin landmark positions after pneumoperitoneum and with simulation results [13]. Their simulation method uses two parameters which are empirically chosen to fit the in vivo data. With an optimally chosen set of parameters, the minimal mean error obtained was 13.8 mm. Although their simulation provides interesting results, its usefulness is still limited since they do not take the viscera motion into account. Moreover the parameters of their model have no biomechanical meaning and cannot be adapted for a particular patient shape.

Shekhar et al. [9] propose a method using low-dose CT acquisitions during the intervention. A high quality acquisition is then registered onto these images, to estimate the movement of abdominal organs. However, a CT-scan in the operating room is unusual due to budgetary reasons. In addition, their method requires numerous patient expositions to radiation. The use of MRI instead of a CT-scan would avoid the radiation issues. Unfortunately, the cost of such an equipment is prohibitive.

To our knowledge, a similar specific simulation of epigastric arteries, has still not been tackled.

1.3   Overview of Our Approach

Our approach is inspired by the work of Hostettler et al. [14]. In the context of radiotherapy or interventional radiology, they present a method to simulate in real-time the movement of the abdominal viscera induced by free-breathing from a preoperative 3D image and a real-time tracking of the skin position. Like them, we consider that skin position can be safely tracked by an optical system (such as the ones presented in [15] or other commercial systems as VisionRT©). Since we only use skin tracking during the intervention, our method does not depend on a medical imaging device such as CT or MRI localised in the operating room. Our method is divided into two parts (Fig. 2).

Preoperative Step. From an abdo-thoracic 3D image, we firstly extract two structures: the first one corresponds to the abdo-thoracic viscera which is considered as a whole entity, and the second one is the abdominal wall (Fig. 1 for anatomical definition). Secondly, a volume mesh is generated for each of these two entities.

Intraoperative Step. The movement of the abdominal wall mesh is simulated by moving its outer surface so that it coincides with the position of the tracked skin after gas injection. During this phase, the abdominal wall movement simulation gives us the widest artery position and the inner surface of its abdominal wall. Using the knowledge of the inner surface position, the new shape of the viscera cavity is then predicted.

The remaining part of the paper is organized as fallows. In Section 2, we explain how we compute a deformation field of the abdominal wall and artery meshes using the knowledge of the skin position, and assumption of the abdominal wall's incompressibility. In Section 3, our method accuracy is evaluated using in vivo pig data and we show that an accurate prediction can be obtained with preoperative 3D CT and skin tracking only. Conclusions are reported in Section 4.

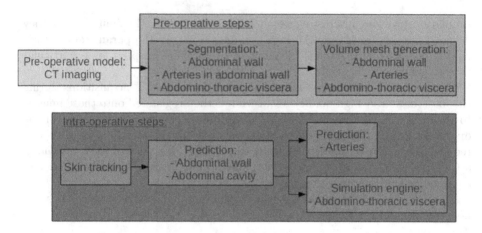

Fig. 2. The workflow of our approach: the upper part describes the preoperative steps and the lower one, shows the intraoperative steps

2   Method

Our method is composed of two preoperative and four intraoperative steps (Fig. 2) to estimate the position of the abdominal wall and its widest arteries. It requires a preoperative image of the patient containing structures of interest (CT with contrast medium).

2.1   Segmentation Step

The bones, the outer and inner surfaces of the abdominal wall and the epigastric arteries of the abdominal wall are segmented in the preoperative image (see Section 3.1 for the acquisition step). The inner surface corresponds to the boundary between the abdominal wall and the abdo-thoracic viscera, and the outer surface to the skin (Fig. 1). These segmentations are done using home-written software with semi-automatic methods (15 min. in average per segmentation). We obtain a binary image for each segmentation and a surface mesh of the arteries.

Then, in each axial slice, we extract the boundary of the segmentation for both inner and outer surface of the abdominal wall: we obtain two curves in each slice (C_{in} and C_{out}).

2.2   Generation of the Abdominal Wall Volume Mesh

A volume mesh of the abdominal wall is built using points located on the curves, C_{in} and C_{out}, of the inner and outer segmentations in each axial slice.

We split each curve into an extensible (C_{exs}) and a non-extensible (C_{nexs}) area. We choose to split the abdominal wall into two areas depending on its extensibility. Each part will behave differently during the simulation step. In fact, rigid anatomical structures like ribs are not deformed during pneumoperitoneum. We assume that these structures locally rigidify the abdominal wall (see Section 3.2 for split position evaluation). One can see in Fig. 3 this separation along the rib extremities, using the image of the bone segmentation.

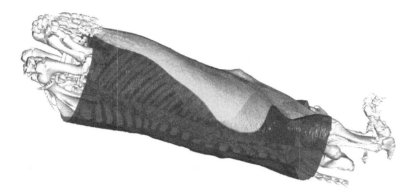

Fig. 3. We identify two parts, one extensible (*blue*) and one non-extensible (*yellow*), on a 3D model of a pig in dorsal supine position according to bone structure position (ribs for the upper part and leg bones for the lower part)

We separate into equal parts the C_{exs} (resp. C_{nexs}) of the outer surface, giving points P^i_{extE} (resp. P^i_{extN}). The inner surface is also divided into the same number of parts, obtaining the P^i_{int} which are matched with the P^i_{ext} (Fig. 4, on the left). A cell of the volume mesh have four vertices in an axial slice and four vertices in an adjacent axial slice. Each set of points is composed of: $P^i_{ext}, P^{i+1}_{ext}, P^i_{int}$ and P^{i+1}_{int}, with P^i_{ext} matched with P^i_{int}. Regarding the choice of the point number by slice, we observed that using more than 200 points does not improve the accuracy results (less than 4% of improvement), whereas using 50 (resp. 100) points leads to 30% (resp. 10%) of supplementary errors.

2.3   Registration of Preoperative Model in the Operative Room

The patient skin will be localized in real-time by an optical system based on structured light projection. To register the preoperative CT image in the frame of the optical system, a iterative closest point registration will be performed [16] before the pneumoperitoneum between the segmented skin in the preoperative image and the skin surface provided by the optical system. Obviously, we assume that the operating table will not be moved during the pneumoperitoneum.

2.4   Deformation Field of the Outer Surface

The previous step allows to register the intraoperative skin position and preoperative model in the same frame. We now try to match each P^i_{ext} of the preoperative

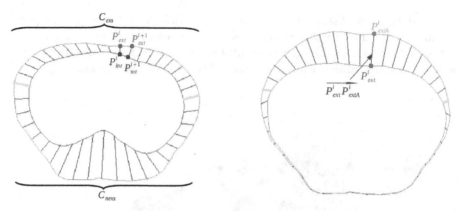

Fig. 4. On the left figure, one can see the matching between the outer (*green*) and the inner (*red*) surface of abdominal wall and, on the right figure, the matching between the skin before pneumoperitoneum (*green*) and acquired with optical tracking system (*orange*). The separation between the extensible and non-extensible area is drawn in yellow.

skin with its physical corresponding point on the tracked skin. We assume that the cranio-caudal movement of the abdominal wall is negligible compared to the antero-posterior. Thus, we seek the corresponding point in the axial slice that contains P_{ext}^i.

Figure 4 shows an axial slice where we can see the preoperative skin and the intraoperative skin positions. We firstly define the boundaries of the intraoperative external curve between the extensible and the non-extensible area: we choose the closest neighbour of the preoperative external curve boundaries. As we assume that the deformation is homogeneous in the extensible area, the intraoperative extensible curve is split into the same number of segments as the preoperative one, thus giving the points P_{extA}^i. The vectors $\overrightarrow{P_{ext}^i P_{extA}^i}$ correspond to the deformation field of the inner and outer surfaces (Fig. 4, on the right).

2.5   Estimation of the Inner Surface Position

To estimate the inner surface position after pneumoperitoneum, we assume that the displacement of a point P_{int}^i is the same as its matched point P_{ext}^i. Then, each point P_{int}^i are moved along the vector $\overrightarrow{P_{ext}^i P_{extA}^i}$ obtaining a P_{estim}^i point. The set of all P_{estim}^i corresponds to our estimation of the inner surface after gas injection (Fig. 5, on the left).

2.6   Prediction of Artery Movement

The abdominal wall artery position is provided by applying the deformation field of the abdominal wall on the artery mesh. We find for each vertex of the mesh, called here P_{art}, a new position according to the P_{ext}^i and P_{int}^i points (Fig. 5, on the right).

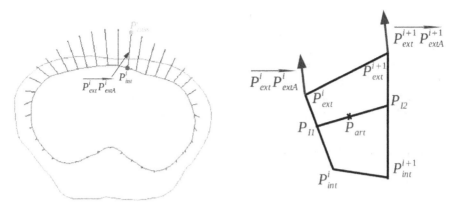

Fig. 5. On the left figure, the estimation of the inner surface position of the abdominal wall after pneumoperitoneum is illustrated. On the right figure, P_{art} motion is computed from P_{I1} and P_{I2} and motion of P_{ext}^i and P_{int}^i.

Firstly, we search, with an iterative method, which quadrilateral $P_{ext}^i P_{ext}^{i+1} P_{int}^{i+1} P_{int}^i$, is the closest to P_{art}. Secondly, we find the line which passes through P_{art} and which intersects the segment $P_{ext}^i P_{int}^i$ in a point P_{I1} and the segment $P_{int}^{i+1} P_{ext}^{i+1}$ in another point P_{I2} such as:

$$\frac{\|\overrightarrow{P_{ext}^i P_{I1}}\|}{\|\overrightarrow{P_{int}^i P_{I1}}\|} = \frac{\|\overrightarrow{P_{ext}^{i+1} P_{I2}}\|}{\|\overrightarrow{P_{int}^{i+1} P_{I2}}\|} \qquad (1)$$

The displacement vector applied to the vertex P_{art} is then:

$$dis(P_{art}) = \frac{\|\overrightarrow{P_{I2} P_{art}}\|}{\|\overrightarrow{P_{I1} P_{I2}}\|} \overrightarrow{P_{ext}^i P_{extA}^i} + \frac{\|\overrightarrow{P_{I1} P_{art}}\|}{\|\overrightarrow{P_{I1} P_{I2}}\|} \overrightarrow{P_{ext}^{i+1} P_{extA}^{i+1}} \qquad (2)$$

3 Experimental Results Using Pig Data

In this section, we evaluate firstly the assumptions related to our model properties, and secondly our model accuracy on the inner surface and artery positions predicted using in vivo pig data.

3.1 Acquisition and Segmentation Steps

Two sets of pig medical images have been used to evaluate the method. These acquisitions were done before and after pneumoperitoneum, using a CT scanner. The pigs were placed in supine position, ventilated mechanically and the pressure after pneumoperitoneum was set to 12 mmHg. A contrast medium was injected to enable blood vessel segmentation.

We extract from both images, with the technique used in Section 2.1, the outer and the inner surfaces and the widest epigastric arteries of the abdominal wall. Unfortunately, a real-time skin tracking of the pig has not been performed during CT acquisitions. Therefore, the evaluation was done using the skin extracted on the CT image after pneumoperitoneum. Thus, our result accuracy is slightly overestimated since it does not take the error due to the registration between the tracked skin and the preoperative model into account.

3.2 Evaluation of Boundary between Tissues

In this part, we evaluate the position of our split method between the extensible and non-extensible areas in the abdominal wall.

We compare the volume of the extensible and the non-extensible areas of the abdominal wall before and after pneumoperitoneum in each axial slice (Fig. 6). To compute those volumes, we count the number of voxels which compose the segmentation of each part. We obtain a difference of 1.6% for the non-extensible part and 1% for the extensible part. These values show that our split position based on the rib positions is well estimated.

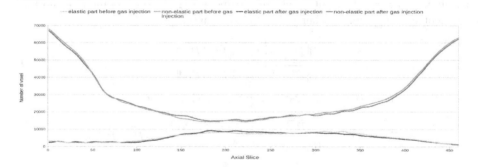

Fig. 6. Measurement of the extensible (*orange-blue*) and non-extensible (*yellow-green*) area volume of the medical image before and after gas injection for each axial slice

3.3 Evaluation of the Inner Surface Position

We illustrate the accuracy of the predicted inner surface of the abdominal wall with a colored mesh, by comparing our prediction with the segmentation from the medical image acquired after gas injection (Fig. 7, on the left).

A surface mesh M_{predi} of our inner surface prediction is generated and a very dense point cloud $M_{gdTruth}$ is extracted from the segmentation of the inner surface in the image after pneumoperitoneum. Each vertex of M_{predi} is then colored according to its distance to its closest neighbour on $M_{gdTruth}$ (Tab. 1).

Table 1. Distance for each vertex sorted into four ranges, each range corresponds to a quartile of the triangle total number

First pig (mm)	Second pig (mm)	Color range
0 - 0.7	0 - 0.6	Blue to Cyan
0.7 - 1.4	0.6 - 1.3	Cyan to Green
1.4 - 2.3	1.3 - 2.6	Green to Yellow
2.3 - 15.3	2.6 - 18.7	Yellow to Red
2.0 ($\pm$ 2.1)	1.9 ($\pm$ 1.8)	Mean ($\pm$ Std.Dev.)

3.4   Evaluation of the Artery Position Prediction

We measure the distance between the artery bifurcations in the artery mesh extracted from the image after pneumoperitoneum and our artery position prediction (Fig. 7, on the right). We find an average distance of these six landmarks equal to 6 mm. By applying a safety margin of one centimeter, we can help surgeons to avoid artery injuries.

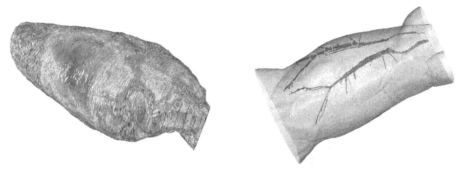

Fig. 7. The mesh on the left figure represents the distance distribution between our estimation and the ground truth using the Tab. 1 color ranges. One can see that the estimation of the inner surface of abdominal wall is accurate enough for our application (90% of the mesh have a distance $<= 3.9mm$). We assume that the largest error area is caused by our model homogeneity. On the right figure, our prediction of artery positions (*red*) and the arteries segmented after pneumoperitoneum (*blue*) are illustrated with the pig skin in transparency.

4   Conclusion

In this paper, we have presented a method to simulate the abdominal wall position after pneumoperitoneum using only a preoperative medical image and a skin tracking during the intervention. From this input data, we firstly compute the abdominal wall outer surface deformation field, which is then used to predict the inner surface position. Finally, the abdominal wall artery positions are estimated using the inner and outer surface positions. The evaluation of our prediction accuracy on two pig data, using images before and after pneumoperitoneum, have shown that our artery prediction accuracy is sufficient to provide practitioners a reasonable safety margin within 1 cm.

This abdominal wall prediction is the first step to simulate the deformation induced by a pneumoperitoneum. In future work, we will focus on the viscera deformation using the inner surface shape of the abdominal wall obtained with our method.

References

1. Sanchez-Margallo, F.M., et al.: Anatomical changes due to pneumoperitoneum analyzed by mri: an experimental study in pigs. Surg. Radiol. Anat. 33(5), 389–396 (2011)
2. Marescaux, J., et al.: Augmented-reality–assisted laparoscopic adrenalectomy. JAMA: the Journal of the American Medical Association 292(18), 2214 (2004)
3. Masamune, K., Sato, I., Liao, H., Dohi, T.: Non-metal Slice Image Overlay Display System Used Inside the Open Type MRI. In: Dohi, T., Sakuma, I., Liao, H. (eds.) MIAR 2008. LNCS, vol. 5128, pp. 385–392. Springer, Heidelberg (2008)
4. Nicolau, S.A., et al.: A cost effective simulator for education of ultrasound image interpretation and probe manipulation. Studies in Health Technology and Informatics 163, 403 (2011)
5. Saber, A.A., et al.: Safety zones for anterior abdominal wall entry during laparoscopy: a ct scan mapping of epigastric vessels. Annals of Surgery 239(2), 182 (2004)
6. Lam, A., et al.: Dealing with complications in laparoscopy. Best Practice & Research Clinical Obstetrics & Gynaecology 23(5), 631–646 (2009)
7. Geraci, G., et al.: Trocar-related abdominal wall bleeding in 200 patients after laparoscopic cholecistectomy: Personal experience. World Journal of Gastroenterology 12(44), 7165 (2006)
8. Kitasaka, T., Mori, K., Hayashi, Y., Suenaga, Y., Hashizume, M., Toriwaki, J.-i.: Virtual Pneumoperitoneum for Generating Virtual Laparoscopic Views Based on Volumetric Deformation. In: Barillot, C., Haynor, D.R., Hellier, P. (eds.) MICCAI 2004, Part II. LNCS, vol. 3217, pp. 559–567. Springer, Heidelberg (2004)
9. Shekhar, R., et al.: Live augmented reality: a new visualization method for laparoscopic surgery using continuous volumetric computed tomography. Surgical Endoscopy, 1–10 (2010)
10. Nakamoto, M., et al.: Recovery of respiratory motion and deformation of the liver using laparoscopic freehand 3d ultrasound system. Medical Image Analysis 11(5), 429–442 (2007)
11. Konishi, K., et al.: Augmented reality navigation system for endoscopic surgery based on three-dimensional ultrasound and computed tomography: Application to 20 clinical cases. International Congress Series, vol. 1281, pp. 537–542. Elsevier (2005)
12. Su, L.M., et al.: Augmented reality during robot-assisted laparoscopic partial nephrectomy: Toward real-time 3d-ct to stereoscopic video registration. Urology 73(4), 896–900 (2009)
13. Oda, M., et al.: Evaluation of deformation accuracy of a virtual pneumoperitoneum method based on clinical trials for patient-specific laparoscopic surgery simulator. In: Proceedings of SPIE, vol. 8316, p. 83160G (2012)

14. Hostettler, A., et al.: Real time simulation of organ motions induced by breathing: First evaluation on patient data. Biomedical Simulation, 9–18 (2006)
15. Nicolau, S.A., et al.: A structured light system to guide percutaneous punctures in interventional radiology. In: Proceedings of SPIE, vol. 7000, p. 700016 (2008)
16. Besl, P.J., et al.: A method for registration of 3-d shapes. IEEE Transactions on Pattern Analysis and Machine Intelligence 14(2), 239–256 (1992)

Appearance Similarity Flow for Quantification of Anatomical Landmark Uncertainty in Medical Images

Yoshitaka Masutani, Mitsutaka Nemoto, Shohei Hanaoka,
Naoto Hayashi, and Kuni Ohtomo

Department of Radiology, The University of Tokyo Hospital
masutani-utrad@umin.org

Abstract. Anatomical landmarks can play key roles in medical image under-standing including segmentation. For example, statistical shape model-based segmentation can be enhanced with landmark information which helps parame-ter initialization such as pose and locations of models. We have been working on local appearance-based landmark detection scheme. When we define land-marks with surrounding appearance in medical images, certain uncertainty is observed depending on local intensity structures around the landmark. It is ob-vious that good landmarks should have low uncertainty and also that uncertain-ty causes difficulty in consistent evaluation of landmark localization error. In this paper, we describe our method for landmark uncertainty quantification based on arrival times of level-set evolution named *appearance similarity flow*, controlled by similarity between landmark appearance and that of the location within whole image. By using 12 clinical CT dataset, the method was evaluated.

1   Introduction

In the general definition, "landmark" means natural or artificial objects which can be recognized easily and can navigate us in unfamiliar environment. Anatomical land-marks in human bodies are also defined and are labeled so that we can make corres-pondence between subjects and can know where we are looking within the part of body. Such landmarks can construct spatial knowledge on biological structures and help us in mapping between the knowledge (often called as "atlas") and a new obser-vation. In the field of morphometrics, landmarks are positively employed to measure shape and sizes of organs.   Bookstein et al. [1] classified three types of landmarks; (I) homologous and salient features such as holes and bifurcations, (II) points with less salient feature but with maximal curvature, and (III) artificial landmark such as gravi-ty center of structures or uniformly located points on profile or surface. While the type (I) is the best for the purpose of landmark, there are not so many within real bio-logical structures. Therefore, the other types; (II) and (III) are often used. Especially, the type (III) is often called as "semi-landmark".

Not only in investigating the real body of biological structures, but also in medical images, anatomical landmarks can play key roles. For example, in image segmentation based on statistical shape models (consisting of semi-landmarks) [2], initialization of

G. Bebis et al. (Eds.): ISVC 2012, Part I, LNCS 7431, pp. 12–21, 2012.

shape, pose and location guided by a few salient landmark can bring higher robustness for model fitting. Also, image acquisition range inference can be done through landmark information. Especially, considering that most of clinical images do not cover whole patient body, landmark detection can be the primary task in image understanding, which performs matching between any structural models and new image data with less coverage of the body than the model. Thus, landmarks can give us important cues for the partial match problems. For such bases, several research groups have been working on models and methods for robust detection of landmarks [3-6]. The state of the art schemes based on various features such as appearance subspace [5] achieved higher than 90 % of detection accuracy depending on the landmark set to be detected.

One of the important problems in landmark detection is uncertainty of several landmarks, which have low saliency and mainly belong to (II) or (III) in the Bookstein's types. When we observe the landmark at the liver top (see Fig.2), similar appearance can be found all over the location, and consequently punctual definition of the landmark is hard in spite of its anatomical importance. Such uncertainty causes difficulty in consistent performance evaluation of landmark detection algorithms. That is, it is less meaningful for such landmarks to investigate error between detected location and gold standard location defined manually. Instead, we need to quantify such landmark uncertainty to be considered in detection error evaluation. Quantified landmark uncertainty can serve as a criterion for choosing good landmark, and furthermore helps us to find new potential landmarks by searching low uncertainty locations in whole image data set. To the best of our knowledge, no other work addresses this problem so far because such uncertain landmarks are generally avoided to use.

The purpose of this study is not to improve or to evaluate landmark detection algorithms, but to develop a method for quantifying uncertainty of landmarks. Therefore, we introduced 3D level-set evolution with a speed function controlled by similarity to original landmark appearance, which is named as *"appearance similarity flow"* [1].

2   Methods

2.1   Appearance Similarity Flow

The idea for our landmark uncertainty quantification is based on a new definition of distance metric between two arbitrary locations in an image, based on the similarity of their local appearance. This seems to be quite natural because we must investigate the extent of the similar intensity structures surrounding the landmark to be evaluated.

We define a point landmark at location $\mathbf{r}_0$ with surrounding appearance Ω within given volumetric image data. The appearance Ω is a local subset of the image data,

---

[1] The term "similarity flow" was originally defined in the physics area by Landau & Lifshitz (Fluid Mechanics, vol. 6 of Course of Theoretical Physics. Pergamon, 2nd ed., 1995). In this paper, however, "appearance similarity flow" stands for level-set evolution controlled by appearance similarity measure as well as "curvature flow" controlled by curvature-dependent speed function.

which consists of N pairs of location vectors relative to r_0 and image signal value at the location. The shape of local domain defined for Ω is often cubic or spherical. We propagate a level-set from r_0 by using a speed function depending on local appearance in the same shape as Ω. For propagating the level-set from the starting point r_0, we compute arrival time $T(r)$ at every location r iteratively. The update of $T(r)$ is written as a discrete formulation of partial derivative equation including Eikonal equation as;

$$T(\mathbf{r}) = T(\mathbf{r}') + \frac{|\mathbf{r} - \mathbf{r}'|}{F(\mathbf{r}')}_.$$

(1)

Here, for our purpose, $F(\mathbf{r})$ is a speed function designed so that level-set proceeds fast if appearance at the location is similar to landmark appearance Ω. We used sum of absolute difference (SAD) as a dissimilarity measure, and defined $F(\mathbf{r})$ as;

$$F(\mathbf{r}) = \frac{1}{1 + \min_{q \in Q} SAD(\mathbf{r}, \Omega(\mathbf{q}))/|\Omega|}$$

(2)

where $\mathbf{q}$ denotes a quaternion and Q shows a set of various quaternions of rotations. That is, we allow appearance rotations for computation of speed function. The dissimilarity measure SAD is;

$$SAD(\mathbf{r}, \Omega) = \sum_{r_\Omega \in R_\Omega} |I(\mathbf{r} + \mathbf{r}_\Omega) - I_\Omega(\mathbf{r}_\Omega)|,$$

(3)

where $I(\mathbf{r})$ shows signal intensity at location $\mathbf{r}$ of given volumetric data, and I_Ω means intensity value within the subset of local appearance so that $I_\Omega(\mathbf{r}_\Omega)$ equals to $I(\mathbf{r}_0 + \mathbf{r}_\Omega)$. R_Ω denotes a set of relative location vectors $\mathbf{r}_\Omega$ within appearance Ω. The update of $T(\mathbf{r})$ is repeated until convergence, then we obtain arrival time map. A toy example is shown in Fig.1.

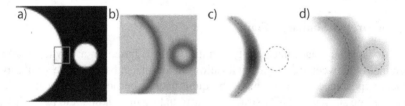

Fig. 1. Toy example of appearance similarity flow. Axial sections of a volumetric image are shown. a) Original data with cubic appearance domain around a landmark at cubic center, b) minimum SAD map, c) and d) displays of arrival time map (in different settings of contrast). Dotted red lines show the surface of the two spherical shapes.

In this simple example, a half of large sphere and an entire shape of small sphere are included in the volumetric data, and have no contact with each other. A landmark is defined at the center of a cubic appearance domain on the surface of large sphere close to small sphere as shown in Fig.1a. First, the minimum dissimilarity map is computed, which determines speed function (Fig.1b). Then, after initialization of arrival time $T(\mathbf{r}_0)=0$ at the starting point and $T(\mathbf{r})=\infty$ at any other locations, $T(\mathbf{r})$ is updated until convergence based on the equation (1) by replacing current time value with the minimum time value brought by neighbors. Implementation details are given in the section 2.4. This is a good sample for landmarks with high uncertainty because appearance identical to that of the landmark can be found everywhere on the surface of larger sphere.

2.2   Tensor Approximation of Arrival Distance Profile and "Landmark-ness"

After acquisition of arrival time map $T(\mathbf{r})$, by setting certain time threshold of $T(\mathbf{r})=\tau$, we can extract an isosurface around the landmark location $\mathbf{r}_0$, of which shape characterizes the local appearance change around the landmark. The isosurface can represent the profile of level-set propagation distance at a time τ. When we compare these profiles of landmarks with fixing τ, the smaller size of profile means that the landmark has lower uncertainty. We can quantify the size of isosurface directly by measuring volume. However, uncertainty could be anisotropic due to orientation-dependent intensity structure at the landmark. For that purpose, it is convenient to approximate the profile by tensor. We define it as "uncertainty tensor" measure at landmark location $\mathbf{r}_0$. The tensor measure $\mathbf{D}$ is a symmetric 2nd order approximation of Euclidean distances from $\mathbf{r}_0$ to the isosurface of time τ in directions $\mathbf{dr}$ ($|\mathbf{dr}|=1$) so that $D(\mathbf{dr}, \tau)=\mathbf{dr}^T\mathbf{D}(\tau)\mathbf{dr}$ as follows.

$$\mathbf{D}(\tau)=\begin{pmatrix} D_{xx} & D_{xy} & D_{zx} \\ D_{xy} & D_{yy} & D_{yz} \\ D_{zx} & D_{yz} & D_{zz} \end{pmatrix} \tag{4}$$

Because we have 6 unknown components in the tensor, we need more than 6 directions of distance measurement. Because this tensor is similar to that of diffusion tensor MR image (DT-MRI), we use the analysis techniques and the metrics for DT-MRI, which are extensively investigated [7]. For example, by using eigenvalues of the tensor; λ_1, λ_2, and λ_3 ($\lambda_1 \geq \lambda_2 \geq \lambda_3 \geq 0$), degrees of linearity, planarity, and sphericity (c_l, c_p, and c_s respectively) are used to quantify anisotropy and are defined as;

$$c_l = \frac{\lambda_1-\lambda_2}{\lambda_1}, \quad c_p = \frac{\lambda_2-\lambda_3}{\lambda_1}, \text{ and } \quad c_s = \frac{\lambda_3}{\lambda_1}. \tag{5}$$

Fractional anisotropy (FA) is also defined in a same way to quantify the direction dependency of local similarity as follows.

$$FA=\sqrt{\frac{3}{2}}\frac{\sqrt{(\lambda_1-\lambda_m)^2+(\lambda_2-\lambda_m)^2+(\lambda_3-\lambda_m)^2}}{\sqrt{\lambda_1^2+\lambda_2^2+\lambda_3^2}}, \tag{6}$$

where λ_m denotes a simple mean of three tensor eigenvalues; $(\lambda_1+\lambda_2+\lambda_3)/3$, which can be used as a general measure of landmark uncertainty in all directions. Alternatively, the mean distance (MD) below is also available.

$$MD = \frac{1}{|\mathbf{E}|}\sum_{\mathbf{dr}\in\mathbf{E}}D(\mathbf{dr},\tau) \tag{7}$$

The $\mathbf{E}$ denotes a set of distance measurement directions. The tensor approximation above assumes point symmetry of the distance profile shape. The profile, however, may have certain asymmetry. The degree of profile asymmetry can be quantified as;

$$ASYM = \frac{1}{|\mathbf{E}|}\sum_{\mathbf{dr}\in\mathbf{E}}\sqrt{\frac{\{D(\mathbf{dr},T)-D(-\mathbf{dr},T)\}^2}{D(\mathbf{dr},T)^2+D(-\mathbf{dr},T)^2}}, \tag{8}$$

In our preliminary experiments, the asymmetry property can be observed only when the time of profile observation τ is rather late. In addition to scalar metrics above, eigenvectors of the tensor measure includes orientations of maximal/minimal decrease of local similarity.

A scalar measure of "landmark uncertainty" or opposite "landmark-ness" can be defined by using the tensor measure. A natural condition of good landmarks is fast decrease of similarity in any directions. Therefore, we defined landmark-ness by;

$$LMness = \frac{C_s}{\lambda_m} = \frac{3\lambda_3}{\lambda_1(\lambda_1+\lambda_2+\lambda_3)}. \tag{9}$$

This definition depends on the uncertainty tensor $\mathbf{D}$ that has a parameter of τ to form isosurface of level-set arrival distance. The time threshold τ must be common when comparing two landmarks or computation of landmark-ness map.

2.3   Implementation

Appearance similarity flow requires computation of SAD in various rotations of landmark appearance. All the rotations are given by sets of quaternion consisting of a rotation axis and a rotation angle. Because rotation axes should be uniformly distributed in all directions, we used an assignment method by Jones, et al. [8]. We used 30 rotation axes and 12 rotation angles. The directions $\mathbf{E}$ for measurement of arrival distance measurement $D(\mathbf{dr},\tau)$ are the 26 directions to the neighbor voxels in the Cartesian coordinates.

The time threshold τ should be short enough so that the profile shape is mostly convex for tensor approximation. We empirically determined the time so that the minimum arrival distance in tensor approximation λ_3 has 5 voxels at least.

Appearance similarity flow is also a sort of level-set evolution and requires even higher computational cost due to consideration of appearance rotation. We employed a parallel computing by GPGPU [9] with the fast marching method by narrow band [10]. With the implementation, we can obtain landmark-ness map in a few hours per CT data set in 128x128x150-170 matrix size.

3   Results

We performed 2 types of experiments for validating our methods. One is quantification of typical three landmarks, of which properties are qualitatively and empirically known. The other is landmark-ness map acquisition to find new landmarks potentially useful for medical image understanding. The landmark-ness map is obtained by computation of the uncertainty tensor at all the voxels within the given volumetric data.

3.1   Quantification of Typical Landmarks

By using clinical X-ray CT images of 12 patients, we computed the landmark uncertainty tensors and derived measures for 3 typical landmarks including; (1) the spinous process of vertebra, (2) the iliac crest top, and (3) the liver top (Fig.2). No contrast agent was used for data acquisition, and the image resolution was reduced in half size. Spherical appearance shape with diameter of 15 mm is used.

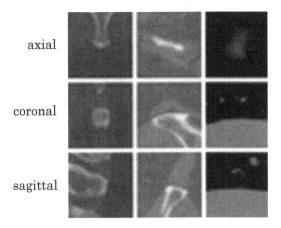

Fig. 2. The three typical landmarks in CT images. From left, (1) spinous process of vertebra, (2) iliac crest top, and (3) liver top. The landmarks are defined at the center of each image.

Table 1 shows the statistics of the uncertainty-related measures. As expected from landmark appearance and our empirical understanding, the three anisotropy measures reflects the characteristics of the landmarks (Fig.3). The spinous process of vertebra has the least uncertainty and its type of anisotropy is spherical due to its prominent shape. The iliac crest top has also similar properties but with more linear anisotropy characteristics. It is natural because it is located on thin ridge on the ilium. The liver top of smooth surface has high planarity. It is revealed that the FA measure may not exactly reflect the characteristics of uncertainty because thin planar tensor may have high FA values than moderately-thin linear tensor. Among the three landmarks, no significant difference was observed in asymmetry measure. Those results show that the uncertainty measures match our natural impression of landmarks including difficulty in punctual determination.

Table 1. Mean and standard deviation values of uncertainty-related measures calculated on the three major anatomical landmarks of 12 patients

	spinous process of vertebra	iliac crest top	liver top
C_l	0.07 ± 0.02	0.19 ± 0.03	0.11 ± 0.07
C_p	0.24 ± 0.04	0.22 ± 0.06	0.68 ± 0.11
C_s	0.60 ± 0.04	0.59 ± 0.05	0.21 ± 0.05
FA	0.18 ± 0.03	0.26 ± 0.04	0.55 ± 0.04
MD	1.28 ± 0.46	1.23 ± 0.16	2.31 ± 0.37
ASYM	0.07 ± 0.02	0.05 ± 0.01	0.05 ± 0.02
LM-ness	0.59 ± 0.17	048 ± 0.08	0.10 ± 0.04

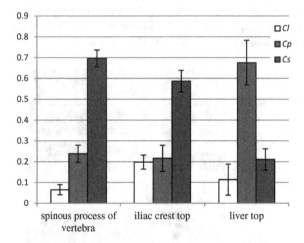

Fig. 3. Statistics of anisotropic type measures (C_s, C_l, C_p) for 3 typical landmarks

3.2   Landmark-Ness Map Acquisition

By computing landmark-ness measure of the equation (9) at all the voxels within volumetric image data, we obtained landmark-ness map for each subject. The image resolution was reduced in quarter size. We used a spherical shape of appearance with diameter of 15 mm. Figure 4 shows coronal and sagittal sections of a landmark-ness map example for a single subject. In the right column, high landmark-ness areas are superimposed in red color on the original CT image. Note that the two sections are in different mapping scales of landmark-ness and the sagittal section is in higher contrast.

In the coronal section, we can find the iliac crest top shown by green arrows as high landmark-ness area. This fact corresponds to our experience that it is a good landmark. Also, the ilium-femur joint, the major vasculature inside the lungs, and the ribs indicate high landmark-ness. Also in the sagittal view, we can find high land-mark-ness areas along the vertebral canal (green arrows), which actually spread to joints between thoracic vertebrae and ribs with saddle-type surface of bones. It is surprising that the areas have even higher landmark-ness than the tips of the spinous process of vertebrae, which are not highlighted with the mapping scale. Therefore, we may be able to define new landmarks in these areas around the vertebral canal. The yellow arrows in the both sections show high landmark-ness areas (osteophyte at the vertebral bodies, and air in the small intestine), which are subject-specific features, and are discussed later.

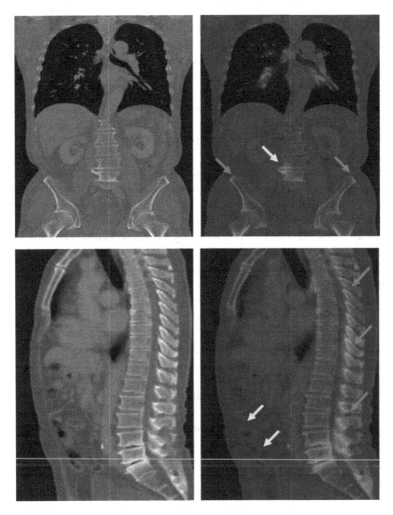

Fig. 4. Example of landmark-ness map in X-ray CT data set of single subject. Left: original CT images, and Right: with superimposed landmark-ness in red color. Top: coronal sections, and Bottom: Sagittal sections. Green arrows show the typical landmark structures such as iliac crest top, while yellow ones show patient-specific features including osteophyte at the vertebral bodies. Different scales are used among the two types of sections for landmark-ness mapping.

4   Discussion and Summary

Through the experiments, the appearance similarity flow is shown to be a powerful tool for quantitative analysis of anatomical landmarks. As mentioned earlier, the tool was introduced not for improvement or evaluation of landmark detection algorithms, but for choosing good landmarks. In other words, the most important contribution of this work is that a systematic method for evaluating characteristics of anatomical landmarks was established. The method can be applied to various applications in medical image analysis. For instance, in our landmark detection method [5,6] treating more than 150 landmarks within body torso, ranking the landmarks by uncertainty and subsequent hierarchical detection scheme (high landmark-ness first) are expected to improve the detection performance because detection errors are often caused by uncertain landmarks.

Analyses similar to our landmark uncertainty quantification seem to be possible through simple extension of feature detectors in computer vision [3, 11]. For example, in our preliminary studies, some types of landmarks showed high correlation between shape index [12] and landmark-ness, while the others not. Among anatomical landmarks, there are several different types of intensity structures such as bifurcation of tree-like objects and prominent structures of bone, on which optimal detector depends. Therefore, we currently choose a unified approach by using local appearance [5]. Comparisons with the other feature detectors are in our scope of future work.

Appearance similarity flow can be also used for definition of non-Euclidean distance metric between two points within image. That is, we can infer whether the two points belong to same structure or not, by using the metric because appearance similarity flow evolves fast only if the similarity is contiguous between the two points. Therefore, faster arrival implies connectivity between the points. This characteristic may be applied in graph-cut segmentation of organ shapes and structures [13].

The landmark uncertainty in our definition is due to internal cause of anatomical structure, while Baka et al. [14] investigated uncertainty due to external factors such as image noise and artifacts. Such external uncertainty of semi-landmarks in shape models critically influences accuracy of regression-based fitting. The external cause should be also considered when we evaluate landmark characteristic because one of the important limitations of similarity flow is sensitivity to noise. With more noise, similarity attenuates faster. In such situations, uncertainty quantification may not yield significant results. Therefore, seeking more robust measure for speed function is in our scope and is included in the future work.

Naturally, one of the important requirements for good landmarks should be that they should be found in common for all subjects except for defect due to pathology or anomaly [6]. Therefore, our next step is to find commonly high landmark-ness area among subjects. This can be simply realized with registration of many samples of landmark-ness map. On the other hand, the subject-specific regions with high landmark-ness are not in the category of good landmakrs. Instead, such subject's own landmarks may contribute to individual identification.

Based on medical expert's advice, we have mainly located landmarks at the tips of bones and other convex structures, which are basically with anatomical labels. However, the landmark-ness map by this study showed us new potential landmarks along the vertebral canal, which are not prominent shape. Generally, human visual perception with the naked eye may tend to focus on convex features more. As known well as the term "figure and ground organization" [15] in the field of psychology, identical structure may bring us different perception when the image intensity is inverted. Thus, with landmark-ness map, we may find new non-convex landmarks still without anatomical labels for computerized medical image understanding.

Acknowledgement. This work is a part of the research project "Computational Anatomy for Computer-aided Diagnosis and therapy: Frontiers of Medical Image Sciences", supported by a grant-in-aid for scientific research on innovative areas MEXT, Japan.

References

1. Bookstein, L.: Morphometric Tools for Landmark Data: Geometry and Biology. Cambridge University Press (1992)
2. Heimann, T., Meinzer, H.: Statistical shape models for 3D medical image segmentation: A review. Medical Image Analysis 13(4), 543–563 (2009)
3. Rohr, K.: Landmark-Based Image Analysis: Using Geometric and Intensity Models. Springer (2001)
4. Criminisi, A., Shotton, J., Robertson, D., Konukoglu, E.: Regression Forests for Efficient Anatomy Detection and Localization in CT Studies. In: Menze, B., Langs, G., Tu, Z., Criminisi, A., et al. (eds.) MICCAI 2010. LNCS, vol. 6533, pp. 106–117. Springer, Heidelberg (2011)
5. Nemoto, M., et al.: A unified framework for concurrent detection of anatomical landmarks for medical image understanding. In: Proc. SPIE, vol. 7962, pp. 79623E–79623E13 (2011)
6. Hanaoka, S., et al.: Probabilistic Modeling of Landmark Distances and Structure for Anomaly-proof Landmark Detection. In: Proc. the Third Int. Workshop on Mathematical Foundations of Computational Anatomy 2011, pp. 159–169 (2011)
7. Westin, C.F., et al.: Processing and visualization of diffusion tensor MRI. Med. Img. Anal. 6(2), 93–108 (2002)
8. Jones, D.K., et al.: Optimal strategies for measuring diffusion in anisotropic systems by magnetic resonance imaging. Magn. Reson. Med. 42, 515–525 (1999)
9. NVIDIA developer zone, http://developer.nvidia.com/
10. Parker, G.J., et al.: Estimating distributed anatomical connectivity using fast marching methods and diffusion tensor imaging. IEEE Trans. Med. Img. 21(5), 505–512 (2002)
11. Tuytelaars, T., Mikolajczyk, K.: Local Invariant Feature Detectors: A Survey. Foundations and Trends in Computer Graphics and Vision 3(3), 177–280 (2007)
12. Koenderink, J.J., van Doorn, A.J.: Surface shape and curvature scales. Image Vision Comput. 10, 557–565 (1992)
13. Boykov, Y., et al.: Fast approximate energy minimization via graph cuts. IEEE Trans. Pattern Anal. Mach. Intell. 23(11), 1222–1239 (2001)
14. Baka, N., Metz, C., Schaap, M., Lelieveldt, B., Niessen, W., de Bruijne, M.: Comparison of Shape Regression Methods under Landmark Position Uncertainty. In: Fichtinger, G., Martel, A., Peters, T. (eds.) MICCAI 2011, Part II. LNCS, vol. 6892, pp. 434–441. Springer, Heidelberg (2011)
15. Rubin, E.: Figure and Ground. In: Yantis, S. (ed.) Visual Perception, pp. 225–229. Psychology Press (2001)

Segmentation of Brain Tumors in CT Images Using Level Sets

Zhenwen Wei[1], Caiming Zhang[1,2,*], Xingqiang Yang[1], and Xiaofeng Zhang[1]

[1] School of Computer Science and Technology, Shandong University, Jinan, China
[2] Shandong Province Key Lab. of Digital Media Technology,
Shandong University of Finance and Economics, Jinan, China

Abstract. This paper proposes an approach based on level sets to segment brain tumors from CT images. Combining edge information with region information dynamically, the novel method introduces a new energy function model, which will make the initial contour evolve towards the desirable boundary while not leak at weak edge positions. In addition, re-initialization of the evolving level set function is avoided by introducing a new simple regularization term, which can eliminate radical changes of level set function(LSF) far away from the contour, and make the LSF prone to be a signed distance function around the contour as well. Experimental results demonstrate that the proposed method performs well on CT images, and can segment brain tumors exactly.

1 Introduction

At present, more and more people are suffering from tumors, which have attracted much attention, especially brain ones. Currently, there are two major medical imaging diagnostic techniques: MRI and CT, and in this paper we will investigate brain tumor segmentation from CT images, for exact segmentation of tumors is very important for doctors to diagnose patients' condition and to perform operations.

As we know, there are many methods for image segmentation, of which level sets is an important one and has been applied widely. In level-set-based methods, the given image is divided into two parts: object and background, and the contour of object is represented by zero level set of a three dimension level set function(LSF). As the LSF evolves in the procedure of minimizing one energy function, the zero level set is also updated[1]. As a result, the energy function plays a crucial role in level-set-based segmentation methods.

Existing level set energy function models can be categorized into two major classes: edge-based models and region-based ones, and there have been some well-known energy function models of the two kinds proposed to solve kinds of problems. In [2], Caselles presented a geodesic active contour model based on gradient information of a given image, in which the energy function was close to the minimal value when the zero level set located at the positions with large

---

* Corresponding author.

G. Bebis et al. (Eds.): ISVC 2012, Part I, LNCS 7431, pp. 22–31, 2012.

gradient. However, periodic re-initialization of LSF to signed distance function is requested in geodesic active contour model. Aiming at this, Li proposed a new level set evolution approach without re-initialization by adding a regularization term, simplifying the level set evolution procedure and improving the computational efficiency[3]. The above two energy functions are edge-based models, yet they perform poorly when the object boundary is too smooth. To solve this problem, Chan and Vese proposed a region-based energy function model without edge information[4], which can detect object with weak boundaries. The model considers the image region information as a whole, but it neglects the influence of local information. As a result, it would not segment images with intensity inhomogeneities. Based on this, Li proposed a new energy function model which emphasizes the local window information[5] and can overcome the limitation in paper [4]. So far, how to avoid re-initialization of the LSF is still a difficult problem in level set evolution, for re-initialization not only raises serious problems as when and how it should be performed, but also affects numeric accuracy and increases computational cost in an undesirable way. In order to solve this problem, Li proposed a distance regularized level set evolution method by introducing a new regularization term to the energy function models[6], which could eliminate radical changes of the LSF far away from the active contour, and make the LSF close to a signed distance function around the contour. Although the methods mentioned above can solve some kinds of segmentation problems, they cannot obtain desirable segmentation results to complex medical images in most of the time, especially for precise segmentation of some organs or tissues. Therefore, the medical image segmentation is still an important issue to be investigated and we will address the segmentation of brain tumors in CT images using level sets.

The rest of this paper is organized as follows. Section 2 proposes a new energy function model combining edge information with region information. Also in this section, we introduce a simplified version of the distance regularized level set evolution, which can avoid re-initialization procedure of the LSF and speed up the level set evolution. In Section 3, we perform the proposed schema to segment brain tumors in a group of CT images, which illustrates the advantages and effectiveness. A brief conclusion is presented in Section 4.

2   Proposed Method

This section will introduce the proposed level-set-based method for brain tumor segmentation, which can be implemented by minimizing one energy functional defined as[7]:

$$E(\phi) = E_{edge}(\phi) + E_{region}(\phi) + R(\phi) \tag{1}$$

in which ϕ is the LSF, and $E_{edge}(\phi), E_{region}(\phi), R(\phi)$ are edge-based energy, region-based energy and regularization term respectively. In the proposed energy function, the three terms play different roles, that is, $E_{edge}(\phi)$ is the main driving force in level set evolution, which will detect object boundaries with large gradient, $E_{region}(\phi)$ will ensure the contour not leak easily at weak edge positions, and $R(\phi)$ will make the LSF evolve more stably and more predictably.

2.1 Edge-Based Energy

In this work the contour C is implicitly indicated by the zero level set, i.e. $C(t) = \{x|\phi(t, x) = 0\}$. For a given image $I(x)$, let Ω represent the whole image domain. Consequently, the zero level set can separate Ω into two regions: Ω_1 for $\phi(x) > 0$ inside C and Ω_2 for $\phi(x) < 0$ outside C.

As we know, the general edge-based energy can be defined as[2, 3]:

$$E_{edge}(\phi) = \mu \int_{\Omega} g\delta(\phi)|\nabla\phi|dx + \nu \int_{\Omega} gH(\phi)dx \qquad (2)$$

where μ and ν are constants, $H(\phi)$ and $\delta(\phi)$ are the Heaviside function and Dirac delta function respectively, and g, the edge-detector, can be defined by

$$g = \frac{1}{1 + |\nabla G_\sigma * I(x)|^2} \qquad (3)$$

in which G_σ is the Gaussian kernel, producing a smooth version of the image $I(x)$.

In (2), the two terms represent the length of the contour and the area of the region Ω_1 inside C with the weighted factor g respectively, and the choice of parameter ν should depend on initial contour position. In other words, when the initial contour is placed outside the object, ν should be assigned positive so that the contour can shrink faster, while negative when inside the object so as to speed up the expansion of the contour. The edge-based energy $E_{edge}(\phi)$ will be close to minimal value when the contour locates at large gradient positions, that is to say, the edge-based energy model can obtain final object contour defined by gradient.

2.2 Region-Based Energy

The region-based energy model can detect objects whose boundaries do not necessarily have large gradient, and also it is robust to noise. Based on previous work[4, 5, 8–10], this paper combines global information inside the contour C with local information around C and defines the region-based energy by

$$E_{region}(\phi) = \alpha \int_{\Omega} log\sigma_1^2 H(\phi)dx + \int_{\Omega} \delta(\phi)\varepsilon_x(\phi)dx \qquad (4)$$

in which

$$\mu_1 = \frac{\int_{\Omega} I(x)H(\phi)dx}{\int_{\Omega} H(\phi)dx} \qquad (5)$$

$$\sigma_1^2 = \frac{\int_{\Omega} |I(x) - \mu_1|^2 H(\phi)dx}{\int_{\Omega} H(\phi)dx} \qquad (6)$$

$$\varepsilon_x(\phi) = \lambda_1 \int_{\Omega_1} K_\sigma(x-y)|I(y) - f_1(x)|^2 dy \\ +\lambda_2 \int_{\Omega_2} K_\sigma(x-y)|I(y) - f_2(x)|^2 dy \tag{7}$$

where $\alpha, \lambda_1, \lambda_2$ are constants, K_σ is a Gaussian kernel with standard deviation σ, which is the scale parameter to control the size of the local window centered at the point x, and y represents those pixels in the local window. $f_1(x)$ and $f_2(x)$ approximate the average intensities inside C and outside C in the local window, which are defined as follows.

$$f_1(x) = \frac{K_\sigma * [I(x)H(\phi)]}{K_\sigma * H(\phi)} \tag{8}$$

$$f_2(x) = \frac{K_\sigma * [I(x)(1 - H(\phi))]}{K_\sigma * (1 - H(\phi))} \tag{9}$$

Commonly, pixels in one object have similar intensities, so there should be a tendency of minimizing the variance in region Ω_1, which can effectively eliminate the possible boundaries leakage due to smooth or discrete edge of the object when we only use edge-based energy defined in Section 2.1. As the boundaries leakage occurs, the pixels' intensities inside the active contour often change drastically, which will obviously increase the value of the variance. This is the main source of the first term in (4). At the same time, in the procedure of brain tumor segmentation from CT images, the pixels around the active contour play a dominant role in detecting the tumor, while the effect of the pixels far away from the contour can be almost neglected. Therefore, local information should be constrained to the pixels only around the contour, which is computed in the second term of (4) by adding $\delta(\phi)$. This is different from the method in [5], in which the local information is retrieved from all pixels of the given image and cannot be applied for the tumor segmentation.

2.3   Regularization Term

In order to avoid re-initialization of the LSF, Li introduced a distance regularization term to the energy function model[6], which can eliminate radical changes of the LSF far away from the active contour, and make the LSF close to a signed distance function around the contour as well. In this paper, a simplified version of the distance regularization term will be proposed and defined as follows.

$$R(\phi) = \rho \int_\Omega P(|\nabla\phi|)dx \tag{10}$$

where $\rho > 0$ is a constant, and $P(|\nabla\phi|)$, the potential function, is defined by

$$P(|\nabla\phi|) = \begin{cases} 0 & |\nabla\phi| < \frac{1}{2} \\ \frac{1}{2}(|\nabla\phi| - 1)^2 & |\nabla\phi| \geq \frac{1}{2} \end{cases} \tag{11}$$

For $|\nabla\phi| \geq 1/2$, the new regularization term forces $|\nabla\phi|$ prone to be 1, while for $|\nabla\phi| < 1/2$, it has no effect on the LSF. In other words, if we use the binary initial method to be described in Section 2.4, the new regularization term will penalize the LSF close to a signed distance function only in the narrow band around the evolving contour[6]. This is another expression of the fact that the pixels around the active contour play the dominant part in the segmentation procedure.

2.4   Implementation

In order to minimize the total energy functional in (1), gradient descent flow is used, which is addressed by

$$\frac{\partial\phi}{\partial t} = -\frac{\partial E}{\partial\phi} \tag{12}$$

Then we can figure out the level set evolution formulation, which will drive the motion of the active contour towards the desirable positions.

$$\frac{\partial\phi}{\partial t} = \delta(\phi)(F_1 + F_2) + F_3 \tag{13}$$

in which

$$F_1 = \mu\,\mathrm{div}(g\frac{\nabla\phi}{|\nabla\phi|}) - \nu g \tag{14}$$

$$F_2 = -\alpha\log\sigma_1^2 - \delta(\phi)(\lambda_1 e_1 - \lambda_2 e_2) \tag{15}$$

$$e_1(x) = \int_\Omega K_\sigma(y - x)|I(x) - f_1(y)|^2 dy \tag{16}$$

$$e_2(x) = \int_\Omega K_\sigma(y - x)|I(x) - f_2(y)|^2 dy \tag{17}$$

$$F_3 = \begin{cases} 0 & |\nabla\phi| < \dfrac{1}{2} \\ \rho[\nabla^2\phi - \mathrm{div}(\dfrac{\nabla\phi}{|\nabla\phi|})] & |\nabla\phi| \geq \dfrac{1}{2} \end{cases} \tag{18}$$

In numerical calculation, the Heaviside function $H(\phi)$ and Dirac delta function $\delta(\phi)$ are defined as[3, 4, 6]:

$$H_\varepsilon(z) = \begin{cases} 1 & z > \varepsilon \\ 0 & z < -\varepsilon \\ \frac{1}{2}[1 + \frac{z}{\varepsilon} + \frac{1}{\pi}\sin(\frac{\pi z}{\varepsilon})] & |z| \leq \varepsilon \end{cases} \tag{19}$$

$$\delta_\varepsilon(z) = \begin{cases} 0 & |z| > \varepsilon \\ \frac{1}{2\varepsilon}[1 + \cos(\frac{\pi z}{\varepsilon})] & |z| \leq \varepsilon \end{cases} \tag{20}$$

where $\delta_\varepsilon(z)$ is the derivative of $H_\varepsilon(z)$, i.e. $\delta_\varepsilon(z) = \dfrac{dH_\varepsilon(z)}{dz}$. The default value of parameter ε is assigned 1.5.

In our method, the LSF $\phi(x)$ is initialized to be not a signed distance function, but a simple binary function, which is defined by[3, 5, 6]:

$$\phi_0(x) = \begin{cases} c_0 & \text{if } x \in \Omega_0 \\ -c_0 & \text{otherwise} \end{cases} \tag{21}$$

where $c_0 > 0$ is a constant, which will influence the width of the narrow band around the evolving contour[6], and Ω_0 is a subregion in the image domain Ω for the initial object contour. In practice, we set $c_0 = 2$, and Ω_0 is defined as rectangles.

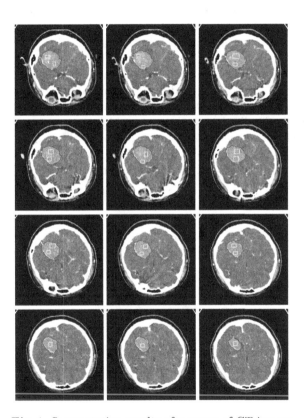

Fig. 1. Segmentation results of a group of CT images

Finally, we get the general steps of the algorithm:

Step 1. Initialize the level set function, i.e. ϕ_n, $n = 0$.

Step 2. Compute $F_1(\phi_n)$, $F_2(\phi_n)$ and $F_3(\phi_n)$ by the formulas (14), (15) and (18) respectively.

Step 3. Solve the partial differential equation (13) which controls the level set evolution, to obtain ϕ_{n+1}.

Step 4. Check whether the solution has converged. If not, $n = n + 1$, goto *Step* 2 and repeat.

Step 5. The zero level set of the function ϕ_{n+1} is the final segmentation result.

3   Experimental Results

The proposed method is used to segment a brain tumor in a group of CT images, which have a resolution of 200×200. The parameters are assigned the same values: $\mu = 300, \alpha = 1.0, \lambda_1 = \lambda_2 = 1.0, \rho = 0.04$ and the time step is set 1. In our experiments, the initial contours are defined inside the brain tumors by one or two rectangles, and initial and final contours are colored in yellow and red respectively. As a result, the parameter ν should be assigned a negative value so that the contours can expand faster to the object boundaries. In subsequent experiments, the parameter ν is set -75 for all images, and encouraging results are retrieved, which will be presented in the following figures.

From *Fig.*1, it can be seen that the precise contours of one brain tumor are obtained in CT slice by slice images, based on which we can reconstruct the three-dimension model, shown in *Fig.*2. In particular, this three-dimension model of the brain tumor is very important for doctors to diagnose the current condition, and then to carry out surgery for the possible removal.

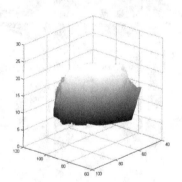

Fig. 2. Three-dimension model of the brain tumor

Additionally, the proposed schema is performed on four other images to segment brain tumors, the results of which are represented in *Fig.*3. In this figure, the first column are the initial contours, the second one are the middle results after several iterations and the third are the final contours respectively.

In order to illustrate the advantages of our method, some comparative experiments have been carried out, and the results are presented in *Fig.*4. As can be seen from this figure, the active contours of the edge-based model in [3] often leak at the positions with weak or discrete edge, while the contours of the local binary fitting model[5] based on window region information are usually redundant, and cannot only locate specified tumors. But our method is able to deal with these problems, and obtains desirable segmentation results. To sum up, this paper aims to propose a robust and accurate algorithm to segment organs or tissues in medical images, especially for brain tumors in CT images.

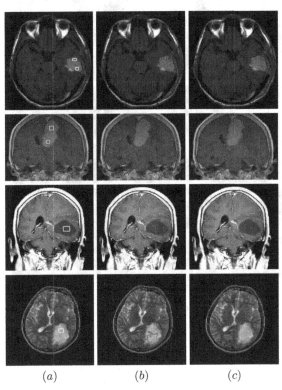

(a) $\qquad$ (b) $\qquad$ (c)

Fig. 3. Four other segmentation results retrieved by the proposed method. Column (a): Initial contours. Column (b): Middle results after several iterations. Column (c): Final contours.

4 Conclusion

This paper proposes a segmentation method of brain tumors from CT images using level sets, which has two advantages: (1)the novel model considers edge information and region information dynamically, which will make the initial contour evolve towards the desirable result while not leak at weak edge positions; (2)a new regularization term is introduced to avoid the re-initialization of the LSF and make the level set evolution more stable and more predictable. In our future work, we will continue to investigate the segmentation method based on level sets, for the results depend on the parameter assignment to some extent.

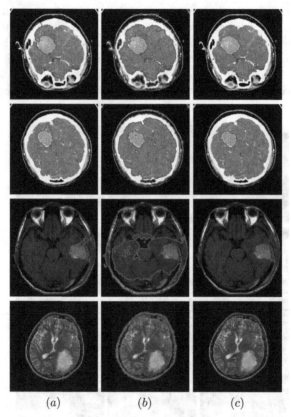

(a) (b) (c)

Fig. 4. Comparison of our method and classical methods. Column (a): Edge-based model in [3]. Column (b): Local binary fitting model in [5]. Column (c): Our method.

Acknowledgements. This work is supported by the National Nature Science Foundation of China (61020106001, 60933008, 61103150) and National Research Foundation for the Doctoral Program of Higher Education of China (20110131130004).

References

1. Osher, S., Fedkiw, R.: Level Set Methods and Dynamic Implicit Surfaces. Springer (2002)
2. Caselles, V., Kimmel, R., Sapiro, G.: Geodesic active contours. Int. J. Comput. Vis. 22, 61–79 (1997)
3. Li, C., Xu, C., Gui, C., Fox, M.: Level set evolution without re-initialization: a new variational formulation. In: Proc. IEEE Conf. Comput. Vis. Pattern Recognit., pp. 430–436 (2005)
4. Chan, T., Vese, L.: Active contours without edges. IEEE Trans. Image Process. 10, 266–277 (2001)

5. Li, C., Kao, C., Gore, J., Ding, Z.: Minimization of region-scalable fitting energy for image segmentation. IEEE Trans. Image Process. 17, 1940–1949 (2008)
6. Li, C., Xu, C., Gui, C., Fox, M.: Distance regularized level set evolution and its application to image segmentation. IEEE Trans. Image Process. 19, 3243–3254 (2010)
7. Chen, S., Radke, R.: Level set segmentation with both shape and intensity priors. In: Proc. IEEE Int. Conf. Comput. Vis., pp. 763–770 (2009)
8. Tsai, A., Yezzi Jr., A., Wells, W., Tempany, C., Tucker, D., Fan, A., Grimson, W., Willsky, A.: A shape-based approach to the segmentation of medical imagery using level sets. IEEE Trans. Med. Imag. 22, 3243–3254 (2003)
9. Cremers, D., Rousson, M., Deriche, R.: A review of statistical approaches to level set segmentation: Integrating color, texture, motion and shape. Int. J. Comput. Vis. 72, 195–215 (2003)
10. Yu, Y., Zhang, C., Wei, Y., Li, X.: Active Contour Method Combining Local Fitting Energy and Global Fitting Energy Dynamically. In: Zhang, D., Sonka, M. (eds.) ICMB 2010. LNCS, vol. 6165, pp. 163–172. Springer, Heidelberg (2010)

Focal Liver Lesion Tracking in CEUS for Characterisation Based on Dynamic Behaviour

Spyridon Bakas[1], Andreas Hoppe[1], Katerina Chatzimichail[2],
Vasileios Galariotis[2], Gordon Hunter[1], and Dimitrios Makris[1]

[1] Digital Imaging Research Centre, School of Computing and Information Systems,
Faculty of Science, Engineering and Computing, Kingston University, Penrhyn Road,
Kingston-upon-Thames, Surrey, KT1 2EE, London, United Kingdom
{s.bakas,a.hoppe,g.hunter,d.makris}@kingston.ac.uk
[2] Radiology & Imaging Research Center, University of Athens, Evgenidion Hospital,
Papadiamantopoulou Street 20, T.K. 115 28, Athens, Greece
katerina@hcsl.com, vgalariotis@hotmail.com

Abstract. This paper presents a methodology for tracking a hypo- or hyper-enhanced focal liver lesion (FLL) and a healthy liver region in a video sequence of a Contrast-Enhanced Ultrasound (CEUS) examination. The outcome allows the differentiation between benign and malignant cases, by characterising FLLs of typical behaviour, according to their Time-Intensity curves. The task is challenging mainly due to intensity changes caused by contrast agents. Initially the ultrasound mask is automatically localised and then the FLL and parenchyma regions are tracked, assuming affine transformations on the image plane, employing the point-based registration technique of Lowe's scale-invariant feature transform (SIFT) keypoints detector. Finally, a quantitative evaluation of the tracking process provides a confidence measure for the characterisation decision.

1 Introduction

Focal liver lesions (FLLs) refer to a particular condition of hepatic disease, which is the fifth largest cause of death in the UK [1]. FLLs are nodules foreign to the liver anatomy that can be either relatively harmless (benignities), or progressively worsening that can potentially result in death (malignancies). There is much interest from clinicians in the potential for the early distinction of a malignant FLL from a benign one as the former, if diagnosed sufficiently early (i.e. in a premature state), can be healed without performing any surgical operation.

Contrast-Enhanced Ultrasound (CEUS) is an attractive imaging modality for the visualisation of FLL candidates as the equipment it requires, when compared with CT or MRI, is relatively inexpensive [2] and portable, allowing its use in any operating room. CEUS has recently gained acceptance for the detection and characterisation of very small FLLs [3]. CEUS requires the intravenous injection of microbubble contrast agents, offering an enhancement to the brightness

G. Bebis et al. (Eds.): ISVC 2012, Part I, LNCS 7431, pp. 32–41, 2012.

Table 1. Signal of FLL VSs during the phases of a CEUS examination. The "+" sign means that the FLL is brighter than the parenchyma and the "-" sign the reverse.

VASCULAR SIGNATURE	Hyper-enhanced		Hypo-enhanced	
	Unipolar. ID:(a)	Bipolar. ID:(b)	Unipolar. ID:(c)	Bipolar. ID:(d)
SIGNAL				
TYPICAL BEHAVIOUR OF	Benign FLL (e.g.Haemangioma, Adenoma, Focal Nodular Hyperplasia)	Malignant FLL (e.g. Hepatocellular Carcinoma, Metastasis)	Benign FLL (e.g. Nodules, Cysts)	Benign FLL

intensity of blood flow in an image. The modality's effectiveness, in terms of diagnostic accuracy for evaluation of malignant FLLs exceeds 95%, according to studies by radiologists [4]. However, few radiologists have been trained to apply this modality and interpret its visual cues.

1.1 CEUS Examination

The examination comprises three phases, whose duration vary depending on the physiology of the patient's liver and heart. The total duration of an examination lasts a maximum of 10 minutes, after which a diagnosis can be made [5]. The first phase is characterised by the enrichment of the FLL and the healthy liver area (parenchyma) due to the inflow of the contrast medium. During the second phase, the flow of the medium is stabilised, resulting in no relative change in brightness. Finally, the third phase is characterised by the outflow of the medium.

There are two main vascular signatures (VSs) of FLLs: the hyper-enhanced and the hypo-enhanced [Table 1]. VSs describe the way that an FLL dynamically behaves in comparison to the parenchyma. During the first phase, the hyper-enhanced FLLs are enriched prior to the parenchyma, while in the hypo-enhanced category the parenchyma is enriched first. FLLs whose VS is hyper-enhanced throughout the three phases are called unipolar hyper-enhanced. In contrast, bipolar hyper-enhanced are those where the contrast medium outflows from the FLL prior to the parenchyma during the third phase. If the VS of an FLL is hyper-enhanced, it is mainly in the third phase where its type (benign or malignant) is identified [6], whereas if the VS is hypo-enhanced then the FLL's type is identified as benign from the first phase [7]. The signal of an FLL's VS describes the FLL's dynamic behaviour. Different signals are linked to different medical conditions. Therefore assigning one of the four different signals [Table 1] to an FLL, is of particular medical importance and assists the radiologist make a reliable diagnosis of an FLL of typical behaviour.

1.2 Challenges of Image Analysis

Accurate and reliable tracking of an FLL poses a very challenging task due to significant changes in its apparent 2D size, shape, intensity and motion over

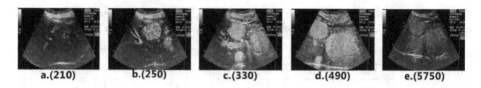

Fig. 1. Significant changes in the appearance of the liver and the FLL during a CEUS examination. In this particular case, the FLL shape is initialised at frame 250 (b).

the course of the examination (Fig.1). In addition, CEUS imagery is often of low quality, due to noise generated by the propagation of the US waves in soft tissue. This low signal-to-noise ratio, as well as the poor definition of boundaries in the US image makes processing it more complex than images acquired with other imaging modalities (e.g. CT, MRI) [8].

Instability of the clinician's hand holding the transducer and the physiological motion of the patient are additional factors that affect the location of the FLL within an image. Physiological motion refers to the combination of cardiac and breathing motion occurring due to the patient's physiopathology (e.g. variability of heart rate, irregular breathing patterns). This, as well as motion of the inner human organs, affect the motion of the devices and consequently the acquisition of the video, resulting in potential out-of-plane movement. During this movement, a radiologist normally attempts the manual stabilisation of the US target's view within the obtained image plane (x, y) by changing the elevation axis (z) of the US probe, but thus introducing the issue of the FLL's dispersion in depth. Furthermore, the continuous irregular repetitiveness of all these disturbances inevitably degrade the quality of the acquired data.

1.3 Related Work

Shiraishi et al. [9] developed a computer-aided diagnostic scheme for distinguishing between only three specific FLL types of hyper-enhanced VS (i.e. Metastasis, Haemangioma and Hepatocellular Carcinoma). Segmentation of the FLLs was obtained by manual annotation of all images processed, by a physician. Then, an Artificial Neural Network was used for the classification of FLLs.

Goertz et al. [6] used the Sonoliver software (Tomtec, Germany) to quantify FLLs. The software allows for semi-automatic motion compensation by combination of manual segmentation of FLLs and parenchyma, background subtraction, and automatic alignment of ROIs, although no specific technical details are given. The outcome of the study [6] implies that the software could only work with data obtained according to particular acquisition standards (e.g. stability of the clinician's hand, no patient's irregular breathing patterns, no FLL dispersion in depth) where motion is minimal.

Bakas et al. [10] developed a histogram-based tracking approach for characterising FLLs of only hyper-enhanced VS. In contrast with [6] and [9], [10] applies a motion tracking approach to the region of interest (ROI) to avoid the need for their manual alignment. This approach is mainly based on intensity histograms of the FLL region and also utilises Lowe's scale-invariant feature transform (SIFT) keypoints detector [11]. However, a histogram-based motion tracking approach may suffer because the intensities of the FLL and parenchyma regions become similar towards the end of the first phase and during the second phase. Instead, an automatic motion tracking method based on affine transformations of salient points is proved to be more reliable (see Section 2.2).

2   Methodology

The proposed system deals with both VSs of FLLs and attempts to distinguish the type of a FLL (benign or malignant) based on the temporal profile (dynamic behaviour) of its brightness intensity in the image sequence. Robust tracking of two ROIs (FLL and parenchyma) is important, in order to accurately obtain intensity information from the exact areas and therefore provide the ability to characterise with confidence the typical behaviour of FLLs from their dynamic behaviour, while taking into consideration the four main types of VS signals. After obtaining the ground truth (GT) and the decision of the system for these ROIs, a quantitative evaluation of the tracking is carried out, in order to demonstrate that the system can be used as a valid second-opinion tool for a radiologist.

The system employs the point-based registration technique of SIFT [11] to track FLL and parenchyma contours within the conical area covering the ultrasonographic image (US mask), whilst overcoming challenges mentioned in Section 1.2. The derived contours are combined with the statistical analysis method of Generalised Procrustes Analysis (GPA) [12] to model the mean shape of the two ROIs. Size and shape information so obtained are then used to localise these two ROIs in an image from the last phase of the examination and obtain their spatial average intensities at this time t. The difference of these average intensities for each frame determine the value of the "signal" of the FLL's VS [Table 1]. Categorising this signal according to pre-defined standard VSs [7] assists the characterisation of the FLL.

2.1   Definition of Workspace

In this section, we propose an automatic approach to define the workspace (US mask) (Fig.3.a), rather than to label it manually as in [10]. Therefore, the intensity values of the pixels of the image in a frame t, are considered as the visual features P_t:

$$P_t = [p_{x_t,y_t}]_{W \times H} \qquad (1)$$

where p_{x_t,y_t} represents the intensity of pixel p with coordinates (x, y) in frame t. In addition, W and H depict the width and the height of the image, respectively.

Subsequently, the maximum and minimum intensity values over time are found for each pixel p_{x_t,y_t} and then the US mask is automatically obtained. The US mask consists of all the pixels satisfying the following criterion:

$$max\,(p_{x_t,y_t}) - min\,(p_{x_t,y_t}) > T_{mask}. \qquad (2)$$

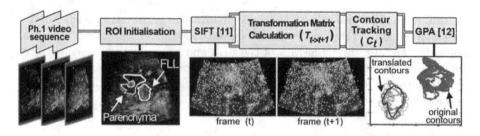

Fig. 2. Phase 1 processing. Where $t \in [1, ..., N_{fr}]$, $N_{fr} = $ numberOfFrames.

where $t \in \{1, ..., N_{fr}\}$, N_{fr} is the total number of frames and T_{mask} is a threshold found iteratively after considering that more than one third of the image should be segmented as the foreground area.

The morphological operation of binary erosion is employed to remove protrusions from objects and thin connections between objects. The connected components algorithm is then used to segment these areas. Finally, the largest-sized area within the resulting image provides the final US mask.

2.2 Motion Tracking under Affine Transformation

After obtaining the US mask, the video sequence of the first phase is processed to obtain the area and shape descriptors of the ROIs (Fig.2). The FLL and the liver contours (described by a matrix C_t) are initialised by a radiologist at frame t_0, based on local texture information, as prior medical knowledge is considered advantageous in correctly spotting an FLL (Fig.3.b,c). C_t is a $2 \times (U + V)$ matrix, comprising $(U + V)$ 2D points, where U and V represent the number of points of the liver and the FLL contour, respectively. The contours are tracked backwards and forwards in time from the frame where they are initialised.

SIFT is used to register salient points (Q_t and Q_{t+1}) within the US mask, in two consecutive frames. Q_t is a $2 \times K_t$ matrix, comprising K_t 2D salient points $q_{\kappa,t}$ at time t, where $\kappa \in [1, K_t]$. A statistical descriptor vector ($D(q_{\kappa,t})$) with 128 dimensions is assigned to each point ($q_{\kappa,t}$) by SIFT [11], in order to characterise it. The correspondence of the registered points between the two frames is estimated by minimising the Nearest Neighbour Distance Ratio (NNDR) between the descriptors of the two frames. Specifically, for every point $q_{\kappa,t}$, another point $q_{\lambda,t+1}$, where $\lambda \in [1, K_{t+1}]$, is found that fulfills the following equation:

$$(q_{\kappa,t}, q_{\lambda,t+1}) = \underset{\substack{q_{\kappa,t} \in Q_t \\ q_{\lambda,t+1} \in Q_{t+1}}}{\mathrm{argmin}} \ [\arccos(D^T(q_{\kappa,t}) \bullet D(q_{\lambda,t+1}))] \qquad (3)$$

Similarly, we find $q_{\lambda',t+1}$, where $\lambda' \in [1, K_{t+1}]$, such that:

$$(q_{\kappa,t}, q_{\lambda',t+1}) = \underset{\substack{q_{\kappa,t} \in Q_t \\ q_{\lambda',t+1} \in Q_{t+1}, \lambda' \neq \lambda}}{\mathrm{argmin}} \ [\arccos(D^T(q_{\kappa,t}) \bullet D(q_{\lambda',t+1}))] \qquad (4)$$

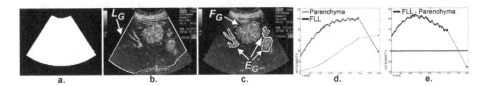

Fig. 3. a.Automatically obtained US mask. b.Liver's GT (L_{G_t}). c.FLL's GT (F_{G_t}) and non-liver areas' GT (E_{G_t}). d.Time-Intensity curves (TIC). e.VS Signal.

We accept that there is a correspondence between points $q_{\kappa,t}$ and $q_{\lambda,t+1}$ in the two frames t and $t+1$, if:

$$\frac{\arccos(D^T(q_{\kappa,t}) \bullet D(q_{\lambda,t+1}))}{\arccos(D^T(q_{\kappa,t}) \bullet D(q_{\lambda',t+1}))} < Z \tag{5}$$

where $Z \in [0,1]$ is a threshold value. If a point $q_{\lambda,t+1}$ is the best match for more than one point $q_{\kappa,t}$, then these correspondences of $q_{\lambda,t+1}$ are rejected. Consequently, we only keep reliable correspondences.

The deformations on the image plane between these correspondences are described by affine transformations (e.g. translation, rotation, shear). Therefore, they are approximated by a transformation matrix $T_{t\to t+1}$ that is computed by the following equation:

$$A_{t+1} = T_{t\to t+1}A_t => T_{t\to t+1} = A_{t+1}A_t^{-1}. \tag{6}$$

where A_t and A_{t+1} are $2 \times R_{t,t+1}$ matrices, comprising $R_{t,t+1}$ 2D salient points at time t and their correspondences at time $t+1$, respectively (Similarly with t and $t-1$ respectively, if tracking backwards). As matrix A_t is not square, A_t^{-1} is its Moore-Penrose inverse [13], computed using the singular value decomposition.

$T_{t\to t+1}$ is then applied to C_t, in order to obtain the corresponding contours in the subsequent frame, estimated as $C_{t+1} = T_{t\to t+1}C_t$. Thus, the contours are tracked in subsequent frames, while their shape and size are also updated based on the affine transformation of subsequent frames, instead of applying the global translation of corresponding SIFT points to the centre of each contour as done in [10].

After all ROIs' contours are obtained in all frames, they are sampled to a set of marked points, in order to model the shape of that ROI. The overall mean shape of the sampled contours is computed by GPA [12] by optimally superimposing the set of all the contours on a single reference orientation. As shown in Fig.2 the FLL's contours originally tracked through the sequence are aligned and then their mean shape is obtained (black-contour). Such a shape is expected to be more accurate than the shape defined manually by the radiologist based on local texture information, as the former takes into account variation across all the frames.

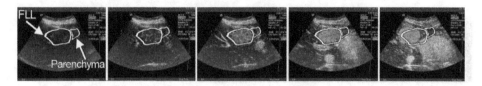

Fig. 4. Motion tracking of the FLL and parenchyma regions with the proposed method.

Finally, the value of the signal of the FLL's VS (Fig.3.e) is determined by the following equation:

$$Signal_{VS} = \frac{\sum_{i=1}^{n_t} p_{x_{t,i}, y_{t,i}}}{n_t} - \frac{\sum_{j=1}^{l_t} p_{x_{t,j}, y_{t,j}}}{l_t} \tag{7}$$

where n_t and l_t are the number of the pixels within the FLL and parenchyma regions respectively, at frame t. The characterisation of the FLL according to the four predefined VSs [7] is then based on this signal throughout the duration of the examination.

3   Results and Validation

3.1   Data Acquisition

The proposed tracking system is applied and evaluated on real clinical data of 14 case studies of patients with similar physical condition. The data acquisition is done by using a Siemens ACUSON Sequoia C512 system with low-frequency 6C2 convex Transducer (2-6MHz) capturing 25fps and following the workflow described in Section 1. Each case includes at least one video sequence from the 1st phase and a static image of the 3rd phase with resolution 768x576 pixels and no compression applied. The contrast medium used was the 2nd-generation contrast medium of sulphur hexafluoride microbubbles (SonoVue, Bracco Diagnostics) that allows for excellent depiction of the FLL vascularity and perfusion [3].

3.2   Ground Truth

To evaluate the accuracy of the tracking method and the level of confidence of the decision on the signal of the FLL's VS, the exact positions that the FLL and the liver areas occupy on each frame were manually annotated, providing their GT as F_{G_t} and L_{G_t} respectively, for every frame t (Fig.3.b,c).

The initialisation of their shapes is performed by the radiologist, whereas their displacements in subsequent frames is done by another operator. The same shape is used for initialising the FLL at frame t_0 (C_{t_0}) and creating $F_{G_{t_0}}$ in the proposed system. Furthermore, the GT of areas that do not belong to, but appear within the liver have been manually labelled on each frame t (E_{G_t}), in order to take these artefacts (i.e. veins and arteries) into account when evaluating the tracking of the parenchyma (Fig.3.c).

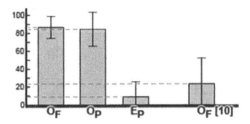

Fig. 5. O_F and O_P depict the overlap of the FLL and the parenchyma, respectively, as explained in Section 3.3. E_P depicts the noise included within O_P in our method.

3.3   Evaluation Metrics

The accuracy of the proposed method is quantitatively evaluated by comparing its decisions with the GT throughout the sequence. The validation of the system is based on considering five regions: the GTs $(F_{G_t}, P_{G_t}, E_{G_t})$, the FLL decision (F_{d_t}) and the parenchyma decision (P_{d_t}). Note that $P_{G_t} = F_{G_t}^C \cap L_{G_t}$, where $F_{G_t}, P_{G_t} \subset L_{G_t}$, and $E_{G_t} \subset P_{G_t}$.

The spatial overlap metric for the FLL regions (O_F) is computed by the equation $O_F = \frac{1}{N_{fr}} \sum_{t=1}^{N_{fr}} O_{F_t}$, where $O_{F_t} = \frac{|F_{G_t} \cap F_{d_t}|}{|F_{G_t} \cup F_{d_t}|}$. This (O_{F_t}) obtains information only from pixels $p_{x_t, y_t} \in F_{G_t} \cap F_{d_t}$ and penalises pixels misclassified as either FLL or non-FLL $(p_{x_t, y_t} \in F_{G_t} \triangle F_{d_t})$.

The spatial overlap metric of the parenchyma regions (O_P) is computed as $O_P = \frac{1}{N_{fr}} \sum_{t=1}^{N_{fr}} O_{P_t}$, where $O_{P_t} = \frac{|P_{G_t} \cap P_{d_t}|}{|P_{d_t}|}$. This (O_{P_t}) obtains information from pixels $p_{x_t, y_t} \in P_{G_t} \cap P_{d_t}$, but its definition differs from that for O_{F_t}, as pixels $p_{x_t, y_t} \in P_{d_t}^C \cap P_{G_t}$, whenever $P_{d_t} \subset P_{G_t}$, should not be penalised. This allows the operator to initialise a smaller region for tracking (P_{d_t}), within P_{G_t}, as correct information will still be provided taking into account the spatial uniformity of the enrichment distribution of P_{G_t}. Last but not least, there is a need to evaluate the percentage of pixels that are non-liver areas but incorrectly classified as parenchyma (i.e. E_{P_t}), which are computed as follows: $E_P = \frac{1}{N_{fr}} \sum_{t=1}^{N_{fr}} E_{P_t}$, where $E_{P_t} = \frac{|E_{G_t} \cap P_{d_t}|}{|P_{d_t}|}$.

3.4   Results

In the cases provided, the third phase is depicted only in static images. Therefore, the FLL ROI is localised in the third phase by employing the exhaustive search method of sliding window and maximising a spatial overlap criterion as in [10]. More specifically, the ROI is localised by maximising the intersection between the automatically segmented areas of the static image and the FLL's mean shape as estimated by GPA in the first phase, allowing translation and rotation of the latter through a sliding window.

Table 2. FLL characterisation. VS ID refers to IDs given in Table 1 (e.g.(a),(b),(c),(d)).

Case studies	1	2	3	4	5	6	7	8	9	10	11	12	13	14
VS ID (GT)	(b)	(b)	(b)	(b)	(a)	(a)	(a)	(b)	(b)	(b)	(b)	(b)	(c)	(c)
VS ID (Proposed System)	(b)	(b)	(b)	(b)	(a)	(a)	(a)	(b)	(b)	(b)	(b)	(b)	(b)	(c)
VS ID of [10]	(b)	(b)	(b)	(b)	(a)	(a)	(a)	(b)	(b)	-	-	-	-	-

An example of the application of the proposed method is shown in Fig.4, demonstrating that the ROIs are tracked even when the video's brightness and contrast are extremely low (e.g. beginning of sequence). The frames show the two ROIs under dramatic appearance changes over time.

The overall average spatial overlaps across our clinical data are above 88% and 86% for O_F and O_P respectively, while for E_P is just above 9% (Fig.5). The mean of O_F and O_P exceeding 87% shows the success of the proposed tracking method, giving confidence that the obtained signal of the VS is truly representative of the FLL, hence providing reliable characterisation of FLLs of typical behaviour. As shown in the first two rows of Table 2 this proved to be valid in 13 out of 14 cases. VS of FLL of case 13 is mischaracterised, as incorrect intensity information is obtained due to much increased dispersion in depth.

The method in [10] is evaluated using the same dataset as ours. Note that [10] is unable to track the parenchyma. The results depict 22.4% overlap (O_F[10] - Fig.5). This low level of overlap, is caused by the histogram-based approach struggling to deal with the intensity variations and therefore fails to properly update the shape of the FLL over time. On the other side, the hereby proposed method considers salient points' deformations of affine transformations for all the regions' updates, which seems more reliable. The final decision of each method is shown in Table 2. Comparison of the results between the two methods depict that [10] is unable to characterise FLLs of hypo-enhanced VS (cases 13,14) as it was not designed for this VS. Also [10] fails if video has low brightness and contrast (cases 10,11,12). In general, tracking in [10] suffers when the relative differences of brightness and contrast between FLL and parenchyma become small (e.g. towards the end of phase 1 and during phase 2). It should be noted that [10] does not characterise an FLL based on its dynamic behaviour, but by observations on the sign change of the signal of hyper-enhanced FLLs, between phase 1 and 3.

4   Conclusion

The proposed system provides accurate tracking and quantitative evaluation of the FLL and the parenchyma areas, whilst characterising the type of an FLL based on the signal of its VS (dynamic behaviour). The tracking of the FLL and the parenchyma is automated by taking into account deformations of the regions as affine transformations on the image plane and only requires the initialisation of the ROIs in one frame as an input by the operator. In our experiments, the

ROIs of 14 clinical cases were tracked with average spatial overlap 87.2% and 13 lesions were correctly characterised by the proposed system.

References

1. BritishLiverTrust: Website,
   http://www.britishlivertrust.org.uk/home/looking-after-your-liver.aspx
   (last accessed December 09, 2011)
2. Sirli, R., Sporea, I., Martie, A., Popescu, A., Danila, M.: Contrast enhanced ultrasound in focal liver lesions–a cost efficiency study. Medical Ultrasonography 12, 280–285 (2010)
3. Wilson, S.R., Burns, P.N.: Microbubble-enhanced us in body imaging: What role? Radiology 257, 24–39 (2010)
4. Strobel, D., Seitz, K., Blank, W., Schuler, A., Dietrich, C.F., von Herbay, A., Friedrich-Rust, M., Bernatik, T.: Tumor-specific vascularization pattern of liver metastasis, hepatocellular carcinoma, hemangioma and focal nodular hyperplasia in the differential diagnosis of 1349 liver lesions in contrast-enhanced ultrasound (ceus). Ultraschall. in Med. 30(4), 376–382 (2009)
5. Albrecht, T., Blomley, M., Bolondi, L., Claudon, M., Correas, J.M., Cosgrove, D., Greiner, L., Jager, K., de Jong, N., Leen, E., Lencioni, R., Lindsell, D., Martegani, A., Solbiati, L., Thorelius, L., Tranquart, F., Weskott, H.P., Whittingham, T.: Guidelines for the use of contrast agents in ultrasound - January 2004. Ultraschall. in Med. 25(4), 249–256 (2004)
6. Goertz, R.S., Bernatik, T., Strobel, D., Hahn, E.G., Haendl, T.: Software-based quantification of contrast-enhanced ultrasound in focal liver lesions - a feasibility study. European Journal of Radiology 75, 22–26 (2010)
7. Rognin, N.G., Mercier, L., Frinking, P., Arditi, M., Perrenoud, G., Anaye, A., Meuwly, J.Y.: Parametric imaging of dynamic vascular patterns of focal liver lesions in contrast-enhanced ultrasound. In: IEEE International Ultrasonics Symposium, (IUS), pp. 1282–1285 (2009)
8. Noble, J.A.: Ultrasound image segmentation and tissue characterisation. Proceedings of the Institution of Mechanical Engineers, Part H: Journal of Engineering in Medicine 224, 307–316 (2010)
9. Shiraishi, J., Sugimoto, K., Moriyasu, F., Kamiyama, N., Doi, K.: Computer-aided diagnosis for the classification of focal liver lesions by use of contrast-enhanced ultrasonography. Medical Physics 35, 1734–1746 (2008)
10. Bakas, S., Chatzimichail, K., Autret, A., Hoppe, A., Galariotis, V., Makris, D.: Localisation and charasterisation of focal liver lesions using contrast-enhanced ultrasonoghaphic visual cues. In: Proceedings of Medical Image Understanding and Analysis (2011)
11. Lowe, D.G.: Distinctive image features from scale-invariant keypoints. International Journal of Computer Vision 60, 91–110 (2004)
12. Gower, J.: Generalized procrustes analysis. Psychometrika 40, 33–51 (1975)
13. Penrose, R.: A generalized inverse for matrices. Mathematical Proceedings of the Cambridge Philosophical Society 51, 406–413 (1955)

Segmentation of the Hippocampus for Detection of Alzheimer's Disease

Maryam Hajiesmaeili, Bashir Bagherinakhjavanlo,
Jamshid Dehmeshki, and Tim Ellis

Medical Imaging International Institutes (QMI3), Faculty of Science,
Engineering and Computing, Kingston University London, Penrhyn Road,
Kingston upon Thames, KT1 2EE, United Kingdom
{m.hajiesmaeili,b.bagheri,J.Dehmeshki,T.Ellis}@Kingston.ac.uk

Abstract. Since hippocampal volume measurement is often used in detection and progression of Alzheimer's disease (AD), segmentation of hippocampus is a significant clinical application. However, it is relatively hard task, due to low signal to noise ratio (SNR), low contrast, indistinct boundary and intensity inhomogeneities. This paper uses Wave Atom shrinkage as an efficient method for enhancing the noisy magnetic resonance images to improve segmentation accuracy followed by a region-scalable active contour model that uses intensity information in local regions. A data fitting energy functional is incorporated into a level set formulation, from which a curve evolution equation is derived for energy minimization. Experimental results of segmenting the hippocampus in T1-weighted MR images yield promising results in the presence of intensity inhomogeneities and low SNR images.

1   Introduction

Magnetic Resonance Imaging (MRI) is a non-invasive medical imaging technology which allows physicians to examine human brain structures. With the rapidly increasing power and difficulties of medical image processing, segmentation of brain structures from MR images has become more feasible. The practical difficulties in segmentation of these datasets can be separated into "internal" and "external" categories. Internal difficulties come up from the natural complexity of brain topology. Although each MR brain image has a similar and recognizable shape, individual brains differ in size, features and other characteristics. External difficulties include various image artifacts, noise, and problems associated with head movement during image acquisition. All these make the design of an ideal segmentation method, which should be simple, accurate, efficient, robust, repeatable, non-invasive and well- tolerated by patients, a difficult goal to achieve.

The hippocampus are two small structures (often linked in shape to a sea horse, and hence the name) that plays a key role in memory and the learning process. Recently, hippocampal atrophy has been proposed as a biomarker for Alzheimer's disease (AD) that can assist with early detection and intervention. The hippocampus is part of the brain's limbic system surrounded by different kinds of tissue and has an

G. Bebis et al. (Eds.): ISVC 2012, Part I, LNCS 7431, pp. 42–50, 2012.
© Springer-Verlag Berlin Heidelberg 2012

ill-defined border with neighboring regions. A large number of studies have shown that hippocampal volume is lower in AD subjects [1]. Hence, the study of MRI image segmentation for the hippocampus has very practical value for clinical application.

Segmentation of the hippocampus from MRI is a challenging task due to their small size, partial volume effects, anatomical variability, low contrast, low signal-to-noise ratio, indistinct boundary and closely proximity to the Amygdaloid body. On the other hand, manual segmentation requires expert operators and is time-consuming. In addition, due to the involvement of human operator in the procedure, manual segmentation is not intrinsically reproducible and automated procedures offer a more objective and consistent solution.

The challenges of, segmenting the hippocampus with conventional methods, e.g., thresholding or region growing does not produce acceptable results [2]. For example [3] proposes a region growing algorithm based on seed which is simple and effective but could fail because of unclear edges of the hippocampus [4]. Atlas-based and registration methods are among the methods which are widely used for human brain segmentation [5] ,[6], [7] do not provide robust results for small and highly variable structures like the hippocampus, due to the limitations in registration and variability to reliable ground truth data. As an automatic method, appearance-based modeling that is a combination of level set shape modeling and active appearance modeling proposed [8] to improve segmentation performance but used multi-contrast images and relies heavily on the intensity normalization.

In recent years, alternative approaches based on active contour models and level set methods have been developed. These methods can be grouped into two classes: region-based models [9], [10], [11], [12], [13] and edge-based models [14], [15], [16]. Edge-based models use edge information for image segmentation and can be applied to images with intensity inhomogeneities, but are quite sensitive to the initial conditions and for weak object boundaries such as the hippocampus are insufficiently sensitive. Region-based models aim to control the motion of the evolving active contour using region properties. However, when applied to MRI images, the algorithm must be able to cope with significant intensity inhomogeneities. Therefore hippocampus segmentation based on this method is difficult. In this paper we propose a region-based active contour model in level set formulation using a local binary fitting model preprocessed using Wave Atom shrinkage to enhance the segmentation results and to increase the image SNR. Section (2) describes the preprocessing technique, the region-based active contour models and the application of level sets. Section (3) compares the performance of the preprocessing algorithm with another widely-used technique (Wavelets) and the results of applying the active contour method to an MRI dataset. The paper is concluded in section (4).

2   Proposed Methodology

2.1   Preprocessing Using Wave Atom Shrinkage

Despite significant improvements in recent years, magnetic resonance (MR) images commonly suffer from low Signal to Noise Ratio (SNR) especially in brain imaging.

The intensity in magnetic resonance images in the presence of noise is described by a Rician distribution that it depends on the data itself. The Wavelet shrinkage [17] is a well known multiresolution analysis tool capable of conveying accurate temporal and spatial information. Wavelets are effective in representing objects with point singularities in 1D and 2D space but fail to deal with singularities along curves in 2D and due to the sparser the representation, the fewer the number of coefficients that are needed to be transmitted is challenging. To overcome the weakness of wavelets, wave atom shrinkage was introduced to improve this method for de-noising Magnetic Resonance Images [18]. Moreover, wave atoms have not only the ability to adapt to arbitrary local directions of a pattern, but the capability to sparsely represent anisotropic patterns aligned with the axes [19].

Two parameters suffice to index many wave packet architectures: α to index whether the decomposition is multiscale ($\alpha = 1$) or not ($\alpha = 0$); and β to indicate whether the basis elements are localized and poorly directional ($\beta = 1$) or, on the contrary, extended and fully directional ($\beta = 0$). Wave Atoms corresponds to $\alpha = \beta = 1/2$, having an aspect ratio $2^{-j/2} \times 2^{-j/2}$ ($j \geq 0$) in space, with oscillations of length~2^{-j}, which obey the parabolic scaling law.

2.2   Segmentation of Hippocampus Using a Region-Based Active Contour

A region-based active contour model is based on intensity information in local regions that must be initialized with the approximate location of the hippocampus, using an initial bounding box based on the approximate size of the hippocampus. In the next step, we define the following local intensity fitting energy in terms of a contour and two fitting functions that locally approximate the image intensities on the two sides of the contour [20]. For a given point $x \in \Omega$

$$\varepsilon^{Fit}(C, f_1(x), f_2(x)) = \sum_{i=1}^{2} \lambda_i \int K(x - y)|I(y) - f_i(x)|^2 \, dy \qquad (1)$$

Where I: $\Omega \to \Re$ is a vector valued image, C is a closed contour in the image domain Ω which separates Ω into two regions: $\Omega_1 = $ outside (C), $\Omega_2 = $ inside (C), λ_1 and λ_2 are positive constants, $f_1(x)$ and $f_2(x)$ are two values that approximate image intensities in Ω_1 and Ω_2, the intensities I(y) that are effectively involved in the above fitting energy are in a local region centered at the point x and K is a Gaussian kernel function

$$K_\sigma(x - y) = \frac{1}{(2\pi)^{n/2} \sigma^n} e^{-|x-y|^2/2\sigma^2} \qquad (2)$$

with a scale parameter $\sigma > 0$.

To obtain the object boundary, we must find a contour C that minimizes the gy ε^{Fit} for all x in the domain. In addition it is necessary to smooth the contour by penalizing its length. Thus, we define the following energy functional:

$$\varepsilon\,(C, f_1(x), f_2(x)) = \int \varepsilon^{Fit}\big(C, f_1(x), f_2(x)\big)\, dx + \upsilon|C| \qquad (3)$$

This energy is then incorporated into a level set formulation to derive curve evolution equation for energy minimization. Due to a kernel function in the data fitting term, intensity information in local regions is extracted to guide the motion of the contour, which thereby enables our model to cope with intensity inhomogeneity. Thus, the energy can be written as

$$\varepsilon_\epsilon\,(\phi, f_1, f_2) = \sum_{i=1}^{2} \lambda_i\ \int(\int K_\sigma(\text{x-y})|I(y) - f_i(x)|^2 M_i^\epsilon(\phi(y))dy\,)dx +$$
$$v \int |\nabla H_\epsilon(\phi(x))|\,dx \tag{4}$$

Where

$$H_\epsilon(x) = \frac{1}{2}\left[1 + \frac{2}{\pi}\tan^{-1}(\frac{x}{\epsilon})\right] \tag{5}$$

$M_1^\epsilon = H_\epsilon(\phi) > 0$ and $M_2^\epsilon = 1 - H_\epsilon(\phi) > 0$.

For accurate computation and stable level set evolution, we define a level set regularization term as

$$\mathcal{P}(\phi) = \int \frac{1}{2}(|\nabla\phi(x)| - 1)^2\,dx \tag{6}$$

Therefore, by keeping $f_1\ and\ f_2$ we propose to minimize the energy functional respect to ϕ as follows

$$\mathcal{F}(\phi, f_1, f_2) + \mu\,\mathcal{P}(\phi) \tag{7}$$

Where μ is a positive constant and by solving the gradient flow equation that can be written as

$$\frac{\partial\phi}{\partial t} = -\delta_\epsilon(\phi)\,(\lambda_1 e_1 - \lambda_2 e_2) + v\delta_\epsilon(\phi)\,\text{div}\,(\frac{\nabla\phi}{|\nabla\phi|}) + \mu\,(\nabla^2\phi - div(\frac{\nabla\phi}{|\nabla\phi|})) \tag{8}$$

δ_ϵ Is the smoothed Dirac delta function and e_1, e_2 are the functions:

$$e_i(x) = \int K_\sigma(\text{y-x})|I(x) - f_i(y)|^2\,dy \qquad \text{i=1, 2} \tag{9}$$

Indeed, the term $-\delta_\epsilon(\phi)\,(\lambda_1 e_1 - \lambda_2 e_2)$ drives the active contour toward the object's boundary and coefficients λ_1 and λ_2 are the weights of the two integrals. The second term has a length shortening (arc length) term and the third term is a regularization term, which maintains the regularity of the level set function.

3   Experimental Results

3.1   MRI Image Acquisitions

Our study is conducted using 3T Vision System image data (GE Medical Systems).Three T1-weighted MRI images scans were downloaded in Dicom format from the Alzheimer's Disease Neuroimaging Initiative (ADNI) data base (http://www.loni.ucla.edu/ADNI) and six additional scans which include ground truth were obtained from the Department of Diagnostic Radiology at Henry Ford Hospital [21]. The data from ADNI are rapid gradient echo (MP-RAGE) with the following specifications: slice thickness = 1.3 mm, matrix = 256*256, TR = 8.6 ms, TE = 3.8 ms, flip angle = 8, TI = 1000 ms and includes data from AD patients. Overall, there are 9 datasets from AD patients.

3.2   Comparison of Wave Atom versus Wavelet Preprocessing

This section gives a detailed analysis of the proposed preprocessing method for hippocampus segmentation. At first the performance of wave atom and wavelet shrinkage method in one AD patient dataset will be shown to choose the best of them as the preprocessing result and then we will compare these two methods based on Peak Signal to Noise Ratio (PSNR) parameter which is defined as

$$\text{PSNR} = 10\log 10\left(\frac{R^2}{MSE}\right) \tag{10}$$

Here R is the maximum fluctuation in the input image data and MSE is mean square error which is given as

$$\text{MSE} = \frac{1}{m*n}\sum_{i=1}^{m}\sum_{j=1}^{n}(N(i,j) - DN(i,j))^2 \tag{11}$$

Where m is the number of rows in the image, N(i,j) is the noisy image and DN(i,j) is the denoised image.

This method applied a threshold to the resulting coefficients of wave Atom transform and similarly to the wavelet, thereby suppressing those coefficients smaller than certain amplitude. Hence determination of the threshold value [22] is an important step and is obtained as follows

$$\sqrt{2 * \ln((maxval) - (minval))^\sigma} \tag{12}$$

where σ is the noise variance which is estimated from the histogram of MR images, maxval is the highest pixel value in the image and the minval is the lowest pixel value in the image [23].

After determining the threshold and estimation of the variance, the forward and then the inverse Wave Atom transform (and similarly Wavelet transform) is applied separately and performance of each are compared using PSNR comparison parameters, seen in table 1.

Table 1. PSNR parameter for Wave Atom and Wavelet transforms for 9 MRI datasets

Image ID	001	002	003	004	005	006	007	008	009
Wave Atom	26.64	26.75	28.04	26.50	26.84	27.97	26.92	28.01	27.89
Wave-let	26.04	26.24	27.38	26.08	26.33	27.28	26.55	27.24	27.36

Table 1 illustrates the comparison of Wave Atom method and Wavelet based on PSNR parameter for nine MR AD images (these values are average over the entire slices for each MRI scan data). In table 1 as can be seen from the results, Wave Atom method provides consistently better PSNR across the entire dataset.

Hence this method will be used to preprocess the image prior to segmentation of the hippocampus. Figure 1 shows the result of applying the de-noising algorithms to a single sagittal slice through an MRI scan (slice 107 out of 210 for scan ID: 003).

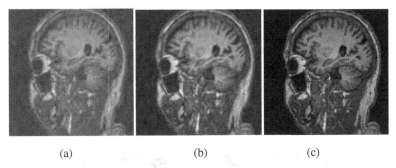

(a)   (b)   (c)

Fig. 1. (a) Original MR image (b) De-noised by Wavelet –PSNR: 26.62(c) De-noised by wave Atom-PSNR:27.27

It is clear from figure (1) that the quality of denoised image using Wave Atom as comparison to the quality of image denoised by Wavelet is better.

3.3   Hippocampus Segmentation Results and Discussion

The proposed method for hippocampus segmentation has been applied after preprocessing using the Wave Atom filter. The initial conditions for the level set algorithm are as follows:

Iteration time step $\Delta t = 0.1$, iteration step is 300 times and the parameters of initial level set function are: $C_0 = 2$, $\sigma = 3$, $\mu = 0.004 * 255 * 255$ and $\lambda_1 = \lambda_2 = 1$. Figure (2) shows the detected boundary of the hippocampus in sagittal images (red curve).

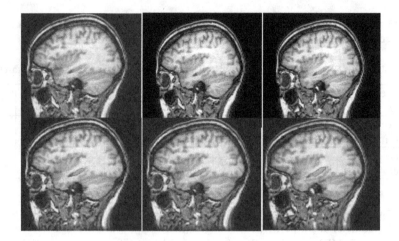

Fig. 2. First row: three slices from one MRI scans, second row: segmentation of the hippocampus result

These results were obtained after pre-processing and implementation of level set.

This method is evaluated in terms of objective results which are obvious based on the hippocampus shape and ground truth. Figure (3) illustrates the evaluation of these results.

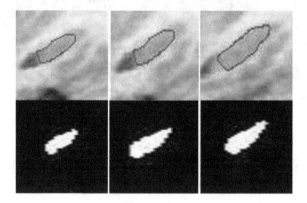

Fig. 3. First row: segmentation of the hippocampus results for three slices of one MRI scans (slice numbers 91-93 for scan ID: 009), second row: ground truth

In order to apply this method for different AD patients, five slices from 5 MRI scans have been chosen which are preprocessed to show the object better and then the region is manually determined as bounding box to implement our methodology. Figure (4) shows these results and each image can be compared with its ground truth.

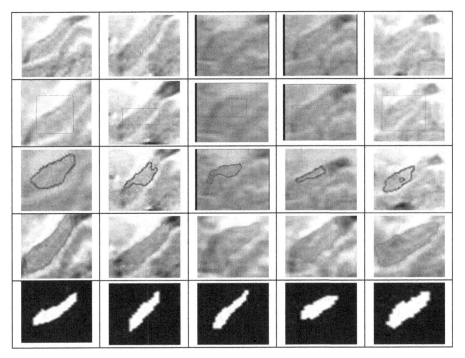

Fig. 4. First row: original MRI image, second row: bounding box, third row: level set results without preprocessing, fourth row: level set results after applying Wave Atom, fifth row: ground truth

4   Conclusion and Future Work

We have proposed the Wave Atom method for noise removal in MRI data as a preprocessing offstage prior to hippocampus segmentation. The Wave Atom was compared to a Wavelet method and shown to offer superior performance; demonstrated higher PSNR. The level set segmentation method, which is based on intensity information in local regions, provides a valuable tool for hippocampus segmentation and its performance at weak object boundaries and low signal to noise levels appears to be good. However this method needs quantitative validation and the accurate estimation of the volume measurement for the hippocampus is the topic of future work.

References

1. Killiany, R.J., Hyman, N., Gomez-Isla, T.: MRI Measures of Entorhinal Cortex vs Hippocampus in Preclinical AD. Neurology 58, 1188–1196 (2002)
2. Claude, I., Daire, J.-L., Sebag, G.: Segmentation and Biometric Analysis of the Posterior Fossa. IEEE Transactions on Biomedical Engineering 51, 617–626 (2004)
3. Wang, Y., Zhu, Y., Gao, W., Fu, Y., Wang, Y., Lin, Z.: The Measurement of the Volume of Stereotaxtic MRI Hippocampal Formation Applying the Region Growth Algorithm Based on Seeds. In: International Conference on Complex Medical Engineering, Harbin, pp. 489–492 (2011)

4. Eindhoven, Philips. Hippocampus Region Segmentation for Alzheimer's Disease Detection. Research, 19–21 (2009)
5. Christensen, G.E., Rabbitt, R.D., Miller, M.I.: Deformable Templates Using Large Deformation kinematics. IEEE Trans. Image, 1435–1447 (1996)
6. Davatzikos, C., Prince, J.L.: Brain Image Registration Based on Curve Mapping. In: Proceedings of IEEE Workshop Biomedical Image, Los Alamitos, CA, pp. 245–254 (1994)
7. Carmichael, O.T., Aizenstein, H.A., Davis, S.W., Becker, J.T., Thompson, P.M., Meltzer, C.C., Liue, Y.: Atlas-Based Hippocampus Segmentation In Alzheimer's Disease and Mild Cognitive Impairment. Neuroimage 27, 979–990 (2005)
8. Hu, S., Coupe, P., Pruessner, J.C., Louis Collins, D.: Appearance-based modeling for segmentation of hippocampus and amygdala using multi-contrast MR imaging. NeuroImage 58, 549–559 (2011)
9. Chan, T.F., Vese, L.A.: Active Contours without Edges. IEEE 10(2), 266–277 (2001)
10. Ronfard, R.: Region-Based Strategies for Active Contour Model. Int. J. Comput. Vis. 13(2), 229–251 (1994)
11. Samson, C., Blanc-Feraud, L., Aubert, G., Zerubia, J.: A Variational Model for Image Classification and Restoration. IEEE 22(5), 460–472 (2000)
12. Tsai, A., Yezzi, A., Willsky, A.S.: Curve Evolution Implementation of the Mumford-Shah Functional for Image Segmentation, Denoising Interpolation, and Magnification. IEEE Trans. Image Process. 10(8), 1169–1186 (2001)
13. Bagheri, B., Ellis, T.J., Dehmeshki, J.: Medical Image Segmentation Using Deformable Models and Local Fitting Binary. In: International Conference on Image, Signal and Vision, pp. 86–89 (2011)
14. Chan, T.F., Vese, L.: A Multiphase Level Set Framework for Image Segmentation Using the Mumford and Shah Model. Int. J. Comput. Vis. 50(3), 271–293 (2002)
15. Caselles, V., Kimmel, R., Sapiro, G.: Geodesic Active Contours. Int. J. Comput. Vis. 22(1), 61–79 (1997)
16. Kichenassamy, S., Kumar, A., Olver, P., Tannenbaum, A., Yezzi, A.: Gradient Flows and Geometric Active Contour Models. In: 5th Int. Conf. Comput. Vis., pp. 810–815 (1995)
17. Taswell, C.: The What, How, and Why of Wavelet Shrinkage Denoising. IEEE Computing in Science and Engineering, 12–19 (2000)
18. Kumar, V., Saini, S., Dhiman, S.: Quality Improvement on MRI Corrupted with Rician Noise using Wave Atom Transform. International Journal of Computer Applications 37(8), 28–32 (2012)
19. Ma, Y.M., Liu, G.J.: Combination of Wave Atoms Shrinkage with Bilateral Filtering for Oscillatory Textural Image Denoising. Advanced Materials Research, 2119–2124 (2011)
20. Li, C., Kao, C.-Y., Gore, J.C., Ding, Z.: Minimization of Region-Scalable Fitting Energy for Image Segmentation. IEEE Trans., Image Processing 17, 1940–1949 (2008)
21. Jafari-Khouzani, K., Elisevich, K., Patel, S., Soltanian-Zadeh, H.: Dataset of Magnetic Resonance Images of Nonepileptic Subjects and Temporal Lobe Epilepsy Patients for Validation of Hippocampal Segmentation Techniques. Neuroinformatics, 335–349 (2011)
22. Rajeesh, Moni, R.S., Palani Kumar, S., Gopalakrishnan, T.: Rician Noise Removal on MRI Using Wave Atom Transform with Histogram Based Noise Variance Estimation. IEEE Communication Control and Computing Technologies, 531–535 (December 2010)
23. Dua, G., Raj, V.: MRI Denoising Using Waveatom Shrinkage. Global Journal of Researches in Engineering Electrical and Electronics Engineering 12(4), 22–27 (2012)

Segmentation of Parasites for High-Content Screening Using Phase Congruency and Grayscale Morphology

Daniel Asarnow[1] and Rahul Singh[2,*]

[1] Department of Chemistry and Biochemistry
[2] Department of Computer Science, San Francisco State University,
San Francisco, CA 94132
rahul@sfsu.edu

Abstract. Schistosomiasis is a parasitic disease with a global health impact second only to malaria. The World Health Organization has determined new therapies for schistosomiasis are urgently needed, however the causative parasite is refractory to high-throughput drug screening due to the need for a human expert to analyze the effects of putative drugs. Currently, there is no vision system capable of relieving this bottleneck with sufficient accuracy for the automated analysis of parasite phenotypes. We presented a region-based method with performance limited primarily by poor edge detection caused by body irregularities, groups of touching parasites and unpredictable effects of drug exposure. Towards ameliorating this difficulty, we propose an edge detector utilizing phase congruency and grayscale thinning. The detector can be used to impose the correct topology on a segmented image – an essential step towards accurate segmentation of parasites.

1 Introduction

1.1 Background

Schistosomiasis is a parasitic disease considered to have global health and socio-economic impacts second only to malaria. Although incidence of the disease in developed countries is extremely low, more than 200 million people are infected worldwide, with an additional 800 million at risk. The chronic illness is caused by infection with one of several species of trematodes, chiefly *Schistosoma mansoni*, *Schistosoma haematobium* and *Schistosoma japonicum*, which are carried to humans through water contaminated with their larvae. Early on, infection is characterized by an inflammatory response to the parasites' eggs, eventually leading to fibrotic granulomas that can occlude the hepatic portal vein and cause hydronephrosis (kidney swelling from urine buildup) and squamous cell bladder cancer. Other effects of schistosomiasis include diarrhea, lesions in the central nervous system and genital sores which enhance the transmission of HIV. The World Health Organization (WHO) has classified schistosomiasis as one of 17 neglected tropical diseases, a set of illnesses grouped together

---

* Corresponding author.

G. Bebis et al. (Eds.): ISVC 2012, Part I, LNCS 7431, pp. 51–60, 2012.
© Springer-Verlag Berlin Heidelberg 2012

because they (1) are proxies for poverty, (2) affect politically disadvantaged populations, (3) do not travel out of the third world, (4) lead to discrimination, especially of women, (5) have serious, widespread health effects, (6) are neglected by research and (7) might be controlled through currently feasible means [1].

Whole-organism drug screens against schistosomiasis have recently been adapted to automated, high-throughput data collection, but the need for an expert observer remains a significant bottleneck. We therefore proposed an image segmentation algorithm [2] for bright-field microscopy images of the juvenile schistosomula, towards the development of a fully automated screen against schistosomiasis. Unfortunately, the separation of touching parasites obtained by this method was often insufficient for precise phenotypic measurements [3].

1.2   Problem Formulation

In order to support the precise measurement of parasite attributes from a video, segmentation must be accurate, robust and able to separate individual parasites. Segmentation of schistosomula in particular raises challenges which are distinct from the segmentation of cells. These include:

- The parasites are all unique individuals and exhibit marked variation in size and shape. In addition, elongation and contraction of the musculature can result in drastic alterations in proportion, shape and orientation. These facts preclude the assumption of an a priori geometric model of shape.
- The proclivity of schistosomula to touch or overlap slightly. This leads to the formation of large groups of parasites in physical contact, which are segmented as a single object due to weak or nonexistent edges.
- The presence of visible anatomical structures within the parasites. These structures create internal edges which do not correspond to parasite boundaries.
- Alterations in each of the above due to the effects of drug exposure, which again differs between individual parasites.

A proposed method must successfully address these difficulties, which are illustrated in Figure 1. Detection of those edges which are most salient to the perceptual separation of individual parasites is essential if the challenges enumerated above are to be overcome.

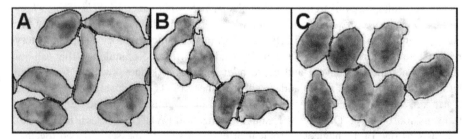

Fig. 1. Phenotypic diversity of schistosomula. (A) Control and exposed to the drugs Praziquantel (B) and Simvastatin (C). Black lines indicate edges found by proposed method.

2   Prior Work

Few computer vision methods targeting parasitic organisms exist in the literature. One image based assay against *T. cruzi*, responsible for Chagas' disease, was designed utilizing the IN Cell Analyzer 1000 (GE Healthcare), a dedicated, high-throughput experimentation and imaging apparatus [4]. The only measurement made is the quantity of parasite in the foreground and no attempt is made to uniquely identify individuals. While not a parasite, *C. elegans* is related to a number of parasite species and is visually similar in some respects to vermiform macroparasites such as Schistosomatidae. Some computer vision studies of *C. elegans* have undertaken to segment multiple touching or overlapping individuals. The approaches taken include articulated models [5] and path searching on probabilistic shape models [6].

In contrast to parasites, segmentation and tracking of cells has been an active research area for some time. Zimmer et al. [7] use a parametric active contour with a repulsive term between regions which enforces the separation of closely touching cells, while Srinivasa et al. [8] devised an "active mask" algorithm in which region masks are iteratively evolved and/or discarded in order to produce a correct labeling of each pixel.

3   Methods

3.1   Overview

In broad terms, the algorithm presented in [2] consists in an initial segmentation using a region-based distributing function adapted from [8], in which touching parasites are typically merged into a single object, followed by edge-based region splitting in which edge information extracted from the image is used to correct border placement and separate erroneously merged parasites. Edges that are irrelevant to the separation of merged parasites are eliminated, giving a pruned set of edges for splitting merged regions. The edge detection component, which is the focus of this paper, originally employed the Canny edge detector. However, the Canny operator proved very susceptible to noise. Furthermore, gradient-based detectors in general are sensitive only to features with a very limited range of phase angles of the spatial frequencies. The remainder of this section further describes the shortcomings of gradient methods and then presents a new edge detector, suitable for localizing perceptual edges between schistosomes – even if they are not well represented by the intensity gradient.

3.2   Edge Detection

Traditional edge detection methods rely on the image gradient. The best known examples of such methods are the Prewitt-Sobel-Roberts family of derivative approximations and the Canny operator [9]. The latter is of particular interest, as it is optimal among the gradient-based approaches and is widely recognized for its efficacy.

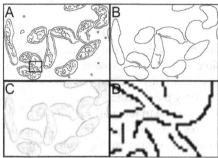

Fig. 2. Canny edges with (A) permissive thresholds, (B) with conservative thresholds, (C) gradient magnitude and (D) close up of problematic result from black box in (A)

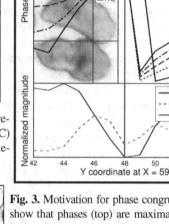

Fig. 3. Motivation for phase congruency; graphs show that phases (top) are maximally congruent just at the perceptual edge indicated by the vertical line, while the gradient (bottom, dashed line) has peaks to either side. Inset shows origin of intensity profile. Note the steep change in phase after the edge is just a 180 degree shift.

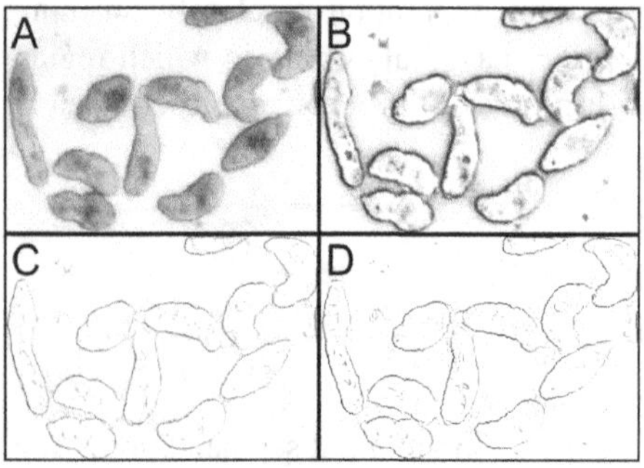

Fig. 4. Original image, (B) phase congruency and ridge detection by (C) grayscale thinning and (D) non-maxima suppression.

Application of the classical formulation of the Canny operator to schistosomula images produces edges which are insufficiently accurate to effectively separate merged parasites without falsely splitting individuals. When permissive thresholds are used, the results are characterized by a high degree of noise and artifacting, while conservative thresholds fail to detect the weak edges which often occur between touching parasites (Figure 2A-B). Furthermore, gradient based edge detection suffers from a "double edge" artifact: edges wide enough to have a small gradient in the center are not identified correctly and in many cases are incapable of separating touching objects regardless of the thresholds used (Figure 2C-D, Figure 3). In order to address these issues, we replace the Canny detector with a novel edge operator aimed at producing accurate edge contours with maximum perceptual salience.

Calculation of valid edge weights is crucial to the performance of the edge detector. Rather than the intensity gradient, we propose using the phase congruency of the grayscale image. Phase congruency (PC) is an approach to feature detection based on the Local Energy Model [10], which holds that perceptually salient features occur where an image's Fourier components are maximally in phase with one another.

PC has a number of qualities which are advantageous in comparison to image gradients. First, it is a dimensionless quantity restricted to the interval [0,1]. Second this notion is illumination and contrast invariant, and can detect and correctly place perceptual elements which do not coincide with steps in the image gradient (such as thick edges). Finally, PC is naturally extensible to include noise cancellation and multi-scale analysis and additionally may be implemented efficiently using fast wavelet transforms [11]. That PC is particularly suitable for images of schistosomula, is demonstrated by Figure 3, which illustrates that congruency of phase coincides with perceptual edges even where the gradient intensity does not.

PC is defined as the ratio between the local energy, or absolute magnitude in frequency space, and the total path length of all frequency component vectors. If F(x) denotes the real components of the spatial frequencies and H(x) the imaginary component, then the local energy is defined by Eq. (1), and the phase congruency itself by Eq. (2). ε is a small constant representing numerical precision.

$$E(x) = \sqrt{F(x)^2 + H(x)^2} \tag{1}$$

$$PC(x) = \frac{E(x)}{\sum_n A_n(x) + \varepsilon} \tag{2}$$

Calculating phase congruency requires a local, phase-preserving frequency analysis. An appropriate method is the wavelet transform, using quadrature pairs of matched even and odd filters. Energy E may then be easily extracted by comparing responses to the symmetric filter f and anti-symmetric filter h, representing real and imaginary parts of Eq. (1). Phase angle φ is computed via Eq. (3).

$$\varphi(x) = atan2(f(x), h(x)) \tag{3}$$

Log-Gabor filters, characterized by a Gaussian transfer function on a logarithmically scaled frequency axis, are chosen because they are psychophysically justified [12], and because they possess zero mean at arbitrary bandwidths. In the Fourier domain, the log-Gabor transfer function with center frequency ω is given by:

$$G(\omega) = e^{\frac{-\ln (\omega/\omega_0)^2}{2\ln (\kappa/\omega_0)}} \tag{4}$$

Using Eq. (4), a log-Gabor filter bank comprised of filters of at different frequency scales is constructed. The implied sense of scale derives not from the spatial extent of a feature itself, but from the spatial extent of its constituent frequency components. Once the filter bank is constructed, analysis at multiple scales can be performed by summing the filter response at each scale. The reader is referred to [11] for a detailed treatment of this approach.

In general, the edge weights obtained via phase congruency (or gradient methods) are, for a given image feature, diffused over a width greater than that needed or desired for edge based region splitting (Figure 4C). In order to accurately localize edges,

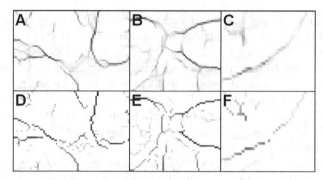

Fig. 5. Comparison of (A)-(C) grayscale thinning versus (D)-(F) non-maxima suppression, demonstrating that thinning is significantly more conservative

as well as reduce the number of pixels under consideration, the intensities of pixels which are not along the center-lines or ridges of edge features must be damped and/or eliminated. Anisotropic non-maxima suppression, using feature orientation estimates, is a well-known method which aligns with topographical sense, in that a ridge point ought to be a peak in the projection along the edge normal. On the other hand, orientation estimates may be subject to noise, and the exact pixel location of a numerical maximum may deviate from the center of the perceptual edge, leading to broken contours. Although the phase angles computed as intermediates in the phase congruency procedure may be used as estimates of feature orientation, in our method an approach to ridge detection is taken which does not rely on any such estimates.

Ridge detection is related to "thinning" of image features, in that the ridge is defined as the high-intensity center of a wider perceptual structure. The operation is therefore analogous to (infinite) morphological thinning of binary images, in which the set theoretic approach of mathematical morphology is used to eliminate pixels whose neighborhoods conform to constraints on their structure. Unlike fundamental morphological operations such as erosion and dilation, the hit-or-miss transform (from which most thinning algorithms are derived) is not well defined for intensity images. Nevertheless, any binary image operation can be extended to grayscale via linear superposition of global thresholding at all intensity values [13]. To carry out a morphological operation by threshold superposition, the intensities of a grayscale image are first quantized by restricting them to N bins. The center of each bin is used as a global threshold to convert the quantized image to binary. A morphological operation is performed on the thresholded images (until convergence if desired), which are then summed to yield a new image under the same quantization. This technique is practically limited to relatively small values of N, due to the exponential proliferation of possible thresholds, and down-sampling may be needed to obtain acceptable performance on available hardware.

Threshold superposition can be used to take advantage of the relationship between ridge detection and skeletonization [14]. In our work, grayscale thinning by threshold superposition is used to thin the phase congruency edge weights. A mathematical formulation of threshold superposition requires the global threshold operator T_t defined over a set of pixels $X=\{x_i\}$ with intensities I_x as given in Eq. (5).

$$T_i = \{x_i \in X | I(x_i) > t \tag{5}$$

Grayscale thinning by superposition is given by the expression in Eq. (6). The notation thin∞ indicates an infinite (binary) morphological thinning operation. Note that the phase congruency PCi is normalized by definition.

$$W_i = \sum_{t=0}^{N-1} thin_\infty \left(T_t\big(round(PC_i \cdot N)\big)\right) \tag{6}$$

The parallel thinning algorithm of Guo and Hall [15], which is simple to implement and obtains good results, is reapplied until convergence at each threshold independently. As shown by Figures 4 and 5, grayscale thinning proves significantly more conservative than non-maxima suppression, preserving the complete ridges that are necessary for effective region splitting.

Once the thinned edge weights are available, a binary edge image is determined using hysteresis thresholding. Hysteresis consists in locating pixels which are above a high threshold, or which are above a low threshold and are 8-connected to a pixel above the high threshold. In terms of binary sub-images given by high and low global threshold operations T_H and T_L, the hysteresis edges may be written as the components of T_L which are supersets of TH. The notation $\prod_i(X)$ denotes the partitioning of a binary image into its connected components.

$$E_i = \prod_i T_L(W_i) \supseteq \prod_i T_H(W_i) \tag{7}$$

The high threshold is taken to be that determined for the (thinned) PC image using Otsu's method; the low is taken as that value times ¼. The hysteresis thresholds are thus reflective of the intensity distribution within each sub-image. The product of hysteresis is finally subjected to infinite (binary) morphological thinning, in order to ensure that edges are one pixel wide in all cases.

Having arrived at a binary edge set, we wish to use it to split connected components found in the initial segmentation. However, the edges will include some which are not relevant to this task; only edges which form closed contours or are connected to the background are capable of internally disconnecting a region. These irrelevant edges are located and eliminated as follows. A marker image is generated by direct subtraction of the edge set from the initial segmentation. Then, each edge pixel is marked as relevant if and only if its 8-neighborhood contains more than one marker region. The presence of multiple markers within the 8-neighborhood is determined by labeling the marker regions and determining if the cardinality of the unique labels within the neighborhood is greater than one. The labeling operation is denoted as $label_8$ and the set of unique elements about a pixel x is denoted as $unique_8$.

$$M_i = label_8(C_i - E_i) \tag{8}$$

$$F_i = \left\{ x \in M_i \big| |unique_8(x)| > 1 \right\} \tag{9}$$

The relevant edges Fi constitute one pixel wide, 8-connected edge contours, which will alter the topology of segmentation upon subtraction (i.e. by splitting regions). Finally, the minimum subset of the relevant edges which reproduces the same segmentation topology is computed using the watershed transform.

4   Experiments

The algorithm we describe is designed to satisfy the criteria for accuracy laid out above, and to mitigate the specific challenges presented by data from HTS of Schistosoma. As discussed previously, it is especially important that accuracy be maintained across the diverse phenotypes exhibited by individual schistosomes, even with this diversity is exacerbated by drug insult. It is therefore necessary to evaluate results for a variety of experimental conditions:

- Control conditions. Parasites exhibiting their natural phenotypes provide an essential baseline for detecting drug induced phenotypic changes. Healthy parasites typically have stronger edges and more regular shapes than those exposed to drugs.
- Extreme phenotypes which occur because of the presence of different compounds. In particular, parasites exposed to the drugs Simvastatin (Sim), Chlorpromazine (Chl) and PZQ provide a diverse phenotypic set.

Volunteers hand-segmented 8 images from these conditions to serve as ground truth for quantitatively evaluating the proposed method versus Canny edge detection. Table 1 summarizes the accuracy of the two methods using two measures: the mean deviation of object boundaries, estimated using a Euclidean distance transform, and correct detection of individual parasites, represented by an excess or deficiency of connected components. The tests show the proposed method is accurate over a wide range of experimental conditions, supplementing results shown in Figure 1. While both methods place edge pixels relatively close to the ground truth, use of Canny results in the loss of 15% of parasites compared to just 3% with the proposed method. As expected, the highest accuracy is obtained for control images. In comparison to the controls, long term exposure to Sim or Chl causes edges between touching parasites to become very weak. For this reason, parasites in these conditions are more difficult to separate when they are in physical contact and are not correctly segmented in all frames. The presence of PZQ evokes a peculiar phenotype in which the parasites tend to "shrivel," adopting very irregular shapes which often bear narrow protrusions from the body. Here, changes in the appearance of anatomical features within the parasites contribute to false edges with the potential to induce the false splitting of single parasites. Success of the proposed method in spite of these difficulties testifies to the high sensitivity of phase congruency to perceptual edge features.

5   Conclusion

One essential component of a fully automated, high-throughput assay for drug discovery against neglected diseases is a computer vision system capable of ameliorating

Table 1. Quantitative Evaluation of the Proposed Method

Experimental conditions	Mean boundary deviation		Number of regions (ratio)	
	Proposed	Canny	Proposed	Canny
Control	0.003	0.002	0.005	-0.067
Sim	4.15	3.20	-0.067	-0.194
Chl	0.596	2.37	-0.077	-0.151
Pzq	2.60	1.08	0.005	-0.199
Overall	1.84	1.92	-0.033	-0.153

the bottleneck created when human experts must be relied upon for data analysis. However, extant methods do not address the distinct challenges found in the segmentation of schistosomes. We presented a segmentation algorithm designed specifically towards the segmentation of standard bright-field microscopy images of schistosomes, which achieved some success without any dependence on proprietary HTS systems. We now improve our method with the development of a novel, high-sensitivity edge operator – the first to combine phase congruency with ridge-detection by grayscale thinning. Quantitative and qualitative analysis demonstrates the power of the algorithm, and paves the way for an effective, high-throughput, phenotypic screen against schistosomiasis. It is hoped that such a system will lead to new drugs which will ameliorate the global health impacts of schistosomiasis.

Acknowledgements. The authors thank Brian Suzuki and Connor Caffrey for generating the image data along with Kristina Finnegan, who hand-segmented the images. This work was funded in part by the NIH grant 1R01AI089896-01 and by NSF grant 0644418.

References

1. Crompton, D.W.T., Daumerie, D., Peters, P., Savioli, L.: Working to overcome the global impact of neglected tropical diseases first WHO report on neglected tropical diseases. World Health Organization. Dept. of Control of Neglected Tropical Diseases, Geneva, Switzerland (2010)
2. Moody-Davis, A., Mennillo, L., Singh, R.: Region-Based Segmentation of Parasites for High-throughput Screening. In: Bebis, G. (ed.) ISVC 2011, Part I. LNCS, vol. 6938, pp. 43–53. Springer, Heidelberg (2011)
3. Lee, H., Moody-Davis, A., Saha, U., Suzuki, B.M., Asarnow, D., Chen, S., Arkin, M., Caffrey, C.R., Singh, R.: Quantification and clustering of phenotypic screening data using time-series analysis for chemotherapy of schistosomiasis. BMC Genomics 13, S4 (2012)
4. Engel, J.C., Ang, K.K.H., Chen, S., Arkin, M.R., McKerrow, J.H., Doyle, P.S.: Image-Based High-Throughput Drug Screening Targeting the Intracellular Stage of Trypanosoma cruzi, the Agent of Chagas' Disease. Antimicrob. Agents Chemother. 54, 3326–3334 (2010)

5. Huang, K.-M., Cosman, P., Schafer, W.: Using Articulated Models for Tracking Multiple C. elegans in Physical Contact. Journal of Signal Processing Systems 55, 113–126 (2009)
6. Wahlby, C., Riklin-Raviv, T., Ljosa, V., Conery, A.L., Golland, P., Ausubel, F.M., Carpenter, A.E.: Resolving clustered worms via probabilistic shape models. In: 2010 IEEE International Symposium on Biomedical Imaging: From Nano to Macro, pp. 552–555. IEEE (2010)
7. Zimmer, C., Labruyere, E., Meas-Yedid, V., Guillen, N., Olivo-Marin, J.-C.: Segmentation and tracking of migrating cells in videomicroscopy with parametric active contours: a tool for cell-based drug testing. IEEE Transactions on Medical Imaging 21, 1212–1221 (2002)
8. Srinivasa, G., Fickus, M.C., Guo, Y., Linstedt, A.D., Kovacevic, J.: Active Mask Segmentation of Fluorescence Microscope Images. IEEE Transactions on Image Processing 18, 1817–1829 (2009)
9. Canny, J.: A Computational Approach to Edge Detection. IEEE Transactions on Pattern Analysis and Machine Intelligence, PAMI-8, 679–698 (1986)
10. Morrone, M.C., Owens, R.A.: Feature detection from local energy. Pattern Recognition Letters 6, 303–313 (1987)
11. Kovesi, P.: Image Features from Phase Congruency. Videre 1, 1–26 (1999)
12. Field, D.J.: Relations between the statistics of natural images and the response properties of cortical cells. J. Opt. Soc. Am. A. 4, 2379–2394 (1987)
13. Maragos, P., Ziff, R.D.: Threshold superposition in morphological image analysis systems. IEEE Transactions on Pattern Analysis and Machine Intelligence 12, 498–504 (1990)
14. Weiss, J.: Grayscale Thinning. Computers and their applications. In: Proceedings of the ISCA 17th International Conference, San Francisco, CA (2002)
15. Guo, Z., Hall, R.W.: Parallel thinning with two-subiteration algorithms. Commun. ACM 32, 359–373 (1989)

Multigrid Narrow Band Surface Reconstruction via Level Set Functions

Jian Ye[1], Igor Yanovsky[2,3], Bin Dong[4], Rima Gandlin[5],
Achi Brandt[1,6], and Stanley Osher[1]

[1] Department of Mathematics, University of California, Los Angeles, CA, USA
[2] Jet Propulsion Laboratory, California Institute of Technology, Pasadena, CA, USA
[3] Joint Institute for Regional Earth System Science and Engineering,
University of California, Los Angeles, CA, USA
[4] Department of Mathematics, The University of Arizona, Tucson, AZ, USA
[5] Department of Mathematics, Carnegie Mellon University, Pittsburgh, PA, USA
[6] Department of Computer Science and Applied Mathematics,
Weizmann Institute of Science, Rehovot, Israel

Abstract. In this paper we propose a novel fast method for implicit surface reconstruction from unorganized point clouds. Our algorithm employs a multigrid solver on a narrow band of the level set function that represents the reconstructed surface, which greatly improves computational efficiency of surface reconstruction. The new model can accurately reconstruct surfaces from noisy unorganized point clouds that also have missing information.

Keywords: Level set, multigrid method, point cloud, surface reconstruction.

1 Introduction

The field of surface reconstruction from scattered point clouds has been developing rapidly in the past decade. It is a challenging problem since point clouds lack ordering information and connectivity, and are usually noisy. There are two ways of representing the reconstructed surfaces: explicit and implicit. Explicit representation usually gives the exact location of a surface in a physical domain, while implicit representation defines the surface as the (zero) level set of some scalar function. Common explicit representations include parametric surfaces [1, 2] and triangulated surfaces [3–7]. Implicit surfaces are most frequently represented using level set functions, typically signed distance functions [8], as well as some more recent ones [9, 10].

One of the traditional approaches for implicit surface reconstruction is the use of radial basis functions representation [11]: $s(\boldsymbol{x}) = p(\boldsymbol{x}) + \sum_{i=1}^{n} \lambda_i \phi(|\boldsymbol{x} - \boldsymbol{x_i}|)$, where p is a polynomial, ϕ is a global smooth function that allows fast summation, e.g. $\phi(r) = r$, and $(\boldsymbol{x_1}, ..., \boldsymbol{x_n})$ is a set that includes the N given surface points and a comparable number of off-surface points at each of which $s(\boldsymbol{x_i})$ is prescribed as an estimated distance from $\boldsymbol{x_i}$ to the surface. The iterative solver

G. Bebis et al. (Eds.): ISVC 2012, Part I, LNCS 7431, pp. 61–70, 2012.

for computation of the coefficients λ_i and the polynomial p requires $C_1 N \log N$ computer operations, where C_1 is a very large constant.

Another well-known method was introduced in [12, 13], where the authors constructed a weighted minimal surface-like model. The reconstructed surface is represented as the zero level set of a scalar function and the proposed energy functional is minimized by solving a nonlinear partial differential equation (PDE) (see [12, 13] for details). However, solving such PDE requires small time steps and hence longer computational time. Furthermore, the energy functional proposed in [12, 13] is nonconvex, which makes the result sensitive to the initialization and noise. Therefore, in order to avoid local minimizers and reduce computation time, one should start from an initial surface which is very close to the given point cloud. More recently, the authors in [14, 15] extended the work in [12, 13], proposing new models and solving the surface reconstruction problem using state-of-the-art optimization algorithms

All the aforementioned level set based surface reconstruction methods aimed at solving some nonsmooth (and often nonconvex) optimization problem. One of the main benefits of solving such nonsmooth problems is to preserve very sharp features, such as edges, of the points clouds in the reconstructed surfaces. However, the nonsmoothness also inevitably increases the difficulty of solving the problems efficiently. In practice, many geometric objects to be reconstructed from their point samples do not have very sharp features. Therefore, for these cases, solving nonsmooth optimization problems is not necessary. In this paper, we propose a differentiable variational model for surface reconstruction problems, which is solved very efficiently by using multigrid techniques on a narrow band of the level set function that represents the reconstructed surface. In addition, important features of the point clouds are also well preserved in the reconstructed surfaces using our proposed algorithm.

2   Proposed Model

Given a data set $\{x_l\}_{l=1,\ldots,N} \subset \mathbb{R}^{\dim}$, i.e. a set of points with dim $= 2$ or 3, we seek a function ϕ that is close to zero on this set and smooth elsewhere (see [16]). For this purpose, we consider the following energy functional:

$$E(\phi) = \int G(\phi(x))dx + \sum_{l=1}^{J} \beta_l (P_l \phi)^2, \tag{1}$$

where the projection operator P_l is some local averaging operator defined as

$$P_l \phi = \int p_l(x)\phi(x)dx, \quad \int p_l(x)dx = 1. \tag{2}$$

The first term in (1) is the regularization term which imposes smoothness on the level set function ϕ. The second term is the fidelity term which forces the zero level set of ϕ align with the data set. If the data is uniformly distributed and the surface is well-resolved in some region, we set

$$G(\phi(x)) = |\nabla \phi(x)|^2. \tag{3}$$

Otherwise, in regions that lack points, more sophisticated regularization is required, for instance, defining $G(\phi(x)) = |\nabla \cdot \nabla \phi(x)|^2$. Furthermore, special regularization is needed for the anisotropic case when the coefficient function varies in different directions. For example in $\mathbb{R}^2$, when the coefficient function $a_1(x, y)$ in the x-direction differs from the coefficient function $a_2(x, y)$ in the y-direction, we set

$$G(\phi(x, y)) = a_{11}|\phi_x|^2 + a_{12}|\phi_x \phi_y| + a_{22}|\phi_y|^2 = \alpha_1|\phi_\xi|^2 + \alpha_2|\phi_\eta|^2$$

with some proper change of variables $(x, y) \to (\xi, \eta)$. In this paper, we will focus on the isotropic regularization (3). However, the proposed method can be easily extended to other cases as well.

The values of the weight function β_l should, in general, depend on the accuracy of data points, the density of data set, and the curvature of surface to be reconstructed. For simplicity, a fixed β is used for all data points in this paper.

We now provide the detailed formulation of the energy $E(\phi)$ and the projection operator P_l in the discrete setting. We first consider the two dimensional case. The discretization of the projection operator (2) takes the following general form:

$$P_l \phi = \sum_{k,n} p^l_{k,n} \phi_{k,n}, \quad \sum_{k,n} p^l_{k,n} = 1. \tag{4}$$

The energy functional (1) is discretized as

$$E(\phi) = \sum_i \sum_j \left(\frac{\phi_{i+1,j} - \phi_{i,j}}{h} \right)^2 + \left(\frac{\phi_{i,j+1} - \phi_{i,j}}{h} \right)^2 + \sum_l \beta_l [P_l \phi]^2. \tag{5}$$

Then at the grid point (i, j), the Euler-Lagrange equation is given by

$$\frac{1}{2} \frac{\delta E}{\delta \phi_{i,j}} = -\frac{\phi_{i+1,j} + \phi_{i-1,j} + \phi_{i,j+1} + \phi_{i,j-1} - 4\phi_{i,j}}{h^2} + \sum_{l=1}^{J} \beta_l p^l_{i,j} P_l \phi = 0, \tag{6}$$

where J is the total number of neighboring points of grid point (i, j). Operator P_l is a linear interpolation operator, and $P_l \phi$ represents the interpolated value of function ϕ at a point x_l. Fig. 1 shows the l-th point inside a grid cell associated to (i, j), with location determined by r_1 and r_2 $(0 \leq r_1, r_2 < h)$. The interpolation coefficients are:

$$p^l_{i,j} = \frac{(h - r_1)(h - r_2)}{h^2}, \quad p^l_{i,j+1} = \frac{(h - r_1)r_2}{h^2}, \tag{7}$$

$$p^l_{i+1,j} = \frac{r_1(h - r_2)}{h^2}, \quad p^l_{i+1,j+1} = \frac{r_1 r_2}{h^2},$$

$$P_l \phi = p^l_{i,j} \phi_{i,j} + p^l_{i,j+1} \phi_{i,j+1} + p^l_{i+1,j} \phi_{i+1,j} + p^l_{i+1,j+1} \phi_{i+1,j+1}.$$

In three dimensions, the corresponding Euler-Lagrange equation at a point (i, j, k) is given as

$$\frac{1}{2} \frac{\delta E}{\delta \phi_{i,j,k}} = -\frac{\phi_{i+1,j,k} + \phi_{i-1,j,k} + \phi_{i,j+1,k} + \phi_{i,j-1,k} - 6\phi_{i,j,k}}{h^3} + \sum_{l=1}^{J} \beta_l p^l_{i,j} P_l \phi = 0. \tag{8}$$

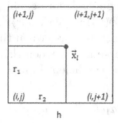

Fig. 1. The illustration of 2D interpolation operator. Here r_1 is the displacement in vertical direction and r_2 is the displacement in horizontal direction.

The natural choice for interpolation operator in three dimensions is trilinear interpolation. In the following formulation, r_1 and r_2 are defined as above, with r_3 denoting the displacement in the z dimension. The interpolation coefficients in three dimensions are given as:

$$
\begin{aligned}
p^l_{i,j,k} &= \frac{(h-r_1)(h-r_2)(h-r_3)}{h^3}, & p^l_{i,j+1,k} &= \frac{(h-r_1)r_2(h-r_3)}{h^3}, \\
p^l_{i+1,j,k} &= \frac{r_1(h-r_2)(h-r_3)}{h^3}, & p^l_{i+1,j+1,k} &= \frac{r_1 r_2 (h-r_3)}{h^3}, \\
p^l_{i,j,k+1} &= \frac{(h-r_1)(h-r_2)r_3}{h^3}, & p^l_{i+1,j+1,k+1} &= \frac{r_1 r_2 r_3}{h^3}, \\
p^l_{i+1,j,k+1} &= \frac{r_1(h-r_2)r_3}{h^3}, & p^l_{i,j+1,k+1} &= \frac{(h-r_1)r_2 r_3}{h^3},
\end{aligned}
\tag{9}
$$

In the next section, we will describe an efficient numerical implementation for solving (6) and (8).

3 Numerical Implementation

In this section, we consider two dimensional case and describe the process of obtaining a numerical solution for equation (6). The treatment for three dimensional case (8) is similar.

3.1 Initialization

In order to solve equation (6) numerically, we define a suitable computational domain. First, given the coordinates of the data $\{x_l\}$, we find the smallest rectangular box that contains the data. Since extra space should be allocated at the boundaries of the computational domain, we extend the rectangular box by some factor ρ. The typical choice for extension is $\rho = 1.2$. In case data set has a large hole (i.e. a large region with missing information), the value ρ is increased.

To make the algorithm efficient, we can restrict our computations within a narrow band containing the data. To construct the narrow band, we first compute the unsigned distance function $d(x)$ to the data set S by solving the following Eikonal equation (See equation (10))

$$|\nabla d(x)| = 1, \text{ on } \Omega \setminus \Gamma \quad \text{and} \quad d(\Gamma) = 0. \tag{10}$$

For the boundary conditions of (10), if the data point is not on the grid, we set the unsigned distance function d to zero at the nearest neighboring grid point to the data point in consideration. We then solve equation (10) using fast sweeping method [17, 18], an efficient algorithm with a computational cost of $O(N)$. The narrow band is obtained by thresholding unsigned distance function at value $\epsilon = \frac{1}{2}m_h h$, where h is the mesh size and m_h is the band width. Here, we denote $\Omega_1 = \{x : d(x) < \epsilon\}$ to be the set of grid points within the narrow band.

Once the narrow band is obtained, we need to find the grid points for the outer and inner boundaries in order to set up the boundary conditions for (6). Outer and inner boundary grid points are denoted as Γ_{in} and Γ_{out}, respectively, and are obtained using the Breadth-First Search (BFS) algorithm. First, we take a thin narrow band Ω_2, which is directly connected to Ω_1, such that the band width is equal to one grid point. Specifically, $\Omega_2 = \{x : \epsilon \leq d(x) \leq \epsilon + h\}$. Second, we select an arbitrary outer boundary grid point $(i, j) \in \Omega_2$. This can be easily done by taking the first grid point in Ω_2 since the outer boundary of the narrow band is usually reached while traversing the grid points. This grid point is then assigned to Γ_{out}. Third, we use BFS algorithm to find all the connected grid points among $(i \pm 1, j), (i, j \pm 1)$ which are assigned to Γ_{out}. The algorithm stops when there are no new grid points to be added into Γ_{out}. The complementary set of Γ_{out} is the inner boundary $\Gamma_{in} = \Omega_2 / \Gamma_{out}$.

If the data has spurious noise, e.g. contains points far from the meaningful data points, the noise and the data set would be disconnected. In this case, all connected components are obtained using the BFS algorithm as described above. We first select one starting point and find the maximally connected component. Then, we select another point in the complementary set and find another maximally connected component. The process is repeated until all data points

Fig. 2. Illustration of inner and outer boundaries

are traversed once. Since spurious noise is usually composed of relatively few data points, we let Γ_{out} and Γ_{in} be the largest and second largest connected components, respectively.

Once exterior and interior boundary grid point sets Γ_{in} and Γ_{out} are obtained, we can specify boundary conditions for $\phi(x)$. For the exterior boundary Γ_{out}, we set $\phi(x) = d(x)$; for the interior boundary Γ_{in}, we set $\phi(x) = -d(x)$; and for the points within the narrow band, we simply set $\phi(x) = 0$.

Fig. 2 shows the initialization of the narrow band associated with the given data set. The red dots and black dots enclose the blue data points with the equal distance $\epsilon = 5h$. The unsigned distance function at the red outer boundary and the black inner boundary points is equal to $5h$ and $-5h$, respectively. The grid points inside the narrow band are all set to zero.

3.2   Multigrid Narrow Band Solver

The problem defined by (6) with boundary conditions is a well-posed elliptic problem and can be efficiently solved using multigrid method. The Full Approximation Scheme (FAS) [19] solves equation (6) on multiple grids, starting with the coarsest and finishing at the finest grid. On coarser levels, a single cell may contain multiple data points. Since high resolution results are not required on coarser grids, mean coordinates of all data points in a given cell can be used to represent all such points for the purpose of calculating the interpolation operator P_l. After the equation is solved on a coarser level using a few Gauss-Seidel relaxations [19], the obtained solution is linearly interpolated onto the finer level. At the finest level, all data points within each cell are considered to be carrying equal weights. The solution obtained on the finest level is the desired final reconstruction. We usually use three-level construction method with the band width $m_h h$. Here, $m_h = 5$, and h is the mesh size for the corresponding level.

We are now ready to give an algorithm for the Multigrid Narrowband Surface Reconstruction:

Algorithm 1. Multigrid Narrow Band Surface Reconstruction

1: Given a set $\{x_l\}_{l=1,\ldots,N}$, solve Eikonal equation (10) to find $d(x)$.
2: Find a narrow band Ω_1 with band width ϵ to enclose the data set.
3: Find the outer boundary Γ_{out} and inner boundary Γ_{in} of the narrow band.
4: Set $\phi(x) = \epsilon$ on the outer boundary and $\phi(x) = -\epsilon$ on the inner boundary.
5: Denote l to be the coarsest level.
6: Solve equation (6) on level l within narrow band. If l is the finest level, then stop.
7: Linearly interpolate the solution from level l to a finer level $l + 1$.
8: Set $l := l+1$ and go to Step 6; or terminate the program if the finest grid is reached.

---

4   Experimental Results

In this section, we show the two and three dimensional results obtained using the proposed Multigrid Narrow Band surface reconstruction model to solve (6)

and (8). In our numerical examples, we found that $\beta = 0.5$ produces desirable results for most data sets. The value of β may change somewhat with changes in noise level and density for a given data set. Fig. 3 shows a two-dimensional point cloud and multigrid surface reconstruction results at four consecutive steps on different levels. The reconstruction captures additional details as the grid is refined, with the finest level having the most information.

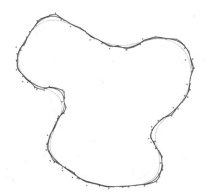

Fig. 3. Contours reconstructed on different levels, from coarsest to finest, are displayed in yellow, green, blue, and black colors

Fig. 4 shows a two-dimensional numerical result for a man-made hand with Gaussian noise. The hand shape has long and thin concave regions which are difficult to reconstruct using traditional methods. However, our method recovers the detailed information. We observe some inaccuracy at the ends of the fingers, which may be attributed to the use of uniform β for high curvature places.

Next we consider some three dimensional examples. Fig. 5 shows 3D multigrid surface reconstruction results for "bunny" point cloud on different levels. The difficulty of reconstructing this point cloud lies in the presence of several holes.

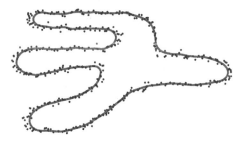

Fig. 4. Contour reconstruction of a man-made hand with Gaussian noise, where the noisy point cloud (in blue) and the reconstructed shape (in red) are displayed

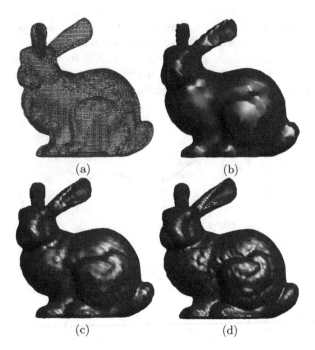

Fig. 5. Three dimensional reconstruction of "bunny". (a) Original point cloud. The results are obtained on (b) 65x65x51, (c) 129x129x101, and (d) 257x257x201 grids.

Table 1. Computational times for 3D data sets

data set	# points	grid size	time (in seconds)
bunny	35,947	257x257x201	6.5
dragon	437,645	301x213x133	7.2
buddha	144,647	149x365x149	6.7

We observe that finer features are recovered as the mesh size becomes smaller, and the holes are filled automatically. On $257 \times 257 \times 201$ mesh, the total computational time to obtain the final result on 2.93 GHz Intel Core 2 Duo CPU is 6.5 seconds (see Table 1).

Fig. 6(a) shows 3D surface reconstruction result from "dragon" point cloud. Since this data set contains numerous regions with missing information (i.e. small holes in the point cloud), it is difficult to reconstruct the corresponding surface. The proposed model is successful in capturing fine details and filling the holes in the surface.

Fig. 6(b) shows 3D surface reconstruction result from Buddha point cloud. Even though this data set contains no holes, it has small bridges. We observe that all fine features are captured well using the proposed surface reconstruction model.

(a)   (b)

Fig. 6. Three dimensional reconstruction of "dragon" (a) and "buddha" (b)

5   Conclusion

The two and three dimensional curve/surface reconstruction results presented in this paper demonstrate that our method is among the fastest surface reconstruction methods. Furthermore, the proposed method is robust to noise and can easily recover surfaces from point clouds with missing information. Important details and features of the point clouds are also well preserved in the reconstructed surfaces.

Acknowledgements. This work was supported in part by NIH GRANT, P20 MH65166, NSF GRANT, DMS-0714807, and NIH GRANT, U54 RR021813. The research of Igor Yanovsky was carried out in part at the University of California, Los Angeles, and in part at the Jet Propulsion Laboratory, California Institute of Technology, under a contract with the National Aeronautics and Space Administration.

References

1. Rogers, D.F.: An Introduction to NURBS. Morgan Kaufmann (2003)
2. Piegl, L., Tiller, W.: The NURBS book. Springer, Berlin (1996)
3. Amenta, N., Bern, M., Eppstein, D.: The crust and the β-skeleton: Combinatorial curve reconstruction. Graphical Models and Image Processing 60, 125–135 (1998)
4. Amenta, N., Bern, M., Kamvysselis, M.: A new Voronoi-based surface reconstruction algorithm. In: Proceedings of the 25th Annual Conference on Computer Graphics and Interactive Techniques, pp. 415–421. ACM, New York (1998)

5. Boissonnat, J.D.: Geometric structures for three dimensional shape reconstruction. ACM Trans. Graphics 3, 266–286 (1984)
6. Edelsbrunner, H.: Shape Reconstruction with Delaunay Complex. In: Lucchesi, C.L., Moura, A.V. (eds.) LATIN 1998. LNCS, vol. 1380, pp. 119–132. Springer, Heidelberg (1998)
7. Edelsbrunner, H., Mucke, E.P.: Three dimensional α shapes. ACM Trans. Graphics 13, 43–72 (1994)
8. Osher, S., Fedkiw, R.P.: Level set methods and dynamic implicit surfaces. Springer (2003)
9. Leung, S., Zhao, H.: A grid based particle method for moving interface problems. Journal of Computational Physics 228, 2993–3024 (2009)
10. Ruuth, S., Merriman, B.: A simple embedding method for solving partial differential equations on surfaces. Journal of Computational Physics (2007)
11. Carr, J.C., Fright, W.R., Beatson, R.K.: Surface interpolation with radial basis functions for medical imaging. IEEE Transactions on Medical Imaging 16, 96–107 (1997)
12. Zhao, H.K., Osher, S., Fedkiw, R.: Fast surface reconstruction using the level set method. In: Proceedings of the IEEE Workshop on Variational and Level Set Methods in Computer Vision 2001, pp. 194–201. IEEE (2002)
13. Zhao, H.K., Osher, S., Merriman, B., Kang, M.: Implicit and non-parametric shape reconstruction from unorganized points using variational level set method. Computer Vision and Image Understanding 80, 295–319 (2000)
14. Goldstein, T., Bresson, X., Osher, S.: Geometric applications of the split bregman method: Segmentation and surface reconstruction. Journal of Scientific Computing 45, 272–293 (2010)
15. Ye, J., Bresson, X., Goldstein, T., Osher, S.: A Fast Variational Method for Surface Reconstruction from Sets of Scattered Points. CAM Report 10-01 (2010)
16. Gandlin, R.: Multigrid solvers for inverse problems. Ph.D. thesis, Department of Computer Science and Applied Mathematics, The Weizmann Institute of Science, Rehovot, Israel (2004)
17. Zhao, H.: A fast sweeping method for eikonal equations. Mathematics of Computation 74, 603–628 (2005)
18. Tsai, Y., Cheng, L., Osher, S., Zhao, H.: Fast sweeping algorithms for a class of Hamilton-Jacobi equations. SIAM Journal on Numerical Analysis 41, 673–694 (2004)
19. Brandt, A.: Multigrid techniques. Ges. für Mathematik u. Datenverarbeitung (1984)

Real-Time Simulation of Ship Motions in Waves

Xiao Chen, Guangming Wang, Ying Zhu, and G. Scott Owen

Department of Computer Science
Georgia State University, Atlanta, USA

Abstract. We propose a new method for simulating ship motions in waves. Although there have been plenty of previous work on physics based fluid-solid simulation, most of these methods are not suitable for real-time applications. In particular, few methods are designed specifically for simulating ship motion in waves. Our method is based on physics theories of ship motion, but with necessary simplifications to ensure real-time performance. Our results show that this method is well suited to simulate sophisticated ship motions in real time applications.

1 Introduction

Ship motion is an important part of water based visual simulation. The most interesting ship motions are the ship oscillations as wave after wave acting upon the hull. In ship motion terms, waves cause the ship to roll, pitch, yaw, heave, sway, and surge. Previous works have not addressed ship motion effectively since most of the previous works are concerned with simulating realistic waves or fluid dynamics [5-7]. Some methods have been developed for simulating solids floating on water [3], [4]. These works often treat the solid as discretized object and apply physics calculation on a node by node base. Although the physics simulation is more accurate, it is not suitable for real time applications. Besides, these methods are not specifically developed for simulating ship motion.

Our method treats ship as a whole and calculates ship oscillations based on the forces generated by waves. The method is based on the physics theories of ship motion, but we made necessary simplifications for real time applications. As a result, our method can simulate realistic ship oscillation in reaction to waves with different directions, frequencies and amplitudes. More importantly, our method runs in real time and therefore is suitable for applications such as games, trainings, and visual simulations.

Compared with previous works, our method is more narrowly designed for simulating ship motion, but it is much more efficient and can present physically realistic ship oscillations. In addition, our method can detect whether a ship will capsize in a particular wave, which is useful for ship and cargo load animation.

The rest of the paper is organized as follows. Section 2 discusses the related work. In section 3 we present our main methodology. Section 4 contains more details our implementation and experiments. Section 5 is the conclusion and future work.

G. Bebis et al. (Eds.): ISVC 2012, Part I, LNCS 7431, pp. 71–80, 2012.

2   Related Work

To simulate the dynamic behaviors of a floating solid, it's important to study the interaction between fluids and solid, especially the effects fluid has on the floating object. There are a lot of research on simulating the interaction between fluid and solid [3-12]. In terms of coupling direction, fluid-solid coupling can be divided into three categories: one-way solid-to-fluid coupling, one-way fluid-to-solid coupling and two-way coupling [8]. The first type of one direction coupling, solid-to-fluid coupling, considers the motion of the rigid body as predetermined and the motion of the fluid is affected by the rigid body. One popular example would be a ball splashing into a pool of liquid [5-7]. We don't discuss more about one-way solid-to-fluid coupling since the motion of the fluid is out of the scope of this paper.

On the contrary, the other type of one direction coupling, one-way fluid-to-solid coupling simulates how fluid affects the motion of rigid bodies without being affected by the rigid bodies. For example, Chen and Lobb [3] simulated objects drifting on fluid as streak-line particles. Foster and Metaxas [4] animated solid object (e.g. tin cans) floating on water. They assumed the solid object is discretized and consists of a set of nodes, and the force applied on each node is calculated according to the pressure and velocity of the fluid. The motion of the solid is determined by applying Lagrange equation.

In two-way coupling [8-12], the solid object and the fluid influence each other's motion. Typically the coupling of fluid and solid is to set the velocity of the solid as boundary condition for the fluid, and use the fluid pressure as the boundary condition for the solid. For example, Genevaux et al. [10] represented the solid with a set of linked point mass. Taking all the forces into account, they calculate the overall force applied to each point and update its position independently. Their method doesn't conserve the torque and therefore couldn't handle the rotation of the solid effectively. Carlson et al. [8] treated the solid as fluid and the motion of the solid is simulated as that of fluid at first. Later on they enforced the rigidity by using a Lagrange multiplier. Batty et al. [9] approximated a J operator to map the pressure of fluid to net force and torque on the solid. However the mass matrix they used to approximate the volume weights is not consistent with those volumes. That becomes an issue when simulating buoyant solid in hydrostatic rest as indicated by the authors in the paper.

In this paper, we focus on simulating ship motion, particularly ship oscillation. Ship oscillation in waves can be considered a special case of one-way fluid-to-solid coupling. However, none of the previous fluid-to-solid and two way coupling methods has dealt with ship motion specifically. It may be argued that some of the previous methods are general enough to handle ship motion. But with the size and complex shape of a ship, it will be difficult to simulate ship motion in real time with the existing methods that treat 3D objects as discretized entities. In fact, there is a more efficient and physically realistic method to simulate ship motion. In this paper, we are exploring this method. Instead of treating ship as a discretized object and calculate forces per node, our method calculates the forces applied to the entire ship based on the theory of ship motion [2]. In other words, most previous methods are based on general physics theory while our method is based on the more specialized ship motion theory. As a result, our method is more efficient and realistic for simulating ship motion.

3   Methodologies

In this section, we describe our method to simulate realistic ship motions in various waves. Our method is based on the physics theories of ship motion but with necessary simplifications for real time simulations. Our method assumes that the fluid is incompressible and the gravity is constant. It also disregards the viscous effect and assumes that the ship has zero forward speed with arbitrary heading. This is because the underlying ship motion theories are also based on the same assumptions.

3.1   General Ship Oscillation Model

Ship motions are caused by multiple forces acting upon the hull. The total force on a ship is a linear combination of hydrostatic forces and hydrodynamic forces. Hydrostatic forces are restoring forces due to gravity and buoyancy. Hydrodynamic forces include first-order wave excitation forces, second-order wave excitation forces, radiation forces, and viscous forces. Here we ignore second-order wave excitation forces and viscous forces due to their complexity of calculation and also because they are insignificant compared to other components.

First-order wave excitation forces can be further divided into Froude-Krylov forces and diffraction forces. Radiation forces can be divided into added-mass forces (with is proportional to wave accelerations) and damping forces (which is proportional to wave velocities).

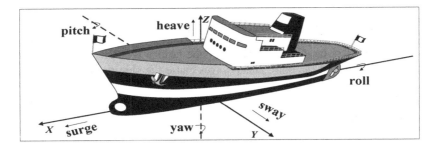

Fig. 1. The six degrees of freedom of a ship

Wave excitation forces cause a ship to oscillate in six degrees of freedom [2]: 1, surge; 2, sway; 3, heave; 4, roll; 5, pitch and 6, yaw (Fig.1). More specifically, let (X, Y, Z) be a right-handed coordinate system aligned with the center of the ship, with Z axis pointing upward. The surge, sway, and heave are the translatory displacements along X, Y and Z axes; the roll, pitch, and yaw are rotations along X, Y, and Z axes.

Based on ship motion theories [1], the total force acting on the wetted surface of the ship is given by:

$$F = (-\rho g \int_S n\zeta dS + \rho g \int_S n\zeta_0 dS) + (\rho \sum_{j=1}^6 \zeta_j^0 \omega^2 \sin \omega t \int_S n\varphi_j dS) +$$

$$(\rho A \sin \omega t \int_S n\varphi_A dS) \tag{3-1}$$

Where ρ is the density of the fluid; g is the gravity; n is the normal of the ship surface; S is the area of wetted surface; ω is the wave frequency; ζ is the submergence under unperturbed surface and ζ_0 is the free surface elevation; ζ_j^0 is oscillation amplitudes, where j ranges from 1 to 6, representing six degrees of freedom; A is wave amplitude; φ_j and φ_A are radiation potential and diffraction potential; t is the instance of time.

In Eq. 3-1, the first term represents the hydrostatic component. The second term is the radiation force, which is comprised of added-mass and damping forces. The third term is the first-order wave exciting force, which is the sum of Froude-Krylov force and diffraction force. Froude-Krylov force F_{FK} is determined by the integration of wave induced pressure as if the ship is fully transparent for incident waves. In Eq. 3-1, radiation and first-order wave exciting forces can be calculated based on the potential flow theories [1], but it is impractical to implement it in real time applications. To simplify the calculation, we assume that the ship is a symmetric object with a slender hull, and a random incident wave can be decomposed into a combination of head waves and transverse waves. Both assumptions are reasonable as they are adopted in ship motion theories [2]. As a result, Eq. 3-1 can be replaced by simplified equations Eq. 3-2 and Eq. 3-7.

A random incident wave can be decomposed into head waves and transverse waves. Therefore, we calculate ship transformations based on the forces generated by the head waves and transverse waves. The combined transformations produce the ship oscillations in the random incident wave.

3.2   Simulating Ship Motions in Head Waves

This section considers the force generated by the head waves. Head waves move in the opposite direction of the ship. Head waves cause the ship to pitch around the Y axis and heave around the Z axis.

The force from small head waves acting on the ship F_{Head} can be calculated by the following equation [1]:

$$F_{Head} = A_H \int_L (\rho g a - \omega_H^2(\rho g + A_{33})) \cos k\eta \, d\eta \cdot \sin \omega_H t +$$

$$\omega_H A_H \int_L B_{33} \cos k\eta \, d\eta \cdot \cos \omega_H t \tag{3-2}$$

Where ρ, g, and t are the same parameters as in Eq. 3-1. A_H and ω_H are head wave amplitude and frequency, respectively; L is the length of the hull; η is the distance from the force acting point to the midship; a and k are constants. B_{33} and A_{33} are damping and added-mass coefficients in heave, respectively. Generally, B_{ij} is the damping coefficient in the i-th direction when the ship oscillates in j-th motion, where i and j range from 1 to 6, representing the six degrees of freedom. Damping coefficient is proportional to the wave velocity. A_{ij} is an added-mass coefficient, which is proportional to wave acceleration. The damping coefficient B_{ij} and added-mass A_{ij} can be computed by:

$$B_{ij} = \int_{-L/2}^{L/2} \frac{\rho g}{\omega_H^3} \cdot \left(\frac{A_H}{\zeta_j^0}\right)^2 dL \tag{3-3}$$

$$A_{ij} = b \cdot B_{ij} \tag{3-4}$$

Where ρ, g, and ζ_i^0 are the same as those in Eq. 3-1 and A_H and ω_H are same as in Eq. 3-2. L is the length of hull. For fast computation, the value of A_{ij} is calculated by Eq.3-4 where b is a constant. Similar to Eq. 3-1, Eq. 3-2 is comprised of the hydrostatic restoring force, the Froude-Krylov force and the radiation force. We neglect the diffraction force since it is insignificant compared with the Froude-Krylov force. Solving Eq. 3-2, Eq. 3-3 and Eq. 3-4, and we can calculate the amplitude of pitch oscillation α, and heave oscillations β, as follows:

$$\alpha = \frac{F_{Head} \frac{1}{m+A_{33}}}{\sqrt{(\frac{\rho g}{m+A_{33}} - \omega_H^2)^2 + 4(\frac{\rho g}{m+B_{33}})^2 \omega_H^2}} \tag{3-5}$$

$$\beta = \frac{F_{Head} \frac{1}{m+A_{55}}}{\sqrt{(\frac{\rho g}{m+A_{55}} - \omega_H^2)^2 + 4(\frac{\rho g}{m+B_{55}})^2 \omega_H^2}} \tag{3-6}$$

Where m is the mass of the ship and B_{55} and A_{55} are damping and added-mass coefficients in pitch. B_{33} and A_{33} are damping and added-mass coefficients in heave.

3.3   Simulating Ship Motions in Transverse Waves

This section considers the forces generated by transverse waves. Transverse waves move perpendicularly toward the hull of the ship. Transverse waves cause the ship to roll around the X axis and heave along the Z axis. According to ship motion theories [1], the force generated by transverse waves acting along the ship's hull can be calculated by the following equation:

$$F_{Transverse} = (\rho g A_{wp} - \omega_T^2 A_{33}) A_T \sin \omega t + B_{33} \omega_T A_T \cos \omega_T t \tag{3-7}$$

Where ρ, g and t are the same as those in Eq. 3-1 and A_{wp} is the water-plane area; A_T and ω_T are transverse wave amplitude and frequency, respectively. A_{33} and B_{33} are added-mass coefficients and damping coefficients, respectively (see Eq. 3-3 and Eq. 3-4). The total force in Eq. 3-7 is the linear combination of Froude-Krylov force and the added-mass and damping forces. Solving Eq. 3-7, Eq. 3-3 and Eq. 3-4, then we can calculate the amplitude of roll oscillation γ and heave oscillation δ as follows:

$$\gamma = \frac{F_{Transverse} \frac{m}{m+A_{44}}}{\sqrt{(\frac{\rho g}{m+A_{44}} - \omega_T^2)^2 + 4(\frac{\rho g}{m+B_{44}})^2 \omega_T^2}} \tag{3-8}$$

$$\delta = \frac{F_{Transverse}\frac{m}{m+A_{33}}}{\sqrt{(\frac{\rho g}{m+A_{33}}-\omega_T^2)^2+4(\frac{\rho g}{m+B_{33}})^2\omega_T^2}} \tag{3-9}$$

Where m is the mass of the ship, B_{44} and A_{44} are damping and added-mass coefficients in roll and B_{33} and A_{33} are damping and added-mass coefficients in heave (see Eq. 3-3 and Eq. 3-4). The results obtained from Eq. 3-8 and Eq. 3-9 determine how the ship rolls and heaves in transverse waves.

The transverse wave-induced force causes the ship to roll around X axis, while the hydrostatic force tries to restore the ship to its rest position. If the rolling angle $\theta >=$ 72 degree [2] around the X axis, the ship will capsize. Therefore, to keep the ship afloat, the hydrostatic restoring force F_R should be proportional to θ: $F_R = \rho g\ \vartheta$ $F_{Transverse}/72$. The interactions between transverse wave-induced force and hydrostatic restoring force produce the lateral ship oscillations.

3.4   Simulating Ship Motions in Random Incident Waves

A random incident wave can be decomposed into a linear combination of a head wave and a transverse wave. This process can be regarded as multiple layers of wave combined to represent the random incident wave (Fig. 2), and is generally accepted in shop motion theories. Therefore, the ship motions in random incident waves can be simulated by combining the oscillations caused by the decomposed transverse waves and head waves. For simplicity, in this method a random wave is decomposed on the horizontal X-Y plane.

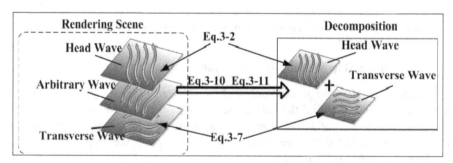

Fig. 2. The six degrees of freedom of a ship

A random incident wave is identified by its frequency $\omega_{Arbitrary}$ and amplitude $A_{Arbitrary}$. In order to calculate forces based on Eq. 3-2 and Eq. 3-7, we decompose the frequency and amplitude of a random incident wave to a transverse wave component and head wave component. Assume that the angle between an arbitrary incident wave and the ship is α, then the frequencies and amplitudes of head and transverse waves ω_H, ω_T, A_H and A_T can be calculated by:

$$\omega_H = \omega_{Arbitrary} \cdot \cos \alpha, \ \omega_T = \omega_{Arbitrary} \cdot \sin \alpha \tag{3-10}$$

$$A_H = A_{Arbitrary} \cdot \cos \alpha, \ A_T = A_{Arbitrary} \cdot \sin \alpha \tag{3-11}$$

By feeding ω_H, ω_T, A_H and A_T into Eq. 3-2 and Eq. 3-7, we can proceed to calculate the amplitude of pitch, roll, and heave. More details will be discussed in the next section.

4   Implementation and Experiment Results

In this section, we discuss our implementation and show the rendering results obtained from our simulation. Our simulation is rendered in real time with an average frame rate of 46 fps at a resolution of 1280×720. The simulation is built with the Unity3D game engine [16] on a PC with Intel core i7 Q7401.73GHz, 4GB RAM, and NVIDIA GeForce 425m GPU with 1GB RAM.

The waves in our simulation are modeled by combining multiple sinusoids. Each wave is defined by its frequency, amplitude, and direction. Our simulated environment is an ocean, which has a large scope and no boundary. In addition to the waves and ship motions, other visual effects, such as reflection and refraction are implemented as well.

Rendering Process

1: First, the values of the parameters in Eq.3-2 to Eq. 3-11 are specified. A sample of the main parameters is shown in Table.1.

2:   The system automatically generates random waves. Each wave is decomposed into a head wave and a transverse wave (see section 3.4).

3: Calculate the forces induced by the head waves and transverse waves, and then calculate the amplitude of pitch, roll, and heave (see section 3.2 to 3.3).

The calculation of surge, sway, and yaw are simplified for performance. Specifically, the amplitude of surge is a simple linear function of the amplitude of pitch.

The amplitude of sway is a linear function of the amplitude of roll. The amplitude of yaw is also a linear function of the amplitude of roll.

All the transformations are applied to the mass center of the ship to produce ship oscillation.

4: Repeat steps 2 to 3 to simulate continuous ship oscillations. .

Table 1. Simulation Parameters and Results

Value of Constants				
Mass of Ship	**Length of Hull**	**a**	**k**	**b**
6,000*tons*	70*meters*	15	2.5	1.76
Samples of Simulation Parameters and Results				
Wave Frequency	**Wave Amplitude**	**Pitch Amplitude**	**Heave Amplitude**	**Roll Amplitude**
2	0.95 *meters*	1.2 *degree*	0.12 *meters*	2.8 *degree*
4	0.95 *meters*	2.8 *degree*	0.26 *meters*	3.9 *degree*
6	0.95 *meters*	3.7 *degree*	0.45 *meters*	5.4 *degree*
8	0.95 *meters*	5.6 *degree*	0.54 *meters*	6.7 *degree*

In Table 1, we show the values of mass of ship, length of hull and constants of a, k and b that we used in the simulation. This table includes the parameters for generating waves as well.

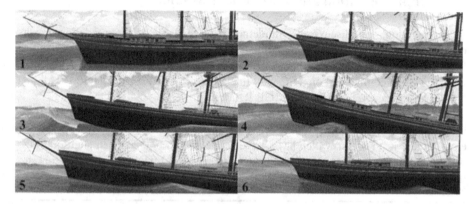

Fig. 3. Screenshots of ship oscillation in head waves

Fig. 3 shows a cycle of ship oscillation in head waves, where the ship first heaved and pitched (Fig.3-1 to Fig.3-4) due to the head wave and then was restored (Fig.3-5 to Fig.3-6) by hydrostatic forces. The force generated by the head wave is calculated with Eq. 3-2, and it causes the ship to pitch and heave.

Fig. 4. Screenshots of ship oscillation in transverse waves

Fig.4 illustrates a cycle of ship oscillation caused by transverse waves. The ship was in the buoyant equilibrium at the beginning of the simulation. Once transverse waves act on the hull, it begins to roll and heave. Hydrostatic forces then try to restore the ship to its initial position.

Fig. 5. Screen Shots of Ship Oscillations in Random Waves

Fig.5 shows the ship oscillations in random waves.

5   Conclusion and Future Works

We have presented a novel method to simulate ship motion. A typical ship has six degrees of freedom: pitch, roll, yaw, heave, sway, and surge. To calculate the ship oscillations, we first decompose a random incident wave into head waves and transverse waves. The forces are calculated for the head wave and transverse wave, respectively. From the head wave force we calculate the amplitude of pitch and heave. Surge is calculated in proportion to the pitch. From the transverse wave force, we calculate the roll and heave. Sway and yaw are calculated in proportion to the roll amplitude. By combining the above transformations the ship motion is produced. Our implementation demonstrates that our algorithm simulates smooth ship oscillations in real time and is visually realistic.

This method is based a number of assumptions and uses a number of simplified physics models. In the future, we plan to improve on these models to produce more physically realistic ship oscillation. For example, in this study we assume that the ship has zero forward speed. In the future, we will integrate forward speed into the model. Our current method also assumes that the waves are regular waves. The next step is to extend our models to handle irregular waves.

References

1. Kornev, N.: Ship Dynamics in Waves. Technical Report, University of Rostock (2011)
2. Salvesen, N., Tuck, E.O., Faltinsen, O.: Ship motions and sea loads. In: Annual Meeting of The Society of Naval Architectures and Marine Engineers, New York (1970)
3. Chen, J.X., Lobo, N.D.V.: Toward interactive-rate simulation of fluids with moving obstacles using Navier-Stokes equations. Graphical Models and Image Processing 57(2), 107–116 (1995)

4. Foster, N., Metaxas, D.: Realistic animation of liquids. Graphical Models and Image Processing 58(5), 471–483 (1996)
5. Foster, N., Metaxas, D.: Controlling fluid animation. In: Proceedings of Computer Graphics International, pp. 178–188 (1997)
6. Foster, N., Fedkiw, R.: Practical animation of liquids. In: Proceedings of ACM SIGGRAPH, pp. 23–30 (2001)
7. Enright, D.P., Marschner, S.R., Fedkiw, R.P.: Animation and rendering of complex water surfaces. In: Proceedings of ACM SIGGRAPH, pp. 736–744 (2002)
8. Carlson, M., Mucha, P.J., Turk, G.: Rigid fluid: animating the interplay between rigid bodies and fluid. In: Proceedings of ACM SIGGRAPH, pp. 377–384 (2004)
9. Batty, C., Bertails, F., Bridson, R.: A fast variational framework for accurate solid-fluid coupling. In: Proceedings of ACM SIGGRAPH (2007)
10. Génevaux, O., Habibi, A., Dischler, J. M.: Simulating fluid-solid interaction. In: Proceedings of Graphics Interface, pp. 31–38 (2003)
11. Robinson-Mosher, A., Shinar, T., Gretarsson, J., Su, J., Fedkiw, R.: Two-way coupling of fluids to rigid and deformable solids and shells. In: Proceedings of ACM SIGGRAPH (2008)
12. Keiser, R., Adams, B., Gasser, D., Bazzi, P., Dutre, P., Gross, M.: A unified Lagrangian approach to solid-fluid animation. In: Proceedings of Eurographics/IEEE VGTC Symposium Point-Based Graphics, pp. 125–148 (2005)
13. Salvesen, N., Smith, W.E.: Comparison of Ship-Motion Theory and Experiment for Mariner Hull and Destroyer with Modified Bow. National Shipbuilding Research Documentation Center, Washingtown, D. C., Report 3337 (1970)
14. Vugts, J.H.: Cylinder Motions in Beam Waves. Netherlands Ship Research Center TNO Report No. 115S (1968)
15. Lautrup, B.: Physics of Continuous Matter: Exotic and Everyday Phenomena in the Macroscopic World, 2nd edn. Taylor & Francis (2011)
16. Unity - 3D Game Engine, http://unity3d.com/

Adaptive Spectral Mapping for Real-Time Dispersive Refraction

Damon Blanchette and Emmanuel Agu

Worcester Polytechnic Institute

Abstract. Spectral rendering, or image synthesis utilizing the constituent wavelengths of white light, enables the rendering of iridescent colors caused by phenomena such as dispersion, diffraction, interference and scattering. Dispersion creates the rainbow of colors when white light shines through a prism. Caustics, the focusing and de-focusing of light through a refractive medium, can be interpreted as a special case of dispersion where all the wavelengths travel along the same paths. In this paper we extend Adaptive Caustic Mapping (ACM), a previously proposed caustics mapping algorithm, to handle physically-based dispersion. Our proposed method runs in screen-space, and is fast enough to display plausible dispersion phenomena at real-time frame rates.

1 Introduction

This paper focuses on rendering physically accurate dispersive refraction at real-time frame rates. Figure 1 shows four examples of dispersive refraction, i.e. different wavelengths of light refracting at different angles, rendered with our technique.

The key difference between dispersion and caustics is that while the wavelengths of light are refracted along different paths in dispersion, all wavelengths are refracted along the same paths to generate caustics. By drawing on these similarities, we have extended a real-time caustics algorithm to render dispersive refraction in real time.

Specifically, we extended the Adaptive Caustic Mapping (ACM) algorithm [20] to perform real-time spectral dispersion. ACM is a real-time image-space method of generating refractive caustics on programmable graphics hardware. Our method, which we call Adaptive Spectral Mapping (ASM), begins with the ACM algorithm

Fig. 1. Four images generated with our algorithm. All four scenes are using seven wavelength samples, and perform between 15 and 30 frames per second.

G. Bebis et al. (Eds.): ISVC 2012, Part I, LNCS 7431, pp. 81–91, 2012.

but adds spectral refraction calculations at object surfaces. To create spectral maps, we simulate external dispersion by refracting seven wavelengths at the surface of refractive objects. In a separate deferred rendering pass, we also calculate internal dispersion, which occurs when white light is split into component colors inside a refractive object such as the colors seen inside diamonds.

The rest of the paper is as follows. Section 2 describes related work. Section 3 gives some background on caustics rendering. Section 4 describes our technique for rendering spectral dispersion using Adaptive Spectral Maps (ASMs). Section 5 describes our implementation. Section 6 describes our results and section 7 is our conclusion and future work.

2   Related Work

Initially, spectral rendering was described in the context of ray or path tracing. Cook and Torrance [2] presented a method for rendering materials that takes into account light wavelengths and spectral energy distribution. Thomas [17] and Musgrave [13] presented specifically on dispersion using ray tracing methods.

Most recent real-time spectral rendering research uses the GPU to perform wavelength calculations. Guy and Soler [7] presented on the real-time rendering of dispersion inside gemstones. Kanamori et al. [12] published on the physically accurate display of rainbows under different atmospheric conditions. Ďurikovič et al. [3] presented an entire spectrally-based framework for interactive image synthesis that could display multilayered thin-film interference.

The work most similar to ours is by Sikachev et al. [16], in which they present spectral dispersion through gems. They differ in that their algorithm can only project dispersion onto planes as opposed to arbitrary surfaces as this paper presents.

3   Background

3.1   Caustics Rendering

Since our proposed technique is an extension of Adaptive Caustic Mapping, we now review the literature on caustics rendering. Kajiya [11] and Shirley [15] both described caustics generation using ray and path tracing algorithms. Jensen's photon mapping algorithm [10] can also generate caustic effects, using a special "caustics photon map" where extra photons are sent in order to create higher resolution data.

The idea for "caustic mapping," a far better performance solution in which a special texture is created containing caustic data that is projected onto a scene similar to shadow mapping, began with the Shah et al. image-space technique [14]. Wyman et al. [18] presented a caustic mapping algorithm similar to Shah's that also operated in image-space. He then extended his own algorithm with a hierarchical caustics generation method [19] that used mipmaps and a reduced resolution version of the scene to increase algorithm speed. Wyman et al. later improved this hierarchical method to yield Adaptive Caustic Mapping [20], which is described in detail in the next section since this work extends it.

3.2   Adaptive Caustic Mapping

We chose to extend Adaptive Caustic Mapping in particular because it solved several issues inherent in other caustic mapping algorithms: notably aliasing due to insufficient sampling and excessive temporal noise due to sampling variations. ACM uses an importance-based adaptive photon sampling algorithm that increases quality while also speeding up the rendering of caustics beyond other methods of similar quality, and in addition they utilize a deferred rendering process that displays refractive objects more quickly than other methods.

For a thorough description of ACM, we refer the reader to Wyman's original paper [20]. However, to aid in understanding, we will summarize the important points of the algorithm here and then describe our spectral dispersion extension.

ACM differs from other caustic mapping algorithms in its photon emission and refractive object "locating" phase. ACM starts with a reduced resolution view of the scene from a light source using mipmaps, and emits only a few regularly spaced photons into that image. In a loop, moving up one mipmap level at a time, each photon that actually hits a refractive object is subdivided into four new photons, increasing photon density and thus the resolution of the caustics. Photons that do not hit a refractive object are simply discarded and never processed. When this phase is completed, the photon buffer contains a high-resolution set of points that all intersect the surface of the refractive object. The photons are then refracted through the object using the normal values at each pixel and splatted onto the spectral map.

The display of the refractive objects in the scene is completed in a separate deferred pass at the end. Pixels that lie on a refractive object's surface are treated as photons, as with the caustics calculations. The photon is refracted once using the front-facing normal at the current pixel's location, and then a second time at the back-facing surface. It is then projected out to the background geometry, where a texture fetch is performed to get the color for that pixel.

4   Spectral Dispersion Using Adaptive Spectral Maps

To create our spectral maps, extensions were made to both the caustic generation algorithm and the deferred refraction algorithm. We chose to sample seven wavelengths that are evenly distributed through the visible spectrum.

Refraction angles of light between two mediums with different refractive indices can be calculated using Snell's Law [8]. Taking into account wavelength, the refractive index can be calculated using Cauchy's equation [1].

For our purposes, it is sufficient to use the following two-term form of Cauchy's equation, with λ representing wavelength, initially used by Musgrave [13]:

$$n(\lambda) = A + \frac{B}{\lambda^2} .$$

$$(1)$$

The A and B coefficients are based on physical measurements and can be found in tables in various sources such as physics textbooks and the Internet [5].

Just before each photon refracts at the front surface of the object, it is split into seven separate photons, and each new one is refracted according to the index of refraction generated by Cauchy's equation. Each of the seven photons is then refracted a second time on the back-facing surface of the object, after which its final position is calculated for splatting into the spectral map. Figure 2 illustrates this process.

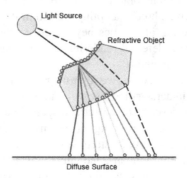

Fig. 2. Refraction using seven samples. The dotted line indicates the original ACM algorithm, and the solid lines are our extension.

In Figure 2, the small yellow circles are the photons. Until they are splatted into the spectral map, they are regarded only as wavelengths. Only just before splatting are they converted from a wavelength to an RGB value.

At this point the spectral map is complete and ready to be projected into the scene. The spectral map is a texture that contains the final locations of photons that have been refracted through the specular object according to their wavelengths and converted to RGB colors.

Once the spectral map has been created and projected into the scene like a shadow map, dispersion within the transparent object is calculated in a completely separate pass at the end. In ACM the color of each pixel on the surface of the refractive object is calculated using a single background color texture fetch. However, with ASM, we perform seven texture fetches – one for each wavelength sample. The location of the texture fetch on the background is calculated by following a ray of light as it passes through the refractive object and intersects the background texture.

The color of the texel chosen from the background texture is altered by the color of the wavelength that hits it, so if all wavelengths arrive at the same location or the same color, then the final color of the pixel on the refractive object is exactly the same as the background. The wavelength is converted to a RGB value here, when it is calculated based on the color of the background texture.

4.1   Filling the Gaps

One of the major issues with spectral rendering using discrete sampling of the spectrum is that gaps or empty portions occur in the resulting spectral map, as shown in figure 3.

We experimented with the simplest brute-force method of fixing gaps by testing our algorithm with 21 samples instead of seven. This indeed reduced the problem to a degree, but gaps still showed up where dispersion between colors was large. In addition, there was a 75% drop in frame rate with 21 samples.

Fig. 3. Problems with discontinuous caustics when using seven samples. Each sample color is clearly visible, with gaps between the colors.

Sikachev et al. proposed interpolating colors between the caustics that do exist in order to solve the color gap problem [16]. They integrate the interpolation results for each point by performing additive blending, and use a given step size which is taken in the view space coordinates.

We propose a similar method using texel marching: for each texel in the caustic map that is not already illuminated, a step is performed one pixel at a time, horizontally and vertically from it. If a colored texel is found, its color is mixed with the current texel's color, akin to interpolating the colors around a gap to fill it in.

Figure 4 illustrates the idea behind our gap-filling algorithm.

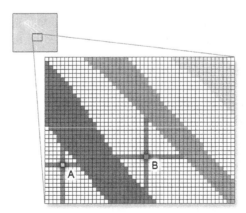

Fig. 4. A diagram showing how our filling algorithm works. Each grid square is one texel in the spectral map.

The gray pixels emanating horizontally and vertically from A and B represent the texel marching step. Texel A is set to a light red color because its vertical and horizontal neighbors are either from the red band or from nothing at all. Pixel B is set to a mixture of the red and orange color bands due to its proximity to both colors in the spectral map. Figure 5 shows images of the spectral map before and after our gap-filling procedure.

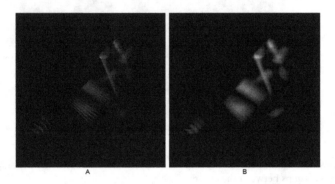

Fig. 5. The spectral map texture before (A) and after (B) filling in the gaps

The chosen number of steps is a tradeoff between better filling with blurry dispersion or sharper dispersed colors with the possibility of gaps not completely filled in. We found that a step size of 20 is a good balance between gap filling and blurriness.

5   Implementation

We implemented using C/C++ and OpenGL 4.2, with vertex, geometry, and fragment shaders written in GLSL. The video card utilized was an NVIDIA GeForce GTX480 in a Windows 7 environment.

We began by implementing Wyman's Adaptive Caustic Mapping algorithm [20].

The photon splatting shaders were extensively modified by the insertion of a new geometry shader to perform photon splitting into seven samples and to handle refraction for each wavelength. The fragment shader was altered to convert the wavelength values to RGB.

Figure 6 shows the entire pipeline for this project from beginning to final image, and each box describes a separate render pass. The yellow, blue, and green boxes in the background show how those passes are being rendered – whether it is from the light's view, from the camera's view, or in image space (on a full-screen quad). This diagram also compares our ASM algorithm with the original ACMs: each white box is unaltered from the original ACM algorithm, light red boxes are altered from the original ACMs, and pass 5, the dark red box, is a completely new pass.

Pass 4's alterations are shown in the pseudocode in figure 7. Both parts 2 and 3 were edited to use our Cauchy equation-calculated refractive indices per wavelength.

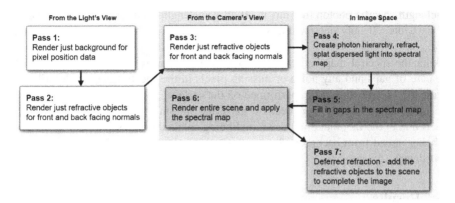

Fig. 6. Our rendering pipeline from beginning to end, completed each frame. This diagram shows both our ASM algorithm and the original ACM algorithm: the red boxes are passes altered from ACM, and pass 5, with the dark red background, is a completely new pass.

```
1. for each photon wavelength
2.    refract at front surface;
3.    refract at back surface;
4.    emit photon;
5.    compute gaussian intensity;
6.    convert wavelength to RGB;
7.    splat into spectral map;
```

Fig. 7. Photon splat pseudocode for pass 4 from figure 6

Part 4 is a geometry shader requirement, just there to emit the photon/vertex after refraction. Part 5 is identical to the ACM author's original code. In our newly created part 6, we convert the wavelength for that photon into an RGB value, described in the next paragraphs. Part 7 is a simple call to gl_FragData, required for all fragment shaders.

Since it is known exactly which wavelengths are being sent to the splat shader, a constant red, green, and blue value can be set for each one. As mentioned previously, these values must be carefully chosen to make sure they sum to white [13]. The color approximations are based on using the CIE color matching functions to get the relative contributions of light from wavelength, and converting them to XYZ color space coordinates [4]. The color matching functions can be described by the integral:

$$X = \int_0^\infty I(\lambda)\overline{x}(\lambda)d\lambda. \tag{2}$$

Where $I(\lambda)$ is the spectral power distribution, $\overline{x}$ is the color-matching function, and λ is the wavelength in nanometers. The Y and Z components are calculated in the same way. From the XYZ coordinates, it is possible to get RGB values using the CIE color space. In this way, a particular pixel's color in the scene is summed for each wavelength-specific photon that hits it. If all wavelengths end up on the same pixel, it will be white. If only one photon wavelength hits a pixel, the pixel will only be that color.

After the spectral map is created, the next pass performs our gap filling shader to take care of gaps and any noise or missing pixels in the spectral map as described in section 4.1. Gap-filling is not performed on the surface of the refractive object as it is with the spectral caustic map because gaps and missing pixels do not occur there. This is due to the fact that photons are not being splatted into a separate map – colors are being pulled from background geometry, which always exists (or is black if nothing is there).

6   Results

Table 1 contains performance data for our algorithm using both seven and 21 samples compared to Adaptive Caustic Mapping. As can be seen in the table, the extra photons needed for dispersion reduces performance in some scenes, and increasing wavelength samples severely reduced speed in all scenes. The sphere, at least with seven samples, still performs at the same speed as with ACM, most likely due to its simplicity and the small size of its footprint from the view of the light. The gem, being composed of far fewer faces than all the other objects, still performs slower than the sphere since it is rendered onto more fragments due to its size, which is an image-space algorithm issue discussed in the following paragraphs. We believe the glass on the table also performed so poorly because of its size in the light's view.

Table 1. Frame rates for our test objects

Object	Faces	ACM	ASM, 7 samples	ASM, 21 samples
Sphere	5120	60	60	19
Ring	65536	27	20	9
Gem	24	60	40	10
Bunny	138902	12	10	6
Glass on Table	12137	60	16	5

Table 2. Table showing relative number of pixels taken up by a refractive sphere as seen from the light source and a frame rate comparison

Frame Rate (frames per second)	Percentage of total pixels covered by object
60	2.5% (sphere on "floor" of Cornell box)
40	4%
30	7%
20	13%
10	33%

Since this algorithm runs in image space, the number of pixels covered by the refractive object from the light's view has an impact on frame rate. The closer the object is to the light, the more pixels involved in caustic calculations, and the slower the performance. On the other hand, this results in better the quality. The relationship between this number of pixels and frame rate is closely tied - table 2 shows what happens as the sphere is moved closer to the light source.

Figure 8 shows a comparison between a screenshot of our software and a ground truth image, which is the same scene rendered with the offline engine LuxRender. Part A was performing at 40 frames per second, and part B took one hour to render. A few things to note: first is that the large caustic directly under the gem is very similar in shape, size, and color in both images. However, there are a couple discrepancies, one of which can easily be explained.

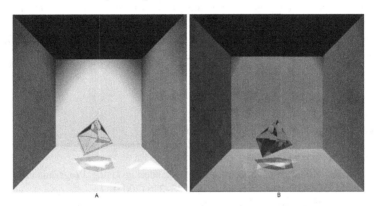

Fig. 8. A comparison between our algorithm, A, and a ground-truth render of the scene, B

The caustics on the walls in the ground truth image were created by reflective caustics, which our algorithm, and indeed ACM, does not simulate. A reason for this is that reflective caustics are more difficult to simulate due to possible extreme changes in a light path's direction, though this could be overcome by using a cube map for the spectral map [14]. The extra caustics on the floor of our screen capture that do not appear in the ground truth are possibly there as a result of imperfect sampling of the refractive object for photon placement.

As for the gem itself, looking closely near the top one can see there is a yellowish tinge above the red color, and on the right there is a green color at the intersection of the walls. These color shifts are a result of utilizing spectral calculations, and would not be present using an algorithm that does not account for the wave nature of light.

There are other dissimilarities between the images as well - the most obvious being the color of the gem to the bottom left. The refraction appears to be correct however, because the lines and positioning of the walls are quite similar.

7   Conclusion and Future Work

We have presented Adaptive Spectral Mapping, a spectral dispersion extension to the proposed algorithm Adaptive Caustic Mapping. Our algorithm displays a plausible approximation of the dispersion phenomenon of light, and does so at interactive and real-time frame rates. Our ASM algorithm is one of the first of its kind, bringing spectral rendering one step closer to being fully displayed in real-time contexts such as games.

There are some limitations to our algorithm, however. The gap-filling procedure creates horizontal and vertical lines in some situations due to our sampling process. This could be ameliorated with a more random sampling method, which might introduce a temporal cohesion issue, but there would be fewer vertical and horizontal lines.

Many opportunities exist for future directions of research. The first could be to extend ASMs to simulate reflective caustics. Others include extending ASMs to display other spectral phenomena that require wavelength calculations, such as diffraction and thin-film interference. Integrating a fast volumetric caustics algorithm with ours would produce beautiful images, along the lines of recent research such as [9]. A distance-aware blurring algorithm such as Screen-Space Soft Shadows [6] could be modified to work with our spectral maps to make them more physically accurate, and might also help with more distant gaps as well.

References

1. Chartier, G.: Introduction to Optics. Springer (2005)
2. Cook, R., Torrance, K.: A Reflectance Model for Computer Graphics. ACM Transactions on Graphics 1(1), 7–24 (1982)
3. Ďurikovič, R., Kimura, R.: Spectrum-Based Rendering Using Programmable Graphics Hardware. In: SCCG 2005 Proceedings, pp. 233–236 (2005)
4. Evans, G.F., McCool, M.D.: Stratified Wavelength Clusters for Efficient Spectral Monte Carlo Rendering. In: Graphics Interface (1999)
5. Gooch, J.W.: Encyclopedic Dictionary of Polymers, vol. 1. Springer (2010)
6. Gumbau, J., Chover, M., Sbert, M.: Screen Space Soft Shadows. In: GPU Pro., pp. 477-490. AK Peters (2010)
7. Guy, S., Soler, C.: Graphics Gems Revisited: Fast and Physically-Based Rendering of Gemstones. ACM Transactions on Graphics 23(3), 231–238 (2004)
8. Hecht, E.: Optics, 4th edn. Addison Wesley (2001)
9. Hu, W., Dong, Z., Ihrke, I., Grosch, T., Yuan, G., Seidel, H.-P.: Interactive Volume Caustics in Single-Scattering Media. In: Proc. 2010 ACM SIGGRAPH Symp. Interactive 3D Graphics and Games, I3D 2010, pp. 109–117 (2010)
10. Jensen, H.W.: Global Illumination Using Photon Maps. In: Rendering Techniques 1996, pp. 21-30 (1996)
11. Kajiya, J.: The Rendering Equation. In: SIGGRAPH 1986 Proceedings, vol. 20(4), pp. 143–150 (1986)
12. Kanamori, S., Fujiwara, K., Yoshinobu, T., Raytchev, B., Tamaki, T., Kaneda, K.: Physically-Based Rendering of Rainbows Under Various Atmospheric Conditions. Computer Graphics and Applications (PG), 39–45 (2010)
13. Kenton Musgrave, F.: Prisms and Rainbows: A Dispersion Model for Computer Graphics. In: Proc. of Graphics Interface 1989, pp. 227–234 (1989)
14. Shah, M., Konttinen, J., Pattanaik, S.: Caustics Mapping: An Image-Space Technique for Real-Time Caustics. IEEE Transaction on Visualization and Computer Graphics (2005)

15. Shirley, P.: A Ray Tracing Method for Illumination Calculation in Diffuse-Specular Scenes. In: Proceedings on Graphics Interface, pp. 205–212 (1990)
16. Sikachev, P., Tisevich, I., Ignatenko, A.: Rendering Smooth Spectrum Caustics on Plane for Refractive Polyhedrons. In: 18th International Conference on Computer Graphics Graphicon 2008, pp. 172–176 (2008)
17. Thomas, S.: Dispersive Refraction in Ray Tracing. The Visual Computer 2(1), 3–8 (1986)
18. Wyman, C., Davis, S.: Interactive Image-Space Techniques for Approximating Caustics. In: ACM Symposium on Interactive 3D Graphics and Games, pp. 153–160 (2006)
19. Wyman, C.: Hierarchical Caustic Maps. In: ACM Symposium on Interactive 3D Graphics and Games, pp. 163–171 (2008)
20. Wyman, C., Nichols, G.: Adaptive Caustic Maps Using Deferred Shading. Computer Graphics Forum 28(2), 309–318 (2009)

A Dual Method for Constructing
Multi-material Solids from Ray-Reps

Powei Feng and Joe Warren

Rice University
{pfeng,jwarren}@rice.edu

Abstract. A ray-rep is an approximation of a solid using a grid of parallel rays; intervals along the ray denote various materials comprising the solid. Ray-reps have a variety of applications in computer graphics and mechanical engineering. Given a ray-rep, this paper focuses on the problem of generating a single polygonal mesh that bounds the materials forming the solid. Most investigations of this problem has focused on the two-material case. These primal methods typically build an approximating volume by connecting intersection points on the ray-rep to form curves and surfaces. In this paper, we describe a new, more flexible method that constructs polygonal approximations using a dual method. In the two-material case with normal data, this dual method typically produces better approximations than existing primal methods. In contrast to existing methods, our dual method also generalizes to three or more materials and produces reasonable polygonal approximations from ray-reps.

1 Introduction

Techniques for reconstructing 3D shape from grid-based samples such as contouring [1,2] are well known. Techniques for reconstructing 3D shape from a sequence of 2D cross sections such as [3] and [4] are also well-studied. One reconstruction problem that is not as well-studied is that of reconstructing a 3D shape from a sequence of intervals attached to rays in a 2D grid of parallel rays. Such 2D grids of rays tagged with intervals are known as a *ray-reps*.

In contrast to a regular 3D gridded volume, ray-reps have higher precision and generally lower memory footprint since they may be stored as a sequence of 2D depth images. In addition to storing intersection points on the boundary of the 3D shape, it is also possible to store auxiliary information with each intersection point, such as a normal or a material identifier. In this work, we derive a single, unified method that allows the accurate reconstruction of multi-material solids from a ray-rep tagged with normals and material identifiers.

Ray-reps are also closely related to two other representations: layered-depth images (LDI) in graphics [5] and dexels in mechanical engineering [6]. In graphics, LDIs have been applied in meshless renderings and CSG renderings [7,8]. In mechanical engineering, ray-reps/dexels/LDIs have been found to be useful for

G. Bebis et al. (Eds.): ISVC 2012, Part I, LNCS 7431, pp. 92–103, 2012.
© Springer-Verlag Berlin Heidelberg 2012

high-precision, interactive solid modeling [9,10]. It is powerful due to the fact that CSG operations are simple to implement under ray-reps.

Most previous work on reconstructing a solid representation from a ray-rep has focused mainly on three orthogonal ray-reps; known approaches form uniform grid interpretation of the ray-reps and apply grid-based contouring methods such as Marching Cubes and Dual Contouring on the uniform grid [11,10].

In contrast, we focus on ray-reps from a single direction, and we propose a simple method that extracts the surface directly from the ray-rep samples without an intermediate grid volume representation. One-direction ray-reps free our reconstruction method from the problem of inconsistency that can arise when three orthogonal ray-reps disagree at their common grid points due to numerical error. Generating one-direction ray-rep from a surface mesh is also faster than generating a three-direction ray-rep, making it more attractive to interactive applications or dynamically deforming meshes.

1.1 Related Work

Ray representations have been studied in both graphics and engineering communities in different contexts. In graphics, the original LDI paper [5] considered generating LDIs from multi-angled reference images to create simulated depth field for 2D photography. LDIs are extended to 3-directions (called layered depth cubes (LDC) [12]) to create a view-independent scene for image-based rendering. Pfister et al.'s work explored using LDC with additional information at each sample point for meshless rendering [7]. More recently, Trapp and Dollner. looked at using LDI to perform real-time volumetric tests for mesh models [8]. These previous works in graphics have not dealt with generating a surface mesh from LDI, which is the focus of our work.

In mechanical engineering, the community has first studied ray-reps in the more specific context of numerical-controlled (NC) milling [13]. Van Hook first noted that ray-rep (dexel) can be used in interactive milling display. In solid modeling, Menon et al. proposed ray-rep as a powerful alternative to boundary and CSG representations [9]. Benouamer and Michelucci proposed using a triple ray representation to allow accurate CSG modeling without using B-reps or CSG data structures [11].

Our work is closely related to the work by Zhang et al [14]. Their work also focus on surface reconstruction from one-direction ray-reps. We will review their method for 2D, two-material reconstruction in the next section. However, we also provide extension of the 2D algorithm into 3D and multi-material domains, which are not available in their work. Furthermore, we will present a dual reconstruction that can be more flexible than their primal method. In another closely related work, Liu et al. present a reconstruction method for planar cross-sections [4]. We do not focus on this general case but instead study the problem of 2D grids of 1D intersections. We will further discuss both of these works and contrast with our works in the sections to follow.

Contributions

Our method builds on a 2D two-material reconstruction method of Zhang et al. [14]. As presented, this method uses a set of rules to generate edges connecting intersection points on adjacent rays in a ray-rep. Starting from this method,

- We re-formulate the method to yield a very simple 2D algorithm.
- We naturally extend the 2D algorithm to 3D.
- For ray-reps with normals, we describe a simple dual algorithm that uses normals to build a better approximation.
- For multiple material ray-reps, we extend the two-material dual algorithm to reconstruct multi-material geometry.

2   Reconstruction from Two-Material 2D Ray-Reps

We begin our investigation by considering the simplest case: a 2D ray-rep consisting of two materials. We will present a simple primal method (connecting intersection points on the rays) that is equivalent to Zhang et al.'s method and then present a dual method based on this primal method.

Given a closed 2D shape M, a ray-rep is usually generated by choosing a scan direction and intersecting a row of evenly-spaced *primary rays* parallel to the scan direction with M. Figure 1(a) shows an example of a ray-rep.

2.1   Primal Reconstruction

To aid in understanding our construction, we first form a rectangular approximation to M using an idea from Liu et al [4]. In particular, we will associate with each interval in the ray-rep a rectangle bounded horizontally by lines through its endpoints and vertically by lines midway between the rays containing the interval and its neighbors. Figure 1(b) shows such a rectangular approximation for the ray-rep of figure 1(a). If we assign the material attached to each interval to its corresponding rectangle, the result is a partition of the area into a collection of rectangles whose union approximates M. This rectangular approximation has two obvious drawbacks: the edges in the approximation are all horizontal or vertical and adjacent rectangles of the same material share a common, redundant edge.

Fig. 1. A ray-rep computed from 2D shape (orange intervals) (a), its rectangular approximation (b), its primal reconstruction (c), and its dual reconstruction (d).

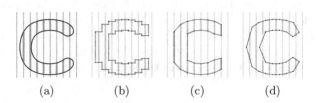

(a)   (b)   (c)   (d)

If each pair of adjacent primary rays bounds a *dual column* in the ray-rep, our goal is to generate a sequence of edges connecting the intersection points on this dual column that bounded M. Zhang et al. tackles this particular problem by independently considering each dual column and using a set of moderately complex rules to pair intersection points. In the following, we will present an equivalent but simpler version of the Zhang et al.'s algorithm based on the rectangular approximation of Liu et al. We choose this particular presentation since it can be extended to both the dual case and the multi-material case in the later sections. Given a ray-rep whose rays are oriented vertically, the algorithm is as follows:

1. Horizontally project the intersection points on adjacent primal rays onto the centerline of the dual column and connect consecutive vertices on this centerline by vertical edges (Figure 2(a)).
2. Remove redundant vertical edges whose corresponding intervals on each primal ray are tagged with the same material (Figure 2(b)).
3. Construct edges connecting pairs of intersection points on the primary rays connected by vertical edges on the centerline (Figure 2(c)).

Note that step one constructs the restriction of Liu et al.'s rectangular approximation to the centerline of the dual column (Figure 2(a)). Step two of the method then deletes redundant vertical edges separating rectangles of the same material (Figure 2(b)). The output of step three are edges connecting intersection points of the original ray-rep (Figure 2(c)). These edges are the ones generated by Zhang et al.'s method. Figure 1(c) shows an example of this method applied to the ray-rep of Figure 1(a). Note that there are two types of edges produced by this method; *regular edges* that connect intersection points on adjacent primal rays and *silhouette edges* that connect intersection points lying on the same primal ray.

In the two-material case, the vertical edges generated by step one have an interesting property. *When vertically ordered, these edges strictly alternate between separating the two distinct materials and separating the same material.* This observation follows since each intersection point on a primary ray causes an alternation between the two materials. Thus, the effect of edge deletion of step two and the edge creation of step three can be summarized in the following elegant rule for directly generating edges:

> *Order the intersection points of each dual column vertically; generate edges connecting the first vertex to the second, the third to the fourth, and generally, $2n - 1^{st}$ to $2n^{th}$.*

Independently applying this method to each dual column of the ray-rep yields a set of edges forming a collection of closed curves.

2.2   Dual Reconstruction

One approach to improving the modeling capabilities of ray-reps is to augment intersection points by their associated surface normals [10]. We next describe a

Fig. 2. Edge generation for a
dual column. Rectangular approximation (a), removing redundant vertical edges (b),
connecting intersection points
linked by a vertical edge (c), and
merging a vertical edge to form
a dual vertex (d) and (e).

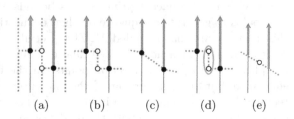

(a)   (b)   (c)   (d)   (e)

method for constructing dual meshes from ray-reps with normals that is similar
in spirit to Dual Contouring [2]. This dual algorithm replaces each primal edge
by a dual vertex positioned according to the normal data attached to endpoints
of the primal edge and connects two dual vertices sharing a common primal
vertex.

We can formulate this dual method directly in terms of the rectangular approximation used in our primal method. After step two of our primal construction
(Figure 2(b)), we can topologically merge each vertical edge into a single dual
vertex (Figures 2(d) and (e)). Observe that the effect of this merge is to remove
all vertical edges from the rectangular approximation and create a dual mesh
formed by deformed horizontal edges. Since each merged edge is the projection
of two intersection points, we can position the resulting dual vertex to minimize
the distance to the tangent lines as defined by the points and normals. (In 3D,
the solution of this problem requires minimizing a quadratic functional [2].) Figure 1(d) shows the dual mesh for the c-shape. When compared to the primal
mesh, the dual mesh provides a better approximation near the silhouette of the
shape.

3   Reconstruction from Two-Material 3D Ray-Reps

The structure of a three-dimensional ray-rep follows that of a 2D ray-rep with
the difference being that the 1D sequence of primary rays is replaced by a two-
dimensional rectangular grid of primary rays. In 3D, Zhang et al. use contour
tiling to reconstruct a 3D surface from a set of parallel 2D curves. In contrast,
we present a simpler method to extract a 3D surface by applying our 2D method
to pairs of adjacent rays in the ray-rep.

3.1   Primal Reconstruction

As in 2D, our method independently processes each dual column (the area
bounded by four face-adjacent primary rays). The method connects the intersection points on these four adjacent primary rays into a set of faces with the
property that the union of these faces over all dual columns forms a closed mesh
that bounds the underlying shape. For each dual column, its associated faces
will correspond to cycles formed by the edges lying on the two-dimensional faces
of the column. After computing the edges on each face using our 2D method,

we form the graph consisting of the union of these edges. To form faces, we then compute the edge cycles in this graph. This search consists of simple iterations through linear chains of edges that terminate each time a cycle is detected.

Figure 3 shows an example of four common cases processed by the method. The leftmost case shows two quad faces, each bounded by four regular edges. The middle left case shows a hexagonal face bounded by four regular edges and two silhouette edges. The middle right case shows another quad face bounded by two regular edges and two silhouette edges, and the rightmost case shows a degenerate two-sided face bounded by two silhouette edges. Other cases involving higher valence faces are possible, but we have not observed arbitrarily large face valences in practice.

The key to the behavior of this face generation method is to observe that two dual columns sharing a common face also share the same edges on that face. This observation follows since these edges are determined solely by the intersection points on the two primary rays bounding the common face. Thus, every regular edge in the resulting mesh is shared by exactly two faces of the mesh and the resulting mesh is guaranteed to be manifold across these edges. For silhouette edges, we observe that the primary ray containing this edge lies on four faces separating the four dual columns sharing that ray. On some faces, our 2D method generates this silhouette edges. On the others, the edge does not exist. For each of the faces containing this edge, our 3D method always generates two faces in the mesh (one in each dual column) that contain this edge. Thus, the edge valence of silhouette edges is always even, and the overall mesh is guaranteed to be closed (but not necessarily manifold). Observe that this argument includes any degenerate 2-sided faces (as shown on the right of Figure 3) as being part of the topological mesh.

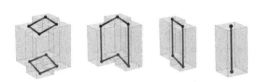

Fig. 3. Four common types of faces in 3D ray-rep meshes. Orange intervals lie on the four rays bounding a dual column. The black edges form faces on the ray-rep mesh.

3.2 Dual Reconstruction

We next consider a method for constructing two-material dual meshes in 3D. Given a 3D primal mesh generated by our method, note that every intersection point on the mesh lies on exactly four faces. Therefore, the corresponding dual mesh consists exclusively of quads. Our method for creating this dual meshes is as follow:

1. Construct edge cycles corresponding to faces in the primal 3D mesh.
2. For the points bounding each face, construct a single dual vertex that minimizes the distance from the tangent planes defined by these points and their normals.

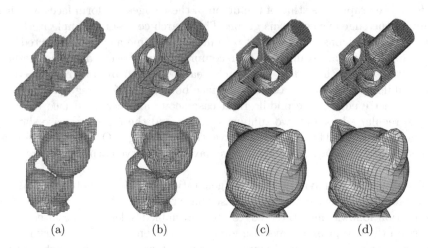

(a)   (b)   (c)   (d)

Fig. 4. Column (a) shows two examples of two-material primal reconstructions from ray-reps. (b) shows the dual reconstruction from the same models. (c) shows a side (silhouette) view of the meshes in (a). (d) shows the same side view for the dual meshes in (b). For ray-reps, sampling on the silhouette is typically poor, and the primal reconstruction reflects this lack of precision. In contrast, the dual meshes are able to reconstruct the sharp features.

3. Since each original intersection point on the ray-rep lies on four primal faces, generate a quad from its four associated dual vertices.

This simple construction requires applying the primal 3D mesh algorithm and identifying cycles of primal vertices in each dual column. Figure 4 shows two examples of the dual meshes.

We consider the dual mesh in terms of the 3D generalization of Liu et al.'s rectangular approximation. For each interval on a vertical primary ray, we construct a rectangular prism that is bounded above and below by horizontal square faces that contain the endpoints of the interval. The prism's vertical sides are bounded by four rectangles whose vertical edges lie on the centerlines of the four dual columns associated with the primary ray. The vertices of the prism all lie on the centerlines of dual columns. In particular, the prism vertices on the centerline of a dual column are the horizontal projections of the intersection points lying on the four primary rays bounding the dual column. As in 2D, the union of these rectangular prism form a prism approximation to the two materials encoded by the ray-rep. Figure 5(left) shows an example of this rectangular approximation for four rays forming a single dual column.

The quad faces in the final dual mesh correspond to the deformed horizontal faces in the 3D prism approximation. To form a closed mesh from these horizontal faces, sets of prism vertices lying on the centerline of dual column are merged by applying our 2D dual method to the faces of the dual column. In particular, each set of prism vertices corresponding to a primal face is merged to form a single dual vertex. For example, Figure 5(right) shows the four prism vertices (white vertices) corresponding to a primal quad merged into a single dual vertex (red vertex). Note that the vertical prism faces shared by different materials have their vertical edges collapsed to vertices by our 2D method causing the face to collapse to a line segment.

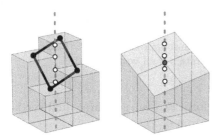

Fig. 5. The two-material dual construction using rectangular approximation. The black primal vertices are projected onto the centerline, denoted by white vertices (left). The four prism vertices on the centerline are used to create a single dual vertex (red vertex) (right).

4   Reconstruction from Multi-material Ray-Reps

We now consider the general case of the ray-rep reconstruction where intervals are tagged with three or more materials. In 2D, straightforward generalizations of our 2D primal and dual methods are available. In 3D, the primal method of tracing edge cycles to form faces leads to ambiguities in the multi-material case. To avoid this problem, we generalize our 3D dual method for two-materials to the multi-material case.

4.1   Multi-material Meshes in 2D

In 2D, the primal version of our multi-material algorithm is almost identical to the two-material case. As before, we focus on one dual column at a time and apply our three-step primal method. The first step forms the rectangular approximation of Liu et al (Figure 6(a)). The second step deletes vertical edges that separate rectangles of the same material (Figure 6(b)). The third step generates edges connecting intersection point on the primary rays whose corresponding projected (white) vertices share a vertical edge (Figure 6(c)).

Our 2D dual method uses the same two initial steps as our primal method, but also topologically collapses sets of white vertices on the centerline of the dual column to form a dual mesh partitioning the materials (Figure 6(d)). In the two-material case, these remaining vertical edges had the elegant property that each projected vertex was incident on exactly one vertical edge allowing us to collapse these edges into a single valence dual vertex. In the multi-material case, a projected vertex can be shared by two vertical edges leading to connected chains of two or more consecutive vertical edges. Our rule for merging these chains of edges is simple: topologically merge the vertices lying on the first and last edges of the chain. In particular, merge a chain of two vertical edges into a single valence-three dual vertex (top of Figure 6(e)). For a chain of three vertical

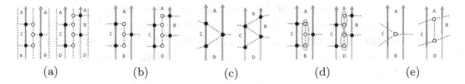

Fig. 6. Two examples of multi-material contouring algorithm. In the top row is an example with three materials, and in the bottom row is an example with four materials. (a) shows the rectangular approximation of the input ray-reps. (b) shows the removal of the redundant vertical edges. (c) shows the primal approximation of the input. (d) shows the merging of the prism vertices to generate a dual vertex in (e).

edges, we merge the first and last edge to form two valence-three dual vertices that connect to form a single dual edge (bottom of Figure 6(e)).

Note that this method preserves the topology of the rectangular approximation: if two materials border each other (either on an edge or vertex) in the rectangular approximation, then the two materials will border each other (at least on a vertex) in the final dual mesh. This property follows from the fact that merging first vertex in a chain to the second vertex in the chain does not affect the topology of the solid. Figure 7 shows an example of 2D multi-material ray-rep.

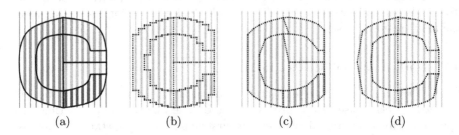

Fig. 7. A 2D multi-material ray-rep (a), its rectangular approximation (b), its primal reconstruction (c), and its dual reconstruction (d). Notice that the primal reconstruction can zig-zag near the boundary between materials. In contrast, the dual reconstruction preserves the straight boundaries.

4.2   Multi-material Meshes in 3D

For two materials, the generalization of our primal method from 2D to 3D involves applying the 2D method to the faces of a single 3D dual column and tracing edge cycles to form faces associated with dual column. For three or more materials, this method produces a network of edges separating the various materials on the faces of the dual column. However, the problem of generating faces from this edge network is much more difficult due to the absence of easily identifiable cycles that form faces.

Instead, we focus our attention on constructing a 3D dual method for three or more materials that generalizes our two-material method. To this end, we begin with a solid approximation of a vertically-oriented ray-rep using rectangular prisms. The vertices of this prism approximation lie on the centerlines of 3D dual columns. In particular, the vertices on the centerline of a single dual column are the horizontal projections of the intersection points on each of the primary rays bounding the column.

In the two-material case, we applied our dual method to each face of the dual column and merge pairs of vertices on the centerline on these faces. These merge operations were then applied to the projections of these vertices onto the centerline of the dual column. As seen in Figure 5, the four merges on the faces of the dual column lead to a single merged vertex on the centerline of the dual column. We take a similar approach in the multi-material case. We compute vertex merges on each face of the dual column using an extension of our 2D multi-material algorithm. We use a distance-based heuristic for merging vertices. Two vertices can be merged if they are joined by an edge, and the distance between the two vertices is below a certain threshold.

In the two-material case, the topological connectivity of these merged vertices was provided solely by the horizontal faces of the original prism approximation. All of the vertical faces in the prism approximation were collapsed into line segments due to vertex merges. In the multi-material case, these horizontal faces are also part of the final surface network used to partition the various materials. However, as opposed to the two-material case, this final surface network also includes some vertical faces inherited from the rectangular approximation. In particular, any vertical edges remaining after applying our extension of 2D dual method to a face of the 3D dual column corresponds to a vertical face separating two rectangular prism of different materials. These vertical faces appear in the final topology of the surface network (subject to vertex merges). The resulting merges on the faces of the dual column corresponds to merging vertices on the

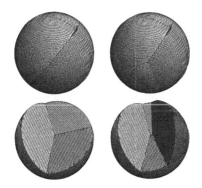

| model | primal | primal | dual | dual |
(resolution)	time(s)	tris	time(s)	tris
fandisk (50^2)	.025	5044	.024	4866
mechpart (100^2)	.093	18116	.0784	17322
mechpart (150^2)	.198	40836	.178	39590
kitten (200^2)	.337	64312	.277	63736
kitten (250^2)	.530	100416	.466	99716
kitten (300^2)	.753	144672	.605	143764

Fig. 8. 3D multi-material reconstructions with 2, 3, 4, and 5 materials sampled at the resolution of 80×80

Fig. 9. Timing results for ray-rep primal/dual mesh generation for the various models

vertical faces in the prism approximation. Each vertical face in the prism approximation is triangulated for rendering, and each pair of merged vertices in the prism approximation correspond to removing a triangle in the triangulation of the vertical face. Each final merged vertex on the centerline is positioned using the position and normal data associated with its corresponding intersection points. Figure 8 shows a few examples of 3D multi-material meshes.

5   Results and Implementation

We implemented the two-material 3D meshing methods in C++ on an Intel Xeon 2.8GHz machine. Table 9 shows the timings for the various examples. (We compute 3D ray-reps from mesh models using depth-peeling [15].) Interestingly, the dual reconstruction is generally faster than the primal reconstruction. There are two reasons for this: first, the primal method generates more triangles than the dual. Since each dual vertex corresponds to a primal face, which consists of at least four vertices, the number of dual triangles should be less than the primal. Second, the primal method requires special care in triangulation; note that we are generating edge cycles in the primal method. For cycles with more than three vertices, we need to determine a good triangulation of the vertices, which requires more computation. Hence, the dual method is typically faster than the primal. Since all of the process takes place for each dual columns independently, we believe that our method can easily be parallelized using a language like CUDA.

6   Conclusion

In this work, we have presented algorithms for building surfaces for two-material and multi-material ray-reps. We have shown that the 2D primal method of Zhang et al [14]. has a very elegant description and can be easily extended to a primal 3D method. Furthermore, we use the rectangular approximation of ray-reps (described by Liu et al. [4]) to examine the dual reconstruction methods for 2D and 3D. We also used the same rectangular approximation to generalize the two-material reconstructions to multiple materials. As demonstrated through examples, the dual mesh is more flexible than the primal in cases where normal data is available. For future work, we will examine the possibility extending our work for three-direction ray-reps.

Acknowledgements. We thank Tao Ju, Scott Schaefer, and the reviewers for their helpful comments. The *fandisk*, *mechpart*, and *kitten* were obtained from Aim@Shape Project.

References

1. Lorensen, W.E., Cline, H.E.: Marching cubes: A high resolution 3D surface construction algorithm. In: Proceedings of the 14th Annual Conference on Computer Graphics and Interactive Techniques, SIGGRAPH 1987, pp. 163–169. ACM, New York (1987)

2. Ju, T., Losasso, F., Schaefer, S., Warren, J.: Dual contouring of Hermite data. In: Proceedings of the 29th Annual Conference on Computer Graphics and Interactive Techniques, SIGGRAPH 2002, pp. 339–346. ACM, New York (2002)

3. Boissonnat, J.D.: Shape reconstruction from planar cross sections. Comput. Vision Graph. Image Process. 44, 1–29 (1988)

4. Liu, L., Bajaj, C., Deasy, J., Low, D., Ju, T.:Surface reconstruction from non-parallel curve networks. In: Computer Graphics Forum Proceedings of Eurographics, vol 27, pp. 155–163. John Wiley & Sons (2008)

5. Shade, J., Gortler, S., He, L.-w., Szeliski, R.: Layered depth images. In: Proceedings of the 25th Annual Conference on Computer Graphics and Interactive Techniques, SIGGRAPH 1998, pp. 231–242. ACM, New York (1998)

6. Huang, Y., Oliver, J.H.: NC milling error assessment and tool path correction. In: Proceedings of the 21st Annual Conference on Computer Graphics and Interactive Techniques, SIGGRAPH 1994, pp. 287–294. ACM, New York (1994)

7. Pfister, H., Zwicker, M., van Baar, J., Gross, M.: Surfels: surface elements as rendering primitives. In: Proceedings of the 27th Annual Conference on Computer Graphics And Interactive Techniques, SIGGRAPH 2000, pp. 335–342. ACM Press/Addison-Wesley Publishing Co., New York (2000)

8. Trapp, M., Dollner, J.: Real-Time Volumetric Tests Using Layered Depth Images. In: Eurographics 2008 Shortpaper, pp. 235–238 (2008)

9. Menon, J., Marisa, R., Zagajac, J.: More powerful solid modeling through ray representations. IEEE Computer Graphics and Applications 14, 22–35 (1994)

10. Wang, C.C.L., Leung, Y.S., Chen, Y.: Solid modeling of polyhedral objects by layered depth-normal images on the GPU. Comput. Aided Des. 42, 535–544 (2010)

11. Benouamer, M.O., Michelucci, D.: Bridging the gap between CSG and Brep via a triple ray representation. In: Proceedings of the Fourth ACM Symposium on Solid Modeling and Applications, SMA 1997, pp. 68–79. ACM, New York (1997)

12. Lischinski, B., Rappoport, A.: Image-based rendering for non-diffuse synthetic scenes. In: Proceedings of the Eurographics Workshop on Rendering Techniques 1998, pp. 301–314. Springer (1998)

13. Van Hook, T.: Real-time shaded NC milling display. In: Proceedings of the 13th Annual Conference on Computer Graphics and Interactive Techniques, SIGGRAPH 1986, pp. 15–20. ACM, New York (1986)

14. Zhang, W., Peng, X., Leu, M.C., Zhang, W.: A novel contour generation algorithm for surface reconstruction from dexel data. Journal of Computing and Information Science in Engineering 7, 203–210 (2007)

15. Everitt, C.: Interactive order-independent transparency. White Paper, nVIDIA 2, 7 (2001)

User Driven 3D Reconstruction Environment

David Sedlacek and Jiri Zara

Czech Technical University in Prague, Faculty of Electrical Engineering
david.sedlacek@fel.cvut.cz

Abstract. An intuitive image-based 3D reconstruction tool based on inaccurate user strokes is presented in this paper. The combination of fast image segmentation method together with user knowledge about reconstructed scene forms a novel low-polygonal editor suitable for architecture reconstruction. The user interaction is minimized thanks to propagation of strokes among input photographs. The final model geometry is created by innovative algorithm. The input to the tool is a set of calibrated photographs together with a sparse pointcloud. The output is a structured low-poly 3D model.

1 Introduction

Inspired by well known 3D editors for low-polygonal image-based modelling (like ImageModeler, PhotoModeler) we propose a novel editor reducing amount of user interaction. We present a combination of state-of-the-art techniques and intuitive interaction methods for geometry construction. Our approach opens new ways for 3D reconstruction focused on low-poly output which is well suited for internet presentation. The user interaction brings also advantages for 3D reconstruction in low-textured or occluded areas where automatic methods often fail. This reconstruction tool is suitable for architecture reconstruction or other

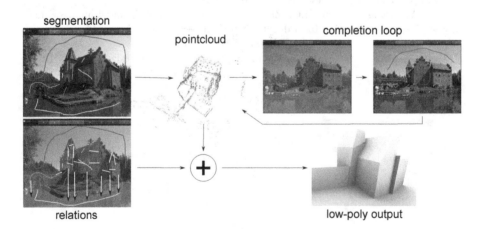

Fig. 1. Program workflow overview

G. Bebis et al. (Eds.): ISVC 2012, Part I, LNCS 7431, pp. 104–114, 2012.

areas where planar structures dominate. Calibrated photos (known camera positions) and sparse pointcloud (reconstructed during camera calibration phase) constitute the input of the method. We have successfully used both Bundler [1] and APERO [2] project as suitable calibration engines.

The core of our approach is geometry primitives fitting based on source photo segmentation, see Fig. 1. The structure of the paper corresponds to individual algorithm steps. The graph-cut image segmentation is driven by inaccurate user strokes (Section 3). The image segmentation is then used for labelling of sparse pointcloud points and is also propagated to close photographs (Section 4). Various geometry primitives are then fitted on labelled points (Section 5). Thanks to user-defined relations between adjacent geometrical structures (Section 6), final polygonal geometry is computed (Section 7).

2   Related Work

The image-based 3D modeling is studied over last 15 years. Several systems were developed over this time, and all of them have common properties. The user works with a set of photographs and selects interesting parts of objects for the reconstruction.

One of the first work was Facade by Debevec at al. [3] which later gave rise to a commercial product called Canoma. The system provided several 3D primitives like prisms, cuboids and pyramids. The user placed the primitives into photographs and improved positions of the primitives in other views. Since the scene was continuously calibrated by user added objects, the system tended to be unstable. Later, other commercial products were inspired by Canoma software, like PhotoModeler by Eos Systems, ImageModeler by RealViz (now Autodesk) and the PhotoMatch component of Google SketchUP. The calibration of input photographs was shifted to user in those systems.

Other systems estimate the camera calibration using automatic methods based on SFM. Some programs have developed their own calibration core (VideoTrace by van den Hengel et al. [4], system for architectural modelling by Sinha et al. [5]). Other works benefit from external SFM calibration tools (like Match-Mover by RealViz, Boujou by 2d3 or Bundler) and suppose calibrated scene as input (Habbecke et al. [6], Paczkowski at al. [7]).

Other works try to get as much information from one photo as possible [8–10]. The one photo calibration is based on finding vanishing points constraints and on the fact that all architectural buildings have parallel lines and the building blocks are build in Manhattan layout [11]. Vanishing points are used even in modelling from more photographs like Sinha et al. [5], Cipolla et al. [12] and Wilczkowiak et al. [13].

Sinha et al. [5] shows that the combination of unordered photo set, detected vanishing points and vanish lines and auto-calibration using SFM leads to user friendly free-polygonal modelling. The user draws lines to the image. The lines are snapped to the direction of vanishing lines and then the lines are extruded to the face in the direction of plane perpendicular to the line in user selected

Fig. 2. User inputs and geometry output: (a) Sketch based segmentation interface; (b) Geometry reconstructed from pointcloud and informations defined by a user in the left photograph

direction. They use RANSAC algorithm for the best plane fitting constrained by vanishing points. The plane fitting algorithm found the best fitting plane in sparse pointcloud restricted by points lying within a face selected by user. They call their user modelling interface sketch-based but it is restricted to placing lines near the real line, in other cases the line is snapped to the wrong vanishing line. Sketch-based interfaces are also applied in works of El-hakim at al. [14] and Xiao at al. [15] where they provide tools for selection, marking, and other specific operations.

The sketch-based interface is mostly related to creation of new 3D models or annotations, see [16] for wide overview. Interesting work is combination of calibrated scene with creation of new architectural design in site, see Paczkowski at al. [7].

Our approach is similar to Sinha et al. [5] however we intensively utilize a sketch-based interface in order to define a higher logic and geometrical structure within 3D reconstructed scene.

3   Image Labelling

In our approach we benefit from user knowledge about the scene to be reconstructed. We suppose that the user recognizes basic scene primitives (a wall, a pillar, etc.) and the connections between them (adjacent or not). For this we have designed an intuitive user interface for photo labelling based on graph-cut algorithm, see Fig. 2a.

Similarly as Boykov at al. [17], we formulate labelling problem as an energy function minimization. As an input we use a gray-scale image P where each pixel $p \in P$ is connected to 4 neighbours (N). The goal is to find labelling l, i.e. to assign each pixel p a label in finite label set $L_p \in L$ while the energy function is minimized. We express the energy function as:

$$E(l) = \sum_{\{p,q\} \in N} V_{p,q}(L_p, L_q) + \sum_{p \in P} D_p(L_p) \qquad (1)$$

Where $V_{p,q}$ is smoothness term representing the energy of intensity discontinuity between two neighbouring pixels and data term D_p is the energy of assigning the pixel to label. The $L_{p,q}$ stands for pixel labelling. We define the smoothness term as:

$$V_{p,q}(L_p, L_q) = 1 + (K * e^{\frac{-(I_q - I_p)^2}{\sigma^2}}) \qquad (2)$$

Where $I_{p,q}$ are pixels intensities, constant K is scaling factor from float to integer values $\langle 0; K \rangle$ and constant σ controls exponent function behaviour. These constants are estimated based on image size, which is mentioned in following paragraphs. Similarly as Sykora at al. [18] avoid segmentation discontinuities using non-zero smoothness term we map the final values of smoothnes term to the $\langle 1; K \rangle$ interval (additive constant in eq. 2). The K and σ constants are set with respect to image size at the end of this section.

The data term in many segmentation methods is usually set to some image-based likelihood such as pattern or color similarity. Due to the fact that our segmentation completely relies on user inputs, the data term is set only to pixels selected by the user and does not take into account any specific image properties.

$$D_p(L_p) = \begin{cases} 0, & p \in L \\ \infty, & p \notin L \end{cases} \qquad (3)$$

Since the smoothness term relies only on pixel intensity and not on the labelling, our energy function can be minimized by solving multiway cut problem on undirect graph [19]. Our graph $G = \{V, E\}$ has the same topology as described in [18], i.e. each image pixel correspond to one vertex P and is connected to 4 neighbours by edges E_p with capacity equal to smoothness term $w_{p,q} = V_{p,q}$ from eq. 2. The edges E_c between pixels P and terminals C has capacity $w_{p,c} = K - D_p$. Considering eq. 3 we can simplify terminal capacity to $w_{p,c} = K$. Recall that only pixels where user made hard strokes are connected to terminals.

Inspired by LazyBrush [18] we solve the multiway cut algorithm by greedy algorithm where one label is randomly chosen as S terminal and all others are connected to the T terminal. Thereafter max-flow/min-cut problem is solved using [17]. All pixels labelled as S are disconnected from graph and new label is selected as terminal S from the set of all remaining labels previously connected to T. This procedure is repeated until T is empty.

The image segmentation is used for labelling sparse pointcloud. Each 3D point of sparse pointcloud contains the list of cameras where it was visible and 2D point representing the image of 3D point in the corresponding camera. Thanks to this information, we are able to propagate the image segmentation to the 3D pointcloud which is then used for 3D reconstruction in section 5.

To achieve best results from calibration phase, we prefer to use photos in original size. Unfortunately the image size of several mega-pixels causes very long segmentation times (Table 1). That is why we internally down-scale images to the size of 640px for longer edge. The computed segmentation is then up-scaled to original images. Thanks to low distribution of underlying pointcloud we do not need to use any of segmentation refinements in higher resolution, for example like [20]. Currently a faster single-core implementation of Boykov algorithm has been introduced by Jamriska et al. [21] but we have not integrated it yet.

Table 1. Processing times of graph-cut algorithm depending on image maximal edge size and labels count. Intel Core2 Quad, 2.5GHz, single core.

size	2 labels	5 labels	10 labels	15 labels
640px	402ms	723ms	921ms	1110ms
800px	800ms	1252ms	1650ms	2043ms
1024px	1342ms	2328ms	2970ms	3882ms

With respect to chosen image size, the edge capacity constants are set as: $K = 2 * (h + w)$ and $\sigma^2 = 32$. To get maximal interactivity during the labelling and reconstruction process, the image segmentation is performed immediately after each user stroke. If the user considers the delay after each stroke too long she can switch to on demand segmentation process where all strokes are given in advance followed by final segmentation evaluation.

4   Propagation of Segmentation

As stated in previous section, the image segmentation is used only for finding labelling of sparse pointcloud. After processing the first photograph, the labelling of the whole model is not complete. Thus other photos are to be used to finish the labelling process. The automatic propagation of labelling is implemented to achieve faster reconstruction and also for the reason of user load reduction. The labelling suggestion is based on the following procedure.

For each feature point in image we get corresponding 3D point of pointcloud. If this 3D point is already labelled we transfer this labelling into next image. In related works the suggested labelling is transferred into segmentation graph as "Soft scribble". It can be interpreted as a connection of node to the corresponding terminal with much less capacity than used for the user input (hard scribble). The tests with real data shown successful propagation of labelling to new photo but brought problems in case of subsequent user corrections. When new user strokes were added the segmentation became less stable and disintegrated into small unconnected regions, requiring additional corrections, see Figure 3b. For this reason we decided to transfer labelling as new user strokes (hard scribbles) which can be easily deleted or over-painted by the user. New hard scribbles are generated as small circles with center at the position of image feature point, see Figure 3c.

5   Fitting of Geometry Primitives

The geometry primitives are fitted to the segmented pointcloud. Each label corresponds to one geometry primitive type (e.g. plane, sphere, box) and primitive

Fig. 3. The segmentation propagated from photo in Fig. 2a using different scribble types. (a) propagated segmentation using soft scribbles; (b) correction of soft scribbles segmentation, notice isolated parts in the closer view; (c) propagated segmentation using hard scribbles; (d) correction of hard scribbles segmentation.

occurrence. It means, that each primitive should be segmented with new label stroke of corresponding type. The RANSAC [22] algorithm is used for geometry model parameter estimation. Thanks to user points pre-selection there is a high probability that all points belongs to the given primitive. For this reason, we do not use any special candidates sampling strategy, for example as in [23], but all label points are sampled randomly.

The output of this algorithm step is not the final geometric representation of found primitives but only the algebraic parameters of geometry model (e.g. a,b,c,d parameters for plane or *center* and *radius* for sphere). The description of final geometry creation is described in section 7. Currently we support fitting of planes and spheres only, however our labelling framework is prepared for other geometry types as well.

6 Relations

In standard image-based modelling software (like ImageModeler, PhotoModeler, Sketchup), the reconstructed model shape is built bottom-up starting from the smallest primitives (points) through edges up to faces and meshes. We introduce another approach (top-down), where the mesh geometry is defined by geometrical primitives and relations among them. In case of planes, a relation of two non-parallel planes defines an edge, similarly three planes form a point.

We recognize two relation types between planes, see Fig. 4. The first one is **cross** relation where two planes form an edge and both planes are split on two half-planes while only two of half-planes forms a mesh. This relation represents a mesh edge. We assume that this relation is enough if and only if all mesh planes are visible and they can be reconstructed. Unfortunately in real scenes a lot of mesh sides are hidden behind some bigger objects, for example a book lying on a table. The bottom side of the book is not visible but we need some reference for restriction of other four book sides. For this reason we introduce second relation - the **restriction**. The restriction relation naturally corresponds to the definition of two different objects and is interpreted like the first plane (book side) is split by the second one (table plane) while the second plane remains unchanged.

The relations are defined in the same interface as labelling by a stroke connecting areas with two different labels, see Fig. 4d.

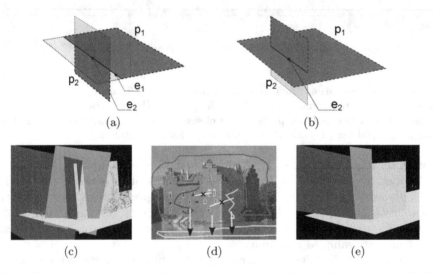

Fig. 4. Two types of relations shown on a schematic view and a real example. In schematic view two planes $p_{1,2}$ are split with respect to the relation type: the cross relation (a) and the restriction relation (b) (T-junction). The plane parts filled by solid color remain. In the real example (d), the cross relation is depicted by line with $\times$ letter in the middle, the restriction relation is depicted by an arrow. Stroke colors correspond to plane colors. Original planes (c) are unrestricted and they intersect mutually. In (e), three upper planes create edges defined by the cross relation. They are all restricted by the bottom yellow plane which is not cropped by other planes.

7   Finding Output Geometry

The final geometry is constructed from outputs of fitting phase in combination with relations. The overall process is similar to modelling boundary representations of solids with boolean operations. The difference is that our primitives are not closed solids and that crossing planes do not determine inner or outer subspace unambiguously. To properly solve the problem we incorporate a knowledge of feature points belonging to fitted geometry. For the sake of simplicity we describe a situation in which planes are considered only.

Each plane defined by a user is initially treated as unbound and the mesh geometry is found as an intersection of related planes. The task is to find bounds for each plane what forms a polygon. Inspired by Bernstein at al. [24] we define convex polygon p as a couple of supporting plane s and counter clockwise ordered list of bounding planes $\{b_i\}_{i \in Z_n}$, see Fig. 5a. The polygon vertices are then given by intersection of three planes: $v_i = (s, b_{i-1}, b_i)$. We construct the final, possibly non-convex polygon from a set of small convex components, see Fig. 5b. All the details connected to polygons like numeric and geometric basis, and BSP operations are described in [24].

To make the process more robust, we artificially add several bounding planes in a far distance. This solves a problem of incompletely specified set of crossing planes.

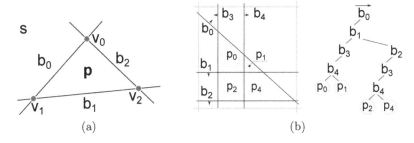

(a)   (b)

Fig. 5. Polygons: (a) convex polygon representation. Supporting plane s is bounded by three boundary planes b_i. Corresponding vertices are shown as dots; (b) Supporting plane is split by related planes into convex polygons. Artificial planes in far distance are illustrated as dashed lines and border polygons (i.e. adjacent to dashed lines) are not labelled for simplicity. The corresponding BSP (sub)tree on the right shows only the leaves containing inner convex polygons. The final polygon consists of three elementary polygons p_0, p_1, and p_2.

We find a final polygon defined by supporting plane s in two steps. Firstly we get a set of all convex polygons defined by plane s and all planes related to it as a result of BSP tree creation. Note that the elementary convex polygons are not treated as geometrical objects only but we keep a list of corresponding pointcloud points attached to them. This allows us to collect all contributing elementary polygons by solving min-cut problem in the second step. The following algorithm describes the first step, its output (BSP tree and a set of convex polygons) is depicted in the Figure 5b.

Routine for finding all possible convex polygons

```
getPolygons (Plane s) {
    R = getRelatedPlanes(s); //get all s-related planes
    p = new Polygon(s); //create polygon bounded by distant planes
    BSP = new BSP(p); //BSP structure for polygon intersections
    for(Plane r : R) {
        BSP.addSplitPlane(r);
    }
    return BSP.getLeaves();
}
```

The second step is a selection of the subset of convex polygon participating in the final polygon. We are looking for binary labelling (IN, OUT) of elementary polygons based on information previously defined by a user. We state out the differences in relation to already described graph-cut algorithm in Section 3. The nodes of graph $G = (V, E)$ are now the leaves of the BSP tree P_i connected with neighbour leaves by edges E_p with capacity equal to $w_{p,q}$, see eq. 4. Some nodes are connected by edges with capacity $w_{p,s}$ and $w_{p,t}$ to S-source and T-sink terminals respectively, see Fig. 6b.

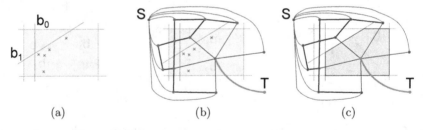

Fig. 6. Graph-cut: (a) An input plane bounded by distant planes (gray rectangle with dashed boundary), two related planes and few plane points. (b) Input plane extended by graph for solving min-cut problem. (c) Labelled graph with depicted final polygon. Point out that minimal-cut correspond to polygon boundary edges. The gray filled polygon is labelled IN, the other three white filled polygons are labelled OUT.

$$
\begin{aligned}
w_{p,q} &= 1 + (K * e^{\frac{-(Q_{points} - P_{points})^2}{\sigma^2}}) \\
w_{p,s} &= K, & if(P_{polygons} = 0) \\
w_{p,t} &= K * P_{points}, & if(P_{points}/points > 0.6)
\end{aligned}
\tag{4}
$$

The terminals are connected to nodes only if the given condition is satisfied. The symbols $P_{points}, P_{polygons}$ stand for a number of pointcloud points or polygons in current node respectively. The symbol *points* is an overall number of points contained in the supporting plane. Notice that only the tree leaves representing the space out of artificial bounding planes have zero polygons. The constants K and σ^2 were experimentally set to $K = 100$ and $\sigma^2 = 1000$. The output of this algorithm is a set of elementary polygons contained in nodes labelled as T (i.e. IN) and forming the final mesh geometry, see Fig. 6c.

8   Conclusion

We have presented a novel interactive 3D reconstruction tool for calibrated photographs. It is composed of several well established techniques extended by new ideas and algorithms. Especially the algorithm presented in Section 7 that is able to construct non-convex polygons from imperfectly specified input planes using graph-cut algorithm represents a promising way.

The reconstruction is driven by inaccurate strokes and is well prepared for use on touch screens. The output is a structured low-poly model suited for online presentations.

Acknowledgment. This work has been partially supported by the Grant Agency of the Czech Republic under research program P202/11/1883, and the Grant agency of the CTU Prague, grant No. SGS10/291/OHK3/3T/13.

References

1. Snavely, N., Seitz, S.M., Szeliski, R.: Photo tourism: exploring photo collections in 3d. In: SIGGRAPH 2006: ACM SIGGRAPH 2006 Papers, pp. 835–846. ACM, New York (2006)
2. Deseilligny, M.P., Clery, I.: Apero, an open source bundle adjusment software for automatic calibration and orientation of set of images. In: Proceedings of the ISPRS Symposium, 3DARCH11 2011 (2011)
3. Debevec, P.E., Taylor, C.J., Malik, J.: Modeling and rendering architecture from photographs: a hybrid geometry- and image-based approach. In: SIGGRAPH 1996: Proceedings of the 23rd Annual Conference on Computer Graphics and Interactive Techniques, pp. 11–20. ACM, New York (1996)
4. van den Hengel, A., Dick, A., Thormählen, T., Ward, B., Torr, P.H.S.: Video-trace: rapid interactive scene modelling from video. In: SIGGRAPH 2007: ACM SIGGRAPH 2007 Papers, p. 86. ACM, New York (2007)
5. Sinha, S.N., Steedly, D., Szeliski, R., Agrawala, M., Pollefeys, M.: Interactive 3d architectural modeling from unordered photo collections. In: SIGGRAPH Asia 2008: ACM SIGGRAPH Asia 2008 Papers, pp. 1–10. ACM, New York (2008)
6. Habbecke, M., Kobbelt, L.: An intuitive interface for interactive high quality image-based modeling. Comput. Graph. Forum 28, 1765–1772 (2009)
7. Paczkowski, P., Kim, M.H., Morvan, Y., Dorsey, J., Rushmeier, H., O'Sullivan, C.: Insitu: sketching architectural designs in context. ACM Trans. Graph. 30, 182:1–182:10 (2011)

8. Jiang, N., Tan, P., Cheong, L.F.: Symmetric architecture modeling with a single image. In: SIGGRAPH Asia 2009: ACM SIGGRAPH Asia 2009 Papers, pp. 1–8. ACM, New York (2009)
9. Guillaume, L.Z., Zhang, L., Dugas-phocion, G.: Single view modeling of free-form scenes. In: Proc. of CVPR, pp. 990–997 (2002)
10. Zheng, Y., Chen, X., Cheng, M.M., Zhou, K., Hu, S.M., Mitra, N.J.: Interactive images: Cuboid proxies for smart image manipulation. ACM Transactions on Graphics 31 (2012)
11. Duan, W., Allinson, N.M.: Vanishing points detection and line grouping for complex building facade. In: Proc. of WSCG 2010 (2010)
12. Cipolla, R., Robertson, D.: 3d models of architectural scenes from uncalibrated images and vanishing points. In: Proceedings. International Conference on Image Analysis and Processing, pp. 824–829 (1999)
13. Wilczkowiak, M., Sturm, P., Boyer, E.: Using geometric constraints through parallelepipeds for calibration and 3d modeling. IEEE Transactions on Pattern Analysis and Machine Intelligence 27, 194–207 (2005)
14. El-hakim, S., Whiting, E., Gonzo, L.: 3d modelling with reusable and integrated building blocks. In: 7th Conference on Optical 3-D Measurement Techniques, pp. 3–5 (2005)
15. Xiao, J., Fang, T., Tan, P., Zhao, P., Ofek, E., Quan, L.: Image-based façade modeling. In: SIGGRAPH Asia 2008: ACM SIGGRAPH Asia 2008 Papers, pp. 1–10. ACM, New York (2008)
16. Jorge, J., Samavati, F.F. (eds.): Sketch-based Interfaces and Modeling, 1st edn. Springer (2011)
17. Boykov, Y., Veksler, O., Zabih, R.: Fast approximate energy minimization via graph cuts. IEEE Trans. Pattern Anal. Mach. Intell. 23, 1222–1239 (2001)
18. Sýkora, D., Dingliana, J., Collins, S.: Lazybrush: Flexible painting tool for hand-drawn cartoons. In: Computer Graphics Forum (Proceedings of Eurographics 2009), vol. 28, pp. 599–608 (2009)
19. Boykov, Y., Veksler, O., Zabih, R.: Markov random fields with efficient approximations. In: Proceedings of the IEEE Computer Society Conference on Computer Vision and Pattern Recognition, CVPR 1998, p. 648. IEEE Computer Society Press, Washington, DC (1998)
20. Lombaert, H., Sun, Y., Grady, L., Xu, C.: A multilevel banded graph cuts method for fast image segmentation. In: Tenth IEEE International Conference on Computer Vision, ICCV 2005, vol. 1, pp. 259–265 (2005)
21. Jamriška, O., Sýkora, D., Hornung, A.: Cache-efficient graph cuts on structured grids. In: Proceedings of IEEE Conference on Computer Vision and Pattern Recognition (2012)
22. Fischler, M.A., Bolles, R.C.: Random sample consensus: A paradigm for model fitting with applications to image analysis and automated cartography. Communications of the ACM 24, 381–395 (1981)
23. Schnabel, R., Wahl, R., Klein, R.: Efficient ransac for point-cloud shape detection. Computer Graphics Forum 26, 214–226 (2007)
24. Bernstein, G., Fussell, D.: Fast, exact, linear booleans. In: Proceedings of the Symposium on Geometry Processing, SGP 2009, Aire-la-Ville, Switzerland, Eurographics Association, pp. 1269–1278 (2009)

Methods for Approximating Loop Subdivision Using Tessellation Enabled GPUs

Ashish Amresh, John Femiani, and Christoph Fünfzig

Department of Engineering, College of Technology and Innovation,
Arizona State University, USA
Laboratoire Electronique, Informatique et Image (LE2I), Université de Dijon, France
{amresh,john.femiani}@asu.edu, c.fuenfzig@gmx.de

Abstract. Subdivision surfaces provide a powerful alternative to polygonal rendering. The availability of tessellation supported hardware presents an opportunity to develop algorithms that can render subdivision surfaces in realtime. We discuss the performance of approximating Loop Subdivision surfaces using tessellation-enabled GPUs in terms of speed and quality of rendering for these methods as well as the implementation strategy. We also propose a novel one pass unified rendering setup for all three methods. Subdivision using the Loop method supports arbitrary triangle meshes and provides for easy transition from polygonal rendering of triangles to the parametric domain. Majority of graphics software applications, especially game engines, render polygons as triangles. The objectives of this paper are to evaluate the performance of smooth rendering algorithms developed to take advantage of tessellator enabled GPUs, provide an easy transition from polygonal to parametric rendering and propose an optimal way to achieve multi-level rendering dependent on performance and visual needs of the application.

1 Introduction

Fast rendering of polygons has been an established process for over a decade and it has been primarily driven by the processing power of the GPU. The common graphics libraries DirectX and OpenGL expose this power and developers reconfigure their applications to deliver increased realism by implementing rendering techniques [1–3] like normal mapping, bloom lighting, shadows, and animation techniques [4, 5] like rigging for character and facial animation. In most cases the polygons are rendered as triangles and animation is performed using a skeleton, that transforms the vertices of the triangle in realtime. With the addition of a hardware tessellator unit to the graphics pipeline in Direct3D11 and OpenGL 4.0, the asset production pipeline can be simplified to a greater detail by having the content creators model both the cut-scenes and application level assets using subdivision surfaces. The tessellator unit adds two new programmable stages , the hull and domain in Direct3D11 and control and evaluation in OpenGL 4.0, to the graphics pipeline. The methods described in this paper use OpenGL 4.0 and our language reflects this usage. Rendering parametric surfaces using tessellation enabled GPUs as been discussed in detail in both Direct3D and OpenGL

G. Bebis et al. (Eds.): ISVC 2012, Part I, LNCS 7431, pp. 115–125, 2012.

developer documentation. For the purposes of this paper we will be looking at various methods for approximate rendering of the Loop subdivision surface [6]. The main factors that limit the performance of the GPU is the number of control points in the input mesh and number of calculations (shader lines of code) performed in the control shader. This is mainly because the GPU is bound by its memory bandwidth and its speed is directly dependent on the number of memory fetches. Our research, presented in Section 5, indicates that algorithms that use many instructions to derive the control points in the control shader can severely limit performance even though their initial memory foot print is low. We therefore choose three methods based on these observations. The Point-Normal (PN) method [7] has 6 control points and 13 operations in the control shader, the Walton-Meek (WM) method [8] has 6 control points and 70 operations in the control shader and our proposed Gregory method has 15 control points and 0 operations in the control shader. A new method for approximating Loop subdivision surfaces with triangular Gregory patches is proposed. The methods are implemented using hardware tessellation. We render various objects using the three methods and describe the performance in terms of their visual quality and speed. We find that the speed is also dependent on the number of models in the scene invoking a particular method and therefore the methods need to be carefully chosen in order to optimize the scene. For this purpose we propose an unified rendering framework, that can combine all three methods at run time. Our goal is to provide application developers the information necessary for migrating from polygonal domain to smooth rendering of triangles. By applying the unified rendering framework, described in Section 4, developers can perform automatic level-of-detail(LOD) calculations for their meshes. The number of control points in the input mesh then becomes synonymous with the LOD level. This gives the ability to have a two step system for LOD, one computes the control points of the input mesh and the other computes the edge tessellation factors.

2   Background

The idea of rendering parametric triangular surfaces in realtime dates back to the introduction of PN-triangles [7], developed to improve the surface quality of low polygonal meshes. Quadratic approximation surface (QAS) [9, 10] used the idea of rendering a PN-patch for limit Loop points and normals. With PNG1-triangles [11] this approach was further improved to satisfy G^1 continuity across each edge. Similarly [8] provides a method for G^1 continuous surface when the boundary curves are known. The idea of approximating Catmull-Clark subdivision surfaces using bi-cubic Bézier patches was introduced in [12] and then further refined in [13] for hardware tessellation using Gregory Patches. These methods are suitable for quad-dominant meshes and do not work well for triangular patches. The above research forms the primary motivation for developing a list of methods for rendering an approximate Loop subdivision surface by varying the number of control points for the input mesh. The Loop subdivision scheme is a face split based on triangular splines [14] and at every subdivision step it calculates a new vertex for each existing one and a new vertex for each edge. We can calculate the limit point and limit tangents for the Loop control mesh directly as shown in [15]. A parametric triangular Bézier patch is defined in [16] and a triangular Gregory patch is defined in [17, 18] and is a modified form of a quartic Bezier patch where the boundary curves are cubic and the interior surface is quartic.However, the inner control points of the patch are dependent on the $(u, v, w = 1-u-v)$ parameter values of the domain. This leads to the formulation of unique inner points for each parameter value in the domain. This construction was introduced by Gregory to ensure that a pair of patches meeting at a shared edge are tangent plane continuous across that edge. The patch is evaluated by the following equation:

$$T(u, v, w) = u^3 \boldsymbol{p}_0 + v^3 \boldsymbol{p}_1 + w^3 \boldsymbol{p}_2 +$$
$$3uv(u + v)(u\boldsymbol{e}_0^+ + v\boldsymbol{e}_1^-) +$$
$$3vw(v + w)(v\boldsymbol{e}_1^+ + w\boldsymbol{e}_2^-) +$$
$$3wu(w + u)(w\boldsymbol{e}_2^+ + u\boldsymbol{e}_0^-) +$$
$$12uvw(u\boldsymbol{F}_0 + v\boldsymbol{F}_1 + w\boldsymbol{F}_2)$$

where

$$\boldsymbol{F}_0 = \frac{w\boldsymbol{f}_0^- + v\boldsymbol{f}_0^+}{v + w},$$
$$\boldsymbol{F}_1 = \frac{u\boldsymbol{f}_1^- + w\boldsymbol{f}_1^+}{w + u},$$
$$\boldsymbol{F}_2 = \frac{v\boldsymbol{f}_2^- + u\boldsymbol{f}_2^+}{u + v} \qquad (1)$$

Figure 1 shows the labeling of control points for the Gregory patch, and how it maps to a quartic Bézier triangle.

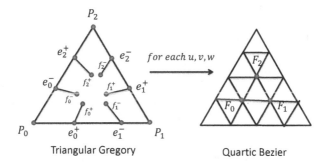

for each u, v, w

Triangular Gregory   Quartic Bezier

Fig. 1. Control points for triangular Gregory and its relation with quartic Bézier

3   Approximation Methods

In this section we will illustrate how to approximate the Loop subdivision surface by using the points, normals and tangents obtained in Section 2. We start by introducing methods that require less input control points and then move on to those that require more. We also discuss implementation details for the control shader and provide performance trade-offs.

3.1   PN-Triangles

Vlachos et al. [7] propose *curved PN triangles* for interpolating a triangle mesh by a parametric, piecewise cubic surface. This established technique generates a C^0-continuous surface which stays close to the triangle mesh. At first, the PN scheme places the intermediate control points $\bar{b}_{ijk}$ at the positions $(ib_{300}+jb_{030}+kb_{003})/3$, $i+j+k=3$, leaving the three corner points unchanged. Then, each b_{ijk} on the border is constructed by projecting the intermediate control point $\bar{b}_{ijk}$ into the plane defined by the nearest corner point and its normal. Finally, the central control point b_{111} is constructed by moving the point $\bar{b}_{111}$ halfway in the direction $m - \bar{b}_{111}$ i.e $(b_{111} = 0.5*(m+\bar{b}_{111}))$ where m is the average of the six control points computed on the borders as described above. The construction uses only the three triangle vertices and its normals. This makes it especially suitable for a triangle rendering pipeline. By using the Loop limit points and normals derived in [15], we get a Loop approximation using PN-triangles.

3.2   Walton-Meek Triangles

Walton and Meek [8] proposed the WM method for achieving G^1 patch from the boundary curves of a triangular control net. Both the WM method and the Gregory Method, described in the next section, construct a triangular Gregory patch to evaluate the surface. The boundary curves are obtained in the same manner as PN-triangles and the inner points are calculated for each u, v, w, this ensures that any two neighboring triangles have a common tangent plane along their shared boundary. The main difference between the WM and Gregory method is based on where and how the calculations for the tangent plane continuity occur, the WM performs this inside the control shader while the Gregory processes this information and sends it to the control shader. We pass the limit Loop points (3) and normals (3)a total of 6 control points for this method. The WM method constructs the inner points based only on the boundary control points as the 1-ring neighborhood is not available inside the control shader. A simple explanation of the WM method is provided below, for details refer to [8]:

- Degree elevate the cubic boundary curve, formed by the limit Loop points and normals, to a quartic and get the control points of the curve.
- For each boundary curve, determine the two interior control points associated with this curve using only the data available on the boundary. Altogether, the 6 interior Gregory control points result.

– For each u,v,w, evaluate the patch by blending the the six Gregory points and obtaining the interior control points of the quartic patch as shown in Equation (1).

3.3 Gregory Triangles

As shown in Figure 1, we need to calculate 15 control points for evaluating a triangular Gregory patch. For each u, v, w parameter value the boundary control points are degree elevated from cubic to quartic using the two corner and two edge points and the the inner quartic points F_0, F_1, F_2 are calculated from the inner six Gregory control points as shown in Equation (1). So an approximation of the Loop surface is possible by calculating these 15 control points, 3 corner, 6 edge and 6 inner for the Gregory triangle.

The corner points are set as the limit loop control points and the edge points are calculated using the limit loop tangents, however we need to choose the right length for these tangents to get the edge points and we know that the derivative at the end points of a Gregory patch is $3(e_0^+ - p_0)$ we can solve e_0^+ by e_0^+ by

$$e_0^+ = p_0 + \frac{2}{3}t_{0,1}\lambda \qquad (2)$$

where $t_{0,1}$ is the Loop limit tangent from p_0 to p_1 and λ is chosen to be the subdominant eigenvalue of the Loop surface and is given by

$$\lambda = \frac{3}{8} + \frac{1}{4}\cos\left(\frac{2\pi}{n}\right) \qquad (3)$$

We now have to construct the inner points so that the surface is tangent plane continuous across the edge. By using the method described in [13], we find that setting the traversal vector r to be equal to the Loop surface cross tangents, see Figure 2, provides the best results. The traversal vector r is calculated by performing two levels of Loop subdivision on the original control mesh and evaluating the cross tangents at $1/4, 1/2, 3/4$ along each edge and then linearly interpolating them to find the values at $1/3, 2/3$.

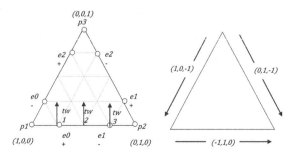

Fig. 2. Loop Cross tangents at $1/4$, $1/2$ and $1/3$ shown by tw1, tw2 and tw3

The equation for the inner points is given by

$$f_0^+ = \frac{1}{4}\left(c_1 p_0 + (4 - 2c_0 - c_1)\,e_0^+ + 2c_0 e_1^- + r\right) \tag{4}$$

where $c_0 = \frac{2\pi}{n_0}$ and $c_1 = \frac{2\pi}{n_1}$, n_0 and n_1 are the valence at p_0 and p_1

4   Unified Rendering Framework

We propose a unified rendering framework that involves three stages, preprocessing, CPU switching and rendering.During the preprocessing stage all the control point calculations for each triangle in the input mesh is done irrespective of the method chosen. Therefore for each triangle, 3 limit points, 3 limit normals, 6 edge points and 6 inner points, resulting in a total of 15 control points and 3 normals are calculated as shown in Figure 3. The application can then dynamically change this data at run time on a case by case basis, giving the developer an automatic one-pass algorithm to switch the methods at run time to manage quality vs. performance or connect it into existing LOD structures. As shown in Figure 4, there are three main stages for the unified rendering framework. In the first stage, control polygon from asset production packages like Maya is read by a custom patch generation tool that creates for each base mesh triangle, the 18 control points. In the second stage the application creates custom draw calls for all the three methods and performs CPU based decisions to choose the appropriate method. The last stage involves creating the effect file using the CGFX file format with the methods implemented as separate techniques. Control and Evaluation shaders need to be written and we provide CGFX files that implement the methods discussed.

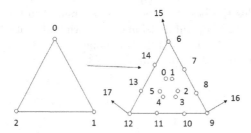

Fig. 3. Indexing of the 18 control points calculated for every input triangle

4.1   Control and Evaluation Shader

The control shader takes the input control mesh based on the method chosen, and performs tessellation calculations to determine each edge tessellation factor. Also depending on the case it can perform other calculations before sending the

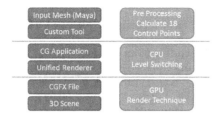

Fig. 4. Various stages of the unified rendering framework

output to the evaluation shader. For example, in the PN method the 10 control points for the cubic Bézier triangle and the 6 normals for quadratic interpolation are calculated in the control shader.

The Evaluation shader receives the u, v, w values from the tessellator and the output from the control shader and calculates the surface point and normal at that parameter set. The PN method evaluates the point using [15] and the normal by quadratic interpolation. The WM method evaluates the point by Equation (1), and the normal by quadratic interpolation. The Gregory method transforms Equation (1) into a quartic Bézier by degree elevating the boundary cubic curves and calculating $F0, F1, F2$ as described in Section 3. This allows for simultaneous calculation of the point and normal using the de Casteljau algorithm. Table 1 shows the the number of arithmetic calculations (shader lines of code) in the control and evaluation shaders for the three methods.

Table 1. Shader properties for the three methods

Method Name	Control Shader	Evaluation Shader	Control Points
PN	13	10	6
WM	70	10	6
Gregory	0	20	15

Table 2. Frame rate comparison for the three methods

Model Name	Base Mesh Tris/Verts	Tess Level/ Object Count	PN FPS	WM FPS	Gregory FPS
Face	102/200	5/1	1850	1180	1780
Bunny	502/1000	5/1	1297	822	1190
Big Guy	1754/2900	5/1	940	635	870
Face	102/200	5/50	479	379	266
Bunny	502/1000	5/50	63	52	32
Big Guy	1754/2900	5/50	53	39	22

Table 3. Geometric error reported by Metro for the three methods

Method Name	Max error	mean error	RMS error
PN	0.5618	0.019	0.030
WM	0.3913	0.013	0.022
Gregory	0.1746	0.009	0.013

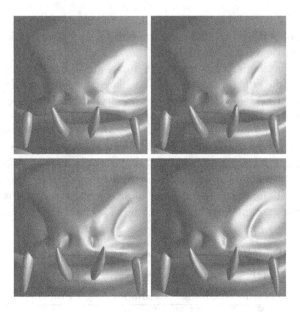

Fig. 5. Monster frog rendered using PN (top-left), WM (top-right), Gregory (bottom-left) and Original Loop (bottom-right) methods

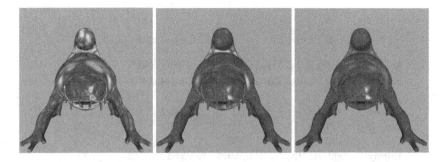

Fig. 6. Geometric approximation error calculated by Metro for PN, WM and Gregory methods

Method	Reflection Lines	Normals

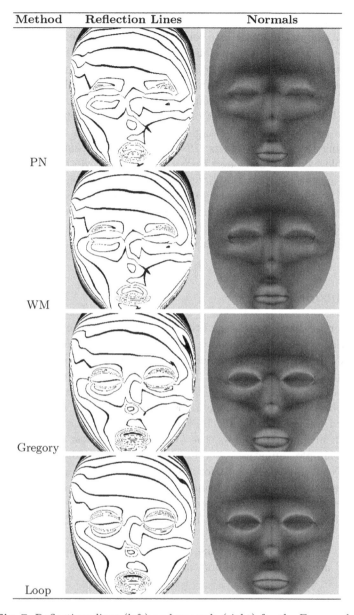

PN

WM

Gregory

Loop

Fig. 7. Reflections lines (left) and normals (right) for the Face model

5 Results

In this section we measure the speed and quality of the surface for the three methods described in this paper. With all conditions being the same we test the frame rate at various tessellation levels. We also render the reflection lines as shown in Figure 7 for each method to test the quality of the surface. Our tests were performed on PC with Windows 7 32bit and NVIDIA Quadro5000 graphics card. Table 2, shows the frame rate comparisons for the face, bunny and big guy models at tessellation level 5 and also compares the numbers by changing the number of models drawn from 1 to 50. It is seen that PN performs best under all situations and Gregory outperforms the WM method only when few models use it. This confirms that performance depends on the number of operations in the hull shader and the number of control points that need to be stored in video memory for a base mesh triangle. Figure 5 shows the monster frog model rendered using the three methods, PN, WM, Gregory and the original Loop method. Table 3 shows the geometric error calculated by Metro [19] by comparing the triangle meshes generated by Loop subdivision to the ones generated by the three methods. Figure 6 shows this error mapped to vertex colors and we can observe that except at extreme extraordinary points (valence greater than 7) the error produced by the Gregory method is minimal. To compensate for this limitation the original control mesh would need to avoid having such points at model time.

6   Conclusion

In this paper we have presented three methods for approximating Loop subdivision surfaces using tessellation enabled hardware. Developers can unify the asset production pipeline and provide automatic switching between these methods at runtime. The maximum number of control points required per patch is 18, the PN and WM methods use a subset of 6 while the Gregory uses a subset of 15. Our observation leads to the following conclusions. The Gregory method is best suited for characters and fluid assets that incorporate complex animations as seen in game characters, the WM method is best suited for dynamic objects with simple animations used in weapons, vehicles and breakable objects, while the PN method works best for static objects like trees, environments and terrain. This observation stems from the information in Table 2 based on the performance of the methods at single and multiple instances as well as at lower and higher tessellations. The unified rendering setup, shown for triangles, can similarly be applied for quads using the Catmull-Clark subdivision method. In future work, we plan to improve the unified rendering setup to include meshes consisting of arbitrary polygons and not just triangles.

References

1. Luebke, D., Humphreys, G.: How gpus work. Computer 40, 96–100 (2007)
2. Kautz, J.: Hardware lighting and shading: A survey. In: Computer Graphics Forum, vol. 23, pp. 85–112. Wiley Online Library (2004)
3. Akenine-Moller, T., Haines, E.: Real-time rendering. AK (2002)
4. Kry, P., James, D., Pai, D.: Eigenskin: real time large deformation character skinning in hardware. In: Proceedings of the 2002 ACM SIGGRAPH/Eurographics Symposium on Computer Animation, pp. 153–159. ACM (2002)

5. Baran, I., Popović, J.: Automatic rigging and animation of 3d characters. In: ACM SIGGRAPH Papers, p. 72. ACM (2007)
6. Loop, C.: Smooth Subdivision Surfaces Based on Triangles. Master's Thesis, University of Utah 1, 1–74 (1987)
7. Vlachos, A., Peters, J., Boyd, C., Mitchell, J.: Curved PN triangles. In: Proceedings of the 2001 Symposium on Interactive 3D Graphics, pp. 159–166. ACM (2001)
8. Walton, D.J., Meek, D.S.: A triangular G1 patch from boundary curves. Computer-Aided Design 28, 113–123 (1996)
9. Boubekeur, T., Schlick, C.: Approximation of subdivision surfaces for interactive applications. In: ACM Siggraph Sketch Program (2007)
10. Boubekeur, T., Schlick, C.: Qas: Real-time quadratic approximation of subdivision surfaces (2007)
11. Fünfzig, C., Müller, K., Hansford, D., Farin, G.: PNG1 triangles for tangent plane continuous surfaces on the GPU. In: GI 2008: Proceedings of Graphics Interface 2008, pp. 219–226. Canadian Information Processing Society, Toronto (2008)
12. Loop, C., Schaefer, S.: Approximating Catmull-Clark subdivision surfaces with bicubic patches. ACM Transactions on Graphics (TOG) 27, 1–11 (2008)
13. Loop, C., Schaefer, S., Ni, T., Castaño, I.: Approximating subdivision surfaces with gregory patches for hardware tessellation. ACM Transactions on Graphics (TOG) 28, 1–9 (2009)
14. Seidel, H.: Polar forms and triangular B-spline surfaces. Computing in Euclidean Geometry, 235–286 (1992)
15. Li, G., Ren, C., Zhang, J., Ma, W.: Approximation of Loop Subdivision Surfaces for Fast Rendering. IEEE Transactions on Visualization and Computer Graphics (2010)
16. Farin, G.: Curves and Surfaces for Computer-Aided Geometric Design — A Practical Guide, 5th edn. The Morgan Kaufmann Series in Computer Graphics and Geometric Modeling, 499 pages. Morgan Kaufmann Publishers, Academic Press (2002)
17. Gregory, J.A.: Smooth interpolation without twist constraints. Computer Aided Geometric Design, 71–87 (1974)
18. Chiyokura, H., Takamura, T., Konno, K., Harada, T.: G1 surface interpolation over irregular meshes with rational curves. NURBS for Curve and Surface Design, 15–34 (1990)
19. Cignoni, P., Rocchini, C., Scopigno, R.: Metro: measuring error on simplified surfaces. In: Computer Graphics Forum, vol. 17, pp. 167–174. Wiley Online Library (1998)

Bundle Adjustment Constrained Smoothing for Multi-view Point Cloud Data

Kun Liu and Rhaleb Zayer

Inria, Villers-lès-Nancy, F-54600, France

Abstract. Direct use of denoising and mesh reconstruction algorithms on point clouds originating from multi-view images is often oblivious to the reprojection error. This can be a severe limitation in applications which require accurate point tracking, e.g., metrology.

In this paper, we propose a method for improving the quality of such data without forfeiting the original matches. We formulate the problem as a robust smoothness cost function constrained by a bounded reprojection error. The arising optimization problem is addressed as a sequence of unconstrained optimization problems by virtue of the barrier method. Substantiated experiments on synthetic and acquired data compare our approach to alternative techniques.

1 Introduction

Over the last decade, *bundle adjustment* (BA) has become one of the key steps in multi-view reconstruction. It intervenes as a single nonlinear optimization which simultaneously fine-tunes the 3D structure and the viewing parameter estimates [1]. BA requires a set of feature correspondences which can be sparse, quasi-dense or dense in order to control the reprojection error and yields a refined visual reconstruction. In an ideal setting the resulting point cloud data would reflect the exact geometry of the original object. In practice however, several factors such as ill-textured objects, spatial discretization, structured noise, and lighting conditions contribute toward matching errors. These errors cannot be fully fixed by BA and the point cloud generally exhibits noise to varying degrees. A commonly adopted solution is the construction of an approximating surface using existing meshing algorithms, e.g. [2, 3]. These geometric algorithms operate mainly in the three dimensional domain and do not necessarily maintain correspondences between the scene and image features. As a result, the cross-image correspondences are lost and can only be approximated by reprojection on the surface. Although recent approaches in multiview reconstruction, e.g. [4], can improve the visual appearance tremendously, they cannot be readily used in applications such as a metrology or non-contact shape and deformation measurement where an accurate and consistent tracking of surface points over time is crucial for gathering information such as strain or parameter estimation. Furthermore, we are not aware of geometry processing methods which enforce bounds on the reprojection error in the literature.

G. Bebis et al. (Eds.): ISVC 2012, Part I, LNCS 7431, pp. 126–137, 2012.

In this paper, we propose a point cloud smoothing approach tailored for multi-view point cloud data. We formulate the problem as the minimization of a smoothness measure constrained by a bound on the reprojection error. For the former, we propose a measure which favors local flatness of the point cloud data and for the latter, we adopt a formulation similar to standard sparse bundle adjustment. Both measures are combined into a constrained nonlinear optimization formulation. A barrier approach is used to drive the numerical optimization towards a smooth point cloud where the bounds on the reprojection error are enforced. In order to overcome numerical problems related to the densely populated nature of the arising matrix equations, we take advantage of the Sherman-Morisson formula. This allows for addressing relatively large data sets while keeping reasonable memory requirements.

We evaluate the quality of resulting point cloud data by means of ground truth data generated synthetically. Tests on real data acquired and reconstructed using existing methods e.g. [5] confirm the quality of our results. We demonstrate the robustness of our approach to irregular data sampling, to sharp features and to shrinkage. Our approach does not make any assumption on the nature of the noise in the data and does not require any additional input, e.g., visual hulls. The only assumption made is the geometric smoothness, which is often a property of the original model. Although, we do not perform any further matching computations on the underlying images, experiments on synthetic data sets suggest that our approach moves existing matches closer towards to the exact matches.

In summary, this paper makes the following contributions:

- Formulate a constrained optimization for smoothing multi-view point clouds with bounded reprojection error
- Develop robust and efficient numerical solution procedure

Our approach can be regarded as a post-processing tool and could be used in conjunction with existing reconstruction algorithms. The rest of this paper is organized as follows: Section 2 covers the most related work, section 3 lays out the general setup and the notation, and section 4 reviews bundle adjustment and discusses using the Laplacian operator as smoothing regularizer. In section 5, we introduce our smoothing cost function and show how to set up the constrained optimization. Numerical aspects of our approach are discussed in section 6 and the results are summarized in section 7.

2   Related Work

The prior art on multi-view model acquisition is extensive. In order to keep this exposition succinct, we restrict ourselves to the most related work and refer the reader to [6, 7] for a general overview.

Despite its long history, most of research effort on BA has been dedicated to numerical optimization strategies [1]. Subjects such as fusing it or enhancing it with additional input has been studied less. For instance, [8, 9] introduce model-based constraints as a regularizer within the bundle adjustment formulation in the context of head reconstruction. The authors of [10] propose using GPS (global positioning system) data as a penalty for the reprojection error and optimize the problem in the least squares sense. More closely related to our work is the approach proposed in [11], where GPS and Structure-from-Motion data are fused within a constrained optimization formulation. The approach is applied to a setting which combines monocular image sequence with GPS data. In our case, we are more concerned about the geometric smoothness of the multi-view data and we cannot take advantage of readily available sensor data.

Smoothing surface meshes [12–15] and point clouds [16, 17], are well studied topics in geometry processing. Traditionally, these approaches are tailored for input data obtained from scanners. Unfortunately, many of these methods do not perform well on point clouds originating from multi-view stereo reconstruction. Furthermore they enforce the correspondences between 3D structure and image feature points.

3   Problem Setting and Notation

In the following, we assume the input data consists of a point cloud originating from standard multi-view acquisition [6] along with the camera parameters. The data can stem from dense [18] or quasi dense [5] matching approaches.

The point cloud will be represented as $x = (x_1^\mathsf{T}, ..., x_n^\mathsf{T})$, where x_i represent the three dimensional coordinates of the i-th point. For a set of m views, the camera matrices can be conveniently assembled as $p = (p_1, ..., p_m)$ where p_j be the vector of parameter for camera j. The point corresponding to x_i on an image j will be denoted a_{ij}.

4   Bundle Adjustment with Smoothing Regularization

The overall goal of this work is to determine a smooth geometric model and the configuration of cameras that are maximally consistent with the observations. A good starting point is the bundle adjustment.

4.1   Bundle Adjustment

BA is a nonlinear least square problem [6] where the cost function penalizes reprojection error with respect to 3D structure and viewing parameters.

Concretely, this can be formulated as:

$$h(x, p) = \sum_i^n \sum_j^m \delta_{ij} \|Q(p_j, x_i) - a_{ij}\|^2 \tag{1}$$

where $Q(p_j, x_i)$ is the predicted projection of point i on image j, and δ_{ij} flags 1 if point i is visible in image j and 0 otherwise.

The best known algorithm to solve this kind of nonlinear least-squares problems is the the Levenberg-Marquardt (LM) algorithm [19].

4.2   Bundle Adjustment with Smoothing Regularization

As we seek to smooth point cloud data while minimizing BA, it seems natural to consider reformulating the problem as a minimization of the following objective function

$$h(p, x) + \alpha \|L(x)\|^2 \tag{2}$$

where h is the BA function from equation 1, L is the Laplacian operator discretized locally using the k-nearest neighbors, and α is a weighting parameter. This formulation blends naturally within the standard Levenberg-Marquardt algorithm as the Laplacian operator can be simply considered as regularizer.

Unfortunately such an approach would suffer from over-smoothing as well as limitations known to Laplacian operator especially with respect to sharp features. Additionally due to the sparsity of the point cloud and the discretization based on nearest neighbors, shrinking effects can appear around holes or areas where data is missing as illustrated in figure 1. In the following section, we propose an alternative approach which remedies such artifacts as illustrated in the aforementioned figure.

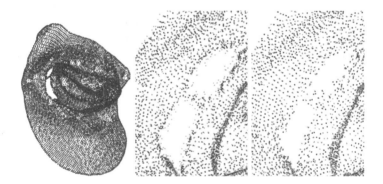

Fig. 1. A zoom on the ear model (left) illustrates the shrinking effect of Laplacian regularization (middle, blue). Constrained smoothing (right, blue) is more robust to such artifacts. In both results, the original data is shown in orange.

5   Bundle Adjustment Constrained Smoothing

We regard the problem as searching for a smooth surface such that reprojection error is minimal. In order to account for possible errors in the cross-image feature correspondences, we can allow the matches to evolve in a small disk around their initial location as illustrated in figure 2. This would allow searching for a smooth surface while maintaining image feature correspondences in the vicinity of their initial positions.

In the context of an optimization for the whole point cloud data, we do not need to enforce the radius constraint for the individual points. Instead, we enforce

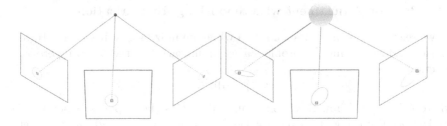

Fig. 2. Starting from a converged bundle adjustment, our approach (left) searches for new spatial position of the 3d point while guaranteeing that the reprojection error is bounded i.e. the matches are maintained within a disk around the input matches. On the other hand, constraining the smoothing within a ball around the initial spatial position (right) can lead to larger reprojection errors as the shape of the corresponding projection (planar ellipses) is not taken into account (please refer to the text for details).

it as a global constraint. In the following, we define our smoothness measure and we show how it can combined with the reprojection error.

It would be possible to tailor a smoothing approach which restricts the displacements within a small ball around the initial spatial point locations. This kind of smoothing however, does not take into consideration the reprojection into image space and can lead to large errors. This is illustrated in figure (2-right) which shows the spatial search domain (ball around the initial point) and its counter part image space (planar ellipses). As the planar ellipses can be elongated, the reprojected point position can lay far from the initial match and hence such an approach would corrupt the initial matching results. This effect can be further amplified when dealing with wide base-line views. In contrast, we formulate the constraints in image space. The spatial position is then forced to lay at the intersection of the fat bundle-lines (small cylinders around the bundle lines) and thus a tight bound on reprojection error is guaranteed.

5.1   Smoothness Measure

In order to define smoothness for point cloud data, we endow the points with local adjacency relations. We use k-nearest neighbor algorithm to construct a directed graph $\mathcal{G} = (\mathcal{V}, \mathcal{E})$, where $\mathcal{V}$ is the point set we add an edge (x_i, x_j) in $\mathcal{E}$ if x_i is one of k-nearest neighbor of x_j. In all our experiments k is set to 10, we note that using higher values hardly changes the results. Additionally, we endow each point with a normal direction. We estimate the point normal n_i ($\|n_i\| = 1$) for every point i by the method proposed in [20], which uses principal component analysis on the local neighborhoods.

Given computed normal directions, we define the local planarity for each edge (x_i, x_j) in $\mathcal{E}$ as

$$(n_{ij} \cdot (x_i - x_j))^2, \tag{3}$$

where n_{ij} is the average normal associated with the mid-point of edge(x_i, x_j). It penalizes the deviation of the points from the average plane defined by the midpoint and its normal. We sum the contribution of all edges in $\mathcal{E}$ to define the global cost function

$$f(x) = \sum_{(x_i,x_j)\in\mathcal{E}} (n_{ij} \cdot (x_i - x_j))^2 \tag{4}$$

This function acts in two ways, it tends to improve the local flatness by minimizing the scalar product and second, since the edge vector is not normalized, it tends to pull neighboring points together. This local flatness measure is commonly used in the context of mesh simplification [25].

A similar formulation in the $\mathbf{L}^1$-norm has been proposed in [21]. In order to avoid differentiability difficulties raised by lower order norms we use the $\mathbf{L}^2$ in view of coupling this measure with the reprojection error. In general, $\mathbf{L}^2$ responds strongly to outliers. We avoid this shortcoming by means of a robust norm

$$\Psi(s^2) = \sqrt{s^2 + \epsilon^2}; \ \epsilon = 1e^{-6} \tag{5}$$

This function can be regarded as a differentiable norm of the absolute norm function and its impact is illustrated in figure 3.

Fig. 3. The result of our approach on a noisy input data using $\mathbf{L}^2$ norm (left) and the robust norm (right). The color coding shows the Hausdorff distance to the ground truth point cloud. (Best viewed in the electronic version).

5.2   Reprojection Error Constraint

Let's assume we apply BA to the input point cloud data and let ϵ_0 be the residual reprojection. We can then define the our reprojection constraint as

$$h(x) < \lambda\epsilon_0 \tag{6}$$

with relaxing parameter $\lambda > 1$. The constraint defined above caps the reprojection error. It also resolves cases where minima of the cost function $f(x)$ is not unique by restricting the search within a very close range.

6 Constrained Optimization

At this stage, we have all the ingredients necessary for the problem setup and we can formulate the smoothing procedure as the following constrained optimization

$$\begin{aligned} minimize \quad & f(x) \\ subject\ to \quad & g(x) \leq 0 \end{aligned} \tag{7}$$

where $g(x) = h(p, x) - \lambda \epsilon_0$. The constraint function $g(x)$ depends only on the structure as we keep the camera parameters fixed in what follows.

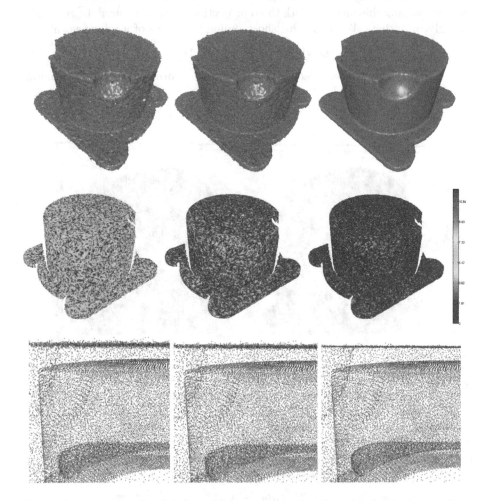

Fig. 4. A noisy point cloud (left-top) is processed using BA with Laplacian regularization (middle-top) smoothing and BA constrained smoothing (right-top), all views are shown in splating mode. The middle row shows the reprojection error for the same view. The bottom row shows a zoom on the corresponding point cloud data.

This problem is more intricate than BA alone as the cost function and the constraint are both nonlinear. We address it using the barrier method [22], which is a procedure for approximating constrained optimization problems by unconstrained ones. The smoothing problem can be solved then as a series of nonlinear minimization problems of the form

$$f(x) + B(g(x), c) \tag{8}$$

where $B(g(x), c)$ is the barrier function and c is a positive constant. The barrier function operates by introducing a singularity along the constraint boundary. Probably, the popular choice is the logarithmic barrier function which tends to infinity at the constraint boundary. Furthermore, Its simplicity w.r.t subsequent derivative computations and its satisfaction of the self-concordance criterion makes it an attractive choice [19]. In the current paper, we use a logarithmic function

$$B(g(x), c) = -c \cdot \ln(-g(x)) \tag{9}$$

Concretely, solving the problem in equation. 7 amount to minimizing a series of function in the form of Eq. 8 with different constant $c = c_k$, where $\{c_k\}$ is a decreasing sequence tending to 0, i.e. for each k, $c_k \geq 0$, $c_{k+1} < c_k$.

Each nonlinear minimization Eq.8 is solved iteratively using Newton's method. The associated Hessian matrix is

$$H = H_{f(x)} + B' H_{g(x)} + B'' \nabla g \nabla g^{\mathsf{T}} \tag{10}$$

where $H_{f(x)}$ and $H_{g(x)}$ are the Hessian matrices of $f(x)$ and $g(x)$ respectively, B' and B'' are the the first and second derivatives, ∇g is the gradient of $g(x)$. In our implementation we approximate $H_{g(x)}$ by $J^{\mathsf{T}} J$, where J is the Jacobian of g.

The last term in equation 10 is a densely populated matrix and turns out to be problematic when solving the linear systems involved at each Newton iteration. We avoid this issue by means of the Sherman-Morrison formula [23] which reads

$$(A + uv^{\mathsf{T}})^{-1} = (I - \frac{A^{-1} uv^{\mathsf{T}}}{1 + v^{\mathsf{T}} A^{-1} u}) A^{-1} \tag{11}$$

and holds for arbitrary invertible square matrix A and vectors u and v such that $1 + v^{\mathsf{T}} A^{-1} u$) is non zero.

Writing $\hat{H} = H_{f(x)} + B' H_{g(x)}$ and $\hat{g} = \sqrt{B''} \nabla g$, we have by virtue of equation 11

$$H^{-1} = (I - \frac{\hat{H}^{-1} \hat{g} \hat{g}^{\mathsf{T}}}{1 + \hat{g}^{\mathsf{T}} \hat{H}^{-1} \hat{g}}) \hat{H}^{-1} \tag{12}$$

Therefore, linear systems of the form $(\hat{H} + \hat{g} \hat{g}^{\mathsf{T}}) x = b$ can be converted into

$$x = (I - \frac{\hat{H}^{-1} \hat{g} \hat{g}^{\mathsf{T}}}{1 + \hat{g}^{\mathsf{T}} \hat{H}^{-1} \hat{g}}) \hat{H}^{-1} b \tag{13}$$

Since $\hat{H}$ is sparse and not densely populated, the system can be handled using standard linear solvers. It is imperative to note that the inverse of $\hat{H}$ need not be computed. Instead, equation 13 is split into two subsystems $\hat{H}y = b$ and $\hat{H}z = \hat{g}$. The results are then plugged back into equation 13.

7   Results

We tested the proposed method on a set of synthetic and real world data. For synthetic data, camera captures were generated using existing 3D models with the help of Blender [24]. In this way, we have all the ground truth data necessary for evaluation. Noise was added in two different ways. In the first scenario, noise was directly added to the image correspondences (figures 3, 5). In the second scenario, noise was added to the original model and the image correspondences were obtained as weighted average of the noiseless and noisy projections (figure 4). In both scenarios, a gaussian function was used to generate the noise. Figure 4 shows a a comparison of Laplacian regularized BA of section 4 and the BA constrained smoothing of section 5. Although the former approach is easier to implement, it under-performs in comparison to latter approach. Figure 5 shows the performance of our approach on a large data set where more than 50 views were used. A typical result of our approach on real world data is shown in figure 6. In this example, 6 views were combined using the quasi-dense propagation approach of [5] to generate the initial point cloud. Our approach reduces

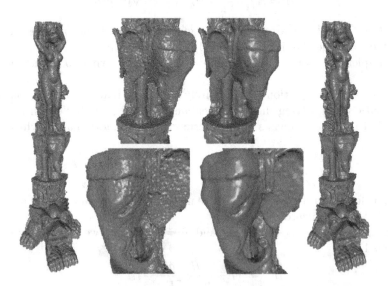

Fig. 5. Illustration of our method on a large data set (200K points). Image correspondences across 56 views were perturbed by a gaussian noise with a unit variance and a peak of 3 which yields the noisy reconstruction (left). The result of our approach is shown to the right. Middle image show a zoom on the elephant head. All views are shown in splating mode.

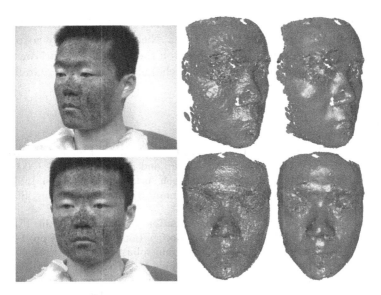

Fig. 6. Sample images (left) out of a set of 6 wide base-line images were used to generate a quasi-dense point cloud (middle) using the propagation approach in [5]. Our result (right) shows an overall quality improvement of the point cloud. Point clouds are shown in splating mode.

the asperities in the point cloud and yields a smoother result. Even in regions such the eye area which are generally difficult to construct, our approach helps smoothing out the noise to a great extent.

Limitations and Discussion: Our approach requires a good initial estimate e.g. results from bundle adjustment and also a sufficient sampling of the data in order to yield optimal results. However, when sampling points are missing in some region, as illustrated in figure 1, our approach does not suffer from shrinkage and still produces coherent results. When the initial data suffers from large noise, our approach can improve the results but only within the limits allowed by the reprojection error control. In this scenario, the noise reflects extensive errors in the matching, projection matrices or both and therefore enforcing reprojection errors based on such corrupt data might not be a viable goal. Nevertheless, relaxing the reprojection error bound would induce smoother geometric results.

8   Conclusion

In this paper, we studied the effect of combining a smoothness measure with bundle adjustment. To overcome limitations of using simple strategies such as Laplacian regularization or constrained spatial smoothing, we developed a robust and efficient approach based on containing the reprojection error while enforcing the smoothness of the point cloud data. In future work, we plan to investigate the use of our approach on time-dependent data.

Acknowledgments and Credits: The author would like thank, Alejandro Galindo for help with acquiring the face data, Nicolas Ray, Dmitry Sokolov, and the anonymous reviewers for their feedback on the paper. This work was funded by the ANR (Agence Nationale de la Recherche) under grant (PhysiGrafix ANR-09-CEXC-014-01).

References

1. Triggs, B., McLauchlan, P., Hartley, R., Fitzgibbon, A.: Bundle adjustment – a modern synthesis. Vision algorithms: Theory and Practice, 153–177 (2000)
2. Kazhdan, M., Bolitho, M., Hoppe, H.: Poisson surface reconstruction. In: Proceedings of the Fourth Eurographics Symposium on Geometry Processing, SGP 2006. Eurographics Association, pp. 61–70 (2006)
3. Lempitsky, V., Boykov, Y.: Global optimization for shape fitting. In: IEEE Conference on Computer Vision and Pattern Recognition, CVPR 2007, pp. 1–8 (2007)
4. Furukawa, Y., Ponce, J.: Accurate, dense, and robust multiview stereopsis. IEEE Transactions on Pattern Analysis and Machine Intelligence 32, 1362–1376 (2010)
5. Lhuillier, M., Quan, L.: Match propogation for image-based modeling and rendering. IEEE Trans. Pattern Anal. Mach. Intell. 24, 1140–1146 (2002)
6. Hartley, R., Zisserman, A.: Multiple view geometry in computer vision, vol. 2. Cambridge Univ. Press (2000)
7. Seitz, S.M., Curless, B., Diebel, J., Scharstein, D., Szeliski, R.: A comparison and evaluation of multi-view stereo reconstruction algorithms. In: Proceedings of the 2006 IEEE Computer Society Conference on Computer Vision and Pattern Recognition, CVPR 2006, pp. 519–528 (2006)
8. Fua, P.: Using model-driven bundle-adjustment to model heads from raw video sequences. In: The Proceedings of the Seventh IEEE International Conference on Computer Vision, vol. 1, pp. 46–53 (1999)
9. Fua, P.: Regularized bundle-adjustment to model heads from image sequences without calibration data. International Journal of Computer Vision 38, 153–171 (2000)
10. Kume, H., Taketomi, T., Sato, T., Yokoya, N.: Extrinsic camera parameter estimation using video images and gps considering gps positioning accuracy. In: Proceedings of the 2010 20th International Conference on Pattern Recognition, ICPR 2010, pp. 3923–3926. IEEE Computer Society (2010)
11. Lhuillier, M.: Fusion of gps and structure-from-motion using constrained bundle adjustments. In: 2011 IEEE Conference on Computer Vision and Pattern Recognition, CVPR, pp. 3025–3032. IEEE (2011)
12. Taubin, G.: A signal processing approach to fair surface design. In: Proceedings of the 22nd Annual Conference on Computer Graphics and Interactive Techniques, pp. 351–358. ACM (1995)
13. Desbrun, M., Meyer, M., Schröder, P., Barr, A.: Implicit fairing of irregular meshes using diffusion and curvature flow. In: Proceedings of the 26th Annual Conference on Computer Graphics and Interactive Techniques, pp. 317–324 (1999)
14. Fleishman, S., Drori, I., Cohen-Or, D.: Bilateral mesh denoising. ACM Transactions on Graphics (TOG) 22, 950–953 (2003)
15. Jones, T.R., Durand, F., Desbrun, M.: Non-iterative, feature-preserving mesh smoothing. ACM Transactions on Graphics (TOG) 22, 943–949 (2003)
16. Weyrich, T., Pauly, M., Keiser, R., Heinzle, S., Scandella, S., Gross, M.: Post-processing of scanned 3d surface data. In: Symposium on Point-Based Graphics, vol. 1 (2004)

17. Schall, O., Belyaev, A., Seidel, H.: Robust filtering of noisy scattered point data. In: IEEE Eurographics/IEEE VGTC Symposium Proceedings, pp. 71–144 (2005)
18. Ferrari, V., Tuytelaars, T., Van Gool, L.: Simultaneous Object Recognition and Segmentation by Image Exploration. In: Pajdla, T., Matas, J(G.) (eds.) ECCV 2004. LNCS, vol. 3021, pp. 40–54. Springer, Heidelberg (2004)
19. Nocedal, J., Wright, S.J.: Numerical optimization. Springer (1999)
20. Hoppe, H., DeRose, T., Duchamp, T., McDonald, J., Stuetzle, W.: Surface reconstruction from unorganized points. SIGGRAPH Comput. Graph. 26, 71–78 (1992)
21. Avron, H., Sharf, A., Greif, C., Cohen-Or, D.: l^1-sparse reconstruction of sharp point set surfaces. ACM Transactions on Graphics (TOG) 29, 135:1–135:12 (2010)
22. Luenberger, D.G., Ye, Y.: Linear and nonlinear programming, vol. 116. Springer (2008)
23. Press, W.H., Teukolsky, S.A., Vetterling, W.T., Flannery, B.P.: Numerical Recipes: The Art of Scientific Computing, 3rd edn. Cambridge University Press (2007)
24. Blender, http://www.blender.org
25. Heckbert, P.S., Garland, M.: Optimal triangulation and quadric-based surface simplification. Comput. Geom. Theory Appl. 14(1-3), 49–65 (1999)

Guided Sampling in Multiple View Robust Motion Estimation Using Regression Diagnostics

Houman Rastgar[1], Eric Dubois[1], and Liang Zhang[2]

[1] School of Information Technology and Engineering
University of Ottawa, Ottawa, Ontario, K1N 6N5 Canada
[2] Communications Research Center Canada, 3701 Carling Avenue
Ottawa, Ontario, K2H 8S2 Canada

Abstract. Random Sampling Consensus (RANSAC) is a standard tool in various estimation problems in computer vision. This paper presents an improvement to RANSAC where regression diagnostics information is incorporated in RANSAC in order to improve accuracy and speed. The improvement in the sampling stage speeds up the estimation by taking advantage of *a priori* information driven from regression diagnostics. Also, a new stopping criterion is presented that enables a much higher speed in the estimation process and guarantees that the iterative estimation is stopped as soon as the best model is found. In addition, the proposed method can easily be combined with existing randomized hypothesis verification techniques to achieve an even higher accuracy and speed. The algorithm is tested extensively on synthetic data and the results show a marked improvement in the accuracy of the estimation and a speed up of more than three times over the traditional RANSAC.

1 Introduction

Robust parameter estimation is a general tool in computer vision and 3D video processing. It is widely used for the task of model fitting in multiple view geometry, such as the estimation of the fundamental matrix [1, 2]. The use of a robust method is essential in order to find an accurate estimation of parameters when the input data are corrupted by non-Gaussian noise. Since the fundamental matrix is estimated from image matches, or point correspondences, its accurate estimation depends on using a robust method. This is due to the fact that image matches are often highly noisy and contain a large number of gross outliers. Moreover, the accurate and fast estimation of the fundamental matrix is of utmost importance since it is central to many computer vision tasks such as 3D reconstruction [3], image stitching [4] and augmented reality [5].

RANSAC has proven to be the most reliable robust estimator in the estimation of the fundamental matrix since it was presented [6]. RANSAC is a hypothesis-and-verify algorithm that proceeds by repeatedly drawing minimal subsets of data (in this case, image correspondences) and fitting a model to these points. Following this, a score is assigned to the parameters estimated using this minimal set with respect to the data. At the end, the best set of parameters that has achieved the highest score is selected.

G. Bebis et al. (Eds.): ISVC 2012, Part I, LNCS 7431, pp. 138–147, 2012.

There have been many improvements presented in the literature for speeding up and/or improving the accuracy of the estimation of the fundamental matrix using RANSAC. These generally fall within one of three main categories. The first category of such methods is those that improve the sampling by using prior information [7–10]. These methods address the "random" part in RANSAC by using information to give higher priority to certain samples rather than choosing them randomly. The second category of improvements are those that offer a better matching score for the models [11], providing a better measure of the "consensus". The third are those that address the hypothesis verification stage by using various methods of improving the efficiency in assessing the quality of the generated samples [12, 13].

The goal of this work is to follow the approach offered by methods that estimate *a priori* probabilities for the samples in order to avoid the time consuming process of choosing samples at random. In order to find meaningful *a priori* values for the sampling process, a new measure is devised for discriminating between inliers and outliers. This method operates without resorting to domain knowledge such as keypoint matching scores. In fact, the proposed *a priori* values are derived solely from regression diagnostics that are readily available during the iterative RANSAC process. In addition, a new stopping criterion for RANSAC is proposed that is consistent with the new sampling method and which guarantees that the iterations are stopped as soon as the best possible model is found, thus avoiding redundant iterations. The result is an algorithm that is more accurate and faster than RANSAC. In fact, the proposed method has an almost constant runtime regardless of the noise ratio, making it a viable tool in real-time vision applications, whereas most existing methods' performance drops exponentially as the noise levels increase as shown in the experimental results.

In addition, the method is not mutually exclusive with various methods of improving hypothesis verification such as Sequential Probability Ratio Test [12] and $T_{d,d}$ test [13]. In fact, results from combining these methods with the proposed one are presented in the experimental results and show the effectiveness of combining the two together.

2   Existing Sampling Methods

Given a set of point correspondences, RANSAC randomly samples a minimal subset of points to calculate the parameters of the model. The quality of these parameters are then evaluated with respect to all the correspondences and their score is calculated based on the total number of inliers, where inliers are defined as data having error less than some threshold T. This process is repeated until the termination criterion is met. The choice of which points are chosen is highly relevant to the final outcome of RANSAC. If no *a priori* information is provided on the correspondences, as is the case in the original RANSAC algorithm, points are chosen with equal probability in the sampling process. This presents a worst case scenario since all points have the same likelihood of being chosen, regardless of being an outlier or inlier. The solution to alleviating this problem is to use non-uniform sampling to give higher priority to points that are more likely to be inliers. Several methods have been proposed towards achieving this end.

GUIDED MLESAC [7] uses the correlation value between grey-level patches around the match points to find a measure of the validity of a correspondence. In other words, the correlation scores become the *a priori* probability of a data point. One of the problems with this approach is that repetitive textures or occlusions cause this validity measure to give erroneous information. In addition, this approach fails when there are multiple motions present in the scene. For example, point matches belonging to local motions in the scene can be perfectly matched but such matches are also outliers with respect to the fundamental matrix. As will be shown, the proposed method can be applied to a scene containing any arbitrary number of structures since it does not rely on correlation or other matching scores.

PROSAC [14] orders the correspondences by their similarity score and operates on progressively increasing correspondence sets to generate hypotheses. Although this method leads to higher quality sample sets, it is also only applicable to cases with a single motion present in the data. It is also prone to degeneracies since points with high correlation scores are often on the same fronto-parallel structures and a hypothesis generated from a set of correspondences on a plane lead to a degenerate motion model [1]. Another method, NAPSAC [9] uses points that are closer to each other in samples. By biasing the random selection towards clusters in multi-dimensional space, it achieves a higher quality of samples than random sampling. The argument is that points that cluster together are more likely to be inliers. Again, with this method, the argument only applies to cases where there is only one structure present in the data and does not extend well to cases of multiple motion. In addition, points that are nearby are often on the same object and form nearly planar structures which also leads to possible degeneracies.

Two existing algorithms where *a priori* information is calculated without resorting to match scores or match consistency information are RES-RANSAC [8] and the Ensemble method presented in [15]. Both methods rely purely on residual information from the random trials to make distinctions between inliers and outliers. In [15] the problem of robust estimation is however reformulated as a classification problem and based on statistical properties of the residuals over a fixed number of random trials. This method is computationally expensive since it runs through a fixed number of iterations regardless of the outlier ratio. Also it is not able to accommodate cases with more than 65% outliers. Moreover, the work presented in [8] solely relies on residuals which are insufficient and often not accurate enough to be used as *a priori* information for the image correspondence points.

3   Proposed Algorithm

As argued, there is readily available information from the iterations of the RANSAC process that can be used to estimate *a priori* information that will guide the sampling process. The proposed method is based on a more statistically meaningful measure than pure residuals and a more general one than matching scores. In addition, the method works well independently of the number of motions present in the scene since it does not rely on matching scores.

3.1 Regression Diagnostics

In the proposed algorithm, the value of the *a priori* probabilities are derived from the measure of each correspondence's *influence*. In other words, the amount by which a single correspondence influences the estimation of the fundamental matrix is used as a measure of its probability to be an inlier. Influential points tend to "pull" the regression coefficients in their direction and are often outliers. It has long been considered that pure residuals are not a good measure of a data point's influence [16]. Before the concept of regression diagnostics is introduced, the linear model of the fundamental matrix will be reviewed.

In classic parameter fitting, the model of a linear system is defined as: $\mathbf{y} = \mathbf{X}\beta + \mathbf{u}$ where $\mathbf{y}$ defines the response values and $\mathbf{X}$ is the matrix of known predictor values. β and $\mathbf{u}$ denote the unknown parameters and the random errors affecting the data respectively. The fundamental matrix is also a linear model that can be formulated as such an equation. In order to assess the influence of a single data point, one of the most widely used metrics is the measure of "leverage". The leverage of the jth data point in the linear model described above can be found from the jth diagonal element of the Hat matrix defined by: $\mathbf{H} = \mathbf{X}(\mathbf{X}'\mathbf{X})^{-1}\mathbf{X}'$ [16]. The value of the leverage gives the amount that a given data point can influence the regression parameters. In addition to the value of leverage, the residuals also do have an influence on the effects of a data point on the estimated parameters. An often used measure of overall influence that combines the value of leverage and residuals is the Cook distance [16]. Cook's distance is defined as:

$$D_j = \frac{r_j^2}{s^2} \frac{h_j}{p^*(1-h_j)^2} \tag{1}$$

where r_j is the residual and h_j the leverage of the jth point. p^* is defined as $RANK(\mathbf{X})$ and $s^2 = \frac{1}{n-p^*}\sum_{i=1}^{n} r_j^2$ and n is the number of data points. In the proposed algorithm, Cook's distance is iteratively updated and the *a priori* probabilities are estimated based on this distance. Since Cook's distance is more meaningful than mere residuals, the algorithm is able to discern between inliers and outliers more effectively, as shown by the results.

3.2 Sampling with Regression Information

In order to denote the *a priori* probabilities of the point correspondences, an indicator variable v_j is used meaning that point j is an inlier and $\bar{v}_j$ if it is an outlier. Therefor, $p(v_j)$ is the *a priori* information for point j, expressing the probability that this point is an inlier. In order to use Cook's distance as a measure of a probability, a density function has to be defined that maps Cook's distance to a probability value. In this case a probability density function is chosen to be a Gaussian function defined by: $p(D_j) = \frac{1}{\sqrt{2\pi\sigma^2}} e^{-\frac{(D_j)^2}{2\sigma^2}}$. The mean of the residuals are assumed to be zero, and the standard deviation is calculated using the median absolute deviation [16]. In other words, a smooth function of the Cook distance measures is used to denote *a priori* probabilities, therefor: $p(v_j) \leftarrow p(D_j)$.

3.3   Termination Criterion

Normally the number of iterations that the RANSAC algorithm has to run through is calculated statistically [11]. This number is based on the number of iterations that it takes to find a set of all inliers with a given confidence level. This can be expressed mathematically as:

$$I_{max} = \frac{\log(1 - \eta)}{\log(1 - p(v)^m)} \tag{2}$$

Where η is the user desired confidence level, $p(v)$ is the inlier ratio and m is the number of points in a minimal subset, i.e., 7 in the case of the fundamental matrix, and I is the number of iterations. As shown in [7], this formula underestimates the number of required iterations. We propose a new stopping criterion where the iterations are stopped when the *a priori* probabilities converge. The convergence is detected when the mean absolute difference between the *a priori* values between two consecutive updates is below a threshold, K. Similar to the traditional method of stopping the RANSAC iterations, the user has to specify this threshold. In other words, the iterations of the proposed algorithm are terminated when:

$$\sum_{j=0}^{n} |p^I(v_j) - p^{I-1}(v_j)| < K \tag{3}$$

where $p^I(v_j)$ is the *a priori* probability assigned to jth data point during the Ith update, and K is set to 0.05 in the experiments. This value was found experimentally; higher threshold values will sacrifice accuracy for speed and lower values will achieve a lower error at a higher computation cost.

3.4   Overall Algorithm

Algorithm 1 outlines the various stages of the proposed method. Here, I denotes the number of the current iteration, h_j and r_j denote the values of leverage and error residual for the jth point. $p(v_j)$ denotes the measure of the validity of point j calculated using Cook's distance. F_I denotes the fundamental matrix that is estimated at iteration I using the minimal sample that is found in this iteration, and F^* denotes the best model found so far. C_I denotes the score at iteration I and C^* denotes the best score found so far. Note that the score used in the proposed algorithm is the MSAC score [11] due to its robustness and simplicity.

Algorithm 1. Guided Sampling with Regression Diagnostics

1: Detect image matches
2: Initialize: $p(v_j) \leftarrow 1$, $I = 0$, $C^* = 0$
3: **while** $I < I_{max}$ **do**
4:     take minimal sample S_I using $p(v_j)$
5:     fit model F_I, find errors, r_j w.r.t. F_I, find score C_I
6:     **if** $C_I > C^*$ **then**
7:         update: $C^* = C_I$ and $F^* = F_I$
8:         form inlier set $S = \{j : r_j < T\}$
9:         update leverage of inliers from design matrix: X_S
10:         find Cook distance D_j and $p(v_j)$ for all the data
11:         **if** $\sum_{j=0}^{n} |p^I(v_j) - p^{I-1}(v_j)| < K$ **then**
12:             Terminate and Return F^*
13:         **end if**
14:     **end if**
15: **end while**
16: **return** F^*

Also the error residual r_j is based on the Sampson error of correspondence j with respect to the hypothesized fundamental matrix [1].

The algorithm proceeds by updating the probabilities when a best new sample is found using the new values of leverage and the new residuals calculated from this best model. From this point on, the sampling process is guided based on the new probabilities. Using a Monte Carlo sampling process, data points with higher values of $p(v_j)$ are more frequently chosen [7]. The probabilities, $p(v_j)$ are updated only when a new best model is found. When this occurs, the set of inliers with respect to the new model are found based on some threshold, T where T is determined to be $1.96 * \sigma$ and where σ is often taken to be 1 pixel [2].

Finally the algorithm ends when one of two termination criteria is met. The first termination criterion is the traditional one used in RANSAC (2) and the second is the proposed criterion (3). The algorithm stops when either one of these criteria is met. However, almost in all cases, the proposed criterion is met first since the presented guided sampling improves the quality of the samples and so almost always a lower number of iterations is needed than plain RANSAC.

4   Experimental Results

In order to test the proposed algorithm, a synthetic correspondence generation framework is devised. For every single test, a random camera pair set up is generated where the two cameras have an arbitrary rotation and translation (non-convergent setups are rejected). Once this setup is created the camera intrinsic parameters are set to some initial values (image size is 512x512, focal length is 700). Then a set of random points in space that are seen by the two cameras are projected into both images and the set of image correspondences are created. This data is then corrupted by two separate noise processes. The locations of the feature matches are affected by a Gaussian noise with a one pixel standard deviation to mimic the process of feature detection and its inherent localization errors. The outliers are also generated by adding an arbitrary value from a uniform distribution to the locations of matches in one image. For any given outlier ratio, 200 such tests are created and the results are averaged over all the tests. Each test configuration consists of 1200 points projected in each of the cameras. The comparisons were made between MSAC, MLESAC [11], RANSAC, RES-RANSAC [8] and the proposed method. Guided MLESAC was omitted due to absence of correlation values since the tests are synthetic and so no texture information exists.

The first set of tests results, as shown in Figure 1(a) presents the comparison of the mean squared error averaged over the number of tests for varying levels of noise contamination. In other words, for every contamination level, 200 tests are performed. The results of a given test for each algorithm is a final estimation of the fundamental matrix. Therefore, the mean squared error for every algorithm is defined as the mean squared error of its hypothesized fundamental matrix with respect to the ground truth matches, averaged over the number of trials. As it can be seen, the error of the fundamental matrices found using the proposed method is significantly lower than other algorithms except for higher outlier ratios compared with RES-RANSAC. This performance drop is negligible when considering the fact that the proposed method finds this solution in a significantly lesser amount of time. The second set of test results, as shown in Figure 1(b), presents the computation time comparison between the tested algorithms. Note that MLESAC has a much higher computation time because of the expectation maximization step [11]. It is clear that the proposed method has a much lower computation time than the competing algorithms. Even though the leverage values need to be calculated, this only happens when a best model is encountered which is only $\log(k)$ times where k is the number of iterations [12]. This additional cost incurred for calculating the leverage is well compensated by the lower number of iterations due to the higher quality samples.

The third and fourth sets of results demonstrate the effectiveness of combining the proposed method with two existing hypothesis verification methods. Randomized hypothesis verification methods accelerate the iterative RANSAC algorithm by avoiding the costly process of finding the error of every hypothesized fundamental matrix with respect to every match point. Since the proposed algorithm only addresses issues of sampling and termination criteria, it is easily combined with existing hypothesis verification methods. Using the same testing framework as in the previous set of results, the experiments assess the accuracy and the computation time of combining the proposed method with various randomized verification techniques on synthetic data. This combined algorithm presents an even lower computational time than the basic version of the proposed without the added accelerated hypothesis verification technique while having a higher accuracy. Figure 2(a) shows the error and time comparison between the $T_{d,d}$ [13] method with the proposed method combined with $T_{d,d}$. The

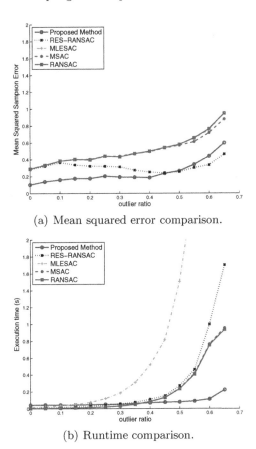

(a) Mean squared error comparison.

(b) Runtime comparison.

Fig. 1. Accuracy and speed comparison on synthetic data

combination of the proposed method with randomized verification shows a lower error for all outlier ratios due to the higher quality of the samples. Although the computation time of the plain $T_{d,d}$ is better for lower error ratios, the combined method maintains its steady computation time for all outlier ratios, whereas $T_{d,d}$ degrades quickly as the contamination levels increase.

Finally, the last set of experiments, as shown in Figure 2(b) show the accuracy and timing comparisons of Randomized RANSAC with Sequential Probability Ratio Test (SPRT) [12] versus the proposed method combined with SPRT. Similar to the $T_{d,d}$ case, this combination yields an even faster method than the pure SPRT case. Here again, the combination of SPRT with the proposed method has a lower estimation error for all outlier ratios. In the case of the computation time, the proposed method performs worse than plain SPRT in lower ratios, but the timing remains steady in the higher error ratios, whereas plain SPRT's speed degrades very quickly in these higher error ratios.

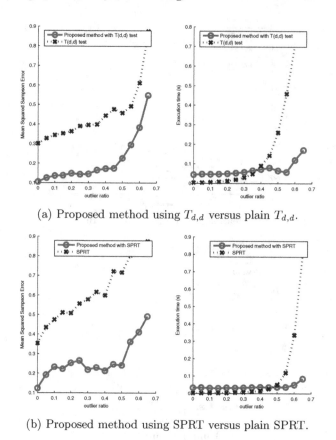

(a) Proposed method using $T_{d,d}$ versus plain $T_{d,d}$.

(b) Proposed method using SPRT versus plain SPRT.

Fig. 2. Combining proposed method with randomized hypothesis verification

5   Conclusion

A new sampling method and a modified termination criterion within the framework of random sampling and consensus is presented. The resulting algorithm is tested for the estimation of the fundamental matrix. This sampling strategy takes advantage of readily available regression diagnostics information. The proposed robust estimator, unlike the existing methods, relies on a meaningful measure of *a priori* probabilities as opposed to using correlation values or other matching scores or error residuals that might not necessarily reflect the relationship of the data point with respect to the manifold that is being estimated. This improved sampling leads to a much faster robust method due to the improved quality of the samples. In addition, a new termination criterion is proposed that does not rely on the existing statistical guess which often understates the number of required iterations. The algorithm is tested on synthetic data and results show the improved accuracy and speed of the proposed method. The error of the estimation is almost half that of RANSAC while performing 7 times faster than

RANSAC and 9 times faster than RES-RANSAC at 60% error ratio. In addition, the speed of the algorithm increases 2.6 times from 10% to 60% error ratio, whereas the speed of RANSAC increases over 50 times in the same interval, thus making the proposed algorithm highly suitable for real-time vision applications. In addition, the algorithm can easily be combined with randomized hypothesis verification methods, resulting in an even faster and more accurate estimation technique.

References

1. Hartley, R., Zisserman, A.: Multiple view geometry. Cambridge University Press (2000)
2. Torr, P., Murray, D.: The development and comparison of robust methods for estimating the fundamental matrix. IJCV 24, 271–300 (1997)
3. Snavely, N., Seitz, S., Szeliski, R.: Photo tourism: exploring photo collections in 3d. ACM Trans. on Graphics 25, 835–846 (2006)
4. Szeliski, R.: Image alignment and stitching: a tutorial. Found. Trends. Comput. Graph. Vis. 2, 1–104 (2006)
5. Cornelis, K., Pollefeys, M., Vergauwen, M., Van Gool, L.: Augmented Reality Using Uncalibrated Video Sequences. In: Pollefeys, M., Van Gool, L., Zisserman, A., Fitzgibbon, A.W. (eds.) SMILE 2000. LNCS, vol. 2018, pp. 144–160. Springer, Heidelberg (2001)
6. Fischler, M., Bolles, R.: Random sample consensus: a paradigm for model fitting with applications to image analysis and automated cartography. Communications of the ACM 24, 381–395 (1981)
7. Tordoff, B., Murray, D.: Guided-MLESAC: Faster image transform estimation by using mathcing priors. IEEE PAMI 27, 1523–1535 (2005)
8. Rastgar, H., Zhang, L., Wang, D., Dubois, E.: Validation of correspondences in MLESAC robust estimation. In: ICPR, pp. 1–4 (2008)
9. Myatt, D.R., Torr, P., Slawomir, N., Bishop, J., Craddock, R.: NAPSAC: High noise, high dimensional robust estimation - it's in the bag. In: BMVC (2002)
10. Sattler, T., Leibe, B., Kobbelt, L.: SCRAMSAC: Improving RANSAC's efficiency with a spatial consistency filter. In: ICCV (2009)
11. Torr, P., Zisserman, A.: MLESAC: A new robust estimator with application to estimating image geometry. CVIU 78, 138–156 (2000)
12. Matas, J., Chum, O.: Randomized RANSAC with sequential probability ratio test. In: ICCV, vol. 2, pp. 1727–1732 (2005)
13. Matas, J., Chum, O.: Randomized RANSAC with $T_{d,d}$ test. Image and Vision Computing 22, 837–842 (2004)
14. Chum, O., Matas, J.: Matching with PROSAC - progressive sample consensus. In: CVPR 2005, vol. 1, pp. 220–226 (2005)
15. Zhang, W., Kosecka, J.: Ensemble method for robust motion estimation. In: 25 Years of RANSAC, Workshop in Conjunction with CVPR (2006)
16. Maronna, R., Martin, R., Yohai, V.: Robust statistics. Wiley (2006)

Hand Shape and 3D Pose Estimation Using Depth Data from a Single Cluttered Frame

Paul Doliotis[1,2], Vassilis Athitsos[1],
Dimitrios Kosmopoulos[2], and Stavros Perantonis[2]

[1] Computer Science and Engineering Department
University of Texas at Arlington Arlington, Texas USA
[2] Institute of Informatics and Telecommunications
NCSR Demokritos Athens, Greece

Abstract. This paper describes a method that, given an input image of a person signing a gesture in a cluttered scene, locates the gesturing arm, automatically detects and segments the hand and finally creates a ranked list of possible shape class, 3D pose orientation and full hand configuration parameters. The clutter-tolerant hand segmentation algorithm is based on depth data from a single image captured with a commercially available depth sensor, namely the $Kinect^{TM}$. Shape and 3D pose estimation is formulated as an image database retrieval method where given a segmented hand the best matches are extracted from a large database of synthetically generated hand images. Contrary to previous approaches this clutter-tolerant method is all-together: user-independent, automatically detects and segments the hand from a single image (no multi-view or motion cues employed) and provides estimation not only for the 3D pose orientation but also for the full hand articulation parameters. The performance of this approach is quantitatively and qualitatively evaluated on a dataset of real images of American Sign Language (ASL) handshapes.

1   Introduction

Gesture recognition has become an essential component for many natural user interface (NUI) systems. It provides humans the ability to interact with machines naturally without the use of any cumbersome mechanical devices. A gesture can be defined by any bodily motion, however hand gestures are more commonly used and can be found in a wide range of applications such as: sign language recognition, robot learning by demonstration and gaming environments, just to name a few. Recognizing hand gestures is a very challenging task and requires solving several sub-problems like automatic hand detection and segmentation, 3D hand pose estimation, hand shape classification and in some cases estimation of the full hand configuration parameters.

G. Bebis et al. (Eds.): ISVC 2012, Part I, LNCS 7431, pp. 148–158, 2012.

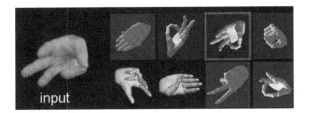

Fig. 1. System input and output. Given the input image, the system goes through the database of synthetic images in order to identify the ones that are the most similar to the input image. Eight examples of database images are shown here, and the most similar one is enclosed in a red square. The database currently used contains more than 100,000 images.

In this work we specifically address the problem of 3D hand pose and shape estimation. Towards developing an effective solution several challenges may arise and some of the main ones are listed bellow:

- High dimensionality of the problem
- Noisy hand segmentation due to cluttered backgrounds
- Increased pose variability and self-occlusions that frequently occur when a hand is in motion

Hand pose estimation is formulated here as an image database retrieval problem. The closest matches for an input hand image are retrieved from a large database of synthetic hand images. The ground truth labels of the retrieved matches are used as hand pose estimates from the input (Figure 1). This paper is motivated by the work presented in [1]. However, one limitation of that method was that it required manual segmentation in order to define a bounding box for the gesturing hand. We propose an automatic hand segmentation method that relies on depth data acquired from the Microsoft $Kinect^{TM}$ device [2]. Another contribution is that we achieve improved performance under clutter by using a similarity measure which is also based on the depth data. A main assumption we make is that the gesturing arm is the closest object to the camera and so it can easily be segmented from the rest of the body and other objects based on depth. To measure the effectiveness of this new method we have collected a dataset of American Sign Language(ASL) handshapes.

In the following Section 2 of this paper we briefly review previous work on 3D hand tracking and pose estimation. Afterwards, in Section 3 we describe in detail our automatic hand segmentation method, the result of which is given as input to our 3D hand pose estimation algorithm (Section 4). Finally we present experimental results (Section 5) and we conclude with a discussion on ideas for future-work (Section 6).

2   Related Work

Some successful early works require specialized hardware or the use of cumbersome mechanical devices. In [3] Schneider and Stevens use a motion capture system while in [4] Wang and Popović employ visual markers with a color glove. Unfortunately such methods impede the user's natural interaction in the scene and they require a costly and complex experimental setup.

Nowadays research is more focused on purely vision-based methods that are non-invasive and are more suitable for Natural User Interface(NUI) systems. The most recent review on vision-based hand pose estimation methods has been published by Erol et al. [5]. They define a taxonomy where initially these approaches are divided in two main categories: "partial pose estimation" and "full DOF pose estimation". "Partial pose estimation" methods can be viewed as extensions of appearance-based systems. They usually take as input image features and map them a small discrete set of hand model kinematic parameters. A main disadvantage is that they require a large amount of training data and hence are not scalable. Appearance-based methods for hand pose recognition, like [6,7,8,9], can tolerate clutter, but they are limited to estimating 2D hand pose from a limited number of viewpoints. Our method can handle arbitrary viewpoints.

"Full DOF pose estimation" approaches are not limited to a small, discrete set of hand model configurations. They target all the kinematic parameters (i.e., joint angles, hand position or orientation) of the skeleton of the hand, leading to a full reconstruction of hand motion. These approaches can be further divided into two other categories: (1) "Model-based tracking" and (2) "Single frame pose estimation".

"Model-based methods" [10,11,12,13] typically match visual observations to instances of a predefined hand model. Formally this is expressed as an optimization problem where an objective function is required in order to measure similarity between actual visual observations and model hypotheses. The main drawback is increased computational complexity due to the high dimensionality of the model's parameter space. On the other hand they require less training and are easily scalable.

"Single frame pose estimation methods" try to solve the hand pose estimation problem without relying on temporal information. The lack of temporal information increases the difficulty of the problem. However successful approaches can tackle the negative effect of motion blur and can also be employed to initialize tracking-based systems. Athitsos et al. [1] have proposed such a single pose estimation method by creating a large database of synthetic hand poses using an articulated model and retrieve the best match from this database. However they require manual segmentation of the test data.

Most recently, due to the advent of commercially available depth sensors, there is an increased interest in methods relying on depth data [14,15,10,16]. Keskin et al. [14] train Random decision forests (RDF) on depth images and then use them to perform per pixel classification and assign each pixel a hand part. Then, they apply the mean shift algorithm to estimate the centers of hand parts to form a hand skeleton. However they don't explicitly address the automatic hand segmentation problem.

According to the aforementioned taxonomy, this paper describes an "appearance-based method" aiming at 3D orientation estimation using features from a single frame. This work builds on top of the work described in [1], where hand pose is estimated from a single cluttered image. The key advantages of the method described here over [1] are that we integrate an automatic hand segmentation method and we use a similarity measure that is based on depth data.

3   Hand Segmentation

As a first step we need to perform a rough segmentation by thresholding the depth data in order to obtain the gesturing arm. Given the assumption that the hand is the closest object to the camera we can automatically find the lower depth threshold. As an upper threshold we take an initial rough estimation, since at this point we are only interested at segmenting the arm. A more precise thresholding is needed however if we need to further segment the hand. To find the palm cutoff point we need to perform the following steps:

1. Compute the axis of elongation for the gesturing arm.
2. Create a sequence of widths.
3. Perform a gradient descent on the sequence of widths in order to identify the local (or global) minimum, at which the palm cutoff point is located.

3.1   Finding the Axis of Elongation

The result of the initial rough segmentation is a blob representing the gesturing arm. The boundary of that blob is essentially a set S of m points in 2 dimensional space: $S = \{x_1, x_2, \ldots, x_m\} \in \mathbb{R}^2$. Noisy smaller groups of pixels are usually part of S. To remove them, we morphologically open the binary image by eliminating all connected components (objects) that have fewer than 20 pixels, considering an 8-connected neighborhood. The remaining boundary pixels will belong to a new set $S' = \{x'_1, x'_2, \ldots, x'_k\} \in \mathbb{R}^2$. In order to define the elongation axis of the gesturing arm we will compute the *Minimum Enclosing Ellipsoid(MEE)* for the boundary pixels $x'_i \in S'$. The major axis of the *MEE* coincides with the arm's axis of elongation (Figure 2).

An ellipsoid in center form can be given by the following equation:

$$\mathcal{E} = \left\{ x' \in \mathbb{R}^2 | (x' - c)^T E(x' - c) \le 1 \right\} \tag{1}$$

where $c \in \mathbb{R}^2$ is the center of the ellipse $\mathcal{E}$ and E is a 2×2 positive definite symmetric matrix, $E \in \mathbb{S}^2_{++}$.*

So finding the Minimum Enclosing Ellipsoid can be formulated as an optimization problem as follows:

$$\begin{aligned} \underset{E,c}{\text{minimize}} \quad & det(E^{-1}) \\ \text{subject to} \quad & (x'_i - c)^T E(x'_i - c) \le 1, \ i = 1, \ldots, k \\ & E \succ 0 \end{aligned} \tag{2}$$

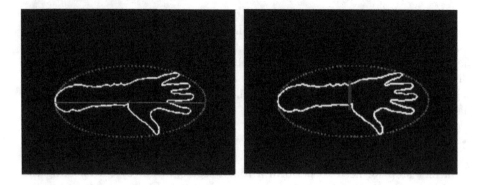

Fig. 2. To the left is an example of a *Minimum Enclosing Ellipsoid(MEE)* along with the major axis. To the right we visually demonstrate the desired palm cutoff location(blue line).

An implementation of a solver based on the Khachiyan Algorithm [17] can be found at the web [18]. The major axis for the arm boundary pixels will coincide with the major axis of the Minimum Enclosing Ellipsoid.

3.2   Creating the Sequence of Widths

After the major (or elongation) axis is obtained we can easily create a sequence of widths. In the discrete domain, the elongation axis is comprised of a set of pixels $\mathcal{P} = \{p_1, p_2, \ldots, p_m\}$. For each p_i we compute the maximum distance of arm pixels belonging to the line that goes through p_i and it's direction is perpendicular to the direction of the elongation axis. The main idea is that at the palm cutoff point the sequence will reach a global or local minimum. Since the contour of the segmented arm is very noisy our method could be prone to other local minima. To alleviate this effect we apply a smoothing on our 2D contour.

Smoothing 2D contours is achieved by using *Local Regression Lines*. Because of the linear nature of fitting it might be possible to loose important information in special cases like corners (or fingertips). To tackle this issue we opt to fit locally the line by employing Weighted Orthogonal Least Squares. The weights are generated from a Gaussian distribution. To be able to calculate the local regression lines we must define an order for the all pixels x_i' that $\in \mathcal{S}' = \{x_1', x_2', \ldots, x_m'\}$. Such an order can be defined with many techniques such as boundary tracing and chain codes. An implementation of this smoothing technique can be found at the web [19].

3.3   Gradient Descent

After the original contour is smoothed the sequence of widths is further filtered with a 1D horizontal mask of ones and of size 5. The next step is to perform

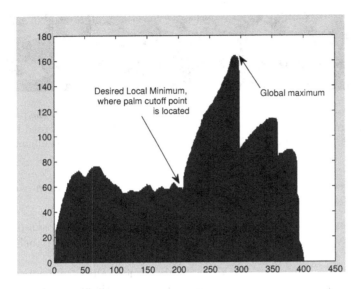

Fig. 3. This a plot of the sequence of widths. The desired local minimum is highlighted indicating the position for the palm cutoff point.

a gradient descent in order to identify the local minimum at which the palm cutoff point will lie. As a starting point we choose the global maximum (i.e. the highest width) which will always reside in the hand area. Then we move towards the end of the arm until we reach our local minimum. In Figure 3 we can see a plot of the sequence of widths along with the desired local minimum where the palm cutoff point is located.

4   Framework for Hand Pose Estimation

We model the hand as an articulated object, consisting of 16 links: the palm and 15 links corresponding to finger parts. Each finger has three links (Figure 4). There are 15 joints, that have a total of 20 degrees of freedom (DOFs). For the 20-dimensional vector of joint angles we use synonymously the terms "hand shape" and "hand configuration."

The appearance of a hand shape also depends on the camera parameters. For simplicity, we consider only the camera viewing direction (two DOFs), and image plane orientation. We use the terms "camera parameters," "viewing parameters" and "3D orientation" synonymously to denote the three-dimensional vector describing viewing direction and camera orientation. Given a hand configuration vector $C_h = (c_1, ..., c_{20})$ and a viewing parameter vector $V_h = (v_1, v_2, v_3)$, we define the hand pose vector P_h to be the 23-dimensional concatenation of C_h and V_h: $P_h = (c_1, ..., c_{20}, v_1, v_2, v_3)$.

Fig. 4. Synthetic images of hands. Left: the articulated hand model. The palm and 15 finger links are shown in different colors. Middle: the 20 basic shapes used to generate model images in our database. Right: four 3D orientations of the same hand shape.

Using these definitions, our framework for hand pose estimation can be summarized as follows:

1. Preprocessing step: create a database containing a uniform sampling of all possible views of the hand shapes that we want to recognize. Label each view with the hand pose parameters that generated it.
2. Given an input image, retrieve the database views that are the most similar. Use the parameters of the most similar views as estimates for the image. The most similar views (Figure 1) are retrieved according to a similarity measure (e.g. Euclidean distance, chamfer distance)

4.1 Database

Our database contains right-hand images of 20 hand shape prototypes (Figure 4). Each prototype is rendered from 86 different viewpoints (Figure 4), sampled approximately uniformly from the surface of the viewing sphere. The rendering is done using a hand model and a computer graphics software ([20]). To accommodate rotation-variant similarity measures (like the chamfer distance), 48 images are generated from each viewpoint, corresponding to 48 uniformly sampled rotations of the image plane. Overall, the database includes 4128 views of each hand shape prototype and 82,560 images overall. We refer to those images using the terms "database images," "model images," or "synthetic images."

4.2 Similarity Measures

Chamfer Distance. The chamfer distance [21] is a well-known method to measure the distance between two edge images. Edge images are represented as sets of points, corresponding to edge pixel locations. The X-to-Y directed chamfer distance $c(X, Y)$ is defined as

$$c(X, Y) = \frac{1}{|X|} \sum_{x \in X} \min_{y \in Y} \|x - y\| , \qquad (3)$$

where $\|a - b\|$ denotes the Euclidean distance between two pixel locations a and b. The undirected chamfer distance $C(X, Y)$ is

$$C(X, Y) = c(X, Y) + c(Y, X) . \qquad (4)$$

We will use the abbreviations *DCD* to stand for "directed chamfer distance" and *UCD* for "undirected chamfer distance."

Depth Matching. Test data have been captured via the $Kinect^{TM}$ device which offers two synchronized streams, one RGB and one depth stream. Frame resolution is 640 × 480. A depth image is a gray-scale image, where each pixel is assigned an intensity value according to how far or close it is located from the camera. We have also managed to create depth-maps for our synthetically generated database images using a 3D modeling and animation software [20]. Some examples of our depth-maps are depicted in Figure 5. Both, model and test depth images are normalized in order to achieve translation invariance for the z-axis. All depth values are in the range from 0 to 1. The depth similarity measure between two images is defined as the total sum of their pixel-wise Euclidean distances. Through the rest of this paper we will refer to the depth matching similarity measure as *depthSM*.

Weighted Depth Matching and Chamfer Distance. Another similarity measure is defined by combining the first two similarity measures. In the following equation the *WeightedSM* is a weighted sum of *depthSM* and *UCD*:

$$WeightedSM = l_1 \times depthSM + l_2 \times UCD \qquad (5)$$

In our experiments $l_1 = l_2 = 0.5$.

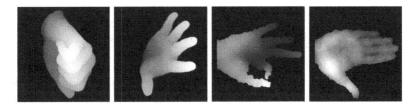

Fig. 5. The two depth-maps at the left side are "database depth-maps" and have been rendered with a 3D modeling software. The two depth-maps at the right side are "test depth-maps" and they have been captured by the $Kinect^{TM}$ device.

5   Experiments

We have tested our system on a challenging dataset of 94 right hand images of American Sign Language(ASL) handshapes. The images are captured with the $Kinect^{TM}$ device in a highly cluttered scene where the hand is not cleanly pre-segmented. We will refer to this dataset as *KinectHandshapes*. These 94 test images are provided in both formats, RGB and depth. We have manually established pseudo-ground truth for each test image, by labeling it with the corresponding shape prototype and using the rendering software to find the viewing parameters under which the shape prototype looked the most similar to the test image. This way of estimating viewpoint parameters is not very exact; we found that manual estimates by different people varied by 10-30 degrees. Model views cannot be aligned perfectly because the anthropometric parameters (like finger lengths and widths) of hands in test images do not match those of the model, and because the hand shapes in the real images are not exact replications of the 26 shape prototypes.

We consider a database view V to be a *correct match* for a test image I if the shape prototype with which we label I is the one used in generating V, and the manually estimated viewing parameters of I are within 30 degrees of those of V ([22]). On average, there are 30.4 correct matches for each test image in the database. Our measure of retrieval accuracy for a given test image I is the rank of the *highest-ranking correct match* that was retrieved for I. 1 is the highest (best) possible rank.

Table 1. Results for KinectHandshapes dataset. For every method used, we show the percentage of test images for which the highest ranking correct match was within each range. DCD stands for "image-to-model directed chamfer distance". UCD depth contours, is undirected chamfer distance between hand contours from depth images and full edges from model images. In UCD color edges, skin color segmentation has been employed to extract the full edges.

Method used	1	1-4	1-16	1-64	1-256
UCD depth contours	14.74	20.0	32.63	40.0	62.11
UCD color edges	9.57	14.89	21.28	32.98	55.32
$depthSM$	34.04	44.68	61.70	76.60	87.23
$weightedSM$	34.04	44.68	61.70	76.60	87.23

From the experiments on the *KinectHandshapes* dataset it is evident that using depth information from the depth-maps enhances the discrimination power of our method, even in highly uncontrolled and cluttered environments. It outperforms even the case "UCD color edges" where full edges of the hand are available due to color information.

We have also evaluated quantitatively our automatic hand segmentation method. To this end we compared for all of our test images the distance between the automatically generated *palm cut-off points* and the manually specified ones. For the 80.85% of our images the distance (measured in pixels) is negligible. The distance is less than 15 pixels for the 85.11% and less than 25 for the 92.55% of the images (Table 2).

6   Discussion and Future Work

We have presented a new method for hand pose estimation from a single depth image. Our method combines a novel hand segmentation method and a

Table 2. Results for our Hand segmentation method

range (measured in pixels)	0	0-15	0-25	0-45	0-68
percentage of test images	80.85	85.11	92.55	96.81	100

similarity measure ($depthSM$) based on depth information from depth images. The similarity measure is used to retrieve the best matches from an image database, thus providing estimates for the 3D hand pose and hand configuration. Depth information increases the discrimination power of our method, according to the experiments conducted. Retrieval accuracy is still too low for the system to be used as a stand-alone module for 3D hand pose estimation. However estimating hand pose from a single image can be useful in automatically initializing hand trackers. Our system currently doesn't achieve real-time performance. In order to do so and since our method is inherently parallel, we are planning to take advantage of the GPU's processing power. Additional future work will be to define more sophisticated similarity measures further exploiting depth information.

Acknowledgments. This work was partially supported by NSF grants IIS-0812601, IIS-1055062, CNS-1059235, ECCS-1128296, CNS-0923494, and CNS-1035913

References

1. Athitsos, V., Sclaroff, S.: Estimating 3d hand pose from a cluttered image. In: Proceedings of 2003 IEEE Computer Society Conference on Computer Vision and Pattern Recognition, vol. 2, pp. II432–II439 (2003)
2. Microsoft Corp. Redmond WA.: Kinect Xbox 360, http://www.xbox.com/kinect
3. Schneider, M., Stevens, C.: Development and testing of a new magnetic-tracking device for image guidance. In: Proceedings of SPIE, vol. 7035, pp. 65090I–65090I-11 (2007)
4. Wang, R.Y., Popović, J.: Real-time hand-tracking with a color glove. ACM Trans. Graph. 28, 63:1–63:8 (2009)
5. Erol, A., Bebis, G., Nicolescu, M., Boyle, R.D., Twombly, X.: Vision-based hand pose estimation: A review. Computer Vision and Image Understanding 108, 52–73 (2007)
6. Moghaddam, B., Pentland, A.: Probabilistic visual learning for object detection. Technical Report 326, MIT (1995)
7. Triesch, J., von der Malsburg, C.: Robotic Gesture Recognition. In: Wachsmuth, I., Fröhlich, M. (eds.) GW 1997. LNCS (LNAI), vol. 1371, pp. 233–244. Springer, Heidelberg (1998)
8. Freeman, W.T., Roth, M.: Computer vision for computer games. In: Automatic Face and Gesture Recognition, pp. 100–105 (1996)
9. Wu, Y., Huang, T.: View-independent recognition of hand postures, vol. 2, pp. 88–94 (2000)

10. Oikonomidis, I., Kyriazis, N., Argyros, A.A.: Full dof tracking of a hand interacting with an object by modeling occlusions and physical constraints. In: 2011 IEEE International Conference on Computer Vision, ICCV, pp. 2088–2095 (2011)
11. de La Gorce, M., Fleet, D.J., Paragios, N.: Model-based 3d hand pose estimation from monocular video. IEEE Trans. Pattern Anal. Mach. Intell. 33, 1793–1805 (2011)
12. Oikonomidis, I., Kyriazis, N., Argyros, A.A.: Markerless and Efficient 26-DOF Hand Pose Recovery. In: Kimmel, R., Klette, R., Sugimoto, A. (eds.) ACCV 2010, Part III. LNCS, vol. 6494, pp. 744–757. Springer, Heidelberg (2011)
13. Rehg, J.M., Kanade, T.: Model-based tracking of self-occluding articulated objects. In: IEEE International Conference on Computer Vision, p. 612 (1995)
14. Keskin, C., Kiraç, F., Kara, Y.E., Akarun, L.: Real time hand pose estimation using depth sensors. In: ICCV Workshops, pp. 1228–1234 (2011)
15. Pugeault, N., Bowden, R.: Spelling it out: Real-time asl fingerspelling recognition. In: ICCV Workshops, pp. 1114–1119 (2011)
16. Mo, Z., Neumann, U.: Real-time hand pose recognition using low-resolution depth images. In: 2006 IEEE Computer Society Conference on Computer Vision and Pattern Recognition, CVPR 2006, vol. 2, pp. 1499–1505 (2006)
17. Khachiyan, L.G.: Rounding of polytopes in the real number model of computation. Math. Oper. Res. 21, 307–320 (1996)
18. Minimum Volume Enclosing Ellipsoid: Matlab Central,
    http://www.mathworks.com/matlabcentral/fileexchange/9542-minimum-volume-enclosing-ellipsoid
19. Smoothing 2D Contours Using Local Regression Lines: Matlab Central,
    http://www.mathworks.com/matlabcentral/fileexchange/30793-smoothing-2d-contours-using-local-regression-lines
20. Smith Micro, Aliso Viejo, CA: Poser 8,
    http://poser.smithmicro.com/poser.html
21. Barrow, H.G., Tenenbaum, J.M., Bolles, R.C., Wolf, H.C.: Parametric correspondence and chamfer matching: two new techniques for image matching. In: Proceedings of the 5th International Joint Conference on Artificial Intelligence, IJCAI 1977, vol. 2, pp. 659–663. Morgan Kaufmann Publishers Inc., San Francisco (1977)
22. Athitsos, V., Sclaroff, S.: An appearance-based framework for 3D hand shape classification and camera viewpoint estimation. In: Automatic Face and Gesture Recognition (2002)

Fusing Low-Resolution Depth Maps into High-Resolution Stereo Matching

Billy Ray Fortenbury and Gutemberg Guerra-Filho

The University of Texas at Arlington

Abstract. This paper introduces a fusion algorithm to integrate low-resolution depth maps and stereo matching algorithms. Our fusion algorithm is a divide-and-conquer approach that produces a number of disparity maps over many small sections, named stripes, of the stereo images. The algorithm then uses statistical methods to evaluate each of these sections in order to produce an accurate high-resolution disparity map for a given scene.

1 Introduction

Two major approaches to 3D reconstruction are stereo matching algorithms and 3D sensing devices. Individually, these 3D reconstruction methods are capable of producing depth maps of their environment with fair accuracy. However, stereo matching results can be time consuming and 3D sensing maps are often low resolution and noisy. Therefore, combining the time performance of 3D sensing with the high-resolution of a stereo matching algorithm can produce results that are potentially far more accurate than either in isolation.

This fusion of low-resolution depth maps and stereo vision algorithms using high resolution CCD cameras is achieved by our fusion algorithm using a divide-and-conquer approach. The algorithm first divides the stereo images into a constant size left stripe consisting of windows and a variable size right stripe (see

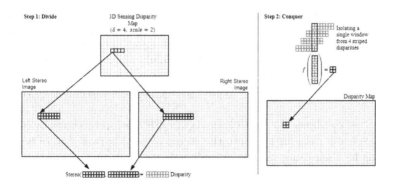

Fig. 1. A conceptual overview of the stereo-flight fusion algorithm

G. Bebis et al. (Eds.): ISVC 2012, Part I, LNCS 7431, pp. 159–168, 2012.

step 1 in Figure 1). The right stripe is computed using the low-resolution disparity map to find the windows that correspond to the left stripe. The algorithm then performs the conquer step which involves using one of several statistical methods in order to combine the windows created for each pixel into a single high-resolution disparity map (see step 2 in Figure 1).

We performed several experiments to evaluate our method on the Middlebury stereo datasets [8]. Our method showed accuracy increases ranging from 15% to 40%. In addition, our fusion algorithm is able to improve accuracy without adding to the time and space complexity of the base stereo matching algorithm.

The primary contributions of this paper are a novel method for the integration of low-resolution depth maps and stereo vision algorithms. This approach may be applied to the fusion of any 3D sensing technology such as structured lighting and time-of-flight cameras with any stereo matching algorithm.

This paper will begin by discussing the related works in Section 2. We will then discuss our fusion algorithm in Section 3. We perform a complexity analysis of our fusion algorithm in Section 4. We present our experimental results in Section 5. Finally, the conclusions drawn from this work will be discussed in Section 6.

2   Related Work

Multi-sensor fusion has many advantages over using a technology in isolation. In general, two major approaches have been used: a priori methods, which fuse the data obtained from the sensor into the stereo algorithm itself [1,5,6,9,11]; and a posteriori methods, which focus on fusing the output of the stereo algorithm and the 3D sensing device in order to obtain more precise results [4,7,10].

A posteriori methods tend to focus on using evidence grids [10] or on separating objects from the scene and performing stereo at the object level [4,7] to obtain high accuracy stereo maps. While these methods work well at producing stereo maps, they tend to lose information that can be gleaned from the stereo maps during the stereo matching process.

In contrast, a priori methods tend to focus on Bayesian modeling [5,9,11] with some methods focusing on volumetric approaches [6] and interpolation methodologies [1]. Bayesian modelling and volumetric approaches tend to produce fairly accurate results. However, they are also very computationally expensive and as such are not suited to real-time depth systems such as motion capture. Interpolation methodologies produce fairly accurate results at a reasonable time trade-off. However, they tend to be tightly coupled to one particular technology (*e.g.*, LI-DAR).

Our method combines the best elements of both a priori methods by integrating the 3D sensing data directly into the stereo algorithm itself and of a posteriori methods by then using various probabilistic methods to choose amongst the output stripes to produce a final output that is accurate in real-time.

3   Stereo Fusion

Given a left reference image I_L, a right image I_R, and a low resolution depth map D_l, our fusion algorithm finds a high-resolution depth map D_h for the scene projected in the images I_L and I_R. Calibration must be performed on both the camera sensors and the 3D sensor. We use a checkerboard pattern to perform this calibration along with the Caltech Camera Calibration Toolbox for Matlab [2]. The calibration procedure finds the necessary correlation between the two stereo images and the 3D sensor.

An important parameter of our fusion algorithm is the number of windows (δ) in the left stripes. Each left stripe corresponds to a rectangular area of $1 \times \delta$ pixels in the low-resolution depth map. Therefore, a left stripe consist of $1 \times \delta$ windows in the left image. Each window is a rectangular region of size $\lfloor s_h \rfloor \times \lfloor s_w \rfloor$, where s_h is the ratio of the height of a stereo image and the height of the low-resolution depth map, and s_w is the ratio of the width of a stereo image and the width of the low-resolution depth map. Therefore, the size of a left stripe is $\lfloor s_h \rfloor \times \delta \lfloor s_w \rfloor$.

The fusion algorithm consists of a divide step and a conquer step. In the divide step, the left and right stereo images are divided into corresponding pairs of stripe regions using the low-resolution depth map. The base stereo matching algorithm is performed on each of these pairs of left and right stripe regions. The conquer step of our fusion algorithm integrates the depth maps obtained for all pairs of corresponding stripe regions into a single depth map.

We propose several methods for the integration of the depth maps obtained for all pairs of corresponding stripe regions. Some of these methods are stochastic and use a normalized probability distribution B. The probability distribution is computed during a pre-processing step from a training set of stereo pair images, along with their corresponding high-resolution depth map ground truth, D_h, and the low-resolution depth map, D_l, acquired from a 3D sensor device (*e.g.*, from structured lighting or time-of-flight camera).

3.1   Probability Distribution Calculation

The probability distribution is used in stochastic methods to integrate the depth maps obtained for all pairs of corresponding stripe regions. The probability distribution is represented by a matrix that has the same size of a left image stripe region, which is the region in the left stereo image that corresponds to a $1 \times \delta$ pixel region in the low-resolution depth map. Each element of this matrix contains the probability that the corresponding element of a depth map obtained for a pair of corresponding stripe regions is an error.

The probability distribution itself represents the normalized error over a sample of stripes from a number of representative images. Thus, the error for each stripe is calculated, and is then added into the total probability distribution which is normalized such that each element sums to 1 across the stripe.

Formally, the probability distribution is used as weights in the conquer step to integrate the depth maps for δ corresponding pairs of windows obtained for each pixel of the low-resolution disparity map.

To calculate the probability distribution, we first create a matrix B equivalent in size to the left stripe region. Then, for each row r and each column c in the low-resolution depth map of size $h \times w$, we get the stripe region T_L from the left image, the stripe region T_R from the right image, and the ground truth T_G for the depth map obtained from T_L and T_R. We use the base stereo matching algorithm to find a disparity map S for T_L as the left image and T_R as the right image. This should give a disparity map equivalent in size to T_L.

Note that the disparity in the low-resolution disparity map at the current location (r, c) should be added to the disparity map for the current pair of stripe regions. This is done due to the fact that the stripes are chosen such that they are already close to being aligned with one another. This means that there is already a shift in the disparity values calculated that must be accounted for by adding in the disparity at the beginning of each stripe.

The disparity error is then computed as the difference between the disparity map S and the ground truth T_G. The error is then added into the matrix B. The error accumulation is performed for each pair of corresponding stripe regions associated with elements (r, c) of the low-resolution depth map. The probability distribution should then be split into δ windows. Finally, for each corresponding element across these windows, the result should be normalized such that the sum for a given element across all those windows is $\lfloor s_w \rfloor$.

3.2   Divide Step: Correspondence in Stripes

The left stereo image I_L was chosen to be the reference image for our fusion algorithm. The size of a strip region in the left stereo image is related to the scales s_h and s_w such that the stripe region corresponds to a $1 \times \delta$ window in the low-resolution depth map D_l. Thus, since each pixel in the low-resolution depth map corresponds to a window of size $\lfloor s_h \rfloor \times \lfloor s_w \rfloor$ in the left stereo image, there are $1 \times \delta$ windows in a left stripe region and, consequently, the size of a stripe region in the left stereo image is $\lfloor s_h \rfloor \times \delta \lfloor s_w \rfloor$.

The top of the left stripe region T_L is the current row times the height scale ($\lceil s_h r \rceil$), the bottom of T_L is the top of the next row of stripe regions minus 1 ($\lceil s_h (r + 1) \rceil - 1$), the left bound of T_L is the current column times the width scale ($\lceil s_w c \rceil$), and the right bound of T_L is the left bound of the next column after δ columns of stripe regions minus 1 ($\lceil \lceil s_w (c + \delta) \rceil - 1 \rceil$).

The top and bottom of the right stripe region is the same as the top and bottom of the left stripe region, respectively. The left and right bounds of the right stripe are calculated similarly. The left bound is the current column c minus the disparity at the current location (r, c) with this result multiplied by the width scale ($\lceil s_w (c - D_l(r, c)) \rceil$).

The right bound of the right stripe region is the next column minus one as above ($\lceil s_w (c + \delta - D_l(r, c + \delta - 1) - 1) \rceil$). However, due to the stereo camera configuration, all objects in the right image must be at the same position or to the left of the object in the left image. Thus, the right bound of the right stripe region (r_R) must be greater than the left bound of the right stripe (l_R) and if this is not

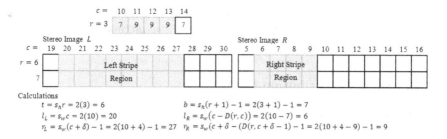

Fig. 2. A figure illustrating the manner in which the left and right stripe regions are chosen

the case, the result must compensate by moving the bound until it is occupying a δ pixel window in the right image ($\lceil r_R + s_w(D_l(r, c + \delta - 1) - D_l(r, c)) \rceil$).

Finally, to calculate the probability distribution, the ground truth depth map for the corresponding pair of stripe regions must be retrieved in the same location as the left stripe region using the ground truth D_h.

Given a left image I_L, a right image I_R, and a low-resolution disparity map D_l, the fusion algorithm constructs a matrix F the same size as the left image which represents the integrated disparity map for the left and right images. Each element of this matrix should be a matrix itself of the same size as the left stripe region T_L. An illustration of the divide step process can be found in Figure 2.

Once the stripe regions T_L and T_R are retrieved from the left and right images respectively, the stereo matching algorithm finds the disparity map S for T_L as the left image and T_R as the right image. The disparity at the current cell (r, c) must be added to S and the result is stored in the matrix F.

3.3   Conquer Step: Disparity Map Integration

Since the divide step produces δ disparity windows per pixel in the low-resolution disparity map, a single disparity window must be determined for each pixel from the corresponding set W of stripes. Once the set of disparities is built for a particular pixel (r, c), we use an integration function to determine the single disparity for pixel (r, c) in the final disparity map I. For this purpose, 7 different integration functions are proposed: maximum, minimum, median, mean, naïve, vote, and weight.

The maximum method chooses the maximum value of all the disparities produced for a given pixel (r, c). Likewise, the minimum chooses the minimum value of all the disparities produced for a given pixel (r, c). The mean method produces an average of all the disparities for a given pixel (r, c) ignoring values outside the range of values a pixel should legitimately be able to take ($0 < pixel_{value} \leq 255$ in our case). The median method takes the median disparity for a given pixel (r, c). The naïve method uses the probability distribution to select the best match. The vote method uses the probability distribution to vote on a best value and then selects the pixel with the most votes. Finally, the weighted method uses the probability distribution as a weight, multiplying each pixel by its corresponding probability distribution value. The three probabilistic methods are discussed in more detail in Section 3.4.

3.4 Probability Distribution Methods

The probability distribution represents the confidence that the algorithm has the correct disparity for a given pixel in a stripe region. The probability distribution can be used in a number of ways. The first is to use it as a best match and simply choose the disparity associated with the best probability value. The second is to use probabilities as votes for disparities and choose the disparity that has the most votes. Finally, it can be used as weights that multiply each disparity value by its associated probability.

All three of the probability distribution methods assume the existence of two arrays of size $h_S \times w_s \times \delta$ called S_c and B_c. The matrix S_c is compiled from the results of the divide step by iterating through each stripe and adding its results into $S_c(i, j, k)$ such that i corresponds to the row of the high-resolution disparity map, j corresponds to the column of the high-resolution disparity map, and k represents the kth element computed from the divide step $(0 < k \leq \delta)$ for that given element. Likewise, B_c is the equivalent map produced from the given probability distribution where for a given row r and a given column c in the low-resolution disparity map, $B_c(i, j, k) = B(i - r, j - c)$.

To use the distribution method (naïve), we iterate through each disparity and check the associated probability value in the corresponding point of the distribution, choosing the value stored in the position of greatest confidence.

To use the probability distribution as a voting system (vote), create a matrix V of a size capable of representing all the possible values that a given pixel can take $(1 \times 255$ in our case). Iterate through each value c in a given window and add the corresponding probability value in the distribution to the location $V(\lfloor c \rfloor)$. The final disparity for a given pixel is the index of the maximum number of votes in V.

Since each window will have an associated window of equal size in the probability distribution, using the probability distribution as a series of weights requires only that each window corresponding to a given pixel be multiplied by its equivalent window in the probability distribution. The final result is then the sum of all windows at a given location.

4   Complexity Analysis

Since our fusion algorithm can use any stereo matching algorithm, the complexity of our fusion algorithm will be stated in terms of the complexity of the base stereo matching algorithm.

Let $n = h \times w$, where h and w are the height and width respectively of the stereo images, and let $m = h_l \times w_l$, where h_l and w_l are the height and width respectively of the low-resolution disparity map. The fusion algorithm

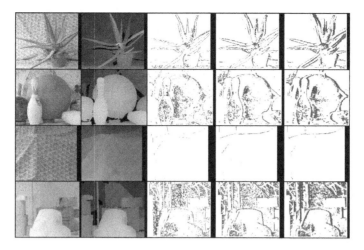

Fig. 3. An image illustrating the original image, the ground truth, and the results of the stereo fusion algorithm for $\delta = 15$, $\delta = 12$, and $\delta = 9$

executes the base stereo matching algorithm for each pixel of the disparity map for a number of iterations. Therefore, the complexity of the fusion algorithm is dependent upon the complexity of the base stereo matching algorithm used. The base stereo matching method used will be performed on stripes of size q equivalent to δ windows such that $q = \lfloor s_h \rfloor \times \delta \lfloor s_w \rfloor$. Thus, we denote the time complexity of the base algorithm as $O(f(q))$, where f may be linear, quadratic, or any other function. Hence, the resultant complexity of the fusion algorithm is $O(mf(q))$. Each iteration of the base algorithm produces a disparity map for δ windows. Thus, the fusion algorithm will produce a total of δm windows. Hence, the resultant space complexity is $O(\delta m) = O(n)$.

Essentially, any base stereo matching algorithm that is more complex than linear will result in fewer operations than running the base stereo matching algorithm by itself in the entire stereo images due to the fact that the actual sub-images being used are much smaller than the image in general.

The time complexity of our actual implementation is $O(n)$. The reason for this is that the base algorithm used (Optimal Dense Stereo Matching - ODSM [3]) is a linear time algorithm. Since each call to the ODSM algorithm results in traversing a region of size q, and because this algorithm is called δm times by the fusion algorithm with δ being constant for any given run of the algorithm, the resultant time complexity is $O(n)$.

5   Experimental Results

The experimental results for our fusion algorithm are broken down into several sections: the scaling variation experiment that shows how varying the scale affects the error; the delta variation experiment which shows how the error behaves

Table 1. The error observed using our fusion algorithm (probability distribution as votes method using left traversal only), the base stereo algorithm by itself (left traversal only), and the top two methods obtained from the Middlebury StereoMatcher program

Image	Stereo Fusion	ODSM	Dynamic Prog	SSD
Aloe	7.3%	47.2%	29.1%	27.74%
Baby1	8.4%	49.5%	20.0%	22.33%
Baby2	13.4%	48.5%	6.3%	5.88%
Baby3	12.9%	44.6%	58.3%	63.40%
Bowling1	21.3%	74.2%	26.6%	27.53%
Bowling2	13.3%	51.5%	15.8%	14.43%
Cloth1	0.4%	43.4%	39.9%	47.35%
Cloth2	8.8%	56.1%	49.8%	51.23%
Cloth3	3.1%	37.4%	17.0%	17.68%
Cloth4	4.7%	46.8%	52.5%	60.46%
Flowerpots	23.2%	55.6%	13.9%	15.71%
Lampshade1	20.3%	56.1%	53.0%	48.44%
Lampshade2	23.4%	70.8%	15.7%	15.21%
Midd1	22.5%	70.9%	20.3%	23.81%
Midd2	27.1%	70.8%	15.3%	16.80%
Monopoly	22.9%	67.8%	62.1%	60.42%
Plastic	30.0%	86.5%	47.9%	55.65%
Rocks1	5.3%	39.7%	17.3%	17.95%
Rocks2	4.7%	40.6%	68.3%	71.91%
Wood1	7.9%	36.8%	38.1%	34.35%
Wood2	11.4%	59.8%	8.7%	8.82%
Mean Error	13.9%	55.0%	32.2%	33.67%

as δ is varied; the integration experiment which shows how the various integration methods perform; and the real data experiment which shows how the process works with real world data. We used the standard two-view Middlebury dataset [8] with the Optimal-Dense Stereo Matching algorithm [3] for the δ and integration experiment. In our implementation, we use the following constant settings: a window size of 7 and a scaling ratio of 1.025. For the real data experiment, we used a Kinect® device to produce a low resolution sparse disparity map. We then performed our fusion algorithm to produce a high-resolution depth map. Since the disparity map was sparse, the δ value varied and was found by finding two contiguous non-zero disparities in the sparse disparity map. The δ value was then set to the distance between those two pixels for that single stripe.

The scaling experiment shows how the scale between the left image and the disparity map affects the error. Specifically, uniform scales in the range from 1:1.1 to 1:10 were used by scaling the full sized image using the nearest neighbor method of scaling in order to observe how the error behaved as the scale between the full sized image and the simulated depth map went from the largest in size to the smallest in size. The primary results from the scaling variation experiment show that error increases linearly as the scale increases.

The probability distribution experiment showed that the probability distribution behaved exponentially such that the error was always highest towards the

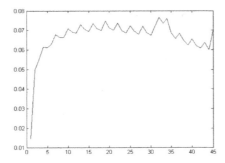

Fig. 4. An example probability distribution. Note that the correctness is lowest towards the beginning of a stripe and increases towards the end.

beginning of each row and dropped towards the end of each row. The probability distribution also showed a notable increase in accuracy around the columns 30 to 35 which then decreased back to the original curve like behavior suggesting that a δ value of 12 will produce the best results (see Figure 1).

The δ value variation experiment shows the behavior of the algorithm as δ increases. The experiment used δ values between 2 and 100 to evaluate the disparity error as the number of windows used increased. The δ variation experiment showed that the results and processing time increased as δ was increased. However, there was one significant exception in that the best results were obtained with a δ equal to 11 for every image in the data set (see Figure 5).

The integration experiment shows how the various integration methods affect the disparity error. We integrate disparity maps using one of the following statistical methods: Minimum of all windows; maximum of all windows; the median value of all windows; the mean value of all windows; the best match according to the probability distribution; the best match determined by votes using the probability distribution; and the weighted addition of all results using the probability distribution as a series of weights. In general, the integration experiment showed that the selective mean algorithm outperformed even the probabilistic measures. In addition, the fusion algorithm showed a significant improvement over using the base stereo matching algorithm by itself with improvements ranging from 29% up to 50% as shown in Table 1 and Figure 3.

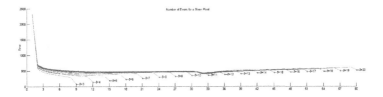

Fig. 5. Results from the delta variation experiment. As δ increases, error decreased along the length of a single stripe.

For the real data experiment, since a ground truth was not available, we measured the planarity of a flat white board in the disparity space in the original Kinect® image and in the disparity map output from our algorithm. This was performed by finding the board of best fit for the data points given, and then calculating the average distance error, in mm, from that plane. The average computered planarity across all the 25 test boards in the original depth maps was 5.92mm. After performing our fusion algorithm on the data, the average planarity was 4.75mm, an improvement of 19.4%.

6   Conclusions

Fusing 3D sensing depth maps with stereo vision algorithms can produce highly detailed and high resolution disparity maps. Our fusion algorithm is one such algorithm that can run generally with any stereo matching algorithm. In addition, it does not add to the complexity of the stereo algorithm and, in the case where the complexity of the stereo algorithm is greater than linear, improves the time complexity of the stereo algorithm without impacting the space complexity.

Our fusion algorithm shows promising results in that there is a significant improvement in accuracy over using the base stereo matching algorithm and a significant increase in resolution over using the low-resolution disparity map.

References

1. Badino, H., Huber, D., Kanade, T.: Integrating LIDAR into Stereo for Fast and Improved Disparity Computation. In: 3DIMPVT (2011)
2. Bouguet, J.Y.: Camera Calibration Toolbox for Matlab
3. Guerra-Filho, G.: An Optimal Time-Space Algorithm for Dense Stereo Matching. The Journal of Real-Time Image Processing (2010)
4. Gurram, P., Rhody, H., Kerekes, J., Lach, S., Saber, E.: 3d Scene Reconstruction Through a Fusion of Passive Video and LIDAR Imagery. In: 36th Applied Imagery Pattern Recognition Workshop (2007)
5. Hahne, U., Alexa, M.: Combining Time-of-Flight Depth and Stereo Images Without Accurate Extrinsic Calibration. International Journal of Intelligent Systems Technologies and Applications (2008)
6. Kim, Y.M., Theobalt, C., Diebel, J., Kosecka, J., Miscusik, B., Thrun, S.: Multi-View Image and TOF Sensor Fusion for Dense 3D Reconstruction. In: 3DIM (2009)
7. Kuhnert, K., Stommel, M.: Fusion of Stereo-Camera and PMD-Camera Data for Real-Time suited precise 3d environment reconstruction. In: IROS (2006)
8. Scharstein, D., Szeliski, R.: A Taxonomy and Evaluation of Dense Two-Frame Stereo Correspondence Algorithms. International Journal of Computer Vision 4 (April-June 2002)
9. Mutto, C., Zanuttigh, P., Cortelazzo, G.: A probabilistic approach to tof and stereo data fusion. In: 3DPVT10 (2010)
10. Nickels, K., Castano, A., Cianci, C.M.: Fusion of lidar and stereo range for mobile robots. In: International Conference on Advanced Robotics (2003)
11. Zhu, J., Wang, L., Yang, R., Davis, J.: Fusion of timeof-fight depth and stereo for high accuracy depth maps. In: CVPR (2008)

Auto-Calibration of Pan-Tilt Cameras Including Radial Distortion and Zoom

Ricardo Galego, Alexandre Bernardino, and José Gaspar

Institute for Systems and Robotics, Instituto Superior Técnico / UTL, Lisboa, Portugal
{rgalego,alex,jag}@isr.ist.utl.pt

Abstract. Although there are intrinsic advantages of using pan-tilt-zoom cameras their application in automatic surveillance systems is still scarce. The difficulty of creating background models for moving cameras and the difficulty of keeping fitted pose and optical geometrical projection models are key reasons for the limited use of pan-tilt-zoom cameras. Geometric calibration is a useful tool to overcome these difficulties.

In this work we propose a method for PTZ camera auto-calibration over the camera's zoom range. A new approach based on a *division distortion model*, which allows designing linear algorithms, is followed to solve the radial distortion when images are captured in different zooms. Results obtained over both synthetic and real data show that a full zoom range, complete field of view, pan-tilt-zoom camera calibration is possible.

1 Introduction

Despite the high versatility and potential of pan-tilt-zoom cameras, their use in current surveillance systems is still much less frequent than the use of fixed cameras. The limited dissemination of pan-tilt-zoom cameras can be justified by the slightly higher costs of the hardware and the higher risk of failure due to the mechanical components. This is however just a partial justification, since the hardware costs and failures can decrease significantly with mass production. Other, more compelling, justifications arise from the operation of the cameras and the difficulty of developing surveillance methodologies. Aiming to have mostly high quality employment implies that installations with numerous pan-tilt-zoom cameras cannot involve many human operators. This motivates developing automatic control and surveillance methodologies in opposition to the currently utilized manual control. The difficulty of creating background models for moving cameras and the difficulty of keeping fitted pose and optical geometrical projection models are also key reasons for the limited availability of automatic surveillance methodologies provided by the industry. Geometric calibration is a useful tool to overcome these difficulties.

There are several methods documented in the literature for calibrating cameras. The method proposed by Bouguet [1] is nowadays one of the most used calibration methods. Bouguet's method allows estimating intrinsic and radial distortion parameters based on imaging a planar chess pattern placed at various orientations. In the case of mobile cameras, in particular the pan-tilt-zoom cameras, there has been shown that they can be auto-calibrated using natural features of a static scenario. Hartley [2] presented an

G. Bebis et al. (Eds.): ISVC 2012, Part I, LNCS 7431, pp. 169–178, 2012.

auto-calibration method for rotating cameras and later Agapito et al. [3,4] introduced a self-calibration method for rotating and zooming cameras. These works showed that geometric calibration of intrinsic and extrinsic parameters can be achieved without requiring non-linear optimization. Estimating of the radial distortion still implied non-linear optimization. Sinha and Pollefeys [5] also recurred to the non-linear minimization of reprojection errors in order to consider the estimation of radial distortion in pan-tilt-zoom cameras. Their approach is based on estimating intrinsic and radial distortion coefficients using imagery taken within a small range of the camera's FOV. In all methods presented till here, there is no approach which achieves an accurate estimation of both intrinsic and radial distortion parameters with low (time) computational costs.

One of the key aspects justifying the complexity of including the estimation of radial distortion in the calibration process is that the distortion model cannot be easily inverted. To overcome this aspect, Fitzgibbon [6] proposed the *division distortion model*, which directly maps the radially distorted points into undistorted points. This direct mapping allows estimating the radial distortion and other calibration parameters in a manner very similar to the conventional calibration of pan-tilt cameras [2]. Steele and Jaynes [7] improved Fitzgibbon's algorithm by making it more stable. Kukelova and Pajdla [8] solve the same problem with the use of the fundamental matrix.

In this work, we propose using a linear algorithm [3] for the calibration of the intrinsic parameters of a pan-tilt-zoom camera, including the estimation of radial distortion. A new approach is followed for solving the radial distortion when images are captured at different zooms.

This paper is organized as follows: Section 2 introduces the pan-tilt-zoom camera model, Section 3 presents several calibrations methodologies, Section 4 proposes a methodology for the calibration of radial distortion when in presence of zoom, Section 5 shows some results with real and synthetic data, and finally, in Section 6 some conclusions are drawn and the future work is stated.

2   Pan-Tilt-Zoom Camera Model

The pin-hole camera model for the perspective pan-tilt-zoom camera consists of a mapping from 3D projective space to 2D projective space. This is represented by a 3x4 rank-3 perspective matrix, P. The mapping from 3D to the image plane takes a point $X = [X\ Y\ Z\ 1]^T$ to a point $u = PX$ in homogeneous coordinates. The projection matrix is usually decomposed as:

$$P = K^z[R\ t] \qquad (1)$$

where t is a 3x1 vector that represents the camera location, R is a 3x3 rotation matrix that represents the orientation of the camera with respect to an absolute coordinate frame and K^z is a 3x3 upper triangular matrix encompassing the intrinsic parameters of the camera:

$$K^z = \begin{bmatrix} k_u & s & u_0 \\ 0 & k_v & v_0 \\ 0 & 0 & 1 \end{bmatrix} \qquad (2)$$

where k_u and k_v are the magnifications in the respective u and v directions, u_0 and v_0 are the coordinates of the principal point of the camera and s is a skew parameter (in this work we assume $s = 0$).

2.1 Specific Aspects of Pan-Tilt-Zoom Cameras

Contrarily to cameras with fixed optics, pan-tilt-zoom cameras allow zoom in/out, and thus allow varying the K^z matrix.

In addition, note that the pan and tilt movements are simply rotations about the projective center $O = -R^{-1}t$, which is usually chosen to be the world origin and thus $t = [0\ 0\ 0]^T$. The pan and tilt movements are included in R, and thus R is also a time varying matrix.

2.2 Radial Distortion

Most cameras deviate from the pin-hole model due to radial distortion. This effect decreases with increasing focal length. Due to radial distortion a 3D point X is projected to a point $x_d = [x_d\ y_d]^T$. This point is deviated from the point $x = [x\ y]^T$ according to the radial distortion function, $\Re^z$:

$$[x_d\ y_d]^T = \Re^z \left([x\ y]^T;\ k_1,\ k_2\right) = L(r)[x\ y]^T \qquad (3)$$

where $L(r) = (1 + k_1 r^2 + k_2 r^4)[x\ y]^T$ and $r = \sqrt{x^2 + y^2}$. This radial distortion model corresponds to a simplified two coefficient version of the model proposed by Heikkila [9] where r is the radial distance (distance from point x to the center of distortion (x_c, y_c)), $L(r)$ is a radially symmetric distortion factor and k_1 and k_2 are the two radial distortion coefficients considered. For every zoom level z, $\Re^z$ is parameterized by $(x_c^z, y_c^z, k_1^z, k_2^z)$. In this work we assume that the principal point (u_0, v_0) is constrained to be the center of distortion and therefore the radial distortion function is only parameterized by coefficients k_1^z and k_2^z.

In summary, the goal of the calibration process of a pan-tilt-zoom camera is to estimate the unknown parameters of a model $(K^z, R$ and $\Re^z)$ that provides the intrinsic parameters for any pan, tilt and zoom admissible configuration. Since R is likely to change continuously due to the operation of the camera, in general the calibration of pan-tilt-zoom cameras refer just to estimating K^z and $\Re^z$.

3 Calibration Methodologies

This section contains a comprehensive review of auto-calibration methodologies for pan-tilt-zoom cameras.

3.1 Calibration Method by Agapito et al.

Agapito et al. in [10] introduced an image based auto-calibration method that estimates the intrinsic parameters of a rotating camera. Given a 3D world point X, it has different projections on different images, $x_i = P_i X$. Since X is the same on all the images and $P = KR$, we can calculate the relation between the images

$$x_i = P_i P_j^{-1} x_j. \tag{4}$$

The relation between images is a homography (projective transform) $H_{ji} = P_i P_j^{-1} = K_i R_{ij} K_j^{-1}$. Since R is a rotation matrix, it satisfies $R = R^{-T}$, and thus

$$H_{ji} K_j K_j^T H_{ji}^T = K_i K_i^T. \tag{5}$$

In the cases that $K_i = K_j$, i.e. images with constant zoom, Eq. 5 can be solved by a set of linear equations in the entries of KK^{-T}.

3.2 Bundle Adjustment, Sinha and Pollefeys' Calibration Method

Sinha and Pollefeys [5] proposed an algorithm to calibrate simultaneously the intrinsic parameters of a camera as well its radial distortion. Their method is based on multiple bundle adjustments. Given a set of images, their algorithm estimates the radial distortion, $\Re$, and the homographies among the images, T, by minimizing the reprojection error, D, on the surface of a cube:

$$min_{T_i, \Re(z_{min}), X^j} \sum_{j=1}^{m} \sum_{i=1}^{n} D(x_i^j, \Re(T_i X^j))^2. \tag{6}$$

Having an initial estimate of the radial distortion and of the homographies, the algorithm uses a method like Hartley and Agapito's to solve for the intrinsic parameters. Afterwards the method encompasses another bundle adjustment to refine the radial distortion parameters and the intrinsic parameters:

$$min_{K(z_{min}), R_i, \Re(z_{min}), X^j} \sum_{j=1}^{m} \sum_{i=1}^{n} D(x_i^j, K(\Re(R_i X^j)))^2 \tag{7}$$

As it has several bundles this method can be slow, depending on the number of images and corresponding points.

3.3 Division Distortion Model, Fitzgibbon's Calibration Method

Fitzgibbon proposed an auto-calibration algorithm that finds the homographies among images taken at various pan and tilt poses and, simultaneously, finds the radial distortion characterizing the camera at a fixed zoom level [6]. In order to estimate simultaneously the homographies and the radial distortion, the radial distortion is represented in the so called *one-parameter division distortion model*:

$$x = \frac{1}{1 + \lambda \|x_d\|^2} x_d \tag{8}$$

where x is the undistorted pixel position, x_d is the distorted pixel position and λ is the radial distortion parameter. The calibration methodology estimates the radial distortion from a pair of images. From Eq. 4 one has that x_i is equal to Hx_j, which leads to:

$$x_i \otimes Hx_j = 0 \tag{9}$$

where $\otimes$ denotes the Kronecker product. If x is replaced using Eq.8 then one obtains:

$$(x_i + \lambda z_i) \otimes H(x_j + \lambda z_j) = 0 \tag{10}$$

where $x_i = [u_i \; v_i \; 1]^T$ and $z_i = [0 \; 0 \; u_i^2 + v_i^2]^T$. The expansion of Eq. 10, based on distorted points directly observed on the images, leads to a quadratic eigenvalue problem (QEP):

$$(D_1 + \lambda D_2 + \lambda^2 D_3)h = 0 \tag{11}$$

where the D_1, D_2 and D_3 matrices are composed from the data, and h is the homography, H vectorized. The solution proposed by Fitzgibbon involves solving a number of times Eq. 11, each one for a pair of images, and then computing the median of the solutions. In addition, Eq. 11 is modified to be written with squared matrices. The matrices in Eq. 11 are made square by multiplying all the terms of the equation by D_1^T:

$$(D_1^T D_1 + \lambda D_1^T D_2 + \lambda^2 D_1^T D_3)h = 0. \tag{12}$$

The solution of the problem described by Eq.12, i.e. λ and h, can be obtained numerically using e.g. the function `polyeig` of Matlab.

3.4 Division Distortion Model, Steele and Jaynes' Calibration Method

Steele and Jaynes [7] improved Fitzgibbon's algorithm by making it lesser biased by the noise existing in the correspondences found among images. Their method takes advantage of an QEP algorithm working directly on rectangular matrices, and thus eliminates the need to enforce square matrices as Fitzgibbon proposed. In their method they add a new variable to Eq. 11:

$$D_1 h + \lambda(D_2 h + D_3 u) = 0 \tag{13}$$

$$u - \lambda h = 0 \tag{14}$$

Solving the equation for h and u one gets $(A - \lambda B)v = 0$ with:

$$A = \begin{bmatrix} D_1 & 0 \\ 0 & I \end{bmatrix}; \; B = \begin{bmatrix} -D_2 & -D_3 \\ I & 0 \end{bmatrix}; \; v = \begin{bmatrix} h \\ u \end{bmatrix} \tag{15}$$

The solution to this problem is obtained iteratively. The algorithm starts with $\lambda = 0$ and then v is computed through single value decomposition. The update to λ is done by solving the quadratic equation

$$v^T(B^T + \lambda A^T)(A - \lambda B)v = 0 \tag{16}$$

and choosing the positive root[1].

---

[1] For the quadratic equation $ax^2 + bx + c = 0$ the positive root is given by $(-b + \sqrt{b^2 - 4ac})/(2a)$

3.5 Comments

The pioneer work by Agapito et al. already focuses on the geometric auto-calibration of pan-tilt-zoom cameras. However, it aims solely to estimate efficiently the intrinsic parameters, not including radial distortion. Sinha and Pollefeys's approach to auto-calibrate these cameras, addresses estimation of both intrinsic and radial distortion parameters, but relies on multiple bundle adjustment processes which make the process computationally complex (time). The methods of Fitzgibbon, and Steele and Jaynes, allow a simple computation of both radial distortion and homographies, but do not provide calibration for multiple zoom levels.

Table 1 summarizes the features of the calibration methodologies presented in this section. One observes that no single methodology has all the presented three features, namely estimating radial distortion, allowing multiple zoom levels, and having at the same time low computational demands (more precisely, low computational time).

Table 1. Comparing calibration methodologies

Calibration methodology	Multiple zoom levels	Radial distortion	Low Comp. resources
Agapito et al. [10]	yes		yes
Sinha and Pollefeys [5]	yes	yes	
Fitzgibbon [6]		yes	yes
Steele and Jaynes [7]		yes	yes

4   Calibration Considering Both Radial Distortion and Multiple Zoom Levels

In this section we propose an auto-calibration methodology that estimates radial distortion, allows multiple zoom levels, and can be implemented efficiently (processing time). This methodology builds on the methodologies available in the literature that were described in the previous section. Before presenting the complete calibration algorithm, we will firstly introduce the calibration procedure for one zoom level given the calibration of the preceding zoom level.

Considering that the radial distortion varies with the zoom level, one has to modify Eq. 10 to represent both distortion levels, λ_1 and λ_2:

$$(x_i + \lambda_1 z_i) \otimes H(x_j + \lambda_2 z_j) = 0 \tag{17}$$

Expanding Eq. 17 one obtains:

$$x_i \otimes Hx_j + \lambda_1 z_i \otimes Hx_j + \lambda_2(x_i \otimes Hz_j + \lambda_1 z_i \otimes Hz_j) = 0 \tag{18}$$

here factorized for λ_2. The equation can be further expanded to factorize the homography as a column vector, $h = [h_{11} \ h_{21} \ \cdots h_{33}]^T$, pre-multiplied by matrices D_1 and D_2:

$$(D_1 + \lambda_2 D_2)h = 0 \tag{19}$$

where

$$D_1 = \begin{bmatrix} 0 & 0 & 0 & -x' - \lambda_1 r'^2 x & -y' - \lambda_1 r'^2 y & -1 - \lambda_1 r'^2 & yx' & yy' & y \\ x' + \lambda_1 r'^2 x & y + \lambda_1 r'^2 y' & 1 + \lambda_1 r'^2 & 0 & 0 & 0 & -xx' & -xy' & -x \end{bmatrix} \tag{20}$$

and

$$D_2 = \begin{bmatrix} 0 \; 0 & 0 & 0 \; 0 & -r^2 - \lambda_1 r^2 r'^2 \; 0 \; 0 & y' r^2 \\ 0 \; 0 & \lambda_1 r^2 r'^2 + r^2 \; 0 \; 0 & 0 & 0 \; 0 & -x' r^2 \end{bmatrix}. \tag{21}$$

Note that Eq. 19 can be once more solved using the `polyeig` function of Matlab, or using the iterative method proposed by Steele and Jaynes (see equations 15 and 16).

Now that one has a methodology to calibrate a zoom level given the calibration of another zoom level, it is possible to design a complete calibration algorithm. Our proposal of a complete calibration methodology encompasses the following main steps:

1. Calibration at minimum zoom using the methodology of Steele and Jaynes (section 3.4)
2. Calibration of one zoom level after another using the methodology introduced in this section (Eq. 19).

Doing calibration at minimum zoom (i.e. maximum *zoom out*) involves moving (panning and tilting) the camera with a fixed zoom, detecting and matching features. In our case we use SIFT features [11]. The calibration for various zoom levels involves fixating the pan and tilt angles while increasing the zoom level. Between each pair of zoom levels one has to find once more corresponding SIFT features.

5 Experiments and Results

This section describes two experiments: (i) comparing the performance of Fitzgibbon's calibration algorithm against the Steele and Jaynes' in a pan-tilt sequence created from a synthetic scenario, and comparing the performance of our zoom algorithm combined with both algorithms; (ii) qualitatively assessment of the performance of our methodology on real data, namely mosaics built from pan-and-tilt and from zoom sequences.

5.1 Noise Analysis

This experiment is based on synthetic data, generated from random $3D$ points imaged by a virtual pan-tilt-zoom camera with radial distortion. White Gaussian noise is added to the position of the features observed by the camera. The noise varies from zero pixels of standard deviation up to two pixels.

The first part of the experiment involves comparing the performance of Fitzgibbon's and Steele and Jaynes' algorithms in estimating the radial distortion from a set of images with constant zoom. The Fitzgibbon's algorithm is tested with a set of 50 pair of images and the Steele and Jaynes' algorithm is tested with just 6 pairs of images. Both algorithms performed well with no noise, as they both found the correct radial distortion parameter $\lambda = 0.1278$ (see fig. 1, green line). However as the noise increases the Fitzgibbon algorithm (left plot, A) tends to go away from the real value of the parameter. On the other hand Steele and Jaynes' algorithm (left plot, B) stays close to the real solution even in a presence of a considerable noise. Despite using a lesser number of image pairs, Steele and Janeys' algorithm performs better.

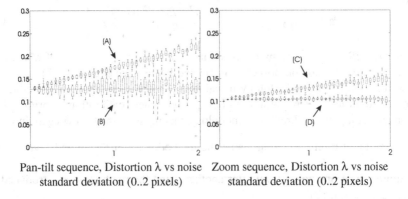

Pan-tilt sequence, Distortion λ vs noise   Zoom sequence, Distortion λ vs noise
standard deviation (0..2 pixels)   standard deviation (0..2 pixels)

Fig. 1. Estimated distortion vs noise in the data. Comparing two estimation methods in a pan and tilt sequence (left plot) and in a zoom sequence (right plot). The green line represents the true distortion value in both plots. Noise is white Gaussian noise varying from 0 to 2 pixels (standard deviation).

In the second part of the first experiment we assess our methodology for calibrating the radial distortion given images taken at different zoom levels. More precisely, this experiment involves generating random $3D$ points and capturing two images with different zooms. One of the two images has known radial distortion. The imaged points, matched between the two images, are corrupted with white Gaussian noise with 0 to 2 pixels pixels of standard deviation. Figure 1(right plot) compares using `polyeig` once (C) with using `SVD` iteratively (D), as proposed by Steele and Jaynes' for the constant zoom case (the horizontal green line denotes the ground truth value). One observes that the iterative algorithm is beneficial for estimating the radial distortion in the case of multiple zoom levels, as it was in the single zoom level.

5.2 Mosaics

The second experiment involves testing the proposed calibration methodologies with real data acquired with an Axis 215 PTZ camera. The calibration of the intrinsic parameters and the radial distortion is performed using a set of 16 images, captured at different orientations, with the same zoom level. First the radial distortion is estimated using Steel and Jaynes' algorithm. After correcting the images the algorithm of Agapito et al. is used to estimate the intrinsic parameters, K^z. Figure 2 (a) shows that both algorithms work well together since the texture is qualitatively consistent at stitching image seams.

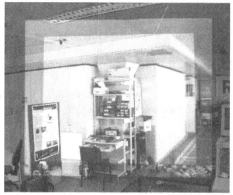

(a) Pan and tilt mosaic   (b) Zooming on mosaic

Fig. 2. (a) Mosaic build from sequence of 16 pan and tilt images. (b) Mosaic from a sequence 4 images with different levels of zoom.

The experiment on real data continues with a sequence of images with varying zoom. Four images were acquired with different zoom levels. The radial distortion parameters are known (estimated in the previews step) just for the first image. The parameters of the other images are estimated using our methodology. The result of the radial distortion correction and pasting of the images is shown on fig. 2 (b).

6   Conclusions and Future Work

In this paper we approach the problem of auto-calibrating pan-tilt-zoom cameras. The auto-calibration comprises estimating intrinsic and radial distortion parameters over the camera's full zoom range.

Fitzgibbon [6] proposed a method for calibrating the radial distortion without knowing anything besides matching features between images. Steele and Jaynes [7] improved the method by making it more robust. In both works the zoom level was considered to be fixed, and thus the radial distortion was constant for all images acquired by the camera. In our methodology we show that auto-calibration can be generalized to the case of multiple zoom levels while still effectively including the estimation of radial distortion. We propose solving the problem of different zoom levels by estimating the radial distortion at one zoom level given the radial distortion of another level.

Future work will focus on investigating even more complex modelings of pan-tilt-zoom cameras, while keeping computationally time-efficient solutions. Recently work by Kukelova and Pajdla [8] is already pointing towards this direction.

Acknowledgments. This work has been supported by the Portuguese Government - FCT project PEst-OE / EEI / LA0009 / 2011, and FCT project PTDC / EEA-CRO / 105413 / 2008 DCCAL.

References

1. Bouguet, J.Y.: Camera calibration toolbox for matlab (2008),
   http://www.vision.caltech.edu/bouguetj
2. Hartley, R.I.: Self-calibration of stationary cameras. International Journal of Computer Science 22, 5–23 (1997)
3. Agapito, L.D., Hayman, E., Reid, I.: Self-calibration of a rotating camera with varying intrinsic parameters. In: Proc. 9th British Machine Vision Conf., Southampton, pp. 105–114 (1998)
4. Agapito, L.D., Hayman, E., Reid, I.: Self-calibration of rotating and zooming cameras. Int. J. Comput. Vision 52, 107–127 (2001)
5. Sinha, S.N., Pollefeys, M.: Pan-tilt-zoom camera calibration and high-resolution mosaic generation. Comput. Vis. Image Underst. 103, 170–183 (2006)
6. Fitzgibbon, A.W.: Simultaneous linear estimation of multiple view geometry and lens distortion (2001)
7. Steele, R.M., Jaynes, C.: Overconstrained Linear Estimation of Radial Distortion and Multiview Geometry. In: Leonardis, A., Bischof, H., Pinz, A. (eds.) ECCV 2006. LNCS, vol. 3951, pp. 253–264. Springer, Heidelberg (2006)
8. Kukelova, Z., Pajdla, T.: A minimal solution to radial distortion autocalibration. IEEE Transactions on Pattern Analysis and Machine Intelligence 33 (2011)
9. Heikkila, J., Silven, O.: A four-step camera calibration procedure with implicit image correction. In: 1997 IEEE Computer Society Conference on Computer Vision and Pattern Recognition, CVPR 1997 (1997)
10. Agapito, L., Hartley, R., Hayman, E.: Linear calibration of a rotating and zooming camera. In: Proc. of the IEEE Conf. on Computer Vision and Pattern Recognition, pp. 15–21 (1999)
11. Lowe, D.G.: Object recognition from local scale-invariant features. In: The Proceedings of the Seventh IEEE International Conference on Computer Vision, vol. 2, pp. 1150–1157 (1999)

Robust 2D/3D Calibration Using RANSAC Registration

Billy Ray Fortenbury and Gutemberg Guerra-Filho

The University of Texas at Arlington

Abstract. An area of increasing interest in computer vision is the fusion of 2D images with depth maps from 3D sensing devices to obtain more robust 3D information about the scene. Before this can be achieved, one must have an accurate method for the calibration of the 3D sensing devices and the pinhole cameras. In this paper, we introduce a robust method for registering depth maps from 3D sensing devices into point clouds reconstructed from 2D images. Our new calibration method explores RANSAC registration to take into account the high-noise nature of current 3D sensing technologies. We solve this by using a novel application of the RANSAC algorithm to robustly register two point clouds obtained from the 3D sensing device and the pinhole camera. The reprojection error after registration using our algorithm is less than 0.3%.

1 Introduction

An area of increasing interest in computer vision is the fusion of 2D images from cameras with depth maps from 3D sensing devices (*e.g.*, time-of-flight cameras or structured lighting) in order to obtain more robust 3D information about the scene. However, before this can be achieved, one must have an accurate calibration between the 3D sensing device and 2D cameras which involves the registration of the 3D depth map into the 2D images.

The primary challenge for solving a problem of this nature lies in the registration of a depth map with high noise content to a point cloud reconstructed from images. The main contributions of this paper are a method for the registration of clouds of 3D points and the calibration of a 3D sensing device with a pair of stereo cameras. The root of this method relies on the RANSAC algorithm [9,4] to obtain the respective transformations between each of the two cameras and the 3D sensing device. Given a known set of 3D points (step 1, 2, and 3 in figure 1), we register the points obtained from the 3D sensing device into the points obtained by 3D reconstruction from 2D images (step 4 in figure 1) and then back project the points into the pinhole cameras image plane (step 5 in figure 1. The evaluation of our registration method is achieved by transforming the 3D point cloud into the respective camera frames and projecting the points into the image planes. Our experimental results show an average accuracy of 2.71 pixels within an image of size 1280×960 pixels using our method with the Kinect® structured lighting device developed by Microsoft®.

G. Bebis et al. (Eds.): ISVC 2012, Part I, LNCS 7431, pp. 179–188, 2012.
© Springer-Verlag Berlin Heidelberg 2012

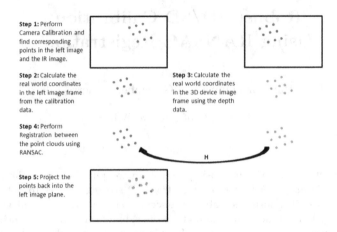

Fig. 1. A conceptual overview of the 2D/3D calibration algorithm

The remaining of this paper is organized as follows. In Section 2, we discuss previous work related to 2D/3D calibration and explain how our approach differs from these other methods. We then introduce our method in Section 3 to register a cloud of 3D points into the points reconstructed from 2D images of a stereo pair of cameras using a modified RANSAC algorithm. Section 4 discusses the experimental results obtained from applying our method to the calibration of a Kinect® device with two CCD cameras. Section 5 contains our conclusions and a discussion on future work involving this methodology.

2   Related Work

Probabilistic methods such as Bayesian Modelling [8] have been used to solve image registration. These methods find a best match based on a probability distribution and are useful when there is a non-rigid transformation between the two 3D point clouds given. Usually, probabilistic methods are not suitable due to the processing power necessary to produce accurate results. Our algorithm addresses registration in the context of image calibration and mostly uses rigid motion transformations. Our algorithm address the issue of processing time by using a constant time algorithm to limit processing necessary for accuracy to a fixed amount of time.

Geometric approaches for the registration of two sets of points include Iterative Closest Point [5,11], Disparity Gradient Limits [10], and Least Squares Fitting [2]. These methods are computationally intensive as they mostly rely on checking each point against all other points in the 3D point clouds to be registered. Our algorithm is fast and robust since it does not need to perform this point to point matching.

In Fontanelly *et al.* [5], they estimate the position of a robot equipped with a laser scanner (LIDAR) and construct the trajectory of the robot over time using an Extended Kalman Filter. To address this localization problem, they solve the point registration of two sensed point clouds. Similar to our approach, they solved the registration problem with an algorithm based on the RANSAC algorithm. However, in order to compute the absolute orientation in a 2D measurement space of the LIDAR device, they assume the plane of measurements is constantly parallel to the plane of motion of the LIDAR. In this special case, the rigid transformation has fewer degrees of freedom and requires only two pairs of corresponding points to be determined. In our approach, we compute a more general rigid transformation (with 6 degrees of freedom) directly in 3D space using three pairs of corresponding points. We also consider affine transformations that include scaling, but our empirical results show that scale may usually be ignored in practice for the 3D sensing device used in our experiments.

3 Robust 2D/3D Calibration

The problem that we address in this paper concerns the calibration of a 3D sensing device with a pair of stereo cameras. More specifically, given a cloud of points in 3D space obtained from the 3D sensing device and their projections into the image planes of the stereo pair of cameras, first we need to find the rigid transformation that best aligns the 3D point cloud with a point cloud reconstructed from the stereo pair of cameras. A major challenge to find this transformation is the significant noise in the 3D point cloud from the 3D sensing device. In order to overcome this issue, we propose a robust approach based on the RANSAC algorithm.

Our method registers a set of 3D points reconstructed from a calibrated pair of stereo cameras and the corresponding set of 3D points given by the 3D sensing device. Our approach uses a novel algorithm based on the RANSAC method. Once a transformation that best aligns the stereo cameras and the 3D sensing device has been found, all other points from the 3D device are then transformed into the 2D camera's coordinate system and projected back onto the image plane.

The 2D/3D calibration process is particularly useful to find corresponding points in the 2D images of the stereo pair. Given a set of points from the 3D sensing device, our method finds the transformations necessary to project each 3D point in this set into the 2D cameras' image plane such that for any particular 3D point, the corresponding 2D point in all 2D cameras is found.

3.1 Camera Calibration

Our approach requires calibration data from a calibration device such as a checker board. The calibration data consists of the left camera images, I_L, the right camera images, I_R, the infrared images, I_F, and the depth maps, I_D, obtained from a setup where a Kinect device is placed in the middle of two stereo CCD cameras. The infrared images, I_F, are captured via an infrared camera in the same 3D sensor device that gives the depth maps, I_D, and, consequently, I_F is aligned with I_D. Figure 2 shows two samples (*i.e.*, top row and bottom row) of our calibration data.

Fig. 2. Two samples of the calibration data. From left to right, the images shown are the left camera image, the infrared camera image, the right camera image, and an image representation of the depth map. The images shown are not all at the same scale as the infrared image and depth map are half the size of the left and right images in both height and width.

For each calibration image, we identify a set of corresponding 2D points in all three image spaces: left image plane, right image plane, and infrared image plane. More specifically, we obtain a set $P_L^{(2)}$ of 2D points in I_L, a set $P_R^{(2)}$ of 2D points in I_R, and a set $P_F^{(2)}$ of 2D points in I_F such that the corresponding points in the three sets $P_L^{(2)}$, $P_R^{(2)}$, and $P_F^{(2)}$ are associated with the same 3D point.

From these points, we use the Caltech Camera Calibration Toolbox [3] to obtain the intrinsic calibration matrix K_L for the left camera, the intrinsic calibration matrix K_R for the right camera, and the intrinsic calibration matrix K_F for the infrared camera. These matrices contain the principal point, aspect ratio, and the focal length of each camera.

3.2   RANSAC Point Registration

After the intrinsic camera calibration has been completed, the extrinsic calibration is performed to obtain a rigid transformation (*i.e.*, rotation and translation) that relates the left camera and the right camera. Using the extrinsic calibration, the sets $P_L^{(2)}$ and $P_R^{(2)}$ of 2D points are triangulated to obtain a set of 3D points, $P_L^{(3)}$, in the left camera frame. Similarly, we obtain the point cloud, $P_R^{(3)}$, in the right camera frame.

A cloud of 3D points, $P_F^{(3)}$, is also obtained for the 3D sensing device by tracing rays travelling from the focal point and through each point in $P_F^{(2)}$ on the image plane of the infrared camera. The distance between the focal point and a 3D point is equal to the depth given by the 3D sensing device for that particular 2D point in the depth map I_D.

Fig. 3. Three example boards showing the original image from the left CCD camera and the 3D reconstruction as a result of our registration method. These images were obtained by using the transformation obtained in Section 3.2 and converting the points from the Kinect® device into the left image frame, plotted using the color of the pixel in the left image at the reprojected pixel location. They were then rotated to show results from different viewpoints. These images were chosen to show depth features (*e.g., a human*) *other than a plane*.

Once two 3D point clouds $P_L^{(3)}$ and $P_F^{(3)}$ have been obtained, where $P_L^{(3)}$ was reconstructed using the stereo pair of cameras and $P_F^{(3)}$ was retrieved from the 3D sensing device, our algorithm based on RANSAC finds a transformation between the two clouds of 3D points that discounts outliers from noise.

For the left camera, our algorithm performs a registration step for a constant number n of iterations. The registration step uses three random 3D points in $P_L^{(3)}$ and their corresponding points in $P_F^{(3)}$. A rotation T_r, a translation T_t, and a scaling factor T_s are then obtained to register these two sets of three corresponding points. The combination of these three transformations results in the 3D transformation $H_{F \to L}$ from points in the 3D sensing device frame to points in the left camera frame.

Once the two sets of three points are registered, all the points in $P_F^{(3)}$ are transformed into the left camera frame using $H_{F \to L}$. We compute the Euclidean distance between each point in $H_{F \to L} P_F^{(3)}$ and the corresponding point in $P_L^{(3)}$. The number c of inliers is then calculated as the number of pairs of corresponding points whose distance is less than a threshold value t. This process is repeated n times and the transformation that maximizes c is selected as the transformation that registers the two sets of 3D points.

The same process is repeated for the 3D sensing device and the right camera. The result of this process will be the transformation $H_{F \to R}$ between the 3D sensing device and the right camera. The pseudocode for our registration method, Ransac-Registration, can be found below along with a listing for the computation of transformations, Get-Transformation.

Ransac-Registration takes in two sets of 3D points, $PF3$ and $PL3$, a number of iterations to perform, n, and an inlier threshold, t. It then chooses three random corresponding points from $PF3$ and $PL3$, finds a transformation between those three points, and then tests the transformation on all the points in $PF3$ and $PL3$ using a Euclidean distance measure. The transformation which causes the maximum number of points to fall below the threshold, t, is then returned.

Get-Transformation takes in two sets of three points from two separate spaces and finds a transformation between them. It makes use of two functions which form specific types of homogeneous transformation matrices: Translation-Matrix which forms a homogeneous translation matrix from a translation vector, and Rotation-Matrix which forms a homogeneous rotation matrix from a non-homogeneous rotation matrix. It does this as follows: 1) translate to the origin using O; 2) translating from point sets B to A; 3) choose two of the points and make this the x-axis of the system; 4) perform the cross product between the x-axis and a vector formed from two other points to get the y-axis; 5) perform the cross product between the x and y-axis to obtain the z-axis; 6) perform the dot product between the matrix $[A_X A_Y A_Z]$ and $[B_X B_Y B_Z]^T$ to get the rotation between the axis formed from A and the axis formed from B. The final transformation is then the translation N from the origin times the rotation R times the translation T times the translation O back to the origin.

```
Algorithm Ransac-Registration(PF3, PL3, n, t)
  var
    cbest, i, p1, p2, p3, c: Integer;
    A, B, H, HFL: RealMatrix;
  begin
    cbest := -inf;
    for i = 1 to n
      p1 := rand() * (length(PF3) - 1);
      p2 := rand() * (length(PF3) - 1);
      p3 := rand() * (length(PF3) - 1);
      A := {PL3[p1], PL3[p2], PL3[p3]};
      B := {PF3[p1], PF3[p2], PF3[p3]};
      H := Get-Transformation(A, B);
```

```
      c := SUM(Euclidean-Distance(PL3, H * PF3) < t);
      if c > cbest
         HFL := H;
         cbest := c;
      endif
   endfor
   return HFL;
end.


Algorithm Get-Transformation(A,B)
  var
    N, O, T, R, H: RealMatrix;
    AX, AY, AZ, BX, BY, BZ: RealVector;
  begin
    N := Translation-Matrix(A[1]);
    O := Translation-Matrix(-A[1]);
    T := Translation-Matrix(A[1] - B[1]);
    A := O * A;
    B := T * O * B;
    AX := (A[2] - A[1]) / Norm(A[2] - A[1]);
    BX := (B[2] - B[1]) / Norm(B[2] - B[1]);
    AY := Cross(AX, (A[3] - A[1]) / Norm(A[3] - A[1]));
    BY := Cross(BX, (B[3] - B[1]) / Norm(B[3] - B[1]));
    AZ := Cross(AX, AY);
    BZ := Cross(BX, BY);
    R := Rotation-Matrix([AX AY AZ] * Transpose([BX, BY, BZ]));
    H := N * R * T * O;
    return H;
  end.
```

3.3   Back Projection

After finding the two transformations from the 3D sensing device's frame to the left camera frame and to the right camera frame, we are able to calculate the location of all 3D points given by depth maps from the 3D sensing device. This is achieved using a ray tracing procedure described above in Subsection 3.1. However, instead of using only the known calibration point cloud, this is performed for all points in a depth map given by the 3D sensing device resulting in a dense 3D point cloud, $P_D^{(3)}$.

Using the 3D points in the cloud $P_D^{(3)}$, obtaining the corresponding 2D projections into the image planes of the left and right cameras is a simple matter of using the calibration matrices and the transformation obtained above. The 3D point cloud for the left camera is $P_{LD}^{(3)}$ which is obtained from $H_{F \to L} P_D^{(3)}$ and the 3D point cloud for the right camera is $P_{RD}^{(3)}$ which is obtained from $H_{F \to R} P_D^{(3)}$.

Fig. 4. A scene in world coordinate space which depicts the results of registration with the calculated points from the Kinect in blue and the known points from the 2D camera in red

Once the point clouds $P_{LD}^{(3)}$ and $P_{RD}^{(3)}$ are obtained, we trace a line between each point in $P_{LD}^{(3)}$ and the left camera focal point. Analogously, we trace a line between each point in $P_{RD}^{(3)}$ and the right camera focal point. We then find the intersection between each of these lines and the respective image planes. This yields the registered correspondence between 2D points in the left image and 2D points in the right image. This ray tracing for back projection is actually achieved by applying the respective intrinsic calibration matrices.

After projecting the points back onto their respective image planes, un-distorting and warping the stereo images and both point clouds using the calibration from Subsection 3.1 should result in rectified image pairs with the 3D sensing device registered into the 2D camera. This allows, for instance, stereo matching algorithms to be performed with the assistance of the 3D sensing device.

Once the 2D/3D calibration between stereo cameras and the 3D sensing device is found, provided the same camera setup is used, we can transform any 3D point from the 3D sensing device into the camera coordinate system and back project it into the stereo images since the resulting transformation is invariant when the cameras are stationary.

4   Experimental Results

Our experimental camera setup uses a stereo pair of CCD cameras with the Kinect® at the middle in between them. When capturing calibration data, the calibration board was placed in the scene in a stationary position. We then captured data from the three cameras of interest (left, right, and infrared) along with the depth maps from Kinect®.

To perform our experiments, a set of calibration images was taken using the two CCD cameras along with the raw infrared camera and depth images from a Kinect® 3D sensing device. The calibration and registration was performed on a subset of these images. The results from this registration were then used to transform and reproject the depth points for all boards and compared to the set of known points for all boards. The transformed points were then compared against the actual points calculated from calibration in both 3D space using an

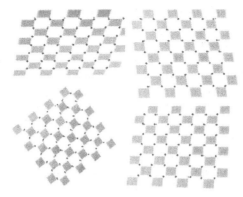

Fig. 5. Four example boards showing the known correct points in red with the reprojected points in green. Since the reprojected points are 3D points calculated from the depth of a Kinect device, they do not form a perfect grid. However, they still are reasonably accurate after registration.

Euclidean distance (*i.e.*, a measure in mm of the difference between actual and transformed points), and in 2D space after reprojecting the points back into the image plane using also an Euclidean distance (*i.e.*, a measure in pixels between the center of the actual pixel and the real projected point).

For the 2D/3D calibration algorithm, we used 49 points per board for a total of 5 boards. For RANSAC, we used 10,000 iterations with a threshold of 5mm. Once the 2D/3D calibration was obtained, we evaluated our method by reprojecting a known set of 3D points on 9 boards obtained from the 3D sensing device into the stereo cameras.

In 3D world coordinates, the average distance between the reconstructed coordinates, which were calculated through calibration of the stereo cameras, and the sensed coordinates, which were obtained as a result of transforming the 3D sensing devices' data into the left camera reference frame, was 7.13mm for all 441 test points used (see figures 4 and 3).

The average reprojection error from performing this experiment 30 times was 2.71 pixels in the 2D camera space of 1280 by 960 pixels (see Figure 5). This corresponds to an error of less than 0.3% in both height and width dimension for the image. Given the high noise content of the Kinect® device, these results show the accuracy and robustness of our approach even in the presence of high noise content.

5   Conclusion

Achieving accurate 3D calibration and registration between a stereo pair of CCD cameras and a 3D sensing device is an important area concerning the fusion of heterogeneous sensing devices. Despite the difficulty of registering a cloud of points through a rigid transformation, robust methods are needed to solve problems related to the 2D/3D calibration of different devices towards their fusion.

In this paper, we proposed a method for 2D/3D calibration with a reprojection error of only 2.71 pixels in an image space covering 1,228,800 pixels which corresponds to an error of less than 0.3% if both the height and width dimension. This was achieved through a three step process of calibrating the images within the 2D image space using a set of known points to obtain a set of reconstructed 3D points, registering the sets of known points using a modified RANSAC algorithm to obtain a transformation between the two 2D camera spaces and the 3D device space, and then projecting the resulting clouds back into the image planes of the 2D devices using the aforementioned transformation for all unknown points.

The proposed method allows for further work to be done in the area of fusion between 2D and 3D devices such as the development of high resolution 3D imaging and motion capture systems.

References

1. Arunm, K.S., et al.: Least-Squares Fitting of Two 3-D Point Sets. IEEE Transaction on Pattern Analysis and Machine Intelligence 5, 698–700 (1987)
2. Besl, P.J., McKay, N.D.: A method for registration of 3-d shapes. IEEE Transaction on Pattern Analysis and Machine Intelligence 14, 239–254 (2001)
3. Bouguet, J.Y.: Camera Calibration Toolbox for Matlab, http://www.vision.caltech.edu/bouguetj/calib_doc/index.html
4. Fischler, M.A., Bolles, R.C.: Random Sample Consensus: A Paradigm for Model Fitting with Applications to Image Analysis and Automated Cartography. Communications of the ACM 24, 381–395 (1981)
5. Fontanelli, D., et al.: A Fast RANSAC Based Registration Algorithm for Accurate Localization in Unknown Environments using LIDAR Measurements. In: IEEE Conference on Automation Science and Engineering (2007)
6. Fukai, H., Xu, G.: Fast and Robust Registration of Multiple 3D Point Clouds. In: 20th IEEE International Symposium on Robot and Human Interactive Communication (2011)
7. Pollard, S.B., et al.: PMF: A Stereo Correspondence Algorithm Using a Disparity Gradient Limit. Perception 14, 449–470 (1985)
8. Szeliski, R.: Bayesian modeling of uncertainty in low-level vision. Journal of Computer Vision 5, 271–301 (1990)
9. Torr, P.H.S., Murray, D.W.: The Development and Comparison of Robust Methods for Estimating the Fundamental Matrix. International Journal of Computer Vision 24, 271–300 (1997)
10. Zhang, Z.: Iterative Point Matching for Registration of Free-form Curves (1992)
11. Zitova, B., Flusser, J.: Image registration methods: a survey. Image and Vision Computing 21, 977–1000 (2003)

Keypoint Detection Based on the Unimodality Test of HOGs*

M.A. Cataño[1] and J. Climent[2]

[1] Pontificia Universidad Catolica del Peru, Peru
mcatano@pucp.edu.pe
[2] Universitat Politecnica de Catalunya, Barcelona Tech, Spain
juan.climent@upc.edu

Abstract. We present a new method for keypoint detection. The main drawback of existing methods is their lack of robustness to image distortions. Small variations of the image lead to big differences in keypoint localizations.

The present work shows a way of determining singular points in an image using histograms of oriented gradients (HOGs). Although HOGs are commonly used as keypoint descriptors, they have not been used in the detection stage before. We show that the unimodality of HOGs can be used as a measure of significance of the interest points. We show that keypoints detected using HOGs present higher robustness to image distortions, and we compare the results with existing methods, using the repeatability criterion.

Keywords: HOG, salient feature, keypoint detection, repeatability, unimodality test.

1 Introduction

Keypoints and their descriptors have been widely used in many computer vision applications, such as correspondence between images, content based retrieval, object detection and tracking, etc. The standard process consists of extracting relevant points from an image, and then obtain robust descriptors that can be compared with other keypoints from different images. The correspondence between keypoints from different images makes possible to locate and identify objects.

Keypoint detection is a low level image process where each pixel is checked in order to determine its significance. Once the most relevant pixels are selected, their information is described by means of a feature vector. These feature vectors are known as keypoint descriptors and they are used to compare keypoints extracted from different images, or from an object database. Several authors have presented different keypoint descriptors (SIFT [1], PCA-SIFT [2], SURF

---

* This research was partially supported by Consolider Ingenio 2010, project (CSD2007-00018) and CICYT project DPI2010-17112.

G. Bebis et al. (Eds.): ISVC 2012, Part I, LNCS 7431, pp. 189–198, 2012.

[3]) that show good properties in robustness, invariance, discriminability, compactness, or computational efficiency. A reference survey about the performance of different descriptors can be found in [4].

In this work, we focus on keypoint detection. Since keypoint detection is a mandatory step prior to obtaining the descriptors, the global process cannot be as much robust as the keypoint detection is. Keypoint descriptors are useless if the detection process is not robust.

A good keypoint detector must be robust. It means that small variations of the image structure may not cause big variations in the keypoint detection. The localization of the keypoints must be accurate as well.

In this work we present a new keypoint detector that improves the robustness of existing methods. It is based on histograms of orientations. Orientation analysis gives invariance to illumination changes, while the use of histograms gives invariance to translation.

Histograms of oriented gradients (HOGs) have been previously used as keypoint descriptors (i.e. SIFT [1]), or for object detection [5][6] , but the contribution of this work is to use HOGs in the detection stage instead of using them as descriptors.

In the results section we make a comparison with existing keypoint detectors in order to show that HOGs make possible a significantly more robust keypoint detection. As it could be expected, the use of gradient orientation makes the detection more robust to illumination changes, but we show that the detection is also more robust to orientation and perspective changes.

2   Keypoint Detection

A wide variety of keypoint detectors exists in the literature. Moravec [7] developed one of the first interest point detectors. His detector is based on the auto-correlation function of the signal. It computes the greyvalue differences between windows shifted in several directions. The detector of Beaudet [8] is based on the determinant of the Hessian matrix and is related to the Gaussian curvature of the signal. Harris and Stephens [9] improve the approach of Moravec by using the auto-correlation matrix. Interest points are detected if the auto-correlation matrix has two significant eigenvalues. Smith and Brady [10] compare the brightness of each pixel in a circular mask to the center pixel to define an area that has a similar brightness to the center. This method is known as SUSAN (Smallest Univalue Assimilating Nucleus). The Harris-Laplace detector [11] uses a scale-adapted Harris function to localize points in space and then selects the points for which the Laplacian of Gaussian attains a maximum over scale. The Harris-Laplace approach provides a compact and representative set of points which are characteristic in the image and also in the scale dimension. The Hessian-Laplace method [12] chooses interest points where the trace and determinant of the Hessian are maxima. An hybrid operator between Laplacian and Hessian determinant is used to localize points in space at the local maxima of the Hessian determinant and in scale at the local maxima of the Laplacian of Gaussian. The detector known as FAST (Features from Accelerated Segment Test) [13] considers the pixels under a circle of radius r around the interest point. The original detector classifies a pixel as a keypoint if there exists a set of n contiguous pixels in the circle which are all brighter or darker than the intensity of the candidate pixel.

Among the methods that present keypoint detectors along with their descriptors, probably the most popular is the Scale-invariant Feature Transform (SIFT) [14]. Although it consists of both a Feature Detector and a Feature Descriptor it is not usual to see them separately. The detection part of the algorithm works by identifying scale-invariant features looking for locations that are extrema of a difference of Gaussian (DoG) function. The SURF detector [3], also known as the Fast-Hessian Detector, is based on the determinant of the Hessian matrix. It uses box filters that approximate second order Gaussian derivatives and can be evaluated very fast using integral images.

A comprehensive evaluation of keypoint detectors can be found in [15]. Interested readers can also find an evaluation of descriptors in [4].

3   HOG Computation

Our keypoint detector is based on HOGs. The HOGs defined in [5] are blocks built from histograms of orientations obtained from overlapping cells, and they are used to detect people in scenes. We work just with the histograms of gradient orientation of single cells, there is no need to work with blocks since the objective is not to detect objects but simple keypoints. Keypoint detection is useful in generic applications, where no a priori information about the objects to be detected is available. The SIFT descriptors are also based on HOGs, but our histograms are constructed differently too. We compute the histograms at a single scale and without dominant orientation alignment. Therefore, computing the histograms in our application becomes an easy task. We first compute the gradient module and orientation images. For every pixel in the orientation image, we construct a histogram of orientations over local spatial windows called cells. Each pixel in the cell votes for its corresponding orientation value. The weighted votes are accumulated into the orientation bins.

To reduce aliasing, each pixel contributes to the histogram according to a Gaussian distribution (instead of voting to a single HOG bin). The standard deviation, σ, of this distribution is determined by the error of the computed orientation. This deviation is computed using the error propagation error law (eq. 1).

$$\sigma = \sqrt{(\frac{\partial \alpha}{\partial G_x}\sigma_{G_x})^2 + (\frac{\partial \alpha}{\partial G_y}\sigma_{G_y})^2} = \frac{\sigma_G}{\nabla f} \tag{1}$$

Where σ_{G_x} and σ_{G_y} are the standard deviations of the computed gradient errors, and can be considered to have the same value (σ_G) since gradient operators are symmetric. In this work, gradients are computed using simple convolution masks, therefore, σ_G is the standard deviation of the image noise.

Histograms computation can be speeded up using the integral histograms presented in [16]. The integral histogram is computed in linear time. Once we have the integral histogram, a HOG is computed for each cell with just 4 additions/subtractions.

Figure 1b shows the gradient orientation of figure 1a. Figures 1c-1f show the HOGS corresponding to the pixels pointed in figure 1b.

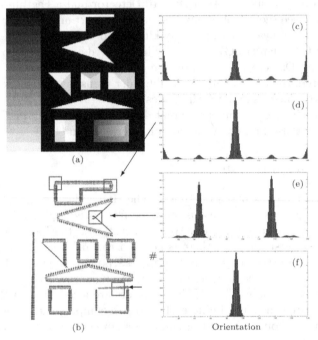

Fig. 1. (a) Synthetic image. (b) Gradient orientation. (c)-(f) Histograms of Orientations.

Each pixel of the image is represented by its corresponding cell HOG. HOGs with high or low entropy, usually unimodal histograms, never correspond to discriminative keypoints. This is due to the fact they are too similar to their neighbor HOGs. The higher the unimodality of the HOG, the less discriminative will be the keypoint. We use the test of unimodality presented by Hartigan and Hartigan in [17] to determine the significance of keypoint candidates depending on the unimodality of their HOGs. Details are given in next section.

4   The Unimodality Test

The most widely used measure of unimodality is the DIP Test [17]. The DIP statistic is defined as the maximum difference between an empirical distribution function and the unimodal distribution function that minimizes that maximum difference. The DIP of a distribution function F is defined by $D(F) = \rho(F, \zeta)$. Being ζ the class of unimodal distribution functions, and $\rho(F, \zeta) = \inf_{G \in \zeta} \sup_x |F(x) - G(x)|$.

To test the null hypothesis that F has a unimodal density Hartigan and Hartigan proposed the statistic $D(F_n)$, where F_n is the empirical distribution function of a random sample of size n. The distribution of $D(F_n)$ is compared with the distribution of $D(F)$, where F is the uniform distribution on $[0; 1]$. They showed that the DIP is asymptotically larger for the uniform distribution than for any other unimodal distribution.

There are other unimodality tests that can be found in literature. In our research we have tried the Variance Reduction Score and the Weighted Variance Reduction Score [18], Kurtosis [19], the Bimodality Index [20], and the Likelihood Ratio Test [21]. They are more sensitive to noise than DIP, and present discontinuities and sudden changes in their responses. It is not recommended to use this tests for measuring the unimodality of HOGs.

We compute the DIP test for every HOG in the image. The DIP test does not work with circular vectors, thus, HOGs are shifted up to a local minima. The unimodality measure is not affected by this rotation. The unimodality measure is not affected by a normalization of histogram amplitude either, therefore, there is no need of normalization as in [5].

We assign the unimodality score to the center pixel of each cell, and build a unimodality image. Figure 2 shows the unimodality image of the synthetic image 1a (Higher values correspond to lower unimodalities). The unimodality image has been computed using the DIP test algorithm published in [22].

Fig. 2. 3D representation of the unimodality image

Peaks of the unimodality image are candidates for being keypoints. We select the keypoints locating the relative maxima of the unimodality image. The unimodality image makes possible to establish a hierarchy of keypoints based on their saliency. The contrast extinction values [23] of the obtained local maxima have shown to be an excellent criterion for evaluating their significance.

5   Results

Keypoints detected in one image should also be detected at approximately corresponding positions in the subsequent image transformations. The most commonly used criterion that proves that the keypoint detection is stable to image transformation is repeatability [24].

The repeatability rate $r(I, \psi(I))$ is defined as the number of points present in both the original and the transformed image, respect to the total number of detected points. Keypoints that are not present in both images, decrease the repeatability score. Furthermore, the repeatability measure has to take into account the uncertainty of the point localization. A keypoint is in general not detected exactly at the same position in the transformed image, but rather in some neighborhood nearby. The size of this neighborhood is denoted by ϵ.

We have computed the repeatability $r(\psi(keyp(I)), keyp(\psi(I)))$ for all image transforms ψ. $keyp(I)$ is an image containing the keypoints of image I; fig. 3 shows a block diagram of the process. This process is repeated for all different keypoint detectors.

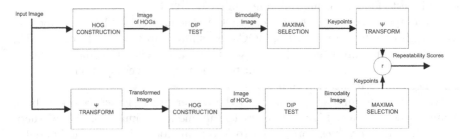

Fig. 3. Process for measuring repeatability

The repeatability rates of our keypoint detector are compared with the ones obtained from the most popular keypoint detectors. Concretely, we have compared our keypoint detector based on HOGs with the cited implementations of Harris, Harris-Laplacian, Hessian-Laplacian (http://www.robots.ox.ac.uk/~vgg/research/affine/detectors.html#binaries), FAST (http://www.edwardrosten.com/work/fast.html), SURF (http://www.vision.ee.ethz.ch/~surf/), and SIFT (http://www.vlfeat.org/~vedaldi/code/sift.html). Results using SUSAN detector are significantly worse than all others methods, and we have omitted them from this section. For a fair evaluation, there should not be a significant discrepancy among the amount of keypoints detected by the different algorithms. For this purpose, we have fixed the parameters of all algorithms to give a constant amount of 100 keypoints. Therefore, there is no need to raise the issue about considering absolute or relative repeatability. Furthermore, the parameters of all keypoint detectors have been tuned to achieve the highest repeatability for every tested set of transformed images. Our algorithm works with unsigned gradient, 180 bin histograms, and 21x21 pixels cells. Repeatability for all methods has been computed using an $\epsilon = 4$.

We show the results for the synthetic image (fig. 1a), the block image (fig. 4a) and the house image (fig. 4b). All images have progressively been transformed with rotations (fig. 4c) and projective transforms (fig. 4d).

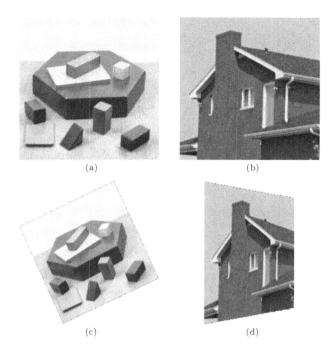

Fig. 4. (a) Blocks image, (b) House image, (c) Block image rotated 25°, and (d) house image projected 50°

Table 1. Mean repeatability (%) for image rotation and projection

	Rotation			Projection		
Detector	Synthetic	Blocks	House	Synthetic	Blocks	House
Harris	97.6	96.3	97.5	80.8	82.8	85.2
Har-Lapl	95.4	87.7	88.8	75.9	76.0	76.0
Hes-Lapl	91.2	92.7	89.5	74.1	79.9	76.9
FAST	98.8	85.2	88.1	85.1	84.4	81.8
SURF	45.4	58.8	63.3	58.6	67.6	61.6
SIFT	83.4	82.1	85.0	62.9	67.7	68.4
HOG	100	93.7	97.8	89.4	88.8	89.7

Figure 5 shows the repeatability score of all mentioned keypoint detectors for rotations of the image from 0° to 90°. The repeatability obtained using HOGs is higher than most of the other methods for most of the rotation angles. Table 1 shows that the mean repeatability of our method is higher than all others methods. Figure 6 shows the repeatability of all methods for projective transformations of the image under different viewpoints from 0° to 70°. Table 1 shows that our method based in HOGs, obtains the highest mean repeatability too.

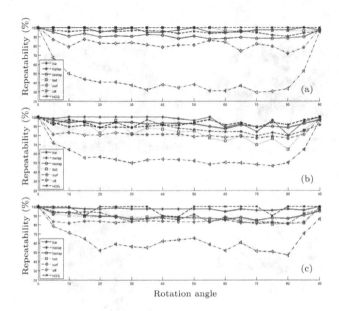

Fig. 5. Repeatability for (a)synthetic, (b)blocks and (c)house images. Images are rotated from 0° to 90°.

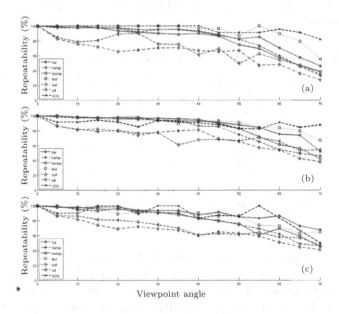

Fig. 6. Repeatability for (a)synthetic, (b)blocks and (c)house images. Images are transformed with a projection from 0° to 70°.

6   Conclusions and Further Work

In this paper we have introduced the use of histograms of oriented gradients in the keypoint detection stage. HOGs have been widely used as keypoint descriptors, but we have shown that their use gives encouraging results for keypoint detection too. The limitations of existing solutions make the keypoint detection too dependent on image variations, and their localization inaccurate. The method presented has shown to be more robust to rotation and viewpoint changes than the other keypoint detectors tested.

The unimodality DIP test of the histograms makes an assessment of the saliency of all possible keypoints. From the unimodality image obtained, we find the local maxima having the highest extinction values to select the most robust keypoints among all singular points.

So far, we have considered only fixed-size cells to compute the histograms, however, cell size has a determinant effect on the image scale where keypoints are detected. In this work we have not considered working at different scales. The research for the extension of this method to the scale space is currently in process.

References

1. Lowe, D.G.: Distinctive image features from scale-invariant keypoints. Int. J. Comput. Vision 60, 91–110 (2004)
2. Ke, Y., Sukthankar, R.: PCA-SIFT: A More Distinctive Representation for Local Image Descriptors. In: Proceedings of IEEE Conference on Computer Vision and Pattern Recognition, vol. 2, pp. 506–513 (2004)
3. Bay, H.: From Wide-baseline Point and Line Correspondences to 3D. PhD thesis, Swiss Federal Institute of Technology (2006)
4. Mikolajczyk, K., Schmid, C.: A performance evaluation of local descriptors. IEEE Transactions on Pattern Analysis and Machine Intelligence 27, 1615–1630 (2005)
5. Dalal, N., Triggs, B.: Histograms of oriented gradients for human detection. In: IEEE Computer Society Conference on Computer Vision and Pattern Recognition, CVPR 2005, vol. 1, pp. 886–893 (2005)
6. Felzenszwalb, P., Girshick, R., McAllester, D., Ramanan, D.: Object detection with discriminatively trained part-based models. IEEE Transactions on Pattern Analysis and Machine Intelligence 32, 1627–1645 (2010)
7. Moravec, H.P.: Obstacle avoidance and navigation in the real world by a seeing robot rover. PhD thesis, Stanford, CA, USA, AAI8024717 (1980)
8. Beaudet, P.R.: Rotationally invariant image operators. In: Proceedings of the International Joint Conference on Pattern Recognition, vol. 579, pp. 579–583 (1978)
9. Harris, C., Stephens, M.I.: A combined corner and edge detector, Manchester, UK, vol. 15, pp. 147–151 (1988)
10. Smith, S.M., Brady, J.M.: SUSAN - a new approach to low level image processing. Int. Journal of Computer Vision 23, 45–78 (1997)
11. Mikolajczyk, K., Schmid, C.: Indexing based on scale invariant interest points. In: Proceedings of Eighth IEEE International Conference on Computer Vision, ICCV 2001, vol. 1, pp. 525–531 (2001)

12. Mikolajczyk, K., Schmid, C.: Scale and Affine Invariant Interest Point Detectors. International Journal of Computer Vision 1, 63–86 (2004)
13. Rosten, E., Porter, R., Drummond, T.: Faster and better: A machine learning approach to corner detection. IEEE Trans. Pattern Analysis and Machine Intelligence 32, 105–119 (2010)
14. Lowe, D.G.: Object Recognition from Local Scale-Invariant Features. In: IEEE International Conference on Computer Vision, vol. 2, pp. 1150–1157 (1999)
15. Schmid, C., Mohr, R., Bauckhage, C.: Evaluation of Interest Point Detectors. International Journal of Computer Vision 37, 151–172 (2000)
16. Porikli, F.: Integral histogram: A fast way to extract histograms in cartesian spaces. In: Proc. IEEE Conf. on Computer Vision and Pattern Recognition, pp. 829–836 (2005)
17. Hartigan, J.A., Hartigan, P.M.: The dip test of unimodality. The Annals of Statistics 13, 70–84 (1985)
18. Hellwig, B., Hengstler, J., Schmidt, M., Gehrmann, M., Schormann, W., Rahnenfuhrer, J.: Comparison of scores for bimodality of gene expression distributions and genome-wide evaluation of the prognostic relevance of high-scoring genes. BMC Bioinformatics 11, 276 (2010)
19. DeCarlo, L.T.: On the meaning and use of kurtosis. Psychological Methods 2, 292–307 (1997)
20. Wang, J., Wen, S., Symmans, W.F., Pusztai, L., Coombes, K.R.: The bimodality index: a criterion for discovering and ranking bimodal signatures from cancer gene expression profiling data. Cancer Informatics 7, 199–216 (2009)
21. Wolfe, J.H.: Pattern Clustering by Multivariate Mixture Analysis. Multivariate Behavioral Research 5, 329–350 (1970)
22. Hartigan, P.M.: Algorithm as 217: Computation of the dip statistic to test for unimodality. Journal of the Royal Statistical Society. Series C (Applied Statistics) 34, 320–325 (1985)
23. Vachier, C., Meyer, F.: Extinction value: a new measurement of persistence. In: IEEE Workshop on Nonlinear Signal and Image Processing, vol. 1, pp. 254–257 (1995)
24. Schmid, C., Mohr, R., Bauckhage, C.: Comparing and evaluating interest points. In: Sixth International Conference on Computer Vision, pp. 230–235 (1998)

Non-rigid and Partial 3D Model Retrieval Using Hybrid Shape Descriptor and Meta Similarity

Bo Li[1], Afzal Godil[1], and Henry Johan[2]

[1] National Institute of Standards and Technology, Gaithersburg, USA
{bo.li,afzal.godil}@nist.gov
[2] School of Computer Engineering, Nanyang Technological University, Singapore
henryjohan@ntu.edu.sg

Abstract. Non-rigid and partial 3D model retrieval are two significant and challenging research directions in the field of 3D model retrieval. Little work has been done in proposing a hybrid shape descriptor that works for both retrieval scenarios, let alone the integration of the component features of the hybrid shape descriptor in an adaptive way. In this paper, we propose a hybrid shape descriptor that integrates both geodesic distance-based global features and curvature-based local features. We also develop an adaptive algorithm to generate meta similarity resulting from different component features of the hybrid shape descriptor based on Particle Swarm Optimization. Experimental results demonstrate the effectiveness and advantages of our framework. It is general and can be applied to similar approaches that integrate more features for the development of a single algorithm for both non-rigid and partial 3D model retrieval.

1 Introduction

Non-rigid 3D model retrieval is a challenging research direction for the community of 3D model retrieval. Compared to generic 3D model retrieval, partial similarity 3D model retrieval is also more difficult and much less studied. Geodesic distance-based global features have intrinsic advantages in characterizing non-rigid 3D models and also have shown their superiority in recognizing deformable models, which has been demonstrated by Smeets et al. [1] [2]. On the other hand, employing local features and Bag-of-Words [3] framework has demonstrated its apparent advantages in dealing with partial similarity retrieval, such as [4] [5]. Curvature is an important local feature and it is the basis of several other important local features, such as Shape Index [6] and Curvedness [6]. Motivated by this, our target is to utilize both geodesic distance-based global features and curvature-based local features together with the Bag-of-Words framework to develop a 3D shape retrieval algorithm that can be used for both non-rigid and partial similarity retrieval. Geodesic distance-based and curvature-based features show different properties and retrieval performances in recognizing non-rigid or partial 3D model retrieval. To adaptively combine these two features, a meta similarity based on Particle Swarm Optimization (PSO) [7] has been proposed to fuse their distance matrices. This framework is general and can be extended to integrate different or more features to develop other similar unified retrieval algorithms for both non-rigid and partial 3D model retrieval.

G. Bebis et al. (Eds.): ISVC 2012, Part I, LNCS 7431, pp. 199–209, 2012.
© Springer-Verlag Berlin Heidelberg 2012

The paper is organized as follows. We briefly discuss the related work in Section 2. Section 3 introduces the hybrid 3D shape descriptor. Section 4 presents our 3D model retrieval algorithm, together with the method of weight assignment for the meta similarity based on Particle Swarm Optimization. We give in detail the experiments in Section 5 and conclude the paper and list the future work in Section 6.

2   Related Work

During the past few years, geodesic distance-based, and local feature together with Bag-of-Words framework based approaches have received much attention, especially in dealing with the non-rigid and partial 3D model retrieval. Combining and integrating heterogeneous features is also an important issue if we employ a hybrid shape descriptor comprising several features. We give a brief review for these four topics as follows.

Geodesic Distance-Based Descriptors. Geodesic distance is an inelastic deformation invariant distance metric, thus popular for the analysis and recognition of non-rigid objects. Typically, the extracted geodesic distance-based feature for 3D is a geodesic distance matrix (GDM) measuring the distances among a set of points sampled on the surface of a 3D object. To deal with deformable 3D model retrieval, Smeets et al. [1] proposed a modal representation method based on the Singular Value Decomposition (SVD) of the GDM of a 3D model. They utilized several largest eigenvalues of a GDM as the shape descriptor. In SHREC 2011 Non-rigid watertight shape retrieval track [2], Smeets et al. further proposed a method by combing GDM and another method called Scale Invariant Feature Transform (SIFT) for meshes (meshSIFT) and they achieved the best retrieval performance among the nine participants.

Local Shape Descriptors. Paul et al. [8] presented a comparative evaluation of several local shape descriptors. Koenderink and Doorn [6] proposed a curvature-based local feature named Shape Index which measures the local topological/convexity geometry, such as ridge, saddle, cup and cap and another local feature called Curvedness which measures the amount of curvature. 3D shape spectrum [9] based on Shape Index distribution was also proposed as the MPEG 3D shape feature standard.

Bag-of-Words Framework. Recently, the Bag-of-Words (BoW) framework has been successfully applied into 3D model retrieval. It has demonstrated successful applications in either view-based (e.g. [10]) or geometry-based (e.g. [4], [5]) 3D model retrieval and apparent advantages in partial similarity 3D model retrieval (e.g. [4], [5]), as well. To reduce the computational cost for distance computation, Ohbuchi et al. [10] encoded the SIFT features of a set of depth views of a 3D model into a histogram by utilizing the BoW approach. Toldo et al. [4] extended the BoW framework from 2D to 3D to represent 3D components. 3D subparts resulting from segmentation are clustered to define a 3D vocabulary comparable to the 2D codewords. Lavoué [5] applied the BoW framework to the Laplace-Beltrami spectrum features of a set of uniformly sampled points on the surface of a 3D model by projecting the geometry onto the eigenvectors of the Laplace-Beltrami operator and also achieved superior partial retrieval performance.

Meta Similarity. Employing several features together in 3D shape retrieval needs a solution of integrating them properly to make them compliment each other to achieve the optimal performance. In the field of 3D model retrieval, compared to new shape descriptors, this topic has received less attention and is also less studied, let alone the adaptive approaches of weight assignment to generate the meta similarity. We can merge several feature vectors directly or merge the distances resulting from different features, as well. Akbar et al. [11] combined features extracted from surface and volume by assigning the weights based on the properties of the two features and they tested on both merging schemes. Unfortunately, the retrieval performance improvement is not apparent. Daras et al. [12] investigated several factors that affect retrieval performance, such as feature selection, dissimilarity metric, feature combination and weight optimization and they suggested that more focus should be given to the efficient combination of low-level descriptors rather than the investigation of the optimal 3D shape descriptor.

3   Hybrid 3D Shape Descriptor

In this section, to represent a 3D model we propose a hybrid shape descriptor composed of a curvature-based local feature vector V_C proposed by us and a geodesic-based global feature vector V_G, described as follows.

3.1   Curvature-Based Local Feature Vector: V_C

Extracting local features are important for partial similarity 3D model retrieval. First, we propose a curvature-based combined local shape descriptor for each vertex of a 3D model and after that we apply the Bag-of-Words framework to generate the local shape descriptor distribution as our proposed local feature vector V_C. To extract the local shape descriptor, we need to define its two basic components: local support region and local features. We regard the adjacent vertices of a vertex as its local support region and consider the following first three curvature-based local features.

(1) Curvature Index Feature. Curvature is an important feature to characterize the local geometry. Based on curvature, Koenderink and Doorn [6] proposed Shape Index and Curvedness. Curvature Index [8] further maps Curvedness values into a reasonable range using a log function. For a vertex p, its Curvature Index CI is computed as follows,

$$CI = \frac{2}{\pi} log(\sqrt{\frac{K_1^2 + K_2^2}{2}}) \tag{1}$$

where K_1 and K_2 are the two principal curvatures in the x and y directions respectively at the point of vertex p.

(2) Curvature Index Deviation Feature. To measure the tendency of the Curvature Index change in a local support region of a vertex, we compute the standard deviation Curvature Index difference of the adjacent vertices of the vertex p,

$$\delta CI = \sqrt{\frac{\sum_{i=1}^{n} (CI_i - \widetilde{CI})}{n}} \tag{2}$$

where CI_1, CI_2,...,CI_n are the Curvature Index values of the adjacent vertices of p and $\widetilde{CI}$ is the mean Curvature Index of all the adjacent vertices.

(3) Shape Index Feature. Shape Index [6] is a feature that has been applied into generic 3D shape retrieval. Here, we utilize it within the Bag-of-Words framework for non-rigid and partial 3D model retrieval. Its definition is as follows,

$$SI = \frac{2}{\pi} \arctan\left(\frac{K_1 + K_2}{|K_1 - K_2|}\right) \tag{3}$$

where K_1 and K_2 are the two principal curvatures in the x and y directions respectively at the point of vertex p. $SI \in [-1,1]$.

(4) Combined Local Shape Descriptor. The three local features described above depict the local properties in different aspects. To more comprehensively measure the local information, a combined local shape descriptor F comprising the above three features is devised,

$$F = (CI, \delta CI, SI) \tag{4}$$

(5) Local Feature Vector Generation: Bag-of-Words. We regard the combined local shape descriptor distribution of all the vertices of a 3D model, with respect to a set of centers, as its local feature vector V_C. Based on the Bag-of-Words framework, the local feature vector generation process includes the following two steps: 1) **Codebook generation**. We cluster the combined local shape descriptors of the vertices of all the 3D models in a 3D dataset into a set of class centers (codewords) $O_1, O_2, \ldots, O_{N_C}$ based on K-means algorithm, where N_C is the number of codewords. In our experiments, L2 distance metric, N_C=500 cluster centers and 100 maximum clustering iteration number are experimentally determined. 2) **Local feature vector formulation**. Based on the generated codebook (cluster centers), we count the distribution V_C of the local shape descriptors of all the vertices of a 3D model with respect to the codewords in terms of maximum similarity,

$$V_C = (h_1, h_2, \cdots, h_{N_C}), \tag{5}$$

where h_i is the percentage of the local shape descriptors whose closest codeword is O_i. To find the closest codeword, Canberra distance metric [13] is utilized to measure the difference between two combined local shape descriptors F_i and F_j: $d_F = \frac{1}{n}\sum_{l=1}^{n} \frac{|F_i(l) - F_j(l)|}{|F_i(l) + F_j(l)|}$, where n is the dimension of F_i and F_j, $d_F \in [0,1]$.

3.2   Geodesic Distance-Based Global Feature Vector: V_G

For non-rigid 3D model retrieval, by utilizing the eigenvalues of global geodesic distance matrix (GDM), Smeets et al. [1] [2] have achieved outstanding retrieval performance. Global GDM considers the geodesic distances among all the sample points on the surface of a 3D model to form a 2D square distance matrix. The eigenvalues of the GDM is comparable to the spectrum of a 3D shape, which shows superior performance when dealing with non-rigid 3D model retrieval. Hybrid approaches by combining global and local features like [14] have been verified to be an effective way to develop a more comprehensive shape descriptor to further improve the retrieval performance. Considering this, we also compute a global geodesic distance matrix-based feature for a 3D model, especially for non-rigid 3D model retrieval.

(1) 3D Model Simplification. To reduce computational cost for geodesic distance-based feature extraction, we simplify each model by adopting the mesh simplification method proposed by Garland and Heckbert [15]. It iteratively contracts vertices pairs under the control of quadric surface error. It is efficient and preserves the most important features. In experiments, we simplify the models to make they contain the same number (e.g. 1000 in our experiments) of vertices.

(2) Geodesic-Based Global Feature Vector Generation. We first compute the geodesic distances based on the method in [16] among all the vertices of a simplified model to form a geodesic distance matrix GDM. Then we decompose the GDM based on Singular Value Decomposition (SVD) and keep the first largest k (e.g. 50 in our experiments) eigenvalues as the global feature vector V_G,

$$V_G = (e_1, e_2, \cdots, e_k) \tag{6}$$

where k is the threshold number of eigenvalues that we are interested in. Similarly, Canberra distance (Section 3.1) is used to measure the distance between two V_G.

4   Non-rigid and Partial 3D Model Retrieval Algorithm Based on a Hybrid Shape Descriptor and Meta Similarity

4.1   Retrieval Algorithm

Given a query 3D model and a target 3D model database, we retrieve relevant models from the target database. Our 3D model algorithm is based on the hybrid shape descriptor presented in Section 3. The complete retrieval algorithm is as follows.

(1) Curvature-based local feature vector V_C and local feature distance matrix M_C computation. For each query and target 3D model, we extract its curvature-based local feature vector V_C as described in Section 3.1. It is very efficient, so we consider all the available vertices and use the original models directly. After that, we compute the Canberra distance (Section 3.1) between the local feature vectors of a query model and a target model to form the curvature-based local feature distance matrix M_C.

(2) Geodesic distance-based global feature vector V_G and global feature distance matrix M_G computation. Based on the algorithms presented in Section 3.2, we simplify each query or target model to make it has 1000 vertices and keep the largest 50 eigenvalues as its global feature vector V_G. Similarly, the Canberra distance between a query and a target model's global feature vectors V_G is computed to form the geodesic distance-based global feature distance matrix M_G.

(3) Meta distance matrix generation and ranking. We adaptively find the weights w_C and w_G for the distance matrices M_C and M_G respectively to generate a meta distance matrix M based on the approach in Section 4.2.

$$M = w_C * M_C + w_G * M_G \tag{7}$$

where w_C and w_G are in the region of $[0,1]$. Finally, we sort all the models in the database in ascending order based on their distances and output the retrieval lists accordingly. The two weights w_C and w_G are needed to be computed only once for each target database which is always available in order to perform a retrieval. If the query database is available, we use it directly as queries to compute the weight values, otherwise, we use the target models as queries. In our experiments, we use the target models directly in Section 5.1 and use the query database in Section 5.2.

4.2   Meta Similarity by Particle Swarm Optimization

The simplest method to find the optimal weights for different features is by performing a brute-force search. We can uniformly sample the values by adopting a fixed step. The drawback of the brute-force search is the high computational cost. For example, in order to find a result with an accuracy of $\Delta\delta$ (e.g. 0.01) for N (e.g. 3) weights, we have to sample at least at a step of $\Delta\delta$ (e.g. 0.01), which means $(\frac{1}{\Delta\delta})^{N-1}$ (e.g. 10000) combinations. As such, the brute-force search is not the ideal method for finding the optimal weights.

To efficiently find the optimal weights, we develop a weight assignment method based on Particle Swarm Optimization (PSO) [7] which is a swarm intelligence optimization technique by imitating the behavior of a flock of birds searching for a piece of food in a region. Each bird learns from its neighboring birds and update itself based on the position of the bird nearest to the food. Our PSO-based weight assignment for the meta distance matrix generation is as follows.

(1) PSO Initialization. We initialize the number N_P and positions of a set of search particles $\{x = (w_C, w_G)\}$ and then compute the private best for each particle and current global best based on all the private bests. In experiments, we uniformly distribute the search particles within its search region of $\{[0,1],[0,1]\}$. We regard the $\lfloor N_P/3 \rfloor$ nearest neighbors of a search particle as its neighborhood, based on which we compute its private best. Finally, we also set the maximum search iterations N_t.

(2) Update Particles. We update the position of each particle by adopting a similar strategy as [17],

$$x(i+1) = x(i) + s \cdot v(i), \tag{8}$$

$$v(i+1) = \omega * v(i) + c_1 \cdot r_1 \cdot (x_p(i) - x(i)) + c_2 \cdot r_2 \cdot (x_g(i) - x(i)). \tag{9}$$

$x(i)$ and $v(i)$ are the position and velocity of a particle; the velocity update step s is inversely proportional to the current iteration number i: $s = \frac{N_t - i}{N_t} + c$, where c is a constant variable and in experiments we choose c to be 0.5. r_1 and r_2 are random variables between 0 and 1; x_p and x_g are the particle positions of private and global bests. c_1 and c_2 are non-negative constants, typically $c_1=c_2=2$ [7]. The inertia-weight ω is a tradeoff between the global and local search abilities. Bigger ω indicates more powerful global search ability and less dependency on the initial locations of the search particles, while smaller ω means finer search within a local area. Similar as [17], we linearly decrease ω from 1.4 to 0 according to the iteration number i: $\omega = \frac{\omega_{min}-\omega_{max}}{N_t} \cdot i + \omega_{max}$, where $\omega_{max}=1.4$ and $\omega_{min}=0$. The new position $x(i+1)$ may be out of the search area, as such we clamp it by subtracting (if larger than 1) or adding (if smaller than 0) 1 .

(3) Search Evaluation. Based on the new position of each particle, we assign the corresponding weights w_C and w_G and compute the meta distance matrix based on Equation (7) and thus the corresponding retrieval performance metrics, such as First Tier (FT) and mean Normalized Discounted Cumulative Gain (NDCG) [18], and regard them as PSO fitness value to evaluate the weight assignment result. After that, we update its private best as well as the global best based on all the private bests.

(4) Result Verification. If the maximum iteration number N_t has been reached, we stop and output the position of the current global best as the optimal weight assignment result and also output the corresponding optimal meta distance matrix M and retrieval performance metrics; otherwise, go to step (2) to continue the search. The complexity of our POS-based weight assignment algorithm is $O(N_P + N_P \cdot N_t)$.

5   Experiments

To investigate the performance of our algorithm in terms of non-rigid and partial 3D model retrieval, we choose to use the following two benchmarks.

(1) **SHREC'11-Non-rigid:** the benchmark for the SHREC 2011 non-rigid 3D watertight models retrieval track [2]. It contains 600 watertight and deformable models, classified into 30 classes, each with 20 models.

(2) **SHREC'07-Partial:** the benchmark used in the SHREC 2007 partial matching track [19]. The target dataset has 400 watertight models, divided into 20 classes, each with 20 models. The query dataset comprises 30 models by combining the parts of two or more models of the target database.

To comprehensively evaluate the non-rigid 3D model retrieval results, we employ six metrics [18] including Precision-Recall (PR), Nearest Neighbor (NN), First Tier (FT), Second Tier (ST), Discounted Cumulative Gain (DCG) and Average Precision (AP). We use the Normalized Discounted Cumulative Gain (NDCG) [18] metric to evaluate the performance of partial retrieval results.

5.1   Non-rigid 3D Model Retrieval

Geodesic distance is invariant to model deformation, which makes it has advantages in non-rigid 3D model retrieval. This means that we should increase its weight during the retrieval. While, adding curvature-based features will probably further improve the retrieval performance. However, it is non-trivial to find an optimal weight assignment

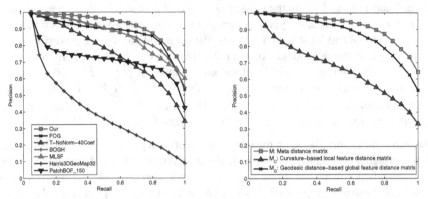

(a) Our retrieval algorithm and other methods   (b) Hybrid shape descriptor and its components

Fig. 1. Precision-Recall performance comparison on the SHREC'11-Non-rigid benchmark

for these two features, let alone for more features. Thus, PSO-based algorithm is utilized to train the weights. We set N_P=10, N_t=20, and select First Tier as the PSO fitness value to evaluate search results. Based on the algorithm in Section 4.2, we find the optimal weights values: w_C=0.349036, w_G=0.650965. Optimal First Tier value is 0.864999.

Table 1. Other performance metrics comparison on the SHREC'11-Non-rigid benchmark

Methods	NN	FT	ST	DCG	AP
Our	**99.7**	**86.5**	**93.1**	**97.1**	**93.4**
FOG	96.8	81.7	90.3	94.4	89.5
T-NoNorm-40Coef	95.5	67.2	80.3	89.7	78.1
BOGH	99.3	81.1	88.4	94.9	89.1
MLSF	98.7	80.9	87.9	94.8	88.2
Harris3DGeoMap32	56.2	32.5	46.6	65.4	43.2
PatchBOF_150	74.8	64.2	83.3	83.7	74.1
M_C	83.7	58.8	77.2	83.7	69.7
M_G	99.3	81.4	88.1	95.3	89.4

We compare with the approaches in the SHREC 2011 Non-rigid track which mainly extract geodesic distance-based features, such as FOG, BOGH and Harris3DGeoMap32; or adopt the Bag-of-Words framework and some other geometric features, like T-No-Norm-40Coef, MLSF and PatchBOF. We also compare with the performances of the two component features of our hybrid shape descriptor, that is, comparing the performances of meta distance matrix M, curvature-based local feature distance matrix M_C, and geodesic distance-based global feature distance matrix M_G. Figure 1 compares their Precision-Recall performances while Table 1 lists their other performance metrics.

As can be seen from Figure 1 (a) and Table 1, our hybrid shape descriptor and meta similarity-based retrieval algorithm outperforms all the six participating approaches which use the features and the Bag-of-Words framework that fall in the same category as our approach. Based on the results shown in Figure 1 (b) and Table 1, we also find that our approach apparently improves the retrieval performances, in terms of all the six metrics, for non-rigid 3D model retrieval.

5.2   Partial Similarity 3D Model Retrieval

Unlike non-rigid model retrieval, in this case curvature-based local features will contribute more for partial 3D model retrieval. Similarly, we optimize their weights based on PSO after computing the curvature-based feature distance matrix M_C and geodesic distance-based feature distance matrix M_G. We set $N_P=10$ and $N_t=20$. Since NDCG is used to evaluate the partial retrieval performance, we use the mean NDCG over all the 400 models to evaluate search results. The optimal weights values are as follows: $w_C=0.397384$, $w_G=0.602614$, while the optimal mean NDCG is 0.613296. Similar as Section 5.1, the NDCG performance comparisons with the participants in the SHREC 2007 partial matching track [19] as well as other approaches mentioned in [18], and the hybrid shape descriptor's components are shown in Figure 2 (a) and (b), respectively. Based on the comparison results in Figure 2, we can draw a similar conclusion as the non-rigid retrieval experiments in Section 5.1 for the partial similarity retrieval.

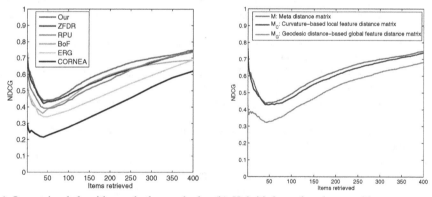

(a) Our retrieval algorithm and other methods   (b) Hybrid shape descriptor and its components

Fig. 2. NDCG performance comparison on the SHREC'07-Partial benchmark

6   Conclusions and Future Work

Non-rigid and partial 3D model retrieval are two important and challenging research directions in the field of 3D model retrieval. While different approaches based on either geodesic distance or some local features have been proposed to deal with either of the above two retrieval problems, little work has been done in developing a hybrid shape descriptor that works for both cases, especially in an adaptive way. We have found that geodesic distance-based global features and curvature-based local features have advantages in non-rigid and partial 3D model retrieval, respectively. To utilize both features and make them compliment each other, we develop a hybrid shape descriptor comprising these two types of features and adaptively combine their feature distance matrices to form a meta distance matrix based on Particle Swarm Optimization.

Experimental results based on a latest non-rigid 3D model retrieval benchmark and a partial 3D model retrieval dataset as well, demonstrate the effectiveness and advantages of our framework. It applies to two different, important and difficult retrieval scenarios and improves the retrieval performances based on an adaptive integration strategy. The idea is general and it can be applied to integrate three or more features for developing a single algorithm for both non-rigid and partial 3D model retrieval, which is also among our future work. Another interesting work is to test the performances of concatenating our global and local feature vectors directly to form a hybrid feature vector by assigning appropriate weights based on our Particle Swarm Optimization algorithm and then perform a comparative evaluation with the retrieval algorithm proposed in the paper.

References

1. Smeets, D., Fabry, T., Hermans, J., Vandermeulen, D., Suetens, P.: Isometric Deformation Modelling for Object Recognition. In: Jiang, X., Petkov, N. (eds.) CAIP 2009. LNCS, vol. 5702, pp. 757–765. Springer, Heidelberg (2009)
2. Lian, Z., Godil, A., Bustos, B., Daoudi, M., Hermans, J., Kawamura, S., Kurita, Y., Lavoué, G., Nguyen, H.V., Ohbuchi, R., Ohkita, Y., Ohishi, Y., Reuter, F.P.M., Sipiran, I., Smeets, D., Suetens, P., Tabia, H., Vandermeulen, D.: SHREC 2011 track: Shape retrieval on non-rigid 3D watertight meshes. In: 3DOR, pp. 79–88 (2011)
3. Li, F.F., Perona, P.: A Bayesian hierarchical model for learning natural scene categories. In: CVPR, vol. 2, pp. 524–531 (2005)
4. Toldo, R., Castellani, U., Fusiello, A.: Visual vocabulary signature for 3D object retrieval and partial matching. In: 3DOR, pp. 21–28 (2009)
5. Lavoué, G.: Bag of words and local spectral descriptor for 3D partial shape retrieval. In: 3DOR, pp. 41–48 (2011)
6. Koenderink, J.J., van Doorn, A.J.: Surface shape and curvature scales. Image Vision Comput. 10(8), 557–564 (1992)
7. Eberhart, R.C., Hu, X.: Human tremor analysis using particle swarm optimization. In: Proc. of the Congress on Evolutionary Computation, pp. 1927–1930 (1999)
8. Heider, P., Pierre-Pierre, A., Li, R., Grimm, C.: Local shape descriptors, a survey and evaluation. In: 3DOR, pp. 49–56 (2011)
9. Bober, M.: MPEG-7 visual shape descriptors. IEEE Trans. Circuits Syst. Video Techn. 11(6), 716–719 (2001)
10. Ohbuchi, R., Osada, K., Furuya, T., Banno, T.: Salient local visual features for shape-based 3D model retrieval. In: Shape Modeling International, pp. 93–102 (2008)
11. Akbar, S., Kueng, J., Wagner, R.: Multi-feature based 3D model similarity retrieval. In: 2006 International Conference on Computing Informatics, ICOCI 2006, pp. 1–6 (2006)
12. Daras, P., Axenopoulos, A., Litos, G.: Investigating the effects of multiple factors towards more accurate 3-D object retrieval. IEEE Transactions on Multimedia 14(2), 374–388 (2012)
13. Laga, H., Nakajima, M.: Supervised Learning of Similarity Measures for Content-Based 3D Model Retrieval. In: Tokunaga, T., Ortega, A. (eds.) LKR 2008. LNCS (LNAI), vol. 4938, pp. 210–225. Springer, Heidelberg (2008)

14. Wu, H.-Y., Zha, H., Luo, T., Wang, X., Ma, S.: Global and local isometry-invariant descriptor for 3D shape comparison and partial matching. In: CVPR, pp. 438–445 (2010)
15. Garland, M., Heckbert, P.S.: Surface simplification using quadric error metrics. In: SIGGRAPH, pp. 209–216 (1997)
16. Kalogerakis, E., Hertzmann, A., Singh, K.: Learning 3D mesh segmentation and labeling. ACM Trans. Graph. 29(4), 102:1–102:12 (2010)
17. Shi, Y., Eberhart, R.: A modified particle swarm optimizer. In: Proc. of IEEE International Conference on Evolutionary Computation, ICEC, pp. 69–73 (1998)
18. Li, B., Johan, H.: 3D model retrieval using hybrid features and class information. Multimedia Tools and Applications, 1–26 (2011) (Online First version)
19. Veltkamp, R.C., ter Haar, F.B.: SHREC 2007 3D Retrieval Contest. Technical Report UU-CS-2007-015, Department of Information and Computing Sciences, Utrecht University (2007)

Large Scale Sketch Based Image Retrieval Using Patch Hashing*

Konstantinos Bozas and Ebroul Izquierdo

School of Electronic Engineering and Computer Science
Queen Mary University of London, Mile End Campus, UK, E1 4NS
{k.bozas,ebroul.izquierdo}@eecs.qmul.ac.uk

Abstract. This paper introduces a hashing based framework that facilitates sketch based image retrieval in large image databases. Instead of exporting a single visual descriptor for every image, an overlapping spatial grid is utilised to generate a pool of patches. We rank similarities between a hand drawn sketch and the natural images in a database through a voting process where near duplicate in terms of shape and structure patches arbitrate for the result. Patch similarity is efficiently estimated with a hashing algorithm. A reverse index structure built on the hashing keys ensures the scalability of our scheme and at the same time allows for real time reranking on query updates. Experiments in a publicly available benchmark dataset demonstrate the superiority of our approach.

1 Introduction

The exponential growth of publicly available digital media during the last two decades highlighted the need for efficient and user friendly techniques to index and retrieve images and videos from large scale multimedia databases. Despite the considerable progress of content-based image retrieval (CBIR) [1], where the goal is to return images similar to a user provided image query, most of the multimedia searches are text-based (e.g. Google Images, YouTube). The latter requires user intervention to tag all the available data and has two main drawbacks: (i) it is time consuming and (ii) most importantly user subjective. It is a common belief that images cannot be succinctly communicated based on words; humans would probably use different words to describe a scene based on their cultural background and experience. It follows from the above that in specific scenarios searching images by text query will return frustrating and dubious results to the user. Sketch based image retrieval (SBIR) emerged as a more expressive and interactive way to perform image search; here the query is formed as a hand-drawn sketch of the imaginary picture. Obviously, a shaded rendition of the query requires great artistic skill [2] and most users will refrain from exercising considerable effort and time to draw it. A more intuitive way is

---

* This work is partially supported by EU project CUBRIK under grant agreement FP7 287704.

G. Bebis et al. (Eds.): ISVC 2012, Part I, LNCS 7431, pp. 210–219, 2012.

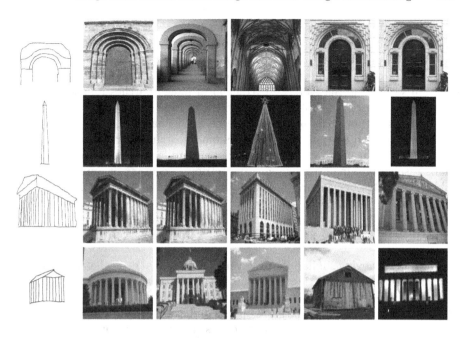

Fig. 1. Top 5 results returned from our system in some queries of [5]

to sketch the main feature lines of a shape or scene, a design choice we follow in this paper, and is supported by recent studies where it has been demonstrated that lines are drawn along contours [3] and line drawings can encode certain shapes almost as well as shaded images [4]. Binary sketches can be easily drawn using the mouse in a personal computer or the touch screen of a modern touch screen device.

Finding similarities between a binary drawing and a database of colored pictures taken under arbitrary conditions consists a challenging problem. Images and binary sketches do not share many common modalities. Images contain rich information in domains such as color and texture, while sketches can be described only by their shape and spatial configuration, therefore traditional CBIR methods relying on texture and color cannot be inherited in SBIR. Furthermore, the vast amount of photos uploaded to social media websites signify the crucial role of *scalability*. We need to be able to search large collections of images in reasonable query times as well as update databases and index files with minimum computational cost.

This paper focuses both on scalability and retrieval quality. We tackle these challenges by retrieving *near duplicate in terms of shape* images to a binary sketch query using a hashing technique. Our approach assumes that users are looking for images spatially consistent with their query, for instance if they draw a sunset scene and the sun has been placed at the top right of the canvas that indicates that images displaying the sun in (approximately) the same spot will be preferred, therefore our system is designed to retrieve near duplicate images to the hand drawn sketches up to small translations. In the core of our retrieval scheme lies min-hash, a set similarity estimation technique originally applied to identify duplicate web pages [6] and later modified for near duplicate image search [7]. Our method differs from [7] where an image is described by a single set of visual words, alternatively we extract local image patches represented with a binary version of the HoG [8] descriptor which allows the utilization of the min-hash algorithm. This provides more detailed image description as well as flexibility during query time since we can omit searching for patches that have not been filled during drawing. An index structure facilitates quick queries and online result updating. We compare the retrieval accuracy of our method in a publicly available benchmark [5] and demonstrate state-of-the-art results.

2   Related Work

The idea of retrieving similar photos given a hand-drawn query has first received attention during the nineties. Hirata and Kato [9] performed visual search by estimating block correlation between digitized oil paintings and a rough sketch illustration of a database painting. Lopresti and Tomkins [10] proposed a stroke matching approach to retrieve hand-drawings. They take advantage of the inherently sequential nature of ink creation in the temporal domain, so they segment a digital pen's input into strokes and export a set of basic strokes types from every drawing. This approach was successfully applied to handwritten text recognition where the stroke order is well defined but struggles with pictorial queries due to the arbitrary order of the strokes. Liang, Sun, and Li [11] decompose sketches into basic geometric primitives by using pen speed and curvature and form a topological graph to quantify similarity. Edge map segmentation, crucial for this technique, fails when applied to natural images. Del Bimbo and Pala [12] presented a template matching technique, where the elastic deformation energy spent is employed to derive a measure of similarity between the sketch and the images in the database and to rank images to be displayed. The method was applied to a pool of 100 images and in spite of its success in that particular database it cannot be employed in large scale due to the computational cost of the elastic deformation calculation. Matusiak *et al.* [13] parametrized the hand-drawn sketch curves in the curvature scale space and match them against a database of eight hundred closed curved shapes. This approach fails when images contain non closed curve shapes, therefore fails to describe the majority of user photos since the presence of a perfect non-occluded shape rarely occurs due to background clutter and other sources of noise.

More recent approaches attempt to extract a visual footprint for each image that satisfies two properties, first similar images are mapped to similar feature vectors and secondly the size of footprint must be kept as small as possible without losing its discrimination power so as to accommodate large scale indexing. Towards this direction, an image is usually partitioned in blocks and a small descriptor is computed for each block, the final footprint is acquired by concatenating all the feature vectors to a longer one. Chalechale, Naghdy, and Mertins

[14] performed angular partitioning in the spatial domain of images and used the magnitude of the Fourier coefficients to obtain a shape description which is rotation invariant but suffers from noisy edge artifacts. The first approach that reported the use of a large scale database for evaluation presented from Eitz *et al.* [15], the authors compute the main gradient orientations for each spatial block achieving robustness to background clutter. Following its success in text retrieval, the well-known bag-of-words (BoW) model has been applied to image search [16], each image is represented with a set of visual words obtained by clustering a large pool of feature vectors. Hu *et al.* [17,18] employed a gradient vector field in an attempt to smooth the divergence between images and hand-drawn sketches, and used the BoW model to facilitate the retrieval, however their experiments were conducted in a small database of 320 images. Cao *et al.* [19] achieved efficient indexing of 2 million images and approximated the Orientated Chamfer Distance to measure database similarities with a binary sketch query. Our work is inspired from ShadowDraw [20], where a min-hash based retrieval mechanism is utilized to acquire image patches similar to the user hand-drawn input, yet our approach employs different patch description process and relying on the patch votes infers a ranking on the database images.

3 Min-Hash Algorithm

In this section, a brief description of the min-hash algorithm is presented. Min-hash is a Local Sensitive Hashing [21] based technique that estimates similarity between sets. Assume a set of tokens S of size $|S|$. The set $D_i \subseteq S$ can be represented by a sequence of size $|S|$ where the presence of a token $s \in S$ is indicated by 1 and the absence by 0. The set overlap similarity (or Jaccard similarity) between two sets D_1 and D_2 is defined as the ratio of their intersection and union and is a number between 0 and 1; it is 0 when the two sets are disjoint, 1 when they are equal, and between 0 and 1 otherwise.

$$sim(D_1, D_2) = \frac{|D_1 \cap D_2|}{|D_1 \cup D_2|} \in [0, 1]. \tag{1}$$

Eq. 1 values equally every member of the set. In [7] an extension proposed to assign to each set member a weight according to its importance. Min-hash approximates the set similarity by creating a collection of hash functions $h_i : S \to \mathbf{R}$ mapping each element of S to a real number. Each hash function h_i infers an ordering on the members of D_i. Min-hash is defined as the smallest element of D_i under ordering induced by h_i and has the property that the probability of two sets having the same min-hash value equals to their Jaccard similarity.

$$P\{h_i(D_1) = h_i(D_2)\} = \frac{|D_1 \cap D_2|}{|D_1 \cup D_2|} = sim(D_1, D_2). \tag{2}$$

An unbiased similarity estimation is obtained by computing for each set D_i multiple independent min-hash functions h_i and counting the occurrences of identical min-hash values for the two sets. To reduce the possibility of false positive retrievals hash functions are organized in s-tuples (often called sketches of min-hash values). Two sets are characterized similar if they share at least one sketch, this follows from the fact that similar sets share many min-hash values while dissimilar share only few. The sketches can be computed fairly fast (linear in the size of D_i) and given two sketches the resemblance of the corresponding sets can be computed in linear time in the size of the sketches. Moreover, each set is represented by k s-tuples to increase the recall of the retrieval. The probability of collision under this scheme is:

$$P\{h(D_1) = h(D_2)\} = 1 - (1 - sim(D_1, D_2)^s)^k . \tag{3}$$

Min-hash has been successfully applied to text [6] and image [7] domains to detect near duplicate instances of a given set. Section 4 describes how we adopt min-hash to sketch based image retrieval.

4   Near Duplicate Patch Retrieval with Hashing

4.1   Method Overview

We retrieve similar images to a sketch query based on a local patch technique. Every image in the database is divided into overlapping patches and for each patch a sequence of min-hash values is extracted. A reverse index is built on the unique min-hash values pointing to the patches containing these values. A sketch query undergoes the same process and for each sketch patch we look into the index to retrieve similar patches. Every index hit contributes a vote to the corresponding image and the final ranking is generated by summing the votes for each image.

Min-hash is employed to estimate the similarity between two patches. Chum et al. [7] proposed to represent an image by an unordered set of visual words acquired by clustering on the feature space. Under this scheme, an image is represented by a single sequence of 0 and 1. Our approach, similarly to [20] adopts a sequence description for every local patch, but instead of utilizing a visual codebook to derive the sequences, we use the non-zero indexes of the binarized patch descriptor. A visual illustration of our method is presented in Fig. 2.

4.2   Preprocessing

Bridging the modality gap between photos and hand-drawn sketches consists the first impediment a SBIR system has to overcome. Ordinarily, this is achieved by extracting edge lines from the images. The well-known Canny algorithm [22] requires in many cases manually tuning of the detection thresholds to return a desired edge map without many erroneous detected edges originating from background clutter. On the other hand, the state-of-the-art edge detector Berkeley Bg operator [23] excels in quality at the expense of computational cost. We found that if we keep the image dimensions reasonably small we can benefit from the

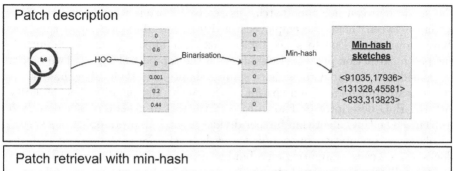

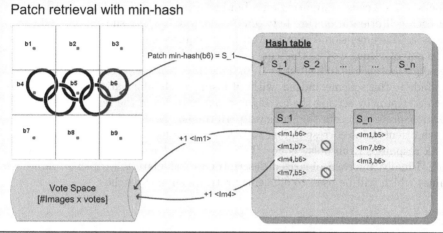

Fig. 2. Patch retrieval framework overview. (Top) Feature extraction for a single patch. The HoG descriptor is computed and then a binarization process is applied to the feature vector. A list of min-hash tuples calculated from the non-zero descriptor indexes is the final patch representation. (Bottom) The patch retrieval mechanism. Every image patch is described individually, for each computed min-hash tuple a look-up in the hash table is performed. We only allow votes originating from neighboring patches.

superior performance of BG detector without burdening much with the computational cost, hence every image is resized to a 200×200 square and edge detection is carried out with the BG detector in less than one second. Weak detected edges are further reduced by thresholding the returned edge map, edgels with edge intensity less than 10 are discarded. Query preprocessing is similar to photos, the edge detection step is substituted with a morphological thinning operation as a means to reduce thick drawn lines to single pixel width followed by downscaling to the same dimensions as photos.

4.3   Feature Extraction

An *overlapping spatial grid* is applied to finely describe the generated edge map and feature vectors are extracted for every patch of the grid. We set the size of spatial grid to 17×17 patches with each patch occupying 40×40 pixels.

The stride between neighbor patches is set to 10 pixels which equals to 75% overlap between them. We employed the Eitz *et al.* database [5] to tune the above parameters. Due to space limitations their impact analysis is omitted. The patch extraction process is visualized in the top part of Fig. 2. Two blocks are considered similar if they share similar shapes, *i.e.* their edges have similar orientation histogram and spatial arrangement. We quantify this similarity with the HoG descriptor [8] known to perform well in general object detection problems. The HoG algorithm further divides a patch in overlapping blocks and calculates an orientation histogram for each block. We configure the HoG to blocks of 2×2 cells, computing an 8-bin histogram of unsigned edge orientations for each cell(orientations are between $0°$ and $180°$). The resulting patch descriptor is a 512-dimensional vector. Min-hash requires each entry to be described by a set of visual words, this can be formed as a binary vector where a word presence/absence is indicated by 1/0. The HoG feature vector can be modified to abide to this scheme, indeed, without losing much information we can binarize the descriptor by setting the $b\%$ highest values to 1 and the rest to 0. The binarization process boosts the retrieval performance as it highlights the strongest patch orientations corresponding to bold continuous contours while eliminating weak responses from noisy edges. The parameter b has been empirically set to 20%. Finally, for each binarized descriptor we calculate k s-tuples of min-hash values which will be used to efficiently retrieve similar patches.

4.4   Index Construction

As discussed in Sec. 4.3, for every patch we compute k sketches of min-hash values. Every value requires 64 bits of memory, hence if we use twenty 2-tuples an image will be described with ~90.3KB of memory, a non-negligible amount especially when the database size grows. Without compression 1 million images require ~90GB of memory, a quantity beyond the memory capacity of a modern computer. The need of an index mechanism is apparent, we take advantage of the sketch collisions that will occur between similar images in a large database and we construct a hash table from the unique min-hash sketches. Every hash table entry can accommodate multiple pairs of type $<image\_id, block\_id>$. Under this scheme, memory is saved by disregarding duplicate sketch values enabling faster search times. Additionally the current patch position in the grid is stored for each entry (*block_id* field), information that can be capitalized during the query process when spatial constraints can be enforced as a means to reject non adjacent patches. Patches containing a small portion of edgels are not taken into account during the index construction.

4.5   Patch Retrieval and Voting

Images similar to a sketch query are returned based on a voting process (Fig. 2 bottom). The pipeline of the query system is as follows: given a binary drawing, features are extracted according to the process illustrated in Sec. 4.3, obviously the edge detection step is omitted. For every patch with adequate number of edgels, k s-tuples are calculated and for each tuple a look-up in the hash table is performed. Retrieved patches located at grid positions above a predefined distance threshold from the current patch center are discarded, the rest vote for their corresponding image. Over-segmentation introduced by the spatial grid leads to multiple votes emanating from one image to the same min-hash tuple. To reduce the bias of this effect an image can cast only one vote for every min-hash tuple extracted from a query patch. The final ranking is generated by counting the votes for each image and sorting them in descending order. The suggested patch based retrieval scheme enhances flexibility since look-ups are taking place only for patches that have been drawn by the user, efficiently reducing query time and facilitating real time result updating when a new stroke is drawn. The query routine can be easily parallelized to enhance scalability even further. Patch queries can be executed independently in different machines and return a vote count for every image. An integration process will then merge all the votes to generate the final ranking.

5   Results

We evaluate our system in the Eitz *et al.* dataset[5] especially designed to measure retrieval performance in a large scale environment. The database consists of 31 user drawn sketch queries outlining objects and sceneries (Fig. 1 depicts some of them). Each sketch query is associated with 40 photos assigned with a value between 1 (similar) and 7 (dissimilar). These 1240 photos are mixed with 100,000 distractor images. A SBIR algorithm must generate a ranking of the database images for each query and retrieve the order of the 40 query related photos, a correlation value between the ground truth and the returned ranking measures the retrieval accuracy. The final benchmark score is the average correlation value across the 31 queries. Eitz *et al.* reported a state-of-the-art score of 0.277 with their tensor descriptor [15] in conjunction with the BoW model. Our approach outperforms [15], we report average correlation value of 0.336. Fig. 1 demonstrates the top 5 returned photos for some of the queries, the reader can verify the spatial coherence between the query and the acquired photos achieved with our patch voting scheme.

Table 1. Results on dataset [5]

Method	Benchmark Score
Eitz *et al.* [5]	0.277
Our method	0.336

We performed the experiments in a desktop machine with 4 cores and 4GB of memory using MATLAB. Twenty 2-tuple min hash values were extracted for each patch. The database outputs ~29.2 million patches, 5.51% of them were empty or with not much information and discarded. The hash table was constructed on the remaining patches outcoming 49,535 unique keys and ~552 million entries. Each entry requires 3 bytes of storage space, therefore the hash table equips ~1.5GB of memory. Hash keys with too many entries were removed from the index achieving two goals, reduction of storage requirements and attenuation of bias caused by very common hash keys (similar to stop words in text retrieval). Building the index takes a considerable amount of time, but needs to be done only once and offline. The code was not parallelized, still a query is executed in less than $2s$. These timings are achieved through the index mechanism which performs a search for a hash key in logarithmic time ($O(log\,n)$) in the size of index.

6 Conclusions

We presented a patch based retrieval technique which can scale to millions of images with the utilization of hashing. Shape and structure information is extracted from a patch with the HoG descriptor and a binarization process further enhances strong continuous contours while facilitating the application of min-hash algorithm. An inverted index created on the unique hash key values allows for sublinear search times and parallelization. Experiments on a large scale database demonstrated the efficiency and scalability of our technique by achieving state-of-the-art results.

Future work includes the optimization of the algorithm in a parallel architecture and a patch description technique that will reduce false positive collisions. Towards this direction a feature descriptor designed to identify and penalize patches with noisy edges could be useful.

References

1. Datta, R., Joshi, D., Li, J., Wang, J.Z.: Image retrieval: Ideas, influences, and trends of the new age. ACM Comput. Surv. 40, 1–60 (2008)
2. Eitz, M., Kristian Hildebrand, T.B., Alexa, M.: A descriptor for large scale image retrieval based on sketched feature lines. In: Eurographics Symposium on Sketch-Based Interfaces and Modeling, pp. 29–38 (2009)
3. Cole, F., Golovinskiy, A., Limpaecher, A., Barros, H.S., Finkelstein, A., Funkhouser, T., Rusinkiewicz, S.: Where do people draw lines? Communications of the ACM 55, 107–115 (2012)
4. Cole, F., Sanik, K., DeCarlo, D., Finkelstein, A., Funkhouser, T., Rusinkiewicz, S., Singh, M.: How well do line drawings depict shape? ACM Transactions on Graphics (Proc. SIGGRAPH) 28 (2009)
5. Eitz, M., Hildebrand, K., Boubekeur, T., Alexa, M.: Sketch-based image retrieval: Benchmark and bag-of-features descriptors. IEEE Transactions on Visualization and Computer Graphics 17, 1624–1636 (2011)
6. Broder, A.Z., Charikar, M., Frieze, A.M., Mitzenmacher, M.: Min-wise independent permutations. Journal of Computer and System Sciences 60, 327–336 (1998)
7. Chum, O., Philbin, J., Zisserman, A.: Near duplicate image detection: min-hash and tf-idf weighting. In: Proceedings of the British Machine Vision Conference (2008)

8. Dalal, N., Triggs, B.: Histograms of oriented gradients for human detection. In: IEEE Computer Society Conference on Computer Vision and Pattern Recognition, CVPR 2005, vol. 1, pp. 886–893 (2005)
9. Hirata, K., Kato, T.: Query by Visual Example. In: Pirotte, A., Delobel, C., Gottlob, G. (eds.) EDBT 1992. LNCS, vol. 580, pp. 56–71. Springer, Heidelberg (1992)
10. Lopresti, D., Tomkins, A.: Computing in the ink domain. In: Yuichiro Anzai, K.O., Mori, H. (eds.) Symbiosis of Human and Artifact - Future Computing and Design for Human-Computer Interaction. Proceedings of the Sixth International Conference on Human-Computer Interaction (HCI International 1995). Advances in Human Factors/Ergonomics, vol. 20, pp. 543–548. Elsevier (1995)
11. Liang, S., Sun, Z., Li, B.: Sketch retrieval based on spatial relations. In: Proc. Int. Computer Graphics, Imaging and Vision: New Trends Conf., pp. 24–29 (2005)
12. Del Bimbo, A., Pala, P.: Visual image retrieval by elastic matching of user sketches. IEEE Transactions on Pattern Analysis and Machine Intelligence 19, 121–132 (1997)
13. Matusiak, S., Daoudi, M., Blu, T., Avaro, O.: Sketch-Based Images Database Retrieval. In: Jajodia, S., Özsu, M.T., Dogac, A. (eds.) MIS 1998. LNCS, vol. 1508, pp. 185–191. Springer, Heidelberg (1998)
14. Chalechale, A., Naghdy, G., Mertins, A.: Sketch-based image matching using angular partitioning. IEEE Transactions on Systems, Man and Cybernetics, Part A: Systems and Humans 35, 28–41 (2005)
15. Eitz, M., Hildebrand, K., Boubekeur, T., Alexa, M.: A descriptor for large scale image retrieval based on sketched feature lines. In: Proceedings of the 6th Eurographics Symposium on Sketch-Based Interfaces and Modeling, SBIM 2009, pp. 29–36. ACM, New York (2009)
16. Sivic, J., Zisserman, A.: Video Google. In: Proceedings of the International Conference on Computer Vision, vol. 2, pp. 1470–1477 (2003)
17. Hu, R., Barnard, M., Collomosse, J.: Gradient field descriptor for sketch based retrieval and localization. In: Proc. 17th IEEE Int. Conf. Image Processing, ICIP, pp. 1025–1028 (2010)
18. Hu, R., Wang, T., Collomosse, J.: A bag-of-regions approach to sketch-based image retrieval. In: 2011 18th IEEE International Conference on Image Processing, ICIP, pp. 3661–3664 (2011)
19. Cao, Y., Wang, C., Zhang, L., Zhang, L.: Edgel index for large-scale sketch-based image search. In: 2011 IEEE Conference on Computer Vision and Pattern Recognition, CVPR, pp. 761–768 (2011)
20. Lee, Y.J., Zitnick, C.L., Cohen, M.F.: Shadowdraw: real-time user guidance for freehand drawing. ACM Trans. Graph. 30, 27 (2011)
21. Indyk, P., Motwani, R.: Approximate nearest neighbors: towards removing the curse of dimensionality. In: Proceedings of the Thirtieth Annual ACM Symposium on Theory of Computing, STOC 1998, pp. 604–613. ACM, New York (1998)
22. Canny, J.: A computational approach to edge detection. IEEE Trans. Pattern Anal. Mach. Intell. 8, 679–698 (1986)
23. Martin, D., Fowlkes, C., Malik, J.: Learning to detect natural image boundaries using local brightness, color, and texture cues. IEEE Transactions on Pattern Analysis and Machine Intelligence 26, 530–549 (2004)

Efficient Scale and Rotation Invariant Object Detection Based on HOGs and Evolutionary Optimization Techniques

Stefanos Stefanou[1,2] and Antonis A. Argyros[1,2]

[1] Institute of Computer Science, FORTH
[2] Brain and Mind Graduate Program, University of Crete
{stevest,argyros}@ics.forth.gr
http://www.ics.forth.gr/cvrl/

Abstract. Object detection and localization in an image can be achieved by representing an object as a Histogram of Oriented Gradients (HOG). HOGs have proven to be robust object descriptors. However, to achieve accurate object localization, one must take a sliding window approach and evaluate the similarity of the descriptor over all possible windows in an image. In case that search should also be scale and rotation invariant, the exhaustive consideration of all possible HOG transformations makes the method impractical due to its computational complexity. In this work, we first propose a variant of an existing rotation invariant HOG-like descriptor. We then formulate object detection and localization as an optimization problem that is solved using the Particle Swarm Optimization (PSO) method. A series of experiments demonstrates that the proposed approach results in very large performance gains without sacrificing object detection and localization accuracy.

1 Introduction

Detecting objects in real-world scenes depends on the availability of local image features and representations that remain largely unaffected by illumination changes, scene clutter and occlusions. A Histogram of Oriented Gradients (HOG) [1] is a descriptor that is computed on a dense grid of uniformly spaced cells and employs overlapping local contrast normalization for improved accuracy. The robustness of HOGs has made them a quite popular image patch/object representation. However, object localization based on HOGs requires the evaluation of the similarity of a reference HOG to the HOG computed in each and every possible placement of a window that slides over the image. Additionally, HOGs are scale and rotation dependent representations. Thus, if one needs to detect and localize objects in a scale and rotation independent way, an explicit and exhaustive consideration of all these search dimensions needs to be performed. This exhaustive search in a multidimensional space becomes computationally prohibitive even for very small image sizes. To overcome this, a variety of methods have emerged [2–4]. Typically, they use heuristics that reduce the number of

G. Bebis et al. (Eds.): ISVC 2012, Part I, LNCS 7431, pp. 220–229, 2012.

HOG similarity evaluations in an image by searching only over a coarse grid of candidate object positions or by using local optimization methods. These methods sacrifice location accuracy to gain speed and, thus, have increased risk of inaccurate localization or even object miss.

In this paper we propose a method to perform accurate object localization in any scale and rotation, avoiding the above drawbacks. We start by proposing a variant of an existing [5], rotationally invariant, HOG-based descriptor. The proposed descriptor relaxes the need of considering rotated versions of it. Furthermore, we formulate object localization as an optimization problem that seeks for the image position and object scale that maximizes the match between the rotationally invariant HOG descriptor and its localization in the image. This optimization problem is solved using the Particle Swarm Optimization (PSO) [6] algorithm. The PSO is a heuristic, evolutionary optimization technique, inspired by search mechanisms employed by certain biological species. Large populations of particles (i.e., candidate solutions) are evolved in iterations called generations to eventually land on the global maximum of the function to be optimized. We demonstrate experimentally that, compared to the sliding window search approach, the proposed approach decreases dramatically the number of descriptor/image similarity evaluations that are needed to localize an object in an image.

2   Related Work

In order to reduce the number of HOG descriptor comparisons required for object localization, many methods have been proposed. Typically, these consist of computing and evaluating the descriptor only over a coarse, limited number of window locations where the object is more likely to be located and over fixed window sizes.

Zhu et al [2] used AdaBoost to select the most relevant windows from an image training set, over 250 random windows per image. In addition, they adopted the integral image representation for a faster formulation of their HOG descriptor variant. This representation of images used in HOG strips the Gaussian mask and trilinear interpolation off the construction of the HOG for each block. In [2], the L2-norm used by Dalal and Triggs [1] is replaced by the L1-norm because it is faster to compute with integral images. Overall, near real-time object localization is obtained but with reduced descriptor robustness. Additionally, the search window locations are heavily depended of the training image set. A similar method [7] uses sparse search at runtime to locate parts of the object in search and then improves the localization by applying a pre-learned Partial Least Squares regression model, followed by an dense search around the approximate locations of the object. Other methods [3, 8–11] employ image pyramids or coarse-to-fine hierarchical schemes. Essentially, detailed searches at higher resolutions are focused on areas where there is evidence for the existence of an object from coarser searches in lower resolutions. This strategy reduces the total number of descriptor evaluations. As an example, Zhang et al [3] applied a multi-resolution pyramid framework on HOGs to produce better performance over the method of [1]. Interestingly, this work demonstrates that the predefined hierarchy performs better compared to the one that is automatically selected by AdaBoost. The method searches each image at one fourth of the original resolution with a constant window size in a dense pattern, identifying regions of the image not containing the reference object. These regions are then excluded from search in finer resolutions and window grids. The resulting method achieves good localization accuracy and faster execution compared to the original HOG. Still, the method does not consider different object orientations and scales, excluding this way a number of interesting search dimensions. The method proposed by Lampert et al [4] uses a branch-and-bound (B&B) search to find the globally maximal region of the search space - the rectangular bounding box enclosing the target - faster than the exhaustive search. This method reduces the computational complexity from $\mathcal{O}(n^4)$ to $\mathcal{O}(n^2)$ for an arbitrary rectangle bounding box, by trading off accuracy for fast convergence.

The PSO based object detection approach proposed in this work exhibits remarkable performance gains over existing sliding window approaches. At the same time, localization accuracy remain largely unaffected. Due to its nature, PSO provides continuous solutions a fact that is particularly important for estimating the true scale and orientation of an object. As a result, objects are localized in subpixel accuracy and at a fraction of the time needed by the other methods. In addition, PSO search operates without any previous knowledge regarding the possible location of objects and requires the adjustment of only very few parameters.

3   The Proposed Method

A HOG is not a rotation invariant representation. Therefore, when used in object detection tasks, it can only handle objects that are observed at a certain orientation. To overcome this limitation, a new variant of the HOG descriptor was recently proposed. The so called Rotation-Invariant Fast Feature (RIFF) descriptor [5] is based on a HOG computed at a circular support area and uses an annular binning to achieve orientation invariance. We study the use of a RIFF like descriptor in object detection in conjunction with the Particle Swarm Optimization (PSO) [6]. Scale invariance is not easily achievable through modifications of the HOG descriptor. Instead, the capability for scale invariant search for objects is delegated to the employed optimization technique. Essentially, the detection of a reference object in an image amounts to searching for the image position and object scale that maximizes the match between the rotationally invariant reference object descriptor and the descriptor computed at that image part. The degree of match is quantified by employing the Quadratic-Chi histogram distance [12] between the reference object and the candidate image area. In PSO terms, the Quadratic-Chi histogram distance between a reference HOG and the HOG computed at an image region constitutes the objective function to be minimized.

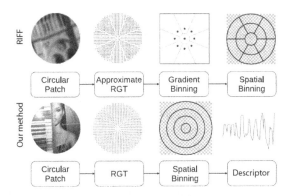

Fig. 1. The RIFF descriptor in comparison with the proposed descriptor variation. The proposed descriptor does not quantize the patch gradients and does not decompose an annulus to sectors.

3.1   Rotation Invariance

The original HOG descriptor performs well with objects that are observed at a certain orientation and scale. In order to handle objects that are presented in arbitrary orientations a rotationally invariant object descriptor is required. There are two prominent techniques for achieving rotation invariance. The first [13] treats rotation as a circular shift and uses the magnitude of the Fourier transform, often not sufficiently robust to view point variations. The second [14] uses steerable filters and computes a descriptor for a number of discrete orientations of the filter.

The Rotation Invariant Fast Features (RIFF) descriptor [5] is a recent approach that leverages on the proven methods of SIFT [15] and HOG [1] and provides robustness and rotation invariance. The RIFF descriptor consists of concentric annular cells, applied to image interest points extracted by the FAST [16] detector. Typically, RIFF descriptors consist of four annular cells with the largest diameter being equal to 40 pixels. In each annulus, the image gradient orientations are computed using the centered derivative mask $[-1, 0, 1]$ and rotated to the proper angle according to the Radial Gradient Transform [5] to achieve rotation invariance. The resulting gradients are quantized with respect to their direction for improved performance. Additionally, at each pixel, a local polar reference frame is created for describing the gradient from the radial and tangential directions of the center of the pixel, relative to the center of the descriptor. The coordinates of the gradient in the local frame of reference are invariant to rotation for the given descriptor center. A binning technique is also employed as in CHOG [17]. Computational performance is further improved based on sparse gradient sampling.

In this paper we use a variant of the RIFF descriptor, adapted for whole object recognition. A single circular descriptor is computed that encloses the reference object. The descriptor for a circular image region is computed by first calculating the edge gradient scale and orientation with a centered derivative mask $[-1, 0, 1]$. We use a signed orientation gradient spanning from $-\pi$ to π. As in the original RIFF, we define four circular, concentric, non-overlapping annuli. The radii of the circles defining the annuli are computed so that the resulting annuli have the same area. To achieve rotation invariance we rotate the gradients according to the Radial Gradient Transform (RGT) [5] without applying any direction quantization. The final descriptor consists of a histogram of 72 discrete bins (4 annuli × 18 gradient directions, each). To avoid boundary effects, bilinear interpolation is used to distribute the value of each gradient sample into adjacent histogram bins. Additionally, each pixel's vote in the histogram is weighted by the edge gradient scale. To account for changes in illumination and contrast, a local normalization is performed between cells using the L2-norm followed by clipping the maximum values by a threshold of 0.2 and re-normalizing as in [15]. The final descriptor is the normalized, concatenated rows of the resulting histogram.

3.2 Descriptor Distance Measure and Matching

Since RIFF is a direct representation of a histogram we can use distance measures that are well suited to histogram comparison. We chose the Quadratic-Chi (Q-Chi) histogram distance [12] in order to reduce the effect of differences caused by bins with large values and because of its performance advantages over the simple χ^2 method. According to [12], let P and Q be two non-negative bounded histograms. Let also A be a non-negative symmetric bounded bin-similarity matrix such that each diagonal element is bigger or equal to every other element in its row. Finally, let $0 \leq m < 1$ be a normalization factor. A Quadratic-Chi histogram distance QC between P and Q is defined as

$$QC^A{}_m(P,Q) = \sqrt{\sum_{ij} \left(\frac{(P_i - Q_i)}{\sum_c (P_c + Q_c)A_{ci}{}^m} \frac{(P_j - Q_j)}{\sum_c (P_c + Q_c)A_{cj}{}^m} A_{ij} \right)}. \quad (1)$$

The normalization factor was set to $m = 0.9$.

Concerning descriptor matching, we experimented in comparing the descriptor produced by the original image with the descriptors produced by sub-sampled instances of the same image in different sizes, using the nearest neighbor sampling method. Using the proposed descriptor design, we concluded that descriptors produced from the same image but at different scales typically differ substantially with respect to their Q-Chi distance. More specifically, the nearest neighbor subsampling of an image gave progressively greater distance as the difference in size was increasing. Using bilinear interpolation to match the size of the sub-sampled instances with the size of the original, higher resolution image resulted in much lower influence from scale difference. Finally, using bi-cubic interpolation instead of bi-linear, improves further the results. So, it turns out that it is of importance to match the resolution of the larger image patch by up-sampling the smaller image patch using bi-cubic interpolation prior to computing the descriptor histogram.

3.3 The PSO Optimization Algorithm

Particle Swarm Optimization (PSO) is an evolutionary technique for the optimization of nonlinear, multidimensional and multimodal functions that is inspired by social interaction. A population of agents, called *particles* is randomly initialized inside the objective function's space. Particles move in search of the function's global maximum for a given number of iterations called *generations*. Each particle is associated with the evaluation of the objective function at its location. Each agent's velocity in the parameter space is determined by three components: a random one, a local one that directs the particle towards its own best position and a global one that directs the particle towards the globally best position. More specifically, the velocity v_i^t for particle i in generation t is given by

$$v_i^{t+1} = K \left(v_i^t + \phi_1 R_1 (\boldsymbol{pb}_i^t - \boldsymbol{x}_i^t) + \phi_2 R_2 (\boldsymbol{gb}^t - \boldsymbol{x}_i^t) \right), \qquad (2)$$

where $\boldsymbol{pb}_i$ is each particle's best position so far, $\boldsymbol{gb}$ is the best position over the whole particles population, $\boldsymbol{x}_i$ the current position of each particle and R_1, R_2 are random numbers in the range $[0..1]$. Additionally, the so called *constriction factor* K is equal to

$$K = \frac{2}{\left| 2 - \psi - \sqrt{\psi^2 - 4 * \psi} \right|}, \psi = c_1 + c_2, \qquad (3)$$

with $c_1 + c_2 = 4.1$ as suggested in [18]. As the swarm evolves, the agents are expected to locate the global maximum of the objective function and keep oscillating around it. The vast percentage of the computational load of PSO is associated to the evaluation of the objective function for each particle in each generation. Thus, the product of the number of generations to the number of particles is a good indication of the computational load for that PSO parameterization.

Although in principle there are no guarantees for convergence, it has been demonstrated that PSO is able to effectively cope with difficult multidimensional optimization problems in various domains, including computer vision [19].

3.4 Employing PSO for HOG-Based Object Detection

Object detection is formulated as a search task across the three-dimensional parameter space formed by all possible 2D translations and scales at which an object might be present in an image. More specifically, the PSO particles are initialized randomly inside this three-dimensional search space. Each particle corresponds to a single 2D position and scale of the descriptor in the image. The boundaries of this space are determined by the minimum and the maximum scale of the window and the size of the image. To account for partially clipped objects near image borders, the image is padded by mirroring its contents near the edges for 10 pixels. PSO seeks to minimize an objective function which, in our case, is the Q-Chi distance (Eq.1) between the reference object descriptor and and a candidate image window.

4   Experimental Results

Several experiments have been performed to assess quantitatively the proposed approach. More specifically, the goal of the experimentation was to evaluate the efficiency and the accuracy of object localization using the proposed technique. The dataset used to evaluate the proposed method is the one used by Tacacs [5] for the evaluation of the RIFF descriptor and consists of images of music CD covers in arbitrary rotation and distance from the camera with partial occlusions and with different backgrounds. The dataset includes 50 different CDs observed in 10 different backgrounds, resulting in a data set of 500 images. For each CD, the clear cover image is provided, based on which the reference descriptor is computed.

In order to evaluate the performance gain of the proposed method, we first produced reference data by locating the position and scale of the window that minimizes the QC distance metric. This was achieved by performing an exhaustive search experiment where all possible object positions and scales were evaluated. To cope with the computational requirements of this exhaustive search experiment, the original images were resized to 320×240 by halving their height and width. With respect to scale, each object has been searched at a minimum window of 60×60 and at a maximum window of 240×240, resulting in 180 different window sizes that were also exhaustively considered.

Next, we ran the proposed method for a variety of PSO parameterizations (number of particles and number of generations). For a particular PSO parameterization our approach reported the position and scale at which an object exists in an image. The localization accuracy of such an experiment was quantified by measuring the F-score (i.e., is the harmonic mean of precision and recall) of the result of our approach and that of the exhaustive experiment. This was repeated for all images. The obtained F-scores for all 500 images are averaged to come up with a single number quantifying the localization accuracy for a certain PSO parameterization. We also measured the accuracy obtained by the sliding window approach where the window displacement step and window size step is equal to $D > 1$ pixels. Essentially, the exhaustive experiment corresponds to $D = 1$ and corresponds to approximately $3,500,000$ objective function evaluations per image. For comparison, running the same experiment with $D = 40$ requires as few as 140 objective function evaluations per image.

Figure 2 summarizes the results obtained from all related experiments. The vertical axis of the plots corresponds to the obtained F-scores. The horizontal axis corresponds to the parameter D. The dashed line corresponds to the average F-score of the sliding window approach as a function of D. As explained earlier, each point in the plot is the average of the F-scores obtained in 500 object searches.

As expected, the exhaustive approach achieves an average F-score of 1 when $D = 1$. As D increases, the average F-score also decreases, reaching the value of 0.25 for $D = 40$. The same plot demonstrates the performance of the proposed approach for a large variety of numbers of particles. For a particular particle count, the number of generations was calculated so that the computational

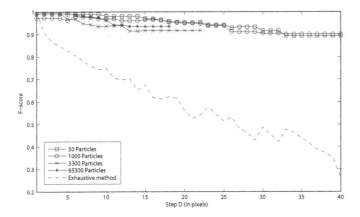

Fig. 2. The mean object localization accuracy with respect to various parameterizations of the proposed algorithm and the exhaustive search approach. The horizontal axis represents the displacement D in location x, y and scale dimensions for the sliding window algorithm. For the proposed method, different plots correspond to different particle numbers. See text for a detailed description.

budget required by our method does not exceed the budget of the corresponding exhaustive approach with displacement D. Thus, the intersection of these plots with e vertical line corresponds to algorithms that have the same computational budget and, therefore, require the same execution time. For some particle counts, the curves do not extend up to $D = 40$ because for F values above a threshold, the above mentioned calculation returns zero PSO generations.

As it can easily be verified, the proposed approach keeps an average F-score above 0.9 for all considered computational budgets. This is true even for a budget as low as 140 objective function evaluations ($D = 40$). Thus, when limited computational resources are devoted to object detection, our approach results in more than 3.5-fold improvement in localization accuracy, compared to the sliding window approach.

We also observed that regardless of the parameterization used, PSO is able to localize the object reasonably well in early generations and then performs only minor improvements. Thus, if localization accuracy can be traded with performance, the proposed approach can result in further performance gains.

Another interesting conclusion that can be derived by studying Fig. 2 is that there is no major difference in the use of more generations over more particles. Despite this general conclusion, in individual images and, especially in those that objects appear at smaller scales, it is preferable to use more particles than generations, so that the parameter space is more densely sampled.

Figure 2 presents representative object localization results obtained by the proposed method. In the first four rows, successful detections are shown (in columns 2 to 6) of the reference objects shown in the first column. It can be verified that the proposed approach manages to localize objects despite significant

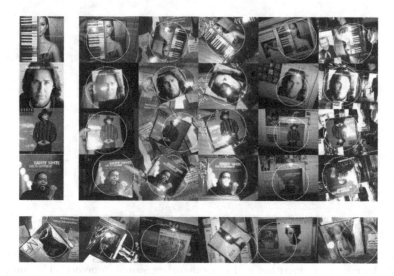

Fig. 3. Representative detection results, for objects of arbitrary rotation and scale in different cluttered backgrounds. The proposed approach exhibits robustness to orientation and scale variations as well as to occlusions and illumination artifacts.

scale and orientation variations as well as partial occlusions, and specular reflections. Interestingly, objects are also accurately localized in the images of the fourth column; these are photos of computer monitors displaying the reference objects. The last row of Fig. 3 shows some of the worst localization results obtained, which we consider as failure cases. In these examples object localization accuracy is small mostly because of the strong specular effects.

5   Discussion

In this paper we formulated object detection as an optimization problem that has been solved with PSO, an evolutionary optimization method. We apply this method to a variant of the HOG descriptor. Experimental results demonstrated that accurate object detection and localization can be achieved at a fraction of the computational cost of the sliding window approach. It is important that PSO has an inherently parallel nature, a fact that can be directly exploited to further reduce the computational time required by employing GPUs. It is also important that the proposed method can very easily be transformed into a tracking framework, which employs object detection at the vicinity of the solution estimated in the previous frame of an image sequence. Current research is considering the employment of PSO in formulations of object detection problem that exhibit even higher dimensionality.

Acknowledgments. This work was partially supported by the IST-FP7-288146 project HOBBIT.

References

1. Dalal, N., Triggs, B.: Histograms of oriented gradients for human detection. In: CVPR, San Diego, USA (2005)
2. Zhu, Q., Avidan, S., Chen Yeh, M., Ting Cheng, K.: Fast human detection using a cascade of histograms of oriented gradients. In: CVPR, pp. 1491–1498 (2006)
3. Zhang, W., Zelinsky, G., Samaras, D.: Real-time accurate object detection using multiple resolutions. In: ICCV (2007)
4. Lampert, C.H., Blaschko, M.B., Hofmann, T.: Beyond sliding windows: Object localization by efficient subwindow search. In: CVPR, pp. 1–8 (2008)
5. Takacs, G., Chandrasekhar, V., Tsai, S., Chen, D., Grzeszczuk, R., Girod, B.: Unified real-time tracking and recognition with rotation-invariant fast features. In: CVPR (2010)
6. Kennedy, J., Eberhart, R.C.: Particle swarm optimization. In: IEEE Int'l Conf. on Neural Networks, pp. 1942–1948 (1995)
7. Wu, J., Wei, C., Huang, K., Tan, T.: Partial least squares based subwindow search for pedestrian detection. In: Macq, B., Schelkens, P. (eds.) ICIP, pp. 3565–3568. IEEE (2011)
8. Epshtein, B., Ullman, S.: Feature hierarchies for object classification. In: ICCV. Springer (2005)
9. Agarwal, A., Triggs, B.: Hyperfeatures – Multilevel Local Coding for Visual Recognition. In: Leonardis, A., Bischof, H., Pinz, A. (eds.) ECCV 2006, Part I. LNCS, vol. 3951, pp. 30–43. Springer, Heidelberg (2006)
10. Amit, Y., Geman, D., Fan, X.: A coarse-to-fine strategy for multi-class shape detection. IEEE Trans. on PAMI 26 (2004)
11. Fleuret, F., Geman, D.: Coarse-to-fine face detection. IJCV 41, 85–107 (2001)
12. Pele, O., Werman, M.: The Quadratic-Chi Histogram Distance Family. In: Daniilidis, K., Maragos, P., Paragios, N. (eds.) ECCV 2010, Part II. LNCS, vol. 6312, pp. 749–762. Springer, Heidelberg (2010)
13. Kingsbury, N.: Rotation-invariant local feature matching with complex wavelets. In: EUSIPCO, pp. 4–8 (2006)
14. Tola, E., Lepetit, V., Fua, P.: A fast local descriptor for dense matching. In: CVPR (2008)
15. Lowe, D.G.: Distinctive image features from scale-invariant keypoints. IJCV 60 (2004)
16. Rosten, E., Drummond, T.W.: Machine Learning for High-Speed Corner Detection. In: Leonardis, A., Bischof, H., Pinz, A. (eds.) ECCV 2006, Part I. LNCS, vol. 3951, pp. 430–443. Springer, Heidelberg (2006)
17. Chandrasekhar, V., Takacs, G., Chen, D.M., Tsai, S.S., Grzeszczuk, R., Girod, B.: Chog: Compressed histogram of gradients a low bit-rate feature descriptor. In: CVPR (2009)
18. Clerc, M., Kennedy, J.: The particle swarm - explosion, stability, and convergence in a multidimensional complex space. IEEE Trans. Evolutionary Computation 6, 58–73 (2002)
19. Oikonomidis, I., Kyriazis, N., Argyros, A.A.: Efficient Model-based 3D Tracking of Hand Articulations using Kinect. In: BMVC (2011)

Neural Network Based Methodology for Automatic Detection of Whale Blows in Infrared Video

Varun Santhaseelan, Saibabu Arigela, and Vijayan K. Asari

University of Dayton, 300 College Park, Dayton OH, USA
{santhaseelanv1,arigelas1,vasari1}@udayton.edu

Abstract. In this paper, we propose a new methodology based on neural networks to detect the presence of whale blows in infrared video. The algorithm is designed based on the spatial and temporal characteristics of whale blows. The first part of the algorithm consists of thresholding techniques that filter out the possible candidates to a group containing whale blows and certain textures on the sea. A novel thresholding technique called grid thresholding is proposed so that the detector is able to detect very small blows while keeping the number of false positives to a minimum. As the final part of the detection algorithm we have used a neural network to differentiate between whale blows and the various textures on the surface of the sea.

1 Introduction

Rapid advancements in the area of computer vision have helped in accelerating the automation process in various research endeavors. In this paper, we propose a new method for automatic detection of whale blows in infrared video. A lot of research is happening in fields related to behavior of mammals and study of the migration patterns of whales has always been a subject of interest for researchers. The first task in the study of whale behavior is to detect the presence of whales. The presence of whales may not be so obvious because of their movement below the surface of the sea. Most of the time, the movement is detected by the presence of whale blows that occur in certain patterns. Researchers that monitor this movement pattern have to analyze hours of video in order to detect blows, count them and then by comparison with the breathing pattern of whales, try to track the whales in video.

It would be of great help to researchers if an automatic process is put in place that can detect the whale blows, estimate the number of whales that are passing through a particular region and track the whales from a distance. In this paper, we tackle this problem by characterizing the behavior of whale blows in the spatial and temporal domain and thus develop an algorithm that can detect whale blows automatically.

The use of infrared imagery to study the behavior of marine mammals was explored by Cuyler et al [1]. An extensive study of gray whale migratory behavior was done by Perryman et al [2]. The paper also explains in detail the different sea conditions that could affect the detection of whale blows and thus affecting the tracking procedure. We have used this paper as a reference to study the performance of our

G. Bebis et al. (Eds.): ISVC 2012, Part I, LNCS 7431, pp. 230–240, 2012.

detection algorithm in various conditions. An extensive study into the detection of whales from thermal imagery was done by Graber et al [3] too. The research focused more on actual detection of whales depending on their shape characteristics rather than the detection of whale blows. Another system for automated whale detection has been implemented by Zitterbart et al [4][5]. Their system was developed to detect the presence of whales in the vicinity of a ship.

Our algorithm focuses on the long range detection of whale blows. The videos are captured from the shore using an infrared camera kept at an elevated position. The primary aim of this detection algorithm is to act as an aid in the study of migratory patterns of whales.

The paper is organized as follows. The characteristics of whale blows are explained in detail in the second section. The third section elaborates on the algorithm that has been developed based on the information available on the behavior of whales. The results that have been obtained are presented in the fourth section. The fifth section summarizes the research findings presented in this paper and also gives an insight into the current research problems that we are addressing.

2   Characteristics of Whale Blows

The problem of whale blow detection is characterized as a change detection problem. In order to detect the change, it is essential to characterize the change in terms of behavior and that forms the crux of this section.

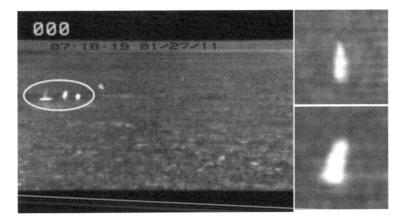

Fig. 1. How whale blows appear on a frame is shown on the left. Two different types of whale blows are shown on the right.

Whale blows appear as distinct high intensity regions in infrared video. The shape of the blow is distinctive when the blow has grown to full size. However, the presence of wind can alter the shape of the spout when it is of full size. This is illustrated in Fig. 1. While the blows have higher intensity in comparison with most of the other elements in the frame, whale blows are not the only high intensity regions in the frame. There could be textures in sea like waves formations that could be of high intensity.

For every whale blow, a rise time and fall time can be defined. The variation of the shape of a blow in a sequence of frames is very distinctive. All the blows will not have the same size when it has reached its peak. This is because the force at which water is blown may vary with whales. For example, the blow from a calf would be considerably smaller than the blow from a mother whale. External characteristics like wind will also cause variations in the shape of the blow when it reaches full size and during the fall time. However, it is to be noted that the growth of the blow is very similar in all situations. This is because of the immense force with which water is blown out by the whale. This forms a characteristic change in the environment and is common for all the whale blows.

Research into whale behavior [6] has shown that the movement of a pod of whales is always in paths that do not intersect. There would be sufficient distance between whales at all times. The inference from this information is that there would always be sufficient distance between two whale blows. This information can be used to differentiate between two whale blows if a pod of whales are present.

3   Algorithm

The algorithm that we propose for whale blow detection is explained in this section. Every frame in the video is scanned for the presence of a whale blow based on the characteristics explained in the previous section. A condensed flow diagram of the implementation model is shown in the Fig. 2.

Fig. 2. Implementation model of the proposed algorithm

The constraints in the second step are defined based on the characteristics of whale blows. Since the scan is being done throughout the entire frame with shifts of a single pixel, multiple detections will be made for a single blow. All the detections for a single blow are fused into one based on the condition that there can only be one base for a whale blow. The final part of the detection algorithm is a neural network that is trained to detect actual whale blows. The neural network was essential because of the presence of textures in the sea that has many characteristics similar to whale blows. The steps prior to classification help in increasing the speed of the algorithm.

3.1   Intensity Thresholding

One of the characteristics that differentiate a blow from many of the other textures on the sea is the high intensity in comparison with its surroundings. Therefore, as the initial step in the algorithm, we apply fixed thresholding to the frames to get a binary image. We also found that adaptive thresholding techniques [7][8] as the first step were not of much help in our case. It is observed that the whale blow appears as a cluster of high intensity and therefore appears as a cluster in the binary thresholded frame as well. An illustration is shown in Fig. 3. It is observed that whale blows are not the only high intensity clusters in the frame. Therefore, clustering high intensity groups alone will not be able to discriminate between whale blows and textures on the sea like waves.

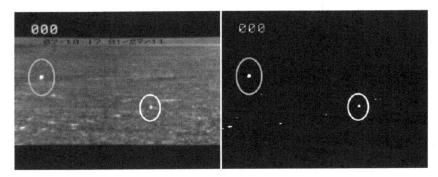

Fig. 3. A frame containing whale blow (in left circle) is shown on left and the corresponding thresholded image on the right. The circle on the right is a non-blow.

However, intensity thresholding is a suitable constraint to filter out many of the non-blows. In our case, we calculate the mean for each window in the frame and check whether it is greater than a particular threshold. The threshold is selected based on the mean for the smallest whale blow in the database.

3.2   Metrics to Capture Shape Variation

The variation of shape of a whale blow over time is illustrated in Fig. 4.

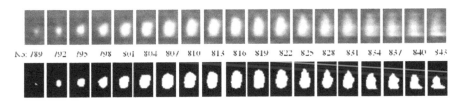

Fig. 4. The growth of whale blow shown at an interval of three frames and the corresponding thresholded images

The size of the blow keeps increasing during rise time and decreases during fall time. The variation is especially clear in the case of the thresholded images. This variation can be visualized as a characteristic volume in three dimensional space. We have utilized two metrics to characterize this shape variation.

Cumulative Absolute Difference: As mentioned in the previous subsection, we make use of the mean of a particular window in the first step to determine whether the window contains a whale blow or not. The mean is calculated on the thresholded image. It is observed that the mean of a particular window keeps on increasing during the rise time of a blow and decreases during fall time. The variation in mean during rise time is shown in Fig. 5. We characterize the variation in size of the blow by defining a threshold for the sum of the absolute differences of mean of a window in successive frames. This metric is called cumulative absolute difference (CAD) as shown in (1).

$$CAD = \sum_{n=1}^{K-1} |mean(n) - mean(n+1)| . \tag{1}$$

The constant K is assigned a value based on the duration of the smallest blow in the database. In the case of a whale blow, the value of CAD would be higher than in the case of non whale blows. In our case, the threshold for CAD has been selected based on the CAD of the smallest blow in the database.

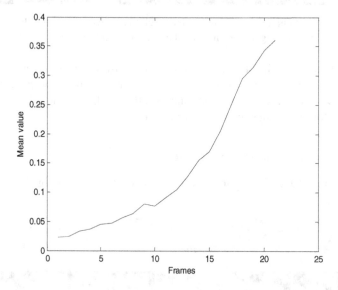

Fig. 5. Variation of mean for a window containing a whale blow

The threshold for CAD provides a good constraint to filter out many of the non-blows. However, there are textures in the sea that have an uneven variation in mean causing the CAD to be high for those windows. This causes an increase in false positives.

Cumulative Difference: Another observation regarding whale blows is the fact that the size of the blow will keep increasing for the initial few frames. This indicates that the slope of the mean curve would be positive. We define another metric called cumulative difference (CD) based on this observation. It is calculated as the sum of the difference of the means for the first few frames as shown in (2).

$$CD = \sum_{n=1}^{\frac{K}{2}} mean(n + 1) - mean(n) \ . \tag{2}$$

The value of K is defined as in the computation of CAD. The number of frames that have been used in our case is the number of frames for the smallest blow. Only those candidate blows are classified as blows if the CD passes a certain threshold. The threshold is decided based on the CD of the smallest blow in the database.

After applying the constraints on CAD and CD, there could still be textures on the surface of the sea that could be classified as whale blows as illustrated in Fig. 6.

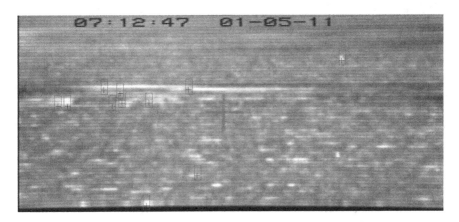

Fig. 6. False detections appear on a frame containing no whale blows after the constraints related to intensity and its variation are applied

3.3   Neural Network for Classification

The metrics defined in the previous section do not capture the variation of the whale blow in its entirety. As mentioned in the second section, the whale blow size variation follows a characteristic pattern and this can be visualized as a volumetric model. All the constraints that we have described so far do not capture the characteristics of this volumetric model. In order to solve this problem, we used a neural network to be trained to recognize patterns of whale blows.

In order to have faster computation, we experimented with Sparse Network of Winnows (SNoW) [9]. However, the network did not work well for our application. This led us to the conclusion that the manifold that can characterize the shape variation has to be non-linear. We used a multi-layer perceptron (MLP) [10][11] as shown in Fig. 7 for our task. Backpropagation [12] algorithm was used for learning. We formed a linear array of a candidate window over a minimum number of frames. The number of input nodes is equal to the size of the linear array. We chose ten hidden nodes and two output nodes.

3.4   Grid Thresholding to Detect the Smallest Blows

It has been mentioned in Sect. 3.1 that the very first step in the algorithm is to do a fixed thresholding to obtain a binary image. It has been observed that the fixed thresholding technique affects the detection of very small blows near the horizon. The difference between a blow closer to the camera and a blow farther toward the horizon is shown in Fig. 8.

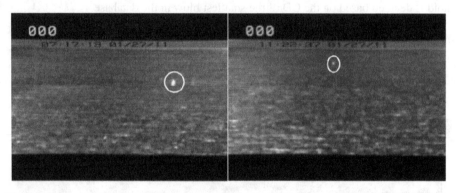

Fig. 7. A blow closer to the camera is shown on the left while a blow very far away is shown on the right. Note that the intensity of the blow region on the right is much lesser.

The difference between a blow closer to the camera and one further away from the camera is the size as well as the intensity in the blow region. It is observed that the intensity in the blow region for faraway blows is not as high as that for nearer blows. The main reason for this change in intensity is atmospheric scattering. This observation necessitated the use of an adaptive thresholding mechanism. We tried decreasing the threshold linearly as distance towards the horizon decreases. However, this continuous variation of threshold increased the number of false detections. The continuous variation of threshold caused undue variations in the binary image that made it similar to a whale blow and this caused the increase in false detections.

In order to overcome this problem, we introduced a technique called grid thresholding. In this case, the frame was divided into horizontal grids. Each grid has a different fixed threshold. The thresholds keep decreasing towards the horizon. In order to decide on the grid boundaries we made use of the Beer-Lambert equation on atmospheric scattering as shown in (3).

$$I = I_0 e^{-\alpha x} \ . \tag{3}$$

where I is the intensity at the observation point, I_0 is the original intensity, α is the absorption constant and x is the distance of observation point from the source.

The grid boundaries were decided based on (4) which was derived from the equation in (3).

$$y = y_0(1 - e^{-n}) \; .$$

(4)

where y_0 is the total frame height and $n = 1,2,3$.

The design of the thresholds for each grid was purely empirical and based on experiments on the videos in the database. However, the general trend was to decrease the threshold as we move close to the horizon.

4   Results

The robustness of the algorithm can be tested by running the algorithm on different kinds of videos. It has been shown in [2] that the detection rate suffers when there are strong winds. Windy conditions cause variation in the blow shape as it grows above a

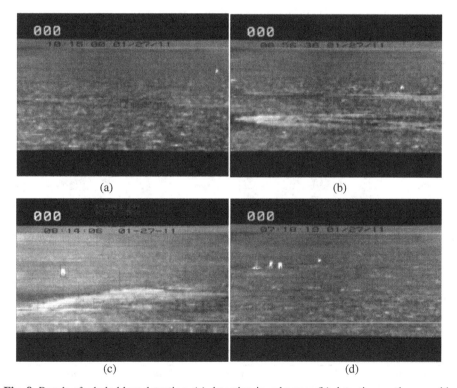

Fig. 8. Result of whale blow detection. (a) detection in calm sea, (b) detection on the sea with some white caps, (c) detection amidst lots of white caps, (d) detection of multiple blows.

particular height. The strength of the blow is also a factor when it comes to the effect of wind on the shape of the blow. There are other objects in the scene as well like birds that can be of concern. However, the movement of birds is very different from that of a whale blow and therefore our algorithm is not affected by those. The most difficult problem is the formation of waves on the sea when there are strong winds. Depending on the strength of the wind, the amount of waves in the sea also varies. The variation in the texture of sea caused due to the turbulence of waves is very similar to the characteristics of the growth of a small whale blow. This is where our algorithm has proven to be very robust in classifying all the waves as non-blows.

We tested the algorithm on different kinds of videos: videos with perfectly calm sea conditions, videos with some white caps on the surface of the sea, videos with lots of white caps on the surface of the sea, videos with multiple blows in a frame and videos with blows close to the horizon (far away from the camera). The videos that were used had a resolution of 640×480 and the thresholds were designed accordingly. The result of the algorithm on different types of videos is shown in Fig. 9.

The comparison of results from adaptive thresholding and grid thresholding for small blows is shown in Fig. 10. It is observed that while adaptive thresholding technique helps in the detection of whale blow, it also creates a lot of false positives. Grid thresholding however overcomes all these problems.

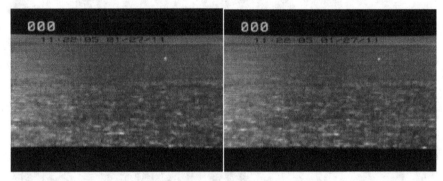

Fig. 9. The result of adaptive thresholding (continuous variation of threshold) is shown on the left and the result of grid thresholding is shown on the right

For the purpose of extensive testing, we ran the algorithm in five hours of video in which there were blows of different sizes and sea textures of different types. Table 1 is a summary of the testing process.

Table 1. Result of testing

Sea Conditions	Visual detections	Automatic detections	False detections	Detection accuracy (%)
Calm sea	65	59	0	90.7
Moderate wind	20	17	3	85
Heavy wind	11	10	3	90.9

The number in visual detections is the ground truth. The data was provided by professionals who study whale migration. The detection accuracy is measured as the ratio of the true positives over the sum of true positives and false positives. The number of false positives is denoted as false detections. It was observed that the false positives occurred at locations where the whale blows were very near to the horizon and the effect of atmospheric noise was at maximum. The noise present in such situations is the presence of haze like conditions. In the presence of heavy winds, long range visibility is affected and that is the reason for the rise in detection accuracy in heavy wind conditions.

5   Conclusion

We have successfully developed an automated mechanism based on a new thresholding technique and neural networks to detect whale blows from infrared video. The proposed technique was able to detect whale blows successfully even when there are a lot of waves in the sea or in the presence of other objects in the scene like birds. The robustness of the algorithm has been tested successfully in videos with different wind conditions. This algorithm will be of great help to researchers in cutting down a lot of time with regard to monitoring video in order to study whale migration. The algorithm is being improved further to incorporate tracking mechanisms based on whale behavior and its breathing patterns. Another short term goal is to design a mechanism to reduce the effect of atmospheric noise in the detection accuracy of the system. An extensive testing of the algorithm will be done by deploying it on a ship that is used for sea exploration.

Acknowledgments. This research has been carried out as part of a collaborative project with Sonalysts Inc. and Whale Acoustics with funding from National Oceanic and Atmospheric Administration.

References

1. Cuyler, L.C., Wiulsrod, R., Oritsland, N.A.: Thermal infrared radiation from free living whales. Marine Mammal Science 8(2), 120–134 (1992)
2. Perryman, W.L., Donahue, M.A., Laake, J.L., Martin, T.E.: Diel variation in migration rates of eastern Pacific gray whales measured with thermal imaging sensors. Marine Mammal Science 15(2), 426–445 (1999)
3. Graber, J., Thomson, J., Polagye, B., Jessup, A.: Land-based infrared imagery for marine mammal detection. In: SPIE Photonics + Optics, San Diego, CA, August 20-25 (2011)
4. Zitterbart, D., Kindermann, L., Burkhardt, E., Boebel, O.: Automated whale detection and local, high resolution ice mapping using a 360° ship-based thermal imager. In: IPY Polar Science Conference, New frontiers, Data Practices and Directions in Polar Research, Oslo 2010 (2010)
5. Zitterbart, D.P., Kindermann, L., Boebel, O.: MAPS: An automated whale detection system for mitigation purposes. SEG Expanded Abstracts 30, 67–71 (2011)

6. Jones, M.L., Swartz, S.L.: Gray Whale, Eschrichtius robustus. In: Perrin, W.F., Wursig, B., Thewissen, J.G.M. (eds.) Encyclopedia of Marine Mammals, pp. 524–536. Academic Press, San Diego (2002)

7. Otsu, N.: A threshold selection method from gray level histograms. IEEE Transactions on Systems Man and Cybernetics 9(1), 62–66 (1979)

8. Sankaran, P., Asari, V.K.: Adaptive Thresholding Based Cell Segmentation for Cell-Destruction Activity Verification. In: 35th IEEE Applied Imagery and Pattern Recognition Workshop (2006)

9. Carlson, A., Cumby, C., Rosen, J., Roth, D.: The SNoW learning architecture. Technical Report UIUCDCS-R-99-2101, UIUC Computer Science Department (May 1999)

10. Rumelhart, D.E., Hinton, G.E., Williams, R.J.: Learning Internal Representations by Error Propagation. In: Parallel Distributed Processing: Explorations in the Microstructure of Cognition (vol. 1: Foundations). MIT Press (1986)

11. Haykin, S.: Neural Networks: A Comprehensive Foundation, 2nd edn. Prentice Hall

12. Rumelhart, D.E., Hinton, G.E., Williams, R.J.: Learning representations by back-propagating errors. Nature 323(6088), 533–536

Gaussian Mixture Background Modelling Optimisation for Micro-controllers

Claudio Salvadori[1], Dimitrios Makris[2], Matteo Petracca[3,1],
Jesus Martinez-del-Rincon[2], and Sergio Velastin[2]

[1] Real-Time Systems Laboratory, Scuola Superiore Sant'Anna, Pisa, Italy
[2] Digital Imaging Research Centre, Kingston University, London, United Kingdom
[3] National Laboratory of Photonic Networks, CNIT, Pisa, Italy

Abstract. This paper proposes an optimisation of the adaptive Gaussian mixture background model that allows the deployment of the method on processors with low memory capacity. The effect of the granularity of the Gaussian mean-value and variance in an integer-based implementation is investigated and novel updating rules of the mixture weights are described. Based on the proposed framework, an implementation for a very low power consumption micro-controller is presented. Results show that the proposed method operates in real time on the micro-controller and has similar performance to the original model.

1 Introduction

Background modelling is a fundamental task for many computer vision applications such as motion detection, object tracking and action recognition. Among the background modelling methods that have been proposed in the literature (see [1], [2], [3] for extended surveys), the Gaussian Mixture Model (GMM) approach [4] has proved very popular. However, its high computational and memory requirements restrict the type of processors in which the GMM algorithm can be deployed. On the other hand, with the increasing demand on the number and cost-effectiveness of video monitoring devices (driven by security concerns but also by applications such as gaming and intelligent spaces), it is highly desirable to explore new computer vision algorithms designed for embedded systems. For this work we have selected a particular micro-controller: the Microchip PIC32 [5], which is typical of the type of low-cost low-power devices used in embedded systems. This family of devices lack a dedicated hardware floating-point processing unit (FPU) and consequently must rely on floating point software libraries or fixed-point/integer algorithmic variants. This limitation represents a major challenge for the implementation of most computer vision and signal processing algorithms.

The goal of this work is to explore ways in which the GMM algorithm could be used on micro-controllers with low power consumption. In particular, an integer precision approach is proposed to reduce both the computational cost and the memory footprint of the algorithm. Nevertheless, this algorithm has to

G. Bebis et al. (Eds.): ISVC 2012, Part I, LNCS 7431, pp. 241–251, 2012.

be able to approximate the original implementation (double precision) without significant loss of robustness under changing light conditions and accuracy for foreground/background segmentation.

Considering micro-controller based architectures, different approaches for background modelling have been proposed in the literature. In [6] and [7] the background is modelled using the *median filter* while [8] use *running averages*, and in [9] a combination of both methods is used. These approaches have low computational requirements, because they use only one parameter to model the background. However, the usage of only one parameter implies also a limitation since it does not explicitly model the foreground and it is likely to underperform in comparison to GMM, specially for complex scenes.

The rest of the paper is organised as follows: in Sec. 2 the methodology to approximate the GMM updating rules is described, given the micro-controller constraints and the operative limits of this approach. In Sec. 3 the implementation constraints imposed by the chosen hardware are discussed. In Sec. 4 the performance evaluation of the implementation using two Gaussians is presented, while conclusions follow in Sec. 5.

2   Methodology

Gaussian Mixture Model is a method to model the background view for a static camera so as to facilitate the separation between foreground and background, on the assumption that over time a pixel is mostly background. One of its most significant characteristic is the capability to adapt to illumination changes at the expense of the inclusion of stationary objects into the background. Each pixel of the background model is separately described as a mixture of G Gaussians, where each one of them is represented by three parameters: mean-value μ, variance σ^2 and a weight w. In Table 1 the ranges of the three parameters are shown. It is particular important to consider the range of the variance: for this parameter a lower bound σ^2_{min} and an upper bound σ^2_{MAX} are defined to prevent the degeneration of the Gaussian distribution into a *Dirac delta* or a *uniform* distribution respectively.

Table 1. Gaussian parameters ranges

μ	σ^2	w
[0, 255]	$[\sigma^2_{min}, \sigma^2_{MAX}]$	[0, 1]

In a mixture, the Gaussians are ordered accordingly to a *sorting rule*, based on the value of ρ [4] (see Eq. (1) below), and splitted into two subsets: the *background sub-set* and the *foreground sub-set*. GMM allows the labelling of every pixel in every image accordingly so as to discriminate the foreground from the background.

$$\rho = \frac{w^2}{\sigma^2} \tag{1}$$

This section presents a general method to deploy the GMM algorithm on micro-controllers with computation and memory constraints (see Sec. 3). Due to these two constraints an integer precision implementation is required to fit the algorithm on a chosen micro-controller. In particular, the updating rules for the Gaussian parameter must be addressed. Inevitably, an integer implementation restricts the range and/or the granularity of the Gaussian parameters. For instance, the number of bits that are allocated to the mean-value and variance of each Gaussian directly affects their range and granularity. The range of learning-rates of GMM algorithm is also affected, however in a more complex way. Specifically, each approximation (mean-value, variance, weight) introduces a lower bound for the learning-rate (see Eq. (2), where α_{min}^w, α_{min}^μ and α_{min}^σ are the lower bounds generated by the updating rules limitation of the weight, the mean-value and the variance, respectively).

$$\bar{\alpha}_{min} > MAX\{\alpha_{min}^w, \alpha_{min}^\mu, \alpha_{min}^\sigma\} \tag{2}$$

In this way, a learning-rate *operating range* is defined as $[\bar{\alpha}_{min}, 0.99]$.

2.1   Mean-Value and Variance Updating

In the GMM algorithm the mean-value and the variance of a given Gaussian of the mixture are updated if and only if the value of the pixel *"belongs"* to that Gaussian: consequently a *belonging criterion* is defined. In this case a pixel, represented by its value $p_{i,j}$, "belongs" to a Gaussian if the condition shown in Eq. (3) is satisfied.

$$(\mu_{g,(i,j)} - p_{i,j})^2 < T\sigma_{g,(i,j)}^2 \tag{3}$$

where $\mu_{g,(i,j)}$ and $\sigma_{g,(i,j)}^2$ are respectively the mean value and the variance of the g-th Gaussian of the mixture and T is a threshold fixed a priori.

To redefine the mean-value and the variance updating rules, and to make them more appropriate for an embedded system, the concepts of *granularity* and *updating step* are introduced. Let assume a sequence of ordered natural numbers $\mathbf{B} = \{b_i \in \mathbb{N} | b_{i+1} - b_i = \gamma\}$, the *granularity* of this set is the value γ. The parameter *updating step* ξ is defined as a function of the *granularity* ($\xi = \mathcal{F}(\gamma)$), so as to characterize a *generalised-round* operation G_ROUND as shown in Eq. (4).

$$b = G\_ROUND(a, \xi, \gamma) = \begin{cases} \left\lfloor \frac{a}{\gamma} \right\rfloor * \gamma + \gamma & \text{if } \left(a - \left\lfloor \frac{a}{\gamma} \right\rfloor * \gamma\right) \geq \xi \\ \left\lfloor \frac{a}{\gamma} \right\rfloor * \gamma & \text{if } \left(a - \left\lfloor \frac{a}{\gamma} \right\rfloor * \gamma\right) < \xi \end{cases} \tag{4}$$

where a is the number to be rounded. In the case considered here, the parameter *updating step* is assumed to be half the *granularity* ($\xi = \frac{\gamma}{2}$). Consequently, the updating rules of the mean-value and the variance are defined as shown respectively on Eq. (5) and Eq. (6).

$$\mu_{i+1} = G\_ROUND(\mu_i + k_i d_i, \xi_\mu, \gamma_\mu) \tag{5}$$
$$\sigma_{i+1}^2 = G\_ROUND(\sigma_i^2 + k_i(D_i - \sigma_i^2), \xi_\sigma, \gamma_\sigma) \tag{6}$$

where $k_i = \frac{\alpha}{w_i}$, p_i is the current value of the pixel, $d_i = p_i - \mu_i$ and $D_i = d_i^2$.

Thus, the mean-value and the variance updating rules introduce lower bound limits on the range of learning-rate values. These limits are required to emulate similar behaviours in both integer and double precision implementations. These lower boundaries are defined by detecting the worst possible case for a given condition. Thus, the worst case for the mean-value updating rule is considered by choosing a background pixel ($w \simeq w_{max} = 1$, consequently $k \simeq k_{min} = \alpha$, and $\sigma \simeq \sigma_{min}$) with its value close to the borders of the interval defined by the *belonging criterion* ($p = \lfloor \mu \pm \sigma_{min}\sqrt{T} \rfloor$). The condition described by Eq. (7) imposes this lower bound limit.

$$\alpha_{min}^\mu = \frac{\xi_\mu}{\left\lceil \sigma_{min}\sqrt{(T)} \right\rceil} \tag{7}$$

On the other hand, the condition reported in Eq. (8) imposes on the variance updating rule the restriction of a lower bound σ_{min}^2 in updating the parameter:

$$\alpha_{min}^\sigma = \frac{\xi_\sigma}{\sigma_{min}^2 + 1} \tag{8}$$

2.2   Weight Updating

Eq. (9) shows the updating rule when the intensity of a pixel k belongs to Gaussian g ($g \in [1, G]$). If the intensity of the pixel does not belong to the Gaussian, the weight value will be decreased as it is shown on Eq. (10).

$$w_k^g(s+1) = (1 - \alpha) * w_k^g(s) + \alpha \tag{9}$$
$$w_k^g(s+1) = (1 - \alpha) * w_k^g(s) \tag{10}$$

where g is the Gaussian index ($g \in [1, G]$), k is the pixel index and s is the counter that indicates how many times the value of the k-th pixel of an image belongs the g-th Gaussian during its temporal evolution, and α is the learning-rate.

Assuming that the intensity of pixel k belongs to Gaussian g for s consecutive frames and solving the iterative Eq. (9) leads to Eq. (11).

$$w_k^g(s) = 1 - (1 - \alpha)^s \tag{11}$$

Finally, solving Eq. (11) and retrieving the value s ($\equiv S(w, \alpha)$), Eq. (12) is obtained. Using this formula it is possible to calculate the number of iterations needed to reach a certain value of weight w.

$$S(w, \alpha) = \frac{\log_{10}(1 - w)}{\log_{10}(1 - \alpha)} \qquad (12)$$

In this situation it is possible to simplify the weight updating rule as an operation of the increase or decrease of an integer counter, which implies a significant gain on performance and a reduction of hardware requirements.

However, Eq. (11) and Eq. (12) can not be directly implemented due to the particular micro-controller features, such as the lack of floating-based capabilities. Consequently two linearly approximated solutions are defined to retrieve the weight value from the integer counter. Fig. 1a and 1b depict both approaches (gray curves) and its comparison with the logarithmic relation of the weight described on Eq. (12) (black curves).

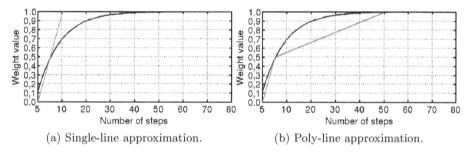

(a) Single-line approximation.   (b) Poly-line approximation.

Fig. 1. The weight trend and its approximations

Similarly to Sec. 2.1, both solutions introduce a lower bound on the learning-rate range. If $\Sigma_1(\alpha)$ is the maximum number of iterations needed by the chosen approximation and learning-rate α ($\Sigma_1(\alpha)$ is the number of steps to reach the value 1, the maximum in the case of weight), the relation shown in Eq. (13) holds, where the *maximum value* L that can be represented is $L = 2^l$, for a given number l of bits required to represent the integer counter.

$$\Sigma_1(\alpha^w) \leq L \qquad (13)$$

Assuming that the function $\Sigma_1(\alpha)$ is known, it is simple to extract the learning-rate lower bound (α_{min}^w) from Eq. (13) (see Eq. (14)).

$$\alpha \geq \alpha_{min}^w \qquad (14)$$

Although this method is general and extensible to any number of Gaussians per pixel, given the hardware constrains of the micro-controller chosen for this work, the number of Gaussians is $G = 2$ and every plot and figure shown in the next sections is related to this case.

Single-Line Approximation. This is based on the approximation of the logarithmic curve described on Eq. (12) by means of a line passing through the origin and the point $P_0 = (S(w_0, \alpha), w_0)$ (with $w_0 = 0.50$) as shown in Fig. 1a and described in Eq. (15), where $m = 2S(w_0, \alpha)$, and the value of steps to increment and to decrement the counters are both equal to 1. The choice of the point P_0 is because around this the swap between the foreground and the background sub-sets happens. Therefore, it is a critical point for achieving a correct emulation of double precision implementation behaviours.

$$w(s) = \frac{s}{m} \tag{15}$$

Consequently, knowing that $\Sigma_1 = 2S(w_0, \alpha)$ the relation shown in Eq. (16) holds.

$$2S(w_0, \alpha_{min}^w) \leq L \tag{16}$$

Finally from Eq. (16) and from Eq. (12), the lower bound of the learning-rate is defined as shown in (17)

$$\alpha_{min}^w = 1 - (1 - w_0)^{\frac{2}{L}} \tag{17}$$

Poly-Line Approximation. Let the two ordered Gaussians be G_1 (the background) and G_2 (the foreground). The relation shown in Eq. (18) and the consequent condition shown in Eq. (19) are valid:

$$w_{max} + w_{min} = 1 \tag{18}$$

$$\begin{cases} w_{max} \geq \frac{1}{2} \\ w_{min} \leq \frac{1}{2} \end{cases} \tag{19}$$

From the condition shown on Eq. (19), it is possible to derive a two-line poly-line approximation of the logarithmic weight trend (see Fig. 1b). The first line of the poly-line (l_1) passes through the origin and the point $P_l = (S(w_l, \alpha), w_l)$ (with $w_l = 0.50$) and maps the lowest weight Gaussian. The second one (l_2) passes through the point P_l and the point $P_h = (S(w_h, \alpha), w_h)$ (with $w_h = 0.99$) and maps the larger weight Gaussian.

Because of Eq. (18), it is possible to represent the two weights (and consequently the two lines) using only one of them. As shown in Fig. 2, line l ($\equiv l_2$) is selected and translated on the origin (line l_0, see Eq. (20), where $m = S(w_h, \alpha) - S(w_l, \alpha) = b - a$, and $\bar{s}$ and $\bar{w}$ are respectively the number of steps and the weight on the translated domain) to simplify the computation.

$$l_0 : \bar{w} = \frac{\bar{s}}{2m} \tag{20}$$

In this configuration, the line l_0, that maps the counter $\bar{s}$, has the same slope of l_2 (the greatest value weight Gaussian), up to some translation constants. Consequently, in the case where the current pixel belongs to the Gaussian related

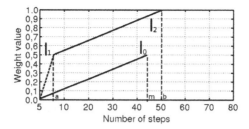

Fig. 2. Poly-line approximation, using a single line

with l_2, the weight updating rule consists on increasing the counter by 1. On the other hand, when the pixel belongs the Gaussian with l_1 the weight updating rule consists on subtracting from $\bar{s}$ a constant derived from the mapping of the line l_1 over l_0. The above mentioned constant is the ratio between the slope of l_1 ($\frac{1}{2a}$) and the slope of l_0 ($\frac{1}{2m}$) (see Eq. 21).

$$STEP = \left\lceil \frac{m}{a} \right\rceil \tag{21}$$

Moreover, from Eq. (20), the relationship between the weight values (w_{max} and w_{min}) and the value $\bar{s}$ (see Eq. (22) and Eq. (23)) can be derived:

$$w_{max} = \frac{\bar{s} + d}{2 * d} \tag{22}$$

$$w_{min} = \frac{d - \bar{s}}{2 * d} \tag{23}$$

Usually the greatest weight Gaussian is labelled as G_1. However, because the sorting rule is based on the values of ρ (see Eq. (1)), in some cases the above mentioned assumption is not valid. For this reason one bit of the weight representation is used as a descriptor for associating the counter correctly to the Gaussian with the largest weight. In this case the learning-rate lower bound is defined by the condition in Eq. (24). Solving Eq. (24), the relation shown in Eq. (25) is retrieved.

$$S_1 = S(w_h, \alpha_{min}^w) - S(w_l, \alpha_{min}^w) \leq L \tag{24}$$

$$\alpha_{min}^w = 1 - \left(\frac{1 - w_h}{1 - w_l} \right)^{\frac{1}{L}} \tag{25}$$

3   Implementation Constraints

Because of the use of micro-controllers such as PIC32MX795F512L [5] (80MIPS, 128 KBytes of RAM), gray-levels images with 8-bits depth and resolution QQ-VGA (160x120) are considered. As described on the previous sections, two integer precision approaches are used, to reduce the computational cost of the algorithm processing (PIC32 family micro-controllers have no FPU and the floating-point numbers algorithms [10] are implemented using software libraries), and the algorithm memory footprint. As we can see in Table 2 both floating-point representations exceed the available memory footprint of the micro-controller (128 KBytes of RAM). However, even the integer representation (uint8_t) uses a large amount of memory and it does not permit the implementation of any other additional image processing algorithm or any communication stack. To address this problem, a two bytes per Gaussian approach is implemented. Consequently using QQ-VGA images and mixtures of two Gaussians, the algorithm memory footprint is 76,800 Bytes.

Table 2. Algorithm footprint (QQ-VGA images and mixture of 2 Gaussians)

Precision	Bytes per parameter	Bytes per Gaussian	Footprint (Bytes)
Double	8	24	921,600
Float	4	12	460,800
uint8_t	1	3	115,200

In Fig. 3 the two bytes per Gaussian representations are shown: the first one, shown in Fig. 3a, is used to fit the single-line approximation approach and the second one, shown in Fig. 3b, to fit the poly-line approximation.

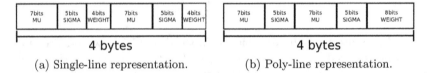

(a) Single-line representation.   (b) Poly-line representation.

Fig. 3. 2 bytes per Gaussian representations

3.1   The Learning-Rate Operating Range

In this section the *learning-rate operating range* is computed taking into account the different learning-rate lower bounds resulting from the constraints related to the integer implementation of the Gaussian parameter updating rules. Because of $T \in [15, 35]$, the worst case is chosen ($T = 15$). Because both implementations differ only on weight representation, the learning-rate lower bounds dictated by the mean-value and the variance are the same (see Table 3).

On the other hand, since the learning-rate lower bounds introduced by the weight depends on which approximation is chosen, the limits introduced by both representations are summarised in Table 4.

Finally, since the learning-rate lower bound is obtained from Eq. 2, *learning-rate operating range* is defined as $[0.15, 0.99]$ and all the performance tests shown on the next section will be evaluated over this range.

Table 3. μ and σ^2 representation constraints

Parameter name	Definition range	N. bits	Granularity γ	Updating step ξ	α_{min} value
μ	[0, 255]	7	2	1	0.125
σ^2	[5, 36]	5	1	0.5	0.083

Table 4. Weight representation constraints

Approximation type	N. bits	Maximum value L	α_{min}^w value
Single-line	4	15	0.088
Poly-line	7	127	0.030

4 Performance Evaluation

The performance of the proposed algorithm is presented in this section. For all the experiments, it is assumed that $T \in [15, 35]$ and $\alpha \in [0.15, 0.75]$.

4.1 Integer Precision vs. Double Precision: Performance Comparison

The comparison between the performance of the integer precision and the double precision is described by means of *precision-recall* diagrams. Particularly, the minimum distance between the P-R curve and the optimal point (the point (1,1)) is computed and plotted. In Fig. 4a and in Fig. 4b this minimum distance is plotted using the two different data-sets: the "IXMAS data-set" [11] (to represent situations with slow movement) and the "Fudan Pedestrian data-set" [12] (to represent situations with fast movement). As we can see in these plots, the performance in the two integer precision approaches is similar and comparable with the results coming from the double precision implementation experiments.

4.2 Integer Precision: Memory Footprint and Processing Time

Both approaches have been implemented in the SEED-EYE [13] board, based on Microchip PIC32MX795F512L micro-controller. In the board, two different firmwares able to acquire, store and process images have been deployed. The memory footprint generated by these applications is the same (both are based on two bytes per Gaussian representation) and use 76% of the micro-controller memory. In Fig. 5 the processing times of the two approaches are shown. The average processing time for single-line approach (Fig.5a) is about 21ms, and for the poly-line one (Fig.5b) is about 28ms. For the type of videos that were considered, both approaches operate in real-time as the achieved frame-rate is more than 30fps (33ms/frame).

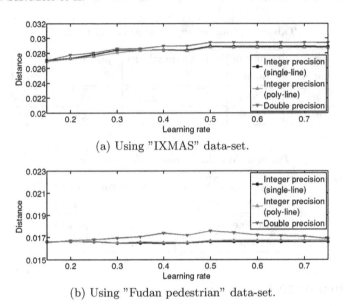

(a) Using "IXMAS" data-set.

(b) Using "Fudan pedestrian" data-set.

Fig. 4. Distance from the P-R to the PR optimal point (1,1)

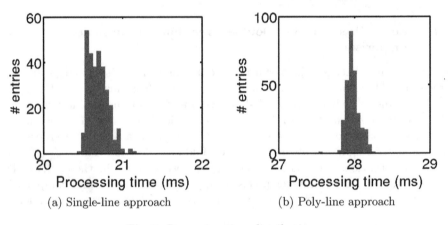

(a) Single-line approach   (b) Poly-line approach

Fig. 5. Processing time distributions

5   Conclusions

In this paper two integer precision embedded-oriented implementations of mixture of Gaussian background modelling have been described. Particularly a *learning-rate operating range* is computed, arising from both an integer implementation and the limited number of bits assigned to each parameter. Instead the retrieved range is $\alpha \in [0.15, 0.99]$. Furthermore, the results coming from the integer precision implementations are compared with the results generated by a double precision one using the P-R curves. From this comparison it is possible to say that the two integer precision approaches have similar performance to the double precision one, in the above-mentioned learning-rate range. Finally, using the SEED-EYE board, the two implementations have a memory footprint of the 76% of the micro-controller memory and a mean processing time of about 21ms and about 28ms for the single-line approach and the poly-line one respectively, both of them less than the real-time constraints of 33ms (30fps).

References

1. Piccardi, M.: Background subtraction techniques: a review. In: IEEE International Conference on Systems, Man and Cybernetics, vol. 4, pp. 3099–3104 (2004)
2. Cheung, S.C.S., Kamath, C.: Robust techniques for background subtraction in urban traffic video. In: Visual Communications and Image Processing, vol. 5308, pp. 881–892 (2004)
3. Radke, R.J., Andra, S., Al-Kofahi, O., Roysam, B.: Image change detection algorithms: A systematic survey. IEEE Transactions on Image Processing 14, 294–307 (2005)
4. Stauffer, C., Grimson, W.E.L.: Adaptive background mixture models for real-time tracking. In: IEEE Computer Society Conference on Computer Vision and Pattern Recognition, vol. 2, pp. 2246–2252 (1999)
5. Microchip Technology Inc.: PIC32MX3XX/4XX Family Data Sheet (2008), ww1.microchip.com/downloads/en/DeviceDoc/61143E.pdf
6. McFarlane, N.J.B., Schofield, C.P.: Segmentation and tracking of piglets in images. Machine Vision and Applications 8, 187–193 (1995)
7. Hung, M.H., Pan, J.S., Hsieh, C.H.: Speed up temporal median filter for background subtraction. In: Proceedings of International Conference on Pervasive Computing, Signal Processing and Applications, pp. 297–300 (2010)
8. Rahimi, M., Baer, R., Iroezi, O.I., Garcia, J.C., Warrior, J., Estrin, D., Srivastava, M.: Cyclops: In situ image sensing and interpretation in wireless sensor networks. In: Proceedings of ACM Conference on Embedded Networked Sensor Systems, pp. 192–204 (2005)
9. Iannizzotto, G., La Rosa, F., Lo Bello, L.: A wireless sensor network for distributed autonomous traffic monitoring. In: Conference on Human System Interactions, pp. 612–619 (2010)
10. IEEE Computer Society: IEEE Standard for Floating-Point Arithmetic. Technical report, Microprocessor Standards Committee of the IEEE Computer Society, 3 Park Avenue, New York, NY 10016-5997, USA (2008)
11. Weinland, D., Ronfard, R., Boyer, E.: Free viewpoint action recognition using motion history volumes. Computer Vision and Image Understanding 104, 249–257 (2006)
12. Tan, B., Zhang, J., Wang, L.: Semi-supervised elastic net for pedestrian counting. Pattern Recognition (2011)
13. Scuola Superiore Sant'Anna and Evidence s.r.l.: SEED-EYE BOARD (2011), http://www.evidence.eu.com/products/seed-eye.html

Automatic Segmentation of Wood Logs by Combining Detection and Segmentation

Enrico Gutzeit and Jörg Voskamp

Fraunhofer Institute for Computer Graphics Research IGD

Abstract. The segmentation of cut surfaces from a stack of wood logs is a challenging task and leads to many problems. Wood logs theoretically have a certain shape and color, which is the main reason to apply object detection methods. But in real world images there are many disturbing factors, such as defects, dirt or non-elliptical logs. In this paper we mainly address the problem of wood and wood log segmentation by combining object detection with a graph-cut segmentation. We introduce an iterative segmentation procedure, which detects the stack of wood, segments foreground and background, and separates the logs. Our novel approach works fully automatically and has no restrictions on the image acquisition other than well visible log cut surfaces. All three steps of our approach are novel and could be applied on similar problems. We implemented and evaluated different methods and show that of these approaches, our methods leads to the best results.

1 Introduction

The automatic and sound segmentation of multiple objects is a known and unresolved problem in computer vision. Many researchers developed methods, which work very well in specific scenarios or for special objects. In the case of segmentation of wood log cut surfaces there exists no approach without restrictions on the image acquisition. In this article we especially address the problem of automatic segmentation of wood logs. For a sound separation of foreground and background we propose using object detection in a pre-step to support the actual segmentation process. Due to variations in shape and color of wood logs an accurate object detection is impossible. Consequently, we additionally address the problem of using uncertain object locations in segmentation.

Number, size and volume of logs in a stack of wood are important data to timber industry. Correct data is needed in different commercial and logistic processes. In our research we aim at developing a fast, cost-effective, mobile applicable and reliable measurement tool by using computer vision methods. We aim at applying those measurement methods on photographs of wood piles. The measurement of the timber itself bases on photographs of the log cut surfaces. Therefor the cut surfaces have to be separated from the background, which leads to the problem of binary image segmentation. Only a robust and accurate automatic segmentation allows for a more precise measurement of wood volume compared to the manual techniques. In the presented article we describe a fully

G. Bebis et al. (Eds.): ISVC 2012, Part I, LNCS 7431, pp. 252–261, 2012.

automatic segmentation method integrating iteratively object detection and image based segmentation. We show that integrating both can improve the final segmentation results and release users from some of the rules when taking the photographs of the log cut surfaces.

2 Related Work

In computer vision our specific application is addressed rarely. The method proposed in [1] is one processing step of a measuring system using a stereo vision system with two cameras. The area of the stack of wood is calculated beforehand by using stereo vision algorithms. The gathered information is pipelined into the proposed method, which detects wood logs in images by using a watershed segmentation on different scales. This method estimates the percentage of wood in the measurement step of wood log piles and works only robust on images without disturbing factors like grass, earth or forest in the background. Fink [2] applies classical vision algorithms and active contours to find the log cut surfaces. The proposed procedure works half automatically only and takes a lot of assumptions about the type of wood, the lighting conditions and the image quality. The evaluation photographs were taken in a controlled environment which makes the proposed algorithms doubtful to work in our scenario. Our approach in [3] addresses the segmentation of wood logs by using a graph-cut segmentation (KD-NN). The segmentation works very well on low quality images. Thereby some general restrictions are set on the image acquisition side. The stack of wood has to be in the center. Furthermore background has to be in the lower and upper part of the image. The outcome of the proposed process is a binary image, which separates only wood and non-wood pixels. In summary there are no existing fully automatic segmentation methods, which segment logs of a stack of wood with an arbitrary position in the image.

In image processing there is a variety of approaches to tackle the segmentation problem [4]. When there is need for a fully automatically working segmentation method, being able to work stable and robust on photographs of different quality and lighting conditions (e.g. different weather conditions), just a few of the proposed methods work well. Graph-cut based segmentation algorithms define the state-of-the-art for separation of foreground and background pixels. The graph-cut algorithm results in a solution of the maximum flow [5]. The minimal cut then describes the partition of the image, in our case log cut surfaces and background. But these kinds of algorithm need prior information about the color and/or the position of the objects of interest. In [6] a rectangle is set manually by the user. After some additional adjustments a very good segmentation is obtained, whereby a Gaussian Mixture Model (GMM) is used to model the color distribution of the fore- and background. The algorithm is derived from [7]. Many region segmentation approaches like watershed [8], pyramid linking [9] or some graph-based methods [10] segment multiple objects, but without prior information they lead to the well known over- or under-segmentation problem.

3   Problem Discussion

Our final aim is to automatically calculate the volume of a stack of wood and the radius for every wood log by applying computer vision methods. We propose a three step algorithm. First, all taken images of the log pile are stitched together, so there is one image showing the whole stack of wood. Then, the resulting image is transformed into a coordinate system suitable for length measurement. Finally, the log cut surfaces are segmented. In this paper we focus on the last step, the log separation in images. Consequently, the aim of this paper is to automatically segment every wood log in the image.

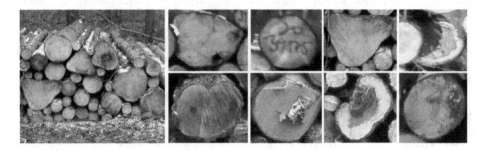

Fig. 1. On the left side a typical input image and on the right side some difficult, but often appearing logs in our input series, are pictured

In our scenario the input images are taken by mobile phones as they are easy to use and easy to transport for the users. The quality of the images heavily rely on the quality of the small lens systems and the compression algorithms. In other words, the quality of the images are quiet low compared to laboratory settings. The images are taken by people of the forestry. The resulting input images to the proposed method could be blurred, over- and under exposed. Images are taken from one frontal side of the pile only. A sample input image can be seen on left side in figure 1.

In our previous research [3] we segment wood pixels by applying the graph cut approach of [5]. To accurately segment wood pixels, we set the graph weights by using a density estimation on a foreground and background model. One model is a set of pixels, which is represented through a KD-Tree. We called the method KD-NN. To ensure correct models, the middle part of the image must be foreground and the upper and lower part background. One of the main problems of the described method was that there were very strict rules to be followed by the users when taking photographs.

To handle an arbitrary location, the stack of wood should be detected in the image before segmentation starts. A wood log cut surface has characteristics in color and shape, whereby object detection methods are appropriate to find a solution. The detection of objects is a wide field of research, whereby an object is mostly detected through a classifier. Specially trained classifiers search in image or feature space to detect the object. Common methods for object detection are template matching [11], eigen-objects or haar-cascades [12]. We found that object detection methods only can not solve our problem, as the detection rates are to low in our scenario and no exact contour will be extracted per log. But these methods are good to give first information about the position and color of the stack of wood. Using the results of object detection in a segmentation method based on gradients and color is an appropriate way for gathering the required data of position and color.

4   Our Approach

Our approach is an iterative segmentation procedure, as illustrated in figure 2. First in Step A, the logs are detected with an object detection method. The detected logs are represented by a set of circles C_{logs}. The set contains log candidates, which means that there are probably some false and not detected logs (true positive in average by 80 %). The set C_{logs} is used in Step B to calculate a tri map I_{tri}, which labels the image pixel with wood, background and unknown. Through a graph-cut segmentation by using I_{tri} the image I_{in} is binary segmented into wood and non wood pixels, which leads to the binary image I_b. Finally, the binary image I_b and the log candidates C_{logs} are together used in step C to separate the logs. In the result, an image I_{logs} is computed, labeling every wood log in the image.

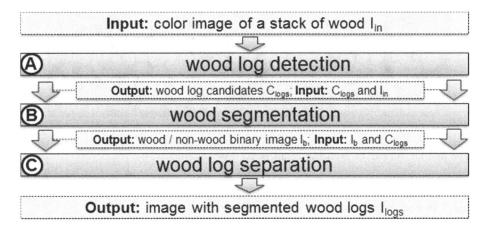

Fig. 2. Our principal methodology and the steps of our automatic wood log segmentation

4.1   Wood Log Detection with PWL-Haarcascades

To detect wood logs, we tested three different classifiers, which are template matching, eigenobjects and haarcascades. In our experiments we received the

best detection rates by applying haarcascades of Viola and Jones [12]. We trained
the classifier with 5729 positive (log images) and 3019 negative samples. The
trained classifier is used to detect wood logs in the image I_{in}. Detection results of
haarcascades are represented through rectangles. To get a better representation,
we first convert all rectangles (size $w * h$) into circles (radius $(w + h)/2$). The
resulting set C_{cand} contains wood log candidates, which are in average only to
65% correct detected. To eliminate some not correct wood logs, the set C_{cand} is
post processed in two steps. In the first step all overlapping logs are removed.
In the most cases a bigger circle is more exact than a smaller one, because
the miss matches often happens inside of defect or dirty logs. Consequently,
we remove all smaller logs, which lie more than 75% into a bigger one. In the
result the set C_{cand} is reduced to the set $C_{cand'}$. Wood logs are piled in common
praxis, whereby every wood log has a neighbor. We use this fact to eliminate
logs (outlier) without a neighbor in the second step. Therefore starting by the
center circle $c_c \in C_{cand'}$, all neighbored circles including c_c are put into a new
set C_{logs}. c_c is the nearest one to the mean position of the elements in $C_{cand'}$. A
circle has a neighbor, if when scaled by g it overlaps with another scaled circle
($g = 2$ is a good value). In the result an improved set of wood log candidates
C_{logs} is available (see figure 3). In summary, our method eliminates overlapping
logs and outlier after a haarcascade based wood log detection. We denote our
approach with PWL-H (Post Processed Wood Log Haarcascades).

Fig. 3. The haarcascade detection result C_{cand} and the PWL-H result C_{logs} are pairwise figured. As can be seen the result is not free of errors.

4.2 Wood Segmentation with KD-NN-A

The wood segmentation is the most important step with a great influence on
the result. The aim is to calculate a binary image I_b, which separates wood and
non-wood pixels. To get a sound binary segmentation we estimate the trimap
I_{tri} by using the log candidates C_{logs}. Image I_{in} is iteratively segmented using
the graph cut based approach KD-NN of [3] extended by the introduction of a
trimap, which leads to a more robust segmentation and lesser restrictions on the
image acquisition. We denote our method KD-NN-A (KD-NN Adaptive). The
two steps are illustrated in figure 4 and the corresponding results in figure 5.

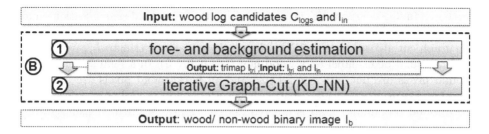

Fig. 4. Abstract illustration of our segmentation process

Fore- and Background Estimation. The object detection result C_{logs} is not free of errors. It is not seldom, that false detected logs occur or logs are missing. Consequently, the fore- and background can only be approximated with C_{logs}. To make sure, that no false fore- and background pixels are marked, we use a trimap I_{tri}, which consists of foreground, background and unknown pixels. Foreground pixels are inside and background pixels are outside of the stack of wood. Our approach works on the binary image I_{bc} created from the set $l \in C_{logs}$. The areas of the circles are set to 1 in I_{bc} and the rest to 0 (see figure 5). To get a good center region of the stack of wood, we first calculate the centroid of I_{bc}. After that, we take the nearest circles to the centroid (25 % of the set of circles). Finally, the region inside the convex hull of the nearest circles are the foreground pixels. The background pixels are calculated through a distance transform. The main point is, that logs are typically clustered. Large empty areas in I_{bc} are supposed to be background. To locate the areas, we use a distance transform on the inverted binary image of I_{bc}. All pixels, which are not fore- or background are set to unknown.

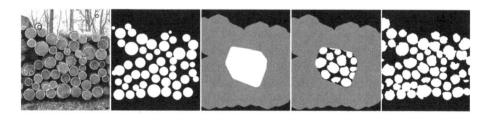

Fig. 5. The in- and outputs of our segmentation steps are illustrated from left to right. The input image I_{in} with the overlaid wood log candidates C_{logs}, the binary log candidate image I_{bc}, the trimap I_{tri}, the refined trimap $I_{tri'}$ and the final segmentation are figured.

Iterative Graph Cut Segmentation. The determined trimap I_{tri} labels every pixel as inside the stack of wood (foreground), outside the stack of wood (background) or as unknown. For an accurate wood segmentation I_{tri} is refined and applied for the segmentation of the image. We refine I_{tri} using the method presented in [3]. We segment the foreground with a threshold segmentation on a modified yellow channel. The result is applied for the training and application of the graph cut algorithm (KD-NN). The result is a refined trimap $I_{tri'}$, whereby all foreground pixels label only wood now (see figure 5). In a final step all foreground pixels are put into one kd-tree and all background pixels are put into an other. By using density estimation for every pixel in the kd-trees a fore- and background probability map is created and the image segmented using graph cut algorithm.

4.3   Wood Log Separation with LSGMC

In the wood log separation step every wood pixel of I_b receives a label. The labels represent a wood log number or background. It is also a segmentation of I_b in circle-like segments. At first, every circle is compared with the corresponding area

Fig. 6. The different steps of our LSGMC are illustrated from left to right. The segmentation result $I_{b'}$ with downscaled $C_{logs'}$, the result of the first GMC, the segmented but undecided wood image I_{bRest}, the result of the second GMC and the final labeled image I_{logs} are pictured.

in I_b. If a certain ratio between wood and non wood pixels is to low, the circle will be removed. The result is a smaller set $C_{logs'}$, where the segmentation and the detection found wood. To reduce holes in the wood logs, we draw all circles of $C_{logs'}$ with a down scaled radius $r * 0.8$ in the binary image I_b. The output is the refined binary image $I_{b'}$. To get the best fitting circles in the binary image $I_{b'}$ a first GMC (Growing Moving Circle) with $C_{logs'}$ is performed. Removing the resulting circles $C_{logs''}$ from $I_{b'}$ leads to a segmented but undecided wood image I_{bRest}. The pixels could be wood or non wood. An extraction of the local maxima after a distance transformation on I_{bRest} leads to a new set of circles C_{rest} (radius is equal to distance value). The union of both sets $C_{logs''} \cup C_{rest}$ is used in a second GMC to find the best fitting circles. Finally all determined logs are projected on the binary image and simultaneously grow until all pixels are labeled in the image I_{logs}. Figure 6 shows intermediate results of our LSGMC.

Growing Moving Circle (GMC). Our algorithm basically searches for the best position and the best radius for every log. The circles grow and/ or move to the position with the highest energy. Energy $e \in [0,1]$ means in our algorithm the ratio between the number of segmented wood pixels inside the circle and the area of the circle. Different directions and sizes are tested. We use an 8 neighborhood and a two pixel testing range (left+1, left+2, leftup+1, leftup+2, radius+1, radius+2, and so on). The circles grow and move until the energy is lower than a certain threshold. After the growing and moving, a post processing step is performed where logs inside another one are removed (maximal overlapping 50%).

5   Results

To provide an evaluation of our algorithm, we compared the segmented images with ground truth data. We manually marked the logs in 71 different wood log images. The marked logs are not connected to each other, so that the ground truth to every log can automatically be extracted by using flood fill. At first, we evaluate the binary wood segmentation (Step B) and thereafter the segmentation results of wood logs (Step C). The wood log detection results (Step A) are used twice in our evaluation. In the wood segmentation the result is seen as binary segmentation and in the wood log segmentation as n-segments. In the evaluation of the binary wood/ non-wood segmentation, we compared our KD-NN-A with the KD-NN from [3]. To get a comparable KD-NN segmentation we fulfill the restrictions in all ground truth images, which means that the stack of wood is in the image center. We determine precision, recall and f-score, which are common in binary segmentation evaluation. We evaluate the segmentation of the wood logs according to the Huang Dom measure [13], which is based on the hamming distance and reasonable to evaluate results with n-segments. The measure interval is $[0,1]$, whereby a value of one represents a perfect segmentation. Furthermore, we simply took the relative error $(nr_{gt} - nr_{seg})/nr_{gt}$ in the number of segmented wood logs, whereby a value of 0 represent no errors. Additionally we compared our algorithms with the common watershed segmentation on a distance image [14], which we call Water-Dis.

Table 1. The upper table shows the segmentation evaluation of wood pixel and the lower of wood logs. All values represent the average of the evaluation, whereby the corresponding standard deviation is in brackets. The asterisks denote our approaches.

wood segmentation method	precision	recall	F-score
Graph-Cut with KD-NN	0,872 (0,089)	0,94 (0,0473)	0,9 (0,0518)
Stamped PWL-Haarcascades*	0,877 (0,044)	0,792 (0,078)	0,83 (0,055)
Graph-Cut with KD-NN-A*	**0,878 (0,08)**	**0,95 (0,0334)**	**0,91 (0,043)**

wood log segmentation method	HD-Measure	log amount error
Water-Dis on the KD-NN-A result	0,918 (0,03)	0,8046 (0,7296)
PWL-Haarcascades*	0,899 (0,021)	0,1908 (0,14)
LSGMC on the KD-NN-A result*	**0,926 (0,03)**	**0,1508 (0,153)**

The binary wood segmentation evaluation in table 1 shows, that stamped PWL-Haarcascades leads to good results, but compared to the others it is the worst method. KD-NN and KD-NN-A produce very good results, whereby our new KD-NN-A lead to the bests. The HD-Measure in wood log segmentation evaluation shows, that Water-Dis is better than PWL-Haarcascades, whereby our LSGMC is still better. However, in the log amount error PWL-Haarcascades are far better than the Water-Dis and the best results we get again with our LSGMC. In summary, our novel approaches lead to the best results applying on wood and log segmentation.

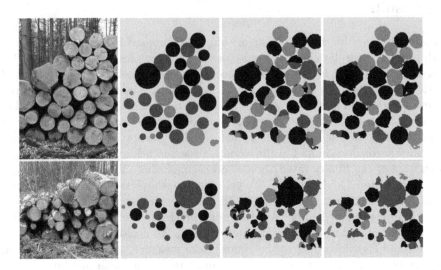

Fig. 7. The images show the different results of the log segmentation. The upper row shows a frequently occurring example and the bottom a challenging one. From left to right the Input Image I_{in}, the result of the PWL-Haarcascades, the Water-Dis and the LSGMC are pictured. The segmented regions are colored only with tree different gray values to make a visual discrimination easier.

6   Conclusion and Future Works

We presented novel approaches to detect and segment wood and/or wood logs from images taken by a mobile phone. At first, we introduced an optimized method to detect stacked wood logs (PWL-Haarcascades). After that, we presented a combined approach for wood/non-wood segmentation, which works very well without hard restrictions on the image acquisition. The novel approach KD-NN-A extends the KD-NN and combines object detection with segmentation. The drawback of the KD-NN approach is, that the stack of wood must be in the center. Otherwise the segmentation get failed. Whereas, the extended approach can handle stack of woods positioned somewhere in the image and additionally leads to better segmentation results. At last, we introduced a approach for wood log segmentation, the LSGMC. The approach use the log detection results to separate the logs from an binary image. Thereby, a new way to separate circle-like segments in a binary image was shown with the GMC algorithm.

In future works we will try to combine disparity maps with our segmentation approach. The stereo map will be created by a stereo mobile phone and should further improve the wood log segmentation. Moreover, we will work on a (LS)GMC extension by using active contours. The main idea is a growing moving contour, which additionally absorb other partly overlapping contours.

References

1. Dahl, A.B., Guo, M., Madsen, K.H.: Scale-space and watershed segmentation for detection of wood logs. In: Vision Day, Informatics and Mathematical Modelling (2006)
2. Fink, F.: Foto-optische erfassung der dimension von nadelrundholzabschnitten unter einsatz digitaler bildverarbeitender methoden. Dissertation, Fakultaet fuer Forst- und Umweltwissenschaften der Albert-Ludwigs-Universitaet Freiburg i. Brsg (2004)
3. Gutzeit, E., Ohl, S., Voskamp, J., Kuijper, A., Urban, B.: Automatic wood log segmentation using graph cuts. In: Richard, P., Braz, J. (eds.) VISIGRAPP 2010. CCIS, vol. 229, pp. 96–109. Springer, Heidelberg (2011)
4. Jaehne, B.: Digital Image Processing, 6th reviewed and extended edn. Springer, Heidelberg (2005)
5. Boykov, Y., Jolly, M.P.: Interactive graph cuts for optimal boundary region segmentation of object in n-d images. In: Int. C. Comput. Vision, pp. 105–112 (2001)
6. Rothar, C., Kolmogorov, V., Blake, A.: Grabcut - interactive forground extraction using iterated graph cuts. In: ACM Transactions on Graphics, pp. 309–314. ACM Press (2004)
7. Orchard, M., Bouman, C.: Color quantization of images. IEEE Transactions on Signal Processing, 2677–2690 (1991)
8. Couprie, M., Najman, L., Bertrand, G.: Quasi-linear algorithms for the topological watershed. Journal of Mathematical Imaging and Vision 22, 231–249 (2005)
9. Zhang, J., Oe, S.: A segmentation method of texture image by using pyramid linking and neural networks. In: Proceedings of the 36th SICE Annual Conference, SICE 1997. International Session Papers, pp. 1267–1272 (1997)
10. Le, T.V., Kulikowski, C.A., Muchnik, I.B.: A Graph-Based Approach for Image Segmentation. In: Bebis, G., Boyle, R., Parvin, B., Koracin, D., Remagnino, P., Porikli, F., Peters, J., Klosowski, J., Arns, L., Chun, Y.K., Rhyne, T.-M., Monroe, L. (eds.) ISVC 2008, Part I. LNCS, vol. 5358, pp. 278–287. Springer, Heidelberg (2008)
11. Brunelli, R.: Template Matching Techniques in Computer Vision: Theory and Practice. Wiley (2009)
12. Viola, P., Jones, M.: Robust real-time object detection. International Journal of Computer Vision (2001)
13. Huang, Q., Dom, B.: Quantitative methods of evaluating image segmentation. In: Proceedings of International Conference on Image Processing, vol. 3, pp. 53–56 (1995)
14. Chen, Q., Yang, X., Petriu, E.M.: Watershed segmentation for binary images with different distance transforms. In: IEEE International Workshop on Haptic, Audio and Visual Environments, HAVE 2004 (2004)

Object Detection from Multiple Images Based on the Graph Cuts

Michael Holuša and Eduard Sojka

Department of Computer Science, FEI, VŠB - Technical University of Ostrava,
17. listopadu 15, 708 33, Ostrava-Poruba
{michael.holusa.st,eduard.sojka}@vsb.cz

Abstract. In this paper, we present a new method for detecting objects
from multiple images. Unlike the general stereo-reconstruction methods,
we propose a method for specific scenes only, which is motivated by
the known fact that, in special circumstances, a specialised method will
probably perform better than a general one. The theoretical model of the
scene expects that the objects are created by the areas placed in a certain
height over the scene base plane. The method is based on minimisation of
an energy function. The function that has been found useful for the model
is presented. For minimisation, the graph cut technique is used, which
is modified by an iterative process in which the original edge weights
are changed on the basis of the residual graph. The method has been
tested on detecting the cars in parking lots; examples of the results are
presented.

1 Introduction

Three-dimensional scene reconstruction is a classic problem in computer vision.
Many methods that find the depth map from multiple images exist. They are
frequently based on energy minimization, e.g., the level-set method [1], belief
propagation [2], dynamic programming [3], and graph cuts [4–6]. The general
methods are usable in various situations, but, on the other hand, they do not
usually give the best possible results for all scenes and images, and they need
not be acceptable in real-time applications. Generally, when solving all image-
processing problems, all a-priori known information should be taken into account
to improve the performance.

The problem that mainly motivated carrying out this work was the problem of
detecting the free or occupied places in the parking lots. The problem was to be
solved by stereo reconstruction. We tried a number of known general methods,
but they reached unsatisfactory results due to their time consumption, and also
due to the big untextured area of the parking lot where it was difficult to find
corresponding points. Therefore, we decided to develop a new method that takes
into account the specific conditions of the mentioned problem. The theoretical
model of the scene expects that the objects of interest are created by the areas
that are placed in a certain height over the scene base plane (Fig. 1). The height
is approximately the same for all the objects. The goal is to detect the objects

G. Bebis et al. (Eds.): ISVC 2012, Part I, LNCS 7431, pp. 262–271, 2012.

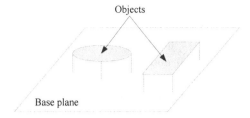

Fig. 1. A scene with a base plane (consisting of *bottom points*) and the objects (consisting of *top points*) that all lie in a certain height above the base plane; the height is approximately the same for all the objects in one scene

and the base plane from a pair of stereo images. Apparently, this model expects only one type of cars (e.g., the passenger cars) since approximately the same height of all objects is required. It was expected that, at the expense of loss of generality, better reliability and computational speed would be achieved.

We propose the method that is based on minimisation of the energy function that is introduced in the paper. For minimisation, the graph cut technique, which was introduced in [11], is utilised. Comparing to the usual use of a single graph cut, we compute the result in an iterative way. When a cut in graph is found, we use the residual capacities of edges to modify the capacities in the original graph and the next cuts are carried out until convergence. This residual-graph based approach differs from previously published iterative graph cuts [8, 9].

The paper is organised as follows. The following section (Section 2) contains a brief description of the graph cut technique. In Section 3, the construction of the energy function for our model is described. The results are presented in Section 4. Section 5 is a conclusion.

2   Graph Cuts in Image Processing

In this section, we recall the fundamentals of graph cuts. Let P be a set of all image pixels, and let L be a set of possible labels. A problem is to assign a label $l \in L$ to each pixel $p \in P$. It can be solved by minimization of the energy function

$$E(f) = E_{data}(f) + E_{smooth}(f) \tag{1}$$

that consists of the data term and the smoothness term, $f : P \to L$ is a labeling. The data term sets individual penalties for assigning a label l to a pixel p. The smoothness term penalizes the discontinuity of labeling between the neighboring pixels. This energy function can be minimized by various techniques. We use the graph cut technique.

A graph $G = (V, E)$ consists of a set of vertices V and a set of edges E. The set $V = P \cup \{s, t\}$ contains all nodes from P and two additional terminal nodes s, t representing the source and the sink, respectively. The set E contains two types of edges: t-links connect all non-terminal nodes with both terminals, and

n-links connect the neighboring non-terminal nodes. Each edge connecting the nodes p, q has a non-negative weight $w(p, q)$.

A cut $C \subset E$ splits the graph G into two disjoint sets S and T. The set S contains the terminal s and all nodes accessible from this terminal, the set T contains the rest of nodes. The size of cut is equal to the sum of weights of edges in C. It has been proven in [10] that the minimum cut problem is equivalent to finding the maximum flow between the source and the sink. The edges whose capacities have been saturated divide the graph into two disjoint sets similarly as the minimum cut. A more detailed description of the graph cut technique can be found in [11, 7].

In our case, the goal is to assign a label $l \in L = \{$'object', 'background'$\}$ to each pixel of input image. Due to this binary labeling, a single graph cut can be used, which is faster than α-expansion method [12], which is usually used in multi-label optimization.

3   The New Method

Firstly, we focus on the geometric transformation without considering occlusions. We suppose that we have two images of a scene taken from different positions. We use the notation u_L and u_R for the brightness function of the left and the right image, respectively. Due to the pinhole model, the images are connected by a projective transformation that is supposed to be known. Let T and B stand for the top and bottom points and let t, b now stand for the object and the background terminal, respectively. Say that, in the left image, we see a certain point whose position is x (its brightness is $u_L(x)$). In the right image, the position of the corresponding point depends on whether it is a top or bottom point. If it is a top point, the position is denoted by $\pi_{LTR}(x)$; in the case of bottom point the notation $\pi_{LBR}(x)$ is used. A similar transformation also works in the opposite direction. A point seen at x (its brightness is $u_R(x)$) in the right image can be found either at $\pi_{RTL}(x)$ or $\pi_{RBL}(x)$ in the left image (top or bottom point). From the geometrical point of view, the point seen at the position of $\pi_{LTR}(x)$ in the right image need not necessarily be only the top point that is seen at x in the left image, but at the same place, a bottom point can also be seen whose position in the left image is $\pi_{RBL}(\pi_{LTR}(x)) \equiv \pi_{LTB}(x)$. Similarly, the point that is seen at $\pi_{LBR}(x)$ in the right image need not only be the bottom point seen at x in the left image, but it can also be a top point that is seen at $\pi_{RTL}(\pi_{LBR}(x)) \equiv \pi_{LBT}(x)$. See Fig. 2 for graphical explanation. For determining the membership of x from the left image to the set of either top or bottom points, we define the function, denoted by $\varphi(x)$. The function is of the form

$$\varphi(x) = [u_R(\pi_{LBR}(x)) - u_R(\pi_{LTR}(x))] - [u_L(x) - u_L(\pi_{LTB}(x))]. \qquad (2)$$

The way how the function works is illustrated in Fig. 3 that depicts the situation if the base plane and the particular objects have constant brightness and if the object is brighter than the base plane. Under these assumptions, $\varphi(x)$ indicates whether the points are the top (object) points (indicated by negative values of

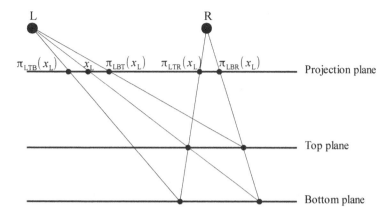

Fig. 2. Projective transformation between the images (without occlusions): If a top point is seen at x_L in the left image, then it should also be seen at $\pi_{LTR}(x_L)$ in the right image; otherwise, if a bottom point is seen at x_L, it should be seen at $\pi_{LBR}(x_L)$. The point at the position of $\pi_{LTR}(x_L)$ in the right image need not necessarily be only the top point seen at x_L; at the same place the bottom point can also be seen whose projection in the left image is at $\pi_{LTB}(x_L)$. Similarly, the point at $\pi_{LBR}(x_L)$ in the right image need not only be the bottom point at x_L; it can also be a top point that, in the left image, is seen at $\pi_{LBT}(x_L)$.

$\varphi(x)$) or bottom (base plane) points (positive values) or whether the membership cannot be decided (zero values).

3.1   Energy Function

The graph-cut algorithm minimizes an energy function that can be written in the form of

$$E(f) = \sum_{p \in P} D_p(f_p) + \alpha \sum_{(p,q) \in E} V_{p,q}(f_p, f_q), \tag{3}$$

where f_p, f_q stand for the labeling of p and q, respectively. Since the second part $V_{p,q}(f_p, f_q)$, called the smoothness term, is realised in an usual way [7], we focus only on the first part, the data term $D_p(f_p)$. Unlike the usual approaches, we propose a new term based on evaluating the mismatches between the stereo images. In addition to this, we also propose a supplementary term that uses the data from the residual graph, which improves the results. The weight parameter α sets a relative importance of the smoothness term versus the data term. The construction of the data term is described in the following paragraphs.

Data Term. The goal here is to find a function that is able, on the basis of the stereo images, to give the information about the locally expected membership of particular points either to the object or to the base plane. For this purpose, we use the function $\varphi(p)$ introduced before (p stands for a point), which means

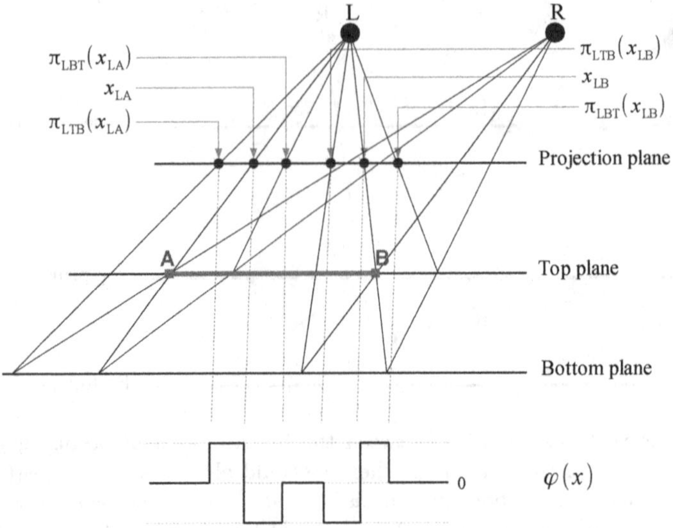

Fig. 3. The graphic explanation of Eq. (2) for the image in which the base plane has a constant brightness of b_B and the object has the brightness of b_T. We suppose that $b_T > b_B$ and we introduce $b_T - b_B = \delta b$. The function $\varphi(x)$ takes the value of $+\delta b$ in the neighbourhood of object, the value of $-\delta b$ inside the object, and the value of zero otherwise; the nonzero values only appear in the neighbourhood of object boundary.

that the labeling is done according to this function everywhere the function can decide. At the remaining points, the work is leaved to other terms, namely to the smoothness term and to the additional part of the data term, which will be discussed later.

Since the values of $\varphi(p)$ vary between positive and negative values that are dependent on the range of brightness in images, we firstly transform the values of $\varphi(p)$ into the interval $\langle 0, 1 \rangle$. It is done by the function

$$F_\varphi(p) = \frac{1}{1 + e^{\eta\varphi(p)+4\xi}}, \qquad (4)$$

whose value can be regarded as a probability $P(p \in T)$ and, at the same time, as the t-link capacity between p and the object terminal. The probability of being a background point is defined as the complement $1 - F_\varphi(p)$, which also defines the capacity of the t-link between p and the background terminal. The η parameter sets the slope of the function; its value is chosen in dependence on the brightness range. It can be easily seen that the negative $\varphi(p)$ values indicating the object points are mapped to higher probabilities, and vice versa. The values of $\varphi(p) = 0$, which indicates the situation that the function cannot decide, should be logically mapped to $F_\varphi(p) = 0.5$. We use the ξ parameter, however, to decrease the probability of the membership to the objects if $\varphi(p)$ cannot decide reliably. It can be easily checked that for $\varphi(p) = 0$, the probability is reduced by ξ in Eq. (4). This initial preferring the background instead of objects reduces the false object detections, i.e., if $\varphi(p)$ cannot decide about an isolated point reliably, we firstly see more probable that it is a background point. On the basis of the situation in the neighbouring points, the probability can be corrected later.

The values of $\varphi(p)$ and, therefore, also $F_\varphi(p)$ are only able to reliably detect the parts of the objects and the back plane near the object boundaries. The remaining work should be done by minimising the energy function from Eq. (3) with the smoothness term. We have found out, however, that the smoothness term itself is not always capable to do all the work sufficiently. It may sometimes happen that the parts of big objects are not merged together or less significant objects are not detected at all. We also recall that the values of probability were reduced in $F_\varphi(p)$ by ξ in order to reduce the false object detections.

Due to these facts, we support the term (4) by an additional process that can increase the reduced probability back, which is now carried out on the basis of distinguishing between the inner object areas and the background. For this purpose, we analysed the results of the graph cut and used them to design an additional part of the data term. It takes into account the data from the residual graph that remains after performing the max-flow algorithm. The data obtained in this way are used to modify the weights of the t-links in the original graph. A new graph cut is carried out afterwards, which makes the method iterative.

Let G_{res} be the residual graph, i.e., the graph that remains after removing the edges that were saturated in the max-flow algorithm. During the computation, at least one t-link is saturated for each $p \in P$; in the second t-link, some residual capacity may remain. The residual capacity of the t-link (p, b) connecting p with b (the terminal corresponding to the base plane) is denoted by $w_r(p, b)$. From the value of $w_r(p, b)$, we can see, how "far" the edge was from saturating. If the value of $w_r(p, b)$ is high, p was marked strongly (and probably correctly) as a bottom point. If, on the other hand, the value of $w_r(p, b)$ is low, the edge was almost saturated, the detection was weak, and a certain chance exists that the point should belong to the object instead to the base plane as was detected. In order to prefer the objects now, we modify the weights of edges in the original graph by increasing the probability of objects.

We have observed that the t-links with low residual weights are situated mainly around the areas convincingly detected as object points by a high value of $F_\varphi(p)$. It is caused by a high flow from the object terminal that is transported mainly to these object points. In order to reach the sink, however, a substantial part of the flow continues through the points neighboring with the object points. The flow to the points outside of the object is eliminated by a low capacity of the edges connecting the points with a higher brightness difference, which occurs between the object and the background. These facts explain why the inner object areas have lower residual capacities $w_r(p, b)$ than the background. We modify the weights of edges in the original graph by increasing the probability of being an object. According to the properties of the residual graph, these probabilities are more increased in the inner object areas. We simply set the maximum increase of probability to ξ, defined in Eq. (4), which compensates for previously preferring the background instead of objects.

The probability of the event that, in the $(k+1)$-st iteration, a point will become an object (top) point, providing that in the k-th iteration it was marked as the back plane (bottom) point is set to the following value

$$F_{\text{res}}^k(p) = P(p \in T^{k+1} | p \in B^k) = \xi e^{\frac{-w_r^k(p,b)}{2\sigma}}. \tag{5}$$

While the probability that p moves from the subset B to T is given by Eq. (5), the move in the opposite direction is not possible since it is assumed that once a point is marked as an object, it is done correctly.

We may now sum up that after introducing the correcting mechanism, the function expressing the capacity for the k-th iteration of the edge connecting p with the object terminal can be formulated as

$$F^k(p) = \min(F_\varphi(p) + F_{\text{res}}^{k-1}(p), 1). \tag{6}$$

In the first iteration, $F_{\text{res}}^0(p) = 0$ for all $p \in P$.

3.2   Graph Cuts Solution

The graph G contains the t-links whose weights are set as $D_p^k(b) = F^k(p)$ and $D_p^k(t) = 1 - F^k(p)$, where $D_p^k(b)$, $D_p^k(t)$ stand for the weights of the edges connecting p with the object and the background terminal, respectively, in the k-th iteration. The weights of the n-links are determined as

$$V_{p,q} = e^{\frac{-(I_p - I_q)^2}{2\sigma^2}}, \tag{7}$$

where I_p, I_q are the intensities of the neighboring pixels in the left input image. For solving the max-flow, the algorithm from [11] was used. Once the flow is found, the weights of the t-links are changed according to the values from the residual graph. The next max-flow iterations then follow until no change in cut is done (according to our experience, two or three iterations are sufficient).

4   Results

In this section, we describe the experiments with our method. Firstly, it has been tested on detecting the cars in a parking lot. This problem is well-studied in the case of detection from a single image [13], but there are still challenges to solve, such as the shadow effects, car occlusions or the luminance variations. Solving the problem from multiple images may overcome some of these difficulties.

The images are provided by a pair of uncalibrated cameras; the lens distortion is not compensated. The transformations $\pi_{\text{LTR}}(x_L)$, $\pi_{\text{LBR}}(x_L)$, $\pi_{\text{LTB}}(x_L)$, and $\pi_{\text{LBT}}(x_L)$ have chosen to be two-value functions that have the form of bicubic polynomials whose coefficients were found by the least squares method from two sets of calibrating points that were determined manually (the exact type of projection and distortion need not be known). The first set of calibrating points

(a) Example 1: 1st & 2nd iteration

(b) Example 2: 1st & 2nd iteration

Fig. 4. Results of our method and a comparison of the results after one and two iterations; the left column contains the results after single graph cut with data term set according to Eq. (4); the right column is after second iteration using Eq. (6)

contained selected corresponding points lying on the ground of parking lot; the second set contained selected corresponding points on the car roofs.

Fig. 4 contains the results from several parking lot situations. Each situation contains the result after one and two iterations, i.e., without and with the F_{res} term. The second iteration marks more pixels, which belong to a vehicle, correctly as object points, some vehicles are detected only after the second iteration. Therefore, the F_{res} term seems to be useful. The detection of bright objects is more successful than of the dark ones, which corresponds to the previously mentioned assumption that the objects are brighter than the background.

We have also compared our results to the stereo matching graph cut algorithm by Kolmogorov and Zabih [5] (K&Z algorithm). The K&Z algorithm finds the depth layers in the image, the layers closer to the camera are depicted as brighter in Fig. 5. For our algorithm, we just mark the detected objects with white color, the rest of the image contains the original image with reduced brightness. For this test, we have prepared two image pairs. The first pair contains a block of paper sheets and a single sheet of paper on a desk. The object we want to detect is the block, the sheet of paper has no height and should not be found. The K&Z algorithm correctly found the block as an object of another height, but also incorrectly detected other non existing depth layers in the background and on the front side of the box, which has to be regarded as a false detection. Our algorithm correctly found only the block.

The second pair of comparisons is detection of cars in a parking lot. The K&Z algorithm finds various levels, but the result contains noise in the background layer and the parts of many vehicles incorrectly fall into many various depths. Our algorithm found all vehicles in the image, but, as expected, the dark vehicles are not labeled as reliably as the bright ones.

A substantial difference between the methods is in their speed. Our algorithm, designed for specific problems only, uses only a small number of iterations of a graph cut algorithm. It makes the detection fast and usable even in the real-time applications. The K&Z algorithm has a wide range of use, but, due to this generality, it is computationally expensive. The speed comparison is in Fig. 5.

(a) Input images: block and one sheet of paper on a desk; a parking lot

(b) Results of the algorithm K&Z

(c) Results of our method

	Kolmogorov & Zabih	Our method
Block on a desk (320 × 180) pixels	92 s	0.3 s
Parking lot (960 × 540) pixels	625 s	2.4 s

Fig. 5. The comparison of the proposed algorithm with the method by Kolmogorov and Zabih [5] on two pairs of images (a). The results of K&Z method (b): Several depth layers incorrectly detected on the block (brighter layers are closer to the camera); noisy detection in the parking lot. The results of our method (c): correctly detected block; the sheet of paper is not detected, which is correct; acceptably detected cars in the parking lot. The table contains the computational time for each example

5   Conclusion

We have proposed a new method for bimodal depth segmentation using the graph cut technique and a new data term of energy function. The term is constructed on the basis of evaluation of mismatches between the pairs of images. The additional part of this term, which utilizes data from the residual graph, is proposed to reach more accurate results. Our method assumes that all objects have similar height and that the transformation between both images is known. Therefore, it is not usable for general stereo reconstruction problems, but it can be used, for example, in traffic surveillance. The method has been tested on detecting the cars in parking lots where it provides fast and promising results if the input conditions are fulfilled.

Acknowledgements. This work was partially supported by the grant SP2012/58 of VŠB - Technical University of Ostrava, Faculty of Electrical Engineering and Computer Science.

Bibliography

[1] Deriche, R., Bouvin, C., Faugeras, O.: Front propagation and level-set approach for geodesic active stereovision. In: Third Asian Conference on Computer Vision, Hong Kong (1998)

[2] Sun, J., Zheng, N.N., Shum, H.Y.: Stereo matching using belief propagation. IEEE Trans. Pattern Anal. Mach. Intell. 25, 787–800 (2003)

[3] Ohta, Y., Kanade, T.: Stereo by Intra- and Inter-Scanline Search Using Dynamic Programming. IEEE Transactions on Pattern Analysis and Machine Intelligence 7, 139–154 (1985)

[4] Roy, S., Cox, I.J.: A maximum-flow formulation of the n-camera stereo correspondence problem. In: ICCV, pp. 492–502 (1998)

[5] Kolmogorov, V., Zabih, R.: Multi-camera Scene Reconstruction via Graph Cuts. In: Heyden, A., Sparr, G., Nielsen, M., Johansen, P. (eds.) ECCV 2002. LNCS, vol. 2352, pp. 82–96. Springer, Heidelberg (2002)

[6] Hong, L., Chen, G.: Segment-based stereo matching using graph cuts. In: IEEE Computer Society Conference on Computer Vision and Pattern Recognition, vol. 1, pp. 74–81 (2004)

[7] Boykov, Y., Funka-Lea, G.: Graph cuts and efficient n-d image segmentation. Int. J. Comput. Vision 70, 109–131 (2006)

[8] Kootstra, G., Bergström, N., Kragic, D.: Fast and automatic detection and segmentation of unknown objects. In: 2010 10th IEEE-RAS International Conference on Humanoid Robots (Humanoids), pp. 442–447 (2010)

[9] Franke, M.: Color Image Segmentation Based on an Iterative Graph Cut Algorithm Using Time-of-Flight Cameras. In: Mester, R., Felsberg, M. (eds.) DAGM 2011. LNCS, vol. 6835, pp. 462–467. Springer, Heidelberg (2011)

[10] Ford, L.R., Fulkerson, D.R.: Flows in Networks. Princeton University Press (1962)

[11] Boykov, Y., Kolmogorov, V.: An experimental comparison of min-cut/max-flow algorithms for energy minimization in vision. IEEE Transactions on Pattern Analysis and Machine Intelligence 26, 359–374 (2001)

[12] Kolmogorov, V., Zabih, R.: What energy functions can be minimized via graph cuts. IEEE Transactions on Pattern Analysis and Machine Intelligence 26, 65–81 (2004)

[13] Huang, C., Wang, S.J.: A hierarchical bayesian generation framework for vacant parking space detection. IEEE Trans. Circuits Syst. Video Techn. 20, 1770–1785 (2010)

Real-Time Semantic Clothing Segmentation

George. A. Cushen and Mark. S. Nixon

University of Southampton, UK
{gc505,msn}@ecs.soton.ac.uk

Abstract. Clothing segmentation is a challenging field of research which is rapidly gaining attention. This paper presents a system for semantic segmentation of primarily monochromatic clothing and printed/stitched textures in single images or live video. This is especially appealing to emerging augmented reality applications such as retexturing sports players' shirts with localized adverts or statistics in TV/internet broadcasting. We initialise points on the upper body clothing by body fiducials rather than by applying distance metrics to a detected face. This helps prevent segmentation of the skin rather than clothing. We take advantage of hue and intensity histograms incorporating spatial priors to develop an efficient segmentation method. Evaluated against ground truth on a dataset of 100 people, mostly in groups, the accuracy has an average F-score of 0.97 with an approach which can be over 88% more efficient than the state of the art.

1 Introduction

Clothing can be considered to be one of the core cues of human appearance and segmentation is one of the most critical tasks in image processing and computer vision. Clothing segmentation is a challenging field of research which few papers have addressed. It can benefit augmented reality [1], human detection [2], recognition for re-identification [3], pose estimation [4], and image retrieval for internet shopping. Although the field has recently been gaining more attention, a real-time clothing segmentation system remains challenging. This is primarily due to the wide diversity of clothing designs, uncontrolled scene lighting, dynamic backgrounds, variation in human pose, and self and third-party occlusions. Secondly, difficult sub-problems such as face detection are usually involved to initialize the segmentation procedure.

In this paper, we present a clothing segmentation method for single images and video which can segment upper body clothing of multiple persons in real-time, as summarized in Figure 1. Our approach is primarily designed to benefit emerging augmented reality applications [1]. These include computer gaming and augmenting localized adverts or statistics onto players' shirts for close-up shots in live TV/internet broadcasting. Shirts worn by sports teams, such as in major league basketball, are often uniformly coloured with text and a logo to indicate the player and team. For this reason, we focus on the case of predominantly monochromatic tops, and attempt to additionally segment any textures on them which can be useful for the purpose of retexturing.

G. Bebis et al. (Eds.): ISVC 2012, Part I, LNCS 7431, pp. 272–281, 2012.
© Springer-Verlag Berlin Heidelberg 2012

Our main contributions include:

1. An efficient automated method for accurate segmentation of multiple persons wearing primarily uniformly coloured upper body clothing which may contain textured regions. Spatial priors are employed and each set of resulting cloth and texture contours are semantically labelled as such and associated to a face. Unlike most previous work which evaluates visually or with respect to applications (such as recognition), we evaluate the segmentation directly and quantitatively against a dataset of 100 people.
2. An initialization scheme where initial points on the cloth are located by estimating skin colour and employing an iterative colour similarity metric to locate the clothing. This can prevent initializing cloth points on the skin in the case of clothing with deep neck lines such as vests and many female tops.

Previous work is described in Section 2. An efficient approach for segmenting clothing is presented in Section 3. Performance is quantitatively assessed in Section 4, and conclusions are drawn in Section 5.

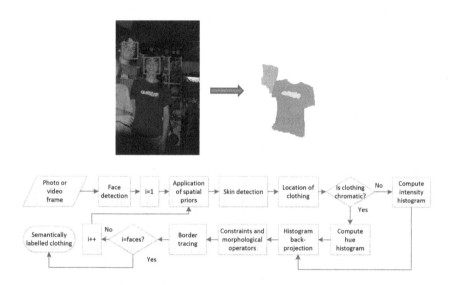

Fig. 1. Segmenting clothing with texture in groups of people. Top: the coloured clothing segments on the right, with texture (logos) depicted in yellow, correspond to the numerically labelled persons on the left. Bottom: system diagram.

2   Previous Work

One of the most popular approaches to clothing segmentation involves a Markov Random Field (MRF) framework based on graph cuts [5,6,3]. Although these approaches have robustness to a diverse range of clothing, they can suffer in accuracy, producing very crude segmentation. This is especially true in cases of occlusions and difficult poses. The MRF has since been reformulated to deal with groups of people [7]. More recently, Wang and Ai [9] introduced a clothing shape model which is learned using Random Forests and self-similarity features, with a blocking model to address person-wise occlusions. The fastest approaches in literature are our previous histogram work [1] (12fps for the overall multithreaded application) and the region growing approach presented in [8]. Although [8] extracts the person including skin pixels, their approach is fast, reporting 16.5ms per detected person for segmenting clothing and 10fps overall (including face detection and a classification application). Their private dataset was captured in a controlled lab setup, featuring a predominantly white background.

Face detection is generally employed in existing approaches to initialize points on the cloth. It should be noted that due to frontal face detection, the segmentation approaches are limited to frontal poses. The initial cloth points are located by applying a (scaled) distance from the bottom of the detected face. In the case of clothing with deep neck lines, such as vests and many female tops, these methods can segment the skin rather than the cloth. In contrast, we design a more complex initialization scheme which attempts to avoid this.

The majority of previous work focusses on segmenting a single image of a single person offline. Contrary to these methods, we attempt to simultaneously process multiple persons and maintain reasonable accuracy whilst increasing computational efficiency to enable real-time image/video processing. Furthermore, no existing approaches yield a semantically labelled segmentation which includes contours for any textured regions, such as logos, on the cloth.

3   Clothing Segmentation

3.1   Pre-processing and Initialization

For pre-processing, the single image or video frame is converted from RGB to the more intuitive and perceptually relevant HSV colour-space. The corresponding illumination channel is then normalized, giving image N. This helps to alleviate, to some extent, the non-uniform effects of uncontrolled scene lighting. Additionally, a 3×3 box blur is performed as a simple denoising measure, yielding image I. We let the H, S, and V channels correspond to I_0, I_1, and I_2 respectively and use the OpenCV HSV intervals $I_0 = [0, 180]$ and $I_{1,2} = [0, 255]$. For our image notation, we also refer to the origin as the top left of the image.

A chromatic/achromatic mask is defined where achromatic pixels are those with illumination extremes or low saturations:

$$\texttt{chrome}(I) = 0 \leq I_0 \leq 180 \wedge 26 \leq I_1 \leq 255 \wedge 26 \leq I_2 \leq 230 \qquad (1)$$

Viola-Jones face detection is performed on image N as a prerequisite for our segmentation approach. This technique is based on a cascade architecture for reasonably fast and accurate classification with OpenCV's popular frontal face trained classifier cascade. We limit the region of interest for object detection to the top half of the image in order to further increase efficiency. For each face detected, the segmentation procedure in the following sections is performed.

3.2   Spatial Priors

To increase robustness against hues/intensities in the background which are similar to those on the clothing, and to increase computational efficiency, we constrain segmentation of each person to a region of interest (ROI). The size of this region is determined by detecting faces in our training dataset (see Section 4) and studying the upper body clothing bounds, given by anatomy and pose, relative to the detected face size and position. As a result of these studies, spatial priors are defined as 5 times the detected face height and 4.5 times the face width and positioned as follows:

$$\text{crop}(I) = \text{Rect}(\text{Point}(F_x - 1.75F_{width}, F_y + 0.75F_{height}),$$
$$\text{Point}(F_x + 2.75F_{width}, F_y + 0.75F_{height} + 5F_{height})) \quad (2)$$

where the F vector for each person is output by face detection. The bounds of the ROI are also clipped to within the image dimensions.

3.3   Locating Points on the Clothing

Points on the clothing are required in order to initialise segmentation. Previous work often employs a scaled distance from a detected face to achieve this. However, this approach is susceptible to initialising clothing points on the skin in the case of clothing with deep neck lines such as vests and many female tops, and hence the segmentation has reduced accuracy. We propose a solution to this problem. The faces detected on the training dataset in the previous section are scaled to within 80×80 pixels, whilst maintaining their aspect ratios. We study the average face and define a region which tends to primarily be skin pixels and avoids occlusion by long hair:

$$\text{FSkin}(I) = \text{Rect}(\text{Point}(F_x + 15s, F_y + 36s), \text{Point}(F_x + 65s, F_y + 56s)) \quad (3)$$

where the scale factor $s = F_{width}/80$. The skin colour α is estimated by computing the mean of the pixels in the $\text{FSkin}(I)$ region.

A sparse iterative procedure is established across the $x = [F_x, F_x + F_{width}]$ and $y = [F_y + F_{height}, F_y + 2F_{height}]$ intervals, shifting a 5×5 pixel window. During each iteration, the mean colour β of the window is computed. The HSV colour similarity between the window's mean β and the estimated skin colour α is calculated. The two cylindrical HSV colour vectors are transformed to Euclidean space using the following formulae:

$$x = \cos(2I_0) \cdot I_1/255 \cdot I_2/255, \quad y = \sin(2I_0) \cdot I_1/255 \cdot I_2/255, \quad z = I_2/255 \quad (4)$$

The Euclidean distance d is then computed between the 3D colour points. If $d \leq 0.35$, we assume the window primarily contains skin pixels. The bottom of the clothing's neck, $Neck_y$, is located as the lowest 'skin window' within the aforementioned x and y intervals. Note that in the case that the subject is wearing clothing which is so similar in colour to their skin that the colour similarity distance remains below the threshold, we establish a cloth sampling window located around the x-coordinate of the face centre at the end of the y-interval. Otherwise, in the typical case, the cloth sampling window is located beneath the garment's neck at:

$$\begin{aligned}\texttt{sample}(I) =&\texttt{Rect}(\texttt{Point}(F_x + 0.25 F_{width}, Neck_y + 1.5\gamma),\\ &\texttt{Point}(F_x + 0.75 F_{width}, Neck_y + 1.5\gamma + 0.25 F_{height}))\end{aligned} \quad (5)$$

where γ refers to the aforementioned window size of 5 pixels.

3.4   Chromatic vs Achromatic

We design a histogram based approach because this is very efficient and can have a high accuracy on segmenting clothing which is primarily monochromatic. In such cases, it can also be suitable for semantic segmentation of printed/stitched textures within the clothing. First, we determine the chromatic ratio of the clothing which is estimated by taking the mean of the binary image $\texttt{chrome}(I)$ (see Equation 1) with the sampling ROI of Equation 5 applied:

$$\text{Chromatic Ratio} = r = \frac{1}{0.5 F_{width} \cdot 0.25 F_{height}} \sum_{x,y \in \texttt{sample}(I)} \texttt{chrome}(I(x,y)) \quad (6)$$

Second, the image plane for segmentation is determined based on whether the clothing is primarily achromatic or chromatic:

$$\text{Segmentation Plane} = S = \begin{cases} I_0 & \text{if } r \geq 0.5 \\ I_2 & \text{otherwise} \end{cases} \quad (7)$$

Based on these two cases, we empirically define some segmentation parameters in Table 1:

Table 1. Segmentation Parameters

Parameter	Segmentation Plane S	
	Hue I_0	Intensity I_2
q	16	15
λ	50	3

3.5   Clothing Segmentation

This section describes our histogram based segmentation routine. A histogram $\{g\}_{i=1...q}$ is computed for image plane S with the ROI $\texttt{sample}(I)$ applied:

$$g_i = \sum_{x,y \in \texttt{sample}(I)} \delta[b(x,y) - i]. \quad (8)$$

where δ is the Dirac delta function and let $b\colon \mathbb{R}^2 \to \{1 \ldots q\}$ be the function which maps the pixel at location $S(x,y)$ to the histogram bin index $b(x,y)$. We empirically choose to quantize to q bins as this provides a good compromise between under-segmentation (due to variation in cloth hue/intensity caused by lighting) and over-segmentation (due to objects with similar hues/intensities which are in direct contact with the clothing). Quantization reduces the computational and space complexity for analysis, clustering similar color values together. The histogram is then normalized to the discrete range of image intensities:

$$h_i = \min\left(\frac{255}{\max(g)} \cdot g_i, 255\right), \forall i \in 1 \ldots q \tag{9}$$

where h is the normalized histogram, g is the initial histogram, and subscripts denote the bin index.

Image S is back-projected to associate the pixel values in the image with the value of the corresponding histogram bin, generating a probability distribution image P where the value of each pixel characterizes the likelihood of it belonging to the clothing (i.e. histogram h). The resulting probability image is thresholded to create a binary image:

$$P(x,y) = \begin{cases} 255 & \text{if } P(x,y) \geq \lambda \\ 0 & \text{otherwise} \end{cases} \tag{10}$$

Scene conditions such as illumination can alter the perceived hue/intensity of the cloth, so we empirically set the λ threshold relatively low (see Table 1).

We further constrain P by considering $\mathtt{chrome}(I)$, the computed chromatic mask. If $S = I_0$, we let $P = P \wedge \mathtt{chrome}(I)$. Otherwise, if $S = I_2$ and $r \leq 0.05$, we constrain with the achromatic mask, letting $P = P \wedge (255 - \mathtt{chrome}(I))$. In the unlikely case that the sampled clothing pixels are mostly achromatic but not entirely (i.e. for $0.05 < r < 50$), we do not constrain P with the achromatic mask as it can exhibit significant holes at the location of coloured cloth pixels.

Morphological closing with a kernel size of 3×3 is employed to remove small holes, followed by opening, with the same kernel, to remove small objects. Experimentally, this has been found to fill small holes in the edges of the clothing caused by harsh lighting, and remove small objects of a similar hue/intensity to the cloth which are in contact with it from the camera's perspective.

Suzuki-Abe border tracing is employed to extract a set T of contours with corresponding tree hierarchy H. We choose to limit the hierarchy to 3 levels deep as this can provide sufficient information for the clothing contour, potential contours for a printed/stitched texture within, and potential holes within the texture contour(s). The top level of the hierarchy is iterated over, computing the bounding box area of each contour. The area is approximated to that of the bounding box for efficiency and experimentally this appears to be acceptable. We define the largest contour T_{max} as the clothing. Therefore, there is robustness to objects in the scene which have a similar hue/intensity as the clothing but are not in contact with it from the camera's perspective. An initial clothing segmentation mask $\tilde{M}$ can be defined by filling T_{max}.

The initial clothing segment $\tilde{M}$ can suffer in accuracy in cases of harsh illumination or patterned clothing. Robustness to these cases can be increased by sequentially performing morphological closing and opening with a large kernel size. The reason for not employing this larger kernel on the first iteration of morphological operations is that this can decrease accuracy if it is not just the clothing segment of interest present, but also other large objects of similar hue/intensity in the background. Since the morphological processes can create additional contours, border tracing is computed again to extract the clothing as one segment M.

3.6   Texture Segmentation

Existing clothing segmentation methods do not purposely attempt to semantically segment printed/stitched textures within clothing masks. Segmentation of any potential printed designs on clothing can be used to make the clothing cue more informative. We hypothesize that this could be useful for the purpose of re-texturing in emerging augmented reality clothing applications.

We iterate through the contours T in the second level of the contour hierarchy H, computing their areas. Unlike the area computation for the cloth contour, we do not approximate by bounding boxes here because textures can have more variation in shape, which may result in inaccurate area estimations. We consider contours with areas above a dynamic empirically defined threshold of $0.25F_{width}F_{height}$ to belong to a printed texture on the clothing. If no contour above the threshold is found, then we assume that there is no texture. Otherwise the extracted contours are filled and the regions of their corresponding hole contours in the third level of the hierarchy are subtracted (if they exist) from this result, yielding the texture mask.

4   Experiments

We study the quality of the clothing segmentation, robustness to noise, and the computational timing. Results are reported in Table 2, alongside a comparison with the state of the art. Two datasets are combined: Soton [10] and Images of Groups [11]. The Soton dataset consists of images of individuals whereas the Images of Groups dataset is very challenging for segmentation, featuring real-world Flickr photos of groups. A testing subset of 100 persons is formed from images featuring predominantly uniformly coloured upper body clothing and does not feature groups where clothing of adjacent persons is of a very similar colour and in direct contact. A training subset of 50 persons is randomly selected from remaining images in the dataset for use in Sections 3.2 and 3.3.

To compare the computational efficiency of our approach to the closest state of the art, we similarly consider static image regions for each person with resolution 200×300. Our approach, excluding pre-processing (Section 3.1), achieves on average 2ms per person using a 2.93GHz CPU core. Thus our method is over 88% more efficient than results reported by [8] under similar conditions, with the

Table 2. Histograms can provide efficient and effective clothing segmentation. The accuracy values are for indicative purposes only as they are not directly comparable.

	Our Method	[8]	[9]	[3]
Timing	**2.0**ms per person on 2.93GHz core	16.5ms per person on 3.16GHz core	Offline	Offline
Accuracy	**0.97** F-score	N/A	92.8%	89.4%

exception that they employ a faster CPU core (3.16GHz). Our overall system, including pre-processing (face detection and simple denoising), is fast, achieving results at an average rate of 25fps (frames per second) for segmenting one person given an input resolution of 480×640 pixels. The equivalent computation time is dissected as 38ms pre-processing per image and 2ms clothing segmentation per person. Face detection is our biggest computational bottleneck, so for high resolutions, the input to face detection could be downscaled. The segmentation procedure could easily be parallelized for each face detected, if using a multi-core CPU and the average number of persons to segment justifies the threading overhead.

Accuracy is reported using the best F-score criterion: $F = 2RP/(P+R)$, where P and R are the precision and recall of pixels in the cloth segment relative to our manually segmented ground truth. We achieve an average F-score over the entire testing dataset of 0.97. Since the F-score reaches its best value at 1 and worst at 0, our approach shows good accuracy. Additionally, by visual inspection of Figures 1 and 2, we can see that our approach can semantically segment clothing of persons in various difficult uncontrolled scenes with some robustness to minor occlusions (Figure 2(c)) and minor patterns (Figure 2(b)). Clothing segmentation literature tends to report accuracy with regards to applications (such as recognition or classification) rather than directly on segmentation. Although not directly comparable, the performance is higher than that reported in [9], using mostly images from the same dataset.

Finally, we consider robustness to one of the most common forms of noise: additive white Gaussian noise. This is caused by random fluctuations in the pixels. Naturally, this could be easily filtered but our aim is to demonstrate robustness. If the input image is represented by I_{input}, and the Gaussian noise by Z, then we can model a noisy image by simply adding the noise: $I_{noisy} = I_{input} + Z$. Z consists of 3 planes which correspond to the RGB planes of I_{input}, and is drawn from a zero-mean normal distribution with standard deviation σ. We study the effects of noise on a randomly selected image of a single person. Figure 3(a) depicts a graph of accuracy (F-score) versus the noise standard deviation (255σ) which shows our approach can handle significant noise. Note that we multiply σ by 255 since we consider integer images, and there is no data point plotted for $\sigma = 0.9$ because face detection mistakenly detects two faces and thus there are two results. Noise can positively affect our segmentation, for example at $\sigma = 0.2$, if noise pixels with hues similar to the cloth are established in dark clothing regions which originally had many unstable hues. At $\sigma = 1.0$, depicted by Figure 3(b), the face detection accuracy continues to decrease; however, the corresponding clothing segmentation in Figure 3(c) remains reasonably accurate. The segmentation fails entirely at $\sigma = 1.1$ since the prerequisite of face detection fails to detect any faces.

Our approach is subject to some limitations. We assume that there are no significant objects of a similar hue/intensity to the chromatic/achromatic clothing which are in direct contact with it from the camera's perspective, and the clothing is predominantly uniformly coloured (i.e. it is not significantly patterned). These limitations should not significantly affect the suggested computer games and broadcasting applications for the purpose of augmented reality.

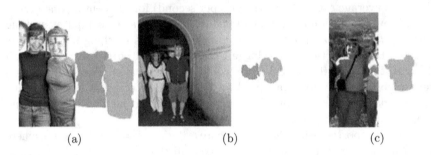

(a)   (b)   (c)

Fig. 2. Further segmentation results. Each pair shows the numerically labelled person(s) on the left with their corresponding colour labelled clothing on the right.

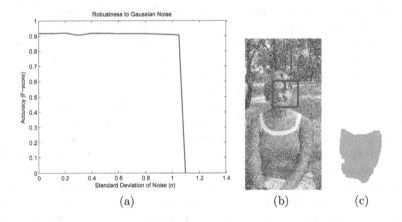

(a)   (b)   (c)

Fig. 3. Robustness to noise: (a) graph of accuracy versus Gaussian noise σ, (b) input with considerable noise ($\sigma = 1.0$), and (c) corresponding clothing segmentation

5   Conclusions

We have presented an algorithm for automatic semantic clothing segmentation of multiple persons. It does so by estimating whether the clothing is chromatic or achromatic and then applying a histogram based approach on the hues or intensities. In order to initialise points on the cloth, we have proposed a method consisting of skin colour estimation and colour similarity to locate the bottom of the garment's neck. We have shown that the proposed framework is able to segment clothing more efficiently than existing state of the art methods, whilst achieving good accuracy and robustness on a difficult dataset. Although our approach is limited to predominantly uniformly coloured clothing (which may contain textured regions), it may be of particular benefit to emerging real-time augmented reality applications such as sports broadcasting and computer gaming.

References

1. Cushen, G., Nixon, M.: Markerless Real-Time garment retexturing from monocular 3D reconstruction. In: IEEE ICSIPA, Malaysia, pp. 88–93 (2011)
2. Sivic, J., Zitnick, C.L., Szeliski, R.: Finding people in repeated shots of the same scene. In: BMVC, vol. 3, pp. 909–918 (2006)
3. Gallagher, A.C., Chen, T.: Clothing cosegmentation for recognizing people. In: CVPR 2008, pp. 1–8. IEEE (2008)
4. Lee, M.W., Cohen, I.: A model-based approach for estimating human 3D poses in static images. IEEE TPAMI, 905–916 (2006)
5. Schnitman, Y., Caspi, Y., Cohen-Or, D., Lischinski, D.: Inducing Semantic Segmentation from an Example. In: Narayanan, P.J., Nayar, S.K., Shum, H.-Y. (eds.) ACCV 2006. LNCS, vol. 3852, pp. 373–384. Springer, Heidelberg (2006)
6. Hu, Z., Yan, H., Lin, X.: Clothing segmentation using foreground and background estimation based on the constrained Delaunay triangulation. Pattern Recognition 41, 1581–1592 (2008)
7. Hasan, B., Hogg, D.: Segmentation using Deformable Spatial Priors with Application to Clothing. In: BMVC, pp. 1–11 (2010)
8. Wang, N., Ai, H.: Who Blocks Who: Simultaneous Clothing Segmentation for Grouping Images. In: ICCV (2011)
9. Yang, M., Yu, K.: Real-time clothing recognition in surveillance videos. In: IEEE ICIP, pp. 2937–2940 (2011)
10. Seely, R.D., Samangooei, S., Lee, M., Carter, J.N., Nixon, M.S.: The University of Southampton Multi-Biometric Tunnel and introducing a novel 3D gait dataset. In: BTAS, pp. 1–6. IEEE (2008)
11. Gallagher, A., Chen, T.: Understanding Images of Groups Of People. In: CVPR (2009)

Detection and Normalization of Blown-Out Illumination Areas in Grey-Scale Images

Karolina Nurzyńska[1] and Ryszard Haraszczuk[2]

[1] Silesian University of Technology, ul. Akademicka 16, 44-100 Gliwice, Poland
Karolina.Nurzynska@polsl.pl
[2] Institute of Medical Technology and Equipment, ul. Roosevelta 118,
41-800 Zabrze, Poland
hrysiek@itam.zabrze.pl

Abstract. This work presents a novel approach for blown-out illumination detection and normalization in grey-scale images. This situation takes place when the source of light is too close to the object or the object has high specularity. The presented algorithm calculates the image histogram and exploits it for further parameters calculation, which become a base for background detection algorithm, too bright pixel range definition, and finally the wrong lighting normalization. Examples of introduced algorithm performance are given.

1 Introduction

Variation of illumination on images causes a big problem for automatic systems for reconstruction of surface from an image, object recognition, and classification. In case of surveillance systems, which aim at face recognition, the non-uniform lighting of the object influences the system efficiency. On the one hand, many systems applied to derive information from the traffic needs to deal with shadows casted by the moving cars. There are also problems when an irregular object's shape generates self-shadowing on the image. On the other hand, when taking a photograph with artificial light, when the distance between light source and object is small or the object surface has high specularity, there are regions with blown-out illumination. For algorithms which try to reconstruct surface from one image, all these problems make this task difficult to achieve.

Shape from shading is a technique which reconstructs 3D object shape from its 2D representation on an image. It derives the information about object shape from shadows and illuminance presented on one image only. It was introduced and developed by Horn [1], who noticed that shadow cannot originate from uniformly illuminated, smooth, continuous surface. Hence, it is assumed that some change in gradient on the image corresponds to the change of the object's shape. There are many works [2], [3], [4], [5], [6], [7], [8] based on this idea, which try to improve the overall quality of the reconstructed surface. Dorou et al. [9] gives an interesting report comparing different approaches presented to solve this problem.

G. Bebis et al. (Eds.): ISVC 2012, Part I, LNCS 7431, pp. 282–291, 2012.

In shape from shadow techniques, the shadows are supposed to determine the surface shape. Yet, it is not always the case. A wrong lighting conditions may generate some additional shadows (cast-shadow, self-shadow), which result in change of the surface shape. Therefore, it would be useful to detect this situation in order to correctly interpret the data on the image.

There is a broad research concerning the problem of shadow detection [10],[11], [12]. For example, Rosin et al. [13] suggest a solution which bases on hysteresis thresholding. More general solution, which is applicable without restriction on the number of light sources, illumination conditions, surface orientations, and object sizes is presented in [14]. Unfortunatelly, most of these techniques works on sequences of images.

On the other hand, there are also techniques which deal with lighting normalization in order to improve face recognition. These methods, however, aim at such image lighting normalization that improves the system efficiency, but does not pay attention on the image quality at all [15],[16].

The contribution of this paper is to describe the technique for detection of parts of images with blown-out illumination on grey scale images. This phenomenon is noticeable on images when the source of light is too close to the object and also on objects with high specularity. This work presents a statistical approach for the wrong illumination area detection and its correction.

2   Blown-Out Illumination Detection

In grey-scale images, which are considered in this research, the information about the surface on the images is reflected by the pixel's value. Pixel's values distribution on the image allows defining the range of grey colours, which in comparison to other shades in the image are too bright. For instance, Fig. 1 presents an exemplary histogram for an image. There are distinguished ranges corresponding to background and object shades. Moreover, using statistical means additional parameters are derived to find the blown-out illumination range.

Pixel's Value Distribution. The distribution of pixel's values on the image is represented as a histogram. The histogram is a compact image representation, which, when understood properly, finds many applications. For instance, one can read from histogram's shape the quality of the image, like its contrast and sharpness. It is also possible to use this information for texture operator definition, which supplies data necessary for image classification and recognition. Next, there are algorithms, which on the basis of image histogram, are able to segment objects on an image.

The goal of the proposed algorithm is to determine with high certainty, the range of grey values that best represent blown-out illumination of objects in an image. As a starting point the image histogram is used. When histogram techniques are applied for object segmentation, the minima of pixel's distribution define the threshold between two separate objects on the image. In presented approach, it is assumed that background and object are these different objects on

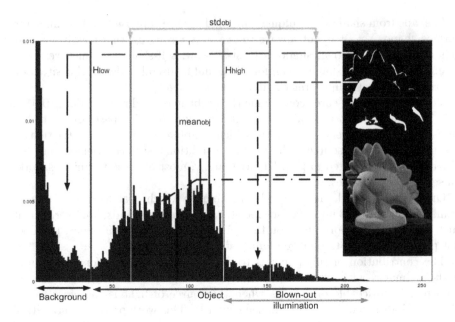

Fig. 1. This image presents histogram calculated for dino image. There are distinguished grey shade ranges corresponding to background and object shades. Additionally, the blown-out illumination range is also marked. The statistical parameters used are explained in the text.

the image. Additionally, the maxima are correlated to object's mean values. The problem is to decide whether each function extremum is global one and can be regarded as described above, or is just a local turbulence and should be neglected. In order to decide, statistical data calculated for the image is exploited. Below the novel algorithm is introduced.

Histogram Extrema Calculation. Lets define pixel's value distribution function H (a histogram) for an image I of resolution MxN as:

$$H(i) = \sum_{x=0}^{M-1} \sum_{y=0}^{N-1} \begin{cases} 1, & I(x,y) = i \\ 0, & I(x,y) \neq i \end{cases} ,$$

where i corresponds to grey level value. Due to the definition, the function is discrete. Moreover, some small jumps between consecutive values are noticeable and may be regarded as an extremum. Therefore, the smoothing operator is applied to the function twice in order to remove local variations. The definition of the smoothing operator is following:

$$H_{smooth}(i) = M_{i-W}^{i+W} H(i) \tag{1}$$

where W defines a window in which a median (M) smoothing takes place; in the implementation this parameter was set to 3.

Having smoothed function, the algorithm is looking for the extrema. Firstly, it labels each i into one of three classes: flat, descending, and ascending. An i is set as flat when $H(i) = H(i + 1)$; it is set descending when $H(i) < H(i + 1)$; it is ascending otherwise. Next, the change from ascending into descending defines the function maximum. In case when the hill is flat on the top, the middle i is set to be a maximum. Similarly, when the function changes from descending to ascending class, the minimum is defined. As a result, a list of global minima L_{min} and a list of global maxima L_{max} are created; both of them store information about extrema placement in histogram (i) and its value $(H(i))$.

Background Information Extraction. In order to better define the object statistical characteristics, it is good to know, which part of the grey scale it may occupy. Therefore, this algorithm assumes input information whether some background is present, otherwise whole histogram data is used. In current version, only dark background is assumed. When user specifies, that there is a background on the image, the algorithm searches the list of global minima for the one with lowest value $(H_{low} = min(L_{min})_{index})$. If it is the first one, the background is defined as follows:

$$Background = \bigvee i < H_{low} \qquad (2)$$

otherwise the list of global minima is searched iteratively comparing two consecutive minima $(L_{min}(k)$, and $L_{min}(k + 1))$ until one of the rules is fulfilled:

$$L_{min}(k)_{value} < 1.10 * L_{min}(k + 1)_{value} \qquad (3)$$

or

$$L_{min}(k)_{index} < max(L_{max})_{index} < L_{min}(k + 1)_{index}. \qquad (4)$$

In both cases H_{low} becomes $L_{min}(k)_{index}$.

Object Statistics. Since background is defined, or there is no background on the image, it is assumed that the rest of the histogram corresponds to the object and for this part statistics are calculated. There are only three statistical parameters which are sufficient for accurate blown-out illumination description. First of them is the mean value of object's histogram given by the formula:

$$mean_{hist} = \frac{\sum_{i=H_{low}}^{256} H(i) * i}{\sum_{i=H_{low}}^{256} \begin{cases} 1, & H(i) > 0 \\ 0, & H(i) = 0 \end{cases}} \qquad (5)$$

and the second and third are object's mean value and standard deviation of object, defined as:

$$mean_{obj} = \frac{\sum_{x=0}^{M-1} \sum_{y=0}^{N-1} I(x, y) * B(x, y)}{D} \qquad (6)$$

$$std_{obj} = \sqrt{\frac{(mean_{obj} - I(x,y))^2}{D}} \qquad (7)$$

where

$$D = \sum_{x=0}^{M-1} \sum_{y=0}^{N-1} B(x,y) \qquad (8)$$

and $B(x,y)$ is a binary image created according to the formula:

$$B(x,y) = \begin{cases} 1, & I(x,y) > H_{low} \\ 0, & otherwise. \end{cases} \qquad (9)$$

Blow-Out Illumination Threshold Calculation. Finally, when all these parameters are known, the algorithm is looking for the H_{high} value, which is the threshold of blown-out illumination in the histogram. The applied algorithm is derived from the idea that the histogram of uniformly lighted object has normal distribution. Therefore, the possibility to distinguish other local maxima and minima suggests that it might be a histogram of many objects, which overlap each other (e.g. uniformly lighted object and its part, which has blown-out illumination).

The definition of blown-out illumination threshold is two-step process: firstly the pixel's distribution function minimum should be found, which index is higher than object's mean value; next the threshold value is tuned in this minimum neighbourhood. Namely, the first step is driven by the formula:

$$j = min(L_{min} > mean_{obj}) \qquad (10)$$

and followed by:

$$H_{high} = min_j^k(H_{smooth}(i) \leq mean_{hist}), \qquad (11)$$

where

$$k = max(L_{max})_{index} < j. \qquad (12)$$

3   Lighting Normalization

Since the blown-out illumination threshold is known, the binary mask which represents too bright pixels is calculated. It is used to determine which pixels should be further transformed in order to diminish the wrong illumination. The pixel colour transformation itself is based on the histogram and image statistics. Generally, the range of shades of too bright pixels is found and the range in which, from statistical point of view, the object pixels should be. Then the shades are scaled linearly.

The range of too bright pixels is enclosed between the H_{high} value and H_{max} the maximal colour recorded in the histogram:

$$H_{max} = max_i(H(i) \neq 0) \qquad (13)$$

On the other hand, the range of the object is defined by the mean object value, $mean_{obj}$, and object standard deviation, std_{obj}. It is assumed in statistics, that when classifing an object to a class described by class's mean value and with known standard deviation, the object, o, with high probability belongs to the class if following rule is fulfilled:

$$mean_{obj} - 3 * std_{obj} < o < mean_{obj} + 3 * std_{obj}. \tag{14}$$

In presented problem it means that a pixel belongs to an object if the difference between pixel shade and object mean colour value is less that 3 standard deviation. However, sometimes the range of the object might be too big and allow the shade to exceed the maximal value of 256. Therefore, in such a case, this rule is narrowed to describe the object colour distribution with only 2 standard deviation range. Having these assumptions in mind, the maximal accepted shade is defined as:

$$H_{acc} = mean_{obj} + k * std_{obj}, \tag{15}$$

where k is set 2 or 3 depending on the circumstances.

Finally, the pixel shade transformation is defined as:

$$I(x,y) = H_{high} + (H_{acc} - H_{high}) \frac{I(x,y) - H_{high}}{H_{max} - H_{high}} \tag{16}$$

and applied only for pixels marked on the binary map.

4 Experiments and Results

This section presents experiments showing the performance of presented algorithm for blown-out illumination detection and normalization. There are examples of images with bright object on dark background and also when it is difficult to define the background. The presented images are also used for performance evaluation in case of following techniques:

- **Beethoven image** - shape from shading object surface reconstruction;
- **Dino dataset** - shape from silhouette object surface reconstruction;
- **Yale database** - face detection and recognition.

Beethoven Image. This is an example of an image presenting an object on the black background. Fig. 2 presents, from right to the left, the original image, the detected area of blown-out illumination, the image after normalization, and in the second row pixel's value distribution. Because the background pixels are the majority on the image, their histograms values were very high, therefore, in order to show the histogram shape in detail, its values have been set to zero. Otherwise, the histogram seems flat.

There are only small regions with too bright pixels on *Beethoven* image, which are depicted on Fig. 2b. One can see on Fig. 2c, the influence of the normalization on overall image quality. Generally, the blown-out illumination is reduced, whereas the smoothness of object surface is kept.

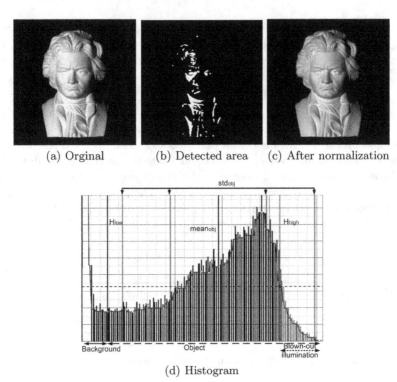

(a) Orginal   (b) Detected area   (c) After normalization

(d) Histogram

Fig. 2. Beethoven image example. Top row presents from right: original image, area with blown-out illumination, and result after normalization. Bottom row depicts image histogram.

Dino Example. Fig. 3a presents an example of a case, when grey-scale object with non-linear lighting was captured on darker background. In this case the background is more complex, however the presented algorithm correctly detects the background threshold. There are also no problems to precisely describe the regions when the lighting changes considerably on the image, as it is presented on Fig. 3b. Finally, Fig. 3c shows a correction of the illumination. It is noticeable that the pixel's values were well corrected in the wrong illuminated area. Yet, the difference in lighting was not completely removed from the image.

Yale Database Example. Fig. 4a is an example of image where it is difficult to define a background. Moreover, it is a very dark image. When one considers detecting a face on such image, there are two problems to address: firstly the non-linear face illumination and secondly the bright pixels, resulting from the window. Presented algorithm, correctly detects both these areas as is depicted on Fig. 4b. Next, the suggested approach normalizes the illumination considerably in the region of interest (see Fig. 4c). What is worth mentioning, the changes do not disturb the image quality, and still the features of face are well maintained.

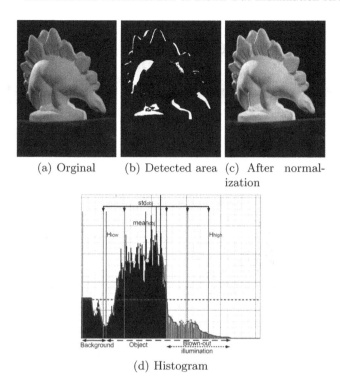

(a) Orginal   (b) Detected area   (c) After normalization

(d) Histogram

Fig. 3. Dino database example. Top row presents from right: original image, area with blown-out illumination, and result after normalization. Bottom row depicts image histogram.

5 Conclusions

This work presents a novel approach for detection of regions with blown-out illumination on the image. The introduced algorithm allows detection of the region of interest on the images and performs its normalization. It works well when there is a bright object on dark background, but it also manages well when an image with non-linear lighting is considered. The presented results show that the algorithm works well and improves the quality of images, as assumed.

This algorithm may found its application in many domains. In this research, it is presented how it improves the information about a shape in the images applied in shape-from-shading surface reconstruction. The example of data used in case of shape-from-silhouette is also considered, as there are approaches which combine the silhouettes with image data. Moreover, it is shown that this algorithm improves the quality of images with non-linear lighting, and may be a useful tool in systems for automatic face detection and recognition. The big advantage of this approach is that only one image is necessary to achieve the lighting normalization, however, using an image sequence probably should improve the performance.

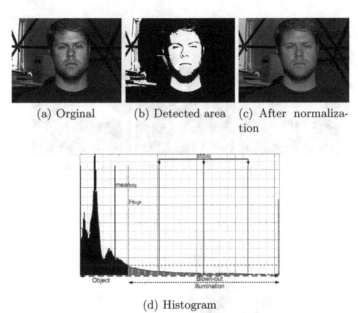

(a) Orginal   (b) Detected area   (c) After normalization

(d) Histogram

Fig. 4. Yale database example. Top row presents from right: original image, area with blown-out illumination, and result after normalization. Bottom row depicts image histogram.

Concluding, it is believed that this approach for blown-out illumination normalization may found a broad application in image processing domain. Although, there are also some problems which need further consideration. One of them is, the problem depicted by dino example, where one might expect, that after normalization the sharp borders between the area with too bright pixels and the rest of the object will disappear. Therefore, the further investigation should consider also the information about pixel's neighbourhood.

Acknowledgements. This work was partially supported by the European Union from the European Social Fund (grant agreement number: UDA-POKL.04.01.01-00-106/09).

References

1. Horn, B.K.P., Szeliski, R.S., Yuille, A.L.: Impossible shaded images. IEEE Trans. Pattern Anal. Mach. Intell. 15, 166–170 (1993)
2. Ascher, U.M., Carter, P.M.: A multigrid method for shape from shading. Technical report, Vancouver, BC, Canada, Canada (1991)
3. Zheng, Q., Chellappa, R.: Estimation of illuminant direction, albedo, and shape from shading. IEEE Transactions on Pattern Analysis and Machine Intelligence 7, 680–702 (1991)

 4. Szeliski, R.: Fast shape from shading. In: Faugeras, O. (ed.) Computer Vision ECCV 1990. LNCS, vol. 427, pp. 359–368. Springer, Heidelberg (1990)
 5. Vogel, O., Breuß, M., Weickert, J.: Perspective Shape from Shading with Non-Lambertian Reflectance. In: Rigoll, G. (ed.) DAGM 2008. LNCS, vol. 5096, pp. 517–526. Springer, Heidelberg (2008)
 6. Vogel, O., Breuß, M., Leichtweis, T., Weickert, J.: Fast shape from shading for phong-type surfaces. In: Proceedings of the Second International Conference on Scale Space and Variational Methods in Computer Vision, SSVM 2009, pp. 733–744. Springer, Heidelberg (2009)
 7. Lei, Y., Jiu-Qiang, H.: A perspective shape-from-shading method using fast sweeping numerical scheme. Optica Applicata 38, 387–398 (2008)
 8. Prados, E., Camilli, F.: A unifying and rigorous shape from shading method adapted to realistic data and applications. Journal of Mathematical Imaging and Vision, 307–328 (2006)
 9. Durou, J.D., Falcone, M., Sagona, M.: Numerical methods for shape-from-shading: A new survey with benchmarks. Computer Vision and Image Understanding 109, 22–43 (2008)
10. Withagen, P., Groen, R., Schutte, K.: Shadow detection using a physical basis. In: IEEE Conference Proceedings Instrumentation and Measurement Technology, pp. 119–124 (2008)
11. Kumar, S., Kaur, A.: Shadow detection and removal in colour images using matlab. International Journal of Engineering Science and Technology 2, 4482–4486 (2010)
12. Guo, R., Dai, Q., Hoiem, D.: Single-image shadow detection and removal using paired regions. In: IEEE Conference on Computer Vision and Pattern Recognition, CVPR 2011, pp. 2033–2040 (2011)
13. Rosin, P., Ellis, T.: Image difference threshold strategies and shadow detection. In: Proc. British Machine Vision Conf., pp. 347–356. BMVA Press (1995)
14. Joshi, A., Atev, S., Masoud, O., Papanikolopoulos, N.: Moving shadow detection with low- and mid-level reasoning. In: IEEE International Conference on Robotics and Automation, pp. 4827–4832 (2007)
15. Xie, X., Lam, K.: An efficient illumination normalization method for face recognition. Pattern Recognition Letters 27, 609–617 (2006)
16. Chen, C.P., Chen, C.S.: Lighting normalization with generic intrinsic illumination subspace for face recognition. In: Tenth IEEE International Conference on Computer Vision, vol. 2, pp. 1089–1096 (2005)

A Synthesis-and-Analysis Approach to Image Based Lighting

Vishnukumar Galigekere and Gutemberg Guerra-Filho

Department of Computer Science and Engineering, University of Texas at Arlington,
Arlington, TX, USA

Abstract. Given a set of sample images of a scene with variable illumination
and the corresponding parameters of the light sources, namely position and/or
intensity, we propose an interpolation based approach to model the variation of
illumination in images. Once our interpolation based model is constructed from
the sample images, we are able to synthesize images under any possible lighting
configuration defined in the parametric space. Moreover, given a query image
of a scene with a known reference object, our method is able to estimate the
lighting parameters of the image. Therefore, our approach allows for both synthesis and analysis of images in different lighting conditions. Our model is
ultimately a compact representation of the set of all images with the lighting
conditions defined within a parametric space. The interpolation model can generate the rendering of different objects in unknown environments and also
perceive unknown environments provided a known object exists in the scene. In
this paper, we show robust image synthesis and analysis with two different datasets: an object image dataset with varying lighting intensity and a face image
dataset with varying light source position.

1 Introduction

In computer graphics, image based modeling techniques are extensively used to synthesize new images of a given scene. Recently, Image Based Modeling approaches
have been proposed to obtain depth maps from stereo images of a real scene and then
to construct a corresponding textured 3D model [9]. The 3D model is re-projected
into a different viewpoint as a synthetic 2D image. An important part of this methodology is referred to as Image Based Lighting (IBL) [4]. IBL considers different lighting conditions retrieved from sample images of an object under several local light
sources and a lighting representation of a different location to synthesize new images
of this object in the different location. Tunwattanapong *et al.* [19] introduced an IBL
approach which allows for the editing of the lighting in terms of intensity and angular
width of local light sources.

In addition to sample images of the target objects under different lighting conditions, a limitation of current IBL approaches is the requirement for a lighting representation of the specific environment to be considered in the synthesis of new images.
This means that the generation of new images requires access to this specific location
for the capture of data that leads to the lighting representation. An example of such a

G. Bebis et al. (Eds.): ISVC 2012, Part I, LNCS 7431, pp. 292–304, 2012.

lighting representation is the environment map or incident light map. Therefore, IBL techniques cannot be applied to the modification of existing images, when access to the actual location where the images were captured is not possible. Furthermore, lighting representations are specific to a particular environment and its corresponding lighting conditions at that time. These representations cannot account for the entire range of possible lighting conditions and, consequently, do not allow the generalization necessary to consider lighting variation.

In this paper, we address the IBL problem to advance the state-of-art towards these two directions: independence of lighting representations and generalization to lighting variation. We propose a novel interpolation based approach for the synthesis and analysis of images under lighting variation. Our approach synthesizes images in different lighting conditions as well as infers the parameters of the scene by perceiving objects (analysis), where the lighting parameters are the positions of the light sources and/or their intensity levels. Formally, given a set S of n training images of an object under different lighting conditions and their corresponding parameters, $S = \{(p_1, i_1), (p_2, i_2), ..., (p_n, i_n)\}$, where $p_j \in P$ is the parameter of image i_j in the parametric space P for $j = 1, ..., n$; our interpolation based approach synthesizes any image i_k for a given parameter p_k within the parametric space such that $(p_k, i_k) \notin S$. Conversely, our approach is also able to estimate the parameter p_q of a given query image i_q such that $(p_q, i_q) \notin S$.

Our approach is motivated by the mirror neuron theory [17] in Neuroscience that essentially places perception and generation under the same foundation. The mirror neuron theory states that the same neurons fire when a person perceives a particular sensory-motor pattern and when the subject generates the same pattern. This theory indicates that both synthesis and analysis are performed according to the same fundamental framework. Inspired by this theory, we propose a framework to perform the synthesis and analysis of object images under general lighting conditions.

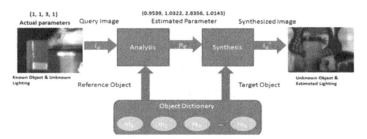

Fig. 1. The query image i_q has unknown parameters and at least one known object. This reference object is used in the analysis component to estimate the imaging parameter p_q. Once the parameter is found, our synthesis component generates the output i_q' as the rendering of an unknown object in the query image according to the estimated lighting parameters.

Our approach enables the editing of existing images and provides an alternative to image based lighting when the construction of lighting models is unfeasible. In other words, we propose a framework where an object can be rendered in any scene (and respective lighting condition) as long as a known object (*i.e.*, an object whose interpolation model is obtained previously) exists in that particular scene. Formally, let us define a dictionary D of k pairs of objects in a known object set O and their corresponding interpolation models in a model set M as follows: $D = \{(o_1, m_1), (o_2, m_2), ..., (o_k, m_k)\}$, where m_j is the interpolation model of a known object o_j for $j = 1, ..., k$. Now, given an image i_q of an unknown scene (under an unknown illumination condition) that contains a known object $o_r \in O$, we can compute the parameter p_q using the model m_r of the reference object o_r. Note that the parameter estimation may use more than one known object in the scene for a more robust outcome. The parameter p_q represents the lighting conditions for the image i_q of the considered scene. This parameter is used to render any known object according to the lighting conditions of the target image i_q. Fig. 1 shows a schematic drawing of this analysis-and-synthesis framework.

Besides eliminating the dependence on lighting representations, our approach is able to generalize the given samples to generate any possible new image associated with a parameter in the parametric space. In this paper, we consider sample images of objects under several local light sources. From the sample data, our method generalizes to light sources at different positions and with different intensities.

We demonstrate that our interpolation based approach performs well in the synthesis from the parametric space to the image space and in the analysis from the image space to the parametric space with images taken under fixed viewpoint and varying illumination. We present experimental results with two image datasets: a *face image dataset* captured with same intensity light sources at different positions and an *object image dataset* captured with fixed pose light sources at different intensities. The synthesized images are compared to ground-truth images using a leave-one-out testing methodology.

2   Previous Work

Image Based Rendering (IBR) techniques either use plenoptic functions [11,12,13] or construct the 3D geometry of the scene from a sparse set of images [20]. Shum *et al.* [18] presented a survey of the IBR techniques. A major issue in most of these techniques is that the rendering is possible only for a fixed viewpoint or a single illumination condition. The geometry based approach is slightly more flexible in terms of rendering images with slightly different views or illumination settings. We will now briefly describe a few popular papers in these areas.

Debevec *et al.* [7] acquire the reflectance field of a human face using a setup called 'light stage'. The light stage allows them to capture still facial data under a small set of viewpoints and dense incident illumination directions. The reflectance function is constructed for each pixel over space of incident illumination directions. Relighting is achieved using mirrored ball images as light sources. This method achieves realistic and visually appealing results. However, the approach requires a light stage setup for data capture and huge amounts of data. Moreover, inclusion of new data or editing existing data requires a significant effort.

Most of the approaches discussed above, are data intensive. Tunwattanapong *et al.* [19] significantly reduce the number of images required to achieve good quality relighting. This method produces visually appealing results despite significantly reducing the data requirements. However, rendering new lighting requires access to the very location. Moreover, having to render a scene with a different lighting setup would require a new environment map. Although changes can be made to an existing image based on the local lights, a capability to render all possible illumination settings given a set of lighting parameters is still missing. Our approach, on the other hand, is capable of rendering any possible illumination settings in the parametric space. The interpolation based model discussed here can work as a standalone application as long as we can have the lighting parameters. If not, it can certainly aid other techniques such as that of [19] to broaden its scope of problems they can address.

Another approach addressing image relighting or retexturing is known as image decomposition [3]. This approach decomposes an image into shading and albedo. Once reflectance and illumination have been decoupled, the scene can simply be relighted with a new illumination map. Bousseau *et al.* [2] decompose a single image into intrinsic images with the aid of user inputs to disambiguate illumination from reflectance. Their approach mostly addresses issues related to user based image editing.

Malzbender *et al.* [14] proposed an image based approach that builds Polynomial Texture Maps (PTM) based on a quadratic polynomial. With regards to the synthesis part of our algorithm, the PTMs are constrained to a quadratic model whereas our approach considers any component function (*e.g.*, polynomials, exponentials, logarithmic, trigonometric) as long as it is invertible to be used in the analysis part of our method. Furthermore, our approach may use any number of component functions. Another major difference is that our method employs the subdivision of the parametric space into neighborhoods. This not only allows for a better modeling of local features but can also significantly speed up the synthesis process. In addition to the image based synthesis, another original contribution of our method concerns the inverse problem where the lighting parameters are inferred from images. This analysis part of our method is what enables the computation of the illumination condition from a sample image. As a consequence, our interpolation based approach may infer the lighting parameters of an unknown scene to render new objects according to the same lighting parameters of this scene.

Drew *et al.* [8] introduced a more robust version of the PTMs described in [14]. The method although much more robust to specular and shadow outliers needs to employ a minimum number of light sources that is more than twice as many observations as the number of variables. On the other hand, the image synthesis part of our approach is a more general model that is versatile in terms of the input parameters and does not mandate a specific number of light sources or illumination levels for that matter.

Matsushita *et al.* [15] interpolate lighting appearance of a scene with sparsely sampled lighting conditions. They use a number of lightfields, each captured under different illumination conditions. The major advantage of this method is the ability to synthesize images with significant realism using a sparse dataset. However, the method requires a precision controlled camera grid to effectively capture the lightfields. Moreover, the synthesized image relies on the performance of the multi-veiw stereo algorithm and the image decomposition method. In contrast, our image synthesis method requires only the position of the light source and/or its illumination level. We are also able to demonstrate good quality synthesis with a reasonable sampling.

3   Interpolation Based Approach

Given a set of n sample images of an object under different given lighting conditions described in terms of a parametric space, we build an interpolation model based on the decomposition of an image matrix I into a kernel matrix K and a parametric matrix C, where K represents the intrinsic features of the object (independent of imaging parameters such as lighting conditions) and C fully embeds the influence of the imaging parameters such as the position of local light sources and their respective intensities. Different from dimensionality reduction techniques such as Component Analysis (*e.g.*, PCA [10], ICA [5]), our decomposition explicitly models the imaging conditions in terms of parameters and, consequently, represents these conditions in a manner that leads to a structure modeling the imaging process more closely.

Since a set of sample images and their corresponding parameters are given, the core of our approach is to learn the kernel matrix K such that when K is multiplied by the parametric matrix C built using the parameters $(p_1, p_2, ..., p_n)$ will result in the corresponding image matrix I, which is composed of the individual images $(i_1, i_2, ..., i_n)$: $KC = I$. Essentially, for a particular image i_q associated with parameter p_q, K is a one-to-one mapping from the parametric space to the image space given by the equation $Kf(p_q) = i_q$, where f is an *interpolation function* used to construct the parametric matrix from individual pairs of images and respective parameters. Here, the interpolation function f can in fact be polynomial, trigonometric, logarithmic, or even a combination of these functions.

As an example of an interpolation function, we have a third degree polynomial equation f_h in h variables, where h is the dimensionality of the parametric space. Consider images taken with four lighting parameters $p_i = (w_i, x_i, y_i, z_i)$, so a third degree polynomial equation in four variables is of the form: $f_4(p_i) = w_i^3 + x_i^3 + y_i^3 + z_i^3 + w_i^2 x_i + w_i^2 y_i + w_i^2 z_i + x_i^2 w_i + x_i^2 y_i + x_i^2 z_i + y_i^2 w_i + y_i^2 x_i + y_i^2 z_i + z_i^2 w_i + z_i^2 x_i + z_i^2 y_i + w_i x_i y_i + w_i x_i z_i + w_i y_i z_i + x_i y_i z_i + w_i^2 + x_i^2 + y_i^2 + z_i^2 + w_i x_i + w_i y_i + w_i z_i + x_i y_i + x_i z_i + y_i z_i + w_i + x_i + y_i + z_i + 1$,   where $p_i = (w_i, x_i, y_i, z_i)$ is a particular lighting parameter such as the illumination level of four local light sources. The equation above has 34 components excluding the constant. We construct a *component matrix* $C_{l \times n}$ using any l individual components of the interpolation function given by $f_h(p_i)$ for $i = 1, ..., n$. The number l of components should be smaller than the number n of training images to avoid a rank deficient system. Formally, if the rank of the component matrix C is r, then l should be in the range $[r, n]$. For example, if we use the four components $\{w_i^3, x_i y_i^2, y_i z_i, w_i\}$ (*i.e.*, $l = 4$), we build the matrix $C_{4 \times n}$ of the form:

$$\begin{bmatrix} w_1^3 & w_2^3 & \ldots & w_n^3 \\ x_1 y_1^2 & x_2 y_2^2 & \ldots & x_n y_n^2 \\ y_1 z_1 & y_2 z_2 & \ldots & y_n z_n \\ w_1 & w_2 & \ldots & w_n \end{bmatrix}$$

To be able to learn the kernel matrix K, we use n training images $(i_1, i_2, i_3, \ldots, i_n)$ and define an *image matrix* $I_{m \times n}$, where $m = r \times c \times d$ (r is the number of rows in the image, c is the number of columns, and d is the number of dimensions, which can be 1 or 3 depending whether the images are treated as grayscale or colored, respetively). Therefore, each image i_j is vectorized as a single column-vector of size $r \times c \times d$ and the image matrix I is the concatenation of n such column vectors.

Given the image matrix I and the component matrix C, we have a linear system of equations that represents the interpolation of n vectorized images of size m using l components of the interpolation function as $K_{m \times l} C_{l \times n} = I_{m \times n}$. Using linear algebra, we can compute the pseudo inverse of the component matrix C and multiply it to both sides of the equation to infer the kernel matrix K. Formally, the kernel matrix is obtained as $K_{m \times l} = I_{m \times n} C_{l \times n}^{-1}$.

Now that the kernel matrix is learnt, the problem of image synthesis reduces to a simple matrix multiplication. That is, for a given query parameter $p_q = (w_q, x_q, y_q, z_q)$ of an image that is not in the training set of the kernel matrix, we plug in the parameters into the interpolation function f_h to get a vector c^q of l components. From our previous example, $c^q = [w_q^3, x_q y_q^2, y_q z_q, w_q]^T$. Multiplying the kernel matrix K by the component vector c^q gives the synthesized image vector i_q. Formally, the synthesis equation is given by $i_{m \times 1}^q = K_{m \times l} c_{l \times 1}^q$.

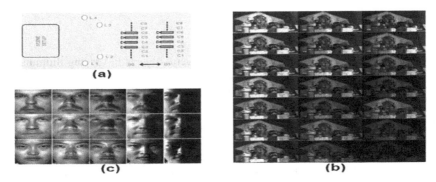

Fig. 2. Camera setup (*a*) for data capture of the object dataset with sample images (*b*) depicting lighting intensity variation. Sample images (*c*) from the face database depicting lighting pose variation.

The analysis part, being the inverse of the synthesis problem, consists of finding the lighting parameter p_q for a given query image i_q. The parameter is obtained by first computing the product of the inverse of the kernel matrix K with the query image vector. That is, the components of the query image are obtained using the matrix equation $c_{l \times 1}^q = K_{l \times m}^{-1} i_{m \times 1}^q$. Having thus obtained the component vector c^q, we find the parameter by solving a linear system of equations. This system is constructed by equating the individual elements of the component vector c^q to the corresponding components in the interpolation function f_h. According to our example above, we have $w_q^3 = c^q(1)$, $x_q y_q^2 = c^q(2)$, $y_q z_q = c^q(3)$, and $w_q = c^q(4)$. We take the logarithm on both sides of these equations and solve for the linear system of equations which results in terms of the logarithm parameter of the query image. Considering the used components of the third degree polynomial in our example, the equations in this system are: $3 \log w_q = \log c^q(1)$, $\log x_q + 2 \log y_q = c^q(2)$, $\log y_q + \log z_q = \log c^q(3)$, and $\log w_q = \log c^q(4)$. Once this system is solved for the logarithm parameter $(\log w_q, \log x_q, \log y_q, \log z_q)$, the actual parameter is found using the exponential function: $\left(w_q, x_q, y_q, z_q \right) = (e^{\log w_q}, e^{\log x_q}, e^{\log y_q}, e^{\log z_q})$.

4   Experimental Results

In this section, we first describe the image datasets we used to demonstrate the image synthesis and analysis capabilities of our interpolation based approach. We constructed an object image dataset to consider the variation of light intensity when local light sources are fixed and we used a face image dataset to consider the variation of the position of light sources under constant intensity. Besides evaluating our approach with regards to the variation of lighting intensity and light source pose, we will also discuss the impact of the number of components and the neighborhood size on the performance of our method. Finally, we demonstrate our approach with regards to synthesis-and-analysis and discuss how they can be used together to render objects into an unknown scene with different lighting given one known object in the scene.

4.1   Image Datasets

We captured images of a complex scene with several objects in varying illumination conditions. We have several objects in the scene to create complex shadows on each other when subjected to different lighting situations.

The camera setup, as illustrated in Fig. 2(a), involved eight CCD cameras mounted with uniform spacing on a straight metal frame. The cameras labeled C1 through C8 were used to capture images of the scene at two depth levels D1 and D2, thereby generating images from sixteen different camera poses.

To generate different lighting conditions, we used four identical incandescent lamps with plain white shades in a fixed spatial configuration as shown in Fig. 2(a). The entire setup was in a dark room to avoid outside lighting conditions from interfering with our lighting setup. Each lamp, namely L1, L2, L3 and L4, could be set at three intensity levels with level-1, level-2, and level-3 representing 100%, 75%, and 50% of the lamp total luminosity, respectively. This setup of four lamps with three levels of lighting leads to $3^4 = 81$ different lighting configurations. All images are in RGB color format with a 1280 x 960 resolution. In Fig. 2(b), we show the lighting variations of the images taken from camera C1 at depth level D2 where the parameters of the lighting range from (1, 1, 1, 1) to (3, 3, 3, 3).

To consider the variation of light source position while the lighting intensity is constant, we used the Extended Yale Face Database B+ [11]. The database contains face images of 38 subjects in 64 different illumination conditions. The illumination variation was achieved by placing the light sources in a set of different azimuth and elevation angles with respect to the camera axis. The face image database [11] is divided into five subsets based on the position of the light source. The subsets 1, 2, 3, and 4 are comprised of images taken with the light source at an angle whose absolute value is less than 12°, between 12° and 25°, between 25° and 50°, and between 50° and 77°, respectively. Subset 5 consisted of images taken with the angle between the light source and the camera axis greater than 77°. The azimuth angles vary from -130° to 130° and the elevation angles from 0° to 90°. This makes subset 5 the most challenging. In our experiments, we have considered all five subsets for testing in order to subject our methodology to an evaluation under extreme lighting conditions. The only images we discarded were that of the eight subjects with corrupted images and, hence, testing was performed with 30 different subjects instead of 38. Moreover, since we are dealing only with illumination variations in this paper, we consider only the frontal pose of the 30 subjects. A set of sample images of three subjects from the 5 subsets are shown in Fig. 2(c). All the frontal face images are cropped and aligned as described in [11]. The images are of dimensions 168 x 192 and in grayscale.

4.2   Image Synthesis and Analysis

For the image synthesis problem, we construct the kernel matrix K by using only the images associated with nearest neighbors (in parametric space) to the query parameter p_q. We also select only a fraction of the original components of the interpolation function that are independent of each other to learn our model.

The performance of the kernel matrix in terms of accurate image synthesis or analysis depends on the following: (a) the way in which neighborhoods are constructed in the parametric space, (b) the size k of the neighborhood and, (c) the number l of components in the interpolation function used to learn the kernel matrix. To investigate the effects of neighborhood construction to our approach, we performed a series of experiments using the k nearest neighbors of the query image parameters according to Euclidian distance in the parametric space. To address the behavior of our approach as the number of used components varies, we consider 27 different numbers of components (values of l) from 1 to 79 incrementing in steps of 3. For each l value, we varied neighborhood size k starting from $l + 1$ to 80, again incrementing in steps of 3. Note that when $l > k$, we have an undetermined system and, hence, we avoid computing the errors in that region.

For each different k and l values, we performed the synthesis of 15 random test images from the total of 81 images corresponding to different illumination parameters. We used the leave-one-out strategy for training, where the test image is left out of the training set used to build the interpolation model. For each of the 15 test images, the synthesis error was computed as the sum of absolute differences between the synthesized image and the test image. The errors thus computed are divided by the

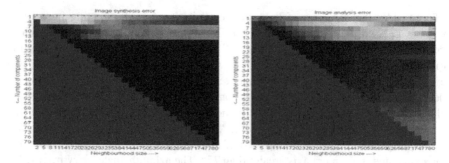

Fig. 3. The average synthesis (left) and analysis (right) error for all possible neighborhood sizes and number of components

dimensions of the image to get the error value in the pixel range [0, 255]. The overall synthesis error for a particular neighborhood size k and number l of used components is the average error for all 15 test images.

The two error matrices for synthesis and analysis are shown in Fig. 3. The error values are depicted using a color scheme where smaller values correspond to colder colors (*i.e.*, blue) and larger values are associated with warmer colors (*i.e.*, red). From this experiment, we learnt that the image synthesis error ranges from 1.154 to 28.673 for all neighborhood size and all possible numbers of components. The image synthesis error is a minimum at 1.154 for a neighborhood size $k = 50$ and number of components $l = 16$. Similarly, the analysis error is computed for the same set of k and l values and for the same 15 testing images. The analysis error is computed as the sum of the absolute differences between the estimated parameters and the known test parameters. The minimum parameter estimation error was a 0.101 for $k = 23$ and $l = 16$. The maximum average error was 1.694. From Fig. 3, we can easily infer that the image synthesis error is better behaved in comparison to the parameter estimation error. However, the parameter estimation error was found to be close to a minimum for most values of k when $l = 16$. For example, the error is 0.124 for $l = 16$ and $k = 50$. Given this situation, we would prefer to set a larger k for a given l to achieve a more compressed model. In other words, choosing $k = 50$ rather than $k = 23$ for $l = 16$ will result in a more compact representation and yet not having compromised much in terms of the analysis accuracy.

In Fig. 4, the first pair of plots on the top shows the average synthesis error (left) and average analysis error (right) of a constant number of components but increasing neighborhood size. The number of components was fixed at 16 for this experiment. In the synthesis error, a clear valley was observed. This shows that over-fitting occurs for smaller neighborhood sizes and then the average error decreases to a minimum of 1.154. After that, the average error increases suggesting the generalization phenomenon. The image analysis error shows a different trend where the average error starts at a small value but increases almost exponentially with increasing neighborhood size.

Similarly, the second pair of plots on the bottom of Fig. 4 shows the behavior of the average error while increasing the number of components with a constant

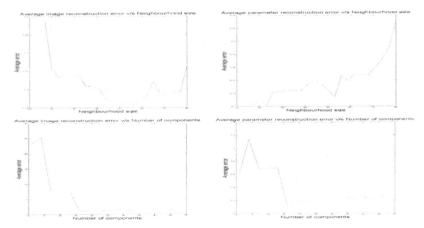

Fig. 4. Average synthesis error and analysis error for constant number of components with increasing neighborhood size (top) and constant neighborhood size with increasing number of components (bottom)

neighborhood size. The neighborhood size in this case was fixed at 50. From the plots for both analysis and synthesis, the average error improves with the increase in the number of components. The error reaches a minimum at 16 components and then shows a saturation which suggests that adding more components further will not produce significant improvement in error. These experiments form the basis for choosing and fine tuning the neighborhood size and an appropriate number of components to address the synthesis and analysis parts for a given dataset.

Fig. 5(a) shows 4 synthesized images with different lighting parameters compared to the corresponding ground-truth images. The synthesized images are almost identical to that of the ground truth. Moreover, on keen observation, one can see that the intricate details of the shadow formations are very well preserved in the generated images. This experiment using the object image dataset was set up with $k = 50$ and $l = 16$ for which the average image synthesis error was 1.154 and the average analysis error was 0.124.

To evaluate our approach with regards to the variation of light source position with constant intensity, we tested our interpolation based approach on the Extended Yale Face Database B+ images [11]. We used the front pose face images with 64 different illumination settings, including some extreme light source angles. We have used our method to address the face recognition problem with an illumination invariant approach. We obtained results with an overall recognition rate of 91.92%. In this paper, we demonstrate the capabilities of our interpolation based technique as a synthesis and analysis tool.

We used 10 independent components of the interpolation function for the generation of face images. A sample set of face images generated from models built based on this dataset are shown in Fig. 5(b). The average image synthesis error was 8.188 for the face image dataset. The face images are reproduced well in terms of shape. However, the specular reflections seem softened. This is the result of the interpolation

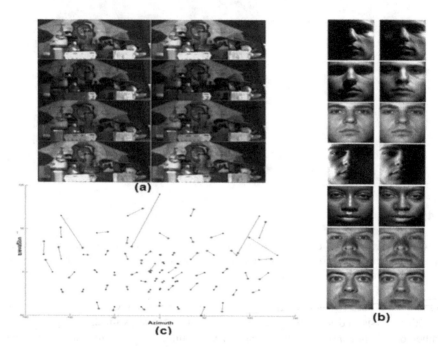

Fig. 5. Image synthesis of object dataset (*a*). The left column is the original images and the right column is the generated images. In (*b*) real face images (left) compared to synthesized images (right). (*c*) shows the analysis of face images for their illumination parameters.

function being only an approximate fit to the sample data. Since specularity is a highly localized lighting effect, a dense sampling is necessary to address this issue while modeling specular objects. However, using a reasonable sampling level, our method performs remarkably well in reproducing the images under different illumination conditions along with the complex self-shadows.

As an inverse process, we demonstrate the models ability in finding the locations of the light sources. Fig. 5(c) shows the estimated lighting parameters for face images of a single subject for all the 64 light sources. The blue and red dots show the actual light source positions and the black asterisk marks show the estimated positions. Red dots mean estimated parameters whose distance to the actual positions is greater than 2.5 degrees while the blue ones indicate that our method estimated the lighting parameters with an error less than the threshold of 2.5 degrees. The average parameter estimation error was 8.836 for a parameter range of [-130°, 130°] in azimuth and [-40°, 90°] in elevation.

4.3   Rendering a Known Object in an Unknown Scene

To demonstrate our method's capability of rendering an object in an unknown scene, we infer the lighting parameters of the unknown scene using the pixels associated with a known object in the scene. Once the lighting parameters are inferred, we can now render a new object according to the illumination conditions of the unknown scene by using the interpolation model of the new object. Fig. 6 depicts the actual rendering of an unknown object rendered into the scene.

Fig. 6. Comparison of an object rendered in an unknown scene with the ground truth

5 Conclusions

We have presented a novel approach based on interpolation that allows for synthesizing a new image given the lighting parameters and also analyzing the lighting parameters given a query image. The model is simple to implement and yet, at the same time, expects a very straightforward and intuitive set of parameters that surprisingly captures the complex nature of lighting variations. The model is not only a compact representation of all possible images in the given parametric space but it is also capable of reconstructing extremely realistic images. Moreover, the same approach can be extended to model geometric variations and hence allowing the synthesis and analysis of objects in different poses. Handling geometric variations of the cameras and/or objects based on this approach is currently part of our ongoing research.

References

1. Adelson, E.H., Bergen, J.R.: The Plenoptic Function and the Elements of Early Vision. Computational Models of Visual Processing, 3–20 (1991)
2. Bousseau, A., Paris, S., Durand, F.: User-Assisted Intrinsic Images. In: Proc. ACM SIGGRAPH, Asia, pp. 16–19 (2009)
3. Barrow, H., Tenenbaum, J.: Recovering Intrinsic Scene Characteristics from Images. Computer Vision Systems (1978)
4. Choudhury, B., Chandran, S., Herder, J.: A Survey of Image-based Relighting Techniques. In: Proceedings of the First International Conference on Computer Graphics Theory and Applications, pp. 176–183 (2006)
5. Comon, P.: Independent Component Analysis, a New Concept? Signal Processing 36(3), 287–314 (1994)
6. Debevec, P.: Rendering Synthetic Objects into Real Scenes: Bridging Traditional and Image-based Graphics with Global Illumination and High Dynamic Range Photography. In: Proc. ACM SIGGRAPH, pp. 189–198 (1998)
7. Debevec, P., Hawkins, T., Tchou, C., Duiker, H.P., Sarokin, W., Sagar, M.: Acquiring the Reflectance Field of a Human Face. In: Proc. ACM SIGGRAPH, pp. 145–156 (2000)

8. Drew, M.S., Hel-or, Y., Malzbender, T., Hajari, N.: Robust Estimation of Surface Properties and Interpolation of Shadow/Specularity Components. Image and Vision Computing (2012)
9. Debevec, P., Taylor, C.J., Malik, J.: Modeling and Rendering Architecture from Photographs: A Hybrid Geometry-and Image-based Approach. In: Proc. ACM SIGGRAPH, pp. 11–20 (1996)
10. Hotelling, H.: Analysis of a Complex of Statistical Variables into Principal Components. Journal of Educational Psychology 24(6), 417–441 (1933)
11. Lee, K.C., Ho, J., Kriegman, D.J.: Acquiring Linear Subspaces for Face Recognition Under Variable Lighting. PAMI 27(5), 684–698 (2005)
12. Levoy, M., Hanrahan, P.: Light Field Rendering. In: Proc. ACM SIGGRAPH, pp. 31–42 (1996)
13. McMillan, L., Bishop, G.: Plenoptic modeling: An Image-based Rendering System. In: Proc. ACM SIGGRAPH, pp. 39–46 (1995)
14. Malzbender, T., Gelb, D., Wolters, H.: Polynomial Texture Maps. In: Proc. ACM SIGGRAPH, pp. 519–528 (2001)
15. Matsushita, Y., Kang, S.B., Lin, S., Shum, H.Y., Tong, X.: Lighting Interpolation by Shadow Morphing Using Intrinsic Lumigraphs. In: 10th Pacific Conf. on Computer Graphics and Applications, pp. 58–65 (2002)
16. Nimeroff, J.S., Simoncelli, E., Dorsey, J.: Efficient Re-rendering of Naturally Illuminated Environments. In: Proc. 5th Erographics Workshop Rendering, pp. 359–373 (1994)
17. Rizzolatti, G., Craighero, L.: The Mirror-Neuron System. Annu. Rev. Neuroscience, 169–192 (2004)
18. Shum, H.Y., Kang, S.B.: A Review of Image-based Rendering Techniques. In: Proc. Int'l Conf. Visual Comm. and Image Processing, vol. 4067, pp. 2–13 (2000)
19. Tunwattanapong, B., Ghosh, A., Debevec, P.: Practical Image-Based Relighting and Editing with Spherical-Harmonics and Local Lights. In: CVMP, pp. 138–147 (2011)
20. Werner, T., Hersch, R.D., Hlavac, V.: Rendering Real-world Objects Using View Interpolation. In: ICCV, pp. 957–962 (1995)

Polynomiography via Ishikawa and Mann Iterations

Wiesław Kotarski, Krzysztof Gdawiec, and Agnieszka Lisowska

Institute of Computer Science, University of Silesia, Poland
{kotarski,kgdawiec,alisow}@ux2.math.us.edu.pl

Abstract. The aim of this paper is to present some modifications of the complex polynomial roots finding visualization process. In this paper Ishikawa and Mann iterations are used instead of the standard Picard iteration. The name polynomiography was introduced by Kalantari for that visualization process and the obtained images are called polynomiographs. Polynomiographs are interesting both from educational and artistic points of view. By the use of different iterations we obtain quite new polynomiographs that look aestheatically pleasing comparing to the ones from standard Picard iteration. As examples we present some polynomiographs for complex polynomial equation $z^3 - 1 = 0$, permutation and doubly stochastic matrices. We believe that the results of this paper can inspire those who may be interested in created automatically aesthetic patterns. They also can be used to increase functionality of the existing polynomiography software.

1 Introduction

Polynomials are objects that can be met in many mathematical fields. They are interesting not only from the theoretical but also from the practical point of view. The problem of polynomial roots finding has a long history. Sumarians 3000 years B.C. then ancient Greeks faced with practical problems which, formulated in modern mathematical language, can be presented as polynomial roots finding. Next, Newton proposed the method of finding polynomial roots approximately. Cayley in 1879 posed the problem related to the behaviour of Newton's method in the complex plane for equation $z^3 - 1 = 0$. Caley's problem was then solved by Julia in 1919 that led directly to Julia set and then in 1970s to Mandelbrot set and fractals [6]. The last interesting discovery in polynomial roots finding history is Kalantari's contribution [4] who introduced to science the so-called polynomiography. Polynomiography defines visualization process in approximation of the zeros of complex polynomial, using fractal and non-fractal images created via the mathematical convergence properties of iteration functions. An individual image is called a polynomiograph. Polynomiography combines both the art and science aspects. Polynomiography, as a method producing nicely looking graphics that could be widely used, was patented by Kalantari in the USA in 2005 [4].

G. Bebis et al. (Eds.): ISVC 2012, Part I, LNCS 7431, pp. 305–313, 2012.
© Springer-Verlag Berlin Heidelberg 2012

Both fractal and polynomiograph are generated by iterations. A shape of fractal is completely defined by the input data, e.g. by the coefficients of an IFS (Iterated Function System), and is rather difficult to control efficiently. Fractal is self-similiar, has complicated and not smooth structure and is not dependent on resolution. Polynomiograph is quite different. Its shape can be controlled and designed in a more predictable way in opposition to typical fractal. Generally, fractals and polynomiographs belong to different classes of graphical objects.

Higher flexibility of polynomiography in comparison to fractals can be explained by taking into account the following arguments. It is known that any complex polynomial:

$$p(z) = a_n z^n + a_{n-1} z^{n-1} + \ldots + a_1 z + a_0 \tag{1}$$

of degree n, according to the Fundamental Theorem of Algebra, has n roots. The polynomial p is well defined by its coefficients $\{a_n, a_{n-1}, \ldots, a_1, a_0\}$ or by its n zeros. So, the degree of polynomial defines the number of basins of attraction. Localization of basins can be controlled by placing roots on the complex plane manually. The chosen roots define the polynomial for which some iteration procedure has to find its zeros. Usually, polynomiographs are coloured based on the number of iterations needed to obtain approximation of some polynomial root with a given accuracy and the iteration method chosen. Description of polynomiography, its theorethical background and artistic applications are described in [3,4].

Summing up, polynomiography can be treated as a theory and visualization tool based on the roots finding process. It has many possible applications in education, math, sciences, art and design.

In this paper we propose to use Mann and Ishikawa iterations instead of Picard iteration to obtain some modifications of the Newton method and iteration methods from Basic Family of Iterations [5]. Earlier, the other types of iterations were used for superfractals [9] and for fractals generated by an IFS [10].

The paper is organised as follows. In Section 2 the theory of Picard, Mann and Ishikawa iterations is presented. Section 3 is devoted to Newton's method of finding polynomial roots and its generalizations, and presents some iteration formulae. In Section 4 the examples of polynomiographs with different types of iterations (Mann, Ishikawa) for complex equation $z^3 - 1 = 0$, permutation and doubly stochastic matrices are presented. The last section, Section 5, describes some conclusions and plans for future work.

2   Iterations

Let $w : X \to X$ be a mapping on a metric space (X, d), where d is a metric. Further, let $u_0 \in X$ be a starting point. Following [1] we recall some popular iterative procedures:

− Picard iteration:

$$u_{n+1} = w(u_n), \quad n = 0, 1, 2, \ldots, \tag{2}$$

– Mann iteration:

$$u_{n+1} = \alpha_n w(u_n) + (1 - \alpha_n)u_n, \quad n = 0, 1, 2, \ldots, \tag{3}$$

where $0 < \alpha_n \le 1$.
– Ishikawa iteration:

$$u_{n+1} = \alpha_n w(v_n) + (1 - \alpha_n)u_n,$$
$$v_n = \beta_n w(u_n) + (1 - \beta_n)u_n, \quad n = 0, 1, 2, \ldots, \tag{4}$$

where $0 < \alpha_n \le 1$ and $0 \le \beta_n \le 1$.

The standard Picard iteration is used in the Banach Fixed Point Theorem [1] to obtain the existence of the fixed point x^* such that $x^* = w(x^*)$ and its approximation under additional assumptions on the space X that should be a Banach one and the mapping w should be contractive. The Mann [7] and Ishikawa [2] iterations allow to weak the assumptions on the mapping w. Further, our considerations will be conducted in the space $X = \mathbb{R}^2$ or $X = \mathbb{C}$ that are obviously Banach ones. We take $u_0 = (x_0, y_0) \in \mathbb{R}^2$ and $\alpha_n = \alpha$, $\beta_n = \beta$, such that $0 < \alpha \le 1$ and $0 \le \beta \le 1$. It is easily seen that the Ishikawa iteration with $\beta = 0$ is Mann iteration, and for $\beta = 0$, $\alpha = 1$ is Picard iteration. The Mann iteration with $\alpha = 1$ is the Picard iteration.

3   Newton Roots Finding Method and Its Generalizations

In this section we recall the well-known Newton method for finding roots of a complex polynomial p. The Newton procedure is given by the formula:

$$z_{n+1} = z_n - \frac{p(z_n)}{\dot{p}(z_n)}, \quad n = 0, 1, 2, \ldots, \tag{5}$$

where $z_0 \in \mathbb{C}$ is a starting point.
   Applying the Mann iteration (3) in (5) we obtain the following formula:

$$z_{n+1} = \alpha \left(z_n - \frac{p(z_n)}{\dot{p}(z_n)} \right) + (1 - \alpha)z_n, \quad n = 0, 1, 2, \ldots, \tag{6}$$

where $0 < \alpha \le 1$.
   Using the Ishikawa iteration (4) in (5) we get:

$$z_{n+1} = \alpha \left(v_n - \frac{p(v_n)}{\dot{p}(v_n)} \right) + (1 - \alpha)z_n,$$
$$v_n = \beta \left(z_n - \frac{p(z_n)}{\dot{p}(z_n)} \right) + (1 - \beta)z_n, \quad n = 0, 1, 2, \ldots, \tag{7}$$

where $0 < \alpha \le 1$ and $0 \le \beta \le 1$.
   The sequence $\{z_n\}_{n=0}^{\infty}$ (or orbit of the point z_0) converges or not to a root of p. If the sequence converges to a root z^* then we say that z_0 is attracted to z^*.

A set of all starting points z_0 for which $\{z_n\}_{n=0}^{\infty}$ converges to z^* is called the basin of attraction of z^*. Boundaries between basins usually are fractals in nature. In [11] some generalizations of the classic Newton formula (5) are discussed. The formulae given above are used in the next section to obtain polynomiographs for complex polynomials that visualize the roots finding process.

Further generalization procedures for roots finding of complex polynomial are given in [4,5]. They are introduced in the following way. First, define $D_0(z) = 1$ and for $m > 0$ let

$$
D_m(z) = \det \begin{bmatrix} \dot{p}(z) & \frac{\ddot{p}}{2!}(z) & \cdots & \frac{p^{m-1}}{(m-1)!}(z) & \frac{p^m}{m!}(z) \\ p(z) & \dot{p}(z) & \cdots & & \frac{p^{m-1}}{(m-1!)}(z) \\ \vdots & \vdots & \ddots & \ddots & \vdots \\ 0 & 0 & \cdots & \dot{p}(z) & \ddot{p}(z) \\ 0 & 0 & \cdots & p(z) & \dot{p}(z) \end{bmatrix}. \tag{8}
$$

The elements of the so-called Basic Family of Iterations are then defined as:

$$
B_n(z) = z - p(z) \frac{D_{m-2}(z)}{D_{m-1}(z)}, \quad n = 2, 3, \ldots. \tag{9}
$$

It is easily seen that B_2 is the Newton method, whereas B_3 is the Halley method. Iterations from the Basic Family can be modified using Mann or Ishikawa iterations because those iterations produce only different orbits in comparison to Picard iteration. Only the character of convergence is different and the basins of attraction to roots of complex polynomial p are looking differently for different kinds of iteration used.

4   Examples of Polynomiographs

In this section we present some polynomiographs for complex polynomial equation $z^3 - 1 = 0$, permutation and doubly stochastic matrices. They are obtained for different parameters α and β via Newton method using Picard, Mann or Ishikawa iterations.

In all examples the colour of each point in the image is determined with the help of Algorithm 1. $I_{\alpha,\beta}$ in the algorithm is the Ishikawa iteration method given by (7), but as we mentioned earlier, for particular values of α and β we obtain Picard or Mann iteration method.

Let us start from equation $z^3 - 1 = 0$ having three roots: 1, $-\frac{1}{2} - \frac{\sqrt{3}}{2}i$, $-\frac{1}{2} + \frac{\sqrt{3}}{2}i$. In Fig.1 nine images with three distinct basins of attraction to the three roots of polynomial $z^3 - 1$ are presented. The colours of different image areas depend on the number of iterations needed to reach a root with the given accuracy $\varepsilon = 0.001$. The upper bound of the number of iterations was fixed as $k = 15$. By changing parameters $\alpha, \beta, \varepsilon$ and k one can obtain infinitely many polynomiographs.

Algorithm 1. Colour determination

Input: $z_0 \in \mathbb{C}$ – starting point, k – maximum number of iterations, ε – accuracy, α, β – parameters of iteration $I_{\alpha,\beta}$

Output: colour c of z_0

1 $i = 0$
2 **while** $i \leq k$ **do**
3 $\quad$ $z_{i+1} = I_{\alpha,\beta}(z_i)$
4 $\quad$ **if** $|z_{i+1} - z_i| < \varepsilon$ **then**
5 $\quad\quad$ **break**
6 $\quad$ $i = i + 1$
7 $c = i$

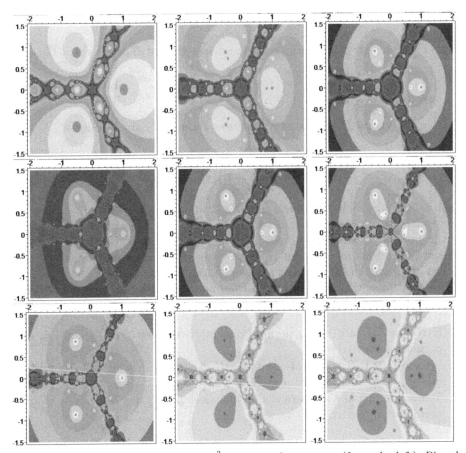

Fig. 1. Polynomiographs of equation $z^3 - 1 = 0$, the top row (from the left): Picard iteration, Mann iterations for $\alpha = 0.8$ and $\alpha = 0.6$, the middle row (from the left): Mann for $\alpha = 0.5$, Ishikawa for $\{\alpha = 0.6, \beta = 0.0\}$, $\{\alpha = 0.6, \beta = 0.1\}$, and the bottom row (from the left): Ishikawa for $\{\alpha = 0.6, \beta = 0.5\}$, $\{\alpha = 1.0, \beta = 0.5\}$, $\{\alpha = 1.0, \beta = 0.7\}$, respectively

Now recall that a $n \times n$ matrix $\Pi = (\pi_{ij})$ is a matrix whose rows and columns form a permutation of the identity matrix. To each matrix Π we can associate a complex polynomial in the following way. To the location (i, j) in Π we set Θ_{ij}:

$$\Theta_{ij} = i + j\mathbf{i}, \tag{10}$$

where $\mathbf{i} = \sqrt{-1}$.

Next, to the matrix Π we further define a $n \times n$ matrix $\overline{\Pi} = (\overline{\pi}_{ij})$ as $\overline{\pi}_{ij} = \pi_{j,(n+1-i)}$. This matrix is analogous to the transpose, except that i-th row of Π corresponds to the i-th column of $\overline{\Pi}$ but written from the bottom up. Finally, for the matrix $\Pi = (\pi_{ij})$ the complex polynomial p_Π can be defined as [5]:

$$p_\Pi(z) = \prod_{\overline{\pi}_{ij}=1} (z - \Theta_{ij}). \tag{11}$$

As an example take 2×2 permutation matrices Π_1 and Π_2 and create $\overline{\Pi}_1$ and $\overline{\Pi}_2$:

$$\Pi_1 = \begin{bmatrix} 1 & 0 \\ 0 & 1 \end{bmatrix}, \Pi_2 = \begin{bmatrix} 0 & 1 \\ 1 & 0 \end{bmatrix}, \overline{\Pi}_1 = \begin{bmatrix} 0 & 1 \\ 1 & 0 \end{bmatrix}, \overline{\Pi}_2 = \begin{bmatrix} 1 & 0 \\ 0 & 1 \end{bmatrix}. \tag{12}$$

Complex polynomials associated to matrices Π_1 and Π_2 are as follows:

$$p_{\Pi_1}(z) = (z - (1 + 2\mathbf{i}))(z - (2 + \mathbf{i})), \tag{13}$$
$$p_{\Pi_2}(z) = (z - (1 + \mathbf{i}))(z - (2 + 2\mathbf{i})). \tag{14}$$

Their polynomiographs are presented in Fig. 2 and Fig. 3, respectively. It is easily seen that localizations of ones in permutation matrices Π_1 and Π_2 correspond to the images of polynomiographs. Polynoniographs obtained via Mann and Ishikawa iterations for different α, β are quite different in comparison to the Picard iteration. All the images have been obtained for $\varepsilon = 0.001$ and $k = 8$.

It is worth mentioning that permutation matrices have the following obvious properties:

1. If Π_1 and Π_2 are $n \times n$ permutation matrices then $\Pi_1\Pi_2$ is also $n \times n$ permutation matrix.
2. Inverse Π^{-1} to a permutation matrix Π exists and it is also a permutation matrix, it is equal to its transpose, i.e. $\Pi^{-1} = \Pi^T$.
3. Tensor product of $n \times n$ permutation matrices Π_1, Π_2, i.e. $\Pi_1 \otimes \Pi_2$, is $n^2 \times n^2$ permutation matrix.

The number of $n \times n$ permutation matrices is huge and equals $n!$. So, very many nice polynomiographs can be generated.

Doubly stochastic matrices have all non-negative elements and the sum of the entries of each row and column equals 1. According to Birkhoff-von Neumann theorem [8] any double stochastic matrix A can be represented as a convex combination of permutation matrices:

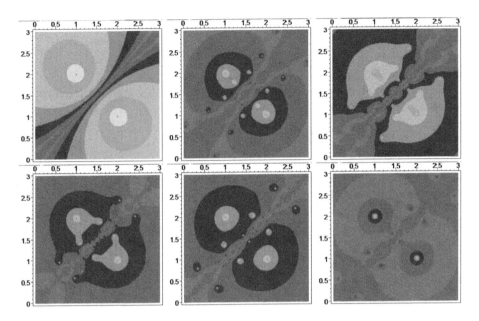

Fig. 2. Polynomiographs of matrix Π_1, the top row (from the left): Picard iteration, Mann iterations for $\alpha = 0.7$ and $\alpha = 0.8$, the bottom row (from the left): Ishikawa iterations for $\{\alpha = 0.7, \beta = 0.4\}$, $\{\alpha = 0.7, \beta = 0.6\}$, $\{\alpha = 0.6, \beta = 0.9\}$, respectively

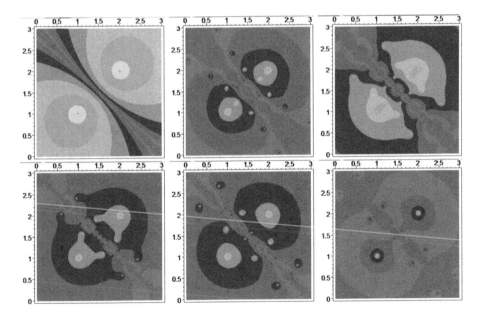

Fig. 3. Polynomiographs of matrix Π_2, the top row (from the left): Picard iteration, Mann iterations for $\alpha = 0.7$ and $\alpha = 0.8$, the bottom row (from the left): Ishikawa iterations for $\{\alpha = 0.7, \beta = 0.4\}$, $\{\alpha = 0.7, \beta = 0.6\}$, $\{\alpha = 0.6, \beta = 0.9\}$, respectively.

$$A = \sum_{i=1}^{k} \alpha_i \Pi_i, \tag{15}$$

where $\sum_{i=1}^{k} \alpha_i = 1$ and $\alpha_i \geq 0$ for $i = 1, \ldots, k$.

The corresponding complex polynomial p_A to a doubly stochastic matrix A can be defined as follows:

$$p_A(z) = \prod_{\overline{a}_{ij} > 0} (z - \overline{a}_{ij} \Theta_{ij}), \tag{16}$$

where matrix $\overline{A}$ to A is constructed in a similar way as matrix $\overline{\Pi}$ to Π.

As an example take the following double stochastic matrix A:

$$A = \begin{bmatrix} \frac{1}{2} & \frac{1}{2} \\ \frac{1}{2} & \frac{1}{2} \end{bmatrix} = \frac{1}{2} \begin{bmatrix} 1 & 0 \\ 0 & 1 \end{bmatrix} + \frac{1}{2} \begin{bmatrix} 0 & 1 \\ 1 & 0 \end{bmatrix} \tag{17}$$

The corresponding complex polynomial p_A to the matrix A has the following form:

$$p_A(z) = \left(z - \frac{1+i}{2}\right)\left(z - \frac{1+2i}{2}\right)\left(z - \frac{2+i}{2}\right)\left(z - \frac{2+2i}{2}\right). \tag{18}$$

In Fig.4 polynomiographs for a double stochastic matrix A are presented.

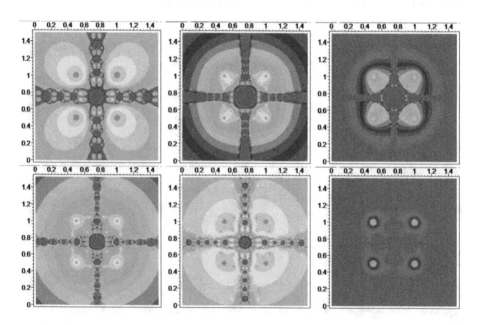

Fig. 4. Polynomiographs of doubly stochastic matrix A, the top row (from the left): Picard iteration, Mann iterations for $\alpha = 0.5$ and $\alpha = 0.3$, the bottom row (from the left): Ishikawa iterations for $\{\alpha = 0.5, \beta = 0.6\}$, $\{\alpha = 0.8, \beta = 0.6\}$, $\{\alpha = 0.2, \beta = 0.7\}$, respectively

5   Conclusions

In this paper we presented some generalizations of the classic Newton method obtained by the use of Mann or Ishikawa iterations instead of the standard Picard iteration. The obtained polynomiographs for complex equation $z^3 - 1 = 0$, permutation and doubly stochastic matrices look quite different in comparison to Picard iteration. Mann and Ishikawa iterations can be used to generalize Basic Family of Iteration. Further experiments will be carried out to check how polynomiographs look after replacing Picard iteration by Mann or Ishikawa iterations. We believe that the results of this paper can be interesting for those whose works or hobbies are related to automatically created nicely looking graphics. We also think that using Mann and Ishikawa iterations can be applied to increase the functionality of the existing polynomiography software.

References

1. Berinde, V.: Iterative Approximation of Fixed Points, 2nd edn. Springer, Heidelberg (2007)
2. Ishikawa, S.: Fixed Points by a New Iteration Method. Proceedings of the American Mathematical Society 44(1), 147–150 (1974)
3. Kalantari, B.: Polynomiography: From the Fundamental Theorem of Algebra to Art. Leonardo 38(3), 233–238 (2005)
4. Kalantari, B.: Polynomial Root-Finding and Polynomiography. World Scientific, Singapore (2009)
5. Kalantari, B.: Alternating Sign Matrices and Polynomiography. The Electronic Journal of Combinatorics 18(2), 1–22 (2011)
6. Mandelbrot, B.: The Fractal Geometry of Nature. W.H. Freeman and Company, New York (1983)
7. Mann, W.R.: Mean value methods in iteration. Proceedings of the American Mathematical Society 4, 506–510 (1953)
8. Minc, H.: Nonnegative Matrices. John Wiley & Sons, New York (1988)
9. Prasad, B., Katiyar, K.: Fractals via Ishikawa Iteration. In: Balasubramaniam, P. (ed.) ICLICC 2011. CCIS, vol. 140, pp. 197–203. Springer, Heidelberg (2011)
10. Singh, S.L., Jain, S., Mishra, S.N.: A New Approach to Superfractals. Chaos, Solitons and Fractals 42(5), 3110–3120 (2009)
11. Susanto, H., Karjanto, N.: Newton's Method's Basins of Attraction Revisited. Applied Mathematics and Computation 215(3), 1084–1090 (2009)

Clustered Deep Shadow Maps for Integrated Polyhedral and Volume Rendering

Alexander Bornik[1], Wolfgang Knecht[2],
Markus Hadwiger[3], and Dieter Schmalstieg[4]

[1] Ludwig Boltzmann Institute – Clinical-Forensic Imaging (CFI), Graz, Austria
alexander.bornik@cfi.lbg.ac.at
[2] Vienna University of Technology, Vienna, Austria
[3] King Abdullah University of Science and Technology, Thuwal, Saudi Arabia
[4] Graz University of Technology, Graz, Austria

Abstract. This paper presents a hardware-accelerated approach for shadow computation in scenes containing both complex volumetric objects and polyhedral models. Our system is the first hardware accelerated complete implementation of deep shadow maps, which unifies the computation of volumetric and geometric shadows. Up to now such unified computation was limited to software-only rendering . Previous hardware accelerated techniques can handle only geometric or only volumetric scenes - both resulting in the loss of important properties of the original concept. Our approach supports interactive rendering of polyhedrally bounded volumetric objects on the GPU based on ray casting. The ray casting can be conveniently used for both the shadow map computation and the rendering. We show how anti-aliased high-quality shadows are feasible in scenes composed of multiple overlapping translucent objects, and how sparse scenes can be handled efficiently using clustered deep shadow maps.

1 Introduction

To address the need for shadows cast by semi-transparent objects, the deep shadow map (DSM) [1] extends the concept of a conventional shadow map by storing a visibility function for each pixel in the shadow image plane. Hadwiger et al. [2] present a GPU implementation of DSM that is suitable for direct volume rendering, but cannot accommodate polygonal or polyhedral geometry, and does not scale well to large, sparse scenes.

Our goal is the integration of dynamic high-quality shadows in a hardware-accelerated rendering framework supporting both polygonal representations and volumetric objects. This has important applications in creating realistic scenes for games and virtual environments, but also in medical applications. We show that a hardware-accelerated DSM implementation on the GPU is feasible based on the Compute Unified Device Architecture (CUDA) [3]. Deep shadow map computation and rendering are both based on ray casting. The ray casting considers segments along each ray, which are homogeneous in terms of object

G. Bebis et al. (Eds.): ISVC 2012, Part I, LNCS 7431, pp. 314–325, 2012.

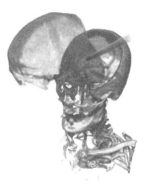

Fig. 1. Examples of deep shadow map rendering in hybrid volumetric/geometric scenes: The left image shows shadow interaction in a complex scene consisting of opaque and translucent geometry with transparent interior material as well as volumetric smoke and a CT dataset. The image on the right shows benefits of deeps shadow maps for depth perception in surgical simulation, showing a polygonal surgery tool (blue rod).

occupancy. Volume rendering acceleration techniques such as early ray termination and empty space skipping allow for interactive frame rates.

This approach yields high-quality anti-aliased shadows and has the following distinguishing properties:

- Arbitrary combinations of *soft shadow casting between polygonal and volumetric objects* are possible. Polygonal objects can be transparent and overlapping with volumetric objects.
- Perspective aliasing of shadows in large, sparse scenes due to insufficient DSM resolution is effectively suppressed through a novel concept, *clustered deep shadow maps*. This approach analyzes the scene structure and adaptively allocates the nodes of the deep shadow map structure where they contribute most. We also present a hardware-friendly chunked memory layout for clustered deep shadow maps.

2   Related Work

The most common methods for computing shadows in geometric scenes are either based on shadow volumes [4] or on shadow mapping [5], both of which are originally limited to opaque geometry and hard shadows.

In contrast to shadow volumes, shadow mapping employs an image-space discretization, which makes the extension toward volume sampling easier. A major problem of shadow mapping methods are several types of aliasing, most of all perspective aliasing. Common approaches reduce perspective aliasing via specific perspective transforms, e.g., in post-projective view space [6] or light space [7], or by using logarithmic transforms [8]. Aliasing artifacts can also be reduced by using adaptive shadow map resolutions like in [9]. Aliasing due to sample positions in the shadow map can be removed entirely using irregular rasterization [10]. Perspective aliasing can also be reduced significantly via cascaded shadow maps [11] or parallel-split shadow maps [12]. However, these approaches are inefficient in sparse scenes. The filtering of shadow maps has been improved by using tailored approaches such as variance shadow maps [13], which can also be layered [14], or convolution shadow maps [15].

Most shadow mapping approaches focus on hard shadows, although soft shadows caused by area light sources can be approximated by superimposing the contribution of multiple point lights. A recent approach for colored stochastic shadow maps of translucent objects is given in [16]. However, approaches working from a surface representation obviously cannot take into account soft shadows caused by continuous absorption of light by participating media or volume data.

In volume rendering, shadow computation is an important topic [17]. The two most common approaches either perform half-angle slicing [18] on-the-fly, or compute an additional volume that stores the amount of light reaching each voxel from the light source [19], which is then used during the actual volume rendering. The latter approach can be combined with ray casting for the volume rendering pass [20], but requires significant additional volume storage. Half-angle slicing does not store any additional volume data, but is restricted to slice-based volume rendering . More recent work has taken on the basic idea for volume ray casting [21], but is still closely linked the regular dataset grid, which prohibits its application to multi-volume-grid scenes and integration with polyhedral objects. An alternative are approaches based on Monte-Carlo sampling [22].

The work in this paper builds on the Deep Shadow Map approach [1], which conceptually unifies the computation of volumetric and geometric shadows. Instead of storing a single depth value per shadow map pixel, a visibility function is computed and stored per pixel.

For volume rendering, DSMs can be implemented efficiently on GPUs [2], which yields real-time frame rates on current hardware even when combined with scattering effects [17]. However, this approach focuses on volume data and does not incorporate geometry.

Several approaches focus on achieving similar functionality for hair geometry with less computational requirements than DSMs [23][24][25][26][27]. Adaptive volumetric shadow maps [28] are an approximate solution to volumetric shadows, which offers fast on the fly lossy compression, efficient lookup, and global shadow support given a fixed map size. However, these algorithmic advantages only apply to a traditional rendering pipeline with unsorted fragments, while we use sorted fragments lists.

3   Basic Algorithm

The implementation of DSM rendering presented in this paper is derived from the ray casting system described in [29]. The whole rendering pipeline including DSM is implemented in CUDA and executes purely on the GPU. It operates on a data structure similar to a volume scene graph, which consists of a tree with volumetric Boolean operations in interior nodes and *polyhedral objects* (topologically closed triangular meshes) in the leaves. First, boundary polygons are rasterized, and all produced fragments that fall onto a given pixel location are depth-sorted. Then a ray is traversed from the frontmost fragment, and visits other boundary fragments for a given location in depth order, while accumulating opacity.

3.1 Deep Shadow Map Generation

The structure of the DSM calculation algorithm is very similar to polyhedral bounded volume rendering. In fact, it reuses a large portion of the kernel code from the ray casting and only differs in the determination of each fragment's 'payload'. Instead of sampling a color and opacity value along a ray, a piecewise linear visibility function $V(z)$ storing the remaining light intensity is calculated from the light's viewpoint as described in [1]. We also perform pre-compression in analogy to [2].

All DSM computation is performed in the per pixel kernel, right after fragment depth sorting. For each homogeneous ray segment, we compute a compressed piecewise linear representation $V'(z)$ of the discrete visibility function $V(z)$, which describes the light attenuation along a ray from the light source through the scene. Its computation is based on opacity samples and represented as a list of nodes storing depth and opacity. The compressed version $V'(z)$ is an approximation of $V(z)$ where

$$|V'(z) - V(z)| \leq \epsilon \tag{1}$$

holds for a maximum error of ϵ [1]. $V'(z)$ is computed on the fly, depending on the types of objects present in a particular ray segment. If the ray is inside of a volumetric object, standard opacity accumulation is used.

Surfaces of polyhedral objects are unconditionally added as nodes to $V'(z)$. In order to accurately represent the step in $V(z)$, two entries with the same z are added, storing the α_{acc} before and after fragment opacity comprehension. In case of more than one light source, omnidirectional light sources or clustering, multiple DSMs are computed concurrently.

3.2 Rendering Using Deep Shadow Maps

During rendering, the precomputed DSM are used to compute the amount of light penetrating to the observed point. The sample location is transformed into light space as for standard shadow mapping. The sample's depth value in light space is used to find the two closest visibility function entries in $V'(z)$, which are linearly interpolated.

Shading is done using the *Phong Illumination Model*. Rendering quality is further improved by bilinear filtering of the four surrounding texel values at lookup time. The contribution of multiple light sources is the sum of all individual contributions, which are computed using the corresponding DSM.

4 Clustering

Using a single shadow map often leads to perspective aliasing artifacts because the map resolution drops with increasing distance of scene objects from the camera. This effect is particularly strong if the current camera location is far from the camera setup used for map calculation. Furthermore, sparse scenes can easily lead to bad shadow map utilization. The above-mentioned problems are also prominent with DSMs.

We attack the aliasing problem with multiple equally-sized DSMs per light source. Each DSM represents a different part of the scene. However, unlike *parallel-split shadow maps* [12], our DSM arrangement is adaptive to the scene content: Instead of splitting the scene along the camera's view vector, scene objects are clustered. Each cluster obtains its own shadow map. Figure 2 gives an overview of the situation.

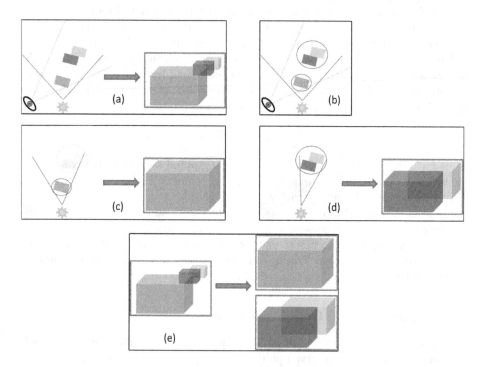

Fig. 2. Illustration of the basic idea of clustering. (a) Overview of the scene and the corresponding single shadow map. (b) Nearby objects are clustered. (c, d) Focusing of the shadow map on the first cluster and on the second cluster. (e) Single shadow map compared to the two focused shadow maps.

Clustered DSM computation is a two stage process. First, clustering groups objects with little shadow contribution into larger clusters, while objects with major shadow contribution are placed in smaller clusters. Every cluster is assigned to one DSM. The rationale is that in order to best represent the shadow with limited resources, the number of DSM entries per object should be proportional to the object's contribution to the shadow.

After cluster assignment, shadow map viewing volumes are fit to the clusters to optimize shadow map usage by adjusting the light camera's orientation, field of view, and near/far planes.

Cluster building is performed in light-space (see [7]). We consider every scene object's bounding box and derive extreme points pp_{min} and pp_{max} as the minimum and maximum projected extents in x and y. The distance between two bounding boxes A and B is defined as follows:

$$d(A, B) = \max |pp_{min}(A) - pp_{min}(B)|, |pp_{max}(A) - pp_{max}(B)| \qquad (2)$$

For omnidirectional lights, $d(A, B)$ is the maximum of the values computed for six different projection directions. The resulting distances are the basis for cluster generation. We use the following heuristic based on a user defined distance threshold d_0, since we want to compute up to m cluster DSMs, where m is the maximum number of DSMs, which fits into pre-allocated shadow map storage (for details on storage management, see Section 5):

1. Sort pairs of objects O_i, O_j with $(i \neq j)$ by $d(O_i, O_j)$
2. Insert objects O_i into existing cluster C_k where $\forall O_j \in C_k : d(O_i, O_j) \leq d_0$. OR insert objects O_i into new cluster C_l if for all clusters C_k: $\exists O_j \in C_k$ with $d(O_i, O_j) > d_0$ and $l \leq m$.
3. Insert remaining objects O_i into the cluster C_k with $\{max(d(O_i, O_j)) \mid O_j \in C_k\} \rightarrow min$.

From the rendering point of view, clustered DSMs are used similarly to DSMs for multiple light sources. Their contributions are appropriately combined along the light ray. Spotlights require an inside/outside illuminated area test per cluster.

Clustering leads to local maps and good resource utilization as well as higher resolution maps for objects farther from the viewer. Re-computation is only required, if the light, the scene or the clustering changes. Note that frame-to-frame coherence means that clustering can often be re-used over multiple frames for a moving camera, and that clustering is computationally cheap compared to DSM computation. An additional advantage of *Clustered Deep Shadow Maps* is the fact that omnidirectional light sources may require fewer than six shadow maps, depending on the scene configuration.

5   Memory Management

The DSM storage scheme used in our implementation is chunked memory. A memory chunk can store a predefined number of L_C depth and opacity values (64 in our case), each pair representing a depth layer for one texel in a DSM. Furthermore each chunk entry stores a link to the global chunk index of the next chunk and the entry number in that chunk. A separate initialization buffer stores the starting chunk number and entry for each DSM texel. Values are stored in a large *float4* texture.

For chunked memory DSM computation, the viewport is split into tiles of 8×8 pixels, corresponding to blocks of 64 parallel threads in CUDA. Each block may reserve a maximum number C_B out of a total number C_{max} of available chunks. The corresponding chunk indices are stored in $b_g[b][i]$, where b is the block index and i the block chunk index. All indices are initialized to N_A (not allocated). Furthermore there is an array $b_l[b]$ used for counting the overall number of samples stored in each block. It is incremented using CUDA atomic operations.

As samples to be added to the DSM arrive, we can calculate $b_c = \frac{b_l[b]}{L_C}$, the current block's chunk number. The current chunk's global index c_d is obtained using an allocation that relies on CUDA atomic operations in order to synchronize access to shared state across threads. Waiting for another thread to exit the transient *pending* state is accomplished by busy waiting since the expected time to wait (if any) is very short (and CUDA does not offer idle wait primitives anyway).

Chunk data is written to separate lists for depth, opacity and indices at index $c_d * (b_l \bmod L_C)$ to take advantage of coalesced writes during computation, while the list elements are copied to the corresponding *float4* texture. The per texel start chunk is initialized with first value obtained when incrementing b_l for a texel and the index of the the corresponding chunk.

Chunk based memory avoids memory wastage, and the number of depth layers per DSM sample is only limited by the amount of GPU memory dedicated to DSM chunks, chunk size L_C, and management entries in $b_g[b]$. However, lack of spatial coherence requires a linear DSM list traversal to find the appropriate depth layer, which results in computational overhead. This overhead is reduced by exploiting fast access to textures for lookup structures and chunk data during rendering.

6   Results

The following results were all obtained on a 2.6GHz Intel i7 system equipped with an NVIDIA GTX 480 GPU running on Windows Vista 64Bit.

6.1   Clustering

The example in Figure 3 depicts the advantages of scene clustering. The region outlined red is far away from the light source. Therefore the shadow map resolution is low in that area, even for a high overall map resolution of 1024×1024, which is visible in the left image. Scene clustering based on the actual viewpoint for $n = 4$ clusters results in a separate DSM for the smoking cup of coffee, which provides much higher local shadow resolution. Note, that the overall memory consumption is the same, since each of the four cluster DSMs is only 512×512.

Figure 4 shows a different view of the same scene and the corresponding clustering, which again guarantees high resolution shadows maps for objects nearby the camera.

Fig. 3. (a) Scene rendered using a single deep shadow map. (b) Same scene rendered using multiple deep shadow maps computed based on scene object clustering. The difference is clearly visible in the magnified highlighted region.

6.2   Performance Analysis

We analyzed the performance of our implementation by varying a number of parameters including DSM resolution, volumetric dataset resolution, number of objects and error tolerance threshold.

In order to study the influence of DSM resolution, we rendered the scene shown in Figure 3 with different DSM resolutions and recorded map computation as well as rendering times with and without filtering. The viewport size was set to 720×480 pixels. The test scene consisted of one spotlight, two volumetric and eight polyhedral objects. Rendering without any shadow mapping took 43ms on the test system. The results with shadows are summarized in Table 1.

Table 1. Computation times in [ms] for varying DSM resolutions and 3D texture based memory management. Numbers in brackets are timings with clustering turned on.

[ms]	256^2 (4×128^2)	512^2 (4×256^2)	1024^2 (4×512^2)
DSM generation	11 (21)	31 (40)	65 (80)
Rendering	46 (54)	46 (55)	46 (57)
Rendering (filt.)	54 (70)	53 (71)	53 (72)

The resolution of volumetric datasets has a significant impact on DSM computation times and even more on rendering. These results – summarized in Table 2 – are in line with ray casting performance in general. A higher dataset resolution also leads to a higher number of depth layers.

As expected, the DSM resolution has a significant impact on performance. Filtering slightly increases rendering times. Clustering timings are compared for four clusters and halved map resolution. Clustering leads to higher map computation and rendering times. The overhead of the n rendering passes and

Fig. 4. Top: Clustering overview for the view from Figure 3 and for a totally different view (bottom). Objects of the same color enter the same DSM.

lookup while rendering is more costly than dealing with a higher resolution map. However, clustering is justified if demands on image quality are high.

Table 3 shows that chunk based memory management decreases DSM computation times, but slightly increases rendering times.

The choice of ϵ as error tolerance threshold for DSM pre-compression influences the number of nodes in the visibility functions as shown in Table 4. The equal maximum number of depth layers per DSM texel for $\epsilon = 0.05$, $\epsilon = 0.1$ and $\epsilon = 0.2$ reflects a scene-dependant lower bound for accurate representation of the visibility function in some region. It can only be lowered by an excessively higher ϵ, resulting in major rendering artifacts.

Rendering times and DSM computing times scale rather linearly with the number of scene objects, regardless of the memory management methods used. Chunked memory management constantly results in faster DSM computation but slightly higher rendering times. However, chunked memory has the additional advantage of lower memory requirements, which can be important for large volumetric scenes. Memory management was tested on a scene with up to 24 similar polyhedral and volumetric objects.

Since memory for DSMs has to be allocated at startup, the 3D texture management size has to be chosen based on the maximum number of depth layers required to store the DSM obtained through pre-compression using a given ϵ.

Table 2. Computation times in [ms] for varying volumetric dataset resolutions. Chunked Memory is used as the memory management method ($4 \times 256 \times 256$ DSM).

	time	Avg. # layers	Max. # layers
DSM generation			
$512 \times 512 \times 96$	74 ms	3.74	31
$256 \times 256 \times 48$	55 ms	3.64	27
$128 \times 128 \times 24$	52 ms	3.64	24
Rendering (filt.)			
$512 \times 512 \times 96$	846 ms		
$256 \times 256 \times 48$	340 ms		
$128 \times 128 \times 24$	130 ms		

Table 3. Computation times in [ms] for varying DSM resolutions and chunk based memory management. Numbers in brackets are timings with clustering turned on.

[ms]	256^2 (4×128^2)	512^2 (4×256^2)	1024^2 (4×512^2)
DSM generation	7 (17)	17 (30)	54 (79)
Rendering	48 (55)	49 (56)	49 (57)
Rendering (filt.)	56 (69)	56 (69)	57 (70)

Table 4. The number of layers / DSM texel depends on the error tolerance ϵ

	$\epsilon = 0.01$	$\epsilon = 0.05$	$\epsilon = 0.1$	$\epsilon = 0.2$
Avg. # layers	4.01	3.32	3.12	3.00
Max. # layers	28	21	21	21

For our test scene (Figure 3), this number (21 for $\epsilon = 0.05$) can be found in Table 4. With chunk based memory, giving an estimated average number of layers is sufficient. Therefore chunk based memory management reduces memory consumption by $\approx 80\%$ for the test scene.

7 Conclusion

We have shown a fully hardware-accelerated implementation of deep shadow maps for scenes combining direct volume rendering with polygonal models. This rendering system yields high quality soft shadows at interactive frame rates. It features a unified handling of volumetric and non-volumetric objects, and can also deal with large, sparse scenes due to its application of multiple deep shadow maps after scene clustering. We believe that our approach is an important step towards the adoption of volumetric shadows in everyday rendering tasks.

We are planning several extensions of the system. Special translucent geometry, in particular hair, can be supported by extending the rasterization stage.

Alternatively, the new CUDA interface in OpenGL 4.0 allows to combine the OpenGL rasterizer with the CUDA raycasting kernel. We have verified in early experiments that this approach increases framerates of typical scenes from 5 to 15-20 fps. Relying the OpenGL rasterizer also makes integration of our approach with conventional game engines much easier. Finally, an extension of our DSM approach to colored shadows inspired by [16] seems straight forward.

References

1. Lokovic, T., Veach, E.: Deep shadow maps. In: SIGGRAPH 2000: Proceedings of the 27th Annual Conference on Computer Graphics and Interactive Techniques, pp. 385–392. ACM Press/Addison-Wesley Publishing Co (2000)
2. Hadwiger, M., Kratz, A., et al.: GPU-accelerated deep shadow maps for direct volume rendering. In: GH 2006: Proc. of the 21th SIGGRAPH/EUROGRAPHICS Symposium on Graphics Hardware, pp. 49–52. ACM (2006)
3. Nickolls, J., Buck, I., Garland, M.: Scalable parallel programming with CUDA. ACM Queue 6, 40–53 (2008)
4. Crow, F.C.: Shadow algorithms for computer graphics. In: SIGGRAPH 1977: Proceedings of the 4th Annual Conference on Computer Graphics and Interactive Techniques, pp. 242–248. ACM (1977)
5. Williams, L.: Casting curved shadows on curved surfaces. In: SIGGRAPH 1978: Proceedings of the 5th Annual Conference on Computer Graphics and Interactive Techniques, pp. 270–274. ACM (1978)
6. Stamminger, M., Drettakis, G.: Perspective shadow maps. In: SIGGRAPH 2002: Proceedings of the 29th Annual Conference on Computer Graphics and Interactive Techniques, pp. 557–562. ACM (2002)
7. Wimmer, M., Scherzer, D., Purgathofer, W.: Light space perspective shadow maps. In: Keller, A., Jensen, H.W. (eds.) Rendering Techniques 2004 (Proceedings Eurographics Symposium on Rendering), Eurographics, pp. 143–151 (2004)
8. Lloyd, D.B.: Logarithmic perspective shadow maps. PhD thesis, Chapel Hill, NC, USA (2007), Adviser-Manocha, Dinesh
9. Lefohn, A., Sengupta, S., Owens, J.: Resolution-matched shadow maps. ACM Trans. Graph. 26, 20 (2007)
10. Aila, T., Laine, S.: Alias-free shadow maps. In: Keller, A., Jensen, H.W. (eds.) Rendering Techniques, Eurographics Association, pp. 161–166 (2004)
11. Dimitrov, R.: Cascaded Shadow Maps. NVIDIA (2007)
12. Zhang, F., Sun, H., Xu, L., Lun, L.K.: Parallel-split shadow maps for large-scale virtual environments. In: VRCIA 2006: Proceedings of the 2006 ACM International Conference on Virtual Reality Continuum and its Applications, pp. 311–318. ACM (2006)
13. Donnelly, W., Lauritzen, A.: Variance shadow maps. In: I3D 2006: Proceedings of the 2006 Symposium on Interactive 3D Graphics and Games, pp. 161–165. ACM (2006)
14. Lauritzen, A., McCool, M.: Layered variance shadow maps. In: GI 2008: Proceedings of Graphics Interface 2008, pp. 139–146. Canad. Information Processing Society (2008)
15. Annen, T., Mertens, T., Bekaert, P., Seidel, H.P., Kautz, J.: Convolution shadow maps. In: Rendering Techniques 2007: Eurographics Symposium on Rendering, Grenoble, France, Eurographics, pp. 51–60 (2007)

16. McGuire, M., Enderton, E.: Colored stochastic shadow maps. In: Proceedings of the ACM Symposium on Interactive 3D Graphics and Games (2011)
17. Rezk-Salama, C., Ropinski, T., Hadwiger, M., Ljung, P.: Advanced Illumination Techniques for GPU-Based Raycasting. ACM SIGGRAPH Course Notes (2009)
18. Kniss, J., Premoze, S., Hansen, C., Ebert, D.: Interactive translucent volume rendering and procedural modeling. In: VIS 2002: Proceedings of the Conference on Visualization 2002, pp. 109–116. IEEE Computer Society, Washington (2002)
19. Behrens, U., Ratering, R.: Adding shadows to a texture-based volume renderer. In: VVS 1998: Proceedings of the 1998 IEEE Symposium on Volume Visualization, pp. 39–46. ACM, New York (1998)
20. Ropinski, T., Döring, C., Rezk-Salama, C.: Advanced Volume Illumination with Unconstrained Light Source Positioning. IEEE Comp. Graph. & Applications 30, 29–41 (2010)
21. Sundén, E., Ynnerman, A., Ropinski, T.: Image plane sweep volume illumination. IEEE Transactions on Visualization and Computer Graphics 17, 2125–2134 (2011)
22. Salama, C.R.: Gpu-based monte-carlo volume raycasting. In: Proceedings of the 15th Pacific Conference on Computer Graphics and Applications, pp. 411–414. IEEE Computer Society (2007)
23. Kim, T.Y., Neumann, U.: Opacity shadow maps. In: Proceedings of the 12th Eurographics Workshop on Rendering Techniques, pp. 177–182. Springer (2001)
24. Mertens, T., Kautz, J., Bekaert, P., Van Reeth, F.: A self-shadow algorithm for dynamic hair using density clustering. In: SIGGRAPH 2004: ACM SIGGRAPH 2004 Sketches, p. 44. ACM (2004)
25. Sintorn, E., Assarsson, U.: Real-time approximate sorting for self shadowing and transparency in hair rendering. In: I3D 2008: Proceedings of the 2008 Symposium on Interactive 3D Graphics and Games, pp. 157–162. ACM (2008)
26. Yuksel, C., Keyser, J.: Deep opacity maps. In: Computer Graphics Forum (Proceedings of EUROGRAPHICS 2008), vol. 27 (2008)
27. Sintorn, E., Assarsson, U.: Hair self shadowing and transparency depth ordering using occupancy maps. In: I3D 2009: Proceedings of the 2009 Symposium on Interactive 3D Graphics and Games, pp. 67–74. ACM (2009)
28. Salvi, M., Vidimče, K., Lauritzen, A., Lefohn, A.: Adaptive volumetric shadow maps. In: Eurographics Symposium on Rendering, pp. 1289–1296 (2010)
29. Kainz, B., Grabner, M., Bornik, A., Hauswiesner, S., Muehl, J., Schmalstieg, D.: Efficient ray casting of volumetric datasets with polyhedral boundaries on many-core gpus. In: SIGGRAPH Asia 2009: ACM SIGGRAPH Asia 2009 papers. ACM, New York (2009)

Bundle Visualization Strategies
for HARDI Characteristics

Diana Röttger[1], Daniela Dudai[1], Dorit Merhof[2], and Stefan Müller[1]

[1] Institute for Computational Visualistics, University of Koblenz-Landau
[2] Visual Computing, University of Konstanz

Abstract. In this paper we present visualization approaches for HARDI-based neuronal pathway representations using fiber encompassing hulls. We introduce novel bundle visualization techniques to indicate characteristics, such as information about tract integrity and multiple intra-voxel diffusion orientations. To accomplish this task, we developed an intra-bundle raycasting approach and use color mappings to encode diffusion characteristics on the bundle's surface. Additionally, we implemented a slicing approach using a plane orthogonal to the centerline of a bundle which reveals intra-bundle diffusion characteristics as well as the local bundle shape. With the presented approaches, we simultaneously reveal features of fiber bundles such as integrity or information about the underlying diffusion profile as well as context information, the shape of a current tract.

1 Introduction

Diffusion imaging is able to characterize organized tissue due to the fact that the movement of water molecules in fibrous material occurs in a larger scale with the fiber course than against it. It poses a large achievement in neuroscience, since it captures information about the organization of neuronal pathways in vivo. High angular resolution diffusion imaging (HARDI), was designed to overcome the limitations of diffusion tensor imaging (DTI) in terms of the ability to model complex intra-voxel diffusion profiles. The output is a probability density function on a sphere, the orientation distribution function (ODF), describing the movement of water molecules. Tractography techniques use the ODF to estimate trajectories and are best known representations of diffusion data. However, clinicians are oftentimes interested in the spatial position and shape of whole bundles: The generation of hulls approximating the reconstructed fibers can be beneficial in neurosurgical planning, for example, to determine the risk of an intervention [1]. Hulls are motivated by the fact that streamlines only approximate the diffusion process within a voxel; the hulls in turn approximate the reconstructed fibers. Conventionally, these hulls are single-colored and do not include information about the underlying diffusion profile. However, this information is essential in neuro-examinations and can be combined with hull visualization to improve tract-related examinations. Therefore, in this paper, we

G. Bebis et al. (Eds.): ISVC 2012, Part I, LNCS 7431, pp. 326–335, 2012.

propose visual exploration approaches for HARDI combining both bundle morphology and intra-voxel fiber characteristics. Main contributions are: *Raycasting-based visualizations of intra-bundle diffusion properties*, using an evaluation of bundle characteristics on a ray from a vertex to the closest bundle centerline point. The information within a bundle is visualized on the geometry's surface using color maps. *Centerline Slicing*, a visual exploration method for diffusion characteristics using a plane, orthogonal to the centerline. The slicing reveals the bundle shape as well as diffusion characteristics for the current cross-section of the bundle. *Visual enhancements* including Phong illumination, ambient occlusion and silhouettes are integrated and can be activated by the user to facilitate depth perception. Therefore, the view does not suffer from visual cluttering and information about the global bundle orientation, shape, integrity as well as intra-voxel orientations is provided. The paper is organized as follows: In Section 2 related work on diffusion data visualization will be discussed. Section 3 will introduce the dataset and applied methods. Novel visualization approaches are proposed in Section 4 and results in Section 5 followed by a conclusion in Section 6.

2   Related Work

Tractography techniques aim to identify white matter tracts in the human brain and use either DTI or HARDI reconstructions [2–4]. These approaches provide essential information about the location and orientation of neuronal pathways, but conventionally do not include any indication about local tract properties such as integrity or fiber configuration. Therefore, a further visualization strategy for diffusion data exists, directly displaying the intra-voxel diffusion pattern using a geometry representation, known as glyphs. Kindlmann [5] presented a glyph representation for DTI, whereas Peeters et al. [6] introduced a fast rendering for HARDI. These approaches are beneficial in terms of analyzing local diffusion profiles precisely. However, the understanding of the global tract configuration, the course of the neuronal fibers, is lost. Therefore, visualizations based on line rendering but additionally including information of the local diffusion pattern arose. Zhang et al. [7] used Streamtubes and Streamsurfaces to visualize the local diffusion profiles of DTI. More precisely, the authors visualized linear diffusion tensor profiles using a tubular geometry whose orientation corresponds with the direction of the principal eigenvector of the diffusion tensor. In addition, the cross-section of a streamtube represents the ellipsis formed by the two remaining eigenvectors of the tensor at the position in the volume dataset. Klein et al. [8] presented an approach for DTI characteristic quantification on fiber bundles. The proposed method uses a resampled fiber bundle and computes an average principal fiber from the bundle. Afterwards, orthogonal planes are generated and used to compute fractional anisotropy (FA) values on the cross-section of the bundle. This approach is beneficial, since the quantification of DTI characteristics is possible for arbitrarily oriented fiber bundles and enables a tract-specific examination of diffusion properties. However, the simple approach presented for centerline reconstruction is not sufficient for complex bundle configurations.

A first step towards visualizing integrity information on bundle surfaces and slicing planes was made by Goldau et al. [9]. In their approach, the FA value is visualized along fiber bundles using either a color-coded slice or directly on the bundle boundary. For bundle color mapping, the authors compute a single FA value out of samples on a slice orthogonal to an average fiber's tangent. Subsequently they interpolate the obtained mean values of two neighboring planes to assign a characteristic to a specific vertex. This comprises many averaging steps and hence, the resulting index value is less representative in case of more complex fiber bundles. In our approach, we compute a ray for each individual vertex, resulting in a more precise representation of diffusion characteristics. In addition, we focus on HARDI fiber bundles and characteristics, compute a more accurate centerline and apply advanced visualizations.

3   Data Acquisition and Precomputaion

In the following, material and methods used for the presented approach will be introduced. Methods were implemented using *MeVisLab*, a development environment for medical image processing and visualization [10].

HARDI Dataset. The human brain dataset was of size $128 \times 128 \times 60$ and was acquired with a voxel size of $1.875 \times 1.875 \times 2$ mm [11]. The applied gradient direction scheme included 200 directions and a b-value of 3000 s/mm^2.

Q-ball Imaging. Q-ball imaging, introduced by Tuch [12] is a method reconstructing the ODF from a HARDI signal and comprises a good balance in terms of acquisition requirements, computation time and a-priori assumptions. Descoteaux et al. [13] introduced an analytical solution using a spherical harmonics basis in combination with a Laplace-Beltrami regularization term. We evaluated the ODF using an icosahedron tessellation of order 3, resulting in 162 directions.

Tractrography. For HARDI-based fiber reconstruction, we used the approach proposed in Röttger et al. [14], a deterministic tractography approach based on the local ODF. The approach includes, amongst other aspects, the computation of distances to white matter boundaries using vectors orthogonal to the current orientation to determine the most adequate propagation.

Voxel Classification. Voxel classification was performed using a local ODF-based index for HARDI introduced by Röttger et al. [15]. It successfully categorizes a diffusion pattern into the following three compartments: isotropic diffusion profiles and anisotropic diffusion into single and multiple fiber populations. The index consists of a computation pipeline, in which at first a separation into white and gray matter takes place, followed by a classification into one and multiple fiber populations by analyzing the ODF.

Hull Generation. The applied hull generation algorithm extends the approach proposed by Merhof et al. [1], consisting of the following four consecutive steps: volume rasterization, volume filtering, surface extraction and surface filtering. A binary fiber volume, the result of the rasterization stage, serves as the basis for isosurface computation. Our approach extends the proposed algorithm in terms of surface filtering: We include a reference volume (the rasterized fiber volume) which controls the adjustment of the geometric hull in the surface filtering stage.

Centerline Extraction. Several approaches of centerline reconstruction for fiber bundles exist in literature [8, 16]. These approaches base on a principal fiber computation representing one centerline for a whole bundle. However, in cases of complex fiber configurations a principle fiber does not characterize a bundle appropriately, which is illustrated in Figure 1. Therefore, we applied a skeletonization approach (MeVisLab module DtfSkeletonization) using the fiber bundle morphology.

4   Visualization Approaches

Using the aforementioned precomputation results we will hereafter introduce the developed visualization techniques, which are illustrated in Figure 2. For enhanced three-dimensional understanding of the bundle shape, we apply a silhouette rendering approach, as well as Phong shading and ambient occlusion.

(a)   (b)

Fig. 1. Schematical illustration of a principle fiber (1a) and a skeleton (1b)

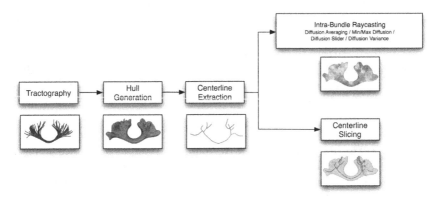

Fig. 2. Computation pipeline for HARDI-characteristic visualization. Preprocessing steps are shown in the left and intra-bundle visualizations in the right part.

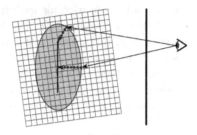

Fig. 3. Ray computation from the current vertex to the nearest centerline point

Silhouette rendering is performed by analyzing variances of neighboring normals in a deferred shading approach as introduced by Saito et al. [17]. Additionally, we implemented a Screen-Space Ambient Occlusion (SSAO) approach, first introduced by Mittring [18], which samples the depth buffer in the neighborhood of each fragment. The number of fragments closer to the viewport than the current one is used to darken the current fragment's color. However, combining color maps with enhanced rendering techniques to facilitate depth perception alters color appearance and can lead to false interpretations. Therefore, in our visualizations the user can interactively switch the rendering techniques on and off. In the following intra-bundle visualizations, we used two different color maps to emphasize the different meaning of both indices. For intra-bundle fiber configuration, red indicates single, yellow multiple and green isotropic diffusion. For variance encoding high variance is displayed in cyan and low variance in blue.

4.1   Intra-bundle Raycasting

The first method introduced in this paper uses a novel intra-bundle raycasting approach, which is realized though a GPU shader pipeline, using the hull geometry as an input. The centerline of the fiber tract is encoded in a 2D texture and transferred to the GPU as well as a 3D texture comprising the diffusion characteristics, in this case the voxel classification volume. The vertex shader computes the texture coordinates for characteristic evaluation, as well as the nearest point of the centerline. Further, a ray is traced from the current hull's vertex to the obtained centerline point, as illustrated in Figure 3. This ray is used for diffusion characteristic evaluation. We developed several visualization strategies for intra-bundle raycasting, motivated by volume rendering and explained in the following.

Diffusion Averaging. For *Diffusion Averaging*, we compute samples along the ray from the vertex to the nearest centerline point using a certain step length and resulting in n distinct positions. These samples are used to evaluate the diffusion characteristics 3D texture and compute the average out of all obtained values. The final fragment color is determined through color mapping and thereby reflects the average diffusion value from a vertex to the centerline.

Min/Max Diffusion. The *Min/Max Diffusion* mode is motivated by the maximum intensity projection (MIP) method, a transfer function, projecting the maximum value along a ray to the image plane. Transferring the idea to our approach results in computing the minimum and maximum characteristic value along the ray and the use of this value for color mapping. Within this visualization, a single characteristic value is displayed.

Diffusion Slider. With the *Diffusion Slider*, the user can interactively examine diffusion characteristics along the ray. In this case, no averaging of diffusion values takes place. Instead, the ray is normalized and sampled at discrete points. The resulting single diffusion values are visualized directly on the bundle's surface. Therefore, visualizations of diffusion characteristics from the hull to the bundle's centerline are feasible.

Diffusion Variance. Using the *Diffusion Variance* mode, diffusion homogeneities as well as inhomogeneities within the bundle, are highlighted. We therefore compute the variance of diffusion characteristics along the traced ray.

4.2   Centerline Slicing

In addition to previously presented approaches, we implemented a second intra-bundle visualization, the *Centerline Slicing*. A plane orthogonal to the tangent of a user specified point on the centerline is generated and utilized for visualizing the color-coded index through texture mapping. The user selects a point through a mouse click on a rendering of the centerline. Afterwards, a shader pipeline uses the precomputed centerline rendered by use of the `GL_LINES` primitive as an input and generates the plane as the output. The coordinates of the user-defined point are transferred to the shaders via uniforms. The geometry shader is designed to generate a plane if the first coordinates of the current primitive are within an interval around the selected centerline point. The tangent of this centerline segment acts as the normal for the plane and is computed by subtracting both coordinates of the primitive. In bifurcations no distinct centerline tangent exists, which can be used to generate the plane. Therefore, in this case we use the view-vector as the plane normal. Thus, the user can interactively examine the diffusion pattern while moving the camera and the plane origin remains unchanged. To enhance the spatial understanding of the bundle, we render the parts of the plane not lying within the bundle transparent. Therefore, the plane not only exhibits information about the diffusivity, but also reveals the shape of the cross-section of the bundle for arbitrarily oriented centerlines.

5   Results and Discussion

Since the presented approaches are designed for challenging pathway configurations, we focus on fibers running in the *centrum semiovale*, a region in the brain, where the following three fiber tracts meet: the *corpus callosum*, the *corticospinal*

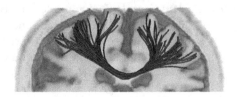

Fig. 4. Streamtube visualization of callosal fibers in the *centrum semiovale*

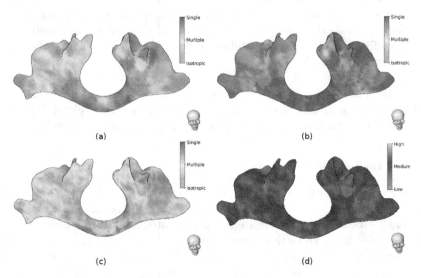

Fig. 5. Color mapping of values along the ray: *Diffusion Averaging* (5a) and *Max Diffusion* (5b) reveal multiple diffusion in the *centrum semiovale*, associated bundle parts appear yellow. Contrary, regions with highest integrity appear red using the *Min Diffusion* approach (5c). The *Diffusion Variance* highlights high diffusion variances (5d).

tract and the *superior longitudinal fasciculus*. Figure 4 shows fibers belonging to the *corpus callosum*, visualized as streamtubes. Results of the *Diffusion Averaging, Max Diffusion, Min Diffusion* and *Diffusion Variance* visualization on the hull without anatomical volume rendering are shown in Figure 5 and the *Diffusion Slider*, the *Centerline Slicing* as well as visual enhancements in Figure 6. Performance evaluation was accomplished on a Core2 Duo, 3.16 GHz with 4 GB RAM and a NVIDIA GeForce GTX 285 graphics card. Following average frames per second (fps) were achieved for callosal fibers shown in this paper. The raycasting-based approach features 11 fps when performing a 360° rotation of the bundle. Ambient occlusion and silhouette visualization reduce the average fps to 6. The *Centerline Slicing* mode achieved 22 fps on average.

Using the presented visualization approaches arbitrary neuronal pathway morphologies can be visualized with significant diffusion encoding color values.

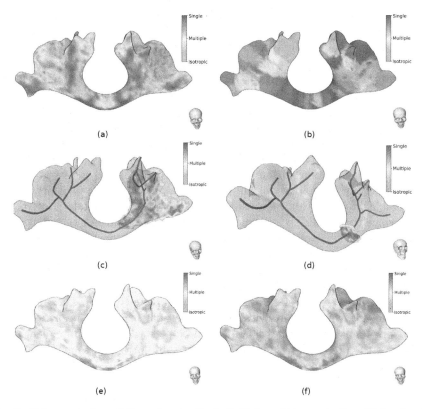

Fig. 6. Color mapping of the value on the hull (6a) and the centerline (6b) using the *Diffusion Slider*. Parts of the hull leaking into gray matter appear in green (isotropic diffusion). High single fiber configuration is obtained in the center of the *corpus callosum*. *Centerline Slicing* within regions of complex diffusion profiles, plane is oriented with the view-vector (6c) and in the center of the *corpus callosum*, plane is orthogonal to the centerline's tangent (6d). Phong illumination (6e) and ambient occlusion (6f) with applied silhouette rendering.

However, our method depends on the bundle's centerline. For complex morphologies, a detailed skeleton is crucial to result in reliable diffusion information, since the ray is determined by the closest centerline point. However, a detailed centerline reduces performance, since more points have to be considered for raycasting. In addition, the amount of diffusion information which is mapped through the ray to a bundle vertex, depends on the centerline's curvature and therefore is not uniformly distributed. In terms of large bundles or a less representative centerline the *Diffusion Averaging* visualization might not be representative, since all values on the ray are taken into consideration. This accounts for inhomogeneous bundles as well, since if the variance of diffusion values along the ray is too high a mean value can lead to false interpretation. In these cases a preliminary examination of the bundle using the *Diffusion Variance* or the *Diffusion Slider* mode and a subsequent adjustment of the centerline enhances the visualization.

Results were additionally discussed with seven neuroscientists to answer the following questions: Whether the approaches provide a better understanding as well as insight into the data, if the introduced perception enhancements facilitate depth impression as well as in which questions the approaches can pose a contribution. Findings resulting from the visualizations correspond with medical knowledge: The center of the *corpus callosum* comprises highest integrity as well as directionality. However, the region of the *centrum semiovale* features more complex fiber distributions. The *Diffusion Averaging* approach was one of the favorite visualization within experts: They stated it provides a good first impression about the underlying diffusion and that it is a straightforward visualization. Diffusion evaluation using the *Diffusion Slider* mode is of special interest, since the bundle skeleton can be considered to represent the whole bundle and visualizations are helpful in tract-based examinations. Furthermore, the *Diffusion Variance* visualization reveals interesting intra-bundle conditions: Regarding the center of the *corpus callosum*, we can identify similar values in both of the *Min/Max Diffusion* visualizations. These regions appear in purple and indicate low variance. Vice versa scenarios can be detected as well. The *Centerline Slicing* mode provides a more detailed examination, which can be beneficial in regions with high variance and therefore can be performed in a subsequent step. Considering the presented visual enhancements, all participants agreed that the combination of ambient occlusion and silhouettes as displayed in Figure 6 is the best depth encoding visualization. Color changes caused by Phong illumination are rated to be more violating than those caused by ambient occlusion. In general, experts think that the provided visualizations are of great interest and can be beneficial in questions related to neurosurgical planning, as well as disorder monitoring, for example in stroke patients.

6   Conclusion and Future Work

In this paper we proposed visualization techniques for intra-bundle diffusion characteristics using HARDI-based fiber reconstruction and diffusion profiles. Today, it is possible to describe intra-voxel diffusion patterns using HARDI classifiers. However an user-friendly and problem-specific visualization had been missing, although, a visual diffusion characteristics exploration in combination with tract morphology has great potential in many clinical questions. With our intra-bundle visualization approach we are taking a step forward towards a combination of global tract morphology and diffusion characteristics for HARDI. Considering HARDI fiber tracking results, an approach managing overlapping bundle regions is needed. Potential approaches include the use of focus and context rendering techniques. From the medical point of view it is interesting to evaluate the application of the presented visualizations to neurosurgical examinations.

References

1. Merhof, D., Meister, M., Bingöl, E., Nimsky, C., Greiner, G.: Isosurface-based generation of hulls encompassing neuronal pathways. Stereotactic and Functional Neurosurgery, 50–60 (2009)
2. Basser, P.J., Pajevic, S., Pierpaoli, C., Duda, J., Aldroubi, A.: In vivo fiber tractography using DT-MRI data. Magnetic Resonance in Medicine 44, 625–632 (2000)
3. Lazar, M., Weinstein, D.M., Tsuruda, J.S., Hasan, K.M., Arfanakis, K., Meyerand, M.E., Badie, B., Rowley, H.A., Haughton, V., Field, A., Alexander, A.L.: White matter tractography using diffusion tensor deflection. Human Brain Mapping 18 (2003)
4. Descoteaux, M., Deriche, R., Knösche, T.R., Anwander, A.: Deterministic and probabilistic tractography based on complex fibre orientation distributions. IEEE Transactions on Medical Imaging 28 (2009)
5. Kindlmann, G.L.: Superquadric tensor glyphs. In: VisSym., pp. 147–154 (2004)
6. Peeters, T.H.J.M., Prckovska, V., van Almsick, M., Vilanova, A., ter Haar Romeny, B.M.: Fast and sleek glyph rendering for interactive HARDI data exploration. In: PacificVis., pp. 153–160 (2009)
7. Zhang, S., Demiralp, Ç., Laidlaw, D.H.: Visualizing diffusion tensor MR images using streamtubes and streamsurfaces. IEEE Transactions on Visualization and Computer Graphics 9, 454–462 (2003)
8. Klein, J., Hermann, S., Konrad, O., Hahn, H.K., Peitgen, H.O.: Automatic quantification of DTI parameters along fiber bundles. In: Bildverarbeitung für die Medizin (2007)
9. Goldau, M., Wiebel, A., Hlawitschka, M., Scheuermann, G., Tittgemeyer, M.: Visualizing DTI parameters on boundary surfaces of white matter fiber bundles. In: Proceedings of the Twelfth IASTED International Conference on Computer Graphics and Imaging, pp. 53–61 (2011)
10. MeVis Medical Solutions AG: MeVisLab, medical image processing and visualization (December 2011), http://www.mevislab.de
11. Poupon, C., Poupon, F., Allirol, L., Mangin, J.F.: A database dedicated to anatomo-functional study of human brain connectivity. In: 12th HBM Neuroimage, Florence, Italie, Number 646 (2006)
12. Tuch, D.S.: Q-ball imaging. Magnetic Resonance in Medicine 52, 1358–1372 (2004)
13. Descoteaux, M., Angelino, E., Fitzgibbons, S., Deriche, R.: Regularized, fast and robust analytical Q-ball imaging. Magnetic Resonance in Medicine 58, 497–510 (2007)
14. Röttger, D., Seib, V., Müller, S.: Distance-based tractography in high angular resolution diffusion MRI. The Visual Computer 27, 729–738 (2011)
15. Röttger, D., Dudai, D., Merhof, D., Müller, S.: ISMI: A classification index for high angular resolution diffusion imaging. In: Proceedings of SPIE Medical Imaging (2012)
16. Chen, W., Zhang, S., Correia, S., Ebert, D.S.: Abstractive representation and exploration of hierarchically clustered diffusion tensor fiber tracts. Computer Graphics Forum 27, 1071–1078 (2008)
17. Saito, T., Takahashi, T.: Comprehensible rendering of 3-D shapes. In: Proceedings of the 17th Annual Conference on Computer Graphics and Interactive Techniques, SIGGRAPH 1990, pp. 197–206. ACM, New York (1990)
18. Mittring, M.: Finding Next Gen - CryEngine 2. In: ACM SIGGRAPH 2007 Courses. SIGGRAPH 2007, pp. 97–121 (2007)

Context-Preserving Volumetric Data Set Exploration Using a 3D Painting Metaphor

L. Faynshteyn and T. McInerney

Dept. of Computer Science, Ryerson University, Toronto, ON, Canada, M5B 2K3

Abstract. This paper presents a technique for fast and intuitive visualization and exploration of unsegmented volumetric data sets. It combines a Maximum Intensity Difference Accumulation (MIDA) visualization algorithm with a 3D interaction model based on a painting metaphor. Under this model the Volume of Interest (VOI) is defined by "painting" an envelope enclosing features of interest. The key advantage of the model is the ability to interactively preview the effect of the painting operation before it is applied, allowing the user to both explore and paint in the volume space at the same time. The result is a flexible and intuitive technique that uses a single, consistent interaction style to achieve desired contextual views and perform several types of volume exploration and editing operations. Examples using several volumetric data sets and the results of a user study are presented to validate the effectiveness, flexibility and ease of use of the technique.

1 Introduction

Despite the significant amount of research in the field of volume rendering, when it comes to providing a user with easily understandable and predictable ways of specifying how the data is to be visualized, there is still much room for improvement. Traditionally this specification has been realized by means of a Transfer Function (TF). However, specifying a TF that achieves the desired visual effect still remains a non-trivial task, even for experienced users. This is primarily due to the fact that TF specification is usually performed in a histogram space and thus doesn't provide the user with a natural feel for the correspondence of features in the histogram with features in the volumetric data set.

The challenge of finding a perfect TF is also compounded by the now common requirement of visualizing areas of interest in the volume using different sets of parameters than the surrounding region, in an effort to preserve the visual context. This secondary goal presents visualization challenges in its own right as navigation in the 3D space of a volume and specification of different VOIs in a 3D data set is often a complex operation. Among the more prominent issues are VOI depth specification and effective depth perception cues, as well as volumetric features separation and occlusion. The traditional approach to the VOI specification, feature separation, and occlusion problems is to use multiple 2D views along with a variety of 2D VOI editing capabilities. While certainly useful in many scenarios, when it comes to volume exploration and VOI specification,

G. Bebis et al. (Eds.): ISVC 2012, Part I, LNCS 7431, pp. 336–347, 2012.

multiple 2D views complicate the user interaction model and it is often difficult to understand the spatial relationships among all the views.

To meet these challenges, we propose a volume visualization and exploration model that combines the Maximum Intensity Difference Accumulation (MIDA) visualization method with an intuitive and flexible volume painting interaction model. The MIDA visualization technique was proposed by Bruckner et al. [1] as a means of quickly visualizing volumetric data sets without the necessity of specifying TFs. The MIDA technique uses a conventional Direct Volume Rendering (DVR) approach and modifies it to exploit inherent data characteristics of the underlying volumetric data set by modulating the accumulated opacity of the image in accordance with changes in the data intensity values. To control which features of a volumetric data set should be emphasized, only a range defined by the minimum and maximum data intensity values needs to be specified.

The painting interaction model supports free-form and constrained painting and editing of envelopes defining VOIs. Features of interest are surrounded by positioning the tip of a paint "brush" at the desired location and depositing a "blob" of paint, extending the envelope by merging with it. The envelope can be thought of as "thick paint", implemented using Metaballs [2] (see section 3.2), and volumetric features inside and outside of the envelope can be visualized using two separate (but functionally identical) sets of MIDA visualization parameters. The key difference between our approach and many other techniques is that it allows the user to preview, in real time, the results of the painting operation before the new blob of paint is deposited. That is, the Metaball-based paint blob acts as a lens that can be moved around to visualize the volume interior before deciding on its final location. This separation of blob positioning and depositing helps the user to interactively guide the painting process and results in a high degree of control over the VOI shape definition.

The combination of the two techniques allows the user to operate within a single 3D data view and employs a minimum of visualization controls. It also enables a variety of possible volume data visualization, exploration and editing operations (e.g. contextual views, volume carving, "Magic Lantern"-like behavior [3] etc.) to be realized using a single, consistent interaction model. In addition, the proposed approach does not depend on any particular visualization method and other DVR techniques can be easily substituted should MIDA prove insufficient in terms of achievable visual results. A number of examples showcasing the visualization and exploration capabilities of the proposed approach are presented, along with the results of a user study validating the ease of use, intuitiveness and effectiveness of the approach.

2   Related Work

Over last two decades, a considerable amount of DVR research has focussed on improving TF specification. Since 1D TFs often suffer from poor volumetric feature separation, more sophisticated higher dimension TFs are now commonly used. Kniss et al. [4] propose a 2D TF specification process based on data intensity and gradient magnitude values that enables better feature separation by representing material boundaries as arches on a 2D histogram. Correa and Kwan-Liu [5] use the relative size of volumetric features to construct a TF with improved feature classification. Sereda et al. [6] construct so called LH histograms that show lower and higher intensities of the materials inside a volume, resulting in a better definition of material boundaries. All these methods, however, force the user to either work in the feature space of 2D histograms or manipulate some other data driven parameters that control the TF specification process. This, in turn, usually results in a time consuming and tedious trial and error specification process.

To simplify the TF specification various automatic and semi-automatic techniques have been proposed. Kindlmann et al. [7] calculate the curvature of volumetric features using iso-surfaces, which is then used as a basis for designing a TF and to generate non-photorealistic images by highlighting the contours of volumetric structures. Prassni et al. [8] introduce a concept of Curve-skeletons, that are used to classify volumetric features inside a pre-segmented VOI using predefined shape descriptors. The resulting shapes are then merged in a process supervised by the user. Haidacher et al. [9] present a method of adaptively estimating statistical properties of the data intensity values in the neighborhood of each sampled point within a volume. These properties are used for TF specification which enables the differentiation of volumetric features. While these and other automatic approaches significantly reduce the complexities of the TF specification, they also require lengthy pre-processing steps.

To better facilitate understanding of the information contained in a volume, it is necessary to be able to explore structures hidden inside it. This can be achieved by specifying a VOI containing the structures of interest and then high-lighting these structures by using some sort of distinctive visual style. At the same time, parts of the volume surrounding the VOI should be preserved for contextual purposes. Most of the approaches that attempt to meet these goals are based on concepts borrowed from the field of traditional illustration. Among these are non-photorealistic rendering with emphasized silhouettes [7], stylized shading, contouring, transparency and haloing [10]. These methods help the user understand the data by making certain features within a volume more prominent. However, highlighting features of interest is often not sufficient since they can still be occluded by other structures.

Several context-preserving techniques addressing the problem of occlusion have been proposed. Inspired by the idea of ghosted views, Bruckner et al. [11] propose a method that uses a shading intensity function for opacity modulation of structures within a volume. Another approach by Bruckner et al. [12] use compositing and masking to combine several pre-rendered layers and draw the user's attention to certain parts of the image while preserving the overall context. The use of masking also enables the effect known as a Magic Lens or Lantern, which is another popular context preserving technique. Monclus et al. [3] use a second TF to visualize a cone-shaped VOI in a way that is distinct from the surrounding material. Various standard and custom magnification effects, within a user defined arbitrarily-shaped 2D region on the screen, are employed in the work by Wang et al. [13]. Bruckner and Gröller [14] use a 3D volumetric painting approach, where the user can define a surface-bound 3D envelope that encloses the features of interest, which are then visualized using an alternative predefined TF to provide contextual data representation.

Another way of dealing with the problem of occlusion is based on the idea of physically removing the occluding parts of the volume or deforming parts of the volume in such a way as to make the features of interest visible. The work presented by Bruckner and Gröller [15] and Ruiz et al. [16] is based on the concept of exploded views. Weiskopf et al. [17] cut away parts of a volume by using various geometric primitives and depth test algorithms. Bernhard et al. [18] use a deformable cutting plane with an adjustable sphere of influence to define the VOI and exclude all other volumetric regions. User-controlled drilling, lasering and peeling operations are used by Chen et al. [19] to remove parts of the surface of a volumetric object to reveal the underlying structures.

Upon considering the complexities and limitations of the TF-based approaches, we decided to focus on algorithms and interaction models with very simple and easily extendible interfaces. The non-TF based MIDA approach combined with the flexible Metaball-based envelope painting forms the basis of our system. Our envelope painting approach shares similarities with many of the techniques mentioned above. In particular, the surface painting described in [14] is similar to our surface painting mode, and the carving and peeling operations presented in [19] can be used to achieve visual results similar to some of ours. However, unlike most of these techniques, we use a higher-level shape representation of the envelope to visualize and define the VOI (i.e. Metaballs). Instead of using sets of independent surface-bound volumetric brushes [14], arbitrarily-shaped depth-fixed peeling regions [19], or shape-restricted envelopes [3], we propose a more general approach that combines and extends all of these visualization and interaction capabilities in a single, flexible and consistent model. The key feature, real time previewing, allows for "magic lens"-like behavior in both the 2D and 3D painting modes and enables precise and easy envelope definition and editing capabilities. The ability to easily control the range of sampled intensity values in the MIDA technique, combined with an unrestricted 3D envelope definition mechanism, enables flexibility in terms of possible types of volume cutting and peeling operations. Furthermore, recognizing that the MIDA visualization technique might not provide the desired visual results in all cases, our system was also designed and implemented in such a way as to be independent of the particular DVR technique used, allowing for easy and straightforward integration with TF-based visualization models.

3 Methodology

3.1 Maximum Intensity Difference Accumulation

MIDA works by altering the monotonically growing opacity accumulation behavior, inherent to the traditional DVR approach, by modulating any previously accumulated opacity values by the amount of positive difference between the currently sampled data intensity value and the previously encountered maximum intensity value. This, in turn, allows volumetric structures represented by higher intensity values (e.g. bones vs. skin) to be given more prominence in the resulting image. A simple linear mapping is used to map data intensity values to gray-scale color and opacity values. The obtained gray-scale values can also be modulated by a user specified color value, resulting in more realistic visualizations. In addition, and rather importantly, a range of mapped data intensity values can be specified. This allows the user to easily control (or filter), with just one set of GUI slider controls, which volumetric features will contribute to the final image and which will not (Fig. 1). As will be shown later, we use this feature extensively both in our exploration work-flow and to provide volume carving and peeling operations.

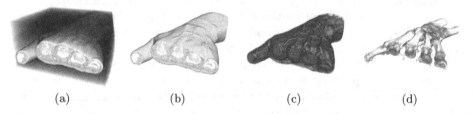

(a)   (b)   (c)   (d)

Fig. 1. By simply adjusting the range of mapped data intensity values and specifying the modulation color, the user is able to control the visualization process: (a) an unfiltered and uncolored volume visualization; (b-d) different volumetric features are visible and colored according to the user's preference

3.2   3D Envelope Painting Overview and Work-Flow

We employ a single interaction model that can be used for volume exploration, VOI definition, and contextual data view generation. It is based on a painting metaphor and realizes a concept of "thick paint" that enables the user to select features of interest within a volume by painting an envelope either in a 2D space of the screen or a 3D space of a volume. The painting process is controlled by moving a virtual brush tip in the 3D space of a scene. The brush is represented by a blob of paint (centered at the brush tip position) that can be smoothly merged with the current envelope painted thus far, allowing for a continuous envelope extrusion. Since everything inside the paint blob (and current envelope) is instantly visualized using a separate MIDA visualization style, the user can move the blob around the current region and preview the volume interior. When the user is satisfied with the results of this previewing operation, they "deposit" the blob with a mouse click, extending the envelope (i.e. spreading the paint)(Fig. 2). This previewing capability can also be used as a tool for volume exploration, acting as a sort of "flash light" or lens, allowing the user to quickly look around the volume interior.

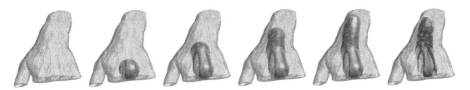

Fig. 2. Envelope painting depicted as a sequence of steps (from left to right) for previewing the results of the painting operation by moving the brush tip to new locations, and spreading the paint (i.e. painting the envelope) by depositing new blobs of paint at desired locations. When the desired features are painted, the envelope is made transparent (rightmost image).

3.3   Painting Modes and Volume Exploration Operation Types

One of the advantages of the Metaball-based painting technique is that the paint blobs can be moved around in the 3D space of the volume, in screen space, or can slide around on the surface of the objects in the volume. To paint in the 3D space using a 2D input device (i.e. a mouse or a touch screen) we allow the user to paint only in the space of a 2D painting plane that is always perpendicular to the current viewing direction of the 3D scene. By changing the position of the mouse (or finger etc.) in the screen space coordinates, the user defines the position of the brush tip as projected onto the painting plane. To add a third dimension to the painting process, the position, orientation and depth of the painting plane relative to the volumetric object can be changed at any time. This constrained interaction of adjusting the orientation and position of the painting plane relative to the volumetric object and of depositing blobs of paint to define arbitrarily-shaped 3D envelopes keeps the 2D input device based user interaction model simple, intuitive and consistent.

3D Surface Painting Mode: In this mode the user can explore a volumetric object by automatically following its surface with a brush tip (Fig. 3b). This is done by overriding the depth of the brush tip, and hence the depth of the painting plane at the current projected position, by the depth value of the object's surface. In addition, an alternative brush type is available in this mode. This type of brush is based on a superquadric metaball (as opposed to a regular quadric metaball) to define cylinder-shaped brushes that are used to paint envelopes that more closely conform to the shape of a volumetric object (e.g. a "thin shell" envelope can be defined that surrounds, for example, only a skin layer) (Fig. 3c). The orientation of the cylinder-shaped brush is based on the surface normal orientation. The depth of the object surface penetration can be controlled by the user.

2D Screen Painting Mode: In this mode the depth of the painting plane is constant and any depth information related to the envelope is discarded. Instead, the outline of the shape of the envelope is used as a stencil, inside of which the secondary MIDA visualization style is applied. Since there is no depth specified, the effect is applied all the way through the volume in the direction of projected rays (Fig. 3d). In addition, unlike the 3D painting modes where the painted envelope always maintains position and orientation relative to the volumetric object, in the 2D screen painting mode the painted envelope stays glued to the screen regardless of the object's orientation in the scene. This allows the user to examine the object from different perspectives using the 2D envelope as a sort of arbitrarily-shaped "X-ray device" (or previously described Magic Lantern), that can highlight volumetric features while the object's orientation is being changed.

Brush and Envelope Related Operations: In all painting modes, the size (radius) (Fig. 3e) and color of the brush can be adjusted as well as the color and transparency of the envelope. Finally, the shape of the envelope can be edited by using an "eraser" brush. This brush simply erases parts of the envelope that fall within its sphere of influence (Fig. 3f). If desired, the envelope can also be completely discarded.

Volume Editing: Another set of volume exploration operations is based on the previously mentioned ability to easily control the range of mapped data intensity values in the MIDA visualization technique. By setting the range to zero, the user can discard (carve or peel) parts of the volume that correspond either to those contained inside of the envelope or outside of it. That is, the paint brush becomes a sculpting tool providing a very simple, yet powerful and flexible method for volume exploration and context-preserving visualization (Fig. 3g,h).

Alternative rendering modes: In cases when MIDA proves to be insufficient in terms of desired visual results, this rendering mode can be easily substituted with alternative DVR methods, while preserving all of the original painting modes and functionality. To showcase this capability, we have implemented additional 1D and 2D TF-based rendering modes that can be used to visualize the parts of the volume both inside and outside of the envelope (Fig. 3i).

3.4   Implementation Details

In the current implementation, the brush and envelope are realized using the technique for defining implicit 3D surfaces known as Metaballs [2]. Metaballs are capable of merging with each other based on their proximity, radius and the chosen field function, creating organic-looking and flowing shapes. We use two types of metaballs: quadric, corresponding to a conventional spherical brush (i.e. a blob), and superquadric, used to define disk or cylinder shaped brushes. The six-degree Wyvill polynomial [20] is used to model the metaball field falloff function. To convert the implicitly defined envelope surface into a 3D polygonal representation, we use the Marching Cubes algorithm. A fast GLSL shader implementation of the algorithm is used to extract the iso-surface and then generate and capture the resulting geometry into a memory vertex buffer (which can be reused later, removing the need to re-compute the geometry for every frame). We use Front-to-Back Depth Peeling to combine polygon-based (i.e. the envelope) and volumetric geometry rendering. The method works by rendering the polygonal geometry in the scene layer by layer, as viewed from the camera position, and filling in the gaps between the layers with the results of the DVR. The results are blended together and there is no limit on the number of rendered layers. Furthermore, as was mentioned earlier, an additional advantage of this approach is its independence of the type of DVR technique used, allowing for

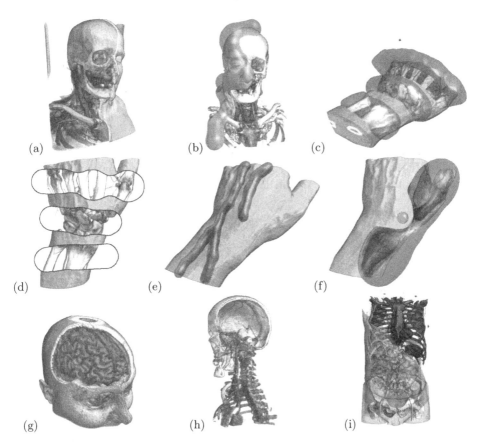

Fig. 3. (a) 3D free-hand painting is used to envelope features deep inside the volume and achieve the desired contextual view; (b,c) 3D surface painting, with spherical and surface aligned cylinder-shaped brushes, slide over the objects' surface; (d) an example of 2D screen painting mode; (e) a small sized spherical brush is used for precise painting on the skin surface; (f) an eraser brush is used to edit the envelope; (g,h) the peeling and carving operations are performed by setting the MIDA mapped data ranges to 0; (i) the MIDA and 2D TF-based DVR techniques are used together to visualize different parts of the same volumetric object

the easy extensibility of the visualization model. Gaussian-smoothed position and normal maps, corresponding to the surface of the visualized object, are generated on-the-fly and are used to control the depth and orientation of the brush in the 3D surface painting mode. For implementation details refer to [21].

4   Results and Validation

In order to evaluate the effectiveness of the proposed exploration technique, as well as to gauge how intuitive and easy to learn and control the painting-based interaction model is, a user study was conducted with ten male subjects and one female subject. Ten of the subjects were undergraduate students and one subject was a graduate student, all of whom were from either the department of computer science or engineering. All subjects were first-time users of the implemented exploration technique and none had any previous experience with any medical software. Approximately half, however, did have limited experience with various 3D modeling programs (i.e. Blender, Maya, etc.). In order to better gauge the level of the participants' proficiency performing navigational tasks in virtual 3D environments, they were also asked to state the number of Hours per Week (HPW) spent using a mouse and playing video games, with the average numbers coming to 25.5HPW and 11.5HPW respectively. The participants were

Fig. 4. Examples of pre-rendered reference images from each set of trials. Left: an example of the 2D painting task, middle: an example of the 3D carving/peeling task and, right: an example of the 3D context-preserving task. The subjects were asked to match these images.

asked to use the implemented software to perform several tasks. A total of nine trials, each based on a different data set, were performed. The goal of each trial was to achieve an approximate visual match to a pre-rendered reference image (4). The trials were further broken down into 3 sets: three 2D painting trials, three 3D volume carving trials and three 3D context-preserving trials. In the course of the trials from both the 3D carving/peeling and 3D context-preserving sets, the users had to use both the free-hand and the surface painting modes. Before each set of trials each participant was given instructions using a separate demonstration data set and was also given time to play with the software to familiarize themselves with the visualization controls and the painting mechanism. The time for each trial was recorded for statistical purposes, although the participants were informed that achieving a visual match in the shortest possible time was not an objective. The average measured trial completion time for these non-expert users was 54 seconds, with results varying from 15 seconds in the case of simple 2D screen painting tasks and up to 2 minutes for more complex 3D painting tasks. Upon completion of all trials the participants were asked to fill out the questionnaire and indicate their level of agreement or disagreement with each statement using a 7-point Likert scale (5).

As is evident from Figure 5, the results were very positive, with a mean average of 6.0 or more. In addition, as can be seen from the spread of the marks,

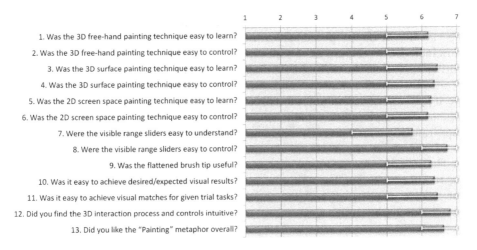

Fig. 5. The results of the user study questionnaire based on a 7-point Likert scale

there was little variability in the results. The only question that received a mean below 6.0 was related to the sliders that control the visible range of values for the MIDA visualization mode. Another common comment from participants was that, even though gauging the depth of the brush tip in relation to the volumetric structures in the scene was usually not a problem, under certain conditions it still could be somewhat confusing, and thus they would prefer to have additional depth cues. We plan to address both of these shortcomings in the near future.

5   Conclusion

As evidenced by ongoing research, providing an intuitive way of visualizing volumetric data sets with predictable visual results is still a challenge. The vast majority of the current techniques rely on a TF mechanism to map volume data characteristics to the visual domain. However, TF specification often remains a complex and time consuming trial and error process. The semi-automatic and automatic approaches address this problem with varying degrees of success and often at the expense of interactivity due to the costly pre-processing operations.

Another challenge lies in providing the user the ability to efficiently navigate the volumetric space and to easily select features of interest while preserving contextual data views. Various techniques based on the ideas of non-photorealistic rendering, a magic lens metaphor, painting, volume sculpting and peeling, and others have been proposed by researchers. To meet these two challenges, we have proposed a volume visualization and exploration framework that combines a non-TF based MIDA DVR method with a simple, yet flexible Metaball-based painting technique. This synthesis enables many of the previously described context-preserving volume exploration and editing techniques to be realized through a single, consistent and intuitive interaction model. The results of a user study as well as feedback the users confirm the validity of these claims and the viability of the proposed solution for relatively noise-free data sets. Future work includes developing more sophisticated undo/redo operations and envelope editing operations by taking advantage of the Metaball-based envelope shape representation. Allowing the user to paint additional VOIs with the ability to specify the visualization parameters for each VOI independently should greatly enhance the exploration capabilities of the system. Finally, and perhaps most importantly, we are currently working on the incorporation of several painting-based interactive segmentation algorithms into the work-flow in order to allow for the visualization and exploration of noisy data sets. Rather than attempting to construct complex TFs, the idea is to retain and augment our simple but flexible painting interaction model and use it to control and constrain the segmentation algorithms.

References

1. Bruckner, S., Gröller, M.: Instant volume visualization using maximum intensity difference accumulation. Computer Graphics Forum 28, 775–782 (2009)
2. Blinn, J.F.: A generalization of algebraic surface drawing. ACM Transactions on Graphics 1, 235–256 (1982)
3. Monclús, E., Díaz, J., Navazo, I., Vázquez, P.: The virtual magic lantern: an interaction metaphor for enhanced medical data inspection. In: ACM Symposium on Virtual Reality Software and Technology, Kyoto, Japan, pp. 119–122 (2009)
4. Kniss, J., Kindlmann, G., Hansen, C.: Multidimensional transfer functions for interactive volume rendering. IEEE Trans. on Visualization and Computer Graphics 8, 270–285 (2002)
5. Correa, C., Kwan-Liu, M.: Size-based transfer functions: A new volume exploration technique. IEEE Trans. on Visualization and Comp. Graphics 14, 1380–1387 (2008)
6. Sereda, P., Bartroli, A., Serlie, I., Gerritsen, F.: Visualization of boundaries in volumetric data sets using lh histograms. IEEE Trans. on Visualization and Computer Graphics 12, 208–218 (2006)
7. Kindlmann, G., Whitaker, R., Tasdizen, T., Moller, T.: Curvature-based transfer functions for direct volume rendering: methods and applications. In: 14th IEEE Conference on Visualization (VIS 2003), Seattle, WA, pp. 513–520 (2003)
8. Prassni, J., Ropinski, T., Mensmann, J., Hinrichs, K.: Shape-based transfer functions for volume visualization. In: IEEE Pacific Visualization Symposium (PacificVis 2010), Taipei, Taiwan, pp. 9–16 (2010)
9. Haidacher, M., Patel, D., Bruckner, S., Kanitsar, A., Gröller, M.: Volume visualization based on statistical transfer-function spaces. In: IEEE Pacific Visualization Symposium (PacificVis 2010), Taipei, Taiwan, pp. 17–24 (2010)
10. Bruckner, S., Gröller, M.: Style transfer functions for illustrative volume rendering. Computer Graphics Forum 26, 715–724 (2007)
11. Bruckner, S., Grimm, S., Kanitsar, A., Gröller, M.: Illustrative context-preserving exploration of volume data. IEEE Transactions on Visualization and Computer Graphics 12, 1559–1569 (2006)
12. Bruckner, S., Rautek, P., Viola, I., Roberts, M., Sousa, M., Gröller, M.: Hybrid visibility compositing and masking for illustrative rendering. Computers & Graphics 34, 361–369 (2010)

13. Wang, L., Zhao, Y., Mueller, K., Kaufman, A.: The magic volume lens: an interactive focus+context technique for volume rendering. In: 16th IEEE Conference on Visualization (VIS 2005), Baltimore, MD, pp. 367–374 (2005)

14. Bruckner, S., Gröller, M.: Volumeshop: an interactive system for direct volume illustration. In: 16th IEEE Conf. on Visualization (VIS 2005), Baltimore, MD, pp. 671–678 (2005)

15. Bruckner, S., Gröller, M.: Exploded views for volume data. IEEE Transactions on Visualization and Computer Graphics 12, 1077–1084 (2006)

16. Ruiz, M., Viola, I., Boada, I., Bruckner, S., Feixas, M., Sbert, M.: Similarity-based exploded views. In: Smart Graphics, Rennes, France, pp. 154–165 (2008)

17. Weiskopf, D., Engel, K., Ertl, T.: Interactive clipping techniques for texture-based volume visualization and volume shading. IEEE Trans. on Visualization and Computer Graphics 9, 298–312 (2003)

18. Bernhard, O.V., Preim, B., Littmann, A.: Virtual resection with a deformable cutting plane. In: Smart Graphics, Magdeburg, Germany, pp. 203–214 (2004)

19. Chen, H., Samavati, F., Sousa, M.: Gpu-based point radiation for interactive volume sculpting and segmentation. Visual Computer 24, 689–698 (2008)

20. Wyvill, G., McPheeters, C., Wyvill, B.: Data structure for soft objects. The Visual Computer 2, 227–234 (1986)

21. Faynshteyn, L.: Context-preserving volumetric data set exploration using a 3D painting metaphor. Master's thesis, Dept. of Computer Science, Ryerson University, Toronto, ON, Canada (2012)

FmFinder: Search and Filter Your Favorite Songs

Tuan Nhon Dang, Anushka Anand, and Leland Wilkinson

University of Illinois at Chicago

Abstract. Choices in music express our taste and personality. Different people have different collections of favorite songs. The explosive growth of digital media makes it easier to access any songs we want. Consequently, finding the songs best fit to our tastes becomes more challenging. Existing solutions record user patterns of listening to music, then make recommendation lists for users. By applying information visualization techniques to this problem, we are able to provide users with a novel way to explore their list of recommendations. Based on that knowledge, users can filter the songs according to their needs and compare the music tastes of different groups of people.

1 Introduction

The growth in both quality and quantity of music-sharing websites makes online entertainment increasingly popular. Mining this huge music collection and making selective recommendations for users is a daunting task. To explore this need, some music websites provide an API to allow read/write access to the full slate of music data resources - albums, artists, playlists, events, users, and more. Last.fm API and Amazon.com's API are the best-known of these sites. These APIs give users the ability to build programs to visualize updated music data.

This paper introduces FmFinder, an interactive web-based application that integrates three visualizations using the Last.fm API to enable linking, brushing and filtering. We use a *Tag Cloud* to present the weekly top 50 tracks, which can be use as input for the main visualization. After selecting a track, we use a *Graph Layout* to present recommended tracks. Finally, we can refine the Graph Layout by filtering tracks on the *Dot Plots* [1], *Bubble Plot*, and *Venn Diagram* [2].

The focus of FmFinder is tracks, not artists. Most likely, people love an artist because of his/her songs, but not the reverse. Moreover, when we like an artist, it does not mean we like all his/her songs. It can also happen that we don't like an artist, but there is a particular song of that artist that we really like. Making recommendations based on artists fails in this case.

The main contributions of this paper:

- Most applications recommend songs similar to a selected song by recoding user's listening habits. In this paper, we propose a song recommending algorithm combining user's listening habits (from the LastFm API) and individual preferences.

G. Bebis et al. (Eds.): ISVC 2012, Part I, LNCS 7431, pp. 348–358, 2012.

– We design an interactive visual interface to drill through the recommendation list to find a subset of tracks that better fit user preferences or to compare favorite song collections of different groups of people (different genders, different ages, and different countries).

The rest of this paper is organized as follows. Section 2 provides a brief overview of prior related work. Section 3 describes components of FmFinder and examples. Section 4 concludes the paper.

2   Related Works

Since 2003, the website Last.fm has been directly recording what songs people listen to. Based on these data, Last.fm created a publicly accessible API. Hundreds of Mashups using the Last.fm API have been created on their website for casual users to view. In this section, we provide a brief overview of prior related work using the Last.fm API.

2.1   Visualizing Listening Histories

In 2006, Lee Byron created a minor sensation in the Last.fm and design communities with his stream graph visualization [3] of his own Last.fm listening history. The visualization shows the changing trends in listening history. However, it is very difficult to view details, such as picking a set of artists out of the huge number of artists in a large fixed stream graph.

Last.fm Explorer [4] resolves this issue through interaction. By interacting with the graph directly, users can view different levels of hierarchy, filter the display by date or text searches, view exact data for particular time points, and rearrange the stacked graph to view changes in a particular tag, artist, or track.

In 2010, Dominikus Baur published LastHistory [5], an interactive visualization for displaying music listening histories, along with contextual information from personal photos and calendar entries. It is aimed toward non-expert users and helps them analyze, reminisce and build stories.

Ya-Xi Chen developed this idea further by combining listening histories of multiple users. The paper [6] presents two interactive visualizations which give users a deeper insight into consent and dissent in their listening behaviors, and help them to compare their musical tastes. HisFlocks shows overlaps in genre and artist in certain time periods and LoomFM illustrates sequential listening patterns.

Recently, Martin Dittus, a former Last.fm employee, grabbed listening data for staff, moderators, and alumni, and visualized 8.7 million scribbles in an calendar heat map [7]. It is like a calendar view, but instead of days, the interest is centered around hours of the day.

2.2   Recommending Artists

In 2005, Frederic Vavrille created Liveplasma [8], a flash-visualization based on Amazon's e-commerce service that explores links between music artists, movies, directors, and actors. It maps and displays music and movie search results with linkages and groupings. Figure 1 shows an example. After the search term is submitted, it is immediately surrounded by other artists. The graph becomes complicated when there are many similar artists/movies. Liveplasma does not offer a way to filter data in this case.

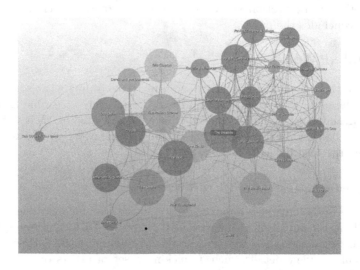

Fig. 1. Liveplasma: Artists related to "The Beatles"

TuneGlue [9] is based on a similar idea. TuneGlue is a force-directed graph exploring relationships between artists based on the Last fm database and ama-zon.co.uk. One can start by inputting an artist name and then the network is built with only one root node. By clicking the root node, one can expand other related artists. One can also explore all albums from each artist on the spot and link out to amazon.com to buy any of them. However, this function does not seem to be working at the moment.

2.3 Recommending Songs

The Genius Sidebar, introduced in iTunes 8, automatically generates a playlist of songs from the user's library which contains songs similar to the selected song. Genius playlists are created by the ratings system and collaborative filtering. An iTunes Store account is required because information about the user's library must first be sent to Apple's database. Algorithms determine which songs to play based on other users' libraries.

Different from the Genius Sidebar, FmFinder presents recommended songs in a graph layout instead of a list view. Moreover, users can filter all similar songs manually on a dot plot, a bubble plot, and/or a venn diagram. No usage history is required. Users can change their preferences at will. The next Section describes components of FmFinder and examples.

3   FmFinder

The FmFinder GUI incorporates three major components as depicted in Figure 2: Input Panel includes Box 1 and Box 2, Output Panel includes Box 3, and Filtering Panel includes Box 4, Box 5, Box 6, and Box 7.

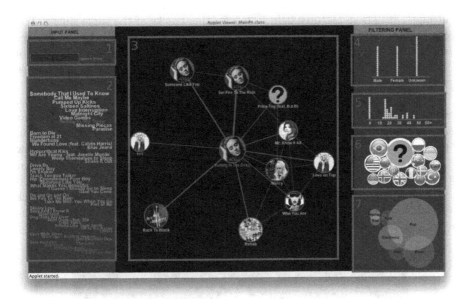

Fig. 2. FmFinder GUI: search box (Box 1), tag cloud (Box 2), graph layout (Box 3), gender dot plot (Box 4), age dot plot (Box 5), country bubble plot (Box 6), and genre venn diagram (Box 7)

Information visualization systems should allow one to perform analysis tasks that largely capture people's activities while employing information visualization tools for understanding data [10]. In the rest of this section, we describe five basic analysis tasks implemented in FmFinder: searching (using the search box), retrieving value (retrieving track information by brushing a node in the graph layout), sorting (sorting tracks by the level of recommendation), characterizing distributions (using Dot Plots, Bubble Plots, and Venn Diagrams to show statistics of the selected song), and filtering (filtering tracks by user preferences).

3.1   Input Panel

The initial layout contains only Input Panel (Output Panel or Filtering Panel are empty). Input Panel enables the user to select a song name. Users can input a song by using the search box (Box 1 in Figure 2) or select a song from the weekly top 50 songs showing in the Tag Cloud (Box 2 in Figure 2). Colors and font sizes in the Tag Cloud are determined by popularity of songs.

3.2   Output Panel

After we have input a song, related songs are displayed in a node-link representation (Box 3 in Figure 2). In particular, we have selected the track "Rolling In The Deep" ranked 8 in the Tag Cloud. The track "Rolling In The Deep" now becomes the root node in the graph (highlighted in red), surrounded by top ten similar tracks. The picture on every node displays the artist of the song labeled right under it. These pictures help in recognizing tracks of the same singer. The sizes of nodes are based on popularity of the track. Additionally, two nodes are linked if they suggest each other. The thickness of the edges indicates the level of recommendation. Edge length is not used to encode any information. The nodes organize themselves to reduce link-crossed.

There are several ways that users can interact with the application. Users can drag nodes to reorganize the graph, use a slider to control the number of similar tracks to be displayed in the graph, brush a node to request detailed information, or play a track. In the example in Figure 3, we have brushed the node "Mr. Know It All". The nodes not directly connected to the brushed node are faded so that we can focus on the relationships of the selected node. Moreover, the brushed track is downloaded and played at the bottom. The details of the brushed track are displayed next to the video. Notably, we can make the brushed track become the root node by a simple click.

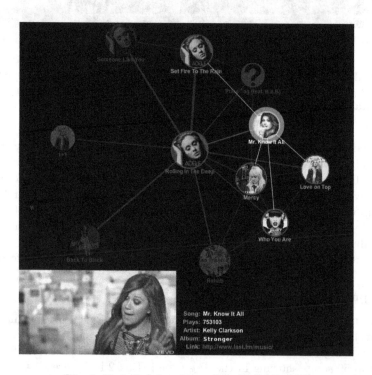

Fig. 3. Node-link brushing: "Mr. Know It All"

3.3   Filtering Panel

Music tastes changes with time, and sometimes depends on one's mood. The songs that I like to listen to today might not be my favorite songs from yesterday. FmFinder allows users to drill through the recommendation list to find a subset of tracks that better fit their current preferences.

Preview the Selected Song. After we have selected a track, the recommended tracks are plotted in the node-link diagram. The statistics of the selected track are also plotted in the Filtering Panel. We use a set of interactive visualization methods to present statistics of the selected track: Dot Plots, Bubble Plot, and Venn Diagram.

Figure 4 shows an example. We have selected the leading track of current week (the first track in the Tag Cloud), "Somebody That I Used To Know". The Dot Plot in Figure 4(a) shows the distribution of the top 50 listeners by gender. Notice that most listeners are female. The Dot Plot in Figure 4(b) shows the density distribution of the top 50 listeners by age. Some listeners not wanting to specify their ages are given the default value 0. Moreover, listeners to this track are divided into 2 groups, from 16 to 24 and from 30 to 36 year olds.

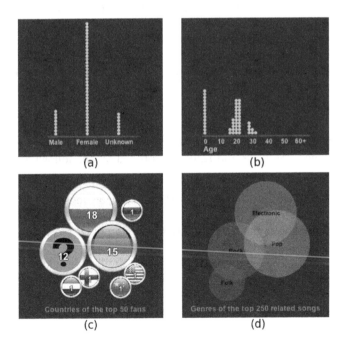

Fig. 4. Filtering Panel, statistics of "Somebody That I Used To Know": (a) Gender Dot Plot of the top 50 fans (b) Age Dot Plot of the top 50 fans (c) Country Bubble Plot of the top 50 fans (d) Genre Venn Diagram of the top 250 related songs

A single dot in the Dot Plots represents a single listener. Dot plots, instead of Histograms or Bar Charts, are used to present statistics of the top 50 listeners because they allow to select every single listener to check out his/her compete profile, including user id, name, age, gender, and nationality. In case there are too many listeners in the result to be displayed as single dots, we can make the dots overlapped vertically to fit the allowed space. This makes a Dot Plot look very similar to a Histogram. However, brushing every single dot is still possible.

Country data is represented in a Bubble Plot which emphasizes their visual appearance. Figure 4(c) shows the distribution of the top 50 listeners by country. Each bubble presents one country. Country flags embedded on top of each bubble facilitate identification. The bubble sizes are decided by the number of listeners from that country. One can request to show the number of listeners on the top of each flag as shown in Figure 4(c). As we notice, there are 12 listeners with non-specified nationality, 18 listeners are from Poland, 15 listeners are from Germany, and the rest are from Russia, Latvia, Finland, US, and China. Overall, 36 over 38 known-nationality listeners are from Europe. Gotye, the singer of "Somebody That I Used To Know," is a Belgian-Australian musician and singer-songwriter, so it is not surprising why the song is popular in Europe.

Figure 4(a), Figure 4(b), and Figure 4(c) are linked. By selecting a country from the Bubble Plot in Figure 4(c), all listeners from that country are highlighted in Figure 4(a) and Figure 4(b).

There are top 250 songs related to the selected song, "Somebody That I Used To Know", returned from LastFm API. We use a Venn Diagram to present the genre distribution of the top 250 related songs. As depicted in Figure 4(d), most of related songs are pop and/or electronic, rock. The genres of a song are retrieved by song tags by listeners. Many songs are mixtures of different genres. Therefore, a Venn Diagram is an appropriate representation for such intersections.

Filter Related Songs. Previewing the selected song provides a guideline to filter related songs. We filter and rank songs based on the following features:

- Number of fans who pass filtering conditions (Let F be the number of fans, F receives a value from 0 to 50. LastFm returns at most 50 top listeners of a selected song).
- Play counts of the songs (Let P be the play counts, P receives a value from 1 to several millions).
- Similarity ranking from Last.fm (Let S be the similarity ranking from Last.fm, S receives a value from 1 to 250. S equals 1 when the song is most similar to the selected song).
- Genres of the songs. These genres are obtained by parsing song tags by listeners.

The weighting function is computed by the following equation:

$$Weight(song) = F * \sqrt[10]{P} - \frac{S}{5} \qquad (1)$$

In Equation 1, we favor the popular songs and LastFm rankings. However, F is the most important factor to decide weight of a song. In particular, the $\sqrt[10]{P}$ mostly produces the values from 3 to 5. For the same S, the least popular song with f fans who pass filtering conditions still get a higher weight than the most popular song with half of f. This is how FmFinder can be used to discover new related songs.

Because LastFm returns at most 250 similar songs ($S \leq 250$), $\frac{S}{5}$ produces the values from 1 to 50. Subtraction from this amount allows more similar songs to get higher weights. Finally, we order the weights to obtain a ranking of the related songs.

More importantly, genres of songs need to be in the set of active genres (selected on the Venn Diagram). Otherwise, $Weight(song)$ is reset 0.

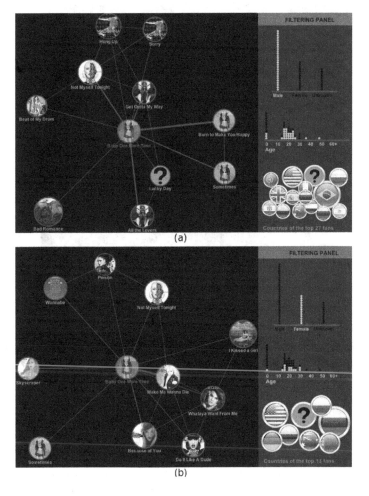

Fig. 5. Filtering recommended songs by gender: a) Male b) Female

Figure 5 shows a filtering example. We have selected "Baby One More Time" of Britney Spears as the root node. The Dot Plots and Bubble Plot now show statistics of the selected song "Baby One More Time". We then apply the filter on gender. Figure 5(a) shows recommended songs for males. Figure 5(b) shows recommended songs for females. The results are different recommended lists. Males like to listen to Madonna, Lady Gaga, and Kylie Minogue while females prefer The Pretty Reckless, Pink, and Kelly Clarkson. Notice that the Dot Plots and Bubble Plot are also updated based on filtering conditions.

Figure 6 shows the distribution of the top 50 listeners by gender of two songs preferred by males and two songs preferred by females regarding the selected song "Baby One More Time" of Britney Spears. The charts help to understand how FmFinder makes decisions resulting in Figure 5.

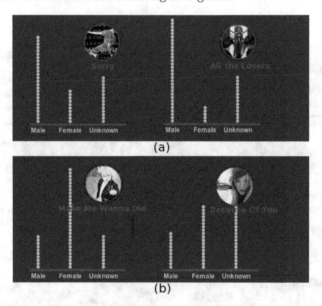

Fig. 6. Distribution of listeners by gender: (a) Songs preferred by males (b) Songs preferred by females

Figure 7 shows another filtering example. We have selected "Baby One More Time" of Britney Spears as the root node. We then apply the filter on genre. The filtering condition is set up on the Venn Diagram (active genres are highlighted, inactive genres are faded). Figure 7(a) shows recommended Rock, Rap, and Hiphop songs . Figure 7(b) shows recommended Pop and Electronic songs. The results are completely different recommended lists.

3.4   Conclusions

FmFinder is an web-based application built in Processing to address the challenges of searching music that fits user preferences. Some distinct features of FmFinder compared to prior related works are:

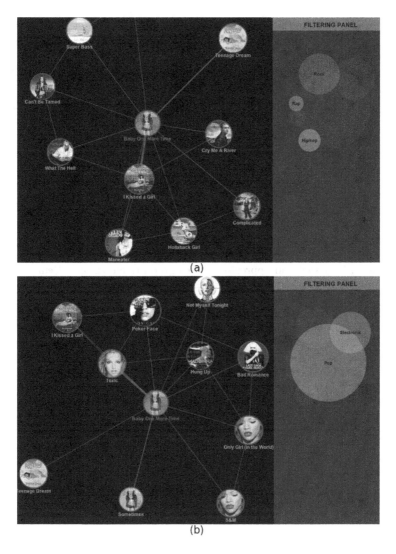

Fig. 7. Filtering recommended songs by genre: a) Rock, Rap, and Hiphop b) Pop and Electronic

- FmFinder works directly on song recommendations, not artist recommendations.
- FmFinder combines multiple visualization techniques in all three processes (Input, Filtering, and Output).
- Users can set their own preferences manually or filter and compare favorite songs of different groups of people.

FmFinder allow users to perform basic analysis tasks required by any information visualization tools:

- Searching an online database (LastFm) to find a song that users require.
- Retrieving track information and/or playing a track by brushing a node in the graph layout.
- Sorting tracks by the level of recommendation which is then encoded by the thickness of edges in the node-link diagrams.
- Characterizing distributions of the top listeners by gender, age, country using Dot Plots, Bubble Plot, and distributions of the related tracks using Venn Diagram.
- Filtering tracks by user preferences.

References

1. Dang, T.N., Wilkinson, L., Anand, A.: Stacking graphic elements to avoid over-plotting. IEEE Transactions on Visualization and Computer Graphics 16, 1044–1052 (2010)
2. Wilkinson, L.: Exact and approximate area-proportional circular Venn and Euler diagrams. IEEE Transactions on Visualization and Computer Graphics (2011) (in press)
3. Byron, L., Wattenberg, M.: Stacked graphs - geometry & aesthetics. IEEE Transactions on Visualization and Computer Graphics 14, 1245–1252 (2008)
4. Pretzlav, M.A.: Last.fm explorer: An interactive visualization of hierarchical time-series data. Tracks A Journal of Artists Writings 7 (2008)
5. Baur, D., Seiffert, F., Sedlmair, M., Boring, S.: The streams of our lives: visualizing listening histories in context. IEEE Transactions on Visualization and Computer Graphics 16, 1119–1128 (2010)
6. Chen, Y.X., Baur, D., Butz, A.: Gaining musical insights: Visualizing multiple listening histories. In: Workshop on Visual Interfaces to the Social and Semantic Web, VISSW 2010 (2010)
7. Dittus, M.: Revealing the periodic listening habits of last.fm users (2011)
8. Vavrille, F.: Liveplasma: Quickly discover similar movies and songs (2005)
9. Team, O.: Music plasma: Music relationship explorer (2006)
10. Amar, R., Eagan, J., Stasko, J.: Low-level components of analytic activity in information visualization. In: Proc. of the IEEE Symposium on Information Visualization, pp. 15–24 (2005)

3D Texture Mapping in Multi-view Reconstruction

Zhaolin Chen, Jun Zhou, Yisong Chen, and Guoping Wang

Graphics and Interactive Technology Lab, Peking University

Abstract. This paper proposes a novel framework for texture mapping of 3D models. Given a reconstructed 3D mesh model and a set of calibrated images, a high-quality texture mosaic of the surface can be created after the process of our method. We focus on avoiding noticeable seams, color inconsistency and ghosting artifacts, which is typically due to such facts as modeling inaccuracy, calibration error and photometric disagreement. We extend the multi-band blending technique in a principled manner and apply it to assemble texture images in different frequency domains elaborately. Meanwhile, self-occlusion and highlight problem is taken into account. Then a novel texture map creating method is employed. Experiments based on our 3D Reconstruction System show the effectiveness of our texturing framework.

1 Introduction

3D Reconstruction is a long-term and challenging problem in computer vision and computer graphics community. This kind of technology aims to reconstruct visual 3D models of real objects or large scale scenes that can be displayed in computer environment. The applications of it include virtual reality, computer animation, online exhibition, etc.

There are a lot of technological ways to perform 3D Reconstruction based on images, video or 3D scanning data. In this paper, we will address the texture mapping problem in the context of image-based modeling with a chessboard-type calibration board (see Fig. 1 below). We assume that we already have an approximate surface model of an object which is represented by a standard triangle mesh encoding. And we also assume that we have a set of images that are taken from different viewpoints and then precisely calibrated. Our focus is how to texture the reconstructed mesh surfaces so as to achieve photorealistic 3D models.

Fig. 1. With 30 images of a stone goldfish, a realistic 3D model with seamless and detailed texture is produced through our method

G. Bebis et al. (Eds.): ISVC 2012, Part I, LNCS 7431, pp. 359–371, 2012.
© Springer-Verlag Berlin Heidelberg 2012

In general, one image cannot provide enough information to texture a 3D model. Given an image sequence, intuitively, we can bind a "best" texture patch to each triangle of the mesh model according to the viewing angle. However, this straightforward method is ineffective in the practical application environment, due to the inaccuracy of shape modeling, calibration error and the inconstancy of lighting and camera conditions. As a result, color discrepancies and lighting discontinuities appear as seams on the rendered surface.

There are mainly three directions that have been explored to remove the texture seams. The first direction is to "blur" the seams by using certain weighted averaging scheme. Second, many methods try to optimize the texture patch layout and therefore reduce the global texture inconsistencies with least details lost. The third direction focuses on the color correction or relighting in the vicinity of texture boundaries or in the global domain.

Adam Baumberg proposes an innovative texturing method [1] which employs a two-band frequency decomposition mechanism and blend images in both frequency domains to implement texture mosaicing. However, it fails to consider the significant connection between the width of transition zones and the size of features in different frequency bands. As a result, this method may produce noticeable artifacts where high frequency features such as lines and edges are broken up across regions that textured (in the high frequency band) from different misregistered images.

J. Digne et al. also make use of the Low/High frequency decomposition scheme to address the problem of high fidelity scan merging [2]. Cedric Allene et al. go a farther way on multi-band blending [3]. They extend the method to more than two frequency bands in a principled manner. Yet they fail to define a reasonable weighting function that can take model shape and viewing angle into account fully.

Victor Lempitsky et al. formulate the texture patch layout problem into a Markov Random Field energy optimization [4]. This method tries to find a balance between texture smoothness and details. Then a seam levelling procedure is applied to remove the residual seams. Ref. [5] extends this work by expanding the combinatorial search to consider local image translations when seeking optimal mosaic. The introduction of MRF indeed reduces many artifacts. However, for sophisticated surface geometries with large number of viewing images, these methods help little in practice. Besides, their seam levelling procedures both fail to achieve perfect smoothness across adjacent patches, seams are visible even in the results demonstrated in papers.

As another interesting work, Bastian Goldluecke proposes a novel approach [6] that allows to recover a high-resolutioon, high-quality texture map even from lower-resolution photographs. This method requires accurate geometry and camera calibration.

In this paper, we develop a more practical, effective and complete texturing framework for 3D modeling. Multi-band blending idea is adopted, clarified and improved. A visibility preprocessing step for solving self-occlusion problem are proposed. Especially, highlight problem, which is always overlooked by previous 3D mapping methods, is handled in a artful way by utilizing the features of our texturing framework. Finally, we describe a novel texture map creating method, which can fully represent the texture data at a low space cost without information loss.

2   Multi-band Blending

A common problem to all applications of photomosaics is how to combine several relative images into a panorama or image mosaic. Due to varying illumination conditions and perspective differences, using some weighted averaging scheme in the overlapping regions to eliminate visible seams is advisable. Apparently, the width of the transition zone plays a crucial role in the blending procedure. If the transition width is small compared to the image features, seams will still be visible. Conversely if it is larger than the image features, features from all images may appear as ghosting within the transition zone. Under some circumstances, deciding an appropriate transition width to achieve color and lighting continuity without introducing ghosting artifacts is very difficult, or even impossible. For instance, if the input images have very large and very small features at the same time, it would be hard to choose an appropriate size as the transition width.

From the viewpoint of the frequency theory, an image can be regarded as a collection of different frequency signals and the sizes of features are proportional to the corresponding wave lengths. If we can decompose the images into a set of band-pass filtered component images and blend them in different frequency domains, then the final image mosaic can be obtained by simply sum all the image mosaics of different bands. This is the basic idea of [7]. Adam Baumberg extends this technique to texturing 3D models by using a two-band decomposition and then fusing the low frequencies while keeping intact the high frequency content [1]. The work is notable, but not enough satisfactory for complex scenes whose features cover a large range of scale.

To remove seams, the transition width should be comparable to the largest feature, and to avoid ghosting, the transition width should not be much larger than the smallest feature. We should "slice" the frequency of the image into more bands to restrict the sizes of features in each band. Then a proper transition width would be easier to achieve.

Following [3, 7], we firstly construct Gaussian pyramids by using a cascade filtering approach. We denote the bth level of the Gaussian pyramid of image I as $G_b(I)$. The first level is just a copy of the original image. Any higher level is generated by convoluting the previous level image with a kernel-constant Gaussian function $g(\sigma)$:

$$G_b(I) = G_{b-1}(I) * g(\sigma) , \tag{1}$$

where σ is proportional to the length of the diagonal of the object's bounding box.

Then we construct Laplacian pyramids by subtracting the adjacent levels of Gaussian pyramids. We denote the xth level of the Laplacian pyramid of image I as $L_b(I)$. The highest level is the same as that of Gaussian pyramid. And the lower levels are calculated as

$$L_b(I) = G_b(I) - G_{b+1}(I) . \tag{2}$$

Apparently, one level of Laplacian pyramid is just one component image of the original one that dominates a limited frequency band. By summing all these levels, the original image can be recovered reversely.

Fig. 2 below shows an example of the pyramid construction procedure.

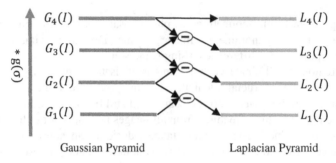

Gaussian Pyramid   Laplacian Pyramid

Fig. 2. Constructing a Laplacian Pyramid with 4 levels

As mentioned above, we need to blend texture in the overlapping regions in different frequency bands. Suppose the Laplacian pyramids have B levels, that is, we have decomposed the images into B bands. The texture patch of each triangle of the mesh surface can be calculated as

$$T(x) = \sum_{b=1}^{B} \sum_{i=1}^{|I|} \frac{W_{b,i}(x) * L_b(I_i)(\Pi_i(x))}{\sum_{i=1}^{|I|} W_{b,i}(x)} , \qquad (3)$$

where I is the viewing image set, x is a triangle of the mesh surface, Π_i is the projection from 3D space to image I_i and W is a weighting function.

To compute $W_{b,i}(x)$, we firstly calculate the angular deviation (denoted as $\alpha_i(x)$) of the viewing vector of source image I_i from the normal vector of mesh triangle x. Then we calculate $W_{b,i}(x)$ as

$$W_{b,i}(x) = \begin{cases} \left(\alpha_i(x) < \frac{\pi}{2} \cdot \frac{b-1}{B-1}\right) ? \ (\frac{\pi}{2} - \alpha_i(x)): 0, & 2 \le b \le B \\ \left(\alpha_i(x) == \max\{\alpha_j(x) | 1 \le j \le |I|\}\right) ? \ (\frac{\pi}{2} - \alpha_i(x)): 0, & b = 1 \end{cases} \qquad (4)$$

This strategy suggests that the higher the frequency band is, the less viewing images of one triangle would be used to blend texture. As a result, the transition zones would be proportional in size to the features represented in each band.

Fig. 3 illustrates the steps of adding the blended texture in each band to a 3D model. In practice, we assign B as 3 or 4 in order to take a balance between algorithm effects and computational cost.

Fig. 3. The steps of adding the blended texture in each band from low to high to a 3D model

3   Visibility Processing

So far, we have not taken self-occlusion and visibility issue into account yet. In practice, quite a lot of objects have self-occlusion problem which gives rise to erroneous texture binding in shadow regions. Obviously, we need a preprocessing step to address it.

The overview of our method is as follows. All triangles of the mesh surface are projected into its viewing images via the corresponding calibration information. Among all projected triangles that can cover it, each pixel of the images records the index of the closest one, while setting the others invisible from this image (see Fig. 4).

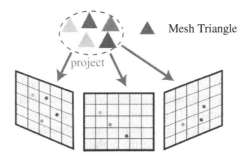

Fig. 4. Each pixel of the images records the nearest triangle that can cover it through projection for now

The details are as follows:

for $i = 1:|I|$

    Build array T and V. $T[m, n]$ records the closest triangle to pixel$[m, n]$. $V[x, i] ==$ true means that triangle x is visible from image I_i, not vice versa.

    Initiate elements of T and V as -1 and true respectively.

    Project each triangle to the current image I_i and then check every pixel that is covered by it.

    Suppose the index of current triangle is x.
      Suppose current pixel is pixel$[m, n]$.
      if $(T[m, n] == -1)$
        $T[m, n] = x;$
      else
        if(dist$(x,$ pixel$[m, n]) <$ dist$(T[m, n],$ pixel$[m, n]))$
          $V[T[m, n], i] =$ false;
          $T[m, n] = x;$
        else
          $V[x, i] =$ false;

4   Highlight Removal

Due to the surface inhomogeneity of real objects, purely Lambertian relection is not possible. In practice, directional illumination often leads to highlights which cause significant variations in the appearances of an object. As presented in Sect. 2, our texture mapping method can handle varying overall illumination conditions very well, however, highlight problem still need one more specific treatment due to its troublesome distinctiveness.

Previous methods for highlight removal can be roughly separated into two categories by the number of images used. The first category methods [8, 9, 10] uses only one single image to remove the undesirable effects of highlights. The key idea of these methods is that the highlight and neighbor pixels contain useful information for guiding the estimation of the underlying diffuse color. However, they only work well with simple colored images. For more general cases, such as multicolored or complex textured images, obtained information may be insufficient to recover shading and texture in highlight regions. To make the problem more tractable, the second category methods utilize more data from multiple images captured under changing conditions, such as different viewpoints [11] or illumination conditions [12, 13]. This type of methods are moderately practical, since their requirements limit their applicability.

We need a specific approach to handle the highlight problem in our texture mapping pipeline. In fact, the context of multi-view reconstruction gives us much convenience to circumvent some limiting assumptions made in previous works. First, each part of an object can be saw in several images in the given sequence. Such redundancy guarantees that the highlight regions in one image can always have highlight-free counterparts in other reference images. Second, the correspondence relationship between pixels of different images are already prepared. Finally, the Multi-band Blending alleviate the difficulty of highlight region replacement.

We develop a simple but very effective approach to remove highlight effects. The highlight regions are detected on the mesh triangle level (remeshing is employed beforehand to guarantee the mesh regularity.), based on the color differences between the corresponding areas in image sequence. After visibility processing, each mesh triangle can be projected into its "visible" images via calibration information. Then we calculate the average colors of these projected regions respectively and find the median. If the average color of some region deviates much from the median value, this region will be identified as a highlight region. Concerning color value deviation, "much" is 20 percent of the R, G, B values in our implementation. Then for each image, we adopt BFS to group the highlight triangle regions and use proportionable circles to cover these groups. So far highlight detection is completed.

A key observance that we can take advantage of is that, due to the relative movement of viewing direction and object, a highlight spot in some image is often out of highlight areas in other images. With calibration information, highlight pixels can be resampled by blending their counterparts in other reference images. After that, Poisson Image Editing [14, 15] is introduced to seamlessly integrate the resampled regions into target images. This measure can change the overall illumination of resampled highlight regions and reduce the lighting inconsistency along the boundaries. However, as mentioned in Sect. 2, any blending strategy (e.g., weighted averaging according to the areas of the projected highlight regions in each image.) inevitably leads to ghosting artifacts. In the context of our texture mapping pipeline, this problem can be circumvented in a very natural and effective way. The key insight is, in the process of Multi-band Blending, we only extract low level frequency information from these resampled highlight regions. For these regions, benefited from the overlapping of image sequence, we can always obtain high level frequency information from other reference images. Fig. 5&6 shows some illustrations of our highlight removal approach.

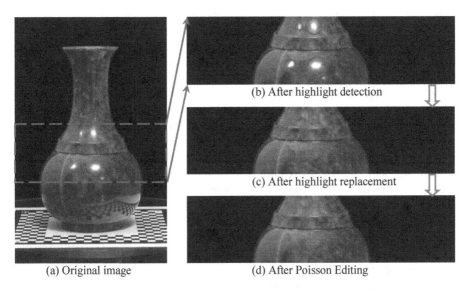

(a) Original image

(b) After highlight detection

(c) After highlight replacement

(d) After Poisson Editing

Fig. 5. The procedure of highlight removal in one image. (b), (c) & (d) show only the ROI.

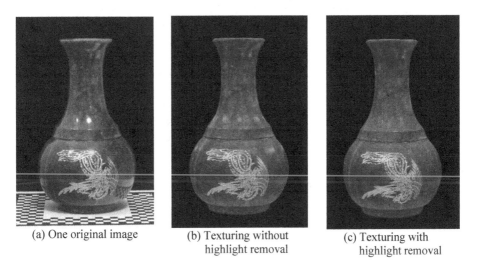

(a) One original image

(b) Texturing without highlight removal

(c) Texturing with highlight removal

Fig. 6. Comparison between texturing without and with highlight removal

5   Texture Map Creation

Finally, we need to pack texture data generated into a single texture map for two reasons. First, blending different (band) images during rendering is inefficient. It's more practical to pre-compute and store the blended texture, which allows the object could be rendered within just one pass. Second, the reconstructed models should be encoded into standard formats in order to be displayed in virtual environments.

The "box" scheme proposed in [1] suffers from two problems. First, it cannot guarantee that every triangle is completely visible in one of the 6 canonical views (top view, front view etc.), which means that the texture data might not be fully represented in the texture map. For example, obviously, the inwall of a cup placed levelly cannot be visible in any canonical view. Second, the combination of the 6 canonical views is not guaranteed to be optimal. There might be another combination of views that can preserve texture data better.

Another reference method comes from [16], which places texture patches of triangles randomly with a surrounding auxiliary area. This method can preserve the full texture of an object, no matter how complex its shape is. However, the randomness of texture patch placement leads to a texture map that is difficult to read for human. In addition, the uniform size of texture triangles somehow causes information loss. Furthermore, this method also suffers from the problem of image space wasting due to a lot of "padding" between texture triangles.

Considering all the advantages and disadvantages of these two strategies, we implement an improved approach of packing texture map.

A preliminary version is as follows: According to the viewing angle, every triangle has a "best viewing image". Based on this "best viewing image" we can easily and naturally partition the surface into several areas. For each area, we can project it to the corresponding "best viewing image", and save its texture data using exactly the same shape of the projected area. Then we sort them in ascending order of the polar angle of area center and pack them together. Finally, for the sake of mipmapping while rendering, we need to pad a three or four pixel width band to enclose each area.

The method presented above works well when the target object has regular shape and the input image number is small. However, if the object is so sophisticated that twenty or more images are needed to cover it, the texture surface would be cut into so many fragmentary areas that it becomes impossible to pack them together without much waste of space (we would need more "padding"). This situation also undermines our intention to make the texture map human-readable, although the texture data is fully represented. So it is necessary to pick out a subset of images which can "see" the object with fairly good viewing angle as a whole.

Let's begin with some definitions:

$\alpha_{i,t}$:   the angle between the view line of image i and the face normal of triangle t.
α_{Thres}:   a threshold of α, such as $\pi/10$.
I: the input image set.
T: the mesh triangle set.
V_t: a conditioned viewing image set of triangle t.
V_t is calculated as: $V_t = \{ i \mid \alpha_{i,t} < \alpha_{Thres} || \alpha_{i,t} == min \{\alpha_{j,t} | 0 \le j \le |I|\} \}$

We transform the problem of selecting an appropriate subset of the input images into a Minimal Hitting Set problem: finding out the minimal subset of I that contains at least one element of each V_t $(0 \leq t \leq |T|)$. As we all know, the Minimal Hitting Problem is NP-Complete. However, the particularity of our context enables us to perform an efficient algorithm which can give an exact solution within tolerable time in most cases. As for other cases in which the generation of an exact solution would cost too much time, an approximate solution is given.

Let we refer the target subset of I as S. Considering that many triangles are visible from only one image, we process them first:

```
for t = 1: |T|
    if (|V_t| == 1)
        S = S ∪ V_t
```

The next step depends on the number of images not in S now. If $|I| - |S| < n$ (in practice, we assign $n = 10$), we perform an exhaustive search to find an exact solution:

```
for num = 1: (|I| − |S|)
    for ∀ I' ⊆ (I − S) && |I'| == num
        if I' hit all remaining traingles
            S = S ∪ I'; break;
```

If $|I| - |S| \geq n$, by controlling α_{Thres}, we can ensure that $|V_t| \leq k$ $(1 \leq t \leq |T|)$, where k is a constant. Then we use the k-approximation of k-hitting algorithm to get an approximate solution:

```
while any un-hit triangle exists
    suppose the current triangle is t
        S = S ∪ V_t
```

α_{Thres} indeed limits the lost of texture information. And S guarantees that most neighborhood relations of the surface triangles are preserved, which means less "padding".

After obtaining S, every triangle has a "best viewing image" among S. The subsequent texture map creating steps are similar to that of the preliminary version described earlier. Fig. 7 shows a result.

In contrast with the reference methods, our texture map creating scheme is obviously outstanding. It not only guarantees that every triangle visible from some image can have its texture data fully represented in the texture map, but also ensures the texture map to be human-readable to some extent.

Fig. 7. Left - a reconstructed model. Right – corresponding texture map

6   Experiments

We demonstrate the effectiveness of our texturing method by comparing it with other four methods:

A. *Best Viewing Image*: For each triangle, choose texture from best viewing image.
B. *Global Blending*:   Weighted averaging used all visible images.
C. *Boundary Blending*:   Blend texture alone boundaries.
D. *Two-band Blending*:   The method proposed in [1].

We conduct hundreds of experiments with different objects. Fig. 8 gives a representative example. This example uses an image sequence consisting of 16 images (3888×2592) of a teapot to test these methods. Note that, due to different surface reconstruction approaches, the mesh models may have minor differences.

(A) Although the Best Viewing Image method can preserve fine texture details very well, it produces noticeable seams due to varied lighting and camera conditions. (B) The Global Blending method can obtain seamless texture with no jumps in color and lighting appearance, however, at the cost of introducing ghosting artifacts. (C) The Boundary Blending method likes a balance or compromise of the two methods above. It tries to relieve the color discrepancies and lighting discontinuities while avoiding ghosting artifacts. Unfortunately, the effort is not enough satisfactory. (D) The overall performance of the Two-band Blending method is quite good. However, due to information loss in the texture map creation step, it fails to remain the features sharp. Besides, some feature lines appear broken up. (E) In contrast, our method not only achieves global texture smoothness but also preserves details very well.

Fig. 9-11 show more example results of our method. Note that the source images are not all displayed.

7   Conclusion

We have presented a novel and practical framework for texture mapping of 3D models in a multi-view setting, where several calibrated views of a object with known surface geometry are available. The basic technique we employ is multi-band blending. We extend it in several respects while taking self-occlusion and highlight problem into

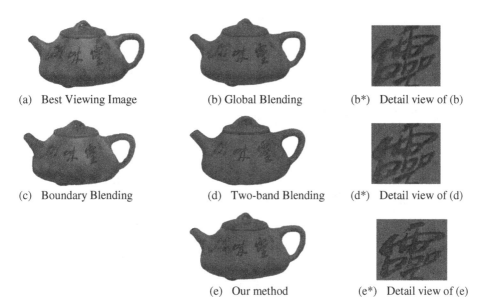

(a) Best Viewing Image   (b) Global Blending   (b*) Detail view of (b)

(c) Boundary Blending   (d) Two-band Blending   (d*) Detail view of (d)

(e) Our method   (e*) Detail view of (e)

Fig. 8. Comparison of different texturing methods

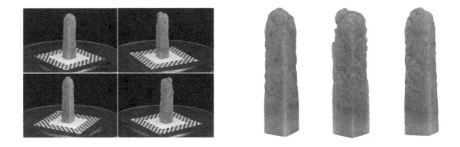

Fig. 9. Signet

Fig. 10. Vase

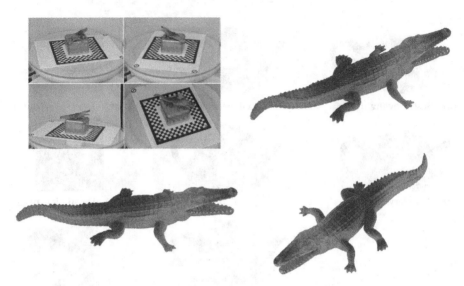

Fig. 11. Crocodile

account. At last, we propose an efficient mechanism to create a single texture map that provides full representation of the texture data.

Our work relies on a modest prerequisite that the images are enough to cover the full range of target object and have relatively wide overlapping regions. Except this, we don't require strong assumption on lighting condition. And we don't need an exact geometry model or perfect calibration data as well. Our method can achieve color continuity and detail preservation while avoiding noticeable seams and ghosting artifacts. On account of the idea's generality, this framework not only can handle indoor objects, but also can be extended to deal with large-scale scenes.

Acknowledgements. This work was supported by National Basic Research Program of China (2010CB328002), National High-tech R&D Program of China (2011AA120301) and National Natural Science Foundation of China (60973052, 61121002, 61173080).

References

1. Baumberg, A.: Blending Images for Texturing 3D Models. In: BMVC (2002)
2. Digne, J., Morel, J.-M., Audfray, N., Lartigue, C.: High Fidelity Scan Merging. Computer Graphics Forum 29(5) (2010)
3. Allene, C., Pons, J.-P., Keriven, R.: Seamless Image-Based Texture Atlases using Multi-band Blending. In: ICPR (2008)
4. Lempitsky, V., Ivanov, D.: Seamless Mosaicing of Image-Based Texture Maps. In: CVPR (2007)
5. Gal, R., Wexler, Y., Ofek, E., Hoppe, H., Cohen-Or, D.: Seamless Montage for texturing Models. In: EUROGRAPHICS (2010)

6. Goldluecke, B., Cremers, D.: Superresolution Texture Maps for Multiview Reconstruction. In: ICCV (2009)
7. Burt, P.J., Adelson, E.H.: A Multiresolution Spline with Application to Image Mosaics. ACM Trans. on Graphics 2(4) (1983)
8. Tan, R.T., Ikeuchi, K.: Separating Reflection Components of Textured Surfaces Using a Single Image. In: ICCV (2003)
9. Tan, P., Lin, S., Quan, L.: Separation of Highlight Reflections on Textured Surfaces. In: CVPR (2006)
10. Yang, Q., Wang, S., Ahuja, N.: Real-Time Specular Highlight Removal Using Bilateral Filtering. In: Daniilidis, K., Maragos, P., Paragios, N. (eds.) ECCV 2010. LNCS, vol. 6314, pp. 87–100. Springer, Heidelberg (2010)
11. Lin, S., Li, Y., Kang, S.B., Tong, X., Shum, H.-Y.: Diffuse-Specular Separation and Depth Recovery from Image Sequences. In: Heyden, A., Sparr, G., Nielsen, M., Johansen, P. (eds.) ECCV 2002. LNCS, vol. 2352, pp. 210–224. Springer, Heidelberg (2002)
12. Petschnigg, G., Agrawala, M., Hoppe, H., Szeliski, R., Cohen, M., Toyama, K.: Digital Photography with Flash and No-Flash Image Pairs. In: SIGGRAPH (2004)
13. Agrawal, A., Raskar, R., Nayar, S.K., Li, Y.: Removing Photography Artifacts using Gradient Projection and Flash-Exposure Sampling. In: SIGGRAPH (2005)
14. Perez, P., Gangnet, M., Blake, A.: Poisson Image Editing. In: SIGGRAPH (2003)
15. Feng, J., Zhou, B.: Automatic Highlight Removal in Visual Hull Rendering by Poisson Editing. In: VRCAI (2008)
16. Maruya, M.: Generating a Texture Map from Object-Surface Texture Data. In: EUROGRAPHICS, vol. 14(3) (1995)

A Novel Locally Adaptive Dynamic Programming Approach for Color Structured Light System

Run Zou, Yu Zhou, Yao Yu, and Sidan Du

NanJing University
School of Electronic Science and Engineering
Hankou Road, Nanjing, JiangSu Province, China, 210093
rzou0623@hotmail.com,
{nackzhou,allanyu,coff128}@nju.edu.cn

Abstract. The authors present a color structured light system by projecting an optimized De Bruijn pattern for 3D model reconstruction under natural light condition. The main focus of this paper is to enhance the accuracy of the correspondence problem by designing a novel locally adaptive dynamic programming algorithm, which adjusts the support-weight of stripe in a given local window. The presented approach performs better in terms of smoothness and accuracy of construction result and is suitable for generating high-quality and real-time scans of moving object.

1 Introduction

The 3D shape acquisition from 2D views has been a longstanding goal in computer vision. Several different groups of techniques have emerged in recent decades, such as stereo vision[1], time of flight[2] and structured light[3]. Structured light techniques simplify the problem by easy image process and can yield high accuracy result. By projecting some designed patterns onto the surface of an object and capturing image(s) with a camera, the numerous correspondence can be found, 3D coordinates of the object can be derived from the distortion of patterns and the surface can be reconstructed by means of triangulation.

In recent years, the main direction of 3D reconstruction is forward to real-time, high accuracy and no limitation about the object and environment. For one-shot structured light system, the final reconstruction result is significantly affected by vital items: designation of projected pattern, the detection of process and the establishment of correspondence.

2 Related Work

The projected pattern has a significant influence on the performance of the 3D shape acquisition. The pattern itself should be easy to detect and to establish correspondence between the projected and captured patterns. Several well

G. Bebis et al. (Eds.): ISVC 2012, Part I, LNCS 7431, pp. 372–381, 2012.

known patterns are evaluated in [4]. Among these methods, colored De Bruijn patterns are powerful in terms of accuracy and resolution. However the shape acquisition result is affected by various of factors, such as object's color, reflectance of object's surface and ambient light. The K-Means and clustering is used to classify the color adaptively, and then high-quality 3D reconstructions are achieved by a one-shot structured light system without need of dark laboratory environments[5]. In this paper, they designed a special De Bruijn sequence colored pattern and use K-Means method to classify the color.

The establishment of correspondence between captured image and projected one also has a significant influence on the quality of 3D reconstruction. Zhang et al.[6] developed a method using multi-pass dynamic programming and edge-based reconstruction, but this approach can not find sub-pixel accurate position of the edge. T.P.Koninckx et al.[7] propose a self-adaptive system for real-time range acquisition, pattern color, geometry, tracking and graph cut are used for solving correspondence problem. However, this method is susceptible to object's color, reflectance of object's surface and ambient light.

In our paper, We develop a adaptive scheme into dynamic programming. Different cost aggregation scheme used in stereo vision, a detailed discussion and evaluation of such methods is found in[8]. Yoon and Kweon[9] present a popular cost aggregation scheme that uses a fixed-size support window with per-pixel adaptive weight for 3D reconstruction through stereo vision, which is computed by color disparity and geometric distance to the center pixel of interest. Satisfying results were achieved without any global optimization. We design the formation of the weight of each stripe in a fixed local window depending on the properties of projected patterns. Our approach has achieved more accurate and complete result.

The reminder of this paper is organized as follows: Section 3 illuminates framework of our system. Section 4 shows the color classification of the stripes. Section 5 presents proposed adaptive dynamic programming algorithm. Section 6 includes experimental results and discussion. Conclusion and future work are stated in Section 7.

3   Framework and Architecture

The hardware used by our framework consists of a DLP projector with a WUXGA resolution(1920×1080) is used to project structured color light, and a camera with a resolution of 5.2 mega pixels(2560×2048) is used to grab the scene. Both the devices are controlled by a PC running the framework. The projector and camera are horizontal and the relative direction angle between the projector and camera is 17 degree, as show in Figure 1. In order to generate the 3D model of the object, the steps are performed: take pictures of the object illuminated with a color stripe pattern; extract stripes and classify color of each stripe; match the prospective stripes with the projected ones; calculate the 3D coordinates of the correspondences and reconstruct the surface.

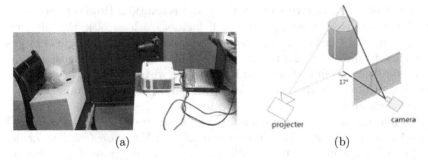

(a)   (b)

Fig. 1. (a)Devices and setting used in our framework. (b)Optical triangulation system.

3.1 Offline Design of Pattern

The pattern projected onto objects should be easy to detect, as described in [10], when high-contrast colors used, the detectability of projected patterns is increased and the ambiguity is reduced. Therefore a set of stripe pattern has been chosen with vertical lines of fully saturated colors with black spaces in between. Red, green, blue, cyan, magenta and yellow are the six colors selected. We use vector $C = \{0, 1, 2, 3, 4, 5\}$ to assign the six colors, as listed in Table 1. P.fechteler B et al.[5] impose a restriction on De Bruijn sequences, that the neighbor stripe should differ at least two channels in (R, G, B). Take element 1 for example, the possible next following colors should only be $0, 2, 4$. Then we achieve a $6 - ary$ De Bruijn sequence with $4 - order$, which has a length of 162.

Table 1. Color assignment for each element

element	color	R	G	B
0	red	255	0	0
1	green	0	255	0
2	blue	0	0	255
3	cyan	0	255	255
4	magenta	255	0	255
5	yellow	255	255	0

4   Color Classification

Accurate peak location needs to be estimated for derivation of 3D coordinates, our previous work[11] develop the M channel method that utilize the information from the rising edge and falling edge, and the calculation can be subpixel accuracy. The output of each stripe is specified by three characters: a scan column index, accurate position and a RGB color value. In RGB color space, the original six types of saturated color should be along the straight axes, such as black-red, black-yellow etc. But experimental result show that projected color, reflected by object and captured by cameras encounter severe distortions. The cross-talk and sensor noise are also detected between camera and projector. The captured color clusters could distribute differently and be roughly identifiable because of vague separation between them, as show in Figure 2, they are approximately shaped along six different direction lines, moreover the visible color distribution varies from one object to another. P.Fechteler et al.[5] presented an excellent adaptive color classification called K-MeansLineFit algorithm without prerequisite measurement of parameters for a model of camera and projector. In RGB space, the 3D color lines which can be written as: $o + xr$ are fitted through these clusters to form prototype of six clusters; In order to determine the individual stripe colors, First step is to calculate the distances $D_{(p_i, g_c)}$ of the pixel p_i to all the phototype lines g_c, and assigning the color of the phototype with the smallest distance. Then the parameters of the each straight lines are updated by the previous classification. Repeating the two steps until no significant changes in labeling process.

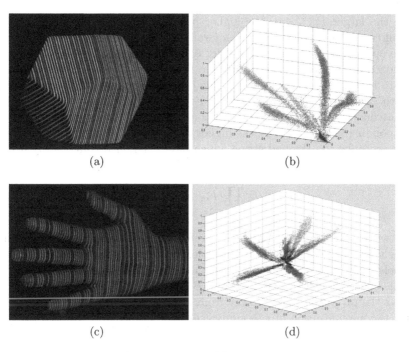

(a)　　　　　　　　　　　　　　　(b)

(c)　　　　　　　　　　　　　　　(d)

Fig. 2. Color distribution of different scene (a)Hexahedron gypsum with structured light pattern. (b)The color distribution of gypsum scene. (c)Hand with structured light pattern. (d)The color distribution of hand scene.

5   Stripe Matching

With previous steps done, each stripe is specified by a location, a classification of being projected with the pattern colors. Now the present assignment is to establish correspondences between projected and detected stripes. This matching is a typical combinatorial optimisation problem finding a combination of correspondences that fits best. We solve this task by setting up an objective function that can be efficiently solved with the dynamic programming method in order to find the best combination. Dynamic programming includes three steps: derive a cost function, build a matrix, and finally trace back the routine.

5.1   Derivation of Cost Function

The quality of a match is computed by assigning a score, measuring the the likelihood of the correspondence between the projected stripe and captured stripe. The specific definition of score can be expressed as

$$P_{i,c} = \frac{1}{D_{(p_i,g_c)}} \tag{1}$$

where the larger $P_{(p_i,g_c)}$ is, the higher possibility the classification is. So this possibility can be a gauge of accuracy rate of classification.

The optimal match can be solved for each horizontal scan column separately. The next step is to set up an objective function that has to be maximised in order to find the best combination. The objective function distinguish between the two cases of stripes being matched($i \in M$) and unmatched($i \notin M$). For the matched stripes, the higher likelihood that this stripe is projected with the corresponding pattern color is, the higher the score is. Additionally, It will assign higher score for the skipped stripe which have a larger distance away from the corresponding pattern color.

$$L = \prod_{\forall i \in M} P_{i,c} \times \prod_{\forall i \notin M} P_{i,\bar{c}} \tag{2}$$

where we adopt a notation in which, $\bar{c}$ refers to the color from the rest of three channels (R, G, B). We take $\bar{r}$ as example, it represents the color from G and B channels, that is cyan; $D_{(i,g_{\bar{r}})}$ represents the distance between the stripe color and the cyan line, which is from G and B channels. It can be a measurement of how far the projected color is away from red line. In order to calculate the cost easily, the maximization of L is equivalent to

$$\max(L) \sim \max(\ln L) \tag{3}$$

which leads to the score function can be as follow

$$Score(i) = \ln \frac{P_{i,c}}{P_{i,\bar{c}}} \tag{4}$$

5.2 Locally Adaptive Support-Weight Computation

There are some constraints in conventional dynamic programming method. For example, when more than one global optimal result exist, the algorithm would choose the first ordering one. In consideration of most part of one object would be locally continuous, a more rational solution is to choose the piecewise smooth path. In order to achieve the goal, it should take the local information into consideration, and incorporate the discontinuity penalty into original cost. We present a new cost aggregation scheme that use a fix-size window with per stripe varying weight to fit our framework.

Choosing the size of support window depends on some property of projected pattern, that is an $6 - ary$ and $4 - order$ De Bruijn sequence, which means every possible $4 - length$ subsequence appears as a sequence of consecutive characters exactly once. The the length of the subsequence less than 4 will appear at least 3 times, which would cause aliasing, while the length equal or bigger than 4 will assign current stripe best score, and cause least aliasing. It is appropriate to choose 4 as the size of the support window, because the length above 4 will sacrifice more margin information.

Next step is to compute the adaptive weight of the stripes in every $4 - length$ window. The adjacent stripes we used differ at least two channels in (R, G, B). We adjusted the weight of the neighboring stripes by using the photometric and geometric relationship with the stripe under consideration. The disparity of color and proximity of distance can be used for computation, which the more disparity between the two corresponding stripes, the larger the weight is, In addition, the closer the stripe is, the larger the weight is.Based on principles described above, local weight representing color disparity can be written as

$$w_d(i,j) = \exp(-\frac{P_{j,c}}{P_{i,c}}) \tag{5}$$

where i is the stripe under consideration and j is the other stripe in local window of i, which can be represented by N_i, Local weight representing proximity of distance can be written as

$$w_p(i,j) = k\frac{\gamma}{|j-i|} \tag{6}$$

where

$$k = \begin{cases} 1 \text{ if } j \in Match \\ 0 \text{ if } j \notin Match \end{cases}$$

where the γ is experimentally determined, $\gamma = 0.4$, and it assigns k good score for stripes being in order with the sequence and assigns bad score for incoherent sequences.So the function of score can be rewritten as

$$Score(i) = \sum_{j \in N_i} w_p(i,j) \ln w_d(i,j) \frac{P_{i,c}}{P_{i,\bar{c}}} \qquad (7)$$

5.3 Build Cost Aggregation Matrix

This section explains the steps for evaluating the DP matrix A, The instructions for accomplishing this are described in pseudocode below. The information stored

---

Algorithm 1. Adaptive DP algorithm

   Upleft = A[i-1][j-1]+Score(i)
   Left = A[i-1][j]
   Up = A[i][j-1]
   A[i][j]=max(Upleft,Left,Up)
   if *A[i][j]==A[i-1][j]* **then**
     M[i][j]=1
   else if *A[i][j]==A[i][j-1]* **then**
     M[i][j]=2
   else if *A[i][j]==A[i-1][j-1]* **then**
     M[i][j]=3
   end if

---

in the $M(i,j)$ can be used to break any ties that could occur in the calculation of $A(i,j)$. The optimum path can be reconstructed directly by traversing through the $M(i,j)$ matrix.

6   Experiment and Discussion

The final result of every correspondence is evaluated by triangulating 3D points cloud. Many experiments have been performed with different objects. Figure 3 shows the 3D points cloud results of three typical scenarios, the objects are hexahedron, doll and hand. Several methods for establishing the correspondences have been worked out and used in 3D reconstruction systems. In order to focus on the amendatory performance of the stripe matching algorithm, we select the region of interest. Two methods presented here: one is locally adaptive DP method(LADP); another is described in [5], which solves such task by combining Markov random field with DP algorithm(MRDP),which just take previous stripe into consideration. These two methods both performs much better than conventional DP Algorithm without considering the local information. The 3D model of objects are calculated and the surfaces are presented in Figure 4. The results show our method performs better in terms of smoothness. This score strategy has

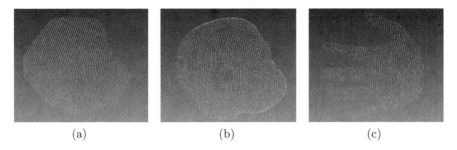

Fig. 3. Points cloud (a)A hexahedron. (b)Doll's head (c)Hand.

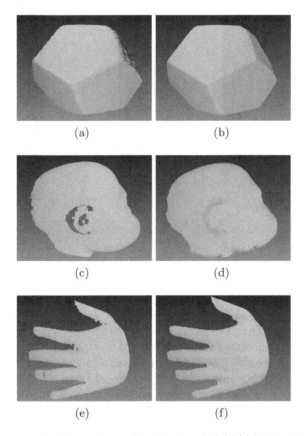

Fig. 4. Surface reconstruction using different methods (a)(c)(e) With MRDP method. (b)(d)(f) With LADP method.

some advantages. Firstly, it can achieve the piecewise smooth matched stripes based on the contextual information within a given local window. Secondly, it can rectify the special stripes inaccurately regarded as outliers derived from color classification. For example, one stripe under consideration is not matched but the others in its local window are all matched, which also will add good cost to the score, So the stripe may not to be skipped. So our method performs higher accuracy. Figure 5 shows a quantitative comparison of the ratio of the valid points to all points, which can be a measurement of the quality of reconstruction. The results are clearly visible that the proposed method drastically improves the number of valid points.

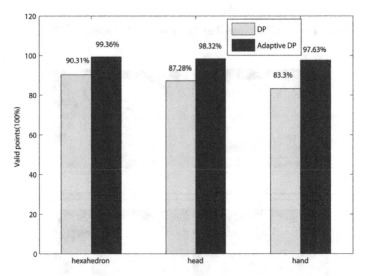

Fig. 5. Quantitative comparison of adaptive DP result

7   Conclusions

In this paper, we proposed a novel locally adaptive dynamic programming method for 3D model reconstruction from one color structured pattern. The presented system is suitable for generating real-time scans of moving object under natural light condition. The adaptive weight of pixel in a given window is computed by measuring the disparity and proximity. The performance show that the proposed method has achieved piecewise smooth matched stripes. The reconstruction of surface has been investigated in terms of accuracy. Experimental results demonstrate that proposed method outperforms other DP method without adaptive local information. The valid points increase at least by 10%. An enhanced accluracy may be achieved by establishing the correspondences with consideration of the whole input image instead of only taking separate scanline into account.

Acknowledgment. This research was supported by The National Natural Science Foundation of China (NSFC No. 61100111)and two Natural Science Foundation of Jiangsu Province Grants(NSFJ No.BK2010391, No.BK2011563).

References

1. Cox, I.J., Hingorani, S.L., Rao, S.B., Maggs, B.M.: A maximum likelihood stereo algorithm. Computer Vision and Image Understanding 63, 542–567 (1996)
2. Yahav, G., Iddan, G., Mandelboum, D.: 3d imaging camera for gaming application. In: International Conference on Consumer Electronics, ICCE 2007. Digest of Technical Papers, pp. 1–2 (2007)
3. Besl, P.J.: Active, optical range imaging sensors. Machine Vision and Applications 1, 127–152 (1988), doi:10.1007/BF01212277
4. Zhang, X., Zhu, L., Chu, L.: Evaluation of coded structured light methods using ground truth. In: 2011 IEEE 5th International Conference on Cybernetics and Intelligent Systems (CIS), pp. 117–123 (2011)
5. Fechteler, P., Eisert, P.: Adaptive colour classification for structured light systems. Computer Vision, IET 3, 49–59 (2009)
6. Zhang, L., Curless, B., Seitz, S.: Rapid shape acquisition using color structured light and multi-pass dynamic programming. In: Proceedings of the First International Symposium on 3D Data Processing Visualization and Transmission, pp. 24–36 (2002)
7. Koninckx, T., Van Gool, L.: Real-time range acquisition by adaptive structured light. IEEE Transactions on Pattern Analysis and Machine Intelligence 28, 432–445 (2006)
8. Tombari, F., Mattoccia, S., Di Stefano, L., Addimanda, E.: Classification and evaluation of cost aggregation methods for stereo correspondence. In: IEEE Conference on Computer Vision and Pattern Recognition, CVPR 2008, pp. 1–8 (2008)
9. Yoon, K.J., Kweon, I.S.: Locally adaptive support-weight approach for visual correspondence search. In: IEEE Computer Society Conference on Computer Vision and Pattern Recognition, CVPR 2005, vol. 2, pp. 924–931 (2005)
10. Je, C., Lee, S.-W., Park, R.-H.: High-Contrast Color-Stripe Pattern for Rapid Structured-Light Range Imaging. In: Pajdla, T., Matas, J(G.) (eds.) ECCV 2004. LNCS, vol. 3021, pp. 95–107. Springer, Heidelberg (2004)
11. Zhao, D., Zhou, Y., Yu, Y., Du, S.: A novel peak detection method of structured light stripes for 3d reconstruction. In: 2011 International Conference on Intelligent Human-Machine Systems and Cybernetics (IHMSC), vol. 2, pp. 43–46 (2011)

Advanced Coincidence Processing
of 3D Laser Radar Data

Alexandru N. Vasile, Luke J. Skelly, Michael E. O'Brien, Dan G. Fouche,
Richard M. Marino, Robert Knowlton, M. Jalal Khan, and Richard M. Heinrichs

Massachusetts Institute of Technology - Lincoln Laboratory, Lexington, MA, USA

Abstract. Data collected by 3D Laser Radar (Lidar) systems, which utilize arrays of avalanche photo-diode detectors operating in either Linear or Geiger mode, may include a large number of false detector counts or noise from temporal and spatial clutter. We present an improved algorithm for noise removal and signal detection, called Multiple-Peak Spatial Coincidence Processing (MPSCP). Field data, collected using an airborne Lidar sensor in support of the 2010 Haiti earthquake operations, were used to test the MPSCP algorithm against current state-of-the-art, Maximum A-posteriori Coincidence Processing (MAPCP). Qualitative and quantitative results are presented to determine how well each algorithm removes image noise while preserving signal and reconstructing the best estimate of the underlying 3D scene. The MPSCP algorithm is shown to have 9x improvement in signal-to-noise ratio, a 2-3x improvement in angular and range resolution, a 21% improvement in ground detection and a 5.9x improvement in computational efficiency compared to MAPCP.

1   Introduction

Three-dimensional Laser Radar (3-D Lidar) sensors output range images, which provide explicit 3-D information about a scene [1][2][3]. MIT Lincoln Laboratory has built a functional airborne 3-D Lidar system, with an array of avalanche photo-diodes (APDs) operating in Geiger mode, that actively illuminates an area using a passively Q-switched micro-chip laser with a short pulse width time [4][5]. On each single laser pulse, light from the laser travels to the target area and some reflects back and is detected by an array of Geiger-mode APDs. Figure 1 captures the 3D Lidar system concept.

Recent field tests using the sensor have produced high-quality 3-D imagery of targets for extremely low signal levels [6][7]. Though there are many advantages to using single-photon sensitive detector technology, the data collected using these Geiger-mode APDs are often noisy with unwanted temporal or spatial clutter. It has been shown in previous publications that by identifying spatial coincidences in data from as few as three laser pulses, we can significantly reduce the probability of false alarms by several orders of magnitude [8][9][10].

The method of finding signal in the presence of noise and clutter by using coincident spatial data is known as coincidence processing. The more 3D points returned at the same spatial location, the more likely that the points came from a real scene

G. Bebis et al. (Eds.): ISVC 2012, Part I, LNCS 7431, pp. 382–393, 2012.

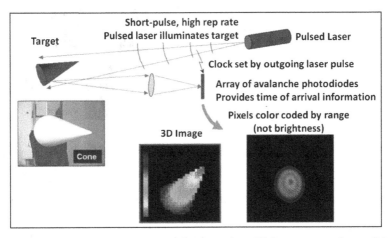

Fig. 1. 3D Lidar system concept. A laser sends out a pulse of light to a target. Some of that light is reflected back and detected by an APD array. The time of flight between the send and receive of the laser pulse is recorded and converted to metric units to create a range image.

surface. In this paper, we discuss the implementation of a novel processing algorithm, known as Multi-Peak Spatial Coincidence Processing (MPSCP), and test it against the current state of the art, Maximum A-posteriori Coincidence Processing (MAPCP) algorithm, [11]. The contributions of this paper are as follows:

1. A set of general methods to address typical 3D Lidar processing challenges that are relevant to most 3D Lidar sensor systems (Linear and Geiger mode).

2. An improved 3D Lidar filtering algorithm that is shown to have a significant improvement over current state-of-the-art, with qualitative and quantitative results shown.

In the remainder of this paper, we first discuss the challenges of processing 3D Lidar data and describe how our algorithm addresses those challenges. Quantitative and qualitative results are shown using data collected over Haiti in support of earthquake rescue operations using the Airborne LIdar Research Test-bed (ALIRT) platform [12].

2   Background

There exists a cause-effect relationship between 3D Lidar system design / data collection methods and the inherent processing challenges that arise and need be subsequently addressed. First we introduce the 3D Lidar system design, where an airborne sensor stares at a pre-designated ground target from multiple perspectives in order to get a more complete measurement of the scene. This concept of operations is shown in Figure 2.

Due to limitations in APD array size (number of pixels), for each viewing perspective the detector array needs to be scanned using a sinusoid pattern in angle-angle

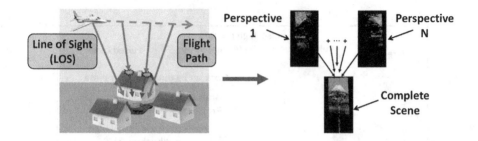

Fig. 2. 3D Lidar concept of operations. An airborne platform, stares at a pre-designated ground target from multiple perspectives in order to get a more complete measurement of the scene.

space to get a higher resolution 3D image of the target. Given the above system design and data collection methods, a list of processing challenges becomes apparent:

1.   *Varying signal and noise levels*: the scanning pattern can lead to large variations in the absolute output level (3D point density). Background light and/or detector thermal excitation can lead to 3D noise points with high spatial coincidence that need to be filtered out. There is a need to know the output level to dynamically determine the statistical significance of coincident returns.

2.   *Photon attenuation*: Obscurants in the range direction might reduce the probability of transmitted photons reaching a ground target. Knowledge of photon attenuation can be used to dynamically adjust statistical significance of coincident returns.

3.   *Detector-specific range attenuation*: due to nature of Geiger-mode APDs, once a pixel is triggered at a closer range, no hits at further ranges are possible as the pixel needs to be reset, leading to output level attenuation in the range direction.

4.   *Laser-detector Point Spread Function* (PSF) can lead to 3D blurring of imaged objects. A method is needed to de-blur the 3D image.

5.   *Platform attitude errors* (GPS/INS) can add further blur to the 3D image.

6.   *Platform motion and signal aggregation from multiple perspectives*: To increase signal-to-noise (SNR) level, a method is needed to evaluate output level from each perspective that contributes to a particular 3D location.

7.   *Automatically determine optimal processing parameters*: Need data-driven method to obtain good, reproducible, single-run results without the need for human intervention.

Before further discussing and addressing each individual challenge, it is crucial to notice that most of these challenges are related by a common denominator: sensor line of sight (LOS). Variations in signal/noise levels are orthogonal to the LOS, while photon/detector signal/noise attenuation are along the LOS. The Laser-Detector PSF is oriented along the LOS direction: error in range due to laser-detector timing jitter is, by definition, aligned to the LOS, while platform attitude errors, such as GPS and inertial navigation errors (INS) can also be readily thought as orthogonal to the LOS.

The crucial insight is that processing in an appropriate line-of-sight coordinate system plays an important role in decoupling the effects of the various processing challenges listed above, so that each challenge can be independently addressed.

Depending on airborne platform velocity, range-to-target and target collection size, a LOS coordinate system can be chosen to approximate the true line-of-sight while avoiding the computational expense of ray tracing each individual APD array LOS vector and storing the information in a 3D volumetric signal map.

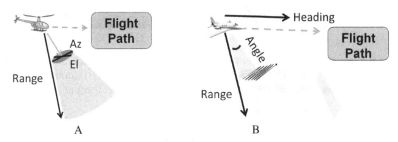

Fig. 3. Line-of-sight (LOS) coordinate systems for various sensor platforms. A) For slow-moving platforms, a spherical coordinate system (angle-angle-range) gives a good approximation of the LOS while B) a skewed-cylindrical coordinate space (heading-angle-range) gives a good approximation of the LOS for fast-moving airborne platforms.

Figure 3 depicts several LOS coordinates systems that might be used. For instance for airborne platforms that are slow-moving in comparison to the range-to-target distance and target area, a sensor-centered spherical coordinate system (angle-angle-range) can best approximate the collection volume which tends to resemble a solid angle. For fast-moving airborne platforms, where target area size is on the same order as platform motion, a skewed-cylindrical coordinate system can best approximate the LOS independent of range-to-target. Another possibility to consider for small target areas is an inverted target-centered spherical coordinate system.

Since the ALIRT system is hosted on a fast-moving airplane and uses the airplane's forward heading motion to scan a target area of the same approximate size, we utilize a skewed-cylindrical coordinate system to best approximate the LOS.

3   Our Approach

The proposed MPSCP algorithm advances current state of the art in 3D Lidar processing by addressing each of the processing challenges noted in Section 2. Figure 4-A/B shows an example of raw 3D data to allow the reader to visually appreciate the large amount of noise and clutter present.

The noisy 3D data is initially stored in Universal Transverse Mercator (UTM) coordinate system, a 3D Cartesian coordinate space [13]. The first processing step of MPSCP is to transform the data from UTM space to an appropriate line of sight space, which for our sensor is a skewed cylindrical coordinate space. Using metadata, such as airborne sensor position, a LOS coordinate basis is created and the data is transformed. We now proceed to explain how utilizing this data-defined LOS coordinate space will lead to improved computational efficiency as well improved 3D filtering results.

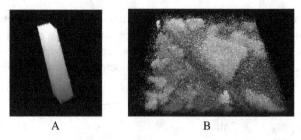

Fig. 4. A) Height color-coded example of raw 3D data (dark grey low altitude, white high altitude). The target area is obscured due to the heavy amount of noise. B) Zoomed-in version of same data set showing the target area with high levels of noise and clutter present.

The next MPSCP processing step is to determine expected 3D output level. MPSCP uses the output level estimate to determine statistical significance of spatially coincident returns. Statistical significance is determined in terms of a maximum likelihood estimator given the expected output level. In this fashion, the MPSCP algorithm can dynamically adjust its internal noise suppression thresholds to work well under most signal conditions.

Variations in output level due to the scanning pattern, photon attenuation as well as detector attenuation, can be accounted accurately using the data-defined LOS coordinate system. We first determine an initial output level, $O_{initial}$, due to variations in scan pattern dwell times. In our LOS coordinate system, $O_{initial}$ varies in only 2 of the 3 dimensions, namely heading and angle, but not range. Compared to output level estimation in a 3D Cartesian coordinates, which would have required computationally expensive 3D ray tracing and storage of a volumetric 3D array of values, the problem of estimating output level reduces to a 2D matrix in heading-angle space using our LOS coordinate system. This leads to increased algorithmic computational efficiency as well as improved implementation with significantly lower memory overhead. An example of the computed $O_{initial}$ output level map is shown in Figure 5, with output level back-projected to 3D from 2D heading-angle space on a per raw 3D point basis.

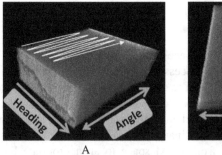
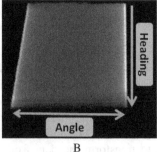

Fig. 5. A) Side-view of a target area, showing a notional scan pattern and the estimate of $O_{initial}$ (dark grey - low value, white - high value) obtained using the LOS space. B) Heading-angle view of the same target area. The LOS is shown to be accurately estimated, with high output levels (white) at the edge of the sinusoid scan pattern due to decreased angular velocity, leading to an increase in 3D point density level.

The output level is also affected by photon attenuation as well as detector attenuation in range. Photon attenuation due to line-of-sight blocking needs to be taken into account when computing the statistical significance of coincident returns. Detector-specific output attenuation in range has a similar effect as photon attenuation, reducing expected output level at further range values along the LOS. To account for these data-dependent effects, the data at a particular heading-angle location (which can be visually represented as a chimney of data in the range direction) is binned into a range histogram H. For each range bin i of histogram H, MPSCP keeps track of returns that have occurred at closer ranges versus returns that have occurred at further ranges to determine an expected output attenuation value. Equation 1 numerically captures the method for determining attenuated output level, $O_{attenuated}$, as a function of range along the LOS, while Figure 6 visually captures photon and detector range attenuation effects.

$$O_{attenuated}(i) = O_{initial}\left[1 - \frac{C_{H(h_1, a_2)}(i)}{C_{H(h_1, a_2)}(N)}\right].$$

(1)

where,
$O_{attenuated}(i)$ is the attenuation-corrected expected output level, $O_{initial}$ is the expected output level at a particular heading-angle location $[h_1, a_2]$ determined solely based on the scan pattern (no attenuation correction), $C_H(i)$ is the cumulative histogram of range histogram H at range bin i and N is the last (furthest) range histogram bin.

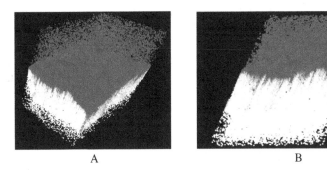

A   B

Fig. 6. A) Side-view and B) view orthogonal to the LOS, showing the effects of photon / detector attenuation as a function of range. Notice how the output attenuation changes from low (dark grey) to high (white) as the line-of-sight passes through obscuration.

Having determined the expected output level, a method is needed to find spatially coincident returns to distinguish signal from noise. Spatial coincidence of points is affected by the laser-detector 3D Point Spread Function (PSF) as well as platform attitude errors, leading to blurring of the 3D image. The laser-detector 3D (angle-angle-range) PSF can be decoupled into the angular response to a step-response in range (such as a ground to building edge), followed by the range response to a flat surface. Figure 7 captures the methodology used to determine the 3D PSF. MPSCP uses the PSF as a 3D matched filter to integrate signal and find 3D locations that have enough returns to be considered statistically significant. Since our LOS coordinate system is already well aligned to the 3D PSF, the 3D matched filter can be efficiently applied to the data. The matched filter is also be used for sub-voxel estimation of the filtered return 3D location, effectively removing the PSF-induced blur.

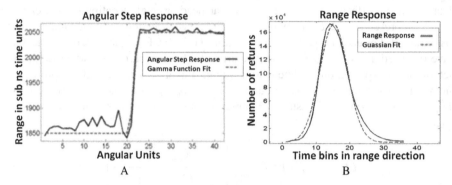

Fig. 7. Computation of laser-detector 3D point spread function. A) Angular response to 3D edge and Gamma-function fit, B) Range response to flat-plate and its associated Gaussian fit.

Another source that affects the spatial coincidence of points is platform attitude errors. These errors occur due to drift in the GPS/INS solution, as well as due to errors induced by the scanning hardware: such errors occur during changes in view-point perspective, which require a sharp step-response in angular space from the scanning hardware. Due to insufficient bandwidth, small angular errors can occur. These angular errors, combined with GPS/INS drift, can lead to blurring that can be several times bigger than the 3D PSF-induced blur. MPSCP corrects for these blurring errors by employing a two-stage filtering process. Figure 8 shows the overall MPSCP processing block diagram. In the first stage filter, data from each single viewpoint is processed independently: starting with a noisy 3D data set per viewpoint, a unique data-defined LOS coordinate system is created and the data is processed along the line of sight to produce a filtered 3D data set per viewpoint. A secondary output is also created, which consists of the original 3D noisy data appended with LOS statistics per point, such as the expected output level value as visualized in Figures 5 and 6. Using the single-viewpoint 3D filtered data sets, we align the data sets using the Iterative Closest Point [14][15] algorithm with six-degrees of freedom (3D rotation and 3D translation), which produces results with sub-pixel error correction. The 6-degree transformation is also applied to the raw 3D point cloud data that has been appended with LOS statistics per point. To detect weak signals that might have been missed when processing data on a single-viewpoint basis, a second-stage filter takes the aggregated, de-blurred, multi-viewpoint data set and processes the data in a similar manner to the first-stage. Since the data is taken from multiple perspectives, MPSCP defaults to using an UTM-aligned 3D Cartesian coordinate space to process the aggregated data. The second-stage coincidence processor uses the expected output level saved on a per-point basis from the first stage filter to determine a statistical noise threshold to filter the multi-viewpoint aggregated data set.

As shown in the block diagram, MPSCP requires a single input parameter: processing resolution in meters. This processing resolution is used to automatically determine a binning size in the LOS coordinate space for the first-stage coincidence processing filter as well as 3D PSF matched filter size. An accurate output level estimate is computed directly from the data, which takes into account photon and detector output attenuation effects, allowing MPSCP to dynamically adjust its noise-suppression thresholds to filter most of the noise while keeping weak signals. In comparison, the MAPCP algorithm, which represents the current state of the art in 3D Lidar processing, does not have this type of automatic, data-dependent parameter tuning, with the user required to manually determine the size of the 3D matched filter and manually choose an optimum threshold. This typically requires multiple runs per data set for an operator to determine a good set of parameters. Furthermore, since MAPCP does not take into account photon or detector output attenuation effects, the algorithm has difficulty in keeping weak signals in low output level regions while at the same time removing noise in high output level regions.

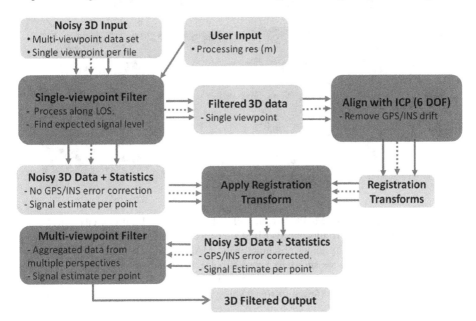

Fig. 8. MPSCP algorithm block diagram. MPSCP has a two-stage filtering process. Data from each viewpoint is first processed independently in its own unique LOS coordinate system. Two outputs are created, namely a filtered 3D data set per viewpoint and the original noisy 3D data with LOS statistics, such as output level on a per point basis. The individual filtered data sets are aligned to remove attitude errors, with the transform applied to the noisy 3D data set. A second filtering stage ingests the aggregated data set to detect weak signals that might have been missed by the first stage filter, leading to a final 3D filtered output.

4   Results and Discussion

The MPSCP algorithm was tested against MAPCP on multiple data sets collected over Port-Au-Prince, Haiti as part of the 2010 earthquake response. The data was used to determine the navigability of streets as well as to quickly respond to population movement into tent-cities that literally sprang out overnight. By accurately counting the number of tents, an accurate assessment could be determined of the quantity of essential supplies for each tent city.

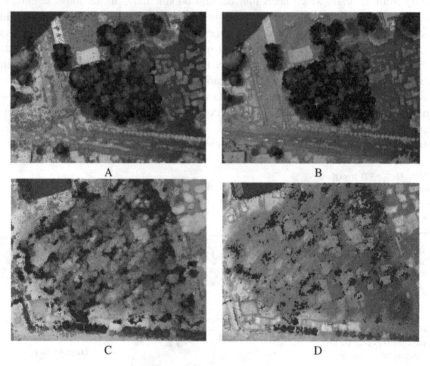

Fig. 9. Visual comparison of MAPCP versus MPSCP results on a Haiti tent city collected in January 2010. A) MAPCP results and B) MPSCP results. C) Zoomed in view to the center of target area showing tent city under obscurant using MAPCP, and D) same view of MPSCP results. The MPSCP results are shown to have less noise, have sharper edges with less blurring on buildings, cars, palm trees lining the street, and have better 3D scene coverage in weak signal areas under obscuration (fewer no-signal voids, shown as black pixels in the image).

4.1   Qualitative Results

Figure 9-A shows height-intensity color-coded MAPCP results for a target-mode data set collected from multiple perspectives. Figure 9-B shows the MPSCP results for visual comparison. From the results, one can visually discern that MPSCP has significantly better angular resolution as well as range resolution compared to MAPCP, with sharp palm tree branches, sharper building edges and car shapes better resolved. In addition, the MPSCP results have almost all the noise removed, while the MAPCP algorithm still has a large amount of noise present (visually seen as salt-and-pepper noise above road, other open areas). In Figure 9-C/D, we are showing the same data set, now zoomed-in and cropped in the z-direction to reveal the presence of tents. The

MPSCP results shown in Figure 9-D, demonstrate improved 3D scene coverage and reconstruction under weak signal conditions compared to MAPCP.

4.2   Quantitative Results

Using metrics developed by Lopez et. al. [11], we quantitatively evaluated the data sets shown in Figure 9. Signal-to-noise (SNR) was measured in a flat area out in open: processed 3D points that fell within a height envelope above and below the ground were considered valid detections; points above or below were considered noise. MPSCP had an SNR of 97x while MAPCP had an SNR of 10.8x. MPSCP has a 9x improvement in SNR, close to an order of magnitude better than MAPCP.

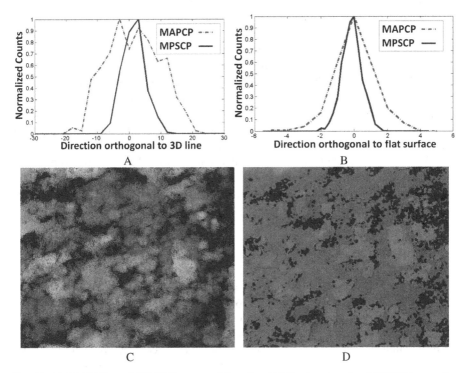

Fig. 10. A) MAPCP vs. MPSCP line spread function (LSF), showing that MPSCP has an improvement of about 3x in angular resolution. B) MAPCP range resolution versus MPSCP range resolution, showing an improvement in the MPSCP result of 2x. C) Ground coverage for MAPCP and D) MPSCP, with voids shown as black pixels. MPSCP recovered 21% more ground cover compared to MAPCP.

Figure 10-A shows the results of a line spread function (LSF) metric to evaluate angular resolution. The LSF results indicate that MPSCP has a 3x improvement in angular resolution. Range resolution was measured by segmenting out the roof-top of a building, followed by slope-bias removal using principal-component analysis to align the plane normal axis to the z, up direction. The resulting MAPCP and MPSCP range histograms are shown in Figure 10-B; MPSCP has 2 times improvement in range resolution. Ground scene reconstruction was also evaluated, as shown in Figures 10-C and 10-D. The 3D data was cropped in the z direction to include only 3D returns on the ground and tents; the data was binned in the x-y directions to create a binary filled vs. empty pixel image. Results indicate that MPSCP found 21% more ground cover compared to MAPCP.

The improved ground signal detection of MPSCP compared to MAPCP, while retaining high-frequency information out in open areas, can be attributed to the use of dynamic thresholding based on an accurate output-level estimate that takes into account photon and detector output attenuation effects due to obscuration. The use of dynamic thresholding allows the MPSCP algorithm to detect weak signals under obscuration, while still removing heavy noise in high output level areas. By contrast, MAPCP does not employ data-driven noise thresholding, leading the algorithm to have difficulty in keeping weak signals in low output areas level while at the same time removing noise in high output level areas.

A timing analysis was run on 4 multi-viewpoint data sets using a 12 core Intel Xeon 3GHz machine. Both MPSCP and MAPCP were run at the same processing resolution with the default processing parameters. The overall conclusion from the timing results is that MPSCP is about 5.9 times faster than MAPCP. Besides extensive testing on 4 multi-viewpoint data sets collected in Haiti, the algorithm has been successfully tested on a large scale 3D map data set covering approximately 30 square km of Port-Au-Prince, Haiti. The MPSCP algorithm produced good, single-run results without the need for parameter tweaking. The removal of the need for human intervention is of tremendous importance for algorithm scalability to the large amounts of 3D Lidar data sets generated in the field.

5   Conclusions

In this paper, we have described a set of general methods to process 3D Lidar data that are relevant to most 3D Lidar sensor systems, with either Linear-mode of Geiger-mode APDs. We have also described in detail a novel 3D Lidar filtering algorithm that is shown to be a significant improvement over the current state of the art. Qualitative results indicate shaper 3D images with building and tree structure better resolved. The algorithm was also able to remove more noise while preserving weak signal areas as visually demonstrated in the form of improved ground coverage under obscuration. The use of automatic, data-driven parameter tuning allows MPSCP to produce good, single-run results without the need of human intervention. The algorithm has been tested on 4 multi-viewpoint data sets, as well as a large scale 30 square km 3D map of Port-Au-Prince, Haiti.

Quantitative metrics run on a multi-viewpoint data set further confirm the qualitative results. MPSCP has close to an order of magnitude improvement in signal-to-noise ratio, a 2-3x improvement in angular and range resolution, and a 21% improvement in weak signal detection. In addition, the algorithmic implementation of MPSCP is much more computationally efficient with run times that are 5.9x faster than MAPCP.

References

1. Gschwendtner, A.G., Keicher, W.E.: Development of Coherent Laser Radar at Lincoln Laboratory. Linc. Lab. J. 12(2), 383–396 (2000)
2. Marino, R.M., Stephens, T., Hatch, R.E., McLaughlin, J.L., Mooney, J.G., O'Brien, M.E., Rowe, G.S., Adams, J.S., Skelly, L., Knowlton, R.C., Forman, S.E., Davis, W.R.: A Compact 3D Imaging Laser Radar System Using Geiger-Mode APD Arrays: System and Measurements. In: SPIE, vol. 5086, pp. 1–15 (2003)
3. Albota, M.A., Aull, B.F., Fouche, D.G., Heinrichs, R.M., Kocher, D.G., Marino, R.M., Mooney, J.G., Newbury, N.R., O'Brien, M.E., Player, B.E., Willard, B.C., Zayhowski, J.J.: Three-Dimensional Imaging Laser Radars with Geiger-Mode Avalanche Photodiode Arrays. Lincoln Laboratory Journal 13(2), 351–370 (2002)
4. Zayhowski, J.J.: Passively Q-Switched Microchip Lasers and Applications. Rev. Laser Eng. 29(12), 841–846 (1988)
5. Zayhowski, J.J.: Microchip Lasers. Lincoln Laboratory Journal 3(3), 427–446 (1990)
6. Heinrichs, R.M., Aull, B.F., Marino, R.M., Fouche, D.G., McIntosh, A.K., Zayhowski, J.J., Stephens, T., O'Brien, M.E., Albota, M.A.: Three-Dimensional Laser Radar with APD Arrays. In: SPIE, vol. 4377, pp. 106–117 (2001)
7. Albota, M.A., Heinrichs, R.M., Kocher, D.G., Fouche, D.G., Player, B.E., O'Brien, M.E., Aull, B.F., Zayhowski, J.J., Mooney, J., Willard, B.C., Carlson, R.R.: Three-Dimensional Imaging Laser Radar with a Photon-Counting Avalanche Photodiode Array and Microchip Laser. Appl. Opt. 41(36), 7671–7678
8. Aull, B.F., Loomis, A.H., Young, D.J., Heinrichs, R.M., Felton, B.J., Daniels, P.J., Landers, D.J.: Geiger-Mode Avalanche Photodiodes for Three-Dimensional Imaging. Linc. Laboratory Journal 13(2), 335–350 (2002)
9. McIntosh, K.A., Donnelly, J.P., Oakley, D.C., Napoleone, A., Calawa, S.D., Mahoney, L.J., Molvar, K.M., Duerr, E.K., Groves, S.H., Shaver, D.C.: InGaAsP/InP Avalanche Photodiodes for Photon Counting at $1.06\,\mu m$. Appl. Phys. Lett. 81, 2505–2507 (2002)
10. Fouche, D.G.: Detection and False-Alarm Probabilities for Laser Radars That Use Geiger-Mode Detectors. Appl. Opt. 42(27), 5388–5398
11. Stevens, J.R., Lopez, N.A., Burton, R.R.: Quantitative Data Quality Metrics for 3D Laser Radar Systems. In: SPIE Proceedings, vol. 8037 (2010)
12. http://www.ll.mit.edu/publications/technotes/TechNote_ALIRT.pdf
13. Karney, C.F.F.: Transverse Mercator with an accuracy of a few nanometers. Journal of Geodesy 85(8), 475–485 (2011)
14. Besl, P., McKay, N.: A method of registration of 3-D shapes. IEEE Trans. Pattern Analysis and Machine Intelligence 12(2), 239–256 (1992)
15. Zhang, Z.: Iterative point matching for registration of free-form curves and surfaces. Int'l Jour. Computer Vision 13(2), 119–152 (1994)

Poisson Reconstruction of Extreme Submersed Environments: The ENDURANCE Exploration of an Under-Ice Antarctic Lake

Alessandro Febretti[2], Kristof Richmond[1], Shilpa Gulati[1], Christopher Flesher[1], Bartholomew P. Hogan[1], Andrew Johnson[2], William C. Stone[1], John Priscu[3], and Peter Doran[4]

[1] Stone Aerospace
3511 Caldwell Lane, Del Valle, TX 78617-3006
[2] Electronic Visualization Laboratory, University of Illinois at Chicago
Dept. of Computer Science, 851 S. Morgan St., Chicago, IL 60607-7053
[3] Dept. of Land Resources and Environmental Sciences, Montana State University
334 Leon Johnson Hall, Bozeman, MT 59717
[4] Dept. of Earth and Environmental Sciences, University of Illinois at Chicago
845 W. Taylor St. Chicago, IL 60607

Abstract. We evaluate the use of Poisson reconstruction to generate a 3D bathymetric model of West Lake Bonney, Antarctica. The source sonar dataset has been collected by the ENDURANCE autonomous vehicle in the course of two Antarctic summer missions. The reconstruction workflow involved processing 200 million datapoints to generate a high resolution model of the lake bottom, Narrows region and underwater glacier face. A novel and flexible toolset has been developed to automate the processing of the Bonney data.

1 Introduction

The McMurdo Dry valleys, located within Victoria Land in Antarctica, are one of the world's most extreme deserts, and represent the largest ice-free region in the continent. The unique conditions in the valleys are in part caused by katabatics, winds reaching speeds of 320 kilometers per hour and capable of evaporating all water, ice and snow in the environment.

Some of the lakes in the Dry Valleys rank among the world's most saline lakes. One of them is Lake Bonney, (figure 1), a perennially ice-covered lake at the end of Taylor Glacier. Anaerobic bacteria whose metabolism is based on iron and sulfur live in sub-freezing temperatures under Taylor Glacier: since the Dry Valleys are one of the terrestrial environments closest to Mars and to some of Jupiter's moons, this is considered an important source of insights into possible forms of extraterrestrial life.

For this reason, NASA funded the Environmentally Non-Disturbing Underwater Robotic ANTarctic Explorer (ENDURANCE) project. ENDURANCE is an autonomous underwater vehicle (AUV) designed to map the geometry, geochemistry and biology of Lake Bonney in three dimensions.

G. Bebis et al. (Eds.): ISVC 2012, Part I, LNCS 7431, pp. 394–403, 2012.
© Springer-Verlag Berlin Heidelberg 2012

ENDURANCE has been specifically designed to minimize impact on the environment it works in. This is primarily to meet strict Antarctic environmental protocols, but will also be a useful feature for planetary protection and improved planetary science in the future: NASA hopes to build upon lessons learned during testing for exploring objects in our solar system known to harbor sizable bodies of water, such as Jupiter's moon, Europa.

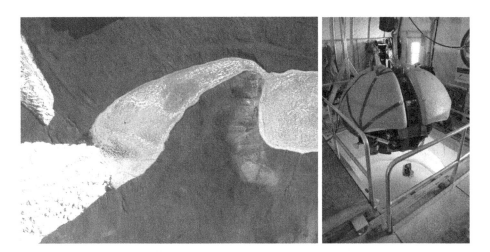

Fig. 1. From the left: A satellite image of West Lake Bonney; The ENDURANCE AUV ready to be deployed

1.1   Challenges

Underwater environments are difficult to navigate, and the simultaneous localization and mapping (SLAM) problem is more challenging to solve, due to the presence of additional sources of uncertainty like water speed, time variant zero-depth levels, and physical properties of the water column influencing the sonar beam geometries.

The sonar mapping at Lake Bonney also required full 3D capabilities. A significant focus of scientific investigation was the degree of influence of the Taylor Glacier at the west end of the lake. To this end, the 3D structure of the glacier face and its interface with the lake bed were to be investigated and mapped in detail. Additionally, the topography of shallow areas around the lake edge, not directly accessible to the vehicle, was of interest.

2   Related Work

2.1   Autonomous Underwater Exploration

3D Mapping with underwater robots ([1–3]) has received relatively low attention when compared to 3D mapping in other environments. Even fewer robotic vehicles have been developed for under-ice operation (like the ALTEX [4], Autosub [5] and SeaBED [6]). Some key differences between these vehicles and ENDURANCE are the very restricted operating volume in Lake Bonney (between 3 and 12 meters depth) and a difficult acoustic environment to operate in, due to smooth ice above the operating volume and a severe halocline below it.

2.2   Surface Reconstruction / 3D Maps

The need for fully 3D, high resolution reconstruction required a careful design of the data processing workflow. The final objective would be to generate a high quality bathymetric and glacier model, while dealing with high noise levels in the data, varying coverage levels, and different sonar configurations from dive to dive.

Delaunay triangulation techniques [7] are affected by data noise so they require pretty aggressive filtering. Also given their plane-fitting nature they do not work well with fully 3D data.

Approaches based on occupancy grid generation plus moving least squares (MLS) have been used successfully in similar works ([8, 9]) but present three major problems relative to our scenario:

- In presence of non-homogeneous data, occupancy grid + MLS approaches leaves holes in the reconstruction. These holes need to be filled by additional postprocessing steps (for instance by expanding neighboring regions)
- For high resolution data, the amount of required voxels can make memory requirements prohibitive (although there are strategies to alleviate this, i.e. the octree grids used in [8])
- The default sonar beam model used to write data to the grid is a cone with negative log-probability volume voxels and positive log-probability base voxels. This works well for isotropic water volumes. But if physical properties change across the water volume (as is the case for the Bonney environment), the beam model needs to be regenerated for each writing step. Although doable from the technical standpoint, this would require a significant amount of work on existing evidence grid solutions.

These considerations (along with preliminary tests on the data) made the occupancy grid + MLS solution less than ideal for our scenario, and led us to consider a reconstruction approach based an implicit surface technique called Poisson reconstruction [10]. Additional details about this technique will be presented in section 4.3. To the best of our knowledge, this is the first time the Poisson technique has been used in the context of sonar-based surface reconstruction. The rest of this paper will present our workflow implementation and discuss several advantages and disadvantages of the chosen approach.

3   Data Collection

ENDURANCE operated during two Antarctic summer seasons (2008 and 2009). Each mission entailed a set of deployments of the AUV, for a total of 45 dives.

For each dive, The AUV operated depending on 3 distinct science objectives: Water chemistry profiling, Bathymetry scanning, and glacier exploration. This work concentrates on processing data from bathymetry and glacier dives, but a full coverage of the mission science objective can be found in [11].

The ENDURANCE AUV was equipped with several independent sonar systems. The primary mapping unit was a 480-point multi-beam sonar with a 120° × 3° field of view. This unit could be mounted in both forward-looking and down-looking configurations to ensure full coverage of the lake geometry.

4   Data Processing

For the purpose of bathymetry reconstruction, the source data consisted of about 200 million distinct sonar range returns, plus navigation data and AUV attitude information at 0.2 second intervals. Figure 2 offers an overview of the data processing steps required to transform the raw range data into the final 3D model of West Lake Bonney. The processing steps required to remove errors from the raw vehicle navigation to arrive at the pose data used for mapping are detailed in [11].

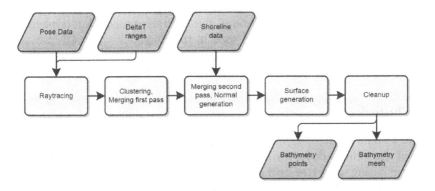

Fig. 2. Overview of the sonar data processing and surface reconstruction workflow. The Pose database contains the AUV position and attitude information. The DeltaT database stores the range return readings from the AUV sonar intrument.

4.1   Raytracing

In section 2.1 we explained how sonar data required corrections dependent on the water column physical properties. The trajectory of sonar beams is influenced by the water density according to Snell's Law. Sound waves get refracted as they travel through the water column. This effect was particularly marked in West Lake Bonney due to the presence of a major halocline at a depth of 16 meters. In order to take this effect into account, we computed a speed-of-sound profile

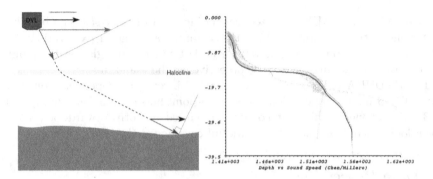

Fig. 3. From the left: geometry of sonar beams in the water column; the sound speed profile for West Lake Bonney: the halocline below the freshwater lens is at a depth of about 16 meters

for West Lake Bonney using the Chen-Millero formula [12] (figure 3) To compute this model, we used the measurements obtained from the ENDURANCE science payload (in particular, the temperature and salinity profiles of the water column).

The sound speed profile was used in conjunction with the sonde telemetry data and sonar configuration to re-trace the beam paths as they traveled across the water column. A modified version of the raytracing algorithm implemented in the MB system software has been used in this step.[1]

Some minor noise filtering was performed during this step to eliminate clear outliers (i.e. using range return timing and beam angle thresholds).

4.2   Clustering

In order to reduce the noise in the raytraced point cloud, we integrated a fairly standard clustering step in our processing pipeline. Range returns were collected in 3D bins, using an octree grid implementation similar to the one presented in [1]. If the number of points in a bin fell under a certain threshold, the entire bin was discarded as noise. Otherwise, points in the bin were averaged in order to generate a single 3D point for each bin.

To process the Bonney dataset, we used a 3D grid with a resolution of 1 meter and a 16 point rejection threshold.

4.3   Surface Reconstruction

As mentioned in section 2.2, we decided to employ an implicit surface technique in our reconstruction step. In [13], Kazhdan et al. show how surface reconstruction from oriented points can be formulated as a Poisson problem. The advantage

---

[1] MB System website:
http://www.ldeo.columbia.edu/res/pi/MB-System/html/mbsystem_home.html

of this formulation is that it considers all the points at once and is quite resilient to noise in the source data.

Like other solutions, Poison reconstruction is based on computing a 3D indicator function χ defined as 1 for points inside the model and 0 otherwise. The gradient of the indicator function $\nabla\chi$ is a zero vector everywhere but for points near the surface, where it equals the inward surface normal. The oriented point samples can therefore be viewed as samples of $\nabla\chi$: the indicator function χ can be defined as the scalar function whose gradient best approximates a vector field V defined by the samples:

$$min_\chi\| \nabla \chi - V \|.$$

By applying the divergence operator, this problem can be transformed into a standard Poisson problem: compute the scalar function whose Laplacian (divergence of the gradient) equals the divergence of a specified vector field:

$$\triangle\chi \equiv \nabla \cdot \nabla\chi = \nabla \cdot V.$$

Due to their interdisciplinary interest, A number of efficient and robust methods have been developed to solve Poisson problems. One additional insight in [13] is that, since we are interested in an accurate solution only near the reconstructed surface, we can use adaptive solvers in order to keep memory and time constraints proportional to the size of the reconstructed surface. Once we have an estimate of the indicator function χ, we can generate the reconstructed surface by extracting an appropriate isosurface. The isovalue is chosen so that the corresponding surface approximates the positions of the input samples. Extraction is then performed using an adapted Marching Cubes technique [14, 15]

Since the Poisson reconstruction technique works on oriented point sets, we used the normal estimation algorithm described in [16] to add normal information to the source points. This method constructs a K-nearest-neighbor-points graph over the input data (a Riemannian graph), and propagates a seed normal orientation using a minimum spanning tree over this graph.

The Poisson reconstruction step was run on the point cloud using a solver depth of 11, in order to produce a final surface with an average vertex distance of 1 meter.

4.4   Postprocessing and Improvements

Cutting. Since the Poisson algorithm solution is a watertight surface, we needed to filter out the surface components that were lying above-water: these were just a low resolution reconstruction artifact generated to close the mesh. This was easily done as a postprocessing step, by placing a cutting plane at the zero water depth level, and eliminating all the geometry above it.

Cleanup. At this point, The reconstructed surface still presented a significant amount of noise, in the form of spherical clusters or blobs disconnected from the main surface. This noise was due to the ENDURANCE science payload deployments, which caused spurious sonar returns every time the AUV stopped to collect data. These undesired sonar returns were coherent enough to pass the cleanup threshold during the clustering step. To address this issue, a second postprocessing identified all the connected components in the mesh below a certain size threshold, and eliminated them from the final surface.

Both the cutting plane and the connected components filter have been implemented as automated scripts running in MeshLab.

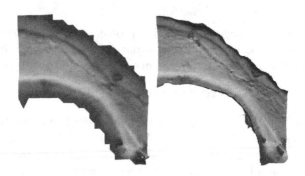

Fig. 4. From the left: detail of Narrows region reconstruction without shore line information; the same region reconstructed with shore line information

Shore Line Data Integration. In order to increase the reconstruction quality along the lake shores, we integrated shore line information generated by satellite imagery to the sonar point cloud. This dramatically reduced reconstruction artifacts for the lake zero-depth contour, as seen in figure 4.

4.5   dttools

The entire workflow has been implemented as a C++ toolkit, consisting of a set of programs implementing the raytracing, clustering and reconstruction steps of the data processing pipeline. All the programs use configuration files to specify their inputs, outputs and parameters. This allowed us to create scripts to run the entire pipeline easily for multiple configurations, and compare results in order to identify the optimal processing settings. The software source code is available online and can easily be adapted to different source datasets. [2]

5   Results

The final reconstruction of West Lake bonney can be seen in figures 5 and 6. The high resolution mesh consists of about 200k vertices and 400k faces. As the

---

[2] dttools website: http://code.google.com/p/dttools/

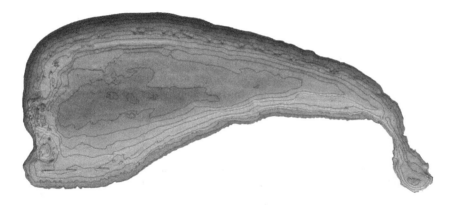

Fig. 5. The new 3D West Lake Bonney bathymetry

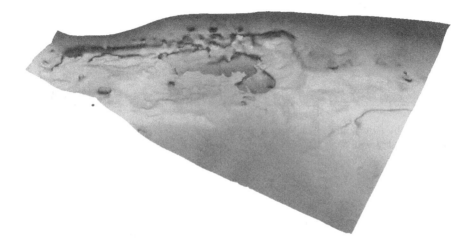

Fig. 6. A detail of the glacier face reconstruction

detail view of the glacier face shows, the reconstruction is fully 3D, while previous bathymetric models from the lake (generated by manually taken samples of the lake depth) were limited to approximate depth contours.

This improved model will allow for precise calculations of the lake water volume, and together with the data collected by the ENDURANCE science payload will allow a better understanding of the chemical and biological features of this extreme environment.

A drawback of the current reconstruction approach is its dependence on clustering and normal estimation parameters. Slight variations of those parameters could lead to incorrectly estimated normals for some regions, which would in turn cause errors in the final reconstruction. Thus, multiple iterations of the reconstruction process had to be run, varying parameters and visually inspecting the result in order to determine the best pipeline configuration. In particular, it was important to balance the parameters in order to get a good reconstruction in all three of the lake regions (the Glacier face, lake bottom, and Narrows). Reconstruction with slight differences had to be compared side by side. Given the high resolution of the models, the use of tiled displays (figure 7) helped in the process. In the future it may be beneficial to expand dttools, integrating tools that help in the iterative refinement of reconstruction parameters and comparison of corresponding results.

Fig. 7. Side-by-side analysis of West Lake Bonney reconstruction results on a high resolution display wall

References

1. Forney, C., Forrester, J., Bagley, B., McVicker, W., White, J., Smith, T., Batryn, J., Gonzalez, A., Lehr, J., Gambin, T., et al.: Surface reconstruction of Maltese cisterns using ROV sonar data for archeological study. Advances in Visual Computing, 461–471 (2011)
2. Fairfield, N., Kantor, G., Wettergreen, D.: Real-Time SLAM with Octree Evidence Grids for Exploration in Underwater Tunnels. Journal of Field Robotics 24, 3–21 (2007)
3. White, C., Hiranandani, D., Olstad, C.S., Buhagiar, K., Gambin, T., Clark, C.M.: The malta cistern mapping project: Underwater robot mapping and localization within ancient tunnel systems. J. Field Robot. 27, 399–411 (2010)
4. McEwen, R., Thomas, H., Weber, D., Psota, F.: Performance of an AUV navigation system at arctic latitudes
5. McPhail, S.: Autosub operations in the Arctic and Antarctic. In: Griffiths, G., Collins, K. (eds.) Proceedings of the Masterclass in AUV Technology for Polar Science, National Oceanography Centre, pp. 27–38. Society for Underwater Technology, Southampton (2006)
6. Jakuba, M.V., Roman, C.N., Singh, H., Murphy, C., Kunz, C., Willis, C., Sato, T., Sohn, R.A.: Long-baseline acoustic navigation for under-ice autonomous underwater vehicle operations. Journal of Field Robotics, 861–879 (2008)

7. Okabe, A., Boots, B., Sugihara, K.: Spatial Tesselations, Concepts and Applications of Voronoi Diagrams. Wiley Series in Probability and Mathematical Statistics. John Wiley & Sons, Chichester (1992)
8. Fairfield, N., Jonak, D., Kantor, G.A., Wettergreen, D.: Field results of the control, navigation and mapping system of a hovering AUV. In: Proceedings of the Unmanned Untethered Submersible Technology Conference (UUST 2007). AUSI, Durham (2007)
9. Papadopoulos, G., Kurniawati, H., Bin Mohd Shariff, A.S., Patrikalakis, N.M.: 3D-surface reconstruction for partially submerged marine structures using an Autonomous Surface Vehicle. In: 2011 IEEE/RSJ International Conference on Intelligent Robots and Systems, pp. 3551–3557 (2011)
10. Kazhdan, M., Bolitho, M., Hoppe, H., Burns, R.: Poisson surface reconstruction. In: Proceedings of the Fourth Eurographics Symposium on Geometry Processing, Eurographics Association, pp. 61–70 (2006)
11. Richmond, K., Febretti, A., Gulati, S., Flesher, C., Hogan, B.P., Murarka, A., Kuhlman, G., Sridharan, M., Johnson, A., Stone, W.C., Priscu, J., Doran, P.: Sub-Ice exploration of an antarctic lake: results from the ENDURANCE project. Arctic (2011)
12. Dushaw, B.D., Worcester, P.F., Cornuelle, B.D., Howe, B.M.: On equations for the speed of sound in seawater. Acoustical Society of America Journal 93, 255–275 (1993)
13. Hoppe, H.: Poisson surface reconstruction and its applications. In: Proceedings of the 2008 ACM Symposium on Solid and Physical Modeling (SPM 2008), p. 10 (2008)
14. Lorensen, W.E., Cline, H.E.: Marching cubes: A high resolution 3D surface construction algorithm. SIGGRAPH Comput. Graph. 21, 163–169 (1987)
15. Wilhelms, J., Van Gelder, A.: Octrees for faster isosurface generation. ACM Trans. Graph. 11, 201–227 (1992)
16. Hoppe, H., DeRose, T., Duchamp, T., McDonald, J., Stuetzle, W.: Surface reconstruction from unorganized points. ACM SIGGRAPH Computer Graphics 26, 71–78 (1992)

Off-road Terrain Mapping Based on Dense Hierarchical Real-Time Stereo Vision*

Thomas Kadiofsky, Johann Weichselbaum, and Christian Zinner

AIT Austrian Institute of Technology GmbH
Donau-City-Str. 1, 1220 Vienna, Austria
{thomas.kadiofsky,johann.weichselbaum,christian.zinner}@ait.ac.at

Abstract. We present a robust and fast method for on-line creation of local maps based on stereo vision for vehicles operating in off-road environments. A 3D vision sensor system with a high accuracy even at long ranges and wide field of view is used as the only sensor input. Due to the hierarchical mode of operation, dense stereo matching on image resolution 1200×525 and a disparity range of 280 becomes feasible at 10 fps. Beside of an achieved speedup factor of 6.47, a significant increase in the density of resulting disparity maps on real-world scenes has been achieved. Multiple captured views are aligned and integrated into a probabilistic elevation map suited for modeling dynamic environments. Efficient computations in the u-disparity-space and a stereo sensor model are at the core of the iterative update process.

Keywords: stereo vision, terrain mapping, real-time, elevation map.

1 Introduction

Mapping the environment of vehicles operating in various kinds of terrain is an enabling technology for advanced driver assistance systems (ADAS), for semi-autonomous operation (auto-pilot), or for fully autonomous operation, depending on the level of technological maturity and legal regulations. Possible platforms range from cars to trucks and various kinds of special mobile machines. The key concept of such local maps is to represent accumulated sensor data of a certain time span or distance up to present and keep this representation up-to-date at a refresh rate of at least several times per second.

Many approaches use laser range scanning sensor technology. However, less work can be found where accurate off-road mapping is solely realised from dense stereo vision depth data. Especially vision based stereo sensor systems can enable cost-efficient solutions with an increased chance for a broad area of application in the near future.

Therefore, we present a system for generating dense depth data of high quality and resolution from a stereo camera setup with at least 10 fps, and a robust

---

* The project has been funded by the Austrian Security Research Program KIRAS - an initiative of the Austrian Federal Ministry for Transport, Innovation and Technology (bmvit).

G. Bebis et al. (Eds.): ISVC 2012, Part I, LNCS 7431, pp. 404–415, 2012.

concept for on-line processing of this depth information to create local elevation maps. In a further step quality maps as a local measure of terrain roughness are created.

The remainder of this paper is outlined as follows. Section 2 gives an overview of relevant stereo vision systems and existing mapping approaches. Section 3 presents a stereo matching method and various enhancements necessary to make the system suitable for the demanding requirements. A detailed description of the robust mapping approach capable to work with stereo depth data and well-suited for efficient real-time implementations is the main part of Sect. 4. Sect. 5 concludes this work.

2   Related Work

Among the large number of existing stereo matching algorithms and techniques (an overview gives [1]), we use the real-time census based stereo matching system according to [2] as a basis for this work.

Recent approaches with adaptive support windows, whose computational complexity is independent from the support window size [3,4], achieve good results against stereo benchmarks and offer potential for real-time capable implementations, at least on GPU platforms. Such methods would be interesting candidates for future evaluations under real-world conditions.

In order to merge multiple views of a scene, the relative motion between the views has to be determined. Feature-based methods like [5] are very popular due to their lower computational requirements. Other approaches like the Iterative Closest Point (ICP) algorithm [6,7] recover these movements based on the geometry of the scene.

Gathering an appropriate description of the environment is a basic requirement for most autonomous mobile platforms. The choice of the representation heavily depends on the area to be modeled. 2D representations like occupancy grids are well suited for indoor applications, where the floor is planar and a binary classification of terrain is sufficient for safe navigation [8,9]. Although there also exist approaches for handling on-road outdoor scenes [10], moving in off-road environments requires more detailed information.

Elevation maps are discretization of a reference plane, where each cell stores the height of the terrain relative to this plane. Although these 2.5D representations suffer from some limitations like their inability to map multiple surface layers, e.g. bridges, this approach is quite popular due to its lower computational complexity compared to full 3D representations [11,12].

A 3D terrain description suited for large scale outdoor environments is discussed in [13]. A major advantage of such tree-based approaches is, that the extent of the modeled area needs not to be known in advance.

3   Hierarchical Dense Real-Time Stereo Vision

According to the concept of the proposed mapping method, our specific stereo vision system shall be the only source of depth information. The starting point of this section and reference system is a high speed census-based stereo matching module [2] with various performance optimizations like those in [14]. We describe special improvements and additional optimizations which have shown to be necessary in order to meet the requirements of this particular application, which are i) sufficient accuracy at long distances, ii) high density of the resulting depth maps, iii) interactive frame rates at high image resolutions.

Since it is our intention to show the feasibility of vision-based terrain mapping on standard computing platforms, we use only standard PC hardware where no GPU processing is employed. The chosen stereo matching algorithm has been shown to be also suitable for embedded DSP platforms – an important advantage for future deployment of such systems on a larger amount of vehicles.

3.1   Accurate Stereo Vision System

It is important that the stereo vision system provides sufficient depth accuracy even at larger distances. According to various considerations, we aim to achieve a depth resolution of about 2 meters at a distance of 100 meters ahead of the vehicle for adequate 3D mapping results under the intended operational conditions. This is a quite challenging requirement for a measurement principle based on triangulation and thus it leads to a design with a stereo baseline of 1.1 meters and the use of gray scale cameras with a resolution of 1600×1200 pixels. Using a focal length of 6 mm on a 2/3" class sensor results in a relatively large horizontal field of view of 72 degrees. Fig. 1 shows the camera assembly. The additional middle camera is also calibrated and it offers color images that can be used for other purposes. The stereo results provided are congruent to the rectified color camera images. The protective lens hoods are shaped to minimize glare and lens flare effects.

Fig. 1. Stereo camera system

Table 1 summarizes the depth resolution characteristics for slightly downscaled input images with a width of 1200 pixels. A moderate subpixel refinement effect of factor three is assumed, which we experienced to be realistic on scenes with an average amount of texture.

Such stereo geometries bring up disadvantages due to the large baseline. One problem that appears is the effect of perspective distortion on the ground level which causes large disparity gradients. Fig. 2(a) illustrates a typical scene from the viewpoints of the left and right camera, respectively. A characteristic figure

Table 1. Depth resolution at different disparities

disparity (pixel)	depth (meter)	depth step (meter)	subpixel step (meter)	lateral pixel size (meter)
280	5.94	0.021	0.007	0.007
160	10.4	0.064	0.021	0.012
75	22.1	0.29	0.098	0.026
33	50.4	1.52	0.50	0.06
20	83.1	4.15	1.38	0.1
17	97.8	5.75	1.91	0.117
16	103.9	6.49	2.16	0.125

describing the severeness of this effect is the ratio between the baseline and the mounting height T/h according to Fig. 2(b). A large T/h quotient makes block-based correlation schemes inefficient, causing the weak results in Fig. 3(c).

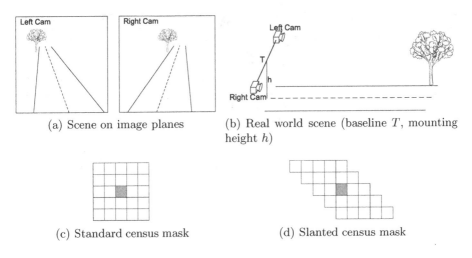

(a) Scene on image planes

(b) Real world scene (baseline T, mounting height h)

(c) Standard census mask

(d) Slanted census mask

Fig. 2. Perspective distortion on ground level and different types of census masks

As a countermeasure, we perform an additional correlation scheme using a slanted census mask (2(d)) for the left input image. The mask is designed to start becoming effective when the "standard" correlation mask (2(c)) becomes more and more ineffective. The stereo matching process has to be performed with both masks, superimposing the results of both to a common disparity map. Fig. 3 shows a distinctive enhancement in the resulting disparity images.

3.2   Fast Hierarchical Stereo Matching

The high image resolution, the large baseline, the resulting large disparity search range and the additional slanted mask correlation cause a computational effort

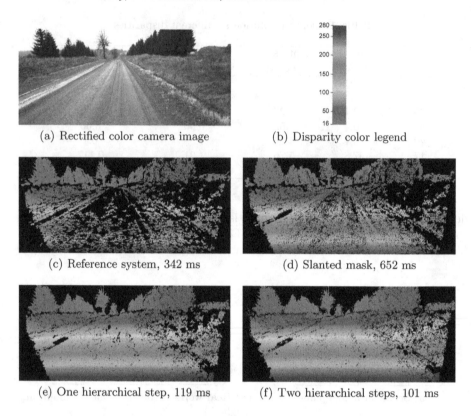

(a) Rectified color camera image

(b) Disparity color legend

(c) Reference system, 342 ms

(d) Slanted mask, 652 ms

(e) One hierarchical step, 119 ms

(f) Two hierarchical steps, 101 ms

Fig. 3. Resulting disparity maps and processing times per frame using standard mask (3(c)), slanted mask (3(d)) and slanted masks combined with hierarchical stereo matching (3(e) and 3(f))

which is too high to achieve interactive frame rates. Hierarchical multi-resolution processing is a well known approach to raise efficiency [15] [16]. We extended our dense stereo matching process by such an approach which yielded a significant performance gain as well as an increased quality of the results.

An initial disparity map is calculated on gaussian-blurred and down-scaled low-resolution input images, which effects also a reduction of the disparity search range. The resulting low-resolution disparity image serves as a hint image for the stereo matching process on the next level of the processing hierarchy. Here, the disparity search range can also be dramatically reduced because the hint image allows a dynamic definition of a search range around the disparity estimate. In Fig. 4(b) the reduced search space for one image line is outlined by areas surrounded with dotted lines. It is important that the refinement strategy operates not only in the disparity dimension but also in the lateral x- and y-dimensions. Otherwise, the effective lateral resolution of the result would not be better than that of the highest level in the hierarchy. To achieve this, the reduced disparity search interval based on a certain pixel of the hint map must be propagated

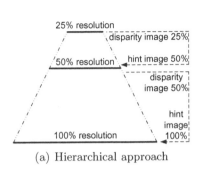

(a) Hierarchical approach

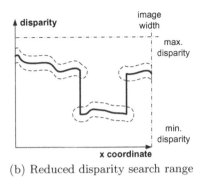

(b) Reduced disparity search range

Fig. 4. Hierarchical stereo matching

to a small lateral neighborhood. For the x-direction this is also illustrated in Fig. 4(b), but in fact, our concept contains an y-propagation too.

Table 2. Computation times of different configurations, varying in mask type (normal/slanted) and number of additional hierarchical steps (0 - 2). All values are in ms per frame

Hierarchical Steps	Reference System	Slanted Mask	Slanted Mask + 1 hierarchical step	Slanted Mask + 2 hierarchical steps
Step 0	341.91	652.20	54.82	35.25
Step 1	-	-	64.43	47.34
Step 2	-	-	-	18.18
Total	**341.91**	**652.20**	**119.25**	**100.77**

Table 2 summarizes the performance figures of selected stereo matching configurations. All configurations were tested on a standard PC with an Intel Xeon W3550@3.07GHz CPU with 4 cores and hyperthreading. We used input images with an image dimension of 1200×525 pixel and evaluated them for a disparity range of 280. The resulting disparity images referred to these configurations can be seen in Fig. 3(c) - Fig. 3(f).

The performance result of the configuration using two additional hierarchical steps shows a computation time of 100.77 ms, which is a final speed up factor of about 6.47 compared to the reference configuration. Using this configuration, we can achieve a frame rate of almost 10 fps without using any GPU capability.

4   Local Mapping

Given the sequence of the most recently captured point clouds and the corresponding stereo disparity images, our approach aims at the iterative fusion of the data into an elevation map describing the actual local vehicle environment. After estimating the motion between the latest measurement and its predecessor, it is integrated into the map.

4.1   Point Cloud Matching

The six degrees of freedom motion aligning a new range image with the previous one is computed with the ICP algorithm [6,17]. One of the ICP algorithm's conviniences is its adaptability to the given scene characteristics, e.g. we prefer the point-to-plane error metric in order to allow sliding planes, like the floor, against each other.

The varying spatial resolution of data obtained by stereo vision negatively effects the result of the registration. Therefore a sampling, e.g. a voxel grid filter, is necessary in order to achieve a uniform distribution of the points. Furthermore, as can be seen in table 1, the depth step is growing with increasing depth. For this reason, lower weights are assigned to correspondences of far distant points.

Hence, using the classification of the six stages of the ICP proposed in [7], our configuration of the ICP is as follows:

1. Points are *selected* in order to achieve a uniform spatial density.
2. Points are *matched* with the closest point in the other cloud.
3. Correspondences are *weighted* according to their depth.
4. Pairs of points are *rejected* if their distance is larger than δ_{dist} or if the angle between their normals is greater than δ_{angle}
5. The point-to-tangent-plane *error metric* is assigned to each correspondence.
6. The sum of squared errors is *minimized.*

In addition, after each iteration a damping is applied to δ_{dist} and δ_{angle} in order to increase the quality of the matches.

The quality of the resulting transformation heavily depends on the initial alignment of the two data sets. Since the motion between two subsequent frames is rather uniform in our application, a simple guess of the translation derived from the velocity and the frame-rate of the sensor is sufficient.

Frame-to-frame registration usually suffers from error accumulation, which leads to the well-known loop closure problem that has to be solved in order to obtain globally consistent maps. The aim of our application is constructing local maps of the vehicle's environment, so this effect can be neglected, because parts lying behind are discarded if they are not needed any more.

Experiments and Results. Since we are lacking any kind of ground truth data for the motion between images captured with our stereo system, an artificial loop $(F_1, F_2, \ldots F_{n-1}, F_n, F_{n-1}, \ldots, F_1)$ is constructed from n frames $(F_1, F_2, \ldots, F_n)$. The final global location after frame-to-frame registration of such a sequence should be equal to the identity. The deviation in the rotation is measured by $20 \cdot \|R - I_3\|$, which is an upper bound for the error of points at a distance of 20 meters. The error in the position is given by the norm of the translation $\|\mathbf{t}\|$. Table 3 shows the resulting rotational and translational error for different

Table 3. Results of the tests with different ICP configurations

Selection	Matching	Weighting	Error Metric	Damping	$20 \cdot \|R - I_3\|$	$\|t\|$ [m]
-	RC	-	P2T	no	3.66	6.22
-	RC	by depth	P2T	no	3.04	4.43
-	RC	by depth	P2T	yes	4.59	6.83
-	RC	by depth	P2P	no	20.79	40.86
-	NN	-	P2T	no	3.72	20.67
USS	NN	-	P2T	no	0.57	1.80
USS	NN	by depth	P2T	no	0.51	1.78
USS	**NN**	**by depth**	**P2T**	**yes**	**0.40**	**1.19**
USS	NN	by depth	P2P	yes	0.23	2.30

USS ... uniform spatial sampling, RC ... reverse calibration [18], NN ... nearest
neighbor search, P2P ... point-to-point, P2T ... point-to-tangent-plane

configurations for a sequence of 80 scans. A comparison of the obtained merged
point cloud together with the resulting trajectories is given in Fig. 5.

4.2   Local Elevation Maps from Stereo Data

An elevation map is a 2.5D grid-based discretisation of an area, that might be
used for navigation and other tasks of autonomous mobile platforms. A reference
plane is tiled in cells of equal size, and each of these cells contains the height
above this plane and possible other attributes.

Elevation Maps in Disparity-Space. Directly converting the 3D points of
a stereo measurement to an elevation map in a Cartesian coordinate system
yields only sparse maps due to the non-uniform spatial resolution of the range
data. Furthermore, complex sensor models have to be applied to describe the
uncertainty of each data point in Cartesian coordinates. Detecting obstacles
and other vertical structures is much more convenient in u-disparity-space [10].
A proper sensor model for correlation based stereo is discussed in [9], where
the region of uncertainty is bounded by ± 0.5 disparities. For this reason, in
our approach we compute a running weighted average of all v-values in a cell
of the u-disparity-plane, as well as the highest elevation in it. Furthermore,
a measurement (u_i, v_i, d_i) provides an upper bound for all cells (u_i, d_j) with
$d_j > d_i$. An example of such a map can be found in fig. 6(a).

Conversion to xyz-Space. Elevation maps with the xz-plane of the sensor
coordinate system as reference plane are generated by reading and interpolating the values at the corresponding positions in the map in u-disparity space.
In order to get a correct representation of the captured area, the elevation of
cells on the ground should be obtained by the averaging values, while vertical
structure should be represented by the maximum elevations. According to [12],

(a) Variant from row five in table 3   (b) Final configuration

Fig. 5. The figure shows samples of the resulting merged point clouds together with the trajectory. The positions resulting from the first and the second part of the artificial loop are marked red and blue.

cells containing upright scene parts can be detected by a high variance of their elevations. The variances are used to assign initial occupancy probabilities $P(n)$, represented by their *logOdds* $L(n)$, to each element n of the grid. An example can be seen in Fig. 6(b).

Map Update. Assuming the latest sensor reading has been successfully aligned with the previous one, it now has to be integrated in the current local elevation map. Each tile in both maps has one of the four states: *floor*, *vertical*, *bounded* and *unknown*. During the merging process, ambiguities may occur. These conflicts are resolved using the updated *logOdds* values

$$L(n|z_{1:t}) \;=\; L(n|z_{1:t-1}) + L(n|z_t)\,. \tag{1}$$

(a) Single elevation map in u-disparity-space (blue/red: low/high v)

(c) Camera Image

(b) Single elevation map in xyz-space (blue/red: low/high elevation)

(d) Meshed surface model

(e) Elevation map(elevation in meter)

(f) Quality map (blue/red: high/low quality; black: vertical obstacles)

Fig. 6. 6(a) and 6(b) are maps computed from a single frame. 6(d), 6(e) and 6(f) represent the local map for the scene shown in 6(c)

If the sign of *logOdds* changes, the old values are replaced by the more recent ones. These are regarded to be more accurate, because of possible changes in the scene or registration errors. All the other cells are updated according to the following rules from [12]:

$$\mu_{1:t} = \frac{\sigma_t^2 \mu_{1:t-1} + \sigma_{1:t-1}^2 \mu_t}{\sigma_{1:t-1}^2 + \sigma_t^2} \qquad \sigma_{1:t}^2 = \frac{\sigma_{1:t-1}^2 \sigma_t^2}{\sigma_{1:t-1}^2 + \sigma_t^2} \qquad (2)$$

In addition, $L(n)$ is reduced for non-empty cells n that are bounded in the new data, in order to let obstacles, that might have moved, vanish. Finally, the variances and occupancy values are used to assign quality values to the grid, that might be used in path planning. The lower the variance, the higher is the confidence and the quality of a region. Tiles representing parts of a road are likely to have a low variation, while the measurements on e.g. grassland are scattered and parts classified as vertical are impassable. An example of an elevation and corresponding quality map is shown in fig. 6(e) and 6(f).

5   Summary and Future Work

In this paper we presented a method that tackles the mapping task for autonomous driving solely based on stereo vision. A camera setup combined with an adapted stereo engine provide measurements satisfying the high requirements concerning accuracy and speed. We consider the achieved results very promising according to a real-time capable implementation within the ROS framework, which actually is an ongoing activity. Both contributed parts, enhanced real-time stereo matching as well as the mapping in the u-disparity space, are designed to run at high frame rates on standard processor platforms. Next steps are an invocation of the third camera. This enables exploitation of an additional smaller stereo baseline suitable for shorter ranges, whose results can be integrated into the hierarchical stereo processing scheme. The mapping module can be extended to accept additional depth sensor inputs simultaneously in order to realize redundant sensor configurations.

References

1. Scharstein, D., Szeliski, R.: A taxonomy and evaluation of dense two-frame stereo correspondence algorithms. International Journal of Computer Vision 47, 7–42 (2002)
2. Humenberger, M., Zinner, C., Weber, M., Kubinger, W., Vincze, M.: A fast stereo matching algorithm suitable for embedded real-time systems. Computer Vision and Image Understanding 114, 1180–1202 (2009)
3. De-Maeztu, L., Mattoccia, S., Villanueva, A., Cabeza, R.: Linear stereo matching. In: Proc. International Conference on Computer Vision, ICCV 2011. IEEE (2011)
4. Rhemann, C., Hosni, A., Bleyer, M., Rother, C., Gelautz, M.: Fast cost-volume filtering for visual correspondence and beyond. In: IEEE Computer Vision and Pattern Recognition, CVPR (2011)

5. Nister, D., Naroditsky, O., Bergen, J.: Visual odometry. In: IEEE Conference on Computer Vision and Pattern Recognition, pp. 652–659 (2004)
6. Besl, P.J., McKay, N.D.: A method for registration of 3-d shapes. IEEE Transactions on Pattern Analysis and Machine Intelligence, 239–256 (1992)
7. Rusinkiewicz, S., Levoy, M.: Efficient variants of the icp algorithm. In: Third International Conference on 3D Digital Imaging and Modeling, pp. 145–152 (2001)
8. Elfes, A.: Occupancy grid: A stochastic spatial representation for active robot perception. In: 6th Conference on Uncertainty in Artificial Intelligence, pp. 136–146 (1990)
9. Murray, D., Little, J.: Using real-time stereo vision for mobile robot navigation. Autonomous Robots 8, 161–171 (2000)
10. Perrollaz, M., Yoder, J.D., Spalanzani, A., Laugier, C.: Using the disparity space to compute occupancy grids from stereo-vision. In: IEEE International Conference on Intelligent Robots and Systems, pp. 2721–2726 (2010)
11. Kweon, I.S., Kanade, T.: High-resolution terrain map from multiple sensor data. IEEE Transactions on Pattern Analysis and Machine Intelligence 14, 278–292 (1992)
12. Pfaff, P., Triebel, R., Burgard, W.: An efficient extension to elevation maps for outdoor terrain mapping and loop closing. International Journal of Robotics Research 26, 217–230 (2007)
13. Wurm, K.M., Hornung, A., Bennewitz, M., Stachniss, C., Burgard, W.: Octomap: A probabilistic, flexible, and compact 3d map representation for robotic systems. In: IEEE International Conference on Robotics and Automation (2010)
14. Zinner, C., Humenberger, M., Ambrosch, K., Kubinger, W.: An Optimized Software-Based Implementation of a Census-Based Stereo Matching Algorithm. In: Bebis, G., Boyle, R., Parvin, B., Koracin, D., Remagnino, P., Porikli, F., Peters, J., Klosowski, J., Arns, L., Chun, Y.K., Rhyne, T.-M., Monroe, L. (eds.) ISVC 2008, Part I. LNCS, vol. 5358, pp. 216–227. Springer, Heidelberg (2008)
15. Hung, Y., Chen, C., Hung, K., Chen, Y., Fuh, C.: Multipass hierarchical stereo matching for generation of digital terrain models from aerial images. In: Machine Vision Applications, pp. 280–291. Springer (1998)
16. Sun, C.: A fast stereo matching method. In: Digital Image Computing: Techniques and Applications (1997)
17. Chen, Y., Medioni, G.: Object modeling by registration of multiple range images. In: 1991 IEEE International Conference on Robotics and Automation, pp. 2724–2729 (1991)
18. Blais, G., Levine, M.D.: Registering multiview range data to create 3d computer objects. IEEE Transactions on Pattern Analysis and Machine Intelligence 17, 820–824 (1995)

Using Synthetic Data for Planning, Development and Evaluation of Shape-from-Silhouette Based Human Motion Capture Methods

Rune Havnung Bakken

Faculty of Informatics and e-Learning,
Sør-Trøndelag University College, Trondheim, Norway
rune.h.bakken@hist.no

Abstract. The shape-from-silhouette approach has been popular in computer vision-based human motion analysis. For the results to be accurate, a certain number of cameras are required and they must be properly synchronised. Several datasets containing multiview image sequences of human motion are publically accessible, but the number of available actions is relatively limited. Furthermore, the ground truth for the location of joints is unknown for most of the datasets, making them less suitable for evaluating and comparing different methods. In this paper a toolset for generating synthetic silhouette data applicable for use in 3D reconstruction is presented. Arbitrary camera configurations are supported, and a set of 2605 motion capture sequences can be used to generate the data. The synthetic data produced by this toolset is intended as a supplement to real data in planning and development, and to fascilitate comparative evaluation of shape-from-silhouette-based motion analysis methods.

1 Introduction

Computer vision-based human motion capture has been a highly active field of research during the last two decades [14]. Both monocular, stereo and multiview approaches have been explored extensively. Multiview approaches put some special demands on the image aquisition process. In order to achieve accurate results the cameras used must be calibrated and properly synchronised, and with a higher number of cameras the complexity of those tasks increase. That complexity and the relatively high cost of the necessary equipment have lead many researchers to rely on publically available datasets in order to perform experiments.

The selection of publically available multiview datasets is limited. Some of the datasets are created with a specific application in mind, for instance evaluation or action recognition, and this can limit their applicability to other application areas because of the camera configuration used, the types of actions performed, the number of subjects, or the length of the sequences. Consequently, it can be problematic for researchers to get access to data suitable for a specific human motion analysis task.

G. Bebis et al. (Eds.): ISVC 2012, Part I, LNCS 7431, pp. 416–426, 2012.

Human motion is governed by the underlying skeletal articulation, and in the context of human motion analysis pose estimation is defined as the process of finding the configuration of that skeletal structure in one or more frames. A possible alternative to relying on datasets with real image data is to use synthetically generated images. A bone hierarchy can be used to control the motion of an avatar, and multiview images of the sequence can be rendered for use in quantitative evaluation of a pose estimation algorithm. Since the underlying skeletal structure is known for these data, it can be used as ground truth in the evaluation. The problem is, however, that evaluation with synthetic data historically have been done in an ad hoc manner, making it difficult to reproduce the results and do comparative evaluation.

Shape-from-silhouette has been a very popular approach for doing 3D reconstruction in multiview human motion analysis since it was introduced by Laurentini [12]. Algorithms based on the shape-from-silhouette paradigm use binary silhouette images as input.

Goal: To make a large number of synthetic image sequences available for use in planning, development and evaluation of shape-from-silhouette-based human motion analysis research.

Contribution: A set of tools that makes it easy to produce silhouette image sequences with arbitrary camera setups by employing the 2605 real motion captures from the CMU motion capture database [6]. Furthermore, the toolset generates a synthetic ground truth that can be used for evaluation of new shape-from-silhouette-based methods. The toolset is made freely available to fascilitate comparative evaluation.

The structure of the paper is as follows: in section 2 we will review some related work on synthetic motion data. Section 3 details the proposed toolset for synthetic silhouette generation, and we explore some possible usage scenarios for the generated data. In section 4 we present the results of using the generated data for evaluating a shape-from-silhouette-based human motion analysis algorithm. Section 5 contains a discussion of the assumptions made when using synthetic data and outlines some further work, and section 6 concludes the paper.

2   Related Work

Using synthetic data for research on human motion analysis is not a novel idea, and is commonly done to alleviate some of the limitations of physical cameras as was demonstrated by Franco et al. [7]. They presented a distributed approach for shape-from-silhouette reconstruction and wanted to test its performance on a higher number of cameras than the four physical cameras they had available, so they used images from between 12 and 64 synthetic viewpoints instead.

Cheung et al. [5] presented a shape-from-silhouette algorithm that considers temporal correspondences when doing 3D reconstruction from image data. The algorithm was applied to both static and articulated objects, and was used for aquiring detailed kinematic models of humans and for markerless human motion tracking. Experiments with synthetic data were done to obtain a quantitative comparison of different aspects of the algorithm.

Mündermann et al. [16] used synthetic data to explore how camera setups influence the performance of shape-from-silhouette methods in biomechanical analysis. Generating silhouette images of a laser scanned human model allowed them to simulate a wide range of different camera setups to study how the accuracy of the reconstruction is affected by the number and positions of cameras, and the location of the subject within the viewing volume.

The quality of the 3D reconstruction from a shape-from-silhouette algorithm is dependent on the amount of noise in the input silhouette images. Landabaso et al. [11] presented a novel algorithm that exibits greater robustness when used with inconsistent silhouettes. In order to produce quantitative results for the algorithm they used a dataset consisting of synthetic images in order to achieve better control of noise levels and occlusion, and for easier comparison with a known ground truth.

Gkalelis et al. [9] used motion capture data from the CMU database to animate ten different avatars performing four different actions. The synthetic sequences were used as a supplement for real data in evaluating their continuous human movement recognition method. The data were rendered using a single viewpoint.

Common for all the aforementioned applications of synthetic data is that the data in question were not drawn from publically available databases. The synthetic images were either generated in an ad hoc manner or came from closed datasets, making it difficult for other researchers to reproduce the results. An effort to standardise the generation of synthetic human silhouettes was made by Piérard and Van Droogenbroeck [17]. They presented a technique for building databases of annotated synthetic silhouettes for use in training learning-based and example-based pose estimation approaches. The generated silhouettes are automatically annotated with pose information and identication of body parts. The images are produced by randomly generated poses of an avatar with Make-Human [3]. Consequently, extra measures must be taken to ensure that the silhouettes are realistic, and only single poses are generated rather than sequences of actual motion patterns.

2.1   Multiview Datasets

Several datasets with multiview images of human motion are publically available. These contain both real and synthetic data.

The HumanEVA [20] datasets are specifically aimed at evaluation of human motion capture methods. The HumanEVA-I dataset consists of 48 image sequences, captured with four greyscale and three colour cameras. Six actions are repeated twice by four different subjects. The HumanEVA-II dataset has two image sequences with one subject each, captured with four synchronised colour cameras.

i3DPost [8] is a dataset with 96 sequences, from eight synchronised colour cameras in a blue-screen studio environment. The data include a wide variety of actions performed by eight different subjects. The CVSSP-3D [22] dataset contains seventeen image sequences, with the same studio setup as for i3DPost. Two subjects each perform a set of actions. The first subject wears different costumes and performs actions related to walking. The second subject performs a set of dance routines. The IXMAS [23] dataset consists of 33 image sequences that were captured with five colour cameras. Eleven subjects perform fourteen everyday actions that are repeated three times. Clean silhouette masks are available for these datasets.

In the MuHAVi [21] dataset there are 119 image sequences. The images were captured with eight colour cameras, but the image streams were not synchronised. There are seven subjects performing seventeen different actions. For a subset of the image sequences silhouettes have been manually annotated (two cameras with five actions and two actors).

The CMU MoBo [10] dataset was originally intended for research on gait recognition, but has been used for other application areas as well. There are 100 image sequences in the dataset. 25 subjects were filmed with six synchronised colour cameras while walking on a treadmill. Four different actions were captured, and noisy silhouette masks are included.

The ViHASi [18] dataset stands out from those described above, in that it consists entirely of synthetic data. Nine virtual subjects performing twenty different actions were animated using a character animation software package. Configurations of 12-40 virtual cameras were placed in two circular formations around the subjects.

A summary of the findings on publically available multiview datasets is shown in Table 1.

Table 1. Publically available multiview datasets. Comparison of number of cameras, subjects, sequences, actions, synchronisation and image types.

Name	Cam.	Sub.	Seq.	Act.	Synch.	Image type
HumanEVA-I	7	4	48	6	Yes	Real
HumanEVA-II	4	2	2	4	Yes	Real
IXMAS	5	11	33	14	Yes	Real w/ silhouettes
i3DPost	8	8	96	12	Yes	Real w/ silhouettes
CVSSP-3D	8	2	17	17	Yes	Real w/ silhouettes
MuHAVi	8	7	119	17	No	Real w/ silhouettes (subset)
CMU MoBo	6	25	100	4	Yes	Real w/ silhouettes
ViHASi	12-40	9	180	20	Yes	Synthetic silhouettes

3   Toolset for Generating Synthetic Silhouettes

The proposed toolset is implemented as a plugin for the open source 3D content creation suite Blender (`http://blender.org`) and a standalone analysis tool.

3.1   Blender Module

This section describes the features of the Blender plugin. The toolset's main interface can be seen in Fig. 1.

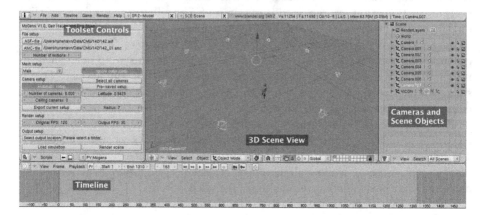

Fig. 1. The main interface of the Blender module

Motions. As mentioned earlier, the CMU motion capture database contains 2605 sequences. These are divided into to six categories (human interaction, interaction with environment, locomotion, physical activities and sport, situations and scenarios, and test motions). The proposed toolbox supports importing motion sequences in the ASF/AMC format. Blender has import plugins for the C3D and BVH motion capture formats, so support for these could be added later. Some of the sequences in the CMU dataset contains noise, particularly around the extremities. Consequently, it is possible to lock the outer joints when importing an action to reduce the effects of the noise.

Multiple motion sequences can be imported with separate avatars. This works both for sequences that contain multiple performers, and for combining sequences with a single performer. Different actions can be stitched together for a single avatar. This motion stitching functionality relies on Blender's interpolation of the sequences, and works best for combinations of actions where the end pose in the first sequence is similar to the start pose in the second sequence.

The standard frame rate for the motion captures in the CMU database is 120 frames per second. The proposed toolset supports downsampling the frame rate when rendering the silhouette images.

For each time step in a rendered sequence the locations of the joints defined by the input motion capture file is calculated, and exported along with the silhouette images. These joint locations can be used as ground truth when evaluating motion analysis techniques that use the silhouette images as input.

Meshes. An unclothed avatar is included with the toolset. This mesh was created with MakeHuman [3], a tool for generating anatomically correct human models with simple controls for setting age, gender, body shape and ethnicity. The mesh used in the proposed toolset is of a fairly athletic, young adult male. It is relatively easy to add new avatars to the toolset. Meshes in the open OBJ format are supported, given that the avatar's pose matches the rest pose in Blender's skeletal animation system. The skeletal structure is automatically skinned with the mesh model using Blender's bone heat method, an implementation of the approach outlined by Baran and Popović [2].

Camera Setup. The toolbox supports simulating arbitrary camera setups. The automatic setup mode will generate a set of cameras arranged in a circle, facing the origin. The user can control the radius of the circle and the latitude of the cameras. It is also possible to add cameras in the ceiling. The default resolution for the virtual cameras is 800×600 pixels. If the automatically generated setup is not satisfactory the user is free to make any necessary modifications. The cameras can be moved and rotated in Blender's 3D view to create an adequate camera setup.

The calibration information for the camera setup is exported along with the rendered silhouette images. The calibration information includes focal lengths, principal points, positions, and rotations for the cameras. An exported camera setup can be loaded and reused for subsequent simulations.

Analysis Tool. Bundled with the toolset is an analysis tool for evaluating human motion capture methods that use the generated silhouette images as input. Estimated joint locations can be compared to the synthetic ground truth data generated by the toolset. The analysis tool can visualise 3D error, as well as errors in the x, y, and z dimensions separately. The analysis tool calculates the mean error and standard deviation for all the joints per time step, and mean error and standard deviation for the entire sequence. Box and whisker plots (following the 1.5 IQR convention) can be generated to examine the dispersion of the data, and identify outliers.

3.2 Usage Scenarios

A number of possible usage scenarios for the toolset can be envisioned.

Planning. The camera equipment required for multiview human motion analysis is expensive, so it can be useful to experiment with different camera setups to find the most suitable configuration for a specific application before making a purchase. The rigging and calibration of cameras is a time-consuming process. Using the proposed toolset to experiment with camera setups is faster, and once a satisfactory configuration has been found it can be replicated in the lab.

Development. Not all researchers have access to a multicamera capture lab and are reliant on using data provided by others. Having a flexible system for generating data like the one presented in this paper can make human motion analysis research more accessible. Coupled with the large action corpus in the CMU database the proposed toolset provides a good starting point for development of algorithms for different applications.

Evaluation. There are few datasets with known ground truth available. Of the datasets examined in Sect. 2, only the HumanEVA datasets are specifically aimed at evaluation and comparison of human motion analysis methods. As Mündermann et al. [15] demonstrated it can be difficult to utilise all the data in the HumanEVA datasets due to problems with background subtraction, further limiting the availability of data for evaluation purposes. The proposed toolset can generate datasets with reproducable results for comparison of different algorithms. The actual data used for evaluation need not be shared, only the calibration parameters and information on which sequences were used. Along with the silhouette images a synthetic ground truth is generated. The ground truth data consists of the locations of the joints in the imported skeleton definition.

4   Case Study: Real-Time Pose Estimation

The proposed toolset was used to evaluate the performance of the real-time pose estimation method described in [1]. The method consists of the following steps. First, a volumetric visual hull is reconstructed from silhouette images. Second, the visual hull is thinned to a skeleton representation. Next, a tree structure is built from the skeleton voxels, and spurious branches are removed. The extremities (hands, feet, and head) are identified, and the tree is segmented into body parts. The final step is to label hands and feet as left or right.

To evaluate the algorithm outlined above, a dance sequence from the CMU database was used (subject 55, trial 1) to generate a test set of silhouette images. The sequence consists of 1806 frames, and has the subject doing a series of pirouettes around the centre of the viewing volume. The ASF and AMC files were loaded into the Blender plugin. This automatically built an armature and skinned it with the chosen avatar mesh. A camera setup of eight cameras was chosen. Seven of the cameras were placed in a circular formation around the subject, and the eighth placed in the ceiling, pointing down. The sequence was downsampled to 30 fps, and silhouettes rendered. Fig. 2 shows the data at different stages of the experiment.

After the outlined algorithm had produced an estimate of the extremity positions for the entire sequence, the analysis tool was used to evaluate those results. The generated ground truth data was loaded into the analysis tool along with the estimates, and a number of graphs were generated, as shown in Fig. 3.

Figure 3a shows the mean error for all extremities per frame of the sequence. The mean error for the entire sequence was 47.8 mm (st. dev. 27.3 mm). Fig. 3b illustrates a problem with the right finger tip. In a considerable number of frames

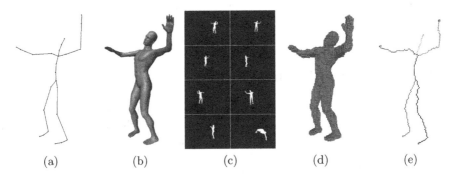

Fig. 2. Frame 37 from the case study. (a) Ground truth skeletal structure with joint positions. (b) Mesh model fitted to skeletal structure. (c) Rendered silhouettes from eight viewpoints. (d) Reconstructed visual hull. (e) Labelled tree extracted from visual hull.

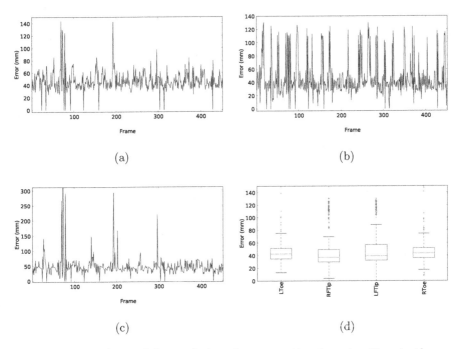

Fig. 3. Four graphs used for analysing the pose estimation algorithm in the case study.(a) Mean error per frame for all extremities. (b) Error for the right finger tip. (c) Error for the right toe tip. (d) Box and whisker plot for all four extremities.

the position estimate is around 120 mm from the ground truth. The cause is that the algorithm chooses the thumb branch as the extremity in those frames. In Fig. 3c a number of spikes can be seen in the estimate of the position of the right toe. This is caused by noise in the reconstructed visual hull between the feet, and the right foot is pointing backwards in these frames. Finally, a box and whisker plot for all four extremities can be seen in Fig. 3d.

5   Discussion and Future Work

The synthetic data produced by the proposed toolset is intended as a supplement to the real datasets available. Naturally there are some differences between real and synthetic data that need to be considered when using synthetic data. There are non-trivial issues with reconstruction and image segmentation when using real data. Depending on the quality of the camera calibration, the reprojection errors in the 3D reconstruction can be problematic. Significant advances are being made on background subtraction algorithms [4], but segmenting the images into binary silhouettes is still a difficult task, that invariably includes some noise.

In comparison the synthetic data have perfectly clean silhouettes and exactly known extrinsic calibration parameters and no lens distortion. It is important to keep in mind that this is an unrealistically ideal situation, and that results produced using synthetic data might not directly transfer to real world applications. However, if the assumption is made that calibration is accurate and background subtraction is robust, then using synthetic silhouettes is a reasonable approximation.

The toolset currently only has one avatar, a hairless and unclothed male mesh. Loose fitting clothing and long hair can have a significant impact on the shape of the reconstructed visual hull, and on the results of further processing. The range of available avatars should be extended with representatives of both genders, with different types of clothing and hair, and with different body shapes.

The automatic skinning method supplied by Blender can in some cases lead to poor results, because it is dependent on the armature and mesh being aligned before skinning. For some sequences this causes the limbs of the mesh to intersect during the motion. Better skinning algorithms should be investigated, for instance the one presented by Rau and Brunnett [19]. The current support for combining motions is rather primitive and relies on Blender's interpolation functionality. This should be improved, for instance by implementing a more robust motion stitching algorithm like the one presented by Li et al. [13].

6   Concluding Remarks

A toolset for generating synthetic silhouette data for use in human motion analysis has been presented. The toolset consists of a Blender plugin for rendering silhouette images of an avatar animated using motion capture sequences. These motion capture sequences are drawn from the CMU database that contains 2605 trials in six categories. The plugin has support for arbitrary camera configurations, basic motion stitching, and ground truth data is exported along with the generated silhouettes. An analysis tool that facilitates evaluation and comparison of shape-from-silhouette-based pose estimation methods is included. The analysis tool can visualise 3D error, errors in each of the x, y, and z dimensions, and generate box and whisker plots. The toolset is available under the BSD license at:

http://code.google.com/p/mogens/

Acknowledgements. The author wishes to thank Geir Hauge, Dag Stuan, and Vegard Løkås for their invaluable assistance with the implementation of the software presented in this paper. The data used in this project was obtained from mocap.cs.cmu.edu. The database was created with funding from NSF EIA-0196217.

References

1. Bakken, R.H., Hilton, A.: Real-Time Pose Estimation Using Tree Structures Built from Skeletonised Volume Sequences. In: Csurka, G., Braz, J. (eds.) Proceedings of the International Conference on Computer Vision Theory and Applications, pp. 181–190. SciTePress (2012)
2. Baran, I., Popović, J.: Automatic Rigging and Animation of 3D Characters. ACM Transactions on Graphics 26(3), 72:1–72:8 (2007)
3. Bastioni, M., Re, S., Misra, S.: Ideas and methods for modeling 3D human figures. In: Proceedings of the 1st Bangalore Annual Compute Conference, pp. 10:1–10:6. ACM (2008)
4. Benezeth, Y., Jodoin, P.M., Emile, B., Laurent, H., Rosenberger, C.: Comparative study of background subtraction algorithms. Journal of Electronic Imaging 19(3), 033003:1–033003:12 (2010)
5. Cheung, K.M.G., Baker, S., Kanade, T.: Shape-From-Silhouette Across Time Part II: Applications to Human Modeling and Markerless Motion Tracking. International Journal of Computer Vision 63(3), 225–245 (2005)
6. CMU: Graphics Lab Motion Capture Database, http://mocap.cs.cmu.edu/
7. Franco, J.S., Menier, C., Boyer, E., Raffin, B.: A Distributed Approach for Real Time 3D Modeling. In: Proceedings of the Conference on Computer Vision and Pattern Recognition Workshop, pp. 31–38. IEEE (2004)
8. Gkalelis, N., Kim, H., Hilton, A., Nikolaidis, N., Pitas, I.: The i3DPost multi-view and 3D human action/interaction database. In: Proceedings of the Conference for Visual Media Production, pp. 159–168. IEEE (2009)
9. Gkalelis, N., Tefas, A., Pitas, I.: Combining Fuzzy Vector Quantization With Linear Discriminant Analysis for Continuous Human Movement Recognition. IEEE Transactions on Circuits and Systems for Video Technology 18(11), 1511–1521 (2008)
10. Gross, R., Shi, J.: The CMU Motion of Body (MoBo) Database. Tech. Rep. 1. Carnegie Mellon University (2001)
11. Landabaso, J., Pardas, M., Casas, J.: Shape from inconsistent silhouette. Computer Vision and Image Understanding 112(2), 210–224 (2008)
12. Laurentini, A.: The Visual Hull Concept for Silhouette-Based Image Understanding. IEEE Transactions on Pattern Analysis and Machine Intelligence 16(2), 150–162 (1994)

13. Li, L., Mccann, J., Faloutsos, C., Pollard, N.: Laziness is a virtue: Motion stitching using effort minimization. In: Proceedings of Eurographics, pp. 87–90 (2008)
14. Moeslund, T.B., Hilton, A., Krüger, V.: A survey of advances in vision-based human motion capture and analysis. Computer Vision and Image Understanding 104, 90–126 (2006)
15. Mündermann, L., Corazza, S., Andriacchi, T.P.: Markerless human motion capture through visual hull and articulated ICP. In: Proceedings of the NIPS Workshop on Evaluation of Articulated Human Motion and Pose Estimation (2006)
16. Mündermann, L., Corazza, S., Chaudhari, A.M., Alexander, E.J., Andriacchi, T.P.: Most favorable camera configuration for a shape-from-silhouette markerless motion capture system for biomechanical analysis. In: Proceedings of SPIE-IS&T Electronic Imaging, vol. 5665, pp. 278–287 (2005)
17. Pierard, S., Van Droogenbroeck, M.: A technique for building databases of annotated and realistic human silhouettes based on an avatar. In: Proceedings of Workshop on Circuits, Systems and Signal Processing, pp. 243–246. Citeseer (2009)
18. Ragheb, H., Velastin, S., Remagnino, P.: ViHASi: Virtual Human Action Silhouette Data for the Performance Evaluation of Silhouette-based Action Recognition Methods. In: Proceeding of the 1st ACM Workshop on Vision Networks for Behaviour Analysis, pp. 77–84. ACM (2008)
19. Rau, C., Brunnett, G.: Anatomically Correct Adaption of Kinematic Skeletons to Virtual Humans. In: Richard, P., Kraus, M., Laramee, R.S., Braz, J. (eds.) Proceedings of the International Conference on Computer Graphics Theory and Applications, pp. 341–346. SciTePress (2012)
20. Sigal, L., Balan, A.O., Black, M.J.: HumanEva: Synchronized Video and Motion Capture Dataset and Baseline Algorithm for Evaluation of Articulated Human Motion. International Journal of Computer Vision 87(1-2), 4–27 (2010)
21. Singh, S., Velastin, S., Ragheb, H.: MuHAVi: A Multicamera Human Action Video Dataset for the Evaluation of Action Recognition Methods. In: Proceedings of the Seventh IEEE International Conference on Advanced Video and Signal Based Surveillance, pp. 48–55 (2010)
22. Starck, J., Hilton, A.: Surface Capture for Performance-Based Animation. IEEE Computer Graphics and Applications 27(3), 21–31 (2007)
23. Weinland, D., Ronfard, R., Boyer, E.: Free viewpoint action recognition using motion history volumes. Computer Vision and Image Understanding 104(2-3), 249–257 (2006)

Moving Object Detection by Robust PCA Solved via a Linearized Symmetric Alternating Direction Method

Charles Guyon, Thierry Bouwmans, and El-Hadi Zahzah

Department of Mathematics, Images and Applications
University of La Rochelle, France
tbouwman@univ-lr.fr

Abstract. Robust Principal Components Analysis (RPCA) gives a suitable framework to separate moving objects from the background. The background sequence is then modeled by a low rank subspace that can gradually change over time, while the moving objects constitute the correlated sparse outliers. RPCA problem can be exactly solved via convex optimization that minimizes a combination of the nuclear norm and the l_1-norm. This convex optimization is commonly solved by an Alternating Direction Method (ADM) that is not applicable in real application, because it is computationally expensive and needs a huge size of memory. In this paper, we propose to use a Linearized Symmetric Alternating Direction Method (LSADM) to achieve RPCA for moving object detection. LSADM in its fast version requires less computational time than ADM. Experimental results on the Wallflower and I2R datasets show the robustness of the proposed approach.

1 Introduction

The detection of moving objects is a key issue in a video surveillance system with static cameras. This detection is commonly done using foreground detection. This basic operation consists of separating the moving objects called "foreground" from the static information called "background" [1]. In 1999, Oliver et al. [2] are the first authors who model the background by Principal Component Analysis (PCA). PCA provides a robust model of the probability distribution function of the background, but not of the moving objects while they do not have a significant contribution to the model. The main limitation of this model [3] is that the size of the foreground object must be small in relation to the size of the image, and don't appear in the same location during a long period in the training sequence, that is the object does not have to be present more than half of the training sequence. Recent research on robust PCA [4][5] can be used to alleviate these limitations. For example, Candes et al. [5] proposed a convex optimization to address the robust PCA problem. The observation matrix is assumed represented as:

$$A = L + S \qquad (1)$$

where L is a low-rank matrix and S must be sparse matrix with a small fraction of nonzero entries. This research seeks to solve for L with the following optimization problem:

$$\min_{L,S} \ ||L||_* + \lambda ||S||_1 \ \ \text{subj} \ \ A = L + S \qquad (2)$$

G. Bebis et al. (Eds.): ISVC 2012, Part I, LNCS 7431, pp. 427–436, 2012.

where $||.||_*$ and $||.||_1$ are the nuclear norm and l_1-norm, respectively, and $\lambda > 0$ is an arbitrary balanced parameter. This approach assumed that all entries of the matrix to be recovered are exactly known via the observation. The other assumption is that the distribution of corruption should be sparse and random enough without noise. Under these minimal assumptions, this approach perfectly recovers the low-rank and the sparse matrices. The optimization in Equation (2) can be solved as a general convex optimization problem by any iterative thresholding techniques [6]. However, the iterative thresholding scheme converges extremely slowly. To alleviate this slow convergence, Lin et al. [7] proposed the accelerated proximal gradient (APG) algorithm and the gradient-ascent algorithm applied to the dual of the problem in the Equation (2). However, these algorithms are all the same to slow for real application. More recently, Lin et al. [8] proposed two algorithms based on augmented Lagrange multipliers (ALM). The first algorithm is called exact ALM (EALM) method that has a Q-linear convergence speed, while the APG is only sub-linear. The second algorithm is an improvement of the EALM that is called inexact ALM (IALM) method, which converges practically as fast as the exact ALM, but the required number of partial SVDs is significantly less. The IALM is at least five times faster than APG, and its precision is also higher. However, the direct application of ALM treats Equation (2) as a generic minimization problem and doesn't take into account its separable structure emerging in both the objective function and the constraint. Hence, the variables S and L are minimized simultaneously. Yuan and Yang [9] proposed to alleviate this problem by the Alternating Direction Method (ADM) which minimizes the variables L and S serially. However, this method is computationally expensive and needs a huge size of memory. Recently, Ma[10] and Goldfarb et al. [11] proposed a Linearized Symmetric Alternating Direction Method (LSADM) for minimizing the sum of two convex functions. This method requires at most $O(1/\epsilon)$ iterations to obtain an ϵ-optimal solution, while its fast version called Fast-LSADM requires at most $O(1/\sqrt{\epsilon})$ with little change in the computational effort required at each iteration. These algorithms [10] have shown encouraging qualitative results on background extraction. These properties allow us to use and evaluate it for moving object detection by using RPCA. The rest of this paper is organized as follows. In Section 2, we present how the background and the moving objects can be separated by using RPCA. Then, the proposed moving object detection method with the LSADM algorithm is presented. Finally, performance evaluation and comparison are given in Section 3.

2 Moving Object Detection by RPCA Solved via LSADM

2.1 Background and Moving Object Separation via RPCA

Denote the training video sequences $D = \{I_1, ...I_N\}$ where I_t is the frame at time t and N is the number of training frames. Let each pixel (x,y) be characterized by its intensity in the grey scale. The decomposition involves the following model:

$$D = L + S \tag{3}$$

where L and S are the low-rank component and sparse component of D, respectively. The matrix L contains the background and the matrix S contains mostly zero columns, with several non-zero ones corresponding to the moving objects. The matrices L and S can be recovered by the convex program based on the Alternating Direction Method (ADM). For this work on moving object detection, we propose to use the Linearized Symmetric Alternating Direction Method (LSADM) proposed in [10][11].

2.2   Linearized Symmetric Alternating Direction Method

This sub-section briefly reminds the principle of LSADM developed in [10][11]. Theorically, the RPCA problem can be initially formulated as a linear constrained convex program [10].

$$min(f(x) + g(y)) \quad subj \quad x - y = 0 \tag{4}$$

Solving this problem by ADM, one operates on the following augmented Lagrangian function:

$$\mathcal{L}_\mu(x, y; \lambda) = f(x) + g(y) + <\lambda, x - y> + \frac{1}{2\mu}|||x - y||^2 \tag{5}$$

with respect to x and y, i.e, it solves the subproblem:

$$(x^k, y^k) = argmin_{x,y} \mathcal{L}_\mu(x, y; \lambda^k) \tag{6}$$

and then the Lagrange multipliers λ^k are updated as follows:

$$\lambda^{k+1} = \lambda^k + \frac{1}{\mu}(x^k - y^k) \tag{7}$$

Minimizing $\mathcal{L}_\mu(x, y; \lambda)$ with respect to x and y alternatingly is often easy. Such an alternating direction method for solving Equation (4) is given below as Algorithm 1.

Algorithm 1: Alternating Direction Method (ADM)

Initialization: Choose μ, λ^0 and $x^0 = y^0$

Line 1   for $k = 0, 1, ...$ do
Line 2 $x^{k+1} = argmin_x \mathcal{L}_\mu(x, y^k; \lambda^k)$
Line 3 $y^{k+1} = argmin_y \mathcal{L}_\mu(x^{k+1}, y; \lambda^k)$
Line 4 $\lambda^{k+1} = \lambda^k + \frac{1}{\mu}(x^{k+1} - y^{k+1})$
Line 5   end

In each iteration of ADM, the Lagrange multiplier λ is updated just once after the augmented Lagrangian is minimized with respect to y (Line 4). For the ADM to be symmetric with respect to x and y, Ma [10] updated also λ after solving the subproblem with respect to x. Such as symmetric ADM is given below as Algorithm 2.

Algorithm 2: Symmetric Alternating Direction Method (SADM)

Initialization: Choose μ, λ^0 and $x^0 = y^0$

Line 1   for $k = 0, 1, ...$ do
Line 2 $x^{k+1} = argmin_x \mathcal{L}_\mu(x, y^k; \lambda^k)$
Line 3 $\lambda^{k+\frac{1}{2}} = \lambda^k + \frac{1}{\mu}(x^{k+1} - y^k)$

Line 4 $y^{k+1} = argmin_y \mathcal{L}_\mu(x^{k+1}, y; \lambda^{k+\frac{1}{2}})$

Line 5 $\lambda^{k+1} = \lambda^{k+\frac{1}{2}} + \frac{1}{\mu}(x^{k+1} - y^{k+1})$

Line 6   **end**

Ma [10] assumed that both $f(x)$ and $g(x)$ are differentiable to linearize the SADM. It follows from the first order optimality conditions that $\lambda^{k+\frac{1}{2}}$=grad $(f(x^{k+1}))$ and λ^{k+1}=-grad $(g(y^{k+1}))$. Then, the Lagrangian functions can be replaced by:

$$Q_f(u,v) = f(u) + g(v) + < grad(f(u)), u - v > + \frac{1}{2\mu}||u - v||^2 \qquad (8)$$

$$Q_g(u,v) = f(u) + g(v) + < grad(g(u)), u - v > + \frac{1}{2\mu}||u - v||_2^2 \qquad (9)$$

Algorithm 3: Linearized Symmetric Alternating Direction Method (LSADM)

Initialization: Choose μ and $x^0 = y^0$

Line 1   **for** $k = 0, 1, ...$ **do**

Line 2 $x^{k+1} = argmin_x Q_g(x, y^k)$

Line 3 $y^{k+1} = argmin_y Q_f(y, x^{k+1})$

Line 4   **end**

In Algorithm 3, the functions f and g are alternatively replaced by their linearizations plus a proximal regularization term to get an approximation to the original function F. The complexity of LSADM is $O(\epsilon)$ for obtaining an ϵ-optimal solution for Equation (4). Finally, to apply the LSADM to the RPCA, the formulation in Equation 3 is firstly restated as the following convex optimization problem:

$$\min_{L,S} \; ||L||_* + \lambda||S||_1 \; subj \; D = L + S \qquad (10)$$

where $||.||_*$ and $||.||_1$ are the nuclear norm (which is the L_1 norm of singular value) and l_1-norm, respectively, and $\lambda > 0$ is an arbitrary balanced parameter. In order to apply the LSADM algorithm to Equation 3, both the nuclear norm $f(L) = ||L||_*$ and the l_1 norm $g(S) = \lambda||S||_1$ are smoothed by using the Nesterov's smoothing technique. A smoothed approximation of $f(L)$ is obtained as follows:

$$f_\sigma(L) = max(< L, W > -\frac{\sigma}{2}||W||_F^2 : ||W||_F \le 1) \qquad (11)$$

where $W_\sigma(L) = U Diag(min(\gamma, 1))V^T$. $U Diag(min(\gamma, 1))V^T$ is the singular value decomposition (SVD) of L/σ. A smoothed approximation to the $l_1$of $g(S) = \rho||S||_1$ is obtained as follows:

$$g_\sigma(S) = max(< S, Z > -\frac{\sigma}{2}||Z||_2^2 : ||Z||_\infty \le \rho) \qquad (12)$$

where $Z_\sigma(S) = min(\rho, max(\frac{S}{\sigma}, -\rho))$. After smoothing f and g, the LSADM algorithm is applied to solve the smoothed problem:

$$min(f_\sigma(L) + g_\sigma(S)) \; subj \; L + S = D \qquad (13)$$

2.3 Moving Object Detection Algorithm

The background and moving object separation can be obtained by solving the Equation (13) with LSADM. Then, the proposed algorithm for moving object detection is as follows:

Algorithm for moving object detection by RPCA solved via LSADM

Input data: Training sequence with N images that is contained in D.

Output data: Background in L and moving objects in S.

Parameters initialization: $L_0 \leftarrow D$; $S_0 \leftarrow O$; $Y \leftarrow O$; $\mu = 30/||sign(D)||_2$; $\rho > 0$, $\sigma > 0$, $k \leftarrow 0$.

$L \leftarrow L_0$; $S \leftarrow S_0$;

Background and Moving Object Separation:

for $k = 1, ...$ to $maxiter$

$\quad X = \mu * Y - S + D$.

$\quad X = UGV^T$ (SVD).

$\quad G \leftarrow Diag(G)$.

$\quad G \leftarrow G - \mu \frac{G}{max(G, \mu + \sigma)}$.

$\quad L_k \leftarrow L$.

$\quad L = U Diag(G) V^T$.

$\quad Y \leftarrow Y - \frac{L+S-D}{\mu}$.

$\quad B \leftarrow Y - \frac{L-D}{\mu}$.

$\quad S_k \leftarrow S$.

$\quad S = \mu B - \mu * min(\rho, max(-\rho, \frac{\mu B}{\sigma + \mu}))$.

$\quad Y \leftarrow Y - \frac{L-S-D}{\mu}$.

$\quad d = ||\frac{L+S-D}{\mu}||_F / ||D||_F$.

$\quad$**if** $d < \epsilon$

$\quad\quad L \leftarrow L_k, S \leftarrow S_k$.

$\quad\quad$goto Moving Object Detection step.

$\quad$**end if**

end for

$L \leftarrow L_k, S \leftarrow S_k$.

Moving Object Detection: Threshold the matrix S to obtain the moving objects.

For the initialization, the low-rank matrix L, the sparse matrix S, and the Lagrange multiplier Y are set respectively to the matrix D, O and O. The parameters $maxiter$ and ϵ are set respectively to 100 and $10e - 7$. The computational cost in the LSADM algorithm is mainly related to the singular value decomposition (SVD). It can be reduced significantly by using a partial SVD because only the first largest few singular values are needed. Practically, we used the implementation available in PROPACK[1]. Fig. 1 shows the original frames 309, 395 et 462 of the sequence from [10] and their decomposition into the low-rank matrix L and sparse matrix S. We can see that L corresponds to the background whereas S corresponds to the moving objects. The fourth image shows the moving object mask obtained by thresholding the matrix S and the fifth image is the ground truth image.

---

[1] http://soi.stanford.edu/rmunk/PROPACK/

Fig. 1. From left to right: Original image, low-rank matrix L (background), sparse matrix S (moving objects), moving object mask, ground truth

3   Implementation

For the implementation, we choose to implement the LSADM algorithm and its fast version called Fast-LSADM which computes an ϵ-optimal solution to the problem in Equation (4) in $O(1/\sqrt{\epsilon})$ iterations [10][11]. It is a successive over-relaxation type algorithm since $(t_k - 1)/t_{k+1} > 0, \forall k \geq 2$ as follows.

Algorithm 4: Fast-LSADM

Initialization: Choose μ, $x^0 = y^0$ and set $t_1 = 1$

Line 1   for $k = 1, 2, \ldots$ do

Line 2 $x^k = argmin_x Q_g(x, z^k)$

Line 3 $y^k = argmin_y Q_f(y, x^k)$

Line 4 $t^{k+1} = (1 + \sqrt{1 + 4 * t_k^2}))/2$

Line 5 $z^{k+1} = y^k + \frac{t_k + 1}{t_{k+1}}(y^k - y^{k+1})$

Line 6   end

The algorithm for moving object detection by RPCA solved via Fast LSADM can be easily derived from the one with LSADM.

4   Experimental Results

We compared LSADM and Fast-LSADM with standard methods (SG [12], MOG [13], PCA [2]) and robust PCA methods (RSL [4], EALM [8], IALM [8]). The experiments were conducted qualitatively and quantitatively on the Wallflower dataset [14] and I2R dataset [15]. The algorithms were implemented in batch mode with matlab. The comparison is essentially conducted against IALM as RSL and EALM are very computational expensive. Furthermore, there was not much difference between the performance of LSADM and that of Fast-LSADM in terms of detection. So, we present only visual results obtained with the Fast-LSADM. Due the 10 page limitation, we present only the visual results obtained by the standard methods on the Wallflower dataset.

4.1 Wallflower Dataset[2]

This dataset consists of seven video sequences, with each sequence presenting one of the difficulties a practical task is likely to encounter: Moved Object (MO), Time of Day (TD), Light Switch (LS), Waving Trees (WT), Camouflage (C), Bootstrapping (B) and Foreground Aperture (F). The images are 160×120 pixels. For each sequence, the ground truth is provided for one image when the algorithm has to show its robustness to a specific change in the scene. Thus, the performance is evaluated against hand-segmented ground truth. The figure 2 shows the qualitative results. For the quantitative evaluation, we used metrics based on the true negative (TN), true positive (TP), false negative (FN), false positive (FP) detections. Then, we computed the detection rate, the precision and the F-measure. The detection rate is given as follows:

Fig. 2. Experimental results on the Wallflower dataset. From top to bottom: original image, ground truth, SG, MOG, PCA, RSL, EALM, IALM, Fast-LSADM. From left to right: MO (985), TD (1850), LS (1865), WT (247), C (251),B (2832), FA (449). More results with other background subtraction methods available on the Background Subtraction Web Site[3].

---

[2] http://research.microsoft.com/en-us/um/people/jckrumm/
wallflower/testimages.htm

[3] http://sites.google.com/site/backgroundsubtraction/
test-image-sequences---results

$$DR = \frac{TP}{TP + FN} \tag{14}$$

The precision is computed as follows:

$$Precision = \frac{TP}{TP + FP} \tag{15}$$

A good performance is obtained when the detection rate is high without altering the precision. This can be measured by the F-measure [16] as follows:

$$F = \frac{2 \times DR \times Precision}{DR + Precision} \tag{16}$$

A good performance is then reached when the F-measure is closed to 1. Table 1 shows in percentage the F-measure for each sequence and its average on the dataset. The F-measure value of MO sequence can't be computed due to the absence of true positives in its ground-truth. We have highlighted when Fast-LSADM outperforms IALM. It is the case on the sequences (MO, TD, LS, WT, C, B, FA). For the sequence LS, the result is still acceptable. RSL and EALM gives the best results on the sequences (TD, WT, C, FA) but the time requirement is very higher than IALM, LASDM and Fast-LSADM. So, the proposed method offers a nice compromise between robustness and computational efficiency. As these encouraging results are obtained by using one ground-truth image one each sequence, we have evaluated the proposed method on a dataset with more ground-truth images in the following sub-section.

4.2   I2R Dataset[4]

This dataset provided by [15] consists of nine video sequences, which each sequence presenting dynamic backgrounds or illumination changes. The size of the images is 176*144 pixels. For each sequence, the ground truth is provided for 20 images. Among this dataset, we have chosen to show results on four representative sequences that are the following ones: airport, shopping mall, water surface and curtain. We ran RSL, IALM, LSADM and Fast-LSADM on these sequences. We skipped EALM since it would take excessive amounts of time due to full SVD calculations. Since all these video clips have more than 1000 frames, we took a part of each clip with 200 frames. Fig. 3 shows the qualitative results. For example, we can see that the LSADM allows to detect the complete silhouette in the sequence called "water surface". Table 2 shows the average F-measure in percentage that is obtained on 20 ground truth images for each sequence and its average on the dataset. We have highlighted when Fast LSADM outperforms IALM. We can see that Fast LSADM outperforms RSL and IALM for each sequence except for the sequence "Airport". The SVDs and CPU time of each algorithm was computed for each sequence. For example, in the case of the sequence "Airport" of resolution 176×144 with 200 images, the CPU time is 40min15s, 3min47s, 4min30s and 1min56s respectively for EALM, IALM, LSADM and Fast-LSADM. The SVDs

---

[4] http://perception.i2r.a-star.edu.sg/

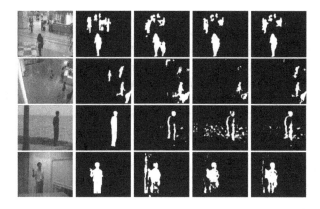

Fig. 3. Experimental results on the I2R dataset. From left to right: original image, ground truth, RSL, IALM, Fast-LSADM. From top to bottom: airport (2926), shopping mall (1980), water surface (1594), curtain (23257).

Table 1. F-measure on the Wallflower dataset

	RSL	EALM	IALM	LSADM	Fast-LSADM
Time of Day	75.73	81.18	80.56	80.24	**80.84**
Light Switch	28.36	70.86	**73.16**	69.11	69.33
Waving Trees	89.69	86.40	40.88	81.25	**81.67**
Camouflage	91.78	75.43	22.02	75.71	**76.01**
Bootstrap	69.38	74.4	73.73	74.11	**74.69**
Foreground Aperture	74.37	72.07	61.92	71.26	**71.56**
Average	71.55	76.73	58.71	75.28	**75.69**

Table 2. F-measure on the I2R dataset

	RSL	IALM	LSADM	Fast-LSADM
Airport	65.26	**74.26**	71.27	71.57
Shopping mall	58.64	50.89	75.24	**75.74**
Water surface	34.42	31.34	45.17	**47.69**
Curtain	70.73	76.59	81.11	**83.33**
Average	57.26	58.27	68.19	**69.58**

times is 550 SVDs, 38 SVDs, 43 SVDs and 6 SVDs respectively for EALM, IALM, LSADM and Fast-LSADM. On these problems of extremely low ranks, the partial SVD technique used in IALM, LSADM and Fast-LSADM becomes quite effective. Even so, the CPU times required by IALM are still about two times of those required by Fast-LSADM. Furthermore, the speed can be improved by a GPU implementation.

5   Conclusion

In this paper, we have presented a moving object detection method based on RPCA that is solved via a linearized alternating direction method. This method allows us to alleviate the contraints of the identities. Furthermore, experiments on video surveillance datasets show that this approach is more robust than RSL and IALM in the presence of dynamic backgrounds and illumination changes. In terms of computational efficiency, the proposed approach has exhibited a significant speed advantage over IALM. Although the proposed method is an offline one, meaning that the processing is done after a large number of frames are acquired and therefore, it is not done in real time. It is possible to extend the concept to incremental, real time processing by adaptively update the low rank component. This will be investigated in details in a future paper.

Acknowledgements. The authors would like to thank Shiqian Ma[5] (Institute for Mathematics and Its Applications, Univ. of Minnesota, USA) who has kindly provided the algorithm LSADM.

References

1. Bouwmans, T.: Recent advanced statistical background modeling for foreground detection: A systematic survey. RPCS 4, 147–176 (2011)
2. Oliver, N., Rosario, B., Pentland, A.: A bayesian computer vision system for modeling human interactions. In: ICVS 1999 (1999)
3. Bouwmans, T.: Subspace learning for background modeling: A survey. RPCS 2, 223–234 (2009)
4. Torre, F.D.L., Black, M.: A framework for robust subspace learning. International Journal on Computer Vision, 117–142 (2003)
5. Candes, E., Li, X., Ma, Y., Wright, J.: Robust principal component analysis? International Journal of ACM 58 (2011)
6. Wright, J., Peng, Y., Ma, Y., Ganesh, A., Rao, S.: Robust principal component analysis: Exact recovery of corrupted low-rank matrices by convex optimization. In: NIPS 2009 (2009)
7. Lin, Z., Ganesh, A., Wright, J., Wu, L., Chen, M., Ma, Y.: Fast convex optimization algorithms for exact recovery of a corrupted low-rank matrix. UIUC Technical Report (2009)
8. Lin, Z., Chen, M., Wu, L., Ma, Y.: The augmented lagrange multiplier method for exact recovery of corrupted low-rank matrices. UIUC Technical Report (2009)
9. Yuan, X., Yang, J.: Sparse and low-rank matrix decomposition via alternating direction methods. Optimization Online (2009)
10. Ma, S.: Algorithms for sparse and low-rank optimization: Convergence, complexity and applications. Thesis (2011)
11. Goldfarb, D., Ma, S., Scheinberg, K.: Fast alternating linearization methods for minimizing the sum of two convex function. Preprint, Mathematical Programming Series A (2010)
12. Wren, C., Azarbayejani, A., Darrell, T., Pentland, A.: Pfinder: Real-time tracking of the human body. IEEE Transactions on PAMI 19, 780–785 (1997)
13. Stauffer, C., Grimson, W.: Adaptive background mixture models for real-time tracking. In: CVPR 1999, pp. 246–252 (1999)
14. Toyama, K., Krumm, J., Brumitt, B., Meyers, B.: Wallflower: Principles and practice of background maintenance. In: ICCV 1999, pp. 255–261 (1999)
15. Li, L., Huang, W., Gu, I., Tian, Q.: Statistical modeling of complex backgrounds for foreground object detection. IEEE T-IP, 1459–1472 (2004)
16. Maddalena, L., Petrosino, A.: A fuzzy spatial coherence-based approach to background foreground separation for moving object detection. Neural Computing and Applications, 1–8 (2010)

---

[5] http://www.columbia.edu/~sm2756/

Tracking Technical Objects in Outdoor Environment Based on CAD Models

Stefan Reinke[1,2], Enrico Gutzeit[1], Benjamin Mesing[1], and Matthias Vahl[1]

[1] Fraunhofer Institute for Computer Graphics Research IGD
{enrico.gutzeit,benjamin.mesing,matthias.vahl}@igd-r.fraunhofer.de
[2] University of Rostock
stefan.reinke@uni-rostock.de

Abstract. Tracking objects under difficult conditions is a challenging topic in many real life and outdoor applications. In particular persistent occlusion of large parts of the model, difficult lighting conditions, and a noisy background result in a failure of many available algorithms. To overcome these difficult problems we significantly extend the existing RAPiD approach, in particular by putting sample control points at random intervals, and providing the possibility for fix points. We show how to automatically extract an edge model that is suitable for tracking from CAD models. The benefit of our approach is presented by means of a real life outdoor application scenario suffering from the mentioned bad conditions. We obtain relative translation and rotation errors that are only moderately higher than those in less challenging setups.

1 Introduction

When tracking objects often the environmental conditions and camera positions cannot be freely chosen. This typically holds within industrial applications, resulting in disadvantageous conditions, like partial occlusion by other objects, difficult lighting conditions and unfavorable backgrounds. Fortunately, for technical structures there is usually a 3D model available. Availability of such a 3D model can significantly simplify the tracking process. The RAPiD algorithm [1] is an edge-based tracking approach which has shown to be suitable for model-based tracking, but has problems with difficult environmental conditions. In fact, currently available research does not provide an approach that addresses all those conditions. We therefore develop a new approach based on the RAPiD algorithm and deal with its shortcomings by making several adaptations. Since CAD-models usually do not directly provide good input models for edge based tracking, we also address the problem of automatically extracting edges from a CAD model suitable for object tracking.

We evaluate our approach on a real life outdoor application scenario suffering from all the aforementioned conditions. For davits (the structures used to launch lifeboats onboard of ships) it is crucial to be able to launch the lifeboats even under extreme conditions like 20° list and 10° tilt. To verify this condition, a

G. Bebis et al. (Eds.): ISVC 2012, Part I, LNCS 7431, pp. 437–446, 2012.
© Springer-Verlag Berlin Heidelberg 2012

multi-body-simulation model simulating the movement of the davit was developed. To validate the simulation the movement of the davit is tracked within a physical test and its movement is compared with the simulation results.

The environmental conditions for our evaluation setup were determined by the weather at the time dynamic testing of the object was performed by the industry partner. For the camera position the best spot available was an elevated position on a nearby crane used to install the structure. The given setup resulted in the following disadvantageous conditions (examples of the input can be seen in Fig. 1):

- during the tracking process large parts of the structure are occluded,
- parts of the structure did almost not distinguish themselves from the background,
- difficult lighting conditions, resulting in many edges to be almost not recognizable.

Fig. 1. On the left: One sample image (1280x1024) of the tracking video; In the middle: Two sub-images with almost imperceptible structure parts; On the right: Two sub images with occlusion

Resulting from the conditions described above we had to counteract (1) drift occurring for structures with one main direction of the visible edges, e.g. with mostly horizontal but only few vertical edges, (2) the tracked object pose being attached to occluding objects when its borders come close to the estimated edges of the tracked object, and (3) many sections of the edges having no detectable correspondences within the image due to occlusion and bad lighting.

In the next sections we first discuss existing approaches. Then we present the RAPiD model and describe our extensions using the CAD model. Next, an evaluation based on the real life outdoor application scenario is given, followed by the conclusions.

2   Related Work

In context of model-based 3D object tracking, the two most widely used approaches are edge-based and feature point-based tracking (see the overview paper by Lepetit and Fua [2]). The first type utilizes an edge model and matches it with strong edges detected within the image. The second type exploits feature-points to locate the object. For technical objects, there is often very little texture and the shape is dominated by long straight edges providing only a limited amount of good feature points. In this case it is advantageous to apply the edge-based approach.

In 1993 Harris published the RAPiD tracking algorithm and with this, laid the foundation for most other model-based tracking approaches using edges [1]. The RAPiD algorithm is based on edges and divided into four main steps. Starting with an initial projection of the 3D model all visible edges are sampled. Those 3D sample points or control-points will be projected into 2D and correspondences are searched in the given image before minimizing the error of the distances using a least-square method.

However, one of RAPiDs major problems is robustness. It performs badly with regard to occlusions, background clutter, shadows or strongly textured objects. Therefore, many modifications were proposed for each step. Drummond and Cipolla used a robust estimator, replaced the least-square estimation with an iterative reweighed least-square method and use a 6-dimensional Lie-Group to calculate the transformation matrix [3]. Armstrong showed a way to use RANSAC to detect outliers and proposed to reject edges completely if the number of control points are below a threshold to avoid false detection by nearby edges [4]. In case of occlusions or previous hidden line segments Yoon et al. suggested to extend the edges to increase the number of usable control points [5]. Due to the usual movement in the tracking process this enables the possibility to find valuable correspondences. It is also possible to use an estimator to handle multiple hypothesis and retain all the maxima as possible correspondents in the pose estimation as demonstrated by Vacchetti et al. [6]. This will exclude false edges and improve the outlier rejection.

Another approach for edge-based or feature point-based tracking is the use of keyframes. Keyframes provide a set of precomputed renderings of the model which can be matched against the current image. Since sole keyframe-based approaches require a huge set of precomputed renderings to be accurate, they are usually used in combination with iterative approaches. Examples for feature point-based hybrid approaches are the ones of Vacchetti et al. [7] and Tordoff et al. [8]. For edge-based approaches Choi and Christensen combine a RAPiD approach with a keyframe approach [9]. They used keyframes for global pose estimation and interpolate between those estimations with an iterative local pose estimation. The local pose estimation is based on the RAPiD extension introduced by Drummond and Cipolla [3]. This combined approach allows for an accurate initialization, reinitalization in the case of tracking failure and additionally a fast tracking process between the keyframes. A disadvantage of all approaches utilizing keyframes is the requirement of an initial set of likely positions for producing the keyframes. This requires a basic set-up step usually with user involvement and limits the usage to objects moving only within a predefined areas.

3   Tracking Approach

In this section we introduce our RAPiD-based tracking approach. We briefly explain the main steps of RAPiD and detail on our adjustments. The tracking process is iterative and requires a few preprocessing steps to be performed. Apart from the extraction of the intrinsic camera calibration, this is also the estimation of an initial pose (position and rotation). Our approach requires a rough manual estimation, after which the iterative tracking process can start. First, the visible edges are extracted from the CAD model, a point not explicitly addressed in most research. Second, the control-points, i.e. sample points on all visible edges, are created. Every control-point is projected into 2D according to the estimated pose. Third, within the image edges are detected. For each control-point a corresponding hit-point is searched on the detected edges. Finally, all correspondences are passed to a minimization process and a new pose is calculated. Due to its edge based nature the approach will not work well for round objects, where only the silhouette edges can be used for tracking purposes.

3.1   Edge Extraction from CAD Model

For edge-based tracking the visible edges need to be extracted from the CAD model. Here, a triangle export of a CAD model is used as input data because this allows utilizing the rendering capabilities and speed of the graphic card. In this section existing approaches for edge extraction along with our approach are presented.

Wuest et al. [10] suggest rendering the object from the estimated viewpoint and detecting discontinuities within the first derivation of the Z buffer at each frame. The approach reliably detects all visible edges but looses correspondence to the model edges, disallowing for enhanced operation within the object space. A similar rendering-based approach is presented by Brüning et. al [11]. The approach additionally creates 3D edges based on the rendering by unprojecting the endpoints of the edges. A reference from the 3D edges to the model edges is created by searching the end-points of the 3D edges within the CAD model. The approach requires additional effort for the search and loses exactness caused by limitations in image and Z-Buffer resolution. Platonov and Langer [12] propose to create a viewpoint independent set of stable edges by rendering the objects multiple times from various locations and under different lighting conditions. Edges detectable under a lot of conditions are collected within the set. The correspondence between the rendering and the edges is preserved by using a color coded rendering step. While within this approach line extraction needs to be performed only once, it cannot be used for hidden line removal.

We combine the above approaches by performing a hidden line removal through rendering while still maintaining the correspondence between image and object-edges. A color coded rendering step is used to detect all visible triangles. They are then compared with each other; adjacent triangles which exceed a certain angle produce strong visible edges and are thus included. Further, each edge without a visible neighbor face is considered to be a silhouette edge and also included. Edges which appear very short in image space are filtered out. Finally parallel edges very close to each other are treated as a single edge. The remaining edges of this basic procedure are the foundation of the tracking process. Fig 2 (middle) illustrates the resulting edges for one pose estimation.

Our approach always retrieves the full edge. For partly occluded edges this leads to parts of the edge with no corresponding visible edges within the image. To remove edges that are mostly hidden, the actually visible pixels of a triangle are compared with the expected number if no occlusion would occur. If large parts of the triangle are covered its edges are not considered in the following steps. The negative impact of the remaining partly occluded edges is relatively small due to outlier rejection as described later.

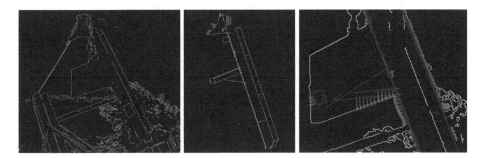

Fig. 2. Filtered image with detected edges through Canny edge (left); visible edges from an estimated pose (middle); problem of edges from occluding objects close to model edges (for illustration purposes with relaxed outlier rejection, but the problem does also occur with stricter outlier rejection) (right)

3.2   Control-Point Creation

The step of creating control-points has a huge impact on the tracking process. To keep the 2D-3D correspondence the lines are sampled in object space, i.e. the model edges are sampled. The amount of sample points for each edge is based on the edge length. Due to the fact that 3D edges can be very short after the projection in 2D, this needed to depend on their projected length. For the projection the previously estimated pose is used.

For the sampling we implemented two different approaches. We started with the basic approach of regularly sampling every line, e.g. at 10 pixels intervals. This results in bad tracking results within our evaluation setup. During the tracking process the model has a huge drift caused by the partly occluded model and background clutter. We identified almost-parallel edges of non-model-objects occluding the original model as one of the main problems. If those edges are close to the model edges, they are not suspect to outlier rejection and negatively influence the tracking result.

To face this behavior we used a randomized sampling approach. Instead of a regularly sampling the edge, the control points are now set at irregular intervals. This has the effect that if only a part of an edge is occluded, in some frames no or only a few control points are put on this part. This is sufficient for the tracker to move the estimated pose to the real pose of the object. The other case, that occluded parts are weighted stronger in other frames does not suffice to mitigate the positive effect. Together with an improved outlier detection this significantly improved tracking results when dealing with occlusion.

Another problem we had to deal with is the shape of the object. Within our evaluation setup there are many long vertical and only a few horizontal edges resulting in horizontal drift. Additionally, the horizontal edges are very small, which results in a small amount of control-points. Given such shapes our approach allows adapting the sampling rate for edges, e.g. favoring small, horizontal lines by using a higher sample rate. Currently the favoring criteria are defined manually. In future an automatic favoring based on the visible edges would be desirable.

3.3   Hit-Point Detection and Outlier Rejection

After the generation of the control-points, correspondences are searched within the current frame. To create an useful edge image, a canny algorithm is run and small edges are filtered out (Fig 2, left). The control points are projected into the image space and starting from those, corresponding edges are searched within the edge image. Usually this search is done in the x- or y-direction of the image to increase the speed of this process [13,2]. Since in our application scenario precision is more important than speed, we use the minimally slower search process along the normal of the edge. Search happens in both directions of the normal, therefore each control-point has zero, one, or two hit-points. Due to the bad lighting conditions, many edges are not visible and only a limited number of hit-point-control-point-correspondences is detected. All gathered control-point-hit-point-correspondences are transferred to the outlier rejection.

For the outlier rejection the following steps are performed in order:

1. All control-points with no hit-points are discarded.
2. For control-points with two hit-points the closer one is chosen as candidate.
3. All hit-points exceeding a certain distance towards the corresponding control-points are discarded.
4. For all edges the corresponding hit-points are discarded if the edge does not have a minimum number of corresponding hit-points, because it cannot be reliably detected if the points belong to an object-edge.
5. A line fitting is performed for all hit-points belonging to the control-points of the same edge. If the angle between this line and the line of the projected control points exceeds a certain value, the hit-points are discarded.

The last step is based on the assumption, that the object rotates only at a slow rate. If the line is not almost parallel to the previous direction, we conclude that those hit-points represent another edge. This especially deals with occlusion, where the original edge is hidden by another object. Once the edge of the occluding object comes close to that of the model, the tracker would otherwise pull the edge to the occluding object (Fig 2, right). It also improves handling of bad lighting conditions. If the original edge is not visible within the image, but the edge of another object close by would otherwise be chosen, it is able to discard this edge. In both cases it obviously only discards false edges with a direction differing from the expected edge.

3.4 Minimization and Estimation of a New Pose

Based on the remaining control-point-hit-point-correspondences the new pose for the object is determined. This step is realized in the same way as presented originally. The new object pose is calculated to minimizes the distance between the projected control- and hit-points. Minimization happens using the least-square based Levenberg-Marquard-Algorithm.

3.5 Fix-Points

When tracking technical objects one can often resort to a few restrictions. Often they can only rotate around one axis or the degrees of freedom are shortened in another way. One way to represent this is by directly representing the kinematic constraints [13,3]. This significantly improves the tracking results but requires the relations to be modeled. We chose to provide a lightweight approach by introducing the idea of fix-points. A fix-point can be defined within the model at a position that is expected not to move, like the center of rotation.

Fix-points are realized as high weighed control points. They make use of the fact that the camera is immobile. The distance of the fix-point is the distance between its two projections, with the old and the new pose estimation. Depending on its weight it is passed multiple time to the minimization process. Despite the name, they are therefore not totally fix. This is important since otherwise they would require a perfect initial pose estimation and would not be able to deal with small unexpected movements of the camera, e.g. through wind. Fix-points can further stabilize the tracked objects.

4   Evaluation

We evaluated our approach using two scenarios and compared the tracked movement of the object with annotated ground truth data. The ground truth data was generated interactively: For selected frames we manually adjusted the model pose until the image and the model overlapped correctly. When the object moved at constant speed, we selected frames at larger intervals, reducing the interval when the movement was accelerated. Because the velocity can be assumed to be almost constant at small time scales, the poses for the remaining frames were calculated by linear interpolation. For recording we used a camera with a resolution of 1280x1024 at 15 fps, 1/3" CMOS and a 12.5 mm television objective.

The first evaluation setup was tracking the movement of a davit arm, a large steel structure used to launch lifeboats. The davit arm rotated around a fixed joint (Fig. 3). The distance of the camera to the steel structure was 14.8 m, the dimension of the structure 4.8 m.

For further evaluation, we did also test the algorithm for tracking a smaller structure. This is a governor with a maximum dimension of 30 cm. This second scenario was especially used to validate robustness of the algorithm for the following conditions: (1) the model did not exactly match the real object, because no CAD data was available but it was remodeled from photos, (2) the camera recorded only parts of the model and additional parts are occluded/shadowed and, (3) the object had a partly reflective surface. The governor was moved along a straight rail without changing its rotation (Fig. 3).

Fig. 3. Sample frames of the first (top) and second (bottom) evaluation scenario; the projection of the tracked 3D model is overlaid; tracked object is partly occluded (top), and only partly visible (bottom)

Two important modifications of the RAPiD algorithm we have introduced were putting control points at random intervals, and introducing fix points. To evaluate the effect, we created test data with random and regular sampling and introduced a fix-point in the first scenario at the rotation center with a weight of 10 (this equals to ∼5% of the overall considered control points). We evaluate the results according to the translation error and rotation error. For the translation we calculate the absolute distance $(\Delta x, \Delta y, \Delta z)^T$ from the tracked position to the ground truth position. Additionally, we calculate the euclidean distance d to the ground truth. For the rotation error, we determine the distance of the roll, pitch and yaw to the ground truth and calculate the average. Table 1 shows the evaluation results.

The comparison of fix-points vs. non-fix-points clearly shows, that fix-points lead to better tracking results for obvious reasons. A look on the sampling methods indicates that the random sampling provides better tracking results especially with regards to the angle. However, due to the random nature it might

Table 1. Mean value ($\bar{x}$) and standard deviation (σ) of difference to ground truth data; results for first scenario within upper part; results for second scenario within lower part

parameters		stat	translation in cm				rotation in degree			
fix p.	sampling		Δx	Δy	Δz	d	roll	pitch	yaw	avg
no	regular	$\bar{x}$	8.9	23.1	10.0	**28.2**	4.54	7.46	1.39	**4.5**
		σ	18.5	13.4	20.8	13.8	9.09	7.67	3.02	–
no	random	$\bar{x}$	15.6	31.5	2.7	**35.8**	2.83	8.49	1.29	**4.2**
		σ	28.0	17.1	5.6	19.5	5.41	12.56	2.73	–
yes	regular	$\bar{x}$	15.9	16.0	7.1	**24.0**	4.91	5.06	1.31	**3.8**
		σ	32.2	3.6	14.7	6.6	9.70	4.24	2.87	–
yes	random	$\bar{x}$	13.6	14.0	5.2	**20.4**	3.94	3.52	1.11	**2.9**
		σ	27.3	3.6	10.4	5.1	8.10	2.59	2.31	–
no	regular	$\bar{x}$	1.60	0.13	1.58	**2.49**	3.58	2.59	4.03	**3.4**
		σ	3.15	0.13	3.38	5.27	5.19	8.15	2.33	–
no	random	$\bar{x}$	1.49	0.12	1.63	**2.49**	3.37	3.40	2.82	**3.2**
		σ	3.12	0.17	3.22	1.83	5.03	6.96	2.35	–

also lead to decreased tracking results in some cases (here the translation in the first setup with no fix-point). The best result of the first scenario has an translation error of $d = 20.4cm$. Taking into account the object-camera distance of $l = 1480cm$, the relative error is $d/l * 100 = 1.4\%$. For the second scenario the relative error is 2.49%

The relative translation error and the rotation error are only moderately higher, than those in less challenging setups [9,5] and the evaluation has shown, that our adaptions improved the tracking results. Comparing with tracking results from current publications, such as Park et al. [14] and considering the difficult environmental conditions within the evaluation setup, one can see that our approach provides very good tracking results.

5   Summary and Outlook

In this paper we have presented an approach for model based object tracking. The approach provides enhances to the RAPiD tracking approach [1]. It especially deals with problems like persistent occlusion of large parts, difficult lighting conditions, inhomogeneous backgrounds and shapes with a dominant direction of their edges. To deal with those conditions, we have introduced the idea of random distance sampling for the control points and a semi-automatic higher weighing of edges. A further step would be, to fully automate the weighting of edges based on criteria like the amount of other edges sharing the same direction. Additionally, we have introduced the concept of fix points, a lightweight approach for making use of rotational constraints. Finally, we have presented a way to automatically derive a visible edge model suitable for tracking from a CAD model.

The evaluation in two industrial scenarios has shown, that the algorithm performs well, considering the difficult environmental conditions. Current limitations of our algorithm are, that the initial pose must be manually estimated and total occlusion for longer periods is not addressed. To this end, a combination with keyframe-based approaches or motion estimation (for the latter) could be implemented as part of future work.

References

1. Harris, C.: Tracking with rigid models. In: Active Vision, pp. 59–73. MIT Press, Cambridge (1993)
2. Lepetit, V., Fua, P.: Monocular model-based 3d tracking of rigid objects. Found. Trends. Comput. Graph. Vis. 1, 1–89 (2005)
3. Drummond, T., Cipolla, R.: Real-time visual tracking of complex structures. IEEE Transactions on Pattern Analysis and Machine Intelligence 24, 932–946 (2002)
4. Armstrong, M., Zisserman, A.: Robust object tracking. In: Asian Conference on Computer Vision, vol. I, pp. 58–61 (1995)
5. Yoon, Y., Kosaka, A., Park, J.B., Kak, A.C.: A new approach to the use of edge extremities for model-based object tracking. In: IEEE International Conference on Robotics and Automation, ICRA 2005, pp. 1871–1877 (2005)
6. Vacchetti, L., Lepetit, V., Fua, P.: Combining edge and texture information for real-time accurate 3D camera tracking. In: Third IEEE and ACM International Symposium on Mixed and Augmented Reality, ISMAR 2004, pp. 48–56 (2004)
7. Vacchetti, L., Lepetit, V., Fua, P.: Fusing online and offline information for stable 3d tracking in real-time. In: Proceedings of the IEEE Conference on Computer Vision and Pattern Recognition, vol. 2, pp. II-241–II-248(2003)
8. Tordoff, B., Mayol, W.W., de Campos, T.E., Murray, D.W.: Head pose estimation for wearable robot control. In: Proceedings of the British Machine Vision Conference, pp. 2–5 (2002)
9. Choi, C., Christensen, H.I.: Real-time 3d model-based tracking using edge and keypoint features for robotic manipulation. In: 2010 IEEE International Conference on Robotics and Automation (ICRA), pp. 4048–4055 (2010)
10. Wuest, H., Stricker, D., Herder, J.: Tracking of industrial objects by using CAD models. Journal of Virtual Reality and Broadcasting 4 (2007)
11. Brüning, A., von Lukas, U., Vahl, M.: Automatic preparation of product data for markerless optical tracking. In: Proceedings of ProSTEP iViP Science Days 2007 (2007)
12. Platonov, J., Langer, M.: Automatic contour model creation out of polygonal cad models for markerless augmented reality. In: Proceedings of the 2007 6th IEEE and ACM International Symposium on Mixed and Augmented Reality, ISMAR 2007, pp. 1–4. IEEE Computer Society, Washington, DC (2007)
13. Comport, A.I., Marchand, E., Chaumette, F.: Object-based visual 3d tracking of articulated objects via kinematic sets. In: Proceedings of the 2004 Conference on Computer Vision and Pattern Recognition Workshop, CVPRW 2004, vol. 1. IEEE Computer Society, Washington, DC (2004)
14. Park, H., Mitsumine, H., Fujii, M.: Adaptive edge detection for robust model-based camera tracking. IEEE Consumer Electronics, 1465–1470 (2011)

Motion Compensated Frame Interpolation with a Symmetric Optical Flow Constraint

Lars Lau Rakêt[1], Lars Roholm[1], Andrés Bruhn[2], and Joachim Weickert[3]

[1] Department of Computer Science, University of Copenhagen
Universitetsparken 5, 2100 Copenhagen, Denmark
[2] Institute for Visualization and Interactive Systems, University of Stuttgart
Universitätsstraße 38, 70569 Stuttgart, Germany
[3] Mathematical Image Analysis Group, Saarland University
Campus E 1.7, 66123 Saarbrücken, Germany

Abstract. We consider the problem of interpolating frames in an image sequence. For this purpose accurate motion estimation can be very helpful. We propose to move the motion estimation from the surrounding frames directly to the unknown frame by parametrizing the optical flow objective function such that the interpolation assumption is directly modeled. This reparametrization is a powerful trick that results in a number of appealing properties, in particular the motion estimation becomes more robust to noise and large displacements, and the computational workload is more than halved compared to usual bidirectional methods. The proposed reparametrization is generic and can be applied to almost every existing algorithm. In this paper we illustrate its advantages by considering the classic TV-L^1 optical flow algorithm as a prototype. We demonstrate that this widely used method can produce results that are competitive with current state-of-the-art methods. Finally we show that the scheme can be implemented on graphics hardware such that it becomes possible to double the frame rate of 640×480 video footage at 30 fps, i.e. to perform frame doubling in realtime.

1 Introduction

Frame interpolation is the process of creating intermediate images in a sequences of known images. The process has many uses, for example video post-processing and restoration, temporal upsampling in HDTVs to enhance viewing experience as well as a number of more technical applications, e.g. in video coding.

In this work we consider optical flow based frame rate upsampling which performs interpolation along the motion trajectories. With this application in mind we propose to reparametrize the optical flow energy such that it fits better to the given problem. The reparametrized energy has a symmetric data fidelity term, that uses both surrounding frames as references. We show that one can improve modern frame interpolation methods substantially by this powerful generic trick, that can be incorporated in existing schemes without requiring major adaptations. We analyze the reparametrization, and show experimentally that it has a great effect on the stability and robustness of the interpolation process.

G. Bebis et al. (Eds.): ISVC 2012, Part I, LNCS 7431, pp. 447–457, 2012.

The idea to symmetrize data matching terms to achieve better results has already established its usefulness in other areas. In image registration Christensen and Johnson [1] explored the benefit of penalizing consistency, by jointly estimating forward and backward transforms, and requiring that they were inverses of one another. A similar idea was applied to the optical flow problem by Alvarez et al. [2], who imposed an additional consistency term. Later that same year Alvarez et al. [3] proposed a reparametrization similar to the one derived in this paper in order to avoid a reference frame, and thereby increase flow consistency. However, they did not use the obtained symmetric flow directly, but interpolated flow values at pixel position of a reference image in order to obtain a flow comparable to the standard asymmetric flow. Recently Chen [4] used a symmetric data term for surface velocity estimation, noting the property that motion vector length is halved, which in turn gives better handling of large displacements.

Apart from being algorithmically different, the difference between the justification given in this paper and the justifications of Alvarez et al. [3] and Chen [4] is that we have chosen the symmetric data fidelity term because it explicitly models the standard interpolation assumption, rather than improves some notion of consistency or better handles large displacements. In turn this also means that we use the estimated flows directly on the unknown frame, and thereby avoid problems of temporal warping. As we will show, the mentioned benefits are clearly reflected in the results. They demonstrate that using a symmetric flow for interpolation is generally better than using either forward or backward flows or both.

The rest of this paper is organized as follows. In the next section we review the estimation process for duality based TV-L^1 optical flow. In Section 3 we discuss a standard method for motion compensated frame interpolation, and in Section 4 we present our reparametrization of the optical flow energy. In Section 5 we consider examples and compare to current state-of-the-art methods, and finally we conclude the paper with discussion and outline future directions in Section 6.

2   Duality Based TV-L^1 Optical Flow

Optical flow estimation concerns the determination of apparent (projected) motion. Given a sequence of temporally indexed images I_t, we want to estimate the optical flow v such that the motion matches the image sequence with respect to some measure. This is often done by computing the flow as the minimizer of an energy of the type

$$E(v) = \lambda F(I, v) + R(v) \qquad (1)$$

where F is a positive functional measuring data fidelity and R is a regularization term. Many energies of this type have been suggested throughout the years, and a large variety of resolution strategies exist. Here we will focus on the TV-L^1

energy, where data fidelity between two frames I_0 and I_1 is measured by the L^1-norm of the difference

$$F(I_0, I_1, v) = \int \|I_1(x + v(x)) - I_0(x)\| \, dx, \qquad (2)$$

and the regularization term R penalize the total variation of the estimated motion

$$R(v) = \int \|\mathscr{D}v(x)\| \, dx, \qquad (3)$$

which, depending on the definition of the operator $\mathscr{D}$ can give different forms of the vectorial total variation [5]. Here we will take $\mathscr{D}$ to be the 1-Jacobian of Goldluecke et al. [6], since this choice of regularizer does not suffer from the channel-smearing of usual definition of vectorial total variation [5]. In order to efficiently minimize E we introduce two relaxations. First we linearize the data fidelity term $I_1(\cdot + v) - I_0 \approx \rho(v)$, where $\cdot$ is a placeholder for the argument of the function (i.e. x),

$$\rho(v) = I_1(\cdot + v_0) - I_0 + J_{I_1}(\cdot + v_0)(v - v_0) \qquad (4)$$

where J_{I_1} is the Jacobian of I_1, and v_0 is the current estimate of v. We further relax E by introducing an auxiliary variable u that splits data fidelity and regularization in two quadratically coupled energies:

$$E_1(v) = \int \lambda \|\rho(v)(x)\| + \frac{1}{2\theta} \|v(x) - u(x)\|^2 \, dx, \qquad (5)$$

$$E_2(u) = \int \frac{1}{2\theta} \|v(x) - u(x)\|^2 + \|\mathscr{D}u(x)\| \, dx. \qquad (6)$$

This relaxation was first proposed by Zach et al. [7], and has a number of advantages, most notably that the first problem can be solved pointwise which makes the solution very easy to implement on massively parallel processors like graphics processing units (GPUs). For completeness we will give the minimizing pointwise solution to (5) in the general case where $\rho(v)(x) = a^\top v + b$, $a \in \mathbb{R}^d$ and $b \in \mathbb{R}$, which is given as

$$v(x) = u(x) - \pi_{\lambda\theta[-a,a]}\left(u + \frac{b}{\|a\|^2}a\right) \qquad (7)$$

where $\pi_{\lambda\theta[-a,a]}$ is the projection onto the line segment joining the vectors $-\lambda\theta a$ and $\lambda\theta a$, which is given by

$$\pi_{\lambda\theta[-a,a]}\left(u + \frac{b}{\|a\|^2}a\right) = \begin{cases} -\lambda\theta a & \text{if} \quad a^\top u + b < -\lambda\theta\|a\|^2 \\ \lambda\theta a & \text{if} \quad a^\top u + b > \lambda\theta\|a\|^2 \\ \frac{a^\top u + b}{\|a\|^2}a & \text{if} \quad |a^\top u + b| \le \lambda\|a\|^2 \end{cases} . \qquad (8)$$

For $a = \nabla I_1(x + v_0)$ and $b = I_1(x + v_0) - I_0(x) - \nabla I_1(x + v_0)^\top v_0$ the above expression reduces to the result of Zach et al. [7]. In the general case of vector valued images, (5) can be minimized by the method presented in [8]. We will not replicate the minimizer of the regularization energy (6) here, but note that it can be minimized effectively following an iterative pointwise Bermùdez-Moreno type algorithm (see Goldluecke et al. [6]). For further implementation details we refer to Section 5.

3   Motion Compensated Frame Interpolation

Given two images I_0 and I_1 and an estimate of the (forward) optical flow $\boldsymbol{v}_f$ we are interested in estimating the in-between image $I_{1/2}$ (the methods are easily extended to any in-between frame I_t, $t \in (0,1)$). A simple approach is to assume that the motion vectors are linear through $I_{1/2}$ and then fill in $I_{1/2}$ using the computed flow. However, since $\boldsymbol{v}_f$ is of sub-pixel accuarcy, the points $\boldsymbol{x}+1/2\boldsymbol{v}_f(\boldsymbol{x})$ that are hit by the motion vectors are generally not pixel positions. This is often solved by warping the flow to the temporal position of the intermediate frame $I_{1/2}$ (see e.g. [9], [10], [11]), in which one defines a new flow $\boldsymbol{v}_f^{1/2}$ from $I_{1/2}$ to I_1

$$\boldsymbol{v}_f^{1/2}(\text{round}(\boldsymbol{x} + 1/2\boldsymbol{v}_f(\boldsymbol{x}))) = 1/2\boldsymbol{v}_f(\boldsymbol{x}), \tag{9}$$

where the round function rounds the argument to nearest pixel value in the domain. There are some drawbacks to this approach. First, if the area around $\boldsymbol{x}$ in I_0 is occluded in I_1, there are likely multiple flow candidates assigned to the point round$(\boldsymbol{x} + 1/2\boldsymbol{v}_f(\boldsymbol{x}))$. In the converse situation, i.e. dis-occlusion from I_0 to I_1 there may be pixels that are not hit by a flow vector, thus leaving holes in the flow.

While the first problem can be solved by choosing the candidate vector with the best data fit, the solution for the problem of dis-occlusions in not that simple. Here we will simply fill the holes in the flow field by an outside-in filling strategy. With a dense flow we can then interpolate $I_{1/2}$ using the *forward* scheme

$$I_{1/2}(\boldsymbol{x}) = \frac{1}{2}\left(I_0(\boldsymbol{x} - \boldsymbol{v}_f^{1/2}(\boldsymbol{x})) + I_1(\boldsymbol{x} + \boldsymbol{v}_f^{1/2}(\boldsymbol{x}))\right), \tag{10}$$

or consider the backward flow $\boldsymbol{v}_b$ (i.e. the flow from I_1 to I_0) and use a backward scheme accordingly. We will in addition consider a *bidirectional* interpolation scheme where the frame is interpolated as the average frames obtained by the forward and backward schemes.

One can sophisticate the interpolation methods by estimating occluded regions and selectively interpolating from the correct frame. We will not pursue any occlusion reasoning here, but refer to [10] and [12] for details.

4   Reparametrizing Optical Flow for Interpolation

The approach presented in the previous section is the standard procedure for frame interpolation and serves as backbone in many algorithms ([9], [13], [14],

[11]). In this section we will reparametrize the original energy functional so the recovered flow is better suited for interpolation purposes. The reparametrization turns out to be beneficial on a number of levels: It makes the temporal warping of the flow superfluous, eliminates the need to calculate flows in both directions, improves handling of large motion, and increase overall robustness.

The original optical flow energy functional take as argument an optical flow v that is defined on a continuous domain. In practice, however, we only observe images at discrete pixels, and the optical flow is typically only estimated at the points corresponding to the pixels in I_0. Since we assume that the intermediate frame $I_{1/2}$ can be obtained from linearly following the flow vectors, we propose to reparametrize the data fidelity functional F using this assumption, so that it is given as

$$\frac{1}{2} \int \|I_1(x + v_s(x)) - I_0(x - v_s(x))\| \, dx. \tag{11}$$

We note that in this parametrization, the coordinates of the optical flow matches those of the intermediate frame $I_{1/2}$, and using this data term will thus eliminate the need for warping of the flow, since interpolation can directly be done similarly to (10). Because the motion vectors of the symmetric flow v_s are only half of the ones of e.g. the forward flow v_f, we need to halve the corresponding λ to keep comparison fair, which is the reason for the factor $1/2$.

Linearizing the data matching term (11) around v_0 gives

$$\rho(v_s) = I_1(\cdot + v_0) - I_0(\cdot - v_0) + (J_{I_1}(\cdot + v_0) + J_{I_0}(\cdot - v_0))(v_s - v_0) \tag{12}$$

which is similar to (4). In 1D the corresponding split energy term (5) is easily minimized using (7), and in general using the L^1-L^2 minimization from [8].

The differences between (12) and (4) are that we now allow sub-pixel matching in both surrounding images, and instead of a single Jacobian we have a sum of two. Thinking of this linearization as a finite difference scheme corresponding to a linearized differential form of the data fidelity term, we see that the temporal derivative is represented by a central finite difference scheme, as opposed to the typical forward differences (4). In addition the sum of the two Jacobians should make the estimation procedure more robust to noise, as the noise amplification caused by derivative estimation is now averaged over two frames. This has previously been used heuristically to improve accuracy in asymmetric flow estimation (see e.g. [15]). Finally we note that the motion vectors will only have half the length of the ones obtained from the regular parametrization. This will make the method better suited to handle large displacements compared to traditional methods that only make use of a one-sided linearization.

5 Results

Motion compensated frame interpolation finds many uses, ranging from the more technical applications such as video coding [13] to disciplines like improving viewing experience [14] or restoration of historic material [11]. For the former type of application the reconstruction quality in terms of quantitative measures is of great importance. For the latter types it is hard to devise specific measures of quality, as the human visual system is very tolerant to some types of errors, while it is unforgiving to other types of errors. In fact, the types of tolerated errors may even depend on the specific sequence.

For the results presented in the following we use a setup where we solve (5) and (6) iteratively in a coarse-to-fine pyramid as illustrated in Algorithm 1. We use $\ell_{max} = 70$ levels and a scale factor of 0.95. On each level $w_{max} = 60$ warps are performed, and within each warp we minimize (5) with linearized data fidelity term (12) using (7), followed by minimization of (6) using using $i_{max} = 5$ inner iterations of a Bermùdez-Moreno type algorithm [6]. We fix the coupling parameter $\theta = 0.2$.

Algorithm 1: Computation of TV-L^1 optical flow

Data: Two images I_0 and I_1
Result: Symmetric optical flow field v_s
for $\ell = \ell_{max}$ *to* 0 **do**
   // Pyramid levels
   Downsample the images I_0 and I_1 to current pyramid level
   for $w = 0$ *to* w_{max} **do**
      // Warping
      Compute v_s pointwise as the minimizer of E_1 (5) with data fidelity (12)
      for $i = 0$ *to* i_{max} **do**
         // Inner iterations
         Compute u_s as the minimizer of E_2 (using methods presented in [6])
      end
      Upscale v and u to next pyramid level
   end
end

As our first experiment we compare the four different types of interpolation suggested in the previous sections, on the four *High-speed camera* training sequences of the Middlebury Optical Flow benchmark. Figure 1 shows the effect of varying the data term weight λ in terms of the mean absolute interpolation error (MAIE). We see that the symmetric flow outperforms the conventional approaches, and that it is typically less sensitive in terms of the choice of λ. In particular we see that the difficult Beanbags sequence which contains large displacements is handled much better by the symmetric scheme. By evaluation on the Middlebury training set it was found that $\lambda = 35$ gave the best overall performance for the symmetric flow, and that $\lambda = 20$ gave the best performance for the other three methods. These λ values will be used in the rest of the experiments presented in this section.

For our second example we consider the results of interpolation under noise. Figure 2 shows the MAIE performance of the four methods on the Beanbags

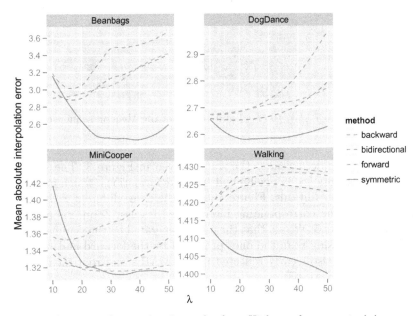

Fig. 1. Performance for varying λ on the four *High-speed camera* training sequences from the Middlebury Optical Flow benchmark [9]

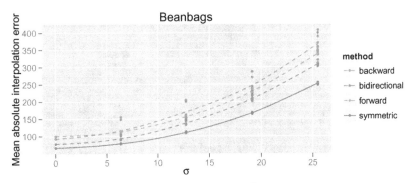

Fig. 2. MAIE performance under additive $\mathcal{N}(0, \sigma^2)$ noise for varying σ. Results are for the Beanbags sequence, and are based on 10 independent replications.

sequence with additive $\mathcal{N}(0, \sigma^2)$ noise. The improved robustness of the symmetric interpolation method is clearly visible from the distances between the MAIEs to the asymmetric methods that increase as the standard deviation of the noise increases. In addition we see that the variance of the MAIEs across the independent replications is significantly lower for the symmetric method compared to the three other methods.

Fig. 3. Frames 7 (I_0), 10 ($I_{1/2}$) and 13 (I_1) of the Mequon sequence

As our third example, consider the frames given in Figure 3. The sequence has large displacements (> 35 pixels) and severe deformations, which makes the estimation of $I_{1/2}$ very difficult. Figure 4 shows the three different flows v_f, v_b and v_s along with the corresponding interpolated frames. Zoom ins of details can be found in Figure 5. We see that the result generated by the symmetric flow is visually more pleasing than the ones produced by the forward and backward flows, a fact that is also clearly reflected in the MAIEs and root mean square interpolation errors (RMSIE).

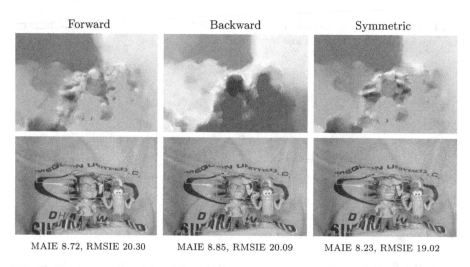

Forward	Backward	Symmetric
MAIE 8.72, RMSIE 20.30	MAIE 8.85, RMSIE 20.09	MAIE 8.23, RMSIE 19.02

Fig. 4. Results for the Mequon sequence. Top row: Color coded optical flows, buttom row: Interpolation results. Zoom ins of details can be found in Figure 5.

Finally let us compare the method to some methods of the current state-of-the-art. Table 1 holds the RMSIEs for six sequences from the Middlebury Optical Flow benchmark and results for a number of methods. While the results cannot fully match the results of Stich et al. [12], which gives significantly better results on 3 of the sequences, our method outperforms all other approaches, including the recent and much more complex methods of Chen and Lorenz [16] and Werlberger et al. [11].

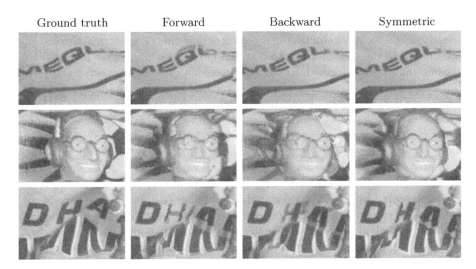

Fig. 5. Details of the interpolated Mequon frame from Figure 4

Table 1. RMSIE for different Middlebury sequences. Bold indicates the best result. [†] Results are taken from [12]. [‡] Marked algorithms have not been implemented by their respective authors, but by the Middlebury authors [9].

Method	Dimetrodon	Venus	Hydrangea	RubberWhale	MiniCooper	Walking
Symmetric TV-L^1	1.93	3.45	3.36	**1.46**	**3.96**	**2.89**
Chen & Lorenz [16]	1.95	3.63	—	—	—	—
Werlberger et al. [11]	1.93	—	—	—	4.55	3.97
Stich et al. [12]	**1.78**	**2.88**	**2.57**	1.59	—	—
Bruhn et al. [17][†,‡]	2.59	3.73	—	—	—	—
Pyramid Lukas-Kanade[†,‡]	2.49	3.67	—	—	—	—

Real-Time Performance

In the presented setup we only have to compute a single flow field between two images and fill in the intermediate frame from the trajectories. The runtime of the interpolation is dominated by the time it takes to compute the flow field, and at a slight cost in accuracy (5 pyramid levels with a scale factor of 2, and 30 warps per level, 1 level of median filtering) the flow fields can be computed in real-time ($\sim$35 fps) for 640×480 images using an NVIDIA Tesla C2050 GPU, which in turn means that we can do real-time frame doubling of 30fps video footage at a resolution of 640×480.

6 Conclusion and Outlook

We have proposed a method for motion compensated frame interpolation that is based on interpolation along the motion vectors. The main contribution is to use the assumption that an in-between frame can be reached by linearly following the motion vectors for reparametrizing the optical flow energy, such that the coordinate system of the flow matches that of the in-between frame. We have showed that one can improve frame interpolation methods substantially using this powerful and generic symmetric parametrization, and that the parametrization can be incorporated in existing frame interpolation methods with only little adaptation. Using a simple TV-L^1 optical flow algorithm as prototype we demonstrated results that are competitive with recent methods that are highly sophisticated. The proposed method can be implemented very efficiently, and using graphics hardware we succeeded in doubling the frame rate of 640×480 video in real time.

The presented work can be extended in a number of directions. The most obvious extension would be to use the symmetric data term with a more advanced optical flow method. If the goal is to improve viewing experience, a spatial regularization of the interpolated frames could probably improve the perceived quality. Spatial regularization could be done by means of total variation (see e.g. [14] and [11]) or by edge enhancing diffusion [18]. The latter has been shown to have very good interpolation properties, and has been successfully used in image compression [19] and for motion compensated deinterlacing [20]. To improve reconstruction quality in terms of e.g. RMSIE, one could do occlusion reasoning and selectively interpolate from the non-occluded frame, or compute motion trajectories over several frames [21] and use this information for interpolation.

References

1. Christensen, G.E., Johnson, H.J.: Consistent image registration. IEEE Transactions on Medical Imaging 20, 568–582 (2001)
2. Alvarez, L., Deriche, R., Papadopoulo, T., Sánchez, J.: Symmetrical dense optical flow estimation with occlusions detection. International Journal of Computer Vision 75, 371–385 (2007)
3. Alvarez, L., Castaño, C.A., García, M., Krissian, K., Mazorra, L., Salgado, A., Sánchez, J.: Symmetric Optical Flow. In: Moreno Díaz, R., Pichler, F., Quesada Arencibia, A. (eds.) EUROCAST 2007. LNCS, vol. 4739, pp. 676–683. Springer, Heidelberg (2007)
4. Chen, W.: Surface velocity estimation from satellite imagery using displaced frame central difference equation. To appear in IEEE Transactions on Geoscience and Remote Sensing (2012)
5. Bresson, X., Chan, T.: Fast dual minimization of the vectorial total variation norm and application to color image processing. Inverse Problems and Imaging 2, 455–484 (2008)
6. Goldluecke, B., Strekalovskiy, E., Cremers, D.: The natural total variation which arises from geometric measure theory. SIAM Journal on Imaging Sciences 5, 537–563 (2012)
7. Zach, C., Pock, T., Bischof, H.: A Duality Based Approach for Realtime TV-L^1 Optical Flow. In: Hamprecht, F.A., Schnörr, C., Jähne, B. (eds.) DAGM 2007. LNCS, vol. 4713, pp. 214–223. Springer, Heidelberg (2007)

8. Rakêt, L.L., Roholm, L., Nielsen, M., Lauze, F.: TV-L^1 Optical Flow for Vector Valued Images. In: Boykov, Y., Kahl, F., Lempitsky, V., Schmidt, F.R. (eds.) EMMCVPR 2011. LNCS, vol. 6819, pp. 329–343. Springer, Heidelberg (2011)
9. Baker, S., Scharstein, D., Lewis, J.P., Roth, S., Black, M.J., Szeliski, R.: A database and evaluation methodology for optical flow. International Journal of Computer Vision 31, 1–31 (2011)
10. Herbst, E., Seitz, S., Baker, S.: Occlusion reasoning for temporal interpolation using optical flow. Technical Report UW-CSE-09-08-01, Department of Computer Science and Engineering, University of Washington (2009)
11. Werlberger, M., Pock, T., Unger, M., Bischof, H.: Optical Flow Guided TV-L^1 Video Interpolation and Restoration. In: Boykov, Y., Kahl, F., Lempitsky, V., Schmidt, F.R. (eds.) EMMCVPR 2011. LNCS, vol. 6819, pp. 273–286. Springer, Heidelberg (2011)
12. Stich, T., Linz, C., Albuquerque, G., Magnor, M.: View and time interpolation in image space. Computer Graphics Forum 27, 1781–1787 (2008)
13. Huang, X., Rakêt, L.L., Luong, H.V., Nielsen, M., Lauze, F., Forchhammer, S.: Multi-hypothesis transform domain Wyner-Ziv video coding including optical flow. In: Multimedia Signal Processing (2011)
14. Keller, S., Lauze, F., Nielsen, M.: Temporal super resolution using variational methods. In: Mrak, M., Grgic, M., Kunt, M. (eds.) High-Quality Visual Experience: Creation, Processing and Interactivity of High-Resolution and High-Dimensional Video Signals. Springer (2010)
15. Wedel, A., Pock, T., Zach, C., Bischof, H., Cremers, D.: An Improved Algorithm for TV-^{1} Optical Flow. In: Cremers, D., Rosenhahn, B., Yuille, A.L., Schmidt, F.R. (eds.) Statistical and Geometrical Approaches to Visual Motion Analysis. LNCS, vol. 5604, pp. 23–45. Springer, Heidelberg (2009)
16. Chen, K., Lorenz, D.: Image sequence interpolation using optimal control. Journal of Mathematical Imaging and Vision 41, 222–238 (2011)
17. Bruhn, A., Weickert, J., Schnörr, C.: Lucas/Kanade meets Horn/Schunck: Combining local and global optic flow methods. International Journal of Computer Vision 61, 211–231 (2005)
18. Weickert, J.: Theoretical foundations of anisotropic diffusion in image processing. In: Kropatsch, W.G., Klette, R., Solina, F. (eds.) Theoretical Foundations of Computer Vision, Computing Supplement, vol. 11, pp. 221–236. Springer (1994)
19. Galić, I., Weickert, J., Welk, M., Bruhn, A., Belyaev, A., Seidel, H.P.: Image compression with anisotropic diffusion. Journal of Mathematical Imaging and Vision 31, 255–269 (2008)
20. Ghodstinat, M., Bruhn, A., Weickert, J.: Deinterlacing with Motion-Compensated Anisotropic Diffusion. In: Cremers, D., Rosenhahn, B., Yuille, A.L., Schmidt, F.R. (eds.) Statistical and Geometrical Approaches to Visual Motion Analysis. LNCS, vol. 5604, pp. 91–106. Springer, Heidelberg (2009)
21. Volz, S., Bruhn, A., Valgaerts, L., Zimmer, H.: Modeling temporal coherence for optical flow. In: Metaxas, D.N., Quan, L., Sanfeliu, A., Gool, L.J.V. (eds.) IEEE International Conference on Computer Vision (ICCV), pp. 1116–1123 (2011)

Ego-Motion Estimation Using Rectified Stereo and Bilateral Transfer Function

Giorgio Panin and Nassir W. Oumer

German Aerospace Center (DLR)
Institute for Robotics and Mechatronics
Münchner Straße 20, 82234 Weßling

Abstract. We describe an ego-motion algorithm based on dense spatio-temporal correspondences, using semi-global stereo matching (SGM) and bilateral image warping in time. The main contribution is an improvement in accuracy and robustness of such techniques, by taking care of speed and numerical stability, while employing twice the structure and data for the motion estimation task, in a symmetric way. In our approach we keep the tasks of structure and motion estimation separated, respectively solved by the SGM and by our pose estimation algorithm. Concerning the latter, we show the benefits introduced by our rectified, bilateral formulation, that provides at the same time more robustness to noise and disparity errors, at the price of a moderate increase in computational complexity, further reduced by an improved Gauss-Newton descent.

1 Introduction

Visual odometry, or ego-motion estimation, is concerned with the estimation of one's own velocities into an unknown, mainly rigid environment, through sequences obtained from one or more cameras fixed on the moving body. In this context, motion is usually constrained to planar (3-dof) or full 6-dof, under the assumption of a rigid scene with a few independently moving items (such as pedestrians, cars) acting as an external disturbance, which are detected and factored out of the estimation procedure.

Model-based approaches, such as [1], use pre-defined models of shape and appearance to be sought in the image, and provide an efficient and robust estimation of absolute pose and motion. However, such methods require an apriori model of the object, which in many scenarios may not be available.

Feature-based methods use a combination of feature detection, matching, tracking, triangulation and pose estimation from corresponding points. Among such techniques, [2] uses RANSAC and iterative pose refinement for stereo and monocular odometry. A similar monocular technique, which minimizes drift using a local bundle adjustment, was presented in [3]. Integration of other sensory modalities, such as GPS or IMU, also allows robustly coping with fast motion as in the real-time 3D modeler [4]. A disadvantage of feature-based techniques is that errors incurred at intermediate processing stages propagate to a higher

G. Bebis et al. (Eds.): ISVC 2012, Part I, LNCS 7431, pp. 458–469, 2012.

level, and that for reliable tracking a sufficient number of features should be available at each frame, which is not always the case.

Direct methods instead use all possible information from the image, including weak gradient regions, to estimate pose and structure of a scene or an object, by minimizing a photometric error rather than a geometric distance between features. As discussed in [5], a major advantage is that feature extraction and matching are not required, while a very large set of measurements are simultaneously available (one per pixel) providing generally more precise and robust performances.

Related techniques, using appearance models and optical flow [6], employ an extended planar pattern for tracking. However, the underlying assumptions about motion (for example a planar homography [7]) or camera model (for example affine cameras [8]) are usually strong, so that they apply to a restricted class of scenes or objects.

A recently developed approach minimizes intensity errors between consecutive image pairs from calibrated stereo sequences [9], and can be considered between model-based and image-based approaches. In this context, dense stereo matching is used to obtain a reference model for motion estimation, that may be built off-line (from a set of key-frames) or updated at each frame. This model consists of a dense point cloud, including color and disparity information. A point transfer function based on the quadrifocal tensor function allows rigid motion estimation directly on the next stereo pair, with full 6-dof. The approach handles arbitrary 3D structures, and improves the convergence domain with respect to planar region-based methods, since the whole image is used for the registration task.

In this paper we improve the above mentioned approach in some important aspects. Firstly, we introduce a symmetric transfer error, simultaneously projecting points forwards and backwards over time, where the Jacobian of the inverse transformation is easily obtained in the Lie algebra setting, and it is computed once per frame using an inverse compositional formulation. Due to the fact that stereo matching on consecutive frames does not provide fully overlapping point clouds, as explained in Section 4.1, this scheme integrates additional information for a more accurate motion estimation. This formulation also respects the symmetry of the problem, since by inverting the image sequence we exactly obtain the inverse motion estimates.

Secondly, instead of trifocal tensors we utilize rectified images, and stereo triangulation in homogeneous coordinates. This results in a simpler formulation, clearly isolating computation into off-line (structure) and on-line (re-projection) terms, while keeping intermediate quantities within a good numerical range. Stereo matching is done here with the semi-global matching (SGM) algorithm [10] using mutual information.

The paper is organized as follows: in Sec. 2 we present the stereo-based estimation framework. The two sub-problems of structure and motion estimation are dealt with in Sec. 3 and 4, respectively. Afterwards, Sec. 5 presents experimental results on simulated and real stereo sequences, compared with ground-truth trajectories. Sec. 6 concludes the paper, mentioning possible improvements to this system.

2   Problem Statement

In the following, we denote by l, r the left and right camera of the stereo rig, and by k the temporal frame index. Given four arbitrary (3×4) camera matrices $P_{l,r}^{k-1,k}$, two corresponding points in homogeneous coordinates $\mathbf{x}_{l,r}^{k-1}$ that satisfy the epipolar constraints (i.e. back-project to the same 3D point) can be transferred forwards in time to the corresponding points $\mathbf{x}_l^k$ and $\mathbf{x}_r^k$, by means of the respective trifocal tensors [9]. When the stereo rig is calibrated, but motion between $k-1$ and k is unknown, both tensors can be parametrized by the rigid transformation $T_l^{k-1,k}$ of the left camera frame, as well as differentiated through the Lie algebra of the Euclidean group.

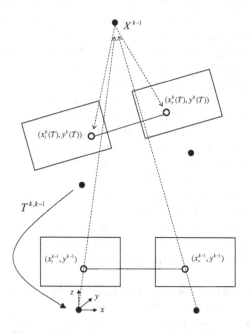

Fig. 1. Forwards re-projection after 3D triangulation

For almost parallel configurations, we simplify the formulation and address normalization issues by considering only rectified images, and casting the trifocal transfer into a mere forwards reprojection (Fig. 1), given by:

1. *(2D-3D)* Structure estimation from the pair at $k-1$, by triangulation in homogenous coordinates

2. *(3D-2D)* Motion estimation, by minimizing photometric error of re-projected points at k

so that the point cloud is computed out of the motion estimation loop.

3   Structure Estimation

As a starting point, we need a rectified stereo pair at time $k - 1$. This can be done using knowledge about the external and internal camera matrices, through an accurate calibration procedure [11, Chap.7][12], for example using a planar chessboard pattern.

Once that camera parameters are known, stereo rectification [11, Chap.11] consists in rotating both cameras around the respective center, until the Cartesian frames are aligned, and the transformation becomes a pure horizontal translation. Furthermore, it requires to align also the projection planes, so that internal parameters become equal.

This is equivalent to apply a planar *homography* to the left and right image

$$H_l = K_l R_l K_l^{-1}; \ H_r = K_r R_r K_r^{-1} \tag{1}$$

where $H_{l,r}$ are functions of the internal camera parameters $K_{l,r}$ and the rotation matrices $R_{l,r}$ that align the two camera frames[1].

In the end, we obtain the following projection matrices:

$$P_l = \begin{bmatrix} f & 0 & p_x & 0 \\ 0 & f & p_y & 0 \\ 0 & 0 & 1 & 0 \end{bmatrix} ; \ P_r = \begin{bmatrix} f & 0 & p_x & -fT_x \\ 0 & f & p_y & 0 \\ 0 & 0 & 1 & 0 \end{bmatrix} \tag{2}$$

where the rectified parameters are given by the common focal length f, the principal point (p_x, p_y), and the horizontal *baseline* T_x, expressed in metric units (e.g. *mm*).

Subsequent triangulation becomes then a trivial task: let $(x_l, y_l), (x_r, y_l)$ be a pair of corresponding poins, where $y_l \equiv y_r$ is the common coordinate, and let $X = (x, y, z, w)$ be the homogeneous coordinates of the corresponding 3D point, referred to the left camera frame. Then, we have

$$x = \frac{x_l - p_x}{f}; \ y = \frac{y_l - p_y}{f}; \ z = 1; \ w = \frac{x_l - x_r}{fT_x} \tag{3}$$

where coordinates are defined up to a scale factor, so we are free to choose $z = 1$, that keeps numerical stability and simplifies the computation of inverse-compositional Jacobians (see eq. 15). This representation also allows points at infinity, given by $w = 0$.

---

[1] The amount of image distortion introduced depends on the convergence angle between optical axes. Therefore, it is best applied to similar and almost-parallel cameras.

In order to perform dense stereo matching we use the state-of-the-art semi-global matching (SGM) algorithm [10],that nicely joins efficiency and robustness properties, through dynamic programming and mutual information. In absence of interpolation, missing disparities will occur over more or less large regions, so that the 3D point cloud will not be 100% dense.

4   Motion Estimation

The point cloud X^{k-1} is now re-projected on the next frame k by

$$\begin{bmatrix} \bar{x}_l^k \\ \bar{y}_l^k \\ \bar{z}_l^k \end{bmatrix} = P_l T X^{k-1}; \quad \begin{bmatrix} \bar{x}_r^k \\ \bar{y}_r^k \\ \bar{z}_r^k \end{bmatrix} = P_r T X^{k-1} \tag{4}$$

where T is the relative motion of left camera[2] and P_l, P_r are the constant matrices defined in (2), followed by normalization

$$x_l^k = \frac{\bar{x}_l^k}{\bar{z}_l^k}; \quad y_l^k = \frac{\bar{y}_l^k}{\bar{z}_l^k} \tag{5}$$

We parametrize motion by using Lie algebras [13]

$$T = \bar{T} \exp \left(\sum_{i=1}^{6} \delta p_i G_i \right) \tag{6}$$

where G_i are the generators, providing a basis for the *tangent space* to the Euclidean group

$$\sum_{i=1}^{6} \delta p_i G_i = \begin{bmatrix} [\omega]_\times & \mathbf{w} \\ \mathbf{0} & 0 \end{bmatrix} \tag{7}$$

using $[\cdot]_\times$ to denote the (3×3) cross-product matrix, and $\delta \mathbf{p} = [\omega, \mathbf{w}]$ the twist velocity, so that the derivatives at $\delta \mathbf{p} = \mathbf{0}$ (i.e. $T = \bar{T}$) are

$$\frac{\partial}{\partial \delta p_i} \begin{bmatrix} x_l^k \\ y_l^k \\ x_r^k \\ y_r^k \end{bmatrix}_{\delta \mathbf{p} = 0} = \begin{bmatrix} J_n\left(\bar{x}_l^k, \bar{y}_l^k, \bar{z}_l^k\right) P_l \\ J_n\left(\bar{x}_r^k, \bar{y}_r^k, \bar{z}_r^k\right) P_r \end{bmatrix} \bar{T} G_i X^{k-1} \tag{8}$$

for $i = 1, \ldots, 6$, where

$$J_n(\bar{x}, \bar{y}, \bar{z}) = \begin{bmatrix} 1/\bar{z} & 0 & -\bar{x}/\bar{z}^2 \\ 0 & 1/\bar{z} & -\bar{y}/\bar{z}^2 \end{bmatrix} \tag{9}$$

is the Jacobian of the normalization (5), evaluated at $\bar{T}$.

---

[2] Notice that in this representation we have $T \equiv T_l^{k,k-1}$, i.e. the transformation from *current* to *previous* left camera frame.

Now we can compute the photometric error of a re-projected pair from $k-1$ to k, under the transformation $\bar{T}$

$$e_l^k = I_l^k\left(x_l^k\left(\bar{T}\right), y_l^k\left(\bar{T}\right)\right) - I_l^{k-1}\left(x_l^{k-1}, y_l^{k-1}\right) \tag{10}$$
$$e_r^k = I_r^k\left(x_r^k\left(\bar{T}\right), y_r^k\left(\bar{T}\right)\right) - I_r^{k-1}\left(x_r^{k-1}, y_r^{k-1}\right)$$

where $(x_l^k, y_l^k, x_r^k, y_r^k)$ are given by eq. (3,4,5).

Derivatives of the residual with respect to local motion parameters δp_i are finally obtained, by taking the image gradients and multiplying them by the screen Jacobians

$$J_{l,i}^k = \nabla I_l^k\left(x_l^k, y_l^k\right) \cdot \frac{\partial}{\partial \delta p_i}\begin{bmatrix} x_l^k \\ y_l^k \end{bmatrix}_{\delta\mathbf{p}=0} \tag{11}$$

$$J_{r,i}^k = \nabla I_r^k\left(x_r^k, y_r^k\right) \cdot \frac{\partial}{\partial \delta p_i}\begin{bmatrix} x_r^k \\ y_r^k \end{bmatrix}_{\delta\mathbf{p}=0}$$

for $i = 1, \ldots, 6$. At non-integer point coordinates (x, y), the corresponding image values and gradients are obtained by bilinear interpolation from the four nearest neighbors. Furthermore, we set a minimum threshold on image gradients in order to avoid uniform regions, that create ambiguities both for stereo matching and motion estimation.

By putting together all of these quantities in vector form $\mathbf{J}^k, \mathbf{e}^k$, where each row of the $(2n_{k-1} \times 6)$ Jacobian is given by the above derivatives, and n_{k-1} is the number of matching pairs at $k-1$, we can write the normal equations for the linearized LSE problem

$$\left(\mathbf{J}^{k,T}\mathbf{J}^k\right)\delta\mathbf{p} = \mathbf{J}^{k,T}\mathbf{e}^k \tag{12}$$

where $\mathbf{H}_{i,j}^k = \mathbf{J}^{k,T}\mathbf{J}^k$ is the Hessian matrix and $\mathbf{g}_i^k = \mathbf{J}^{k,T}\mathbf{e}^k$ the gradient.

All of these quantities are evaluated at the current $\bar{T}$ that, after solving eq. (12), is updated to $\bar{T} \leftarrow \bar{T} \cdot \exp(\sum_i \delta p_i G_i)$.

As a further speed-up, we apply the *inverse-compositional* method [7]: instead of computing the Jacobian $\mathbf{J}^k$ over I^k at each iteration, we rather evaluate it using ∇I^{k-1} at the identity transform $\bar{T} = I$, and call it $\mathbf{J}_0^{k-1}$, so that

$$\delta\mathbf{p} = -(\mathbf{J}_0^{k-1,T}\mathbf{J}_0^{k-1})^{-1}\mathbf{J}_0^{k-1,T}\mathbf{e}^k \tag{13}$$

where the sign also changes, because the linearized residual now depends on $\bar{T}$ through $I_{l,r}^{k-1}$ instead of $I_{l,r}^k$, as explained in [7]. The notation $\mathbf{J}_0^{k-1}$ may create some confusion, however we preferred it in order to underline that this is a quantity related to the previous frame, and not to the current one.

Dropping the 0 subscript, eq. (11) becomes

$$J_{l,i}^{k-1} = \nabla I_l^{k-1} J_{n,l}^{k-1} P_l G_i X^{k-1} \tag{14}$$
$$J_{r,i}^{k-1} = \nabla I_r^{k-1} J_{n,r}^{k-1} P_r G_i X^{k-1}$$

where image gradients are taken at (x^{k-1}, y^{k-1}), and J_n is also evaluated at $\bar{T} = I$ with X given by (3), so that

$$J_{n,l}^{k-1} = \begin{bmatrix} 1 & 0 & -x_l^{k-1} \\ 0 & 1 & -y_l^{k-1} \end{bmatrix} \tag{15}$$

and similarly for $J_{n,r}^{k-1}$.

Also considering the simple structure of P_l, P_r, this provides a very fast computation. In fact, since the set of pairs is changing at each frame, $\mathbf{J}_0$ must be re-computed at each k. However, the cost of doing only one evaluation becomes negligible, with respect to the iterated point transfer and Gauss-Newton updates.

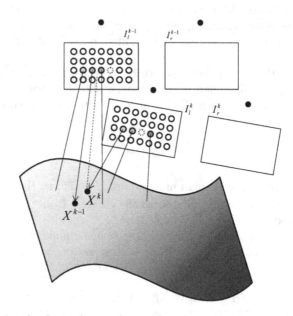

Fig. 2. Two point clouds at adjacent frames do not exactly overlap, neither in space nor in brightness, because of image resolution, noise, stereo matching errors, and occlusions due to motion. Therefore, a symmetric transfer error can improve motion estimation accuracy.

4.1 Symmetric Transfer Error

The previous result applies to one-directional point transfer, where the previous pair $I_{l,r}^{k-1}$ plays the role of a template, reprojected onto $I_{l,r}^k$. Therefore, we may also add a *backward*-transfer term, parametrized by the inverse transformation T^{-1}, where the current stereo pair $I_{l,r}^k$ is the template, reprojected onto $I_{l,r}^{k-1}$. In this way we have a symmetric error, roughly requiring twice the amount of computation for the motion estimation part[3].

The main motivation for that lies in the different point clouds $\{X^{k-1}\}, \{X^k\}$ sampled at consecutive frames (see also Fig. 2): in fact, due to many factors such as image resolution, occlusions on boundaries, brightness changes, shading effects, missing (or mis-matched) stereo disparities etc., the two point clouds will never exactly overlap, neither in spatial nor in brightness values, already for a small inter-frame motion. Therefore, they provide two different sets of measurements for the estimation of T, with a higher accuracy and robustness with respect to the mono-lateral case.

To make an interesting comparison, in a feature-based approach the symmetric formulation could be applied to the *geometric* re-projection error, as mentioned in [11, Chap. 4.2.2]. However, in that case temporal matching has already been performed through optical flow, so that the 4-tuple of corresponding points (at least under ideal conditions of an exact localization) is supposed to come from the *same* 3D point $X^k \equiv X^{k-1}$, and therefore no significant benefit is observed by adding the backwards term.

The backwards re-projection error is then

$$e_l^{k-1} = I_l^{k-1}(\bar{T}^{-1}) - I_l^k \tag{16}$$
$$e_r^{k-1} = I_r^{k-1}(\bar{T}^{-1}) - I_r^k$$

and inverse-compositional Jacobian

$$J_{l,i}^k = -\nabla I_l^k J_{n,l}^k P_l G_i X^k \tag{17}$$
$$J_{r,i}^k = -\nabla I_r^k J_{n,r}^k P_r G_i X^k; \quad i = 1, \ldots, 6$$

The opposite sign of eq. (17) is because the backwards-projection Jacobian is computed on the tangent space at $\bar{T}^{-1}$. In fact, we have $(\bar{T}\delta T)^{-1} = \delta T^{-1}\bar{T}^{-1}$, where the inverse of the exponential matrix is $(e^M)^{-1} = e^{-M}$, with $M = \sum_{i=1}^6 \delta p_i G_i$, and $\bar{T} = I$, so that G_i are simply replaced by $-G_i$.

Finally, the overall Hessian matrix is

$$\mathbf{H}_{i,j} = \sum_{l=1}^{n_{k-1}} J_{l,i}^{k-1} J_{l,j}^{k-1} + \sum_{r=1}^{n_{k-1}} J_{r,i}^{k-1} J_{r,j}^{k-1} + \sum_{l=1}^{n_k} J_{l,i}^k J_{l,j}^k + \sum_{r=1}^{n_k} J_{r,i}^k J_{r,j}^k \tag{18}$$

that is computed only once per frame, while the gradient

$$\mathbf{g}_i = \sum_{l=1}^{n_{k-1}} J_{l,i}^{k-1} e_l^k + \sum_{r=1}^{n_{k-1}} J_{r,i}^{k-1} e_r^k + \sum_{l=1}^{n_k} J_{l,i}^k e_l^{k-1} + \sum_{r=1}^{n_k} J_{r,i}^k e_r^{k-1} \tag{19}$$

is updated at each step.

---

[3] Instead, the triangulated structure at frame k, as well as $\mathbf{J}_0^k$, are re-used for the next forwards transfer, from k to $k+1$.

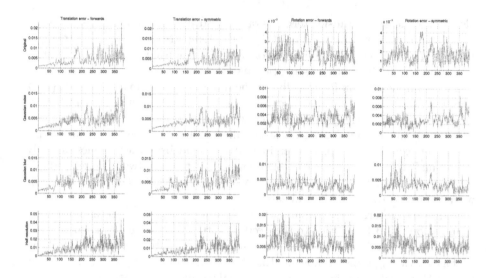

Fig. 3. Frame-to-frame motion estimation errors for different versions of the car sequence, comparing forwards-only with symmetric transfer errors. Translations are given in m, rotation angles in *deg*.

5   Experimental Results

In order to test our formulation, we performed experiments on sequences involving simulated and real camera images.

For each sequence, comparisons are done with ground truth data, obtained respectively by the simulation environment (hence, exact) or by different sensory data, such as GPS/IMU or robot kinematics. The latter also require knowledge of the rigid transform between the left camera and the sensor, obtained through a *hand-eye* calibration procedure [14], to determine the transformation between the robot TCP and the left camera frame Therefore, accurate ground truth is provided by the direct robot kinematics, through the absolute angular measurements from the joints. Since we do pure frame-to-frame estimation, without keeping a constant structure, we only compare incremental motion $T^{k-1,k}$.

The code has been implemented in C++ on a multi-core (Intel Xeon W3530) CPU with 2.8 GHz and 5 GB RAM, however without exploiting parallelization. Concerning processing times, for both sequences (at VGA resolution 640 × 480) we observe an average of 0.5 sec/frame using the symmetric transfer error, that decreases to 0.25 sec/frame for the mono-lateral case, as previously explained. However we emphasize that, on a dual-core platform, the symmetric reprojection function can be easily splitted into the two terms (forwards and backwards) at each LM iteration.

The car-driving simulation has been taken from a public dataset of the University of Auckland[4], also related to the work [15]. This sequence consists of

---

[4] http://www.mi.auckland.ac.nz

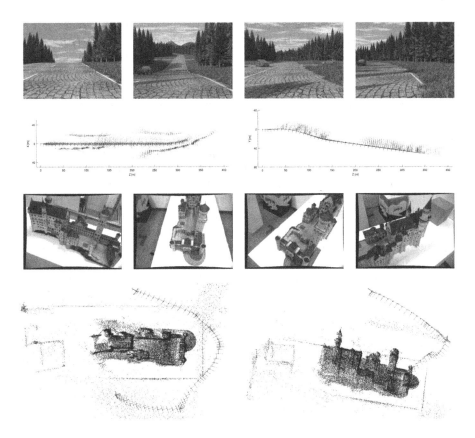

Fig. 4. Reconstructed map (subsets of points, re-projected onto the first camera frame) and left camera trajectories, from two different viewpoints. A few undistorted and rectified frames are also shown.

396 stereo frames, with a baseline of 0.3 m, showing navigation on a road with textured pavement, and forest trees surrounding it. The road goes briefly uphill and downhill, and makes a brief left turn at the end. Several cars appear in the field of view, from crossroads or running along the opposite lane, resulting in outliers that are detected by the algorithm.

The original image sequence is perfecly rectified, and no image noise is present, so that almost an ideal disparity map (dense and accurate) is obtained from the SGM algorithm, limited to a depth range of $z_{min} = 1$m, $z_{max} = 100$m. In this idealized scenario, we obtain good results for both the forwards and symmetric reprojection error (top row of Fig. 3), however we already notice an improvement in accuracy for the translation. The overall frame-to-frame translation error is around 1 cm/frame, for a motion of $50 - 100$ cm/frame. Rotation errors are very low (about 0.005 deg/frame), although the car maintains almost a constant attitude, with the exception of a left turn at the end, of about 0.3 deg/frame.

Subsequently, we tested the same sequence in the presence of image noise, blurring and lower resolution, all of them affecting the estimated disparities and colors. These conditions were obtained, respectively, by adding Gaussian noise with $\sigma = 10$ in the range $[0, 255]$, by Gaussian filtering with $\sigma = 2.5$ pixels, and by sub-sampling to half resolution. As we can see from the related plots in Fig. 3, benefits of the symmetric error become more evident; a combined effect of those disturbances has not yet been tested.

Next, we consider a real camera sequence, recorded by a small-baseline stereo rig (5 cm) mounted onto an industrial manipulator, after performing stereo as well as hand-eye calibration. The sequence consists of 1190 frames, showing a miniature model of the *Neuschwanstein* castle, put on a white table and surrounded by the robot arm, that performs a smooth full 6-dof trajectory. Here we use a depth range of $z_{min} = 0.05$ m, $z_{max} = 2$ m. As a result, the average motion error is about 0.1 mm/frame and 0.02 deg/frame, with the exception of frame 400, where an unpredicted large inter-frame motion caused an error of about 1 cm and 1 deg.

By considering the absolute errors, over a forth-and-back sweep covering about 2 m and 360 deg, using symmetric transfer the estimated trajectory accumulates a maximum drift of 25 mm and 4 deg, both corresponding to roughly 1% final error, which is a good result since no drift reduction (e.g. by means of key-frames selection and wide-baseline matching), is done. Using the forwards-only transfer, the final errors grow up to 35 mm and 5 deg.

6   Conclusions

We presented an efficient and accurate method for visual odometry in rectified stereo cameras, based on symmetric pixel transfer and photometric error minimization, making use of stereo triangulation in homogeneous coordinates at both consecutive frames. This method naturally handles normalization issues, while efficiently splitting computation between structure and motion estimation, the former executed once per frame, the latter in a Levenberg-Marquardt optimization loop, with outlier rejection and multi-resolution matching.

The current system employs a semi-global matching algorithm for computing dense stereo disparities, that can be accelerated by means of existing GPU [16] or FPGA implementations. The same applies to the odometry algorithm, concerning the computation of per-pixel reprojection errors and Jacobians.

Other issues may concern the robustness of the cost function to photometric outliers, due to shading or specularity effects, global brightness and contrast variations, as well as independently moving objects. In this context, apart from the standard rejection scheme of the present work, further improvements may be obtained by introducing more general cost functions, including explicit modeling of local illumination [17], or mutual information [18].

References

1. Comport, A., Marchand, E., Pressigout, M., Chaumette, F.: Real-time markerless tracking for augmented reality: the virtual visual servoing framework. IEEE Trans. on Visualization and Computer Graphics 16, 615–628 (2006)
2. Nistér, D., Naroditsky, O., Bergen, J.R.: Visual odometry. In: CVPR (1), pp. 652–659 (2004)
3. Mouragnon, E., Dekeyser, F., Sayd, P., Lhuillier, M., Dhome, M.: Real-time localization and 3d reconstruction. In: IEEE Conference of Vision and Pattern Recognition, Washington DC, vol. (1)
4. Strobl, K.H., Mair, E., Bodenmüller, T., Kielhöfer, S., Sepp, W., Suppa, M., Burschka, D., Hirzinger, G.: The Self-Referenced DLR 3D-Modeler. In: Proceedings of the IEEE/RSJ International Conference on Intelligent Robots and Systems, St. Louis, MO, USA, pp. 21–28 (2009), best paper finalist
5. Irani, M., Anandan, P.: About direct methods. In: Proc. Workshop Vision Algorithms: Theory Practice, pp. 267–277 (1999)
6. Lucas, B.D., Kanade, T.: An iterative image registration technique with an application to stereo vision (darpa). In: Proceedings of the 1981 DARPA Image Understanding Workshop, pp. 121–130 (1981)
7. Baker, S., Matthews, I.: Equivalence and efficiency of image alignment algorithms. In: IEEE Computer Society Conference on Computer Vision and Pattern Recognition, vol. 1, p. 1090 (2001)
8. Hager, G.D., Belhumeur, P.N.: Efficient region tracking with parametric models of geometry and illumination. IEEE Transactions on Pattern Analysis and Machine Intelligence 20, 1025–1039 (1998)
9. Comport, A.I., Malis, E., Rives, P.: Real-time quadrifocal visual odometry. I. J. Robotic Res. 29, 245–266 (2010)
10. Hirschmüller, H.: Stereo processing by semiglobal matching and mutual information. IEEE Trans. Pattern Anal. Mach. Intell. 30, 328–341 (2008)
11. Hartley, R.I., Zisserman, A.: Multiple View Geometry in Computer Vision, 2nd edn. Cambridge University Press (2004) ISBN: 0521540518
12. Zhang, Z.: A flexible new technique for camera calibration. IEEE Transactions on Pattern Analysis and Machine Intelligence 22, 1330–1334 (2000)
13. Drummond, T., Cipolla, R.: Visual tracking and control using lie algebras. In: CVPR, vol. 02, pp. 2652–2659. IEEE Computer Society, Los Alamitos (1999)
14. Strobl, K.H., Hirzinger, G.: Optimal hand-eye calibration. In: 2006 IEEE/RSJ International Conference on Intelligent Robots and Systems, IROS 2006, Beijing, China, October 9-15, pp. 4647–4653 (2006)
15. Vaudrey, T., Rabe, C., Klette, R., Milburn, J.: Differences between stereo and motion behavior on synthetic and real-world stereo sequences. In: 23rd International Conference of Image and Vision Computing New Zealand (IVCNZ 2008), pp. 1–6 (2008)
16. Ernst, I., Hirschmüller, H.: Mutual Information Based Semi-Global Stereo Matching on the GPU. In: Bebis, G., Boyle, R., Parvin, B., Koracin, D., Remagnino, P., Porikli, F., Peters, J., Klosowski, J., Arns, L., Chun, Y.K., Rhyne, T.-M., Monroe, L. (eds.) ISVC 2008, Part I. LNCS, vol. 5358, pp. 228–239. Springer, Heidelberg (2008)
17. Silveira, G., Malis, E., Rives, P.: An effecient direct approach to visual slam. IEEE Transactions in Robotics 24, 969–979 (2008)
18. Dame, A., Marchand, E.: Mutual information-based visual servoing. IEEE Trans. on Robotics 27 (2011)

Generative 2D and 3D Human Pose Estimation with Vote Distributions

Jürgen Brauer, Wolfgang Hübner, and Michael Arens

Fraunhofer Institute of Optronics, System Technologies and Image Exploitation
Gutleuthausstr. 1, 76275 Ettlingen, Germany
{juergen.brauer,wolfgang.huebner,michael.arens}@iosb.fraunhofer.de

Abstract. We address the problem of 2D and 3D human pose estimation using monocular camera information only. Generative approaches usually consist of two computationally demanding steps. First, different configurations of a complex 3D body model are projected into the image plane. Second, the projected synthetic person images and images of real persons are compared on a feature basis, like silhouettes or edges. In order to lower the computational costs of generative models, we propose to use vote distributions for anatomical landmarks generated by an Implicit Shape Model for each landmark. These vote distributions represent the image evidence in a more compact form and make the use of a simple 3D stick-figure body model possible since projected 3D marker points of the stick-figure can be compared with vote locations directly with negligible computational costs, which allows to consider near to half a million of different 3D poses per second on standard hardware and further to consider a huge set of 3D pose and configuration hypotheses in each frame. The approach is evaluated on the new Utrecht Multi-Person Motion (UMPM) benchmark with the result of an average joint angle reconstruction error of 8.0°.

1 Introduction

Estimating the 2D and 3D articulation of a person is an important step for action recognition and automatic visual scene understanding. There is a huge amount of literature on human pose estimation. Surveys are provided e.g. by Sminchisescu [1], Poppe [2], and Ji and Liu [3]. Work on human pose estimation can be divided into *top-down* (generative) and *bottom-up* (discriminative, conditional, recognition-based) approaches.

Bottom-up approaches try to predict the 2D or 3D pose directly given image features. There are model-free and model-based bottom-up approaches. Model-free approaches do not make use of a body model, but learn the mapping from image features to 2D/3D directly using a huge training set of image and 2D/3D pose pairs. In contrast, model-based bottom-up approaches make explicit use of a body model to map the features to a 2D/3D pose.

Top-down approaches are inherently model-based. Hypothesized 3D poses of a human body model are rendered into the 2D image and compared with image

G. Bebis et al. (Eds.): ISVC 2012, Part I, LNCS 7431, pp. 470–481, 2012.

features. A key issue for any top-down approach is the definition of a robust matching method, which allows to compare a synthetic view and a real image. For modelling the 3D body different geometric primitives as ellipsoids [4], cylinders [5], or super-quadrics [6] are used. The most detailed 3D human body model so far is [7] and consists of a polymesh based shape model with 25,000 polygons. The authors generate 2D silhouette projections of different 3D pose hypotheses and compare these with the current estimated silhouette from the person image to be analyzed. Beside silhouettes, edges are the image features mostly used as basis for the projected model vs. image evidence comparison. [8] compare edges from the projected model with computed edges by searching into the direction of the projected edge normal for an edge extracted from the image. An alternative approach to using fixed geometric primitives for modeling the 3D body model was recently presented [9]. For each limb a model of its projected shape is learnt using Microsoft's Kinect, which provides a easy way to collect example data, since it provides 3D poses and camera images simultaneously. Generative approaches are claimed to be computationally demanding compared to discriminative approaches due to the high computational costs for projecting a huge set of 3D pose candidates into the image and to compare each projection with the person image using low-level image features as edges and silhouettes.

We propose a computationally efficient approach which avoids the need for a complex 3D body model. An Implicit Shape Model (ISM) ([10], [11]) is used to generated a distribution of votes for potential body landmark locations. Thereby we replace the low-level image evidence with a representation that one the hand, still captures the ambiguity inherent in the 2D domain, but on the other hand, allows to use a simple 3D stick-figure body model, since we can compare projected marker points of this 3D body model with the votes directly and with low computational costs.

There are two approaches ([12], [13]) which followed a similar idea: lifting the low-level image evidence to some inter-mediate level. Both works first lift the image evidence to a 2D pose estimate and then use this 2D pose estimate to compare it with projected 3D body model configurations. [13] estimate consistent sequences of 2D poses which are called 'tracklets' using the pictorial structures model and use these tracklets as input for the 3D pose estimation. [12] use non-parametric belief propagation (NBP) to infer probability distributions representing the belief in the 2D pose state of each limb. Nevertheless, in both approaches an early decision about the 2D pose is made by using a 2D kinematic model which restricts the space of possible 2D poses and thereby the space of possible 3D poses that can be detected if the 2D kinematic model does not include all possible projections of all 3D poses. In contrast, we do not use any 2D kinematic model and do not even try to estimate a 2D pose since we believe that 2D pose ambiguities can better be solved in the 3D world. For this, the output of our first stage is a vote distribution for possible marker locations and not a set of possible 2D poses.

This article is structured as follows. In section 2 we explain how we determine potential locations for each 2D body marker, while section 3 describes the 2D and 3D pose estimation step. Section 4 evaluates our approach on the recently published Utrecht Multi-Person Motion (UMPM) benchmark designed especially for quantitative 3D pose estimation evaluations.

2   2D Body Marker Location Estimation

In this section we describe how we transform a set of local image features (we use SURF features) – extracted from a person image – into a set of potential marker locations.

Fig. 1. Small images: sample descriptor regions of extracted SURF features that are assigned to the same visual word. Only the ground-truth marker locations within the corresponding descriptor region are used for training samples of (visual word, marker location) pairs. Ground-truth marker locations as head, neck, shoulders, elbows, hands, knees, and feet are visualized by circles. Big images at left: visual word mean image and corresponding collected marker locations from the samples. The mean image for the visual word is computed by averaging all descriptor region images (resized to fixed mean image size) of the features to the right.

We assume that we already know the approximate location of the persons in the image. State of the art person detectors for this task are sliding detection window based approaches as the HOG detector [14] and part-based approaches as e.g. the Implicit Shape Model (ISM) [10] which uses SIFT features, or the Fastest Pedestrian Detector in the West [15] which uses Integral Channel Features [16]. For a survey on state of the art person detection we refer the reader to [17].

For the body marker location estimation step we adopt the ISM based body part detection approach presented in [11] but introduce three important modifications to the original work. The basic idea of ISM based 2D pose estimation is to learn the spatial distribution of body markers relative to each visual word of a codebook. More precisely, we start with a training step in which we first

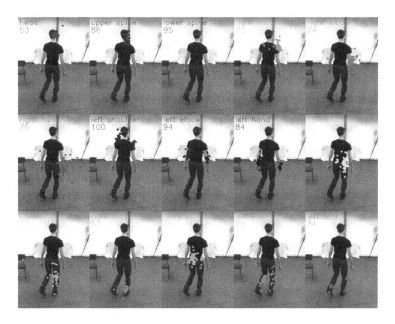

Fig. 2. Body part location ambiguity. For each marker we collect votes casted from visual words. Votes are displayed by small circles. Deciding for an unique location for each marker is not possible at this early stage of pose estimation, since left/right ambiguities cannot be solved just based on identified image structures. Note the left/right ambiguities for nearly all body parts (shoulders, elbows, hands, hips, knees, feet).

compute local image features for a training set of 2D pose ground-truth labeled person images, map each feature to a visual word of a given codebook (computed by clustering local features extracted from a set of person images), and record the location offsets for each body marker relative to this visual word. The result is a list of example locations for each (visual word, marker) pair. These lists are the training result and will be used later to estimate the 2D pose for a new person image. In the detection phase, visual words then cast marker specific votes according to the previously learned example locations.

A first modification to the original ISM approach concerns the collection of (visual word, marker location) observation pairs. The original approach [10] and its extension to 2D pose estimation [11] used all (feature, marker) observations, even if the marker location was not within the descriptor region of the feature. Thus a local image structure e.g. occurring spatially restricted only near to the feet was later allowed to vote for the right shoulder location as well. We observed this idea to be highly problematic, since votes of some few visual words that e.g. appear on the right shoulder are then typically dominated by a huge set of votes stemming from all other visual words, which actually do not have a meaning for the location of the right shoulder. This leads to the new idea of collecting only (feature, marker) observation pairs in the training step for which the marker location is within the descriptor region of the corresponding observed feature (see Fig. 1).

The second modification relates the selection of features that are used in the vote casting phase. The SURF keypoint detector typically detects blob-like structures, thus it can provide also a huge amount of meaningless small scale features e.g. on textured blob-like structures appearing on the person's clothings or small image structures. These small region sized features can appear at many different locations on the human body. In the original ISM approach these small sized features are nevertheless matched against visual words and allowed to cast votes, resulting in many wrong votes, which again can dominate the votes casted by visual words covering large parts of the person image and containing worthwhile information about the marker locations. Therefore we discard all extracted SURF features that have a descriptor region size smaller than 15% of the average descriptor region size of all features extracted from the current person image.

A third important modification concerns the final 2D pose estimate determination and representation. In [11] vote maxima were considered to decide for a final location for each marker, thereby making an early decision about the 2D pose although such a decision is mostly not reasonable, since 2D left/right ambiguities and problems due to missing or wrong estimated marker locations can better be solved in the 3D model domain (see Fig. 2). For this, we use the full body marker location vote distributions as input for the determination of a plausible 3D pose.

The ISMs transform the image representation consisting of a set of extracted SURF features to a vote distribution $\mathcal{E}_m$ for each marker which can be considered as a new image evidence $\mathcal{E} = \{\mathcal{E}_m\}$ $(m = 1, \ldots, M)$, where $\mathcal{E}_m = \{\boldsymbol{v_k} = (x_k, y_k)\}$ $(k = 1, \ldots, N_m)$ is the 2D vote distribution for marker m. N_m is the number of votes for marker m in the current frame. M is the number of markers of the body model used (here: $M = 15$).

3   Pose Estimation

This section explains how we estimate a 3D and a 2D pose based on the estimated marker locations, represented by the vote distributions $\mathcal{E}_m$ for each marker.

The pose estimation problem can be formulated using Bayes' theorem:

$$P(o|\mathcal{E}) \propto P(\mathcal{E}|o)P(o) \tag{1}$$

where o is a 3D pose. The final 3D pose estimate is then the pose that maximizes the posterior probability $P(o|\mathcal{E})$. Many approaches (e.g. [13], [18], [7]) follow the Bayesian formulation. The difference between the approaches is how the prior probability $P(o)$ over the 3D pose space and the observation likelihood $P(\mathcal{E}|o)$ are approximated.

Prior Modeling. For modeling the prior $P(o)$ we use an example-based approach, i.e. use a set of example 3D poses that have prototypical character. For this, we traverse a motion capture database (here: UMPM dataset) and incrementally collect example 3D poses. While traversing we consider each 3D pose, compare it to all 3D poses collected so far and only add it to our example set S, if the average joint angle difference is above some threshold θ. This threshold θ allows to control how different the poses in the example set will be: a large threshold e.g. will result in a small example set with large joint angle differences between the example 3D poses. We use the same prior probability $1/|S|$ for each of the 3D example poses, since the occurrence frequency of a 3D pose in a motion capture database does not need to correspond to its occurrence frequency in real world. Example based approaches can only recognize 3D poses seen before and contained in the example set S, which seems to be very restrictive. It depends on the size of S and whether there are poses in S which are similar to the poses we test on, how restrictive this approach is in practice. On the other hand, a major advantage of such an example-based approach is its robustness. We can put similar example 3D poses into S to the ones we expect to be of interest in our application scenario and we do not need to worry about invalid 3D pose configurations as final pose estimates, which is a major issue in geometric reconstruction based approaches as [19].

Likelihood Estimation. The most important issue concerns the question how to model the likelihood $P(\mathcal{E}|o)$. The key idea is to project each of our example 3D poses $o \in S$ onto the image and to approximate the likelihood by comparing the projected marker locations with the marker specific votes, i.e. the evidence E_m for potential locations of the m-th marker (see Fig. 3).

Since we do not know from which viewpoint we will observe the person, we generate for each pose o a set of 2D example projections while rotating and tilting the pose – encoded by two angles α_{pan} and α_{tilt} or a single rotation matrix R (where we use only 2 DOFs, no camera-up vector). Different focal lengths f are used to handle different perspective limb foreshortening situations. If we do not expect the person images to be recorded from a strong perspective, we use a large focal length to approximate the parallel projection. Each 3D marker point $q = (x, y, z)$ is projected to a 2D image point $q' = (u, v)$ using a perspective camera model, i.e.

$$\begin{bmatrix} x' \\ y' \\ z' \end{bmatrix} = R \begin{bmatrix} x \\ y \\ z \end{bmatrix} + t, \qquad u = f\frac{x'}{z'} + p_x, \qquad v = f\frac{y'}{z'} + p_y \qquad (2)$$

To handle different locations of the person in the image, we further use different principal points $p = (p_x, p_y)$ for the final location of the projected pose. If we know the rough person center (e.g. due to a person bounding box) we can use locations p sampled around this person center. To handle different sizes of the person, we further scale the final projected 2D pose limb lengths with an uniform scaling factor s. A single projection of a pose o can therefore be described by the projection parameters $c = (\alpha_{pan}, \alpha_{tilt}, f, p, s)$. For each 3D pose o we generate a set $T(o) = \{o'\}$ of example projections $o' = \{q'_i = (u_i, v_i)\}$ $(i = 1, \ldots, M)$ which we can compare individually with the image evidence $\mathcal{E}$ by comparing each projected marker location $q'_i = (u_i, v_i)$ with all the votes $v_k \in \mathcal{E}_m$ for this marker.

For this, we compute a weighted sum of all votes which are nearby to the projected marker location, i.e. assess how much evidence we can find that the observed marker location $\mathcal{E}_m$ is near to the projected marker location q'_i:

$$P(\mathcal{E}|o') = \sum_{i=1}^{M} \mathbf{f}(q'_i, \mathcal{E}_i) \tag{3}$$

where $\mathbf{f}$ is some kernel function that weights the votes in $\mathcal{E}_i$ depending on their distance to the projected marker location q'_i. The simplest choice for $\mathbf{f}$ is a flat kernel, i.e. we count the number of votes that are within a circle around the projected marker which represents the kernel center. Since the meaning of *nearby* depends on the overall size of the projected pose o', we should adapt the kernel size proportional to the scale factor s, i.e. $r \propto s$. For a flat kernel, $\mathbf{f}$ is defined as:

$$\mathbf{f}(q'_i, \mathcal{E}_i) = \sum_{v_k \in \mathcal{E}_i} \mathbf{g}(q'_i, v_k) \qquad \mathbf{g}(q'_i, v_k) = \begin{cases} 1 \, , \|q'_i - v_k\| \le r \\ 0 \, , \|q'_i - v_k\| > r \end{cases} \tag{4}$$

While the flat kernel discards all votes having an Euclidean distance greater than r from the projected marker, a Gaussian kernel function $\mathbf{f}$ can be used to consider all votes, but giving the votes different weights depending on their distance to the projected marker location:

$$\mathbf{g}(q'_i, v_k) = \frac{1}{\sigma\sqrt{2\pi}} e^{-\frac{1}{2}(\frac{\|q'_i - v_k\|}{\sigma})^2} \tag{5}$$

where the variance of the Gaussian should be set again according to the scale of the 2D pose, i.e. $\sigma \propto s$.

The overall likelihood for a single 3D pose o is

$$P(\mathcal{E}|o) = \max_{o' \in T(o)} P(\mathcal{E}|o') \tag{6}$$

Since we find the best 3D pose by comparing projections of our example 3D poses with the image evidence under different projection situations, the final result of the pose estimation step is not only (i) a 3D pose o, but (ii) also a 2D pose estimate o' as well, and further (iii) the overall relative orientation of the person to the camera, since we know from which camera view angles α_{pan}, α_{tilt} we projected the 3D pose o to the 2D pose o'.

4   Evaluation

In this section we evaluate the quality of our 3D pose estimation method by comparing ground truth 3D poses with our estimated 3D poses.

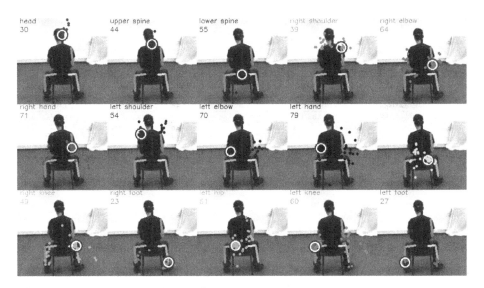

Fig. 3. Observation likelihood computation principle. Example 3D poses o are projected into the image using different projection parameters c. For comparing the projected pose o' with the image evidence $\mathcal{E}$ we count how many votes we can find nearby each projected marker q'_m within some radius (visualized by circle). This is done independently for each marker m using the marker specific vote lists $\mathcal{E}_m$. While the flat kernel does consider only nearby votes, the Gaussian weighting strategy considers all votes, but weighted with their distance to the projected marker location q'_m. The pose o' displayed is the best match found for the vote distribution in this frame.

Test Dataset. The new [20] Utrecht Multi-Person Motion (UMPM) benchmark[1] dataset allows for detailed comparisons of estimated vs. ground-truth poses. It includes synchronized motion capture and video data for 4 cameras and therefore allows to compare an estimated 3D pose with the corresponding ground truth 3D pose. It further provides the extrinsic (camera location + rotation) and intrinsic (focal lengths, principal points, distortion polynomial coefficients) parameters for each of the 4 cameras. Thus we can project each 3D pose into the image to yield the corresponding ground truth 2D pose as well, which we can use for generating training pairs (image, 2D pose) for training the body marker localizer described in section 2.

Error Measures. We evaluate our pose estimates quantitatively using two different error measures: E_1 is used to measure the articulation error and is defined as the average joint angle difference between the estimated and the ground-truth 3D pose. E_2 is used to measure the global orientation angle error. For this, we project the line connecting left and right hip onto the floor (XZ plane in the OpenGL visualizations in Fig.4) and define the angle between this line and the

---

[1] http://www.projects.science.uu.nl/umpm/

Table 1. Experiment definitions and results. The average over all 8 experiments for E_1 is 8.0° (36.3°), while for E_2 it is 27.6° (22.8°).

exp no.	experiment description	Train 2D [frames] / [persons]	$\|S\|$ [poses]	Test [frames]	E1 (var) [°]	E2 (var) [°]
1	calibration + flat + limited S	535 / 1	148	514	8.3 (5.2)	29.6 (33.1)
2	calibration + flat + exhaustive S	535 / 1	514	514	7.7 (3.9)	34.7 (45.5)
3	calibration + Gaussian + limited S	535 / 1	148	514	8.4 (4.8)	39.5 (49.2)
4	calibration + Gaussian + exhaustive S	535 / 1	514	514	7.8 (4.3)	36.6 (43.2)
5	generalization + flat + limited S	213 / 4	245	504	9.6 (4.8)	25.6 (2.3)
6	generalization + flat + exhaustive S	213 / 4	504	504	5.5 (3.2)	17.7 (3.4)
7	generalization + Gaussian + limited S	213 / 4	245	504	11.4 (7.1)	20.3 (3.1)
8	generalization + Gaussian + exhaustive S	213 / 4	504	504	5.5 (3.0)	16.5 (2.2)

camera plane as the global orientation. E_2 measures the relative angle difference between the global orientation of the estimated and the ground truth pose.

Limited vs. Exhaustive Example Poses. Since the smallest reachable pose error E_1 is limited by the similarity between our example 3D poses in our example set S and the actual ground-truth 3D poses we conduct two experiments. In the first setting (limited set of example poses), we use a training set S with prototype 3D poses, which were extracted from motion capture sequences of other person, i.e. different to the ones we test on, using the procedure described in section 3 by collecting example 3D poses ($\theta = 0.05$ radians). This is the realistic setting in which we do not have knowledge about the 3D poses that will occur in the test phase. In the second setting (exhaustive set of example poses), we put all 3D poses of the sequence we test on into the training set. This is an unrealistic setting in terms of a later application, since we know which 3D poses will occur before. Nevertheless, it is interesting to see which quality of 3D pose estimates we can reach if S is perfect, since (i) the body part localization step and (ii) the discrete sampling of projection configurations c and the (iii) search for the right 3D pose within the set of examples poses will introduce additional errors into the overall processing pipeline.

Calibration vs. Generalization. A further important distinction has to be made between scenarios, (i) in which we train the ISM body part localizer on a person P_1 and test the pose estimation on new sequences showing P_1 (experiments 1-4), or (ii) in which we train the ISM body part localizer on some persons $P_1, \ldots, P_N$ and test on a new person P_{N+1} which was never seen before (experiments 5-8). In (i) we will have very similar local features in the test phase as in the training phase, whereas in (ii) train and test local features will differ and therefore the generalization ability of the SURF feature descriptor, the feature to visual word matching, and the ability of the successive pose estimator to compensate for wrong estimated body marker locations is of high importance. Since (i) is nevertheless an important application scenario in which we calibrate the 2D pose estimator on a person using a marker-based approach and can later estimate poses marker-less using monocular camera information only for this person, we test both scenarios.

Flat vs. Gaussian Kernel. We further differ between experiments in which we used the flat kernel vs. the Gaussian kernel (see section 3).

Experiments Conducted. Table 1 shows the experiments conducted and the results regarding E_1 and E_2 (mean and variance of these errors). We further specify the total number of training and test frames used. In experiment 5 e.g. we train the ISM based body part localizer using 213 (image, 2D ground truth pose) pairs of 4 different persons $P_1, \ldots, P_4$, while the set of example poses S contains 245 example 3D poses (from motion capture data from persons $P_1, \ldots, P_4$). Testing (3D pose estimation) is done for a sequence of 504 frames showing a new person P_5 never seen before (image, motion capture data), resulting in an average joint angle error (averaged over all joint angles and 504 frames) of 9.6° (variance of this error: 4.8°), while the global orientation error was 25.6°.

Computing Time. For projecting a single 3D pose onto the 2D image and comparing it with the image evidence using the flat kernel only 0.00222 ms = 2.22 μs are needed in average (straightforward C++ implementation, standard hardware), i.e. we can roughly project and compare 450450 3D pose configurations with the body marker vote distributions per second. For the Gaussian vote weighting strategy we need in average 0.0035 ms = 3.5 μs which allows to test 285714 3D pose configurations per second.

Results. The results show that using an exhaustive example 3D pose set S (exps 2,4,6,8) yields better results compared to the limited example 3D pose set (exps 1,3,5,7), which we expected. Nevertheless, there is a huge difference between the calibration and the generalization scenario. For the calibration scenario there was no significant difference, while for the generalization scenario, the average joint angle error dropped e.g. from 9.6° to 5.5° (exp 5 vs. 6), and from 11.4° to 5.5° (exp 7 vs. 8) for E_1. This could indicate that considering as many 3D poses as possible can dramatically help to improve 3D pose estimation, which in turn underlines the need for a lightweight processing pipeline for generative approaches as presented here, in order to allow for testing several thousands of different 3D poses and configurations per frame. The results for the flat vs. Gaussian kernel were unexpected. We expected the Gaussian kernel to yield better results than the flat kernel since it weights votes nearer to the projected marker locations higher than far distant votes and thereby distinguishing between near and far distant votes. But the opposite was true, which is important to know, since the usage of an Gaussian (exponential) kernel is connected with higher computational costs (3.5 μs vs. 2.22 μs per pose evaluation): for the Gaussian kernel we need to evaluate the exponential function, while for the flat kernel, we only need to consider the distance of a vote to the projected marker location. The average joint angle error – averaged over all joints, frames, and experiments – is 8.0° (E_1), while the average global orientation error for the estimated 3D poses is 27.6° (E_2). The smallest 3D pose articulation error E_1 was reported by Agarwal and Triggs [21] and is 5.91° for single-frame pose estimation and 4.1° for multiple-frame pose estimation. While our average joint angle error is

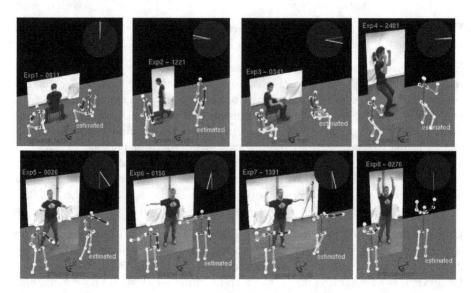

Fig. 4. Qualitative examples of estimated 3D poses. For each of the 8 experiments we show one sample screen-shot of ground-truth and estimated 3D pose (automatically generated from our pose error evaluation tool). Experiment and frame number are rendered in top left corner of the corresponding camera frame image. In each image the left pose is ground-truth, while the right is the estimated pose. In the top right corner we visualize the global orientation of the ground-truth and estimated pose. E_2 is the relative angle between the two lines (averaged over all test frames of the corresponding experiment).

larger, one has to underline, that the results presented in [21] are for artificial 2D training and testing silhouettes, generated using motion capture data.

5   Conclusions

We have presented a new idea for generative human pose estimation that is quite simplistic, but showed to be effective for recovering the 3D pose of a person. Instead of using low-level image features as edges, or silhouettes on the one hand, or high-level 2D pose estimates on the other hand, for comparing with model projections, we propose to use body marker vote distributions from a ISM based body marker localizer. These can be compared with negligible computational costs with a simple 3D stick-figure body model projected to the image plane. A flat kernel, that counts for the number of votes nearby to projected marker locations of this 3D stick-figure, is sufficient as basis for an observation likelihood that allows to filter for the correct projection parameters and 3D example pose from a set of prototype poses. Additionally, we motivated several modification ideas for a recently published ISM based body part localization approach for improving 2D body marker localization.

Our next step will be to incorporate temporal information into this approach.

References

1. Sminchisescu, C.: 3D human Motion Reconstruction in Monocular Video. Techniques and Challenges. In: Human Motion Capture: Modeling, Analysis, Animation, vol. 36. Springer (2007) ISBN 978-1-4020-6692-4
2. Poppe, R.: Vision-based human motion analysis: An overview. CVIU 108, 4–18 (2007)
3. Ji, X., Liu, H.: Advances in view-invariant human motion analysis: A review. IEEE Transactions on Systems, Man, and Cybernetics, Part C 40, 13–24 (2010)
4. Bregler, C., Malik, J.: Tracking people with twists and exponential maps, p. 8. IEEE Computer Society, Los Alamitos (1998)
5. Roth, S., Sigal, L., Black, M.J.: Gibbs likelihoods for bayesian tracking. In: CVPR, pp. 886–893 (2004)
6. Sminchisescu, C., Triggs, B.: Kinematic jump processes for monocular 3d human tracking. In: CVPR, vol. 1, p. 69 (2003)
7. Sigal, L., Balan, A.O., Black, M.J.: Combined discriminative and generative articulated pose and non-rigid shape estimation. In: NIPS (2007)
8. Drummond, T., Cipolla, R.: Real-time tracking of highly articulated structures in the presence of noisy measurements. In: ICCV, pp. 315–320 (2001)
9. Charles, J., Everingham, M.: Learning shape models for monocular human pose estimation from the microsoft xbox kinect. In: ICCV Workshops, pp. 1202–1208. IEEE (2011)
10. Leibe, B., Leonardis, A., Schiele, B.: Robust object detection with interleaved categorization and segmentation. IJCV 77, 259–289 (2008)
11. Müller, J., Arens, M.: Human pose estimation with implicit shape models. In: ACM Artemis, ARTEMIS 2010, pp. 9–14. ACM, New York (2010)
12. Sigal, L., Black, M.J.: Predicting 3D People from 2D Pictures. In: Perales, F.J., Fisher, R.B. (eds.) AMDO 2006. LNCS, vol. 4069, pp. 185–195. Springer, Heidelberg (2006)
13. Andriluka, M., Roth, S., Schiele, B.: Monocular 3d pose estimation and tracking by detection. In: Proc. of CVPR 2010, USA (2010)
14. Dalal, N., Triggs, B.: Histograms of oriented gradients for human detection, vol. 1, pp. 886–893 (2005)
15. Dollár, P., Belongie, S., Perona, P.: The fastest pedestrian detector in the west. In: BMVC, Aberystwyth, UK (2010)
16. Dollár, P., Tu, Z., Perona, P., Belongie, S.: Integral channel features. In: BMVC (2009)
17. Dollár, P., Wojek, C., Schiele, B., Perona, P.: Pedestrian detection: An evaluation of the state of the art. PAMI 99 (2011)
18. Andriluka, M., Roth, S., Schiele, B.: Pictorial structures revisited: People detection and articulated pose estimation. In: CVPR, pp. 1014–1021 (2009)
19. Taylor, C.J.: Reconstruction of articulated objects from point correspondences in a single uncalibrated image. CVIU 80, 349–363 (2000)
20. Aa, N.v.d., Luo, X., Giezeman, G., Tan, R., Veltkamp, R.: Utrecht multi-person motion (umpm) benchmark: a multi-person dataset with synchronized video and motion capture data for evaluation of articulated human motion and interaction. In: HICV Workshop, in Conj. with ICCV (2011)
21. Agarwal, A., Triggs, B.: Recovering 3d human pose from monocular images. PAMI 28, 44–58 (2006)

TV-L1 Optical Flow Estimation with Image Details Recovering Based on Modified Census Transform

Mahmoud A. Mohamed and Baerbel Mertsching

GET Lab, University of Paderborn, 33098 Paderborn, Germany
{mahmoud,mertsching}@get.upb.de
http://getwww.upb.de

Abstract. This paper proposes an improved optical flow estimation approach based on the total variational $L1$ minimization technique with weighted median filter. Furthermore, recovering image details using modified census transform algorithm improves the overall accuracy of estimating large scale displacements optical flow. On the other hand, the use of the Taylor expansion approximation in most of the optical flow approaches limits the ability to estimate movement of fast objects. Hence, a coarse-to-fine scheme is used to overcome such a problem of the cost of losing small details in the interpolation process where initial values are propagated from the coarse level to the fine one. The proposed algorithm improves the accuracy of the estimation process by integrating the correspondence results of the modified census transform into the coarse-to-fine module in order to recover the lost details. The outcome of the proposed approach yields state-of-the-art results on the *Middlebury* optical flow evaluations.

1 Introduction

The most known methods of optical flow estimation were developed by K.P. Horn, B.G. Schunck [1] and B. D. Lucas, T. Kanade [2]. Lukas Kanade used a local method and assumes that small regions of pixels have the same flow. On the other hand, the Horn Schunck model is based on global optimization and proposes a variational approach to optical flow estimation.

$$E = \sum \left[\left(I_1\left(x, y\right) - I_2\left(x + u, y + v\right) \right)^2 + \lambda \left(|\nabla u|^2 + |\nabla v|^2 \right) \right] \qquad (1)$$

where E is the energy error function that has to be minimized and (u, v) are the displacement values in the x and y direction, respectively. $I_1(x, y)$ and $I_2(x, y)$ are the first and second frame and $\nabla = (\partial/\partial_x, \partial/\partial_y)^T$ while λ is the regularization parameter. The optimal solution of equation (1) is calculated using the Euler-Lagrange and the least square minimization within an iterative scheme. In order to enhance the estimation accuracy for large displacements, [3] proposed a new variational approach combining local and global methods named CLG.

G. Bebis et al. (Eds.): ISVC 2012, Part I, LNCS 7431, pp. 482–491, 2012.

However, CLG model used the squared $L2$ norm which results in many drawback such as high sensitivity to noise and over-propagation. Moreover, they are not directionally selective and do not preserve motion boundaries where a Gaussian filter is applied on the image gradient which blurs the edges. [4] proposed a variational model with a parallel numerical scheme. This model used CLG with the total variational $L1 - norm$ instead of $L2 - norm$ and applied a bilateral filter on the data term and used a diffusion filter to limit the propagation of optical flow among the neighbor pixels. The $L1$ total variation minimization equation is:

$$E_{TV-L1} = \sum \left[\lambda | I_1 (x,y) - I_2 (x+u, y+v) | + (||\nabla u|| + ||\nabla v||) \right] \qquad (2)$$

The data term and the smoothness term represent the isotropic total variation, while [5] decomposed the above function into three parts:

$$E_{TV-1} = \sum \left[\lambda | I_1 (x,y) - I_2 (x+u, y+v) | + \frac{1}{2\theta} (u - \hat{u})^2 + \frac{1}{2\theta} (v - \hat{v})^2 \right] \quad (3)$$

$$E_{TV-u} = \sum \left[\frac{1}{2\theta} (u - \hat{u})^2 + ||\nabla u|| \right] \qquad (4)$$

$$E_{TV-v} = \sum \left[\frac{1}{2\theta} (v - \hat{v})^2 + ||\nabla v|| \right] \qquad (5)$$

To solve the above highly nonlinear system, [5] proposed a numerical scheme. The solution was originally developed to solve a denoising problem and it is subjected to convex optimization [5]. Most of the optical flow approaches approximate $(I_1(x,y) - I_2(x+u, y+v))$ by using the Taylor expansion and ignoring all terms of order higher than two, which yields:

$$I_1(x,y) - I_2(x+u, y+v) = I_t + I_x u + I_y v \qquad (6)$$

Where I_x and I_y are the image derivatives in x and y direction and $I_t = I_1(x,y) - I_2(x,y)$. Due to the approximation the minimization is valid only for small values of the displacements u and v. A coarse-to-fine scheme is used to produce large displacements optical flow [6], which propagates the initial values from the coarse level to the fine one. This propagation is done by the interpolation causing the loss of many small image details.

The goal of this paper is to improve the overall accuracy of large and small displacements in the optical flow by preserving the small image details. This can be done by using a large scale displacement optical flow technique with image details recovering technique. Our approach is based on the $L1$ total variation minimization algorithm combined with the CLG approach. Furthermore, the modified census transform corresponds are used for recovering the lost image details during the implementation of the coarse-to-fine levels and the weighted median filter is integrated to prevent the propagation among pixels in different regions.

The organization of the paper is as follow: Section two introduces the related work and section three represents the image details recovering module by using the modified census transform. The proposed model is described in section four, while the experiment results with synthetic and real sequences including a comparison with classical and state-of-the-art methods are represented in section 5. Conclusions and future work are finally given in section six.

2   Related Work

Most of the coarse-to-fine variational optical flow models have high-ranked results in the *Middlebury* optical flow evaluation [7,8]. An optimization approach in [8] is used to reduce the reliance of the flow estimates on their initial values propagated from the coarser level and to recover many motion details in each scale. This approach used the SIFT algorithm for finding point correspondences. In [9], a dense optical flow field is produced by using the SIFT and the discrete optimization algorithms. However, in order to introduce the regularity constraint, [9] considers all the possible matches for the SIFT correspondents which is a very time consuming process. [10] proposed a solution to estimate large motions of small structures by integrating the matched correspondences in a variational approach. However, the addition descriptor matching function improves only the accuracy of estimating large displacements, while the accuracy of estimating small displacements still needs to be enhanced.

In order to prevent the pixels propagation of different regions as in [8], our approach applies the CLG algorithm on the data term with the total variational minimization and use of the weighted median filter on the smooth term. Furthermore, we used the modified census transform to produce the features matches by using the census patches representation of the pixels. The outcome of a modified census transform module is used as initial values for solving the variational minimization equation. In addition, the possible matching correspondences is limited by considering only strong matching results in the calculation of the occurrence of each signature which overcomes the high computational problem in [9]. Moreover, contrary to [10], our approach improves both large and small displacements in the optical flow by integrating the CLG algorithm in the data cost function and by using the matching correspondences to recover the lost image information during the interpolation process, while the image details at the small displacements are preserved by implementing the weighted median filter.

3   Image Details Recovering

To extract image details we used the modified census transform described in [11,12] which produced a high texture density in areas with high structural information.The census transform has been used as the basic primitive feature due to its robustness with respect to outliers, its allowance of a large span of displacement vector lengths and its computational efficiency.

The main idea behind the census transform is to retrieve a set of promising image to image correspondence hypotheses [11]. These correspondences were obtained by applying the census operator on the two images and extract signature vectors for each image. Each signature vector has a fixed length and contains information about all the neighbor pixels. The matching can be done using a structural indexing scheme. In order to reduce the search area limits, the occurrence frequency c for each signature vector is calculated and only the vectors which have $c < c_{max}$ are considered. In our approach, we use a small value for occurrence frequency which yields complexity of $O(c * n)$ where n is the number of pixels.

Zabih and Woodfill[13] introduce the census operator as a binary operator, while [11] defines the census transform as a non-linear transform which maps a local neighborhood surrounding a pixel P to a ternary string representing the set of neighboring pixels. Each census vector $\xi(P, P')$ is defined as:

$$\xi(P, P') = \begin{cases} 0 & P - P' > \varepsilon \\ 1 & |P - P'| \geq \varepsilon \\ 2 & P' - P > \varepsilon \end{cases} \tag{7}$$

The signature vector of length c represents 3^c different patches. In this work, we use the Modified Census Transform (MCT), originally proposed by [14] for the face recognition algorithms. MCT compares the average intensity in a block instead of the center pixel. We used the algorithm in [11] for calculating the correspondences between two images. In order to preserve the output of the image details recovering module, the weights of the implemented median filter has to be calculated based on regions classification.

Although census transform solves the correspondence problem between two images in a very efficient way, the matching correspondence has major drawbacks. First, it produces a sub-pixel output. Second, the outliers of the matching correspondences cause high disturbance in the motion calculation at image discontinuities.

Our goal is to combine the sparse matching correspondence with the variational model in a coarse to fine optimization scheme , which could be used as an initial values for each new level during the coarse to fine optimization process.

4 The Proposed Model

In this section, the proposed approach (TV-L1-MCT) are represented in more details. We integrated the CLG [3] algorithm with the $L1$ total variational algorithm in the energy function to be:

$$E = \sum \left[\psi \left(w^T J_\rho \left(\nabla_3 f \right) w \right) + \lambda_1 \left(\psi \left(\nabla u \right) + \psi \left(\nabla v \right) \right) + \lambda_2 \left(\left(u - \hat{u} \right)^2 + \left(v - \hat{v} \right)^2 \right) \right] \tag{8}$$

where $\psi(x^2) = \sqrt{x^2 + \varepsilon^2}$ and $\varepsilon = 0.001$ to prevent the division by zero. $w = (\hat{u}, \hat{v}, 1)^T$ which $\hat{u}$ and $\hat{v}$ are auxiliary optical flow variables. f is $I(x, y)$ convolved with a Gaussian $K_\sigma(x, y)$ with standard deviation σ, while $\nabla_3 f = (f_x, f_y, f_t)^T$. J_ρ is Gaussian $K_\rho(x, y)$ with a stander deviation ρ and $J_\rho(\nabla_3 f) = K_\rho * (\nabla_3 f \nabla_3 f^T)$. The last term $\lambda_2((u - \hat{u})^2 + (v - \hat{v})^2)$ is used to enforce $\hat{u}$ and

u to be equal. λ_1 and λ_2 are regularization parameters. The function (8) can be decomposed as in [5] into three parts:

$$E_M = \sum \left[\psi \left(w^T J_p \left(\nabla f \right) w \right) + \lambda_2 \left((u - \hat{u})^2 + (v - \hat{v})^2 \right) \right] \tag{9}$$

$$E_u = \sum \left[\lambda_2 \left(u - \hat{u} \right)^2 + \lambda_1 \psi \left(\nabla u \right) \right] \tag{10}$$

$$E_v = \sum \left[\lambda_2 \left(v - \hat{v} \right)^2 + \lambda_1 \psi \left(\nabla v \right) \right] \tag{11}$$

where u , v are constants and E_M has two unknowns $\hat{u}$, $\hat{v}$. The optimal solution for $\hat{u}, \hat{v}$ does not depend on spatial derivatives of $\hat{u}$ and $\hat{v}$ and is calculated point-wise by using the Euler-Lagrange equation then optimized by using least square minimization. Similarly, E_u and E_v have two unknowns u and v, while $\hat{u}, \hat{v}$ are constants. We have used the numerical scheme in [4] to solve E_u and E_v. The Euler-Lagrange equation for E_u is:

$$(u - \hat{u}) - \lambda.div \left[\frac{(\nabla u)}{\psi(\nabla u)} \right] = 0 \tag{12}$$

let $P_u = \nabla u / \psi(\nabla u)$, then we have

$$u = \lambda.div(P_u) + \hat{u}, \tag{13}$$

which can be solved using a fixed-point iteration scheme as in [4]

$$P_u^{n+1} = \frac{P_u^n + \tau.\nabla(div(P_u^n) + \frac{\hat{u}}{\lambda})}{1 + \tau.||\nabla(div(P_u^n) + \frac{\hat{u}}{\lambda})||} \tag{14}$$

Where $\tau \leq 1/8$ is the time step. The same can be applied to get P_v. Equation (8) is isotropic and propagates the flow in all directions regardless of local properties. In order to enhance the propagation, the weighted median filter has been used to solve the isotropic propagation problem as introduced in [15,16]. We applied the algorithm in [7], but we integrate the spatio-temporal image segmentation approach introduced in [17] to calculate the weighted function:

$$\hat{u}_{i,j} = \min \sum \omega_{i,j,i',j'} |\hat{u}_{i,j} - u_{i',j'}| \tag{15}$$

where $(i', j') \in N_{i,j} \cup \{i, j\}$ which $N_{i,j}$ is the $N \times N$ local window, and $\omega \in [0,1]$. The approach in [17] segments the image into three different regions, texture moving, homogeneous moving and static regions based on the spatial and the temporal image derivatives.

$$I(x,y,t) \in \begin{cases} Texture - Moving & SNR \leqslant \tau, |\cos(\delta)| \simeq 1, |\cos(\beta)| \simeq 0 \\ Homogenous - Moving & SNR > \tau, ||\cos(\delta)| \simeq 1 \\ Stationary & Otherwise \end{cases} \tag{16}$$

where SNR is the signal-to-noise ratio of the gradient magnitudes and δ is the angle between the spatio-temporal gradient $(I_x, I_y, I_t)^T$ and a unit vector $(0, 0, 1)^T$. β is another angle which $|cos(\beta)|$ is close to one when the image gradient is very small. If a seed pixel belongs to a textured region while a neighboring pixels belong to a homogeneous region, then the propagation should not be affected by such pixels and thus $\omega = 0$. Similarly, if both pixels belong to the same type of region (homogeneous or textured) but the states are different i.e. one is a moving region and the other is a static region, then $\omega = 0$, otherwise $\omega = 1$. We have used the warping technique as in [6] to improve the coarse-to-fine module. At each level of coarse-to-fine scheme, the solution of equations (13), and (14) are calculated iteratively. the initial values of u, v are propagated from each coarser level into the finer one. The matching correspondences are calculated gives a set of hypothesis points. These points used to refine the propagated values from the coarser level.

At this stage we have two initial sources of data: (a) matching correspondences which neglects regularity and (b) propagated values from the coarser level which neglects image details. Our proposed algorithm integrates both sources of information in the calculation of the initial values. In case that there is no matching correspondence, only the propagated value from the coarser level is considered. Otherwise, a fusing function is used to validate the motion vector values based on the neighborhood pixels information. This decision is done by comparing the mean value of the vector lengths $\bar{d}_p$ of the propagated $(N \times N)$ window values and the vector length d_c of the matching correspondences. d_c is assumed to be an outlier in case that the difference between d_c and $\bar{d}_p$ is bigger than a threshold and then only the propagated value is considered. On the other hand, if the propagated value is similar to the neighbor pixels while its location is not homogeneous then its probably the motion information has been lost in the interpolation process. In such case, we consider only the matching correspondences. The initial values for the first coarse level are set to the values of matching correspondences instead of zeros in other approaches.

5   Evaluation and Experimental Results

In this section, the optical flow results of the Middlebury data sets [18] are represented in figure 1. On the other hand, we have evaluated our approach on the training database available at Middlebury on the end-point error (EPE) and the average angular error(AAE) as shown in table 1. We compare our improved results with the baseline results by using the implemented Matlab code of the algorithm[4] available at http://vision.middlebury.edu/flow/eval/results/. The parameters, that have been used in our results, are as follow: The outer fixed point iteration value was 20 and the inner fixed point iteration was 8 while $\lambda = 8$. The census transform used a windows of $5x5$ to construct the signatures with search windows size of 50 pixels and the maximum frequency occurrence value was 3 and the vector size was 20 digit, while the coarse-to-fine factor was 0.75. To display the results, We use the Middlebury color coding to display the optical flow in all estimated optical flow in this paper.

Fig. 1. Optical flow results of our approach, on a part of the semi-synthetic Middlebury images. The first row shows frame10 from each sequence, while the second row shows the Middlebury color coding for the estimated optical flow.

Our result has been ranked on the *Middlebury* benchmark at the time of publication June, 2012 as 12th position for angular error and 12th for the end point error while [4] was in 37th position for end point error and in 38th position for the average angular error. Figure 2 shows the average end-point and angular errors of our approach comparing with the the selected approaches on the Middlebury benchmark. The key difference between our approach and most of the top ranks approaches is that our approach deals with large scale optical flow. We have evaluated our approach on large scale displacement sequences, unfortunately Middelbury do not have ground-truth for the large scale displacement. We evaluate our approach on real and large scale displacement sequences using the available data at http://i21www.ira.uka.de/image_sequences/, http://people.csail.mit.edu/celiu/OpticalFlow/, and http://lmb.informatik.uni-freiburg.de/resources/datasets/sequences.en.

Figure 3 shows some of our results of our approach comparing with the output of the baseline. The implementation of our approach done by using C++ and opencv library on a PC with Intel(R) Core(TM)2 Duo CPU with 3.33 GHZ. The execution time was about 90 sec for 2 frames. This approach can be implemented using parallel processing and can be easy converted to GPU and work as real time optical flow approach.

Table 1. The average angular and the end-point errors of our approach (TV-L1-MCT) applied on the training data set from Middlebury benchmark comparing with the baseline

	Error	Venus	Dimetrod	Hydrangea	Rubber	Grove2	Grove3	Urban2	Urban3
TV-L1-MCT	AAE	3.320	2.013	1.834	2.472	1.440	5.492	2.071	2.997
	EPE	0.238	0.103	0.156	0.078	0.099	0.509	0.223	0.407
Baseline	AAE	7.060	3.951	2.271	3.332	2.844	7.756	3.407	13.302
	EPE	0.500	0.195	0.187	0.103	0.217	0.894	0.551	1.501

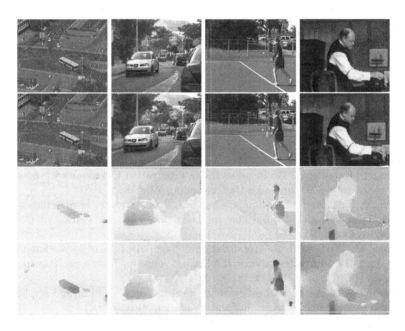

Average end-point error	avg. rank	Army (Hidden texture) GT im0 im1 all disc untext	Mequon (Hidden texture) GT im0 im1 all disc untext	Schefflera (Hidden texture) GT im0 im1 all disc untext	Wooden (Hidden texture) GT im0 im1 all disc untext	Grove (Synthetic) GT im0 im1 all disc untext	Urban (Synthetic) GT im0 im1 all disc untext	Yosemite (Synthetic) GT im0 im1 all disc untext	Teddy (Stereo) GT im0 im1 all disc untext
LSM [41]	13.0	0.08₃ 0.23₆ 0.077	0.22₁₇ 0.73₁₉ 0.15₁₉	0.28₁₀ 0.64₉ 0.19₁₈	0.14₄ 0.70₅ 0.09₁₀	0.66₈ 0.97₇ 0.48₁₀	0.50₂₁ 1.05₇ 0.33₁₅	0.15₃₃ 0.12₅ 0.29₄₆	0.50₁₁ 0.98₉ 0.73₁₆
TC-Flow [48]	13.0	0.07₁ 0.21₂ 0.06₁	0.15₁ 0.56₄ 0.11₁	0.31₁₈ 0.78₂₁ 0.14₁	0.15₁₅ 0.85₂₀ 0.08₅	0.75₁₇ 1.11₁₈ 0.54₁₅	0.42₈ 1.40₂₄ 0.25₂	0.11₉ 0.12₅ 0.20₄₅	0.52₂₂ 1.35₂₃ 0.93₃₄
Ramp [68]	13.2	0.08₃ 0.24₁₂ 0.077	0.21₁₄ 0.72₁₇ 0.18₁₉	0.27₇ 0.62₇ 0.19₁₈	0.15₈ 0.71₇ 0.09₁₀	0.66₈ 0.97₇ 0.49₁₂	0.51₂₃ 1.09₁₀ 0.34₂₀	0.15₃₃ 0.12₅ 0.30₄₉	0.48₅ 0.96₄ 0.72₁₂
Classic+NL [31]	14.7	0.08₃ 0.23₆ 0.077	0.22₁₇ 0.74₂₃ 0.18₁₉	0.29₁₄ 0.65₈ 0.19₁₈	0.15₈ 0.73₁₀ 0.09₁₀	0.64₆ 0.93₆ 0.47₈	0.52₂₄ 1.12₁₄ 0.33₁₅	0.16₄₁ 0.13₁₆ 0.29₄₆	0.48₅ 0.98₄ 0.74₁₉
TV-L1-MCT [70]	15.2	0.08₃ 0.23₆ 0.077	0.24₃₁ 0.77₂₅ 0.19₂₇	0.32₂₀ 0.75₂₀ 0.19₁₈	0.14₄ 0.69₄ 0.09₁₀	0.72₁₄ 1.03₁₂ 0.60₂₀	0.54₂₆ 1.10₁₂ 0.35₂₁	0.11₉ 0.12₅ 0.20₁₈	0.54₁₇ 1.04₁₃ 0.84₂₈
IROF-TV [56]	16.5	0.08₆ 0.25₁₆ 0.08₂₆	0.22₁₇ 0.77₂₅ 0.19₂₇	0.30₁₆ 0.70₁₀ 0.19₁₈	0.18₂₃ 0.93₂₆ 0.11₂₈	0.73₁₄ 1.04₁₃ 0.56₁₈	0.44₁₁ 1.65₃₈ 0.31₈	0.09₃ 0.11₁ 0.12₄	0.50₁₁ 1.08₁₅ 0.73₁₆
MDP-Flow [39]	17.7	0.08₁₄ 0.26₁₆ 0.08₂₆	0.19₈ 0.54₂ 0.18₁₉	0.24₄ 0.55₆ 0.20₂₅	0.16₁₅ 0.91₂₆ 0.09₁₀	0.74₁₅ 1.06₁₅ 0.61₂₂	0.45₁₅ 1.02₆ 0.35₂₁	0.12₁₄ 0.14₂₈ 0.17₁₁	0.78₃₈ 1.68₄₁ 0.97₃₈
OFH [36]	19.2	0.10₂₉ 0.25₁₆ 0.09₂₆	0.19₈ 0.69₁₄ 0.14₆	0.43₂₉ 1.02₂₃ 0.17₆	0.17₂₀ 1.08₃₆ 0.08₅	0.87₂₆ 1.25₂₅ 0.73₂₉	0.43₆ 1.69₃₉ 0.32₁₉	0.10₄ 0.13₁₅ 0.18₁₄	0.59₂₁ 1.40₃₈ 0.74₁₉

Average angle error	avg. rank	Army (Hidden texture) GT im0 im1 all disc untext	Mequon (Hidden texture) GT im0 im1 all disc untext	Schefflera (Hidden texture) GT im0 im1 all disc untext	Wooden (Hidden texture) GT im0 im1 all disc untext	Grove (Synthetic) GT im0 im1 all disc untext	Urban (Synthetic) GT im0 im1 all disc untext	Yosemite (Synthetic) GT im0 im1 all disc untext	Teddy (Stereo) GT im0 im1 all disc untext
TC-Flow [48]	13.2	2.91₂ 8.00₂ 2.34₅	2.18₁ 8.77₉ 1.52₁	3.84₂₀ 10.72₂ 1.49₂	3.13₁₉ 16.8₂₂ 1.46₁₁	2.78₉ 3.73₉ 1.95₂	3.08₄ 11.4₁₇ 2.66₄	1.94₆ 3.43₉ 3.20₄₉	3.05₂₄ 7.04₂₄ 4.08₄₃
Ramp [68]	13.2	3.18₁₁ 8.83₁₂ 2.73₁₁	2.89₁₈ 10.1₁₇ 2.44₂₃	3.27₅ 8.43₇ 2.38₂₀	2.74₁₀ 14.2₉ 1.46₁₁	2.62₇ 3.68₈ 2.28₇	3.37₁₁ 9.31₅ 2.93₁₇	2.62₂₇ 3.36₈ 3.19₄₈	1.54₅ 3.21₅ 2.24₁₃
COFM [64]	14.8	3.17₉ 9.90₂₈ 2.45₈	2.41₄ 8.34₆ 1.92₈	3.77₁₇ 10.5₁₉ 2.54₂₈	2.71₈ 14.9₁₅ 1.19₄	3.08₁₈ 3.92₁₆ 3.25₃₈	3.63₂₂ 10.9₁₃ 3.15₂₃	2.20₁₄ 3.35₅ 2.91₄₁	1.62₈ 2.56₅ 2.09₁₀
Classic+NL [31]	15.0	3.20₁₈ 8.72₆ 2.81₁₇	3.02₂₃ 10.6₂₀ 2.44₂₃	3.46₁₉ 8.84₉ 2.38₂₀	2.78₁₂ 14.3₁₀ 1.46₁₁	2.83₉ 3.66₆ 2.31₈	3.40₁₉ 9.09₃ 2.76₁₄	2.87₃₈ 3.82₂₃ 2.85₃₇	1.67₉ 3.53₉ 2.26₁₅
TV-L1-MCT [70]	15.5	3.16₈ 8.45₅ 2.71₁₀	3.28₃₁ 10.8₂₆ 2.50₃₁	3.95₂₀ 10.5₁₉ 2.38₂₀	2.69₆ 13.9₆ 1.45₈	2.94₁₅ 3.79₁₂ 2.63₂₃	3.50₁₆ 9.75₁₀ 3.05₂₀	2.08₆ 3.35₅ 2.29₂₃	1.95₁₄ 3.89₁₁ 2.71₃₀
SimpleFlow [52]	17.9	3.35₁₉ 9.20₁₅ 2.98₂₂	3.18₂₇ 10.7₂₄ 2.71₃₄	5.05₂₉ 12.6₂₇ 2.70₂₉	2.95₁₅ 15.1₁₆ 1.58₁₆	2.91₁₂ 3.79₁₂ 2.47₁₇	3.59₁₆ 9.49₈ 2.99₁₈	2.39₁₈ 3.46₁₁ 2.24₁₃	1.60₇ 3.46₁₁ 1.57₄
CostFilter [42]	19.6	3.84₂₆ 9.64₂₃ 3.05₂₄	2.55₁₆ 8.09₅ 2.03₉	2.69₉ 6.47₃ 1.88₆	3.65₃₁ 16.8₂₄ 1.88₂₆	2.52₅ 3.34₃ 1.99₄	4.05₂₇ 11.0₁₅ 3.65₃₃	4.16₅₇ 7.18₆₈ 4.66₅₈	1.16₁ 3.35₇ 0.87₁
OF-Mol [48]	19.7	3.19₁₂ 8.76₁₀ 2.77₁₄	3.94₃₈ 14.0₄₁ 2.69₃₃	3.44₁₁ 8.78₈ 2.39₂₄	2.98₁₆ 15.8₁₉ 1.53₁₆	2.95₁₆ 3.89₁₅ 2.34₁₁	3.40₁₃ 9.30₄ 2.73₁₁	2.83₃₄ 3.92₂₆ 2.98₄₄	2.46₁₇ 4.98₁₇ 2.69₂₁

Fig. 2. Snapshot of the web pages of the average angular and end-point errors of the TV-L1-MCT approach comparing with some selected approaches on the Middlebury evaluation web-site at the time of publication June, 2012

Fig. 3. Optical flow results of our approach comparing with the baseline, first row shows frames at time t from Ettlinger-Tor, MIT, tennis, and marple2 sequences, and the second row shows the frames at $(t + 1)$. while the third row shows the estimated optical flow produced by using the baseline and the fourth row is our estimated optical flow.

6   Conclusion

This paper proposes a combined $L1$ total variational and CLG optical flow approach integrating an image details recovering module. A modified census transform algorithm has been used to improve the recovering module and a weighted median filter has been used to prevent pixels propagation among different regions. The proposed approach used the matching corresponds output of the modified census transform module as initial values for solving the variational minimization equation during the coarse-to-fine scheme. In addition, the possible matching correspondences are limited by considering only strong matching in order to improve the overall computational time. Moreover, the matching correspondences has been used, was combined with weighted median filter to recover the lost image information and preserve image details during the interpolation process in order to improve both large and small displacements. The proposed method gives competitive results on the $Middlebury$ benchmark in both endpoint and average angular error.

Acknowledgment. Particular acknowledgment is due to DAAD (German Academic Exchange Service) and Egyptian Ministry of Higher Education for the financial support for Mr. Mohamed during the study period in Germany. We are very grateful to Ing. M. Salah for his technical assistance and for fruitful discussions.

References

1. Horn, B., Schunck, B.: Determining optical flow. Artificial intelligence 17, 185–203 (1981)
2. Lukas, B., Kanade, T.: An iterative image registration technique with an application to stereo vision. In: Image Understanding Workshop (1981)
3. Bruhn, A., Weickert, J., Schnörr, C.: Lucas/kanade meets horn/schunck: Combining local and global optic flow methods. International Journal of Computer Vision 61, 211–231 (2005)
4. Drulea, M., Nedevschi, S.: Total variation regularization of local-global optical flow. In: 2011 14th International IEEE Conference on Intelligent Transportation Systems (ITSC), pp. 318–323. IEEE (2011)
5. Chambolle, A.: An algorithm for total variation minimization and applications. Journal of Mathematical Imaging and Vision 20, 89–97 (2004)
6. Brox, T., Bruhn, A., Papenberg, N., Weickert, J.: High accuracy optical flow estimation based on a theory for warping. In: Computer Vision-ECCV 2004, pp. 25–36 (2004)
7. Sun, D., Roth, S., Black, M.: Secrets of optical flow estimation and their principles. In: 2010 IEEE Conference on Computer Vision and Pattern Recognition (CVPR), pp. 2432–2439. IEEE (2010)
8. Xu, L., Jia, J., Matsushita, Y.: Motion detail preserving optical flow estimation. In: 2010 IEEE Conference on Computer Vision and Pattern Recognition (CVPR), pp. 1293–1300. IEEE (2010)

9. Liu, C., Yuen, J., Torralba, A.: Sift flow: Dense correspondence across scenes and its applications. IEEE Transactions on Pattern Analysis and Machine Intelligence 33, 978–994 (2011)

10. Brox, T., Malik, J.: Large displacement optical flow: descriptor matching in variational motion estimation. IEEE Transactions on Pattern Analysis and Machine Intelligence 33, 500–513 (2011)

11. Stein, F.: Efficient computation of optical flow using the census transform. Pattern Recognition, 79–86 (2004)

12. Puxbaum, P., Ambrosch, K.: Gradient-based modified census transform for optical flow. Advances in Visual Computing, 437–448 (2010)

13. Zabih, R., Woodfill, J.: Non-parametric local transforms for computing visual correspondence. In: Computer Visionâ ECCV 1994, pp. 151–158 (1994)

14. Froba, B., Ernst, A.: Face detection with the modified census transform. In: Proceedings of the Sixth IEEE International Conference on Automatic Face and Gesture Recognition, pp. 91–96. IEEE (2004)

15. Buades, A., Coll, B., Morel, J.: A non-local algorithm for image denoising. In: IEEE Computer Society Conference on Computer Vision and Pattern Recognition, CVPR 2005, vol. 2, pp. 60–65. IEEE (2005)

16. Gilboa, G., Osher, S.: Nonlocal operators with applications to image processing. Multiscale Model. Simul. 7, 1005–1028 (2008)

17. Rashwan, H., Puig, D., Garcia, M.: On improving the robustness of differential optical flow. In: 2011 IEEE International Conference on Computer Vision Workshops (ICCV Workshops), pp. 876–881. IEEE (2011)

18. Baker, S., Scharstein, D., Lewis, J., Roth, S., Black, M., Szeliski, R.: A database and evaluation methodology for optical flow. International Journal of Computer Vision 92, 1–31 (2011)

Automatic Reference Selection for Parametric Color Correction Schemes for Panoramic Video Stitching

Muhammad Twaha Ibrahim[1], Rehan Hafiz[1], Muhammad Murtaza Khan[1],
Yongju Cho[2], and Jihun Cha[2]

[1] National University of Sciences and Technology, Islamabad, Pakistan
[2] Electronics and Telecommunications Research Institute (ETRI), Korea

Abstract. Panoramic views enhance the immersive visual experience by providing seamless high resolution image formed by stitching multiple low resolution images or videos. Color correction is a fundamental step in this process that operates to match the color of individual views with each other. Typically, one arbitrary view is taken as a reference and colors of the remaining views are matched to the reference view. This paper presents a scheme for automated selection of a reference image that results in a high contrast, visually appealing stitched panorama. The scheme is computationally efficient and applicable to a broad range of global parametric color correction schemes.

1 Introduction

Image stitching is the process of seamlessly aligning multiple images into a single image. Video stitching is a similar process where the images to be stitched are either acquired from the time elapsed sequences of the video frames or acquired from a synchronized set of multiple cameras. In the latter case, the goal is to generate a wide angle (high resolution) video using multiple low resolution cameras and we shall term it as Panoramic Video Stitching. With the advent in the display technologies and ever-growing demand for viewing high definition video contents; generation of high definition media content is a highly desirable necessity. By employing Panoramic Video Stitching, one can always go beyond what state-of-the-art video recorders can record. This can be typically achieved by placing multiple video cameras in a circular or planar fashion. Thus Panoramic Video Stitching can be an attractive technology for future real-time tele-broadcasting of, for example live sporting events [15]. There are three main steps in constructing a panorama i.e. geometric registration, photometric correction and finally blending [1]. A lot of work has been done in the domain of image and video stitching algorithms related to geometric correction [16, 17, 20]. Photometric color correction has also received attention recently [1]. When stitching images or video frames from multiple views into a single panorama, typically one of the views is arbitrarily selected as the reference or chosen by the user and the color of the remaining views is matched to the reference [2, 8, 9]. However, if the arbitrarily selected reference image has low contrast, the final output results into a visually un-appealing panorama. Figure-1 compares the effect of selecting a low contrast image as a reference with that of selecting a higher contrast image.

G. Bebis et al. (Eds.): ISVC 2012, Part I, LNCS 7431, pp. 492–501, 2012.

Fig. 1. (Top) Four images used to construct panoramas. (*Middle*) shows the output when the left-most image has been selected as reference. The second panorama (*Bottom*) is the output when third image is taken as reference.

As shown in the figure, the panorama with the higher contrast is more pleasing to look at. This effect is independent of the efficiency of the selected global color matching algorithms since the stitched output depends on the color constituency of the reference image, hence making automatic reference selection a useful task during the process of color correction. In this paper we present a simple and effective scheme that automatically selects an appropriate reference image that results in a visually pleasing panorama. Furthermore, it is shown that the scheme is applicable to a wide range of higher order global parametric color correction schemes. This paper is structured as follows. Section 2 provides a brief survey of color correction schemes. Section 3 describes the proposed scheme; Section 4 discusses the results followed by conclusions in Section 5.

2   Literature Survey

A lot of work has been done in the field of color correction for panorama stitching. As defined by Wei et al. in [1], color correction approaches can be divided into two broad categories: parametric and non-parametric. Parametric approaches assume a relation between the colors of the target image and those of the source image whereas non-parametric methods do not follow any particular model and most of them use some form of a look-up table to record the mapping for the full range of color intensity levels [9, 12]. As stated in [1], while non-parametric approaches provide better color matching results, parametric approaches are more effective in extending the color in non-overlapping regions of the source images without producing grain artifacts.

Parametric approaches are further classified as global schemes that assume a single relation between the reference and the target image, and local correction schemes. Local correction schemes are typically content based [2, 18] and are computationally expensive [19] thus making global color correction schemes a better candidate for the case of Panoramic Video Stitching. A few noteworthy global parametric approaches are reviewed next. Xiong et al. [5] employed diagonal model [3] for color and luminance compensation where the correction coefficients are the ratio of sum of pixel values in adjacent images in the overlap region. In another work [6] by the same authors, they perform gamma correction for luminance component and linear correction for chrominance component by minimizing error functions based on pixel values in the overlapping regions. In a later work [7], the authors use Poisson blending to reduce visibility of image seams produced due to difference in image colors. They compute the difference of the pixel values between the source image and the current panorama on the seam and then distribute and add this color difference to the rest of the image. In [8], the authors use a heuristic to select an image with the most similar means in R, G & B channels as the color reference. However, they do suggest a need for user input to select the best reference. Ha et al. [10] compensates color and luminance by using the linear model in the YCbCr color space, assuming that the objects in the scene have Lambertian reflectance. The gains are computed as the ratio of the average luminance value of adjacent overlapping regions, which are applied on the chrominance components too. In [11], Brown & Lowe use gain compensation for reducing color differences between source images forming a panorama. The gains are computed using an error function, which is the sum of gain normalized intensity errors for all overlapping pixels. However, Tian et al. [3] states that although gain compensation (diagonal model) is simple, it may not be sufficiently accurate. Doutre & Nasiopoulos [13] use a second-order polynomial to render a pixel in one image with the exposure and white balance of the reference image. The polynomial weights can be computed by comparing images in the overlap regions using standard linear least-squares regression. In [14], the blending width between two adjacent images is adjusted according to the color difference of corresponding pixels between seams of two adjacent images forming a panorama.

In almost all the methods mentioned above, the color reference for panorama construction is either selected by the user or chosen arbitrarily. To the best of our knowledge, there have only been two efforts [5,8] to automatically select the reference. Though Xiong et al. [5] suggests searching for an image with the overall best color and luminance distribution in the image sequence and using it as color reference, formal mathematical basis for the proposed scheme is not provided. Furthermore, an image with best global color and luminance distribution can have poor distribution at the image boundaries (due to phenomena such as vignetting). Since image boundaries are almost always part of the overlapping regions in panorama stitching and correction coefficients are computed by comparing the overlapping regions, this can result in the selection of an inappropriate reference. In this paper, a technique for selecting the most suitable color reference image is proposed that compares the global color correction parameters of all the input images evaluated only over the overlapping regions of adjacent images. In particular, we define an image with the most suitable color reference as the one that shall result in a high contrast panoramic image (or video frame for the case of video stitching). The technique requires no user input and is computationally simple.

3   Proposed Methodology

The proposed methodology comprises of three main steps. Firstly, color correction parameters are estimated using an arbitrarily selected image as reference using a global parametric color correction scheme. Next, the most suitable reference image is determined from the pre-calculated color correction parameters. Finally, the correction parameters are updated according to the selected reference image.

3.1   Color Correction Parameter Estimation

Consider a panoramic view being stitched from n different views. Let I_1, I_2 ...I_n represent the images acquired from n different views where I_1 is the left-most image in the panorama. Given an adjacent image pair forming a panorama, the colors of the right image can be matched to the colors of the left image by the following relationship:

$$I^c_{i,(R,G,B)} = T(M_i, \psi_i).$$ (1)

$$\psi_i = \begin{bmatrix} R_i & G_i & B_i \end{bmatrix}^T.$$ (2)

where ψ_i consists of the R, G and B values of image I_i and M_i is a transformation matrix matching the colors of image I_i to a reference image. I^c_i refers to the corrected image i.e. after applying the transformation T. The diagonal plus affine model [3] is employed due to its low computational complexity and thus applicability to Video Stitching, although higher-order transfer functions can also be used. The transformation matrix M for the Diagonal plus Affine model [3] is:

$$M = \begin{bmatrix} \alpha & & \alpha_1 \\ & \beta & \beta_1 \\ & & \gamma & \gamma_1 \end{bmatrix}.$$ (3)

where α, β and γ are channel gains and α_1, β_1 and γ_1 are channel offsets for R, G and B channels respectively. For an image pair $\{I_{i-1}, I_i\}$ and their corresponding overlapping regions $\{I_{i-1,o}, I_{i,o}\}$, the left image i.e. I_{i-1} is used as the color reference. The histogram of $I_{i-1,o}$ is then specified onto $I_{i,o}$.

$$I^h_{i,o} = H(I_{i-1,o}, I_{i,o}).$$ (4)

where $H(a,b)$ represented histogram specification of image a onto image b in RGB domain. The transformation matrix can then be calculated by comparing corresponding pixel values of $I_{i,o}{}^h$ and $I_{i,o}$ [3]. After applying the correction parameters to I_i, the corrected output, i.e. I_i^c is used as a reference for the next pair in the image sequence. In this way, the correction coefficients of all the images in the panorama with respect to the left-most image are determined.

3.2   Automatic Reference Selection

The next step is to automatically determine the reference image. Let $T(M_i, j)_{(k)}$ be the output value of applying the parametric transformation matrix to input gray value j in the kth channel of the ith image. We define the best reference as the one that has the minimum value of Ω, which is the sum of normalized transformed grayscale values over all channels. Thus, for an image I_i:

$$\Omega(i) = \frac{1}{3} \sum_{k \in R,G,B} \frac{1}{255} \sum_{j=1}^{255} \frac{T(M_i, j)_{(k)}}{j}. \tag{5}$$

$$ref = \arg\min\left(\{x : x = \Omega(i), 1 \le i \le n\} \right). \tag{6}$$

The reference image is therefore represented by I_{ref}. The image that requires the least gains and offset as a result of parametric correction is thus selected as the reference image since it will maximize the parameters for all other images, resulting in a high contrast panorama.

3.3   Color Correction Parameter Adjustment

The final step is to modify the correction parameters. This is done by transforming all the color transfer functions such that the transfer function of I_{ref} becomes identity i.e. I_{ref} remains unchanged and the colors of all other images will be matched to I_{ref}. To set the correction coefficients relative to I_{ref}, we use the following equation:

$$T^{new}(M_i, j)_{(k)} := \frac{T(M_i, j)_{(k)}}{T(M_{ref}, j)_{(k)}} \quad \{1 \le i \le n, 0 \le j \le 255, k \in R,G,B\}. \tag{7}$$

where T^{new} represents the updated transfer function. The updated transformation matrix can be extracted by interpolating a function of the desired order through T^{new}.

   The same steps can be applied to higher-order parametric transfer functions since equation (1) represented a generic global parametric color transfer function. As an example, a quadratic transfer function [13] can be expressed in terms of M and ψ_i as:

$$M = [m_{R1} \ m_{R2} \ m_{R3} \quad m_{G1} \ m_{G2} \ m_{G3} \quad m_{B1} \ m_{B2} \ m_{B3}]^T. \tag{8}$$

$$\psi_i = \begin{bmatrix} R_i^2 & R_i & 1 & 0 & 0 & 0 & 0 & 0 & 0 \\ 0 & 0 & 0 & G_i^2 & G_i & 1 & 0 & 0 & 0 \\ 0 & 0 & 0 & 0 & 0 & 0 & B_i^2 & B_i & 1 \end{bmatrix}. \tag{9}$$

Equations (5)-(7) require no manual input and thus completely automate the reference selection procedure.

4   Results and Discussion

The performance of the proposed scheme is reported for four test cases. Fig-2 compares the output panoramas generated using the default reference (I_1) to using the automatically selected reference for both diagonal plus affine models and quadratic models. The pitch, scoreboard and the crowd all have a higher contrast, giving the panorama a pleasing look (Fig-2b & d). Table-1 compares the Ω values for the images of Fig-2. In both cases the Ω value for I_5 is the least and hence selected as reference.

(a)   Diagonal + affine model: (*Top*) Default, (*Bottom*) Auto-Reference

(b) Quadratic model: (*Top*) Default, (*Bottom*) Auto-Reference

Fig. 2. Baseball Stadium Panorama stitched from five images

Table 1. Color Correction Coefficients for Panorama images in Figure 2. I_5 is selected as the reference for both cases.

Image	Diagonal + Affine Ω_i	Quadratic Ω_i
I_1	1	1
I_2	1.104	0.764
I_3	1.018	0.757
I_4	0.939	0.753
I_5	**0.845**	**0.629**

(a)   Fountain Panorama : (*Top*) Default, (*Bottom*) Auto-Reference

(b)   Athletics Stadium Panorama : (*Top*) Default, (*Bottom*) Auto-Reference

Fig. 3. Panoramas stitched using four camera views with automatic color reference selection

In the fountain panoramas (Fig-3a), the top panorama appears dark, especially on the right side, where the details of the hills are difficult to discern. The bottom panorama of Figure-3a is the output when the third image is automatically set as reference resulting in enhanced contrast, making the hills on the right clearer. In the athletics stadium panoramas (Fig-3b), the color of the sky and grass has better color distribution in the automatically selected reference panorama compared to the default reference panorama.

In some cases (Fig-4), the color distribution of the selected reference image is such that the output panorama becomes saturated. This may become undesirable. When I_5 was selected as reference automatically, the overall panorama (Fig-4 *Middle*) became saturated. In particular, the details of the building in the middle are reduced due to clipping of intensity values as a result of saturation. The number of pixels whose intensity is clipped due to saturation depends on the nature of content in a particular view and its transformation T. To prevent this from happening, the increase in the amount of pixels being saturated for each image before and after modifying the correction parameters is determined using the following equation:

$$S = \frac{1}{n \times \sum\limits_{k \in R,G,B} \sum\limits_{j=1}^{255} p(I_{1,k}, j)} \times \left(\sum\limits_{k \in R,G,B} \sum\limits_{i=1}^{n} p(I_{i,k}^c, 255) - p(I_{i,k}, 255) \right). \qquad (10)$$

where $p(I_{i,k}, j)$ returns the number of pixels of gray value j in the kth channel of image I_i. If this increase is beyond a certain threshold (5% in our experiments), the next image with the least Ω value is considered a candidate for the auto-referencing scheme. Table-2 shows the percentage increase in saturation when the respective image is set as reference. Initially I_5 was automatically chosen as reference, but it resulted in the middle panorama of Fig-4. The percentage increase in saturation was greater than 5% and hence, it was removed from the reference selection process. Consequently, image I_4 was selected reference. However, it should be noted that the output panorama's color distribution depends entirely on the color distributions of the source images. In the case where all source images have low contrast, the output panorama will also have low contrast.

Fig. 4. Panorama stitched using five camera views with automatic color reference selection. *(Top)*: Default reference I_1, *(Middle)*: Setting I_5 as reference, *(Bottom)*: Setting I_4 as reference.

The proposed technique works successfully for a number of image sets. It is computationally efficient and equations (5)-(7) took 35.5 msec to execute in MATLAB when run on Intel Core I5 3.3 GHz with 4 GB RAM.

Table 2. Ω and %S values for Panorama image in Figure 4. Negative %S values represent a decrease in the number of saturated pixels.

Image	Ω_i	%S
I_1	1	-0.356
I_2	1.137	-0.3649
I_3	1.100	-0.345
I_4	**0.967**	**3.952**
I_5	**0.814**	**10.799**

5   Conclusion

In this paper an automatic scheme for reference image selection for global parametric color correction techniques was presented. The scheme results in an output panorama with richer colors due to better color distributions of the output color channels. The effectiveness of the scheme is shown using real test images. Furthermore, a method was discussed to avoid making images as reference which result in the panorama becoming oversaturated. The scheme is scalable for higher order parametric color correction and supporting results were also provided. In the future, we would like to extend this technique to non-parametric color correction methods.

Acknowledgements. We would like to thank IT R&D program of MKE/ETRI (12ZI1140, HCI-based UHD Panorama Technology Development) for their generous funding.

References

1. Xu, W., Mulligan, J.: Performance evaluation of color correction approaches for automatic multi-view image and video stitching. In: 2010 IEEE Computer Vision and Pattern Recognition (CVPR), pp. 263–270 (2010)
2. Uyttendaele, M., Eden, A., Skeliski, R.: Eliminating Ghosting and Exposure Artifacts in Image Mosaics. In: 2001 IEEE Computer Vision and Pattern Recognition (CVPR), vol. 2, pp. II-509–II-516 (2001)
3. Tian, G,Y., Gledhill, D., Taylor, D., Clarke, D.: Colour correction for panoramic imaging. In: Proceedings of the Sixth International Conference on Information Visualisation (IV 2002), pp. 483–488 (2002)
4. Xiong, Y., Pulli, K.: Color Correction for Mobile Panorama Imaging. In: Proceedings of the First International Conference on Internet Multimedia Computing and Service (ICIMCS), 2009 ACM SIGMM, pp. 219–226 (2009)

5. Xiong, Y., Pulli, K.: Fast and High-Quality Image Blending on Mobile Phones. In: 2010 7th IEEE Consumer Communications and Networking Conference (CCNC), pp. 1–5 (2010)
6. Xiong, Y., Pulli, K.: Color Matching for High-Quality Panoramic Images on Mobile Phones. IEEE Transactions on Consumer Electronics 56(4), 2592–2600 (2010)
7. Xiong, Y., Pulli, K.: Fast Image Stitching and Editing for Panorama Painting on Mobile Phones. In: 2010 IEEE Computer Vision and Pattern Recognition Workshops (CVPRW), pp. 47–52 (2010)
8. Xiong, Y., Pulli, K.: Fast Panorama Stitching for High-Quality Panoramic Images on Mobile Phones. IEEE Transactions on Consumer Electronics 56(2), 298–306 (2010)
9. Jia, J., Tang, C.-K.: Image Registration with Global and Local Luminance Alignment. In: Proceedings of the Ninth IEEE International Conference on Computer Vision (ICCV 2003), vol. 1, pp. 156–163 (2003)
10. Ha, S.J., Koo, H., Lee, S.H., Cho, N.I., Kim, S.K.: Panorama Mosaic Optimization for Mobile Camera Systems. IEEE Transactions on Consumer Electronics 53(4), 1217–1225 (2007)
11. Brown, M., Lowe, D.G.: Automatic Panoramic Image Stitching using Invariant Features. International Journal of Computer Vision 74(1), 59–73 (2007)
12. Kim, S.J., Pollefeys, M.: Robust Radiometric Calibration and Vignetting Correction. IEEE Transactions on Pattern Analysis and Machine Intelligence 30(4), 562–576 (2008)
13. Doutre, C., Nasiopoulos, P.: Fast Vignetting Correction and Color Matching For Panoramic Image Stitching. In: 16th IEEE International Conference on Image Processing (ICIP), pp. 709–712 (2009)
14. Lee, K.-W., Jung, J.-Y., Jung, S.-W., Morales, A., Ko, S.-J.: Adaptive Blending Algorithm for Panoramic Image Construction in Mobile Multimedia Devices. In: 2011 IEEE International Conference on Consumer Electronics (ICCE), pp. 93–94 (2011)
15. Christoph, F., Christian, W., Ingo, F., Markus, M., Peter, E., Peter, K., Hans, B.: Creation of High-Resolution Video Panoramas of Sport Event. In: Eighth IEEE International Symposium on Multimedia (ISM), pp. 291–298 (2006)
16. Steedly, D., Pal, C., Szeliski, R.: Efficiently Registering Video into Panoramic Mosaics. In: 10th IEEE International Conference on Computer Vision (ICCV 2005), vol. 2, pp. 1300–1307 (2005)
17. Ikeda, S., Sato, T., Yokoya, N.: High-resolution Panoramic Movie Generation from Video Streams Acquired by an Omnidirectional Multi-camera System. In: Proceedings of IEEE International Conference on Multisensor Fusion and Integration for Intelligent Systems (MFI 2003), pp. 155–160 (2003)
18. Tai, Y.-W., Jia, J., Tang, C.-K.: Local color transfer via probabilistic segmenta-tion by expectation-maximization. In: IEEE Computer Vision and Pattern Recognition (CVPR 2005), vol. 1, pp. 747–754 (2005)
19. Xiang, Y., Zou, B., Li, H.: Selective color transfer with multi-source images. Pattern Recognition Letters 30(7), 682–689 (2009)
20. Jia, J., Tang, C.-K.: Image Stitching Using Structure Deformation. IEEE Transactions on Pattern Analysis and Machine Intelligence 30(4), 617–631 (2008)

Asynchronous Occlusion Culling
on Heterogeneous PC Clusters
for Distributed 3D Scenes

Tim Süß[1], Clemens Koch[2], Claudius Jähn[3],
Matthias Fischer[3], and Friedhelm Meyer auf der Heide[3]

[1] Johannes Gutenberg-University Mainz
[2] CC HPC - Fraunhofer ITWM
[3] Heinz Nixdorf Institute, University of Paderborn

Abstract. We present a parallel rendering system for heterogeneous PC clusters to visualize massive models. One single, powerful *visualization node* is supported by a group of *backend nodes* with weak graphics performance. While the visualization node renders the visible objects, the backend nodes asynchronously perform visibility tests and supply the front end with visible scene objects. The visualization node stores only currently visible objects in its memory, while the scene is distributed among the backend nodes' memory without redundancy. To efficiently compute the occlusion tests in spite of that each backend node stores only a fraction of the original geometry, we complete the scene by adding highly simplified versions of the objects stored on other nodes. We test our system with 15 backend nodes. It is able to render a $\approx 350\,M$ polygons ($\approx 8.5\,\text{GiB}$) large aircraft model with 20 to 30 fps and thus allows a walk-through in real-time.

1 Introduction

The CAD-based design produces increasingly complex models. Because the complexity of the models is growing faster than the performance of graphics hardware, the real-time visualization of such models is of the current top ten problems in CAD [1]. However, the development cycles of graphics hardware are so short that new hardware soon after purchase seems to be no longer powerful. If the computers of a PC cluster are updated gradually over a longer period, this can easily result in a heterogeneous system; i.e., the network consists of computers that have varying graphics adapters with partially weak performance. This raises the algorithmic question, how the computational power of such heterogeneous networks can still be exploited for efficient parallel rendering.

We look at a single *visualization node*, equipped with a fast graphics adapter that is supported by a group of *backend nodes* which may have weak graphics performance only. This means that their polygon throughput may be significantly lower than on the visualization node. With our approach, we aim at visualizing scenes with a complexity that can neither be stored in the visualization node's memory, nor be displayed in real-time under common conditions. We must

G. Bebis et al. (Eds.): ISVC 2012, Part I, LNCS 7431, pp. 502–512, 2012.

therefore consider how the slower backend nodes can reduce the visualization node's workload and, additionally, how the scene's objects are stored in the PC cluster.

Our approach is as follows: The backend nodes serve two purposes. First, their accumulated main memory provides an additional memory system for the visualization node. Second, they identify the set of currently visible objects by performing asynchronous, approximate occlusion culling. The asynchronous communication scheme allows exploiting even weak backend nodes for occlusion culling without reducing the framerate on the visualization node. The visualization node receives the data of the visible objects, stores it in its video memory (or in its main memory, if the amount of visible objects exceeds the graphics card's storage) and simply renders it using OpenGL.

As in most parallel rendering systems we have to balance the rendering and the data load to prevent congestion and support efficient occlusion culling. More specifically, we identified the following requirements for our system:

- The scene must be distributed evenly among the backend nodes to achieve a balanced network and computational load. Therefore, we distribute the original objects without redundancy in a pseudo-randomized manner across the backend nodes.
- The backend nodes' response times must be kept to a minimum. To achieve this, only backend nodes' main memory is used to store the original objects' data (together with additional approximations). Although this limits the maximal scene size according to the combined memory of the backend nodes, this removes the need for high latency hard disc accesses.
- While each backend node only has a fraction of the whole scene, it should be capable of performing approximative occlusion culling with a high quality. For this, each backend node stores the scene in a hierarchical data structure that provides tight bounding boxes for improved occlusion test accuracy (based on the *hull tree*[2]). As replacement for the objects that are not stored at a node (but which are necessary as occluders for visibility calculations), memory efficient, simplified approximations are calculated and stored in the data structure.

2   Related Work

We survey some work of parallel rendering for massive data sets [3] as well as parallel occlusion culling, which are related to our approach.

Rendering systems for massive data sets: If a three-dimensional scene does not fit into main memory of a single machine, out of core mechanisms, which store the scene on hard disks, can be used. Such approaches store objects, which contribute to the final image, in primary memory in order to archive real-time rendering. A fast loading of important objects from the hard disk into primary memory is possible if e.g., pre-fetching methods using region information [4] or extended view frustums for selective object loading [5] are used. To prevent memory fragmentation and to reduce data management costs, objects can be stored in equally sized memory blocks[6]. Also the approximate rendering of objects can be utilized [7]. Unsimilar to the approaches above, we remotely store our scene in multiple back end nodes' main memory instead of using hard disks. Thus, our system has a memory size available that goes beyond the typical memory size of a single computer, but does not scale as in the case of a hard disk-based out-of-core system. Since we store the scene in the backend nodes' RAM, we have fast access to all objects and do not require a pre-fetching mechanism.

Some works aim at limiting the amount of data sent over the network: Goswami et al. [8] presented a parallel out-of-core rendering system for very large terrain datasets reducing the data accesses by using equally sized tiles. The data is stored on a network-attached storage. Kendall et al. presented methods to accelerate the sort-last rendering of Radix-K [9]. They reduce the data sent through the network by a run-length encoding. Our system takes a different approach and reduces the network load by filtering occluded scene objects. Potential visible objects are only sent to the visualization node.

Parallel occlusion culling: The parallel occlusion culling system of Naga K. Govindaraju et al. [10] uses one node for rendering and two nodes for occlusion-tests. In every frame one occlusion-test node computes the ids of the visible scene-graph nodes sending it to the other nodes afterwards. At the same time, the rendering node uses the previously received ids to render the image. The second occlusion-test node renders these objects too and uses the image for the nodes' next occlusion test. In contrast to our approach this approach supports only a fixed number of similar backend nodes which are all required to store all objects. Hua Xiong et al. [11] presented different parallel occlusion culling methods partitioning the scene-data into voxels. These voxels are distributed over the different nodes. In the *Sort-First Occlusion Culling* approach, every computer computes the occlusion culling on its own - the culling process is not distributed. Each node of the *Sort-Last Occlusion Culling* computes the visibility for all voxels in its local memory. Afterwards, the geometry of the visible voxels is sent to the nodes assigned to the tile where it has to be displayed.

3   Preprocessing of Randomized Hull Tree and Interior Object Approximations

For the presented technique, the envisaged types of scenes are complex structures exported by CAD systems (consisting of a large number of single objects). In order to prepare a scene for using it in our system, we store the scenes' objects into a hierarchical data structure – a modified version of the hull tree[2] (see below)– in a preprocessing step. We also preprocess the computation of simplified meshes for each object. These serve as additional occluders for the occlusion tests on the backend nodes. These steps are described in more detail in the following sections:

Our Bounding volume hierarchy is based on a hull tree. This data structure is similar to a loose octree where all elements are stored in tree's leaves, but the hull tree's nodes cover the interior geometry tighter than the bounding boxes of an octree.

One important property of a bounding volume hierarchy is, that if the bounding volume of an inner node is evaluated as fully occluded (and the observer position is outside the volume), its complete subtree can immediately be discarded as invisible. If only a fraction of the bounding volume is visible (but no object), the visibility of the nodes in the subtree can not be predetermined and has to be evaluated through additional tests. To increase the accuracy of the occlusion tests, the hull tree replaces the underlying octree's cubic cells by a set of different, tighter volumes (for details, see [2]). The overall complexity of the bounding volumes of hull trees is typically only slightly higher when compared to an ordinary octree ($\approx 0.05\%$ more vertices).

Due to the nature of the applied approximative occlusion culling algorithm, nodes at a deeper level of the tree may need several frames to be identified as visible. To increase the visual quality – especially when the observer performs quick rotations – some objects are placed closer to the root node of the tree, than they ought to be based on their size. If these objects are well chosen, they can help to quickly fill the occurring holes and thereby improve the visual impression during the walk-through, while only marginally influencing the systems overall performance. In analogy to the randomized sample tree[12], the objects which are "'pulled up"' from the tree are chosen randomly; weighted according to their size. Thereby, large objects that are likely to contribute much to the final image are preferred.

Interior object approximation are required for the occlusion culling. The scene's objects are distributed disjunctively over all backend nodes; but each backend node has to evaluate whether its assigned objects are currently occluded by any other object. To substitute the missing objects, the hull tree's interior approximations are used (which are based on the edge collapse algorithm [13] using a special quality function). Each backend node thereby holds the complete set of scene objects; but only fraction of those objects are memory consuming originals and the rest are lightweight, highly simplified approximations. As the interior approximations can be slightly smaller than the original objects, this can lead to an increased number of objects, which are erroneously classified as visible.

4   Data Distribution, Visibility Tests and Rendering

As mentioned, the system aims at visualizing scenes with a complexity exceeding the capacities of a single node's main memory. The data is therefore distributed over the nodes in the following manner:

The **visualization node** stores the data of those objects, which are currently classified as visible in its graphics memory. As soon as an object is identified as occluded, its data is deleted.

Each **backend node** stores a subset of the scene's objects, the precomputed interior approximations of the corresponding remaining objects and the hierarchical data structure in their main memory. The assignment of the objects to the backend nodes is done by using a pseudo-random hash function, which leads to a almost uniformly distribution where each object is stored on exactly one backend node (this could be confirmed by our measurements). The scene is thereby completely stored by the backend nodes and can be efficiently transferred to the visualization node without the need of any low latency hard disk access. Furthermore, every backend node has a simplified representation of the whole scene, to be able to perform the occlusion culling.

The rendering loop of the visualization node is simple and straightforward (see Figure 2). In every frame, it is checked if a backend node has finished its current visibility calculations. If new data is available, the ids of the objects that were newly classified as invisible are received and the corresponding objects are removed. Then the mesh data of the now new objects is received and stored in graphics memory. After this, the backend node is informed of the observer's new position (and viewing direction) in order for it to begin with the next visibility calculation. If the data of more than one backend node is available, its processing is postponed to the next frames to reduce the fluctuations of the frame rate. Finally, all stored objects are rendered and the camera position is updated according to the user's input before the rendering loop starts over.

In the proposed setting, in which a cluster with a high network bandwidth is given, the actual time needed for receiving the updated mesh data does in most situations not influence the overall frame rate decisively. If the network between the visualization node and the backend nodes becomes a bottleneck or if a more stable frame rate is required, this problem could be reduced by receiving the data on the visualization node asynchronously to the rendering process.

Occlusion culling on the backend nodes: As of the low graphic performance of the backend nodes and the high complexity of the scene, we chose to use an occlusion culling algorithm which does determine the visibility of the whole scene adaptively over several frames instead of one frame (similar to the algorithm presented in [2]). During the culling process, the visibility of the hull tree's nodes is determined. Each frame, the level of the tree up to which elements may be classified as visible is thereby adaptively moved by only one level towards the leafs. On the one hand, this leads to a very efficient culling process with almost no pipeline stalls. On the other hand, it may take few frames until all objects are classified as visible after they first came in sight.

The rendering algorithm works in four phases: Initializing the depth buffer and testing formerly visible nodes, testing potentially visible objects, fetching and evaluating the test results and sending the data to the visualization node.

In the first phase, the depth buffer is filled by rendering the objects visible in the last frame. This is done in front to back order and a hardware assisted occlusion query is initiated for each rendered object to identify whether it was occluded by other objects. The rendered objects can be both: original meshes stored on a backend node or the corresponding interior approximations. To prevent a possible pipeline stall caused by retrieving the results of occlusion queries before they passed the whole rendering pipeline, the retrieval is postponed to a later phase.

The second phase is used to test objects of visible nodes and those nodes whose parent in the tree is visible or which have an invisible child by recursively traversing the tree. The set of visible marked nodes can thereby be extended or reduced by one level of the tree per frame. For the nodes' occlusion test, the precomputed bounding volumes are used.

In the third phase the results of all issued occlusion queries are evaluated. For tested inner nodes, just their visibility flag is updated. The visibility flag of every tested object is also updated and all visible objects are collected into the list of visible objects used for the next frame. Those objects whose original mesh is stored on this specific backend node are processed further in the next phase: If they are found visible, but their data is not present on the visualization node, their data is prepared for sending. If they are found invisible and their data is on the visualization node, their id is added to the list of disappeared objects.

In the last phase, the backend node sends the ids of disappearing objects to the visualization node, followed by the mesh-data of the appearing objects. Finally, it receives the new camera parameters before the loop starts over again.

5   Evaluation

In this evaluation we analyze our system's properties and performance. First, we introduce our system environment, the used scene, the properties of the precomputed object approximations, as well as the used camera path. First, we analyze the achieved rendering performance and network load of our system in Section 5.2. Here we present the measured times for rendering the images and receiving data, as well as the amounts of data sent over the network. We show that our rendering performance allows real-time interaction, and that the network is not saturated. In Section 5.3, we evaluate the *visibility delay*, where we measure the time until all visible objects have been send to the visualization node. We compare the visibility delay, when the camera is *teleported* to random scene positions to the delay during walk-through. Finally in Section 5.4, we analyze the influence of a changed hardware environment.

5.1   Benchmark

For our evaluation we used the PC cluster of the *Paderborn Center for Parallel Computing* (PC^2). This cluster consists of one powerful visualization node, equipped with two AMD Opteron 250 processors (2.4 GHz), 8 GiB RAM, and a NVidia GeForce 9800 GX2 (512 MiB). The 15 backend nodes are equipped with two Intel Xeon 3.2 GHz processors, 4 GiB RAM, and a NVidia Quadro NVS-280. All nodes are connected via Infiniband (about 1.8 GiB/sec at full-duplex transmission). CentOS-5.5 is the operating system installed on all cluster nodes. Our application is written in C++, using GCC-4.4, OpenMPI, and OpenGL. For our walk-through tests we use the Boeing-777 model, which requires a storage of about 8.5 GiB. The size of the interior approximations is about 2.2 GiB, the exterior approximation's size is about 320 MiB.

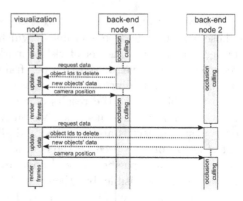

Fig. 1. Visualization of the camera path through the Boeing-777, used for our tests

Fig. 2. Communication scheme between visualization node and backend nodes

For our tests we perform a walk-through as shown in Figure 1, using 15,000 fixed camera positions. The walk-through starts in front of the model, focusing the cockpit. We move the camera straightforward through the airplane. After we leave the plane trough its tail we rotate the camera and walk directly through the turbine. Finally, we return to the starting point.

The speed of the camera during the walk-through has a strong influence on the systems overall behavior. For the evaluation, we chose to limit the frame rate to 30 fps, while always stepping one fixed step forward on the path each frame. On average, this corresponds to the speed of a slowly walking person.

5.2   Rendering Performance

In our first tests we analyzed the performance of the visualization node. The time needed on the visualization node to receive the data from the backend nodes and

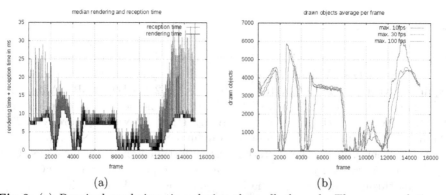

Fig. 3. (a) Required rendering time during the walk-through. The measured times include the reception of updates and the pure rendering. (b): Drawn objects per frame.

to render an image is shown in Figure 3(a). The black base line illustrates the net rendering time, where as the red peaks show the time spent to receive and process updates from the backend nodes. The rendering time for a single frame is the time spent for receiving data plus the net rendering time. As the visualization node does not perform any additional operations for the rendering itself and the test scene's objects are all of similar complexity, the rendering time strongly correlates with the number of drawn objects (see the 30 fps plot in Figure 3(b)). The additional time spent on the visualization node correlates with the amount of received data from the backend nodes. Other measurements showed, that the number of deleted objects which have to be updated on the visualization node, does not influence the overall performance decisively.

In every frame the number of drawn objects is small in relation to the complete number of objects (about 720,848 in total). On our walk-through with 30 fps the number of drawn objects never exceeds 5, 600, which is the reason for why the frame rate is high enough to allow real-time interaction with the system.

5.3   Visibility Delay

As the occlusion culling on the backend nodes is significantly slower than the rendering on the visualization node, the outdated visibility information leads to temporary image errors.

Depending on the speed with which the user moves through the scene, the number of drawn objects changes (see Figure 3(b)). The slower the user moves (e.g., with 10 fps), the more objects render on the visualization node as the backend nodes have more time to adapt their own visibility information closer to the actual position. If the user moves faster (e.g., with a limit of 100 fps), the visibility tests on the backend nodes do not have time to reach deeper levels of the scene tree and only the larger objects (and those which were randomly lifted up) are identified as visible.

The worst-case scenario for the image quality is when the observer does not walk to a position in the scene but immediately teleports to a new position. Figure 4 shows the positions and measurements for such teleports. Depending on the target position, the time needed until the image is complete and no further objects are transmitted ranges from 6 seconds (in the Boeing's cockpit) to 50 seconds (when the complete Boeing is in sight). In general, a user moves more steadily through the scene, allowing to exploit spatial coherence: While the complete reload of the turbine needed 31 seconds after a teleport, it took only 8 seconds during the walk-through to load the complete engine. Another observation is, that in these experiments the different backend nodes still work in a synchronized manner and their updates arrive at similar times. This results in larger popping artifacts where many objects become visible at once. During the walk-through, nodes send their results more evenly distributed over time.

In general, the experiments have shown that the system allows a fluid navigation through a complex scene with a high frame-rate, while achieving a reasonable image quality.

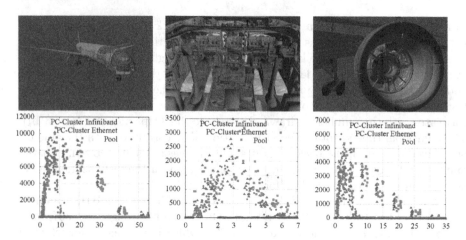

Fig. 4. The upper row shows screenshots of different camera positions, while the lower row shows the required update times on these different scene's locations, when the camera is teleported to those positions. Diagrams' y-axis the received amount of data is plotted in KiB and on the x-axis the time since teleport in seconds.

5.4 Hardware Influences

In our last test we evaluated the influences of improving the graphics card and deteriorating the network on the system. Here we use a visualization node equipped with a NVidia 480 GTX and 16 backend nodes each equipped with a NVidia Quadro FX-3500. The network is reduced to 1 GBit Ethernet. We perform the same tests as presented in Section 5.3. Here the times required for the complete updates have also been decreased to a minimum of 9 sec (see Figure 4; pool-network; blue dots). These tests have shown that these improvements are reasoned due to the used graphics adapters. If the communication in the PC cluster is also reduced to gigabit Ethernet the visibility delay does not change significantly (see Figure 4, green dots).

6   Conclusion

We developed a asynchronous parallel rendering system for massive scenes. Our system requires only a single visualization node with fast hardware. We use weak backend nodes' main memory as secondary storage and to compute visibility tests. In order to enable backend nodes to perform occlusion tests for complex scenes, despite of their limited memory, we use object approximations. We achieved a fairly uniform data distribution and test load balancing for most camera positions by distributing the scene's objects pseudo-randomly on the backend nodes. Due to exploiting spatial coherence, our rendering system copes with large delays of the visibility tests. Because of these properties our rendering system allows real-time interaction with large scenes, while producing images whose error depending on the movement speed.

In order to support a wider range of different backend nodes, one could easily switch the used rendering system on the backend nodes to a software based system like Mesa. While the drawback would be, that the latency of the occlusion queries might increase, the backend nodes then would no longer need any graphics hardware at all.

One possible way to overcome the situation where the amount of actually visible objects exceeds the capacities of the visualization node, is to apply an additional approximative rendering technique. As the randomized sample tree is already internally used as part of the data structure, this would be a good candidate for such an extension of the presented method.

Acknowledgement. This work is supported by the DFG-project DA155/31-1, ME872/11-1, FI1491/1-1 AVIPASIA.

References

1. Kasik, D.J., Buxton, W., Ferguson, D.R.: Ten CAD challenges. IEEE Computer Graphics and Applications 25, 81–92 (2005)
2. Süß, T., Koch, C., Jähn, C., Fischer, M.: Approximative occlusion culling using the hull tree. In: Proc. of Graphics Interface 2011, pp. 79–86. Canadian Human-Computer Communications Society (2011)
3. Kasik, D., Dietrich, A., Gobbetti, E., Marton, F., Manocha, D., Slusallek, P., Stephens, A., Yoon, S.E.: Massive model visualization techniques: course notes. In: ACM SIGGRAPH 2008 Classes, SIGGRAPH 2008, pp. 40:1–40:188. ACM, New York (2008)
4. Aliaga, D., Cohen, J., Wilson, A., Baker, E., Zhang, H., Erikson, C., Hoff, K., Hudson, T., Stuerzlinger, W., Bastos, R., Whitton, M., Brooks, F., Manocha, D.: Mmr: an interactive massive model rendering system using geometric and image-based acceleration. In: Proc. of the 1999 Symposium on Interactive 3D Graphics, I3D 1999, pp. 199–206. ACM, New York (1999)
5. Corrêa, W.T., Klosowski, J.T., Silva, C.T.: Visibility-based prefetching for interactive out-of-core rendering. In: Proc. of the 2003 IEEE Symposium on Parallel and Large-Data Visualization and Graphics, PVG 2003, pp. 1–8 (2003)
6. Sajadi, B., Huang, Y., Diaz-Gutierrez, P., Yoon, S.E., Gopi, M.: A novel page-based data structure for interactive walkthroughs. In: I3D 2009: Proceedings of the 2009 Symposium on Interactive 3D Graphics and Games, pp. 23–29 (2009)
7. Brüderlin, B., Heyer, M., Pfützner, S.: Interviews3d: A platform for interactive handling of massive data sets. IEEE Comput. Graph. Appl. 27, 48–59 (2007)
8. Goswami, P., Makhinya, M., Bösch, J., Pajarola, R.: Scalable parallel out-of-core terrain rendering. In: Proceedings Eurographics Symposium on Parallel Graphics and Visualization, Eurographics Association, pp. 63–71 (2010)
9. Kendall, W., Peterka, T., Huang, J., Shen, H.W., Ross, R.: Accelerating and benchmarking radix-k image compositing at large scale. In: Proceedings Eurographics Symposium on Parallel Graphics and Visualization, Eurographics Association, pp. 101–110 (2010)
10. Govindaraju, N.K., Sud, A., Yoon, S.E., Manocha, D.: Interactive visibility culling in complex environments using occlusion-switches. In: Proc. of the 2003 Symposium on Interactive 3D Graphics, I3D 2003, pp. 103–112. ACM, New York (2003)

11. Xiong, H., Peng, H., Qin, A., Shi, J.: Parallel occlusion culling on gpus cluster. In: Proc. of the 2006 ACM International Conference on Virtual Reality Continuum and its Applications, VRCIA 2006, pp. 19–26. ACM, New York (2006)

12. Klein, J., Krokowski, J., Fischer, M., Wand, M., Wanka, R., Meyer auf der Heide, F.: The randomized sample tree: a data structure for interactive walkthroughs in externally stored virtual environments. In: Proc. of the ACM Symposium on Virtual Reality Software and Technology, VRST 2002, pp. 137–146 (2002)

13. Hoppe, H.: Progressive meshes. In: SIGGRAPH 1996: Proc. Conference on Computer Graphics and Interactive Techniques, pp. 99–108 (1996)

A Novel Color Transfer Algorithm for Impressionistic Paintings

Hochang Lee, Taemin Lee, and Kyunghyun Yoon*

ChungAng University
khyoon@cau.ac.kr

Abstract. Existing color reproduction algorithms achieve image enhancement or visualization by correcting tone and hue. Although these algorithms are good at incorporating a natural feel in an image, they do not accurately represent intended colors that are contrasted. This limitation is particularly evident in the case of impressionist paintings that have a color contrast based on the style of the painter and the conventions of impressionism. In this study, we propose a novel color reproduction algorithm for impressionism, primarily focusing on paintings by Vincent van Gogh. First, the color value that is most obvious to the user is extracted. Based on this color, the color of the main object is transferred. Then, the color contrast information of a sample painting is extracted based on the color harmony theory. Finally, the color of other regions is transferred. Using our proposed algorithm, a more artistic result that matches the contrast style of the artist is generated. This work should be extended further to enhance the simulation of stroke color by a painterly rendering algorithm.

1 Introduction

Impressionism reached the peak of its popularity in the latter half of the 17th century. Impressionist painters attempted to use a new painting style significantly different from previous academism painters who expressed objects with an ideal and realistic depiction. From [1], "The Impressionists developed a style of painting that used vivid color and dynamic brush strokes to capture a moment." According to [2], "Impressionist painters put emphasis on capturing reality and depicting what they saw at a given moment." Impressionists wanted to express their initial impressions of a scene on the canvas. These attempts are well represented in their artwork. They express various scenes by using different styles such as stroke expression, perspective, and color selection, based on their individuality. In this study, we attempt to simulate the color expression of impressionism, using paintings by Vincent van Gogh, which are considered the representative work of impressionism. The characteristics observed in the color expression of the art work by Vincent van Gogh are:

---

* Corresponding author.

G. Bebis et al. (Eds.): ISVC 2012, Part I, LNCS 7431, pp. 513–522, 2012.

Impressionist characteristics

1. He tries to express his first impression of color on the canvas.
2. He does not use black.

Personality characteristics

1. He expresses a scene, not with its original colors, but with new colors according to his style.

Color is a very important factor for the viewer to feel emotional inspiration. Therefore, we might obtain more artistic results with impressive effects if we could extract and represent the features of artistic color. However, artistic theory based color transfer studies are not widely researched in computer graphics and image processing research field.

For this purpose, we propose a color-reproduction algorithm based on color-transfer techniques. This study focuses on the paintings of Van Gogh. The proposed technique deals with two inputs: the sample of a painting by Van Gogh (S), and the target image (T) which is to be converted. Each pixel color in T is updated by a transfer function that considers S. The resultant image (R) is composed of colors that consider the color contrast information in S. We segment the target image according to objects, and then define the main object that is most prominent to the viewer. The main object is transferred based on the color of various "focused points". Other regions are transferred using color contrast information from the source image. Finally, we correct black regions using a luminance correction step. From our algorithm, we can generate artistic results based on the color expressing mechanism in impressionism and the color contrast pattern of Van Gogh.

2   Related Work

A color transfer algorithm modifies the color of the target image based on the color of a reference image or color palette. Research on various color transfer algorithms has generally focused on image enhancement or visualization by correcting tone and hue. Based on the matching method, these techniques are broadly classified into two approaches. The global approach involves direct use of reference images but does not use area segmentation. In the local approach, image segmentation and color transfer are applied to each area. [3,4] are examples of the global approach. In [3], a representative study on color transfer, the author considered the average and standard deviation of Lab color space. In [5] and [4], the authors performed color reproduction based on histogram matching in HSV and Lab color space. These techniques use reference images without manual intervention, and hence, they are efficient. However, it is impossible for these techniques to express artistic effects that have a distinct color in each area. [6,7] use the local approach. These studies performed area segmentation and assigned matching relations for each area. Then, they transferred hue and

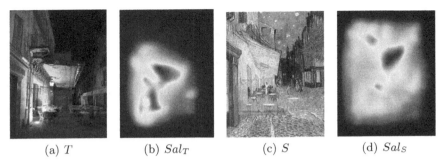

(a) T (b) Sal_T (c) S (d) Sal_S

Fig. 1. Saliency information

tone based on matched areas of reference. Although this approach is good for transferring a natural feel to the target image, it is time-consuming because a matching function must be set for each area. Our algorithm is based on a local approach for expressing distinct color contrast between areas. However, we reduce the time cost because our algorithm matches each area automatically based on color and position information. Further, we focus on the color contrast between regions of the source image rather than the color. Additionally, differing from previous work, we attempt to select a color by considering the "impression" of impressionists and simulating this mechanism.

3   Color Transfer for Impressionist Paintings

3.1   Definition and Analysis of "Impression"

In order to simulate impressionist paintings, the meaning of the word "impression" must be defined. We define "impression" as the region of the painting that attracts the attention of the user. To simulate this, we use image saliency techniques. Image saliency is a measure of the strength of local saliencies in a 2D image, indicating an area that can be expected to be the focus of visual perception. We use Perona's image saliency technique [8]. Fig. 1 shows the saliency information of T and S, which we define as Sal_T and Sal_S, respectively. In these figures, the red regions have higher saliency than the blue regions. Each pixel has a value $0 - 1$; a value closer to 1 signifies higher saliency. We utilize this value in region segmentation and color transfer.

3.2   Painting Analysis

Contrast Analysis of Reference Image. We extract the representative color by a k-means clustering in the hue and saturation channel. First, we perform a color quantization of S with k colors. Next, we use the initial points to determine the distribution groups. In general, we use k value as 3.

Let r_i be the points set of i'th clustering, and let p_i be the hue value with maximum frequency in the hue color wheel. The p_i has a value $1 - 360$

$$Cluster = \{(r_1, p_1), (r_2, p_2), ..., (r_k, p_k)\} \tag{1}$$

We need to select a basis color for analyzing the color contrast between represented colors. We define P_{main} as the basis color;

$$P_{main} = \{p, range\} \tag{2}$$

where p is the p_i with the maximum saliency value in each clustering set. After determining P_{main}, we find the artistic color pattern based on it. For hue contrast, we use the color harmony theory proposed in [9]. The authors of [9] defined 8 types of color patterns that humans recognize as harmonious, as shown in Fig. 2(b). We determine the harmony type of each p_i based on P_{main}, and the difference in degree with P_{main}. We also calculate the average position of the current area, pos, and then define $P_{other(i)}$ as follows:

$$P_{other(i)} = \{Type, d, pos\} \tag{3}$$

where $Type$, d, and pos represent the hue contrast type in [9], the hue distance from P_{main}, and the x and y position of the current region, respectively.

Fig. 2(a) shows the hue distribution of each area of "Cafe Terrace at Night" by Vincent van Gogh and color harmony pattern proposed in [9]. In this figure, P_{other1} represents type I and P_{other2} represents type V in relation to P_{main}. Finally, we save the information for S as follows:

$$S = \{P_{main}, P_{other1}, P_{other2}, ..., P_{otherk-1}\} \tag{4}$$

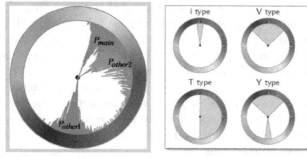

(a) Distribution of color   (b) Harmony pattern

Fig. 2. Distribution of palette color & Harmony pattern

3.3   Color Reproduction Based on Impressionist Style

In the color reproduction step, the color of T is transferred by using the contrast information. To express a more distinct color contrast, T is segmented by mean-shift segmentation, and then, color is transferred according to each segment. In segmentation, T is divided by k'th region. In a manner similar to the method for determining P_{main} from S, the average saliency value of each region is first calculated; then, T_{main}, having the maximum saliency value, is identified. The other regions are defined as $T_{other(i)}$. Further, the maximum color collected point in the hue space at each region is determined, and defined as $C_{main}, C_{other1}, C_{other2}, ..., C_{otherk-1}$. Each $T_{other(i)}$ is defined as follows:

$$T_{other(i)} = \{d, pos\} \tag{5}$$

$$d = |C_{main} - C_{other(i)}| \tag{6}$$

where, d indicates the difference in degree between C_{main} and $C_{other(i)}$, and pos represents the average position of the region. After segmentation, we perform the color transfer step. First, the T_{main} area is transferred; then, the $T_{other(i)}$ areas are transferred using T_{main} and contrast information.

Main Area Color Transfer Based on Saliency. In order to transfer the color of the main area, color pixels with high saliency value in T_{main} are extracted, and $\widetilde{C_{main}}$ is calculated by determining the average color of pixels with high saliency value. Hue correction is performed based on the determined value of $\widetilde{C_{main}}$. C_{main} is moved to $\widetilde{C_{main}}$, and the hue of each pixel is transferred onto an arc whose center value is $\widetilde{C_{main}}$. The arc degree is determined by the range of P_{main}. Fig. 3(a) shows the distribution of the hue range of T_{main}, and the shifting of the hue value based on $\widetilde{C_{main}}$.

Color Reproduction for Remaining Area. After transferring the main area, the other areas are transferred based on the contrast information. In this step, the contrast information between P_{main} and P_{other} is used. In order to represent the contrast information between P_{other}'s and each region, we must perform a corresponding match between regions. In this study, we consider the color and position information of $T_{other(n)}$. The hue and position distance between the current region and each P_{other} are calculated. Then, the matching set with minimum distance is selected as represented in equation (10), and P_{other} is assigned to each region.

$$V1 = \{T_{other1}, T_{other2}, ..., T_{othern}\}, \tag{7}$$

$$V2 = \begin{pmatrix} P_{other1} \\ P_{other2} \\ \vdots \\ P_{othern} \end{pmatrix} \tag{8}$$

Find the set $(V1_i, V2_i)$ which satisfy $Min(V1 \otimes V2)$, $\qquad$ (9)

$$A \otimes B = \sum dis(a_i, b_i), \ A = (a_1, .., a_n), B = (b_1, .., b_n) \tag{10}$$

$$dis(a, b) = ||a(d) - b(d)|| + ||a(pos) - b(pos)|| \tag{11}$$

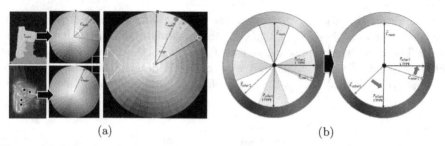

(a)   (b)

Fig. 3. (a) Color transfer step for the main area. The color of main area is transferred by color with high saliency value, (b)Color transfer step for the other region.

where $||\cdot||$, $X(d)$, and $X(pos)$ represent the Euclidean distance, d of X region, and the average position of X, respectively. After determining the relation between $T_{other(n)}$ and $P_{other(n)}$, the hue distribution is transferred based on contrast information. We shift for the hue information to represent color contrast. The method for color shift is the same as the method used for obtaining $\widehat{C_{main}}$. Figure 3(b) shows the color transfer step for other regions.

4   Experimental Results

4.1   Analysis of the Palette of Van Gogh

We used famous paintings by Van Gogh that show definite contrast. Table 1 shows our analysis information. Several common rules can be confirmed from the observations. When the image is segmented into three regions, it is observed that one region has a slightly different hue value compared to the main object, while another region uses a color complementary to the color of the main object. This implies that the paintings have V or L type in similar color, and I type in complementary color. Essentially, each painting does not have the same values. However, if we consider that color expression was limited at the time of his painting, this contrast analysis is reasonable.

Table 1. Pattern type in each painting. (1)"Cafe Terrace at Night"(1888)(2) "The night Cafe in the place Lamartine in arles"(1888) (3)"Sunflowers"(1888), (4)"Wheat Field with Crows"(1890), (5)"La chambre de Van Gogh a Arles"(1889).

Painting	$C_{main}(Hue)$	$C_{other1}(d)$	$C_{other2}(d)$
1	$(45 - 75)$	V(-30)	I(+180)
2	$(90 - 120)$	V(-30)	L(-90)
3	$(40 - 50)$	V(+30)	L(+180)
4	$(0 - 60)$	L(+60)	I(+170)
5	$(35 - 45)$	V(-30)	I(+170)

4.2   Test Results

Fig. 4 shows our results at each step. Figure 4(a) depicts the saliency image and high saliency regions of Fig. 1(a). The segmented image is seen in Fig. 4(b). Based on this, we transfer the color of the main area (Fig. 4(c)); the other area is transferred by considering the contrast information from the main object. Then, we obtain the final result (Fig. 4(d)). Fig. 4(e) shows the result of luminance correction. It can be observed that our result has the color contrast seen in Table 1(1)(V and I type).

(a) saliency image and high   (b) segmented   image   and   (c) main area transfer
saliency regions   pattern of S

(d) other area transfer   (e) luminance correction

Fig. 4. Results at each step(S=Fig. 1(c))

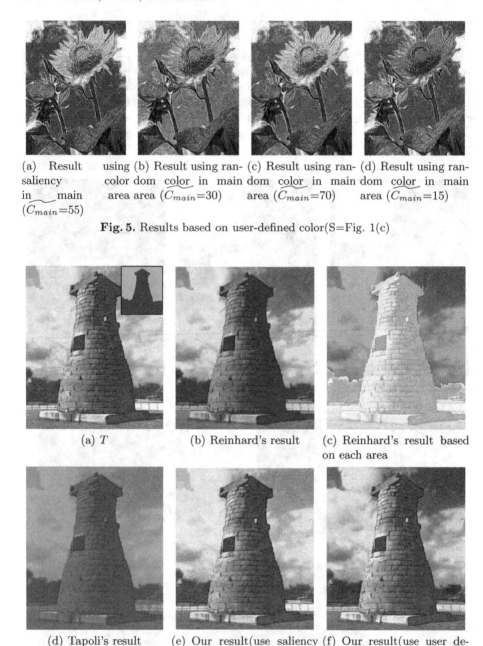

(a) Result using saliency in main $(C_{main}=55)$

(b) Result using random color dom color in main area area $(C_{main}=30)$

(c) Result using random dom color in main area $(C_{main}=70)$

(d) Result using random dom color in main area $(C_{main}=15)$

Fig. 5. Results based on user-defined color(S=Fig. 1(c)

(a) T

(b) Reinhard's result

(c) Reinhard's result based on each area

(d) Tapoli's result

(e) Our result(use saliency color)

(f) Our result(use user defined color)

Fig. 6. Comparison with results from other studies(S=Fig.1(c))

Figure 5 shows the various results obtained when the user manually selects the color of the main object. Figure 5(a) uses saliency color when the color of the main object is transferred. Figures 5(b)-(d) use a random color determined by the user. Impression colors were randomly assigned to the main object, and the colors of other areas were transferred using the contrast information. Based on each main object color, our algorithm can express various results by considering the contrast information. Figure 7 shows comparisons with previous research. Figure 6(a) is the original (T), and Figure 6(b) shows the result obtained from Reinhard's algorithm [3]. Owing to the usage of a global approach, the result cannot represent yellow and red colors and these express an average. Figure 6(c) shows the result of adopting Reinhard's techniques for each area. By assigning a specific color to each area, yellow and blue tones are prominent in the resulting image. However, owing to the usage of averages and standard deviations in this technique, the result appears unnatural. Figure 6(d) shows the results obtained using a local approach [7]. This technique performs well when using natural scenes. However, the result loses texture information when the technique is applied to artistic paintings. Figure 6(e) shows the result from our algorithm. In this result, the color of the high saliency region is gray; hence, there is slight change in the main object while the remaining regions are changed with color contrast. Figure 6(f) shows the result with user-defined colors in the main object. In contrast with previous work, our algorithm can naturally express results while preserving texture and color contrast information when using artistic images.

5   Conclusion and Future Work

In this study, we proposed a method to simulate the color expression mechanism of Vincent van Gogh, which is based on impressionism. First, impression points were extracted by image saliency, and then, a color transfer of the main object was achieved based on colors with high saliency points. The remaining objects were transferred based on the color contrast information of S. We are able to transfer color by impressionist color expression mechanisms, and, based on user studies, we are certain that our algorithm generates an emotional response. The proposed algorithm operates well when sources with definite contrast information, such as a painting by Van Gogh, are used. However, our algorithm does not consider saturation value, and hence, cannot generate good results when an image with low saturation is used. Our algorithm makes several significant contributions to the field. It is a novel color reproduction algorithm for impressionist paintings. Owing to the simple and easy manner in which it transfers color based on artistic paintings, it can be utilized in many applications with other painterly rendering results. We also analyze artistic contrast information from famous paintings, and identify a harmony pattern from color harmony theory. This technique can be utilized in painting analysis. Our future work includes a plan to adapt the concept of impressionism to other painting techniques such as texture and edge. This will be a good step for expressing the effects of other impressionist painters. Next, we plan to adapt the color palette of Van Gogh for color selection. Finally, we intend to analyze more paintings from different artists and periods.

Acknowledgements. This work was supported by a Korean Science and Engineering Foundation(KOSEF) grant funded by the Korean government(MEST)(No.20110018616). And this work was also supported by a Seoul R&BD Program(PA110079M093171).

References

1. Callen, A.: The art of impressionism: Painting technique and the making of modernity. Yale University Press (2000)
2. Impressionism: What is impressionist art? introduction to impressionism, http://www.impressionism.org/
3. Reinhard, E., Ashikhmin, M., Gooch, B.: Color transfer between images. IEEE Computer Graphics and Applications 21, 34–41 (2001)
4. Xiao, X., Ma, L.: Gradient-preserving color transfer. Computer Graphics Forum 28, 1879–1886 (2009)
5. Neumann, L., Neumann, A.: Color style transfer techniques using hue, lightness and saturation histogram matching. In: Proc. CAe 2005, pp. 111–122 (2005)
6. Tai, Y.W., Jia, J., Tang, C.K.: Local color transfer via probabilistic segmentation by expectation maximization. In: Proc. CVPR 2005, pp. 217–226 (2005)
7. Pouli, T., Reinhard, E.: Progressive histogram reshaping for creative color transfer and tone reproduction. In: Symp. Non-Photorealistic Animation and Rendering, NPAR 2010, pp. 83–90 (2010)
8. Onathan Harel, C.K., Perona, P.: Graph-based visual saliency. In: Proceedings of the Conference Advances in Neural Information Processing Systems, pp. 545–552 (2007)
9. Cohen-Or, D., Sorkine, O., Gal, R., Leyvand, T., Xu, Y.Q.: Color harmonization. In: ACM SIGGRAPH 2006 Proceeding, pp. 624–630 (2006)

Gaze-Dependent Ambient Occlusion

Radosław Mantiuk and Sebastian Janus

West Pomeranian University of Technology in Szczecin,
Faculty of Computer Science,
Żołnierska 49, 71-210, Szczecin, Poland

Abstract. In this paper we present a gaze-dependent ambient occlusion technique in which information about the human viewing direction is used to vary accuracy of ambient occlusion computation. The screen region surrounding the observer's gaze position is rendered with maximum precision, decreasing gradually towards the parafoveal regions. The rendering is speeded up by reducing the number of sampling rays with increasing distance to a gaze point captured by the eye tracker. We conduct experimental evaluation of the perceptual visibility of the ambient occlusion shadows outside observer's region-of-interest. We also present our GPU-based implementation of the gaze-dependent ambient occlusion system.

1 Introduction

Ambient occlusion (AO) is technique used in the local illumination shading models to improve realism of renderings through shadowing of the ambient light. In comparison to the full global illumination solutions, AO requires less computations and produces plausible perceptual simulation of the global lighting model. However, ambient occlusion algorithm still needs demanding resources to achieve high quality renderings and cannot be rendered in real time even based on the contemporary high-end GPUs.

In this work we present a gaze-dependent ambient occlusion technique in which information about the human viewing direction is used to varying accuracy of ambient occlusion computations. The screen region surrounding the observer's gaze position is rendered with maximum precision, decreasing gradually towards parafoveal regions. Number of the ambient occlusion sampling rays is decreased with increasing distance to a gaze point what speeds-up the rendering.

The main goal of this work is a perceptual evaluation of the visibility of the ambient occlusion shadowing in the parafoveal regions. We argue that the AO effect can be clearly visible only in high frequency regions. We use the gaze-dependent Contrast Sensitivity Function (CSF) to model observer's region-of-interest (ROI) and we conduct experimental evaluation of the visibility of the AO shadows outside ROI.

G. Bebis et al. (Eds.): ISVC 2012, Part I, LNCS 7431, pp. 523–532, 2012.

Our gaze-dependent AO implementation relies on NVIDIA OptiX library[1], which operates on the CUDA[2] engine. The system renders images with an interactive performance of a few frames per second.

The concept of gaze dependent rendering was thoroughly studied (see [1] for review). However, to the best of our knowledge, a gaze-dependent ambient occlusion is a novel technique and was not so far proposed. The latest progress in GPU performance and eye tracking technologies make this solution practical and interesting from the computer graphics and the vision research point of view.

In Section 2 of this paper we describe the ambient occlusion algorithm and discuss why the full ambient occlusion method has been chosen instead of approximated AO methods like the screen space ambient occlusion. In Section 3 our concept of the gaze-dependent ambient occlusion rendering is presented followed by description of the implementation (Section 3.4). The experimental evaluation and results are discussed in Section 4. The paper ends with conclusions and future work in Section 5.

2   Background

In the Phong reflection model, diffuse and specular reflections are varying due to observer and lights position, but ambient light is constant. Having this assumptions, we miss the inter-reflections between rendered objects. Adding ambient occlusion for varying ambient light creates very convincing soft shadows, that combined with indirect lighting give realistic images [2, Sect. 9.2].

The goal of the AO technique is to take the visibility into account during integration of radiance values at the surface point p. Some directions over the hemisphere Ω are blocked by other objects in the scene, or by other parts of the same object. It is assumed that these directions have zero incoming radiance and this information is taken into account during computation of the occlusion factor k_a:

$$k_a = \frac{1}{\pi} \int_{\Omega} V_p(\boldsymbol{\omega})(N \cdot \boldsymbol{\omega})d\omega, \qquad (1)$$

where $V()$ denotes the binary visibility function from the point p, which returns 0 if a ray cast from p in the direction of ω is blocked, and 1 if it is not (N is a normal vector of a surface at p). The k_a factor is further used instead of the ambient light coefficient in the Phong reflection equation.

In our gaze-dependent technique, we use extension of the AO algorithm proposed in [3], in which the visibility function is replaced with a distance mapping function $\rho()$. $\rho()$ is continuous function ranging from 0 at an intersection distance of 0 to 1 at intersection distance greater than a specified maximum distance. In contrary to the standard method, the distance function gives correct results also for the enclosed geometry.

---

[1] See: http://www.nvidia.com/object/optix.html

[2] See: http://www.nvidia.com/object/cuda_home_new.html

The significant computational complexity of the AO algorithm is caused by a large number of rays (counted in hundreds of millions) that must be cast from an intersection point and traced to approximate the visibility. A few real-time methods of ambient occlusions computation were proposed [4–6]. However, they are based on approximations that introduce visible artefacts like haloing. Objects outside of the screen are not considered during computations resulting in inaccurate values of k_a, computed occlusion is dependent on viewing angle and camera's position, etc. We argue that a fundamental shift to the full AO method is necessary and we do not evaluate simplified techniques (e.g. the screen space ambient occlusion) in this paper.

3 Gaze-Dependent Rendering of Ambient Occlusions

In the gaze-dependent ambient occlusion technique the detailed computations are performed in the region of interest. The further from the gaze point, the less detailed ambient factor is rendered saving computation time. While, the use of eye tracker leaves observer with a feeling that the scene is fully detailed.

3.1 Rendering Engine Architecture

The outline of our gaze-dependent ambient occlusion system is presented in Fig. 1. The viewing direction of an observer is captured by the eye tracker. Sampling of the screen is varied based on location of the gaze point and the shape of the gaze-dependent Contrast Sensitivity Function ([7], see also Sect. 3.2). In general, less AO rays are generated in the parafoveal region to compute the AO factors with decreased precision but significantly faster. In parallel, the Phong based lighting equation is solved to compute ambient, diffuse, and specular components for every pixel. The ambient component is blended with the AO factors also considering the shape of the gaze-dependent CSF, resulting in the final image with gradually decreasing influence of the AO factor in the parafoveal regions.

3.2 Region of Interest Sampling

The ambient occlusion factors have rather subtle influence on the final image appearance, especially when the image is rendered basing on the Phong lighting model. This shadowing of ambient light is visible as a high frequency information in characteristic regions of a scene (e.g. at corners, close to complex objects, etc.). Frequency of AO shadows is higher than variability of the Phong ambient shading. We use this perceptual phenomenon to model the drop-off of the AO rendering quality based on the gaze-dependent CSF. This idea is consistent with drop-off of the visual sensitivity across the visual field modelled by the classic CSF [8]. The number of the AO rays is reduced with increasing distance from the gaze point, therefore the total number of rays for a region is decreased. It

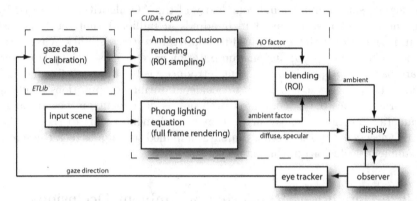

Fig. 1. Gaze-dependent ambient occlusion rendering system

reduces the screen space sampling frequency of the AO shading and decreases visibility of the AO effect.

Large reduction in the number of the AO rays generates adverse noise. This noise is noticeable even in the parafoveal vision and generates strong temporal aliasing (see Figure 2). Therefore, we additionally blend the AO factors with the regular ambient colour component which removes noise from the parafoveal regions (see Fig. 6). This blending is also weighted by the gaze-dependent CSF.

Fig. 2. Gaze-dependent ambient occlusion without blending (visible noise). The same scene rendered with blending is presented in Fig. 6.

In our gaze-dependent AO technique we use the gaze-dependent CSF proposed by Peli et al. [7]:

$$C_t(E, f) = C_t(0, f) * exp(kfE), \qquad (2)$$

where C_t denotes contrast sensitivity for spatial frequency f at an eccentricity E, k determines how fast sensitivity drops off with eccentricity (the k value is ranged from 0.030 to 0.057). Based on the above equation, the cut-off spatial frequency f_c can be modelled as:

$$f_c(E) = min(max\_display\_cpd, 43.1 * E_2/(E_2 + E)), \qquad (3)$$

where E_2 is retinal eccentricity at which the spatial frequency cut-off drops to half its foveal maximum (from 43.1 cpd to 21.55 cpd, $E_2 = 3.118$, see details in [9]). In this equation, we flatten the CSF function with $min(max\_display\_cpd, ...)$ operator to take into consideration the limited resolution of our display (see Sect. 4.1). The shape of the CSF function and the computed region-of-interest mask are presented in Fig. 3.

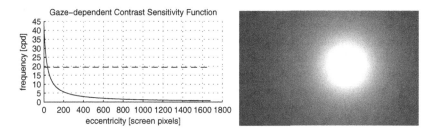

Fig. 3. Left: Gaze-dependent Contrast Sensitivity Function (the dashed line denotes threshold frequency of our display). Right: ROI coefficient computed based on CSF (centre at $(1000, 500)$), lighter areas denotes that AO will be computed with higher accuracy.

3.3 Eye Tracker Fixations

Eye trackers capture two types of eye movement called *saccades* and *smooth pursuit*. A smooth pursuit is active when eyes track moving target and are capable of matching its velocity. A saccade represents a rapid eye movement used to reposition the fovea to a new location, which lasts from 10 ms to 100 ms. The main goal of the gaze tracking is however to capture a single location an observer intents to look at. This process is known as a *visual fixation*. A point of fixation can be estimated as a location where saccades remain stable in space and time [10].

In our work we approximate the mechanism of visual fixation by pooling gaze points and using averaged gaze position for consecutive rendering frames. We argue that this simple technique is the best suitable for our application. There are many fixation algorithms that provide more advanced spatial-temporal analysis of gaze points (see [1, 10] for survey), but generally they decrease the accuracy of the fixation point estimation [11] and are not suitable for the computer graphics applications [12].

3.4 Implementation

Our application is based on the AO example delivered with the OptiX SDK package [13]. We extended this code to the gaze-dependent technique based on ROI and CSF. We also implemented the eye tracker interfaces.

Ambient Occlusion Based on OptiX. The implementation of gaze-dependent AO is divided into a host program (written in C++), and a set of GPU shader programs: *ray generation program* - tracing rays from the viewer towards the scene (the primary rays), *closest hit program* - computing lighting equation, this is main shader which looks for intersections of rays with scene objects and computes the colour based on the Phong lighting equation together with our gaze-dependent AO technique, *any hit program* - calculating any hit occlusion, it is used for tracing rays from the intersection points (pointed by the rays which are calculating radiance), these secondary rays are used to compute the ambient occlusion factor. The main modules of the gaze-dependent AO implementation are located in the *closest hit program* where the ROI-based sampling is performed. This program also computes the final blending of the AO factors and the ambient light.

Eye Tracker Library. We implemented the ETlib library responsible for managing communication between the eye tracker software and our application. ETlib also supports the calibration of eye tracker, which must proceed every eye tracking session [1]. During calibration observer is asked to look at calibration markers displayed one-by-one in different region of the screen. The process is finished after registration of gaze positions for a given number of the markers (5 in our case). Based on the captured data, the mapping from the eye tracker camera coordinates to the display screen coordinates is computed. Precise calibration is extremely important because it affects further accuracy of the captured gaze positions [14].

4 Experimental Evaluation

The main objective of our experimental evaluation was to prove that one can minimise the ambient occlusion sampling in parafoveal area with only slight influence on observers perception of the AO effect. We also analyse a performance boost achieved due to decrease of the rendering computations.

4.1 Stimuli and Hardware Setup

Our experimental scene presents the fixed Stanford Dragon Model[3] (50,008 vertices and 100,005 faces) enclosed in the 5 walls box (see Fig. 4). The scene is

---

[3] http://www.mrbluesummers.com/3572/downloads/stanford-dragon-model

rendered with interactive speed by our AO software based on the OptiX engine (see Sect.3.4).

We use SMI RED250 [15] eye tracker working with the frequency 250 Hz and the accuracy close to 0.5° of the viewing angle, and controlled by the proprietary SMI iViewX software. This software runs on the remote PC and sends gaze data via TCP network to the ETlib driver (see Sect. 3.4). Eye tracker captures location of observer's gaze points that are further used to compute the ROI-based AO factors and to blend the final colour. All computations are performed on PC equipped with 2.8 GHz Intel i7 930 CPU with 8 GB of RAM, Windows 7 64bit OS, and a GPU NVIDIA GeForce 480 GTX 512MB graphics card. The stimuli is displayed on 22 inch Dell, with the 1680x1050 resolution (60 Hz) and dimensions 47.5 cm width x 30 cm height. The observers sit in a 60 cm distance from the display what results in the screen resolution close to 20 cpd (cycles per degree).

We use high quality image rendered with the full frame AO effect (without ROI sampling, with 400 AO rays per pixel, see Fig. 4) as a reference in the quality and in the performance comparisons.

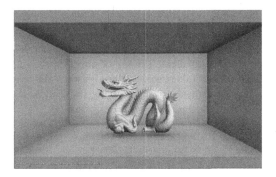

Fig. 4. The reference image with the full frame ambient occlusion

4.2   Quality Evaluation

During quality evaluation observers were asked to assess the perception of AO quality deterioration caused by the reduced sampling in the parafoveal area and by blending with the ambient light. We conducted the double stimulus experiment where observers looked freely at the full frame AO and then at gaze-dependent AO driven by the eye tracker (for 20 seconds each). Next, observers were asked to assess deterioration of the AO quality using a 5-point Likert scale (5-no deterioration, 4-slight, 3-medium, 2-large, 1-extremely large). The procedure was repeated for 5 different camera settings (different view points). During observation, the camera was slowly animated to force constant recomputing of the image. We asked 7 participants to take part in the experiment, 6 males and one female, age between 21 and 23 years, they had normal or corrected to normal visual acuity. All participants passed basic computer graphics course and were familiar with the AO technique.

Figure 5 (left) depicts results of the experiment. The average difference mean opinion score (DMOS) [16] for every observer and every camera setting is equal to 1.31 (sem = 0.11, std = 0.68) which corresponds to judgement between the medium and the slight deterioration. Significantly better results were achieved for the camera setting 5 (DMOS = 0.86, see Fig. 5,right). In this case the AO shadows are away from each other. Observers cannot see opposite region in parafoveal region and their perceptual experiences were close to observation of the full screen AO.

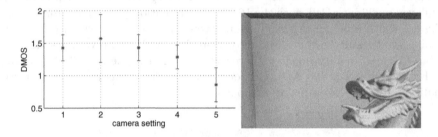

Fig. 5. Left: Result of experiments showing observers assessment of the AO quality deterioration. The circles depict the DMOS for five different camera settings. Error bars show the standard error of mean (SEM) of the DMOS values. DMOS values correspond to difference beween Likert scale values between reference and test images. Right: The rendering for the camera setting 5 giving the best assessment results.

In Fig. 6 two example frames depicting the gaze-dependent renderings are presented (Camera setting 1). It can be seen that shading caused by the AO factor is stronger in the centre of the ROI and weakens with the distance.

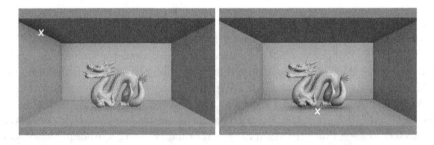

Fig. 6. Images rendered with the gaze-dependent ambient occlusion. The white X indicates the centre of ROI. In the left image there is visible AO effect in the upper corner of the box, and there is no shadow below the dragon model. In contrary, in the right image where the position of the ROI is close to the dragon, these shadows are visible very well, and there is only a slight shadow on the wall behind the dragon.

4.3   Performance Evaluation

To compare timings for the full frame AO and the gaze-dependent AO, we measured the speed of rendering for 3 different camera settings. We achieved 1.26 fps, 0.96 fps, and 0.62 fps for Camera 1, 2, and 3 respectively in the full frame AO mode (for performance reasons, the resolution of renderings was reduced to 840x525 pixels). The same images were rendered using eye tracker and the gaze-dependent AO rendering technique. We measured the rendering speed for different locations of ROIs because the local complexity of a scene affects the rendering time (there are more intersection tests in regions with more triangles). The results are summarised in Table 1. We achieved a 276% performance boost in the best case and average speed-up equal to 140.07%. However, even in that worse case, we get a noticeable rendering speed-up.

Table 1. Rendering speeds for gaze-dependent AO. Regions A-E are located in Camera 1 image, regions F-J in Camera 2, and K-O in Camera 3.

Region	A	B	C	D	E	F	G	H	I	J	K	L	M	N	O
Speed [fps]	4.10	4.21	4.06	1.68	2.11	2.36	2.65	2.19	1.30	1.40	1.82	2.33	2.24	1.00	1.17
Speed-up [%]	201	210	199	24	55	146	176	128	35	46	194	276	261	61	89

Generally, the results show significant performance increase in a regions of a small triangle number (box walls all around the screen). Worse outcome is when we take the hardest to compute environment - middle of the screen (the biggest samples number) and the dragon object (high level of the triangles). It is consistent because when ROI is outside complex object, the computation connected with this object are skipped.

5   Conclusions and Future Work

In this work we propose to vary the accuracy of the ambient occlusion computations based on information about observer's viewing direction. Thanks to reducing a number of the AO rays in the parafoveal regions, we achieved significant rendering speed-up without decreasing the perceivable quality of the AO shadowing.

In the future work we plan to implement a fixation algorithm suitable for graphics rendering systems. It would help to stabilise capturing of the viewing direction and would make the perceptual experiment less prone to interferences. However, further progress in performance of the graphics systems and in accuracy of eye trackers are clue factors projecting the development of gaze-dependent technologies in computer graphics.

Acknowledgements. This work was supported by the Polish Ministry of Science and Higher Education through the grant no. N N516 508539.

References

1. Duchowski, A.T.: Eye Tracking Methodology: Theory and Practice, 2nd edn. Springer, London (2007)
2. Akenine-Möller, T., Haines, E., Hoffman, N.: Real-Time Rendering, 3rd edn. A. K. Peters, Ltd., Natick (2008)
3. Zhukov, S., Iones, A., Kronin, G.: An ambient light illumination model. In: Proceedings of the 9th Eurographics Workshop on Rendering, pp. 45–56 (1998)
4. Mittring, M.: Finding next gen-cryengine 2. In: SIGGRAPH 2007 Advanced Real-Time Rendering in 3D Graphics and Games Course Notes (2007)
5. Bunnell, M.: GPU Gems 2, Dynamic Ambient Occlusion and Indirect Lighting. Addison Wesley (2005)
6. Hegeman, K., Premoze, S., Ashikhmin, M., Drettakis, G.: Approximate ambient occlusion for trees. In: Proceedings of ACM Symposium in Interactive 3D Graphics and Games (I3D 2006), pp. 41–48 (2006)
7. Eli Peli, J.Y., Goldstein, R.B.: Image invariance with changes in size: the role of peripheral contrast thresholds. JOSA A 8(11) (1991)
8. Daly, S.: The visible differences predictor: An algorithm for the assessment of image fidelity. In: Watson, A.B. (ed.) Digital Images and Human Vision. MIT Press (1993)
9. Yang, J., Qi, X., Makous, W.: Zero frequency masking and a model of contrast sensitivity. Vision Research (1995)
10. Salvucci, D.D., Goldberg, J.H.: Identifying fixations and saccades in eye-tracking protocols. In: Proceedings of the 2000 Symposium on Eye Tracking Research & Applications (ETRA), New York, pp. 71–78 (2000)
11. Shic, F., Scassellati, B., Chawarska, K.: The incomplete fixation measure. In: Proceedings of the 2008 Symposium on Eye Tracking Research and Applications, ETRA 2008, pp. 111–114. ACM, New York (2008)
12. Mantiuk, R., Bazyluk, B., Tomaszewska, A.: Gaze-Dependent Depth-of-Field Effect Rendering in Virtual Environments. In: Ma, M., Fradinho Oliveira, M., Madeiras Pereira, J. (eds.) SGDA 2011. LNCS, vol. 6944, pp. 1–12. Springer, Heidelberg (2011)
13. Parker, S.G., Bigler, J., Dietrich, A., Friedrich, H., Hoberock, J., Luebke, D., McAllister, D., McGuire, M., Morley, K., Robison, A., Stich, M.: Optix: A general purpose ray tracing engine. ACM Transactions on Graphics (2010)
14. Mantiuk, R., Kowalik, M., Nowosielski, A., Bazyluk, B.: Do-It-Yourself Eye Tracker: Low-Cost Pupil-Based Eye Tracker for Computer Graphics Applications. In: Schoeffmann, K., Merialdo, B., Hauptmann, A.G., Ngo, C.-W., Andreopoulos, Y., Breiteneder, C. (eds.) MMM 2012. LNCS, vol. 7131, pp. 115–125. Springer, Heidelberg (2012)
15. SMI: RED250 Technical Specification, SensoMotoric Instruments GmbH (2009)
16. Dijk, A.M., Martens, J.B., Watson, A.B.: Quality assessment of coded images using numerical category scaling. In: Proc. SPIE, vol. 2451, pp. 90–101 (1995)

Profile-Based Feature Representation Based on Guide Curve Approximation Using Line and Arc Segments

Jinggao Li and Soonhung Han

School of Mechanical, Aerospace and Systems Engineering, KAIST

Abstract. The profile-based features of a mechanical computer aided design (CAD) system commonly have complicated shapes because of the guide curve, which is free-form type. However, while considering interoperability between different CAD systems, due to this kind of free-form guide curve, it is difficult to represent the corresponding feature shape in another CAD system, for example, ship CAD systems usually use relatively simple shape primitive to represent shape. Thus, in this paper, we propose a straightforward algorithm to represent profile-based features that is based on guide curve approximation using line and arcs segments. In addition, the solid alignment and filling operation are also given to complete the process of solid model reconstruction, and we use several test cases to demonstrate the effectiveness of the proposed method.

1 Introduction

The profile-based features of a mechanical CAD system, for example, sweep or helical features, as shown in Fig.1, commonly have complicated shapes because of the guide curve, which is free-form type. Recently, collaborative design has played a vital role in the product design process since the whole design task is too big for a company to deal with in-house; meanwhile, in order to improve the design efficiency and minimize the design cycle, collaboration between companies and their partners is also needed. Hence, while considering the interoperability between different CAD systems and the diversity caused from the preference of companies and partners, sometimes, it is difficult to represent the corresponding feature shape in other CAD systems due to the free-form guide curve. Therefore, this paper presents a straightforward algorithm to represent profile-based features that uses relatively simple shape primitives, such as line and arc segments, to approximate the complex shape of the guide curve. The proposed algorithm then uses these line or arc segments to generate the relevant solid model segments, and the filling operation is also given to complete the process of solid model reconstruction. The remainder of this paper is organized as follows: Section 2 reviews related research work; Section 3 mainly presents the proposed method, including how to find the optimal reference plane, line, and arc segment interpolation, solid alignment, filling operation, etc.; Section 4 shows the implementation result; and Section 5 ends up with the conclusion.

G. Bebis et al. (Eds.): ISVC 2012, Part I, LNCS 7431, pp. 533–543, 2012.

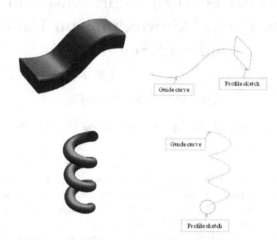

Fig. 1. Example of the sweep and helical features and their elements

2   Related Research Work

A circular arc or a helix curve is widely used in computer graphics, industrial engineering, motion path, etc. Thus, in terms of circular arc approximation, several studies have been conducted in this area using the Bézier curves [1] [2] [3] [4] [5] [6] [7] [8] [9], B-spline curves [10] [11], or other polynomial curves [12]. In addition, a helix segment is a natural generalization of a circular arc in 3D space, and the following is a quick review of the relevant research of helix curve approximation. A helix segment is a kind of geometric curve with a non-vanishing constant curvature and non-vanishing constant torsion [13]. Though a helix segment can be represented accurately with a combination of trigonometric functions and polynomials, there is no exact representation for it by polynomials or rational polynomials [14]. Hence, several studies have been conducted on the helix curve approximation with the rational Bézier curve. Mick and Röschel [15] proposed one method that uses degree 3 and 4 to interpolate the helical patches. Juhász [16] introduced an approach that approximates helix segments by cubic rational Bézier curves, and the error bound is also given, which enables the helix to be approximated within any prescribed tolerance. Seemann [17] discussed the case using rational Bézier curves of degree 4, 5, and 6. Yang [14] proposed a method that uses quintic Bézier curves or quintic rational Bézier curves based on the geometric Hermite interpolation technique to approximate a helix segment. Ahn [18] presented an approach of the approximation of a cylindrical helix by conic and quadratic Bézier curves, and the error bound analysis is also given. Furthermore, Ahn extended this work to the cases of surface [19]. Pu [20] introduced a subdivision scheme for approximating a circular helix with NURBS curve that combines the algorithm for the planar NURBS circular arc and the height function defined by using the knot insertion and the degree elevation for linear NURBS. Lu [21] presented a method for polynomial approximation of circular arcs and helices by expressing the trigonometric functions using the two-point Taylor expansion.

Lee [22] presented an algorithm to approximate a set of unorganized points with a simple curve without self-intersections and proposed an improved moving least squares technique based on the Euclidean minimum spanning tree to thin a given point cloud. Lin [23] proposed a curve reconstruction method based on an interval B-spline curve that focuses on generating a piece of centric curve from a planar strip-shaped point cloud.

However, in this situation, the problem is different from the cases mentioned above because, here, we cannot use an approximation tool like the Bézier curves or B-spline curves due to the request of the receiving system while exchanging CAD data. We have to use line and arc segments to approximate the free form guide curve, such as helix or B-spline. In addition, in the cases mentioned above, it does not discuss the solid model situation, but we have to deal with the solid model reconstruction. Thus, in the first place, it is better to discuss the guide curve approximation, which is based on line and arc segments, then the solid model reconstruction process will be explained in the next section.

3   The Proposed Method

3.1   The Entire Guide Curve Approximation and Model Reconstruction Process

In this section, the solid model reconstruction, which is based on guide curve approximation, will be explained in detail. Actually, it mainly includes two stages: firstly, it explains how to use line and arc segments to approximate the guide curve and, secondly, based on these line and arc segments, it explains how to create the relevant solid models.

In order to more accurately represent the free form guide curve using line and arc segments, we should locally approximate the guide curve. Thus, we use Moving Least Squares (MLS) concept to achieve this goal: firstly, descretize the free form guide curve and divide it into several pieces, and then, for each piece, use line and circular arc segments to approximate the points of each piece in the local reference domain. In fact, the MLS idea was firstly introduced by McLain smoothing and interpolating scattered data [24], and after that, many significant advances were achieved [25] [26] [27] [28] [29] [30] [31] [32] [33]. Furthermore, the MLS method is applied in different domains with a variety of applications. For example, it is used for shape deformation [34] [35], ray tracing [36] [37], finite element analysis [38] [39] etc. In fact, we utilize the MLS concept for the guide curve approximation and solid model reconstruction, which can be applied into interoperability between two different CAD systems.

After the guide curve approximation, we create a corresponding solid model based on a profile sketch, and these line and arc segments then proceed with the filling operation to complete the whole shape approximation. The whole procedure is given as follows:

Step1: From the original model, extract the feature data;

Step2: In order to approximate the guide curve, discretize the guide curve;

Step3: Get the segment number and divide the guide curve into several pieces;

Step4: Find the optimal reference domain;

Step5: Using a line and arc, interpolate the end points to approximate each piece;

Step6: Create a corresponding solid model;

Step7: Proceed with the filling operation;

In Fig. 2, the whole procedure is demonstrated with pictures of the creation of a related model. Among all of the steps, step 4, 5, and 7 will be explained later in detail.

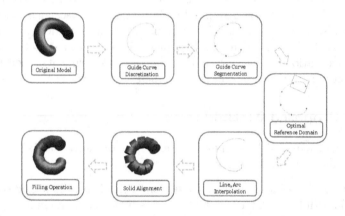

Fig. 2. Model reconstruction procedure based on the approximation technique

3.2   Optimal Reference Plane

First, we explain how to find the optimal reference domain. As mentioned before, here, we refer to the MLS concept to complete the guide curve approximation. In fact, the guide curve is a 3D curve. However, the line and arc should be planar segment; hence, the optimal reference plane for each line or arc segment has to be found first. The basic MLS procedure consists of two steps: first, defining a local reference domain and, second, constructing an MLS approximation with respect to the reference domain [31]. We only use the first stage to find the optimal reference plane, and explain the related procedure briefly. For more a detailed explanation, refer to [26] [27] [31].

Let points $p_i \in \mathrm{R}^3$ and $i \in \{1, ..., N\}$. It is assumed that the points are sampled from a surface S. In order to define a reference domain or a local approximating hyperplane for r which is near S. The local reference plane, which has the following expression $H = \{x | \langle n, x \rangle - D = 0, x \in \mathrm{R}^3\}$, $n \in \mathrm{R}^3$, $\|n\| = 1$, is computed to optimize a local weighted sum of square distances of the points p_i to the plane. Then H is obtained by minimizing the following summation:

$$\sum_{i=1}^{N} (\langle n, p_i \rangle - D)^2 \theta(\|p_i - q\|) \tag{1}$$

where θ is a smooth, radial, and monotone decreasing function. A Gaussian function is proposed in [31]:

$$\theta(d) = e^{-\frac{d^2}{h^2}} \tag{2}$$

where h, known as the radius of the supporting region or bandwidth, is a spacing scalar parameter that can be used to smooth out small features in the data [28] and where q is the projection of r onto H.

In Fig. 3, points p_i are sampled from the guide curve, point c is the average value of the point set of one piece, and the goal is to find optimal normal vector n. The following is the practical computation process:

Firstly, calculate the average value point c of the point set of each piece;
Secondly, calculate the weighted function value $\theta_i = \theta(\|p_i - c\|)$;
Thirdly, obtain the weighted covariances matrix $B = \{b_{jk}\}, B \in \mathrm{R}^{3 \times 3}$;
Next, solve the optimization problem, min $n^T B n, \|n\| = 1$;

Finally, calculate the eigenvector of B that has the smallest eigenvalue, and use it as the normal vector of the optimal reference plane.

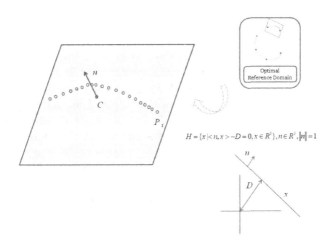

$$H = \{x \mid < n, x > -D = 0, x \in R^3\}, n \in R^3, \|n\| = 1$$

Fig. 3. An optimal local reference plane

3.3   Line and Arc Interpolation

Next, we use line and arc segments to approximate the point set of each piece. As shown in Fig. 4, it has two cases. For the line case, P_i, P_{i+n} are the end points, and we project them onto the local reference plane, which is obtained from the previous step. Then there can be two intersection points V_i, V_{i+n}, and we use these two projected points to approximate the point set with the constraint of interpolation V_i, V_{i+n}. For the arc case, we select four points, two end points P_i, P_{i+n}, and two points P_{i+1}, P_{i+n-1} that are adjacent to end points, and project them onto the local reference plane. Then we obtain the four intersection points $V_i, V_{i+1}, V_{i+n-1}, V_{i+n}$, using these four points to calculate two vectors, $V_i V_{i+1}$ and $V_{i+n} V_{i+n-1}$. Finally, based on these two vectors and two end points, we calculate the arc, which approximates the point set.

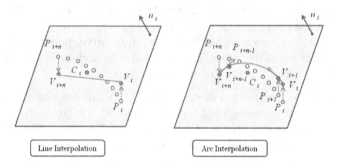

| Line Interpolation | | Arc Interpolation |

Fig. 4. Line and arc segment interpolation

3.4   Filling Operation

In terms of the filling operation, a simple and effective method used to compute the position and size of filling solid model segment is proposed. In Fig. 5, P_i, P_{i+n} are points belonging to the solid models that are created from step 6, and they are also the end point of each line or arc segment that is created from step 5. V_i, V_{i+n} are the projected points of P_i, P_{i+n} for each plane of the adjacent solid model segment. Then the center position Q_i and normal vector n_{qi} are calculated using the formula shown in Fig. 5, and the shape parameter α is used to determine the size of filling solid model segment. At this point, the whole approximation process is completed.

As shown in Fig. 6, we can use different numbers of primitives to approximate the helical feature, and a kind of spring shape with one rotation is created. The number noted in each picture is the segment number that is used to approximate the guide curve. The specific number, from left to right, varies from 10 to 35.

3.5   Approximation Quality Evaluation

The approximation quality can vary according to the number of primitives that are used to approximate the original shape. There are a variety of metrics that are used to measure the approximation quality. Here, we select one of the most typical metrics - distance error - to measure the approximation quality. To measure how well the newly created model approximates the original one, we calculate the distance error between the two models. Hence, we first sample a number of

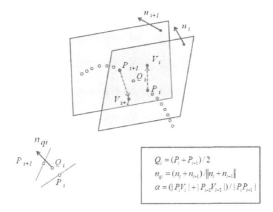

$$Q_i = (P_i + P_{i-1})/2$$
$$n_{qi} = (n_i + n_{i-1})/\|n_i + n_{i-1}\|$$
$$\alpha = (|P_iV_i| + |P_{i-1}V_{i-1}|)/|P_iP_{i-1}|$$

Fig. 5. Filling operation and its calculation process

Fig. 6. Approximated models with different numbers of segments

points on the surface of the approximated model and the original one, and then, find the correspondence between the two point sets that have been sampled from the approximated and the original models. Next, we calculate the distance from one point of the approximated model to its corresponding point in the original model and, finally, calculate the summation of distance of all the sampled points and average it. A chart of distance error is shown in Fig. 7. In the chart, the horizontal axis represents the line segment number and the vertical axis represents the related error value. As shown in the chart, the distance error decreases monotonously when the number of segment increases.

3.6 Guide Curve Segmentation Based on Curvature Value

In terms of the sweep feature, the whole translation procedure is more or less similar with the case of the helical feature; therefore, we just concentrate on the part that is different: the guide curve segmentation. In the sweep feature case, without loss of generality, because the guide curve is usually free form, like B-spline, after discretizing the guide curve, calculate the curvature on each sampling point of the guide curve. Based on the curvature, divide the guide curve into several pieces, and use line and arc segments to fit the dicretized points of the guide curve. Then create a corresponding solid model using these line and arc segments. In the process of the guide curve segmentation, according to a

Ave. Distance Error

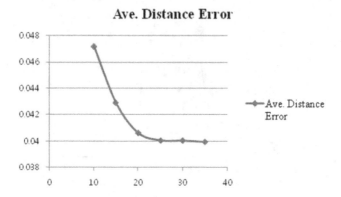

Fig. 7. Distance error with different numbers of segments

threshold of curvature, divide the guide curve, and the threshold used here is the average curvature value of all of the sampling points. The calculation of the curvature on each sampling point and the classification result, which is based on threshold value of the curvature, are shown in Fig. 8. On the left-hand side of the figure, the blue line on the guide curve is the visualization effect of the size of the curvature value. Then, based on these classified sampling points, we use line and circular arc segments to fit these points that are on the guide curve, and finally, we use these line and circular arc segments to create the solid model, which approximates the original model of the sweep feature.

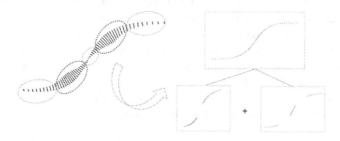

Fig. 8. Calculation of the curvature and classification of the guide curve segments

4   Implementation and Test Cases

In terms of the implementation environment, we use ACIS and Hoops as the geometric modeling kernel and visualization tool, respectively, and also use Eigen as the mathematical library for matrix calculation. Depicted in Fig. 9, on the left-hand side is the original model and on the right-and side is the corresponding model that is obtained through the approximation process.

Fig. 9. Example of the original and approximated solid model of the sweep and helical features

5   Conclusion

In this paper, a straightforward algorithm is proposed that uses line and arc segments to approximate the free-form guide curve of profile-based features. Then the related solid model segments are generated, and the filling operation is also given to complete the whole process of profile-based feature representation. In terms of future work, in the filling operation, there is a shape parameter that is used to determine the size of the solid model segment, especially for the length. Then there could be a more suitable decision process to obtain the shape parameter. While dealing with the sweep feature, for its guide curve segmentation, here, we use the average curvature value of the sampling points as the threshold for dividing the guide curve. More consideration for threshold selection may be needed to achieve a better segmentation result. In addition, the problem of how to optimize the entire approximation process also remains.

Acknowledgement. This research was supported by WCU(World Class University) program through the National Research Foundation of Korea funded by the Ministry of Education, Science and Technology (R31-2008-000-10045-0).

References

1. Ahn, Y., Kim, H.: Approximation of circular arcs by bézier curves. Journal of Computational and Applied Mathematics 81, 145–163 (1997)
2. Ahn, Y., Shin, Y., et al.: Approximation of circular arcs and offset curves by bézier curves of high degree. Journal of Computational and Applied Mathematics 167, 405–416 (2004)

3. De Boor, C., Hollig, K., Sabin, M.: High accuracy geometric hermite interpolation. Computer Aided Geometric Design 4, 269–278 (1987)
4. Dokken, T., Dæhlen, M., Lyche, T., Mørken, K.: Good approximation of circles by curvature-continuous bézier curves. Computer Aided Geometric Design 7, 33–41 (1990)
5. Goldapp, M.: Approximation of circular arcs by cubic polynomials. Computer Aided Geometric Design 8, 227–238 (1991)
6. Hur, S., Kim, T.: The best g1 cubic and g2 quartic bézier approximations of circular arcs. Journal of Computational and Applied Mathematics 236, 1183–1192 (2011)
7. Kim, S., Ahn, Y.: An approximation of circular arcs by quartic bézier curves. Computer-Aided Design 39, 490–493 (2007)
8. Mørken, K.: Best approximation of circle segments by quadratic bézier curves. In: Curves and Surfaces, pp. 331–336. Academic Press Professional, Inc. (1991)
9. Riškus, A.: Approximation of a cubic bézier curve by circular arcs and vice versa. Journal of Information Technology and Control 35, 371–378 (2006)
10. Cripps, R., Lockyer, P.: Circle approximation for cadcam using orthogonal c² cubic b-splines. International Journal of Machine Tools and Manufacture 45, 1222–1229 (2005)
11. Piegl, L., Tiller, W.: Circle approximation using integral b-splines. Computer-Aided Design 35, 601–607 (2003)
12. Fang, L.: Circular arc approximation by quintic polynomial curves. Computer Aided Geometric Design 15, 843–861 (1998)
13. Do Carmo, M.: Differential geometry of curves and surfaces, vol. 2. Prentice-Hall, Englewood Cliffs (1976)
14. Yang, X.: High accuracy approximation of helices by quintic curves. Computer Aided Geometric Design 20, 303–317 (2003)
15. Mick, S., Röschel, O.: Interpolation of helical patches by kinematic rational bézier patches. Computers & Graphics 14, 275–280 (1990)
16. Juhász, I.: Approximating the helix with rational cubic bézier curves. Computer-Aided Design 27, 587–593 (1995)
17. Seemann, G.: Approximating a helix segment with a rational bézier curve. Computer Aided Geometric Design 14, 475–490 (1997)
18. Ahn, Y.: Helix approximations with conic and quadratic bézier curves. Computer Aided Geometric Design 22, 551–565 (2005)
19. Ahn, Y.: Error analysis for approximation of helicoid by bi-conic and bi-quadratic bezier surfaces. Journal of the Korean Society for Industrial and Applied Mathematics 10, 63–70 (2006)
20. Pu, X., Liu, W.: A subdivision scheme for approximating circular helix with nurbs curve. In: IEEE 10th International Conference on Computer-Aided Industrial Design & Conceptual Design, CAID & CD 2009, pp. 620–624. IEEE (2009)
21. Lu, L.: On polynomial approximation of circular arcs and helices. Computers & Mathematics with Applications 63, 1192–1196 (2012)
22. Lee, I.: Curve reconstruction from unorganized points. Computer Aided Geometric Design 17, 161–177 (2000)
23. Lin, H., Chen, W., Wang, G.: Curve reconstruction based on an interval b-spline curve. The Visual Computer 21, 418–427 (2005)
24. McLain, D.: Two dimensional interpolation from random data. The Computer Journal 19, 178–181 (1976)
25. Lancaster, P., Salkauskas, K.: Surfaces generated by moving least squares methods. Mathematics of Computation 37, 141–158 (1981)

26. Alexa, M., Behr, J., Cohen-Or, D., Fleishman, S., Levin, D., Silva, C.: Computing and rendering point set surfaces. IEEE Transactions on Visualization and Computer Graphics, 3–15 (2003)
27. Amenta, N., Kil, Y.: Defining point-set surfaces. In: ACM Transactions on Graphics (TOG), vol. 23, pp. 264–270. ACM (2004)
28. Cheng, Z., Wang, Y., Li, B., Xu, K., Dang, G., Jin, S.: A survey of methods for moving least squares surfaces. In: Symposium on Point-Based Graphics, pp. 9–23 (2008)
29. Fleishman, S., Cohen-Or, D., Silva, C.: Robust moving least-squares fitting with sharp features. In: ACM Transactions on Graphics (TOG), vol. 24, pp. 544–552. ACM (2005)
30. Levin, D.: The approximation power of moving least-squares. Mathematics of Computation 67, 1517–1532 (1998)
31. Levin, D.: Mesh-independent surface interpolation. Geometric Modeling for Scientific Visualization 3, 37–49 (2003)
32. Lipman, Y., Cohen-Or, D., Levin, D.: Data-dependent mls for faithful surface approximation. In: Proceedings of the Fifth Eurographics Symposium on Geometry Processing, Eurographics Association, pp. 59–67 (2007)
33. Mederos, B., Velho, L., De Figueiredo, L.: Moving least squares multiresolution surface approximation. In: XVI Brazilian Symposium on Computer Graphics and Image Processing, SIBGRAPI 2003, pp. 19–26. IEEE (2003)
34. Schaefer, S., McPhail, T., Warren, J.: Image deformation using moving least squares. In: ACM Transactions on Graphics (TOG), vol. 25, pp. 533–540. ACM (2006)
35. Zhu, Y., Gortler, S.: 3d deformation using moving least squares. Technical report, Harvard University, Computer Science (2007)
36. Adams, B., Keiser, R., Pauly, M., Guibas, L., Gross, M., Dutré, P.: Efficient raytracing of deforming point-sampled surfaces. In: Computer Graphics Forum, vol. 24, pp. 677–684. Wiley Online Library (2005)
37. Adamson, A., Alexa, M.: Ray tracing point set surfaces. In: Shape Modeling International, pp. 272–279. IEEE (2003)
38. Liew, K., Huang, Y., Reddy, J.: Moving least squares differential quadrature method and its application to the analysis of shear deformable plates. International Journal for Numerical Methods in Engineering 56, 2331–2351 (2003)
39. Nguyen, V., Rabczuk, T., Bordas, S., Duflot, M.: Meshless methods: a review and computer implementation aspects. Mathematics and Computers in Simulation 79, 763–813 (2008)

Real-Time Illumination for Two-Level Volume Rendering

Andrew Corcoran and John Dingliana

Trinity College Dublin

Abstract. We propose improvements to real-time two-level rendering incorporating ray cast shadows and ambient occlusion of 3D volumetric datasets. Our ambient occlusion calculation utilises the same sampling scheme as standard per-voxel Phong shading thus allowing for an extremely computationally efficient rendering in comparison to other ambient occlusion algorithms. Our ray cast shadows technique requires no pre-processing, does not significantly increase memory requirements and is compatible with ray cast volume renderers. We validate these techniques through a number of user experiments. The results indicate that our ambient occlusion method increases the visual information in an image. Meanwhile, ray cast shadows appear to not provide the hypothesised improvement to image understanding, however this raises some interesting implications regarding the practical relevance of shadows in such two-level volume renderings.

1 Introduction

Due to the highly complex nature of volume datasets, illumination calculations are computationally expensive and difficult to render at interactive frame rates. Researchers have developed a variety of techniques to improve the rendering speed of illumination calculations for volume data but these techniques are either not applicable to ray cast renderings, require a reduction in quality, an increase in memory requirements or a pre-processing step in order to function. The large datasets that are being captured in recent times can no longer fit in memory and thus techniques which increase memory requirements are no longer attractive. Additionally the large body of research that has been directed towards optimising ray cast rendering systems means that texture slicing approaches to volume rendering are no longer competitive in terms of computational speed and illumination techniques which are compatible with ray cast solutions are favoured.

Focus and context techniques are highly prevalent in medical illustration and are used to emphasise certain anatomical objects while peripheral regions are abstracted but sufficiently retained in order to provide cues for spatial reference [1]. Two-level volume rendering is one type of such a technique which allows for selectively using different rendering techniques for different sections of a volume dataset [2, 3]. We propose enhancing two-level volume rendering by integrating real-time illumination techniques.

G. Bebis et al. (Eds.): ISVC 2012, Part I, LNCS 7431, pp. 544–555, 2012.

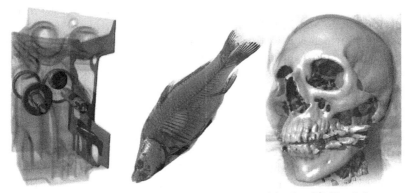

(a) The Engine dataset.   (b) The Carp dataset.   (c) The CT-Head dataset.

Fig. 1. Sample renderings combining ray cast shadows and ambient occlusion

The two-level technique we investigate utilises this paradigm by allowing the user to specify a target isosurface within the volume data. This target surface is rendered with high quality ray cast shadows. The surrounding regions are illuminated in a less pronounced manner with a computationally efficient ambient occlusion calculation which is able to highlight the important structures of the surrounding regions for negligible cost. By splitting a rendering into two levels it is possible to provide high quality, computationally expensive rendering techniques for the target surface while simultaneously enhancing the surrounding context regions with a more computationally efficient calculation which still enhances the final image but for a much lower computational cost. This technique allows for real-time illumination of the entire dataset with no extra memory requirements by trading computational speed for a less detailed illumination solution in the surrounding regions.

In addition we perform a number of user experiments to test if the addition of ray cast shadows and ambient occlusion can enhance users perception of shape, depth and visual information.

2   Related Work

Hauser et al. [2] propose the use of two-level volume rendering, which merges direct volume rendering (DVR) and maximum intensity projection (MIP) techniques in an interactive tool. They argue that such an approach, based around a *focus-and-context* strategy, provides intuitive benefits to users, allowing them to peer inside inner structures, while keeping surrounding objects integrated for spatial reference. A wide variety of focus and context techniques have been developed for volume visualisation [4, 5]. Viola et al. [6] propose a technique where certain features inside a volume are specified as important by the user and these areas are rendered in high quality. Focus and emphasis is also discussed by Preim at al [7], who survey a number of existing areas where various enhancement and non-photorealistic (NPR) techniques can be used to improve tasks in medical visualisation. Kim and Varshney [8] use a visual saliency metric to compute an emphasis field which is used to elicit viewer attention to relevant parts of a volume rendering. Hadwiger et al. [9] perform two-level rendering using multiple rendering modes for segmented volume data. Their technique allows each object to be assigned an individual compositing mode in conjunction with a single global mode. Corcoran et al. [10] combine two-level volume rendering with image space line drawings and validate their work using user studies to show that two-level enhancement can improve perception of shape.

Traditional volume shading techniques have relied on simple shading models such as per-voxel Phong lighting [11] but increasingly researchers are investigating more advanced illumination methods in order to provide more realistic light simulations to improve user perception [12, 13]. Half angle slicing [14] is a slice-based approach to volume illumination which modifies the standard slice based volume rendering technique. Instead of orientating texture slices in the view direction they are instead orientated at the half angle between the view direction and the light position thus allowing both light propagation and the final image to be calculated in parallel. Dynamic ambient occlusion [15] involves a preprocessing step which analyses a local area surrounding each voxel to compute a histogram of the surrounding data which is then subsequently compressed using vector quantisation. This technique is independent of transfer function updates and can provide interactive rendering but does not perform a full global lighting calculation. Ropinski et al. [16] propose generating a three dimensional global illumination volume by performing a fast pre-processing step which propagates the light information using a slice-based rendering approach. They reduce the high memory requirement of the illumination volume by performing chroma subsampling, a compression technique which exploits human perception by downscaling the RGB channels while maintaining a high resolution for the alpha channel. Kronander et al. [17] use spherical harmonics to encode local and global volumetric visibility. This technique allows high quality shadows to be generated using a ray casting approach with only a small recomputation required for transfer function changes but requires a significant increase in memory requirements and is unable to generate hard shadows. Sundén et al. [18] have developed a global illumination technique for ray cast volume rendering which requires no pre-processing but is only able to calculate illumination information for the portion of the volume on screen. Their technique utilises the plane sweep paradigm in order to iteratively generate the final image, thus allowing the illumination volume to be updated in between sweep steps.

3   Pipeline

Our technique utilises high quality ray cast shadows and ambient occlusion in order to generate a final enhanced image of the complete volume. We combine high quality shadow ray casting with computationally efficient ambient occlusion in order to generate a final image which approximates global illumination but assigns the most computation time to areas with the highest importance as determined by the user. The user specifies a focus isosurface which is rendered in high quality with ray cast shadows, the surrounding volume is rendered with ambient occlusion to provide context to the final image.

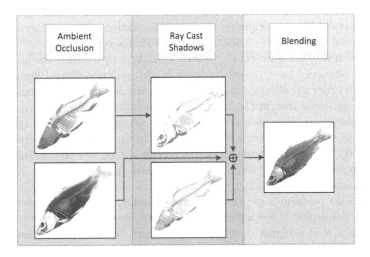

Fig. 2. The rendering pipeline

We have implemented this technique using a two pass graphics processing unit (GPU) shader algorithm. By separating the ambient occlusion and the shadow ray calculation into separate passes we avoid expensive conditional checks in the GPU shader which can dramatically increase computation time.

In the first pass, we iterate along each ray computing the local ambient occlusion for every sample until the isovalue of the target surface is reached. At this point we perform an iterative bisection procedure [19] in order to accurately determine the location of the isosurface and avoid artifacts due to low sampling rates. We then output the accumulated value of the ambient occlusion ray and the location of the isosurface for use in the second pass (Figure 2).

In the second pass, the ray cast shadow contribution is calculated by casting shadow rays from the isosurface location towards the light source utilising adaptive sampling in order to refine the shadow locations.

Lastly, the shadow ray information and the ambient occlusion rendering are blended together to generate the final image.

3.1 Ambient Occlusion

We have modified the ambient occlusion technique used by Ropinski et al. [15] in order to implement our efficient ambient occlusion calculation. Instead of precomputing and storing histograms for each voxel we compute the ambient occlusion contribution on-the-fly during the rendering process therefore eliminating the memory requirement incurred by Ropinski et al. In order to ensure that the ambient occlusion calculation is performed in real-time we are required to reduce the number of samples used by the ambient occlusion calculation to the surrounding six voxels. One would expect that this reduction in sampling size would result in a dramatic decrease in visual quality but as we show in Section 4 this method still provides an increase in visual information (Figure 3). In addition we perform this fast method of ambient occlusion only in the area of the volume classified as peripheral context information and therefore a tradeoff between visual quality and computational speed is acceptable.

We perform this fast ambient occlusion calculation by utilising the information that would normally be accessed during a traditional per-voxel Phong shading approach. In order to calculate the gradient information for Phong shading, most techniques generally use on the fly gradient calculation using finite differences. This gradient calculation method samples the volume data once in each of the six directions surrounding each sampling point in order to generate gradient information. By limiting our ambient occlusion calculation to only the surrounding six directions we utilise the same texture accesses that are required by the gradient calculation thus reducing the computation required. Using the gradient information we perform a dependant texture lookup into the transfer function in order to determine the opacity of the surrounding region. If any of the surrounding area is of a higher opacity than the current voxel we assume that the current voxel is occluded and modify its luminance accordingly.

3.2   Shadow Ray Casting

Standard shadow ray casting techniques terminate the ray when it reaches the light source. This technique works well for typical ray casting scenes where the light source is positioned within the scene. However, for the vast majority of cases in volume rendering, the light source is positioned outside the bounding box of the volume data which results in unneeded shadow ray calculations being performed in areas outside the volume where no occlusion can possibly occur. Using this knowledge it is possible to perform early shadow ray termination by calculating the distance from each voxel to the maximum extents of the volume bounding box and terminating the shadow ray when it exceeds this distance. By taking the minimum of the distance from the voxel to the light source and the distance from the voxel to the extents of the bounding box we can ensure that no extra ray iterations are performed for all possible light source positions both inside and outside the volume.

We use the isosurface location information generated in the first pass as the starting point for shadow rays cast in the direction of the light source. This pass iterates along the shadow rays until either the isosurface is encountered again or the termination criteria for the ray is reached. As the ray nears the isosurface we reduce the sampling rate in order to ensure an accurate intersection, thus increasing the quality of the shadows. If a ray terminates without encountering an isosurface, it is not occluded and the shading calculation is performed normally. If the ray encounters the isosurface, we know it is occluded and the light contribution is reduced accordingly.

4 Rendering Speed

We compared the rendering time of our method to a standard per-voxel Phong shading model by calculating the average frames per second (fps) over 100 frames of a 360° rotation around three datasets as shown in Figure 1. As can be seen from Table 1 our technique is approximately 28% slower for the CT-Head dataset, 19% slower for the Engine dataset and 21% faster for the Carp dataset (we have determined this speed increase is due to the ambient occlusion calculations).

Sundén et al. [18] have proposed Image Plane Sweep Volume Illumination (IPSVI) a state of the art technique for volume illumination. They compare their IPSVI technique to standard Phong shading and show that their technique experiences an 85% reduction in frame rate. This compares favourably to our results which show an average reduction of 9% over the three datasets. Assuming both Phong implementations by Sundén et al. and ourselves incur a similar computational cost we can compare the computationally efficiency of our technique to IPSVI by comparing the relative reduction in speed versus Phong shading. Using this comparison our technique is approximatley five times faster than IPSVI while still providing a reasonable illumination calculation.

Table 1. Average frames per second for a 360° rotation around each dataset

Dataset	Size	Phong	Our Method
CT-Head	256x256x113	34.26 fps	24.49 fps
Engine	256x256x256	19.63 fps	15.75 fps
Carp	256x256x512	14.09 fps	17.16 fps

5 User Experiments

In order to further test our technique we devised a number of user experiments to gauge how participants perception of shape, depth and visual information changed with the addition of ray cast shadows and ambient occlusion. All experiments were conducted on a 24 inch monitor in a well lit office environment.

5.1 Subjective Evaluation

In this experiment participants were shown two images side by side of the same dataset from the same camera position. The only difference between the two images was that one image had our ambient occlusion calculation enabled and the other did not. Participants were asked to select which image they felt had the most visual information. The users used the mouse to select an image by pressing the left mouse button to select the left image and the right mouse button to select the right image.

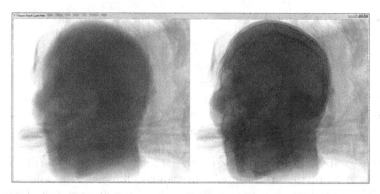

Fig. 3. The evaluation program used by participants in the ambient occlusion user study. The participants were asked to select which of the two images they felt contained the most visual information.

For each dataset, 2 pairs of images (one with and one without ambient occlusion) were generated all from the same viewpoint. One pair of images contained no target isosurface and purely ambient occlusion calculations, the other set of images contained a combination of a target isosurface with ray cast shadows and our ambient occlusion technique. Each pair of images was tested 3 times for a total of 27 trials. The order in which the sets of images were shown and the left/right ordering of each individual set of images was randomised to eliminate any bias. For all trials the light source was positioned to the left of vertical and above the horizontal as this position has been shown by other researchers to be the best location for perception of shape [20–22].

Eighteen naive participants took part in this experiments (12M-6F). The participants ranged in age from 21 to 31 and came from a wide variety of backgrounds and occupations. Twelve participants rated themselves as having a novice knowledge of computer graphics, two participants rated themselves as experienced and four participants rated themselves as experts. Each participant was shown renderings generated from a knee, skull and brain dataset in random order.

In order to evaluate if participants were answering significantly above chance level (i.e., significantly different from 50%), we averaged the results for each isovalue and performed a single sample t-test. The results show that participants selected our ambient occlusion technique significantly above chance ($t(17) = 15.912, p < 0.0001$) thus showing that our computationally efficient ambient occlusion method improves the visual information contained within a dataset for very little extra computation cost.

In addition to testing our technique on its own, we tested it with a large amount of the dataset covered by a target isosurface on which ray cast shadows were displayed. By setting up the images in this way, only a very small amount of the dataset receives our ambient occlusion technique therefore making our technique far less noticeable. We wanted to test if this worst case scenario for our ambient occlusion method resulted in a statistically significant result. Analysis of the results show that for this scenario our ambient occlusion technique is also favoured above chance $(t(17) = 3.387, p = 0.004)$.

An ANOVA analysis shows there is a main effect of isovalue showing that there is a statistically significant difference between how well our ambient occlusion technique performs in this worst case scenario when compared to a more plausible use case $(F(1, 17) = 25.576, p = 0.0001)$. Our ambient occlusion technique was shown to significantly increase the the amount of visual information contained within an image, with a larger effect reported when no isosurface was rendered.

5.2   Shape Perception

In the shape perception experiment, the same 18 participants were shown a series of images from the knee, skull and brain dataset in random order and asked to rotate gauges which overlaid the images to match the surface normal of the isosurface at that point. This evaluation method has been employed by various other researchers [10, 20, 23, 24].

We postulated that the addition of ray cast shadows may cause participants to experience a reduction in shape perception due to the addition of shadowed regions causing distractions and also due to the reduction in luminance that shadowed areas receive. In fact, our results indicate that the addition of ray cast shadows has no statistically significant effect on the participants shape perception $(F(1, 17) = 0.399, p = 0.536)$. This shows that we can add ray cast shadow calculations to a volume rendering scene without reducing a users understanding of the shape of the dataset.

Post-hoc analysis shows that there is a statistically significant difference between the accuracy of participants for gauges placed in shadowed areas $(p < 0.012)$. Interestingly this effect remains when shadows are disabled and gauges are placed in regions where shadows would be cast if shadows were rendered. This shows that the addition of shadows is not the reason for less accurate results, these regions receive statistically significant less accurate results even when no shadows are displayed.

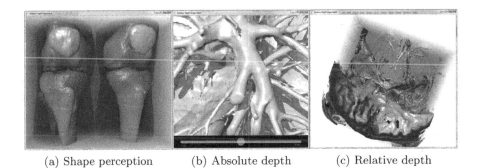

(a) Shape perception   (b) Absolute depth   (c) Relative depth

Fig. 4. The evaluation programs used for the shape and depth experiments

Post-hoc analysis also shows that there was a significant reduction in accuracy for the brain dataset ($p < 0.015$). A number of participants commented that they recognised the knee and skull datasets and could use prior knowledge in order to assist in gauge orientation. In contrast, participants found that they had no prior knowledge of the brain dataset and were relying solely on the images presented to them in order to estimate gauge orientation. We believe this lack of prior knowledge explains the reduction in accuracy.

5.3   Relative Depth Perception

For the relative depth perception task we conducted an experiment where two points were highlighted on a static image and the same 18 participants were asked to estimate relative depth by selecting which point they felt was closest to the camera [13, 24]. Participants were shown images generated from the knee, skull and brain dataset in random order.

This experiment was unable to show any statistically significant increase in depth perception due to the addition of ray cast shadows ($F(1, 17) = 0.265, p = 0.613$). In discussions with our participants we were informed that a large number of users utilised prior knowledge of basic anatomy in order to determine depth information. Analysis of the results show that the average accuracy (89%) for this experiment was very high which may have contributed to the nonsignificant result. Some participants received perfect scores for some images for both ray cast images and standard renderings which due to the binary nature of the experiment may have masked a possible difference between the two renderings.

5.4   Absolute Depth Perception

In an absolute depth perception experiment users were shown an image with a single point highlighted and asked to estimate the absolute depth of the point using a slider in relation to the nearest and furthest point in the scene from the current point of view [13]. This experiment had 14 participants (11M-3F), ranging in age from 16 to 62. Five participants rated themselves as having a novice knowledge of computer graphics, three participants rated themselves as experienced and six participants rated themselves as experts. In this experiment five datasets were shown to participants, a skull, heart, clouds, bonsai tree and a randomly generated sine wave dataset (Figure 5).

This depth experiment showed that there was a main effect of shadows ($F(1, 12) = 8.135, p = 0.015$) with the addition of shadows reducing depth perception. Other researchers [12] have shown that the addition of shadows can produce the sensation of depth but our results show that this perception of depth does not result in increased accuracy at estimating depth.

5.5   Discussion

Our results show that two-level volume rendering can be enhanced by the addition of our computationally efficient ambient occlusion enhancement on the

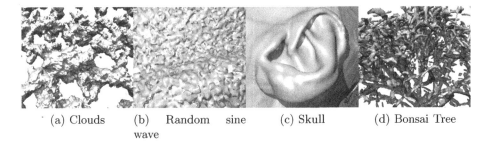

(a) Clouds   (b) Random sine wave   (c) Skull   (d) Bonsai Tree

Fig. 5. Some of the datasets shown to participants

context area of the rendering. However the results of our user study indicate that the addition of ray cast shadows does not improve shape perception, has no effect on relative depth perception and worsens absolute depth perception in two-level volume rendering. Therefore the addition of ray cast shadows, despite their usefulness in other techniques is not strongly recommended for enhancing user understanding of two-level volume rendering.

6 Conclusions and Future Work

We have presented an investigation of real-time illumination for two-level volume rendering which combines high quality ray cast shadows of a target isosurface within a volume dataset while simultaneously providing computationally efficient ambient occlusion shading for the surrounding volume.

User experiments indicate that our ambient occlusion technique can significantly increase the visual information in the image. This potentially increases the structures that users can perceive, and thus provides more context information allowing users to more easily interact with the dataset. Our experiments also reveal that the addition of ray cast shadows to the target surface has a detrimental effect on absolute depth perception and has no benefit to the perception of relative depth or surface shape. This is a somewhat surprising discovery as some previous works have demonstrated the benefit of shadows to depth perception in volume renderings [13, 24]. We suggest that this observed effect may be specific to two-level approaches such as ours that apply detailed shadows to only the focus layer. Based on these results, we propose that a combination of ambient occlusion on the context layer with other types of enhancements to the focus layer may provide the best increase in visual information. For instance, NPR stylisations, have been shown to increase shape perception in two-level renderings [10]. An example of our ambient occlusion combined with such NPR stylisations is shown in Figure 6b.

In future work, we intend to perform a detailed investigation into how techniques such as NPR can be integrated with real-time illumination. In particular, we plan to perform a user study testing how shape and depth perception are affected by combining NPR line drawing with our ambient occlusion technique. In

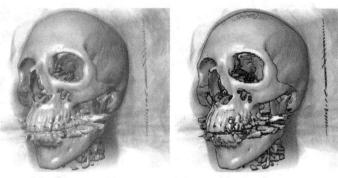

(a) Phong Shading   (b) NPR and Ambient Occlusion

Fig. 6. Comparison between Phong rendering and a combination of NPR Difference of Gaussian line extraction and ambient occlusion. Notice the addition of lines enhance the shape perception of the target surface and the addition of ambient occlusion lighting increases the visual information in the context layer.

addition we plan to compare the quality of our fast ambient occlusion technique to state of the art techniques that require precomputation such as those used by Ropinski et al. [15].

References

1. Svakhine, N., Ebert, D., Andrews, W.: Illustration-Inspired depth enhanced volumetric medical visualization. IEEE Transactions on Visualization and Computer Graphics 15, 77–86 (2009)
2. Hauser, H., Mroz, L., Bischi, G.I., Groller, M.E.: Two-level volume rendering - fusing MIP and DVR. In: Proceedings of Visualization 2000, pp. 211–218. IEEE (2000)
3. Hauser, H., Mroz, L., Italo Bischi, G., Groller, M.: Two-level volume rendering. IEEE Transactions on Visualization and Computer Graphics 7, 242–252 (2001)
4. Bruckner, S., Gröller, M.E., Mueller, K., Preim, B., Silver, D.: Illustrative Focus+Context approaches in interactive volume visualization. Scientific Visualization: Advanced Concepts (2010)
5. Nikolov, S.G., Jones, M.G., Agrafiotis, D., Bull, D.R., Canagarajah, C.N.: Focus+Context visualisation for fusion of volumetric medical images. In: Proceedings of the 4th International Conference on Information Fusion, Montreal, QC, Canada (2001)
6. Viola, I., Gröller, A.K.M.E., Kanitsar, A., Gröller, M.E.: Importance-Driven volume rendering. In: Proceedings of IEEE Visualization, pp. 139–145 (2004)
7. Preim, B., Tietjen, C., Dörge, C.: NPR, focussing and emphasis in medical visualizations. In: Simulation und Visualisierung, pp. 139–152 (2005)
8. Kim, Y., Varshney, A.: Saliency-guided enhancement for volume visualization. IEEE Transactions on Visualization and Computer Graphics 12, 925–932 (2006)

9. Hadwiger, M., Berger, C., Hauser, H.: High-quality two-level volume rendering of segmented data sets on consumer graphics hardware. In: Visualization, VIS 2003, pp. 301–308. IEEE (2003)
10. Corcoran, A., Redmond, N., Dingliana, J.: Perceptual enhancement of two-level volume rendering. Computers & Graphics 34, 388–397 (2010)
11. Levoy, M.: Display of surfaces from volume data. IEEE Computer Graphics and Applications 8, 29–37 (1988)
12. Medina Puerta, A.: The power of shadows: shadow stereopsis. Journal of the Optical Society of America A 6, 309–311 (1989)
13. Lindemann, F., Ropinski, T.: About the influence of illumination models on image comprehension in direct volume rendering. IEEE Transactions on Visualization and Computer Graphics 17, 1922–1931 (2011)
14. Kniss, J., Premoze, S., Hansen, C., Shirley, P., McPherson, A.: A model for volume lighting and modeling. IEEE Transactions on Visualization and Computer Graphics 9, 150–162 (2003)
15. Ropinski, T., Meyer-Spradow, J., Diepenbrock, S., Mensmann, J., Hinrichs, K.: Interactive volume rendering with dynamic ambient occlusion and color bleeding. Computer Graphics Forum 27, 567–576 (2008)
16. Ropinski, T., Doring, C., Rezk-Salama, C.: Interactive volumetric lighting simulating scattering and shadowing. In: 2010 IEEE Pacific Visualization Symposium (PacificVis), pp. 169–176. IEEE (2010)
17. Kronander, J., Jönsson, D., Löw, J., Ljung, P., Ynnerman, A., Unger, J.: Efficient visibility encoding for dynamic illumination in direct volume rendering. IEEE Transactions on Visualization and Computer Graphics 99 (2011)
18. Sundén, E., Ynnerman, A., Ropinski, T.: Image plane sweep volume illumination. IEEE Transactions on Visualization and Computer Graphics 17, 2125–2134 (2011)
19. Hadwiger, M., Sigg, C., Scharsach, H., Bühler, K., Gross, M.: Real-Time Ray-Casting and advanced shading of discrete isosurfaces. Computer Graphics Forum 24, 303–312 (2005)
20. O'Shea, J.P., Banks, M.S., Agrawala, M.: The assumed light direction for perceiving shape from shading. In: Proceedings of the 5th Symposium on Applied Perception in Graphics and Visualization, APGV 2008, pp. 135–142. ACM, New York (2008)
21. Sun, J., Perona, P.: Where is the sun? Nature Neuroscience 1, 183–184 (1998)
22. Langer, M.S., Bülthoff, H.H.: Depth discrimination from shading under diffuse lighting. Perception 29, 649–660 (2000)
23. Cole, F., Sanik, K., DeCarlo, D., Finkelstein, A., Funkhouser, T., Rusinkiewicz, S., Singh, M.: How well do line drawings depict shape? In: ACM SIGGRAPH 2009 Papers, pp. 1–9. ACM, New Orleans (2009)
24. Šoltészová, V., Patel, D., Viola, I.: Chromatic shadows for improved perception. In: Proceedings of the ACM SIGGRAPH/Eurographics Symposium on Non-Photorealistic Animation and Rendering, NPAR 2011, pp. 105–116. ACM, New York (2011)

Spatial Colour Gamut Mapping by Orthogonal Projection of Gradients onto Constant Hue Lines

Ali Alsam and Ivar Farup

Sør-Trndelag University College, Trondheim, Norway
Gjøvik University College, Gjøvik, Norway

Abstract. We present a computationally efficient, artifact-free, spatial gamut mapping algorithm. The proposed algorithm offers a compromise between the colorimetrically optimal gamut clipping and an ideal spatial gamut mapping. This is achieved by the iterative nature of the method: At iteration level zero, the result is identical to gamut clipping. The more we iterate the more we approach an optimal spatial gamut mapping result. Our results show that a low number of iterations, 20-30, is sufficient to produce an output that is as good or better than that achieved in previous, computationally more expensive, methods. More importantly, we introduce a new method to calculate the gradients of a vector valued image by means of a projection operator which guarantees that the hue of the gamut mapped colour vector is identical to the original. Furthermore, the algorithm results in no visible halos in the gamut mapped image a problem which is common in previous spatial methods. Finally, the proposed algorithm is fast- Computational complexity is $O(N)$, N being the number of pixels. Results based on a challenging small destination gamut supports our claims that it is indeed efficient.

1 Introduction

To accurately define a colour three independent variables need to be fixed. In a given three dimensional colour space the colour gamut is the volume enclosing all the colour values that can be reproduced by the reproduction device or present in the image. Colour gamut mapping is the problem of representing the colour values of an image within the gamut of a reproduction device, typically a printer or a monitor. Furthermore, in the general case, when an image gamut is larger than the destination gamut some visual image information will be lost. We therefore redefine gamut mapping as the problem of representing the colour values of an image within the gamut of a reproduction device with minimum loss of visual information, i.e., as visually close as possible.

Unlike single colours, images are represented in a higher dimensional space than three, i.e. knowledge of the exact colour values is not, on its own, sufficient to reproduce an unknown image. In order to fully define an image, the spatial context of each colour pixel needs to be fixed. Based on this, we define two categories of gamut mapping algorithms. In the first, colours are mapped independent of their spatial context [1]. In the second, the mapping is influenced

G. Bebis et al. (Eds.): ISVC 2012, Part I, LNCS 7431, pp. 556–565, 2012.

by the local context of each colour value [2–5]. The latter category is referred to as spatial colour gamut mapping.

Eschbach [6] stated that although the accuracy of mapping of a single colour is well defined, the reproduction accuracy of images isn't. To elucidate this claim, with which we agree, we consider a single colour that is defined by its hue, saturation and lightness. Assuming that such a colour is outside the target gamut, we can modify its components independently. That is to say, if the colour is lighter or more saturated than what can be achieved inside the reproduction gamut, we shift its lightness and saturation to the nearest feasible values. Further, in most cases it is possible to reproduce colours without shifting their hue.

Taking the spatial context of colours into account presents us with the challenge of defining the spatial components of a colour pixel and incorporating this information into the gamut mapping algorithm. Generally speaking, we need to define rules that would result in mapping two colours with identical hue, saturation and lightness to two different magnitudes depending on their context in the image. The main challenge is, thus, defining the spatial context of an image pixel in a manner that results in an improved gamut mapping. By improved we mean that the appearance of the resultant in-gamut image is closer to the original as judged by a human observer. Further, from a practical point of view, the new definition needs to result in an algorithm that is fast and does not result in image artifacts.

It is well understood that the human visual system is more sensitive to spatial ratios than to absolute luminance values [7]. This knowledge is at the heart of all spatial gamut mapping algorithms. A rephrasing of spatial gamut mapping is then the problem of representing the colour values of an image within the gamut of a reproduction device while preserving the spatial ratios between different colour pixels. In an image, spatial ratios are the difference, given some metric, between a pixel and its surround. This can be the difference between one pixel and its adjacent neighbors or pixels far away from it. Thus, we face the problem that spatial ratios are defined in different scales and dependent on the chosen difference metric.

McCann suggested to preserve the spatial gradients at all scales while applying gamut mapping [8]. Meyer and Barth [9] suggested to compress the lightness of the image using a low-pass filter in the Fourier domain. As a second step the high-pass image information is added back to the gamut compressed image. Many spatial gamut mapping algorithms have been based upon this basic idea [2, 10–12, 4].

A completely different approach was taken by Nakauchi et al. [13]. They defined gamut mapping as an optimization problem of finding the image that is perceptually closest to the original and has all pixels inside the gamut. The perceptual difference was calculated by applying band-pass filters to Fourier-transformed CIELAB images and then weighing them according to the human contrast sensitivity function. Thus, the best gamut mapped image is the image having contrast (according to their definition) as close as possible to the original. Kimmel et al. [3] presented a variational approach to spatial gamut mapping where it was shown that the gamut mapping problem leads to a quadratic programming formulation, which is guaranteed to have a unique solution if the gamut of the target device is convex. However, they did not apply their method to colour images.

Finding an adequate description of the surface of the gamut, commonly denoted a gamut boundary descriptors (GBDs) is an important step in any colour gamut mapping algorithm. One of the main challenges is the fact that gamut surfaces are most often concave. Many methods for finding the GBD have been proposed over the years. Recently, Bakke et al. [14] presented an evaluation of the most common method, showing that the modified convex hull algorithm by Balasubramanian and Dalal [15] is generally the most reliable one.

The algorithm presented in this paper adheres to our previously stated definition of spatial gamut mapping in that we aim to preserve the spatial ratios between pixels in the image while preserving hue. The task of preserving three dimensional gradients while maintaining the original hue values is challenging–the correction of the image gradients in a three dimensional colour space results in unavoidable change in the hue. This is true because adding a three-dimensional gradient vector to the colour triplet results in a modified vector that isn't necessarily parallel to the original.

The first contribution of this paper is, thus, to derive a new n-dimensional gradient operator that is, mathematically, guaranteed to results in gradient vectors that are parallel to the original colour. In the literature, there are a number of gradient operators that provide estimations of the three dimensional colour-image-gradients. As an example the first and second eigenvalues of the tensor matrix are used to define gradients. From a spatial gamut mapping point of view, these operators share two drawbacks: The first is that the gradient vector is not in the direction of the original colour while the second is that operators result in the absolute value of the gradient without its orientation, i.e. the gradient is defined along a line not a vector. The operator derived in this paper remedies both these problems. We define the difference between two colour vectors as the norm of the first minus the norm of orthogonally projected component of the second onto the first. In so doing we arrive at an oriented gradient that is in the direction of the first vector.

The second contribution of this paper is the use of a new computationally efficient approach to restrict the gamut of the spatially mapped image to be within the destination gamut. This is achieved by observing that the resultant image pixels are a convex combination of the colour values that are obtained by clipping the gamut to the gamut boundaries and the neutral gray. Thus we start by calculating the gradients of the original image in the CIELab colour space. The image is then gamut mapped by projecting the colour values to the nearest, in gamut, point along hue-constant lines. The difference between the gradients of the gamut mapped image and that of the original is then iteratively minimized with the constraint that the norm of resultant colour is no greater than that of the gamut clipped vector. The scale at which the gradient is preserved is related to the number of iterations and the extent to which we can fit the original gradients into the destination gamut.

The third contribution relates to halos which are a main drawback in previous spatial gamut mappings techniques. We observe that halos are visible in the resultant gamut mapped images at strong lightness or chromatic edges. Furthermore, those edges are generally visible in the gamut clipped image. That is to say that halos are the result of over enhancing visible edges. We avoid this problem by using anisotropic diffusion [16] where the gradients of the gamut mapped image are improved based on their strength. In other words, diffusion is encouraged within regions and prohibited across strong edges thus avoiding the introduction of halos.

Finally, our results show that as few as ten to thirty iterations are sufficient to produce an output that is similar or better than previous methods. Being able to improve upon previous results using such low number of iterations allows us to state that the proposed algorithm is fast.

2   Spatial Gamut Mapping: A Mathematical Definition

Let's say we have an original image with pixel values $\mathbf{p}(x, y)$ (bold face to indicate vector) in CIELab or any similarly structured colour space. A gamut clipped image can be obtained by leaving in-gamut colours untouched, and moving out-of-gamut colours along straight lines towards $\mathbf{g}$, the center of the gamut, G, on the L axis until they hit the gamut surface. Let's denote the gamut clipped image $\mathbf{p}_c(x, y)$. In a previous papers [17], we showed that spatial gamut mapping can be achieved by minimising

$$\min \int ||\nabla \mathbf{p}_s - \nabla \mathbf{p}||^2 \, dA \quad \text{subject to} \quad \mathbf{p}_s \in G. \tag{1}$$

where $\mathbf{p}_s$ is the spatially gamut mapped image that we are solving for. The numerical solution to this problem was found by solving the corresponding Euler–Lagrange equation,

$$\nabla^2(\mathbf{p}_s - \mathbf{p}) = 0 \tag{2}$$

using a finite difference method with Jacobi iteration, subject to the constraint that the resultant colour vectors are inside the gamut boundaries defined by the gamut clipper image.

One of the problems with this approach was that there was a tendency towards the creation of halos near strong edges. In Reference [18], we therefore proposed to exchange the simple diffusion equation with the anisotropic diffusion equation proposed by Perona and Malik [16]:

$$\nabla \cdot (D\nabla(\mathbf{p}_s - \mathbf{p})) = 0. \tag{3}$$

The diffusion constant was chosen in accordance with Perona and Malik: $D = 1/(1 + |\nabla \mathbf{p}/\kappa|^2)$, κ being a regularisation parameter, which resulted in the following equation:

$$\nabla \cdot \left(\frac{\nabla(\mathbf{p}_s - \mathbf{p})}{1 + |\nabla \mathbf{p}/\kappa|^2} \right) = 0. \tag{4}$$

In order to simplify the problem further, we solved this equation for the grayscale versions of the original, p, and gamut mapped images, p_s only. The final colour gamut mapped image was assumed to be a convex linear combination of the original image and the neutral gray color at any pixel position. The main sacrifice by this approach was that we were not able to recover details that were lost in the conversion between the colour gamut mapped image and its grayscale version.

Here, we propose a new way to deal with Equation (4). Instead of working only on the grayscale images, as in Reference [18], we perform the gamut mapping directly in the full three-dimensional colour space. However, in order to ensure hue constancy during the mapping process, we force the changes to occur on lines of constant hue by projecting the gradients of the original image onto the vectors $\mathbf{p} - \mathbf{g}$, i.e. instead of using the gradient of the original image directly, we substitute it with

$$(\nabla \mathbf{p})_{\parallel} = \mathbf{e}(\mathbf{p})(\mathbf{e}(\mathbf{p}) \cdot \nabla \mathbf{p}), \tag{5}$$

where $\mathbf{e}(\mathbf{p}) = (\mathbf{p} - \mathbf{g})/|\mathbf{p} - \mathbf{g}|$ is a unit vector in a direction of constant hue. Thus, the equation we are seeking to solve for $\mathbf{p}_s$ is

$$\nabla \cdot \left(\frac{\nabla \mathbf{p}_s}{1 + |(\nabla \mathbf{p})_{\parallel}/\kappa|^2} \right) = \nabla \cdot \left(\frac{(\nabla \mathbf{p})_{\parallel}}{1 + |(\nabla \mathbf{p})_{\parallel}/\kappa|^2} \right), \tag{6}$$

subject to $\mathbf{p}_s \in G$. This equation is discretised using the finite difference method with homogeneous boundary conditions, and iterated using the steepest decent method. In order to avoid loss of saturation, we imposed a constraint on how much the out-of-gamut colours can be compressed. This constraint is described by the parameter s (suggesting "saturation"). A value of, e.g., $s = 0.7$ constrains the resulting colour of an out-of-gamut pixel to be outside the inner 70% of the target gamut as measured along the line from $\mathbf{g}$ to $\mathbf{p}_c$.

3 Results

Figure 1 shows two original colour images and the results of pure gamut clipping. Clearly, many details are rendered invisible in the clipped images. For the first image, this loss of detail is evident in the left side of the face, the hair and the face paint.

Figures 2 and 3 show the results of running our algorithm for various number of iterations and for different values of the saturation parameter for the two images, respectively. The saturation parameters are $s = 0.65, 0.75, s = 0.85$ are in the three columns from left to right column, and the number of iterations are $N = 5, 10, 20, 50, 100, 500$ from top to bottom. We observe that small details and edges are corrected to match the original better. With more iterations, the local changes are propagated to larger regions in order to maintain the spatial ratios, however, already at ten iterations, the result resemble that presented in [4], which is, according to Dugay et al. [19] a state-of-the-art algorithm. For many of the images tried, an optimum seems to be found around ten to twenty iterations. Thus, the algorithm is very fast, the complexity of each iteration being $O(N)$

Fig. 1. The original colour images (left) and the gamut clipped images (right)

for an image with N pixels. At the bottom of Figures 2 and 3, the result with as many as 400 iterations are shown. Here, we notice that the details are preserved in a fashion that indicates the power of spatial gamut mapping were we observe that the details are as good as those in the original image.

We further notice that improving the details have not resulted in halos around strong edges. This is due to the use of anisotropic diffusion where we limit diffusion over strong edges. Finally, we notice that, with the introduction of the s parameter, the de-saturation of some colours resulting from previous versions of the algorithm [17, 18] is much more controllable.

As part of this work, we have experimented with 20 images which we mapped to a small destination gamut. Our results shows that the proposed algorithm results in improvement in the visualisation of all the images.

4 Conclusion

In this paper, we presented a spatial gamut mapping algorithm that is derived to minimise the difference, in local contrast, between an original image and its in-gamut counterpart. The first contribution of this paper, is the introduction of a gradient operator that results in three dimensional gradients that are in the direction of the original colour. The motivation behind the use of this operator is to maintain the hue of the original colour (subject to the linearity of the hue lines in the colour space). By employing anisotropic diffusion and a constraint

Fig. 2. The first image from Figure 1 gamut mapped by the proposed algorithm for the three values $s = 0.65, 0.75, s = 0.85$ in the three columns from left to right, and for the number of iterations $N = 5, 10, 20, 50, 100, 500$ from top to bottom

Fig. 3. The second image from Figure 1 gamut mapped by the proposed algorithm for the three values $s = 0.65, 0.75, s = 0.85$ in the three columns from left to right, and for the number of iterations $N = 5, 10, 20, 50, 100, 500$ from top to bottom

on the extent to which gradient correction is allowed to result in de-saturation, we were able to achieve results that improve on the state of art algorithm- our results are free from halos, without loss of saturation. Finally, this improvement is achieved to a small computational cost which makes this algorithm suited for practical implementation as part of a colour reproduction pipeline.

References

1. Morovič, J., Luo, M.R.: The fundamentals of gamut mapping: A survey. Journal of Imaging Science and Technology 45, 283–290 (2001)
2. Bala, R., de Queiroz, R., Eschbach, R., Wu, W.: Gamut mapping to preserve spatial luminance variations. Journal of Imaging Science and Technology 45, 436–443 (2001)
3. Kimmel, R., Shaked, D., Elad, M., Sobel, I.: Space-dependent color gamut mapping: A variational approach. IEEE Transactions on Image Processing 14, 796–803 (2005)
4. Farup, I., Gatta, C., Rizzi, A.: A multiscale framework for spatial gamut mapping. IEEE Transactions on Image Processing 16 (2007), doi:10.1109/TIP.2007.904946
5. Giesen, J., Schubert, E., Simon, K., Zolliker, P.: Image-dependent gamut mapping as optimization problem. IEEE Transactions on Image Processing 16, 2401–2410 (2007)
6. Eschbach, R.: Image reproduction: An oxymoron? Colour: Design & Creativity 3, 1–6 (2008)
7. Land, E.H., McCann, J.J.: Lightness and retinex theory. Journal of the Optical Society of America 61, 1–11 (1971)
8. McCann, J.J.: A spatial colour gamut calculation to optimise colour appearance. In: MacDonald, L.W., Luo, M.R. (eds.) Colour Image Science, pp. 213–233. John Wiley & Sons Ltd. (2002)
9. Meyer, J., Barth, B.: Color gamut matching for hard copy. SID Digest, 86–89 (1989)
10. Morovič, J., Wang, Y.: A multi-resolution, full-colour spatial gamut mapping algorithm. In: Proceedings of IS&T and SID's 11th Color Imaging Conference: Color Science and Engineering: Systems, Technologies, Applications, Scottsdale, Arizona, pp. 282–287 (2003)
11. Eschbach, R., Bala, R., de Queiroz, R.: Simple spatial processing for color mappings. Journal of Electronic Imaging 13, 120–125 (2004)
12. Zolliker, P., Simon, K.: Retaining local image information in gamut mapping algorithms. IEEE Transactions on Image Processing 16, 664–672 (2007)
13. Nakauchi, S., Hatanaka, S., Usui, S.: Color gamut mapping based on a perceptual image difference measure. Color Research and Application 24, 280–291 (1999)
14. Bakke, A.M., Farup, I., Hardeberg, J.Y.: Evaluation of algorithms for the determination of color gamut boundaries. Journal of Imaging Science and Technology 54, 050502–050511 (2010)
15. Balasubramanian, R., Dalal, E.: A method for quantifying the color gamut of an output device. In: Proc. SPIE. Color Imaging: Device-Independent Color, Color Hard Copy, and Graphic Arts II, San Jose, CA, vol. 3018 (1997)

16. Perona, P., Malik, J.: Scale-space and edge detection using anisotropic diffusion. IEEE Transactions on Pattern Analysis and Machine Intelligence 12, 629–639 (1990)
17. Alsam, A., Farup, I.: Colour Gamut Mapping as a Constrained Variational Problem. In: Salberg, A.-B., Hardeberg, J.Y., Jenssen, R. (eds.) SCIA 2009. LNCS, vol. 5575, pp. 109–118. Springer, Heidelberg (2009)
18. Alsam, A., Farup, I.: Spatial Colour Gamut Mapping by Means of Anisotropic Diffusion. In: Schettini, R., Tominaga, S., Trémeau, A. (eds.) CCIW 2011. LNCS, vol. 6626, pp. 113–124. Springer, Heidelberg (2011)
19. Dugay, F., Farup, I., Hardeberg, J.Y.: Perceptual evaluation of color gamut mapping algorithms. Color Research and Application 33, 470–476 (2008)

Accelerated Centre-of-Gravity Calculation for Massive Numbers of Image Patches

Andreas Maier

University of Salzburg, CS Department, Salzburg, Austria
Andreas.Maier@sbg.ac.at

Abstract. Some image processing algorithms require centre-of-gravity (CoG) information of a huge number of different regions or patches of an image. The traditional approach of calculating the CoG from image moments is slow in this circumstance and can significantly limit the system performance. Alternative techniques for fast and quasi patch size independent CoG calculations are presented in this paper. The proposed methods are described and run-time figures, constraints on their usage and their respective memory overhead are assessed. This theoretical complexity analysis is verified for example configurations with different patch sizes.

1 Introduction

The image centre-of-gravity (CoG) is a property that describes the brightness distribution in an image[1]. It is conventionally used for binary or grey scale images. For a grey scale image I the image CoG $cg(I)$ with coordinates (x_{cg}, y_{cg}) is defined as

$$cg(I) = (x_{cg}, y_{cg}) = \left(\frac{\sum_{x,y} x I(x,y)}{\sum_{x,y} I(x,y)}, \frac{\sum_{x,y} y I(x,y)}{\sum_{x,y} I(x,y)} \right) \tag{1}$$

where $I(x,y)$ denotes the brightness value of the pixel at coordinates (x,y) of the image while the sums iterate over all pixels of an image, respectively. Frequently the CoG is alternatively defined using zero and first order spacial images moments m as

$$cg(I) = \left(\frac{m_{1,0}}{m_{0,0}}, \frac{m_{0,1}}{m_{0,0}} \right) \quad where \quad m_{p,q} = \sum_{x,y} I(x,y) x^p y^q \tag{2}$$

Due to its algorithmic simplicity the CoG is regularly utilised in many image processing tasks. Some examples of applications are: image registration using the CoG of corresponding image regions[2,3], tracking of featureless object with occlusion[4], accurate object localisation with sub-pixel accuracy[5] or accelerating the processing speed for template matching. The CoG is only a subset of the more expressive and universal (invariant) image moments[6,7], but it is

G. Bebis et al. (Eds.): ISVC 2012, Part I, LNCS 7431, pp. 566–574, 2012.

faster to compute and sufficient for the before mentioned examples. Many applications do not suffer from a slow CoG computation while others are in need for a fast CoG computation (i.e. [8]). This is emphasised by the fact that specialised hardware for fast CoG computation[9] or imaging sensors with support for CoG computation[10] have been built before CPUs have become powerful enough to perform the calculations at real time in software.

For an image of width W and height H the calculation of the CoG requires $2WH$ multiplications, $3WH-1$ additions and two division. (For implementations a check for a zero denominator is also necessary but such details are not considered throughout this paper.) In applications like CoG accelerated template matching the template slides across the image selecting consecutive patches. For each patch the CoG has to be calculated individually, i.e. for a full match of a square $N*N$ pixel image with a square $M*M$ pixel template $(N-M+1)^2$ patches are selected and processed[11]. This gives an overall complexity of approximately $5M^2(N-M+1)^2$ operations for the calculation of the entire supporting CoG information assuming that multiplications and additions are equally expensive. For increasing patch sizes this number can become undesirably large. Apart from the simplicity of this approach another advantage of directly and instantaneously computing the CoG for every patch is that this technique does not need to store additional data, so the memory overhead is zero.

The contribution of this paper is to present and discuss speed up techniques for such situations where massive CoG calculations for image patches occur. Section 2 presents different algorithms to speed up consecutive CoG calculation and gives runtime and memory usage estimates. Section 3 shows results from evaluating the number of operations for the presented algorithms while Section 4 concludes the paper.

2   Speed-Up Techniques for Consecutive CoG Calculation

In the following a number of speed-up techniques are presented and discussed that improve the performance of the CoG calculation for consecutive image patches. Without loss of generality only square images (of $N * N$ pixels) and image patches (of $M * M$ pixels) are considered to simplify notation.

Discrete Fourier Transform: The first class technique to speed up the processing of the naive approach from the previous section is usually to apply Discrete Fourier Transform (DFT). The calculation of the spacial image moments $m_{0,0}$, $m_{1,0}$ and $m_{0,1}$ can be phrased as a correlation of an image patch with specifically designed kernels. The correlation of the whole image with such kernels for calculating the entire supporting CoG information is best handled in Fourier space where the operation is reduced to an elementary multiplication. The kernels used in this case are

$$K_{0,0} = \{K(x,y) = 1 \mid 0 \le x < M, 0 \le y < M\} \quad for \quad m_{0,0} \qquad (3)$$
$$K_{1,0} = \{K(x,y) = x \mid 0 \le x < M, 0 \le y < M\} \quad for \quad m_{1,0} \qquad (4)$$
$$K_{0,1} = \{K(x,y) = y \mid 0 \le x < M, 0 \le y < M\} \quad for \quad m_{0,1} \qquad (5)$$

and are naturally zero extended to the size of the image for the Discrete Fourier Transform. The overall effort for transforming the image and a kernel, multiplying the transforms and applying the inverse transform is approximately $30N^2 log_2(N)$ operations[11]. This has to be done for all three kernels giving slightly less than three times the number of operations of a single correlation because some effort can be saved on transforming the image into Fourier space only once for all three correlations.

The DFT approach has a significant memory overhead because the spacial image moments are calculated and stored for the whole image at once. For the three image moments this adds up to $3N^2$ elements in the general case or to at least $3(N-M+1)^2$ elements if the irrelevant borders are dropped. Furthermore, $2N^2$ elements of storage space is needed for the transformed image and kernel during the correlation process, but this memory space can be deallocated strictly afterwards. Other disadvantages of the DFT approach are the algorithmic complexity of fast DFT implementations and the use of finite precision floating point numbers in the computations.

Integral Images: In the case of consecutive CoG calculation other approaches apart from DFT are possible that increase the computation speed by taking into account that consecutive patches selected from the image do have many pixels in common and the processing of these pixels can be at best avoided completely. These approaches are based on the either explicit or implicit use of integral images[11]. An integral image S of an image I of width W and height H is defined as

$$S = \left\{ S(x,y) = \sum_{q<x,r<y} I(q,r) \,\middle|\, 0 \le x \le W, 0 \le y \le H \right\} \qquad (6)$$

This definition can be extended to two specific forms of weighted integral images S_x and S_y defined as

$$S_x = \left\{ S_x(x,y) = \sum_{q<x,r<y} qI(q,r) \,\middle|\, 0 \le x \le W, 0 \le y \le H \right\} \qquad (7)$$

$$S_y = \left\{ S_y(x,y) = \sum_{q<x,r<y} rI(q,r) \,\middle|\, 0 \le x \le W, 0 \le y \le H \right\} \qquad (8)$$

for which the intensity value of an image pixel is weighted by its horizontal or vertical position, respectively.

The specific structure of the integral images enables the computation of the image moments $m_{0,0}$, $m_{1,0}$ and $m_{0,1}$ for an image patch in constant time. The spacial image moments for a square image patch of size M at position (x,y) are

$$m_{0,0} = S(x+M,y+M) + S(x,y) - S(x,y+M) - S(x+M,y) \qquad (9)$$

$$m_{1,0} = S_x(x+M,y+M) + S_x(x,y) - S_x(x,y+M) - S_x(x+M,y) - x * m_{0,0} \qquad (10)$$

$$m_{0,1} = S_y(x+M,y+M) + S_y(x,y) - S_y(x,y+M) - S_y(x+M,y) - y * m_{0,0} \qquad (11)$$

The computation time is thereby independent of the patch size. This property is useful in template matching scenarios that process an image with templates of different sizes and thus require different patch sizes[12]. The use of integral images in CoG calculations has been proposed before[13] but a detailed analysis of computational complexity and additional memory requirements was not formulated.

The computation of S, S_x and S_y utilises two add operations per pixel per integral image and additionally a multiply operation per pixel for each of the two weighted integral images. For square images the overall complexity of the integral images calculation is therefore approximately $(2 * 3 + 2)(N + 1)^2$. The effort of calculating the spacial moments of an image patch from the integral images is eleven add and two multiply operations. This results in approximately $13(N - M + 1)^2$ operations for the overall CoG computation in addition to the effort of the integral image pre-processing step. Each of the three integral images requires an additional storage space of $(N + 1)^2$ elements while the CoG calculation thereafter does not require additional memory.

Slice Configuration: The memory overhead induced by the use of integral images can be reduced at the expense of the patch size independence and the random patch selection capability. Therefore, all image patches for which the CoG has to be computed have to be ordered according to the addressed columns or rows in the image, i.e. in the row-ordered scenario considered below patch $p(x_1, y_1)$ precedes $p(x_2, y_2)$ if $y_1 < y_2$. To compute the CoG of all patches at the same row y only selected rows of the integral images are required (see Equations 9 - 11).

In a slice configuration only $M + 1$ rows $[y, y+M]$ of the integral images are kept in memory. The elements for the computations of the image moments are directly accessible and the consecutive row $y+M+1$ of the integral images can be computed from row $y+M$ on demand, overriding the no longer required storage location of row y. For this update procedure to work correctly each row has to be addressed one after the other and no rows can be skipped. The computations of the slice configuration are exactly the same as for creating the whole integral images at once but these computations are performed on demand. In always reusing the no longer required rows in the slices the storage requirement of each slice is reduced to $(N + 1)(M + 1)$ elements.

Two-Row Configuration: In a two-row configuration only rows y and $y+M$ of the integral images are stored. This is again enough to compute the necessary image moments but now the computation of consecutive rows is more expensive as not only row $y+M+1$ has to be computed from row $y+M$ but also the (already previously computed but abolished) row $y+1$ has to be computed from row y for the next series of CoG calculations. As nearly every row of the integral images has to be computed twice and no rows can be skipped, the effort increases to $(2 * 3 + 2)((N + 1)^2 + (N - M)^2) + 13(N - M + 1)^2$ but the additional storage space is reduced to $2(N + 1)$ elements per integral image.

One-Row Configuration: For even tighter storage space requirements a one-row configuration is possible where only one row of supplemental information for every image moment used in CoG calculation is stored. Such a configuration places additional constraints on the ordering of the patches of an image. Apart from the row-ordering mentioned above a column-ordering has to be applied, where patches are additionally ordered within a row according to the addressed columns, i.e. patch $p(x_1, y)$ precedes $p(x_2, y)$ if $x_1 < x_2$. The one-row configuration uses a support row for every image moment. The elements of these support rows are no longer part of the integral images but have a different format. Each element contains the (weighted) sum of the pixels of a section of the corresponding column in the image. The section of a column is given by the pixels of the column that are selected from the latest patch that covered the column. The content of the support rows $s_{0,0}$, $s_{1,0}$ and $s_{0,1}$ for the spacial image moments $m_{0,0}$, $m_{1,0}$ and $m_{0,1}$ after processing patches $p(x, y)$ for each x at a constant y is

$$s_{0,0}[x] = \sum_{0 \leq r < M} I(x, y+r) \tag{12}$$

$$s_{1,0}[x] = \sum_{0 \leq r < M} x I(x, y+r) \tag{13}$$

$$s_{0,1}[x] = \sum_{0 \leq r < M} y I(x, y+r) \tag{14}$$

For each row patches address columns in ascending order due to the ordering constraints. Therefore, to compute required image moments for the patch $p(x, y)$ only the support rows elements at position $x + M - 1$ have to be updated because all other involved elements are already updated from processing the patch $p(x-1, y)$. The update procedure for an element of all support rows is to shift the section of the corresponding column to the next position

$$s_{0,0}[x+M-1] := s_{0,0}[x+M-1] + I(x+M-1, y+M-1) - I(x+M-1, y-1) \tag{15}$$

$$s_{1,0}[x+M-1] := s_{1,0}[x+M-1] +$$
$$(x+M-1)(I(x+M-1, y+M-1) - I(x+M-1, y-1)) \tag{16}$$

$$s_{0,1}[x+M-1] := s_{0,1}[x+M-1] +$$
$$(y+M-1)I(x+M-1, y+M-1) - (y-1)I(x+M-1, y-1) \tag{17}$$

Subsequent to the update procedure is the calculation of the resulting image moments for patch $p(x, y)$ from the support rows elements

$$m_{0,0} = \sum_{0 \leq q < M} s_{0,0}[x+q] \tag{18}$$

$$m_{1,0} = \sum_{0 \leq q < M} s_{1,0}[x+q] - x * m_{0,0} \tag{19}$$

$$m_{0,1} = \sum_{0 \leq q < M} s_{0,1}[x+q] - y * m_{0,0} \tag{20}$$

Summing up the support rows elements for patch $p(x, y)$ can likewise be reduced to a small number of operations through the help of temporary variables. If temporary variables $tm_{0,0}$, $tm_{1,0}$ and $tm_{0,1}$ contain the partial sums used in the calculation of $p(x-1, y)$ the incremental update step to $p(x, y)$ is

$$tm_{0,0} := tm_{0,0} + s_{0,0}[x+M-1] - s_{0,0}[x-1] \tag{21}$$

$$tm_{1,0} := tm_{1,0} + s_{1,0}[x+M-1] - s_{1,0}[x-1] \tag{22}$$

$$tm_{0,1} := tm_{0,1} + s_{0,1}[x+M-1] - s_{0,1}[x-1] \tag{23}$$

where the resulting values in the temporary variables can directly replace the sum term in Equations 18 - 20. For all update procedures of this configuration to work correctly each possible patch position has to be addressed in the order specified and no columns or rows can be skipped. The algorithm further more requires an initialisation phase at the start and an intermediate reinitialisation step whenever processing moves to a new row. The initialisation phase at first fills up the support rows with information from the first M rows of an image according to Equations 12 - 14 ($y = 0$). The reinitialisation step has to update the first M elements of the support rows according to Equations 15 - 17 as well as to set up the temporary variables $tm_{0,0}$, $tm_{1,0}$ and $tm_{0,1}$.

The computational complexity of the one-row configuration is 19 operations for the calculation of image moments for each consecutive patch. The complexity of the initialisation and the reinitialisation steps is slightly below these numbers but can be approximated sufficiently well with the upper bound of 19 operations too. The overall CoG computation effort for the whole image is therefore below $19N^2$ operations which is virtually equivalent to the integral image based methods. The storage overhead is determined from the three support rows and adds up to $3N$.

Table 1 summarises the findings of this section. In accordance with the notation used in the section N denotes the width and height of an image while M denotes the width and height of a single image patch.

Table 1. Approximate computation complexity and memory requirements for centre-of-gravity calculation for a $N*N$ pixel image and $M*M$ pixel image patches

Algorithm	Number of operations	Required add. memory
Classic (Eq. 1)	$5M^2(N-M+1)^2$	0
DFT*	$\approx 3 * 30N^2 log_2(N)$	$(3+2)N^2$
Integral Images (pre-processing)	$13(N-M+1)^2 +$ $8(N+1)^2$	$0+$ $3(N+1)^2$
Slice Configuration	$13(N-M+1)^2 + 8(N+1)^2$	$3(N+1)(M+1)$
Two-Row Configuration	$13(N-M+1)^2 +$ $8((N+1)^2 + (N-M)^2)$	$6(N+1)$
One-Row Configuration	$< 19N^2$	$3N$

(*) Approximate number of operations according to [11]

3   Experimental Verification

All algorithms for exhaustive CoG calculations that are presented on a theoretical basis in the previous section have been evaluated for their correctness and performance. The correctness has been verified by comparing the results of each algorithm against the results achieved with the classical algorithm of Equation 1. As expected, all algorithms produce results equivalent to the classical algorithms and thus are considered correct.

For an assessment of the validity of the performance statements the number of operations for each algorithm has been measured. All algorithms have been implemented in the Python programming language and the operations have been counted, except for the Discrete Fourier Transform for which no implementation has been done but the approximation from [11] has been used as a reference value. The operations are subdivided in additions and multiplications but are only counted when they utilise image elements or "helper" data like integral image elements. Program overhead like loops or elements access has been ignored. A more detailed performance examination would require the use of a better suited programming language like C and a complete analysis of machine code.

For an image of size 1024x1024 pixels and three different image patch sizes (3x3, 15x15, 41x41 pixels) the average number of operations per image patch of each algorithm is shown in Table 2.

Table 2. Complexity of centre-of-gravity calculation in operations per image patch for different algorithms and patch sizes on a 1024x1024 pixels image

Algorithm	Average number of operations per image patch					
	3x3 patch		15x15 patch		41x41 patch	
	add/sub	mult	add/sub	mult	add/sub	mult
Classic (Eq. 1)	26.89	17.93	656.67	437.78	4656.71	3104.47
DFT*	≈ 900.00		≈ 900.00		≈ 900.00	
Integral Images	16.96	3.99	16.71	3.95	16.16	3.85
(included pre-processing)	(6.01)	(2.00)	(6.01)	(2.00)	(6.01)	(2.00)
Slice Configuration	16.96	3.99	16.70	3.95	16.16	3.85
Two-Row Configuration	22.94	5.99	22.61	5.92	21.92	5.77
One-Row Configuration	13.97	4.99	13.78	4.93	13.38	4.80

(*) Number of operations according to [11], exact distribution is unknown.

The effort of the classical approach of the CoG computation is already higher than most of the other approaches for even very small patch sizes like 3x3 pixels patches and increases quadratically with the size of the patch, overtaking also the effort for the DFT at a moderate patch size of 15x15 pixels. All other algorithms show a quasi constant effort independent of the image patch size at about twenty to thirty operations per image patch, except for the DFT which has an effort more than a magnitude higher. The observable slight decrease in the numbers at larger patches is due to the decreasing numbers of different patches selectable from an image when the patch size increases. It is interesting to observe that the use of full or reduced integral images is slightly more effort computationally wise than the one-row configuration although they all have more storage overhead than the latter.

4   Conclusion

Algorithms for centre-of-gravity calculation of massive numbers of image patches have been presented and discussed that outperform the classical approach. A subset of these algorithms have been identified that show a small and constant processing time for an image patch independent of the patch size. A clear winner is the one-row configuration which not only requires the smallest number of operations per image patch but also has the lowest memory overhead of all accelerated methods. In case of non-consecutive image patch processing or when different patch sizes are to be processed simultaneously, the integral images approach is beneficial for its flexibility while having a small number of operations but it reveals an important storage overhead.

Acknowledgements. This work has been partially supported by the Austrian Federal Ministry for Transport, Innovation and Technology (FFG Bridge 2 project no. 822682).

References

1. Gonzalez, R., Woods, R.: Digital Image Processing, 2nd edn. Prentice-Hall (2002)
2. Flusser, J., Suk, T.: A moment-based approach to registration of images with affine geometric distortion. IEEE Transactions on Geoscience and Remote Sensing 32, 382–387 (1994)
3. Goshtasby, A., Stockman, G., Page, C.: A region-based approach to digital image registration with subpixel accuracy. IEEE Transactions on Geoscience and Remote Sensing GE-24, 390–399 (1986)
4. Gordon, G.: On the tracking of featureless objects with occlusion. In: Proceedings of Workshop on Visual Motion 1989, pp. 13–20 (1989)
5. van Assen, H., Egmont-Petersen, M., Reiber, J.: Accurate object localization in gray level images using the center of gravity measure: accuracy versus precision. IEEE Transactions on Image Processing 11, 1379–1384 (2002)
6. Hu, M.K.: Visual pattern recognition by moment invariants. IRE Transactions on Information Theory 8, 179–187 (1962)
7. Paschalakis, S., Lee, P.: Pattern recognition in grey level images using moment based invariant features. In: Seventh International Conference on Image Processing and Its Applications (Conf. Publ. No. 465), vol. 1, pp. 245–249 (1999)
8. Maier, A., Uhl, A.: Fast orientation invariant template matching using centre-of-gravity information. In: Proceedings of the 3th Conference on Computational Modeling of Objects Presented in Images: Fundamentals, Methods and Applications, CompIMAGE 2012 (accepted 2012)

9. Andersson, R.: Real-time gray-scale video processing using a moment-generating chip. IEEE Journal of Robotics and Automation 1, 79–85 (1985)
10. Astrom, A., Astrand, E.: Analog sensor processing using exposure control. A new concept for high speed image processing. In: Proceedings Fourth IEEE International Workshop on Computer Architecture for Machine Perception, CAMP 1997, pp. 68–74 (1997)
11. Brunelli, R.: Template matching techniques in computer vision: theory and practice. Wiley (2009)
12. Maier, A., Uhl, A.: Robust automatic indentation localisation and size approximation for vickers microindentation hardness indentations. In: Proceedings of the 7th International Symposium on Image and Signal Processing, ISPA 2011, Dubrovnik, Croatia, pp. 295–300 (2011)
13. Doretto, G., Yao, Y.: Region moments: Fast invariant descriptors for detecting small image structures. In: 2010 IEEE Conference on Computer Vision and Pattern Recognition, CVPR, pp. 3019–3026 (2010)

An Optimization Based Framework for Human Pose Estimation in Monocular Videos

Priyanshu Agarwal[1], Suren Kumar[1], Julian Ryde[2], Jason J. Corso[2], and Venkat N. Krovi[1]

[1] Mechanical and Aerospace Engineering Department
[2] Computer Science and Engineering Department
University at Buffalo, Buffalo, NY, USA
{priyansh,surenkum,jryde,jcorso,vkrovi}@buffalo.edu

Abstract. Human pose estimation using monocular vision is a challenging problem in computer vision. Past work has focused on developing efficient inference algorithms and probabilistic prior models based on captured kinematic/dynamic measurements. However, such algorithms face challenges in generalization beyond the learned dataset.

In this work, we propose a model-based generative approach for estimating the human pose solely from uncalibrated monocular video in unconstrained environments without any prior learning on motion capture/image annotation data. We propose a novel Product of Heading Experts (PoHE) based generalized heading estimation framework by probabilistically-merging heading outputs (probabilistic/ non-probabilistic) from time varying number of estimators to bootstrap a synergistically integrated probabilistic-deterministic sequential optimization framework for robustly estimating human pose. Novel pixel-distance based performance measures are developed to penalize false human detections and ensure identity-maintained human tracking. We tested our framework with varied inputs (silhouette and bounding boxes) to evaluate, compare and benchmark it against ground-truth data (collected using our human annotation tool) for 52 video vignettes in the publicly available DARPA Mind's Eye Year I dataset[1]. Results show robust pose estimates on this challenging dataset of highly diverse activities.

1 Introduction

Estimating and tracking 3D pose of humans in unrestricted environments using monocular vision poses several technical challenges due to high-dimensionality of human pose, self-occlusion, unconstrained motions, variability in human motion and appearance, observation ambiguities (left/right limb ambiguity), ambiguities due to camera viewpoint, motion blur and unconstrained lighting [1]. Efforts at addressing this challenging problem can be broadly classified into: (i) model-based approaches, and (ii) model-less approaches [2]. Sminchisescu [3] alternately categorizes the research into: (i) generative approaches and (ii) discriminative approaches. While generative approaches are highly generalizable, the use of stochastic sampling methods to deal with

---

[1] Available at: www.visint.org

G. Bebis et al. (Eds.): ISVC 2012, Part I, LNCS 7431, pp. 575–586, 2012.

the multimodal posterior/likelihood function increases their computational complexity. On the other hand, discriminative approaches are computationally tractable (for moderate sized training sets) but lack generalizability to unseen exemplars. However, there is always one or more fundamental assumptions involved that there is a priori knowledge about the physical properties (e.g. mass, inertia, limb lengths, ground plane and/or collision geometries), the activity in the scene, calibrated camera, imagery from multiple cameras (often in laboratory settings), availability of similar motion dataset [4,5,6].

No formal studies exist on which methods are employed by the human visual system for its marvelous visual perception. However, studies have constantly shown that humans use motion based cues (the instantaneous retinal optical flow) for instantaneous retino-centric heading (3D translation direction), eye-body rotation, and the relative depth of points in the world [7]. Humans appear to use motion based cues whenever motion is present in the scene and resort to visual cues (color, texture) when no/subtle motion is present in the scene. To the best of our knowledge, no prior work has used motion based cues for the task of explicitly estimating human heading direction.

Our work employs a model-based generative approach for the task of human pose estimation for general human movements in unrestricted environments. Unlike many previous approaches, our framework is fully automatic, without using camera calibration, prior motion (motion capture database), prior activity, appearance, body size information about the scene. Evaluations on a challenging dataset (DARPA Mind's Eye Year I) show the robustness of the presented framework.

Research Contributions

1. *Product of Heading Experts* - We model the heading estimation task independent of features/types of individual estimators using the proposed Product of Heading Experts (PoHE) based generalized heading estimation framework which probabilistically merges heading outputs from time varying number of estimators to produce robust heading estimates under varied conditions in unconstrained scenarios.
2. *Motion Cues Based Heading Estimation* - We propose a novel generative model for estimating heading direction of the subject in the video using motion-based cues thus, significantly reducing the pose search space.
3. *Decoupled Pose Estimation* - We propose a sequential optimization based framework optimizing the uncoupled pose states (camera/body location, body joint angles) separately using a combination of deterministic and probabilistic optimization approaches to leverage the advantages associated with each.
4. *Probabilistic-Deterministic Optimization Scheme* - We achieve faster convergence to the global minima by obtaining initial guesses using population based global optimization technique for deterministic convex optimization scheme.
5. *Identity Maintained Pose Evaluation Metric* - We introduce the notion of pose evaluation for videos with multiple humans by defining identity maintained pose evaluation metrics.

2   Optimization Based Pose Estimation

Fig. 1 provides an overview of the human pose estimation framework. We use background subtracted binary images [8] and point features at the low-level to detect/track

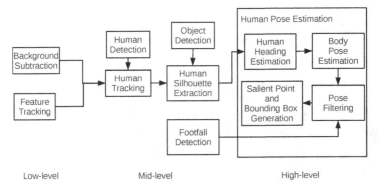

Fig. 1. System diagram for human pose estimation framework

humans/objects in the scene, extract human silhouettes at the mid-level leading to human heading and pose/salient-point estimation/filtering at the high-level. We consider 3 position variables and 5 angular variables to define the pose of a human (Figs. 2(a)-(c)).

2.1 Human Heading Estimation

Knowledge regarding the heading direction can significantly restrict the pose search space and can result in better pose estimates at lower computational costs. In the past, the task of heading estimation is not addressed separately from the actual body pose which significantly increases the complexity of the problem. Furthermore, heading is often modeled as a discrete variable using discriminative approaches with few possible values [9]. Fig. 2(d) illustrates a sequence of frames where invaluable human heading direction information can be inferred from following cues: (i) human silhouette centroid, (ii) human silhouette bounding box centroid, (iii) detected human bounding box centroid, (iv) area of human silhouette, (v) aspect ratio of bounding boxes, (vi) human silhouette/bounding boxes centroid velocity (x and y coordinates), (viii) regression/classification-based estimation of heading direction (Adaboost/Support Vector Machine), and/or (ix) optical flow.

Product of Heading Experts: We use a time evolving Product of Experts (PoE) [10] model to optimally fuse hypothesis from various heading estimators at each instant in time to propose a Product of Heading Experts (PoHE). We consider each estimator $T_1, T_2, ..., T_K$ as experts for predicting the heading direction. Product of experts model for heading ensures that the resulting model for heading is explained by all the experts. Let θ_k be the parameters associated with probability distribution of each expert ($=$ $[\mu^k, \Sigma^k]^T$ in current case). Probability of any direction ϕ to be true heading of a human as explained by all the expert estimators is given by Equation 1.

$$p(\phi|\theta_{T_1}, \theta_{T_2}, ..., \theta_{T_K}) = \frac{\prod_{k=1}^{K} p_k(\phi|\theta_k)}{\int \prod_{k=1}^{K} p_k(\phi|\theta_k)\mathrm{d}\phi} \tag{1}$$

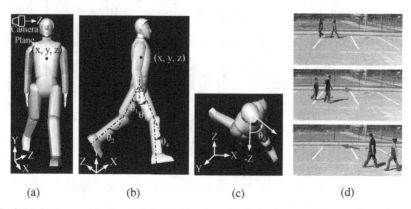

(a)   (b)   (c)   (d)

Fig. 2. Variables used in the model. (a) side view, (b) front view, and (c) top view of the human model (d) frames from a vignette in the DARPA corpus depicting that the motion cues provide significant information regarding the heading direction of a human. The red arrow portrays the direction of motion of the human in the respective frame.

This model results in robust estimation because it allows to incorporate (or leave out) arbitrary number of estimators, even those providing non-probabilistic output, which could also be incorporated using Equation 3.

A wealth of information about the heading direction of the human torso can be inferred solely from information regarding the human motion. In the current implementation, we focus on a PoHE based generative heading estimation method using (i) human silhouette centroid, and (ii) human silhouette bounding box centroid. Once a silhouette corresponding to a detected/tracked human is found in a frame, internal holes/gaps are filled [11] for subsequent use in the pose-estimation process. The silhouette centroid and the silhouette bounding box centroid are then evaluated for every valid frame and any gaps are filled using linear interpolation. We model the 3D heading direction as a continuous variable and approximate it as the 2D heading angle (which is the projected 3D heading angle) which works fairly well as will be evident in results. Fig. 3a depicts two human silhouettes from two different frames (N frames/δt time apart) in a video. The red triangle (solid line) connects the centroid of the two silhouettes ((x_1, y_1) to (x_2, y_2)) and the blue triangle (dashed line) connects the centroid of the two silhouette bounding boxes ((x_{b1}, y_{b1}) to (x_{b2}, y_{b2})). It can be seen that the true silhouette centroid and the silhouette bounding box centroid information are corrupted by the merging of the silhouette due to the shadow in the original human silhouette. In cases where partial silhouette information is obtained, the silhouette centroid tends to be biased towards the region where the foreground pixels are concentrated. However, the bounding box centroid locates the centroid of the region irrespective of the foreground pixel density. By merging information from both the sources we tend to reduce the effect of noise in estimating heading direction. Equation 2 is used to evaluate an estimate of the heading direction given the centroid information for two frames.

$$\mu_k = tan^{-1}\left(\frac{y_2 - y_1}{x_2 - x_1}\right) \qquad (2)$$

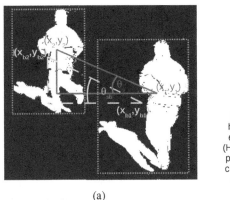

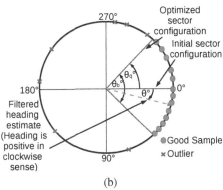

(a)   (b)

Fig. 3. Human heading estimation modeling. (a) Silhouette and bounding box centroid modeling of human heading estimation, and (b) Outlier detection in angular data using optimization based sector positioning.

We build a Gaussian distribution for each heading estimate considering the distribution mean to be situated at the corresponding estimated value and the variance to be equal to the variation in the value from its vector mean in a local temporal window. *Intuitively, we seek to weight the heading direction changes with uncertainty within each temporal window.* Please note that directional statistics [12] is required to deal with the heading angle data.

$$p(\phi_k(t)) = N(\mu_k(t), \sigma_k^2), \qquad (3)$$

where, $\sigma_k^2(t) = \phi_k(t) - \bar{\phi}_k(t)$, $\bar{\phi}_k(t) = Arg(\bar{\rho}_k)$, $k \in \{s, sb\}$, $\bar{\rho}_k = \frac{1}{N}\sum_{n=1}^{N} z_{kn}$, $z_{kn} = \cos\phi_k(t) + i\sin\phi_k(t)$

Outlier Detection in Angular Data: The raw heading estimates obtained are noisy due to noise in silhouettes and so contain outliers which are eliminated using outlier detection. For outlier detection, we use an optimization based sector positioning technique in which the data lying within a sector is considered to be fit for evaluating the heading estimate within a local temporal window (Fig. 3b). The green circles on the main circle represents good samples and the red crosses represents the outliers. The blue sector represents the angular region (of angle $\theta_b = \pi/2$ degrees) samples in which are considered to be good and valid for heading estimation. Initially the sector is aligned with the main quadrant ($\theta_q = 0$) and the sector positioning (θ_q) is determined by solving the optimization problem in (4) which maximizes the number of samples lying in the angular region:

$$\arg\max_{\theta_q}(\max \#\{\theta_q | \theta \in bin(k)\}) \; k = 1...K, K = \frac{2\pi}{\theta_b} \qquad (4)$$

where the symbol '#' stands for angular histogram. The optimization is carried out in local temporal sliding window to remove the outliers and Gaussian filtering is carried out on the filtered data considering the same temporal window. *Intuitively, we rely on the continuity of motion i.e. the human heading direction does not change within a fraction of a second.*

2.2 Optimization Based Body Position Estimation

We formulate the problem of determining the position of the body relative to the camera as two optimization subproblems.

Z Coordinate Estimation: The camera depth (z coordinate) estimation is based on the fact that an actual body with proportional dimensions and similar orientation in space will roughly occupy a similar area in an actual image as that of the model in the synthetic image. We set up an optimization problem based on the difference in the silhouette area in the original image and the model generated image, and minimize the square of this difference as in Fig. 4 (c_z is the z coordinate of the camera in the model coordinate system, A_o and A_m is the silhouette area in the original and model generated image, respectively.). We also specify an upper and lower bound on z coordinate such that the model generates a reasonable area in the synthetic image.

X,Y Coordinate Estimation: The estimation of the x, y coordinate is based on the fact that the centroid of the silhouette in the original image and the model generated image should roughly be the same for model with similar orientation. We setup another optimization problem in which square of the distance between the centroid of the original silhouette and the model generated silhouette is minimized constraining the (x,y) coordinates such that the model silhouette is within the synthetic image as in Fig. 4 ((c_x, c_y) is the (x, y) coordinate of the camera in the model coordinate system, (x_{co}, y_{co}) and (x_{cm}, y_{cm}) is the centroid of the silhouette in the original and model generated image, respectively).

2.3 Optimization Based Pose Estimation

For a given camera position, the difference between original and model generated images is minimum when the correct limb pose is achieved. The absolute subtracted image (of model generated and actual human silhouettes) measures the extent of mismatch and serves as the objective function (Fig. 4 where the subscript i indicates i^{th} joint in the human body model, I_a and I_m denotes the actual and model generated silhouette image, respectively). Limits on the human joint angles are imposed based on the biomechanical constraints set by the human body [13].

3   Optimization Approach

The probabilistic optimization techniques are good at identifying promising areas of the search space (exploration), but slow at fine-tuning the approximation to the minimum (exploitation) [14]. Thus, a much faster convergence to the local minima can be achieved if initial guesses are obtained using population based global optimization technique (Genetic Algorithm (GA) [15]) and then convergence to the global optima is accomplished using convex optimization techniques.

3.1 Convex Optimization

We use Augmented Lagrangian method (ALM) [16] for solving the optimization problem considering its advantage over penalty methods which are less robust due to sensitivity to penalty parameter chosen. In order to solve the ND unconstrained optimization subproblem, we use Powell's conjugate direction method [17] as it requires only the objective function value and is more robust to noise in function evaluation, which is often the case with image based objective functions. For 1D optimization subproblem we employ Golden section with Swann's bounding [18].

3.2 Optimization Framework

The optimization subproblems in Fig. 4 are highly coupled and cannot be solved independently. While a weighted/combined optimization problem may be posed, it suffers from multiple local minima as well as sensitivity to weightage of each objective. Hence, in lieu of this, we adopt a sequential optimization framework as shown in Fig. 4. Once, we have the heading estimates for each frame in the video, we first optimize for the camera parameters (relative location of body with respect to camera) and then for the pose assuming fixed geometries for the human body parts. In order to deal with the well-known problem of pose ambiguity due to symmetric nature of human body and keep the framework computationally feasible, we only resort to GA when either the difference between the joint angles for the the left and the right leg are below a certain threshold or the joint angle limits are exceeded, to obtain good initialization for pose. The presented framework is executed on each frame in the video to estimate two corresponding poses (left leg forward and right leg forward).

4   Experiments

We evaluated the proposed human pose estimation framework on 52 challenging video vignettes in the DARPA Mind's Eye Year I[1] dataset (resolution: 1280×720) of different activities (collide, enter, follow, flee, leave, run, jump, walk, approach, fall, pass, stop, replace, take, turn, throw, kick, go, hold, get) performed by multiple people interacting with other entities (humans/objects) in outdoor scenes.

Inputs: In order to thoroughly test the system performance we test the pose estimation framework on three different types of inputs: (i) Manually Labeled Silhouette (MLS) and Manually Labeled Human Bounding Boxes (MLHBB) for a selected set of the videos (6 in number) where it was possible to get good pose estimates as significant lower limb movement was involved. This trial was carried out to establish the

---

[1] Available at https://sites.google.com/site/poseestimation/

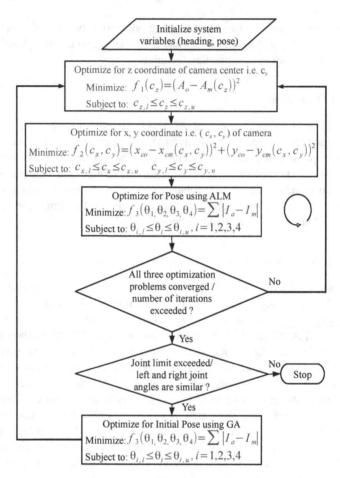

Fig. 4. Summary of optimization framework implemented for pose estimation on each frame

benchmark against which to compare the performance of the algorithm with inputs of varying fidelity; (ii) Background Subtracted Silhouette (BSS), Detected Human Bounding Boxes (DHBB) [19], and Detected Object Bounding Boxes (DOBB) for the entire dataset to establish the system performance over a larger set and all algorithm generated inputs. We observed that the human detection results contains a lot of false positives along with ambiguity in entity identity while tracking; (iii) Background Subtracted Silhouette (BSS), Tracked Human Bounding Boxes (THBB) and Detected Object Bounding Boxes (DOBB) for the entire corpus again to establish the system performance over a large set and more reliable human detections [20].

Pose Evaluation Metrics: Human annotation GUI[2] was developed in order to assess and quantify the performance of the pose estimation algorithm. 13 salient points on

---

[2] Available at https://sites.google.com/site/poseestimation/

human body: head center, right shoulder, right elbow, right hand, left shoulder, left elbow, left hand, right hip, right knee, right foot (ankle), left hip, left knee, left foot (ankle) were manually marked for all videos in the corpus. We build upon the pose error metric proposed in [21] and define the following pose evaluation metrics for each vignette in the corpus: (a) Average error per frame as in (5), (b) Average error per marker per frame (D_{aepmpf}) (average of (5) for number of markers), (c) Average error for different markers per frame as in (6).

$$D_{aepf}(X, \hat{X}) = \frac{1}{N} \left(\sum_{n=1}^{N} \sum_{m=1}^{M=13} ||x_m - \hat{x}_m||_2 \right) \tag{5}$$

$$D_{aedmpf}(X, \hat{X}, m) = \frac{1}{N} \left(\sum_{n=1}^{N} ||x_m - \hat{x}_m||_1 \right) \tag{6}$$

where N is the number of processed frames in the considered vignette. For vignettes with multiple humans, we first associate the estimated pose tracks with the ground truth pose tracks by using the nearest neighbor approach on the entire track, as in (7).

$$j_i = \arg\min_k \sum_{n=1}^{N} \sum_{m=1}^{M=13} ||x_{mk} - \hat{x}_{mi}||_1, E_x = \frac{1}{K} \left(\sum_{n=1}^{P} D_x \right), \text{x} \in \{\text{aepf, aepmpf, aedmpf}\} \tag{7}$$

where $x_{mk}, \hat{x}_{mi}$ are the coordinates of the mth marker in the ground truth data of the kth person and in the estimated data of the ith person, respectively, j_i is the ground truth track associated with the ith detected track, K is the number of humans present in the ground truth and P is the number of detected humans. The error over the entire corpus is the average error obtained considering all the vignettes in the corpus as in 7.

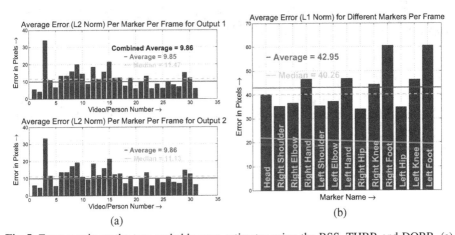

(a)   (b)

Fig. 5. Error metric on the two probable pose estimates using the BSS, THBB and DOBB. (a) Average error (L2 norm) per marker per frame, and (b) Average error distribution across markers (L1 norm) per frame.

Table 1. Error metric comparison for different inputs provided to the developed human pose estimation framework on 52 vignettes from DARPA ARL-RT1 dataset

Input/Error Metric	Average Error Per Frame (L2 Norm) (pixels)	Average Error Per Frame Per Marker (L2 Norm) (pixels)	Average Error for Different Markers Per Frame (L1 Norm) (pixels)
MLS + MHLBB (6 vignettes)	80	6	17
BSS + DHBB + DOBB (52 vignettes)	166	13	65
BSS + THBB + DOBB (52 vignettes)	128	10	43

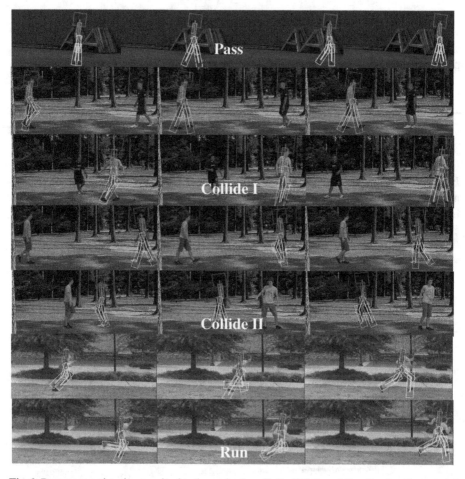

Fig. 6. Raw pose estimation results for the verbs "pass", "collide", and "run" using the system generated inputs. N.B. Identity of the persons is maintained before and after collision for the verb "collide". (Please view in color)

5 Results

Fig. 6 depict the stick figure and bounding boxes superimposed over the original video frame for vignettes corresponding to the verbs "pass", "collide", and "run" in the dataset, respectively. As can be seen the tracking is carried out while maintaining the identity of people in the video. Please note that the presented framework works well with different types of verbs[2] and does not make assumptions regarding the activity in the scene which is an unstated assumption in many state-of-the-art pose trackers.

Fig. 5 shows the error metric obtained for the two probable pose estimates using BBSS and THBB. Table. 1 shows a comparison of the pose evaluation metric for different inputs described in the Section 4. As expected, the average error per frame per marker increased from a value of 6 to 13 when BSS, DHBB, DOBB are provided as input as opposed to MLS, MLHBB. However, the error reduced from 13 to 10 when tracked human bounding box detections are used showing the performance of the pose estimation framework over the entire dataset. Please note that an average human head for the current dataset has a dimension of ≈ 50 pixels (ground truth), so an accuracy of around 10 pixels (L2 norm) and 40 pixels (L1 norm) is fairly good. Since, the current framework does not reliably distinguish between the left and the right leg the error corresponding to the foot and the knee markers is relatively high (Fig. 5b).

6 Discussion

In this work, we propose a Product of Heading Experts (PoHE) based generalized heading estimation framework bootstrapping an integrated probabilistic-deterministic optimization framework for human pose estimation in uncalibrated monocular videos. We benchmarked the standalone performance of the pose estimation framework against ground-truth data for the DARPA video corpus using the proposed pixel-distance based metrics emphasizing identity maintained human tracking and low false human detections. Results showed the robustness and performance of the proposed framework.

Acknowledgements. The authors gratefully acknowledge the support from Defense Advanced Research Projects Agency Mind's Eye Program (W911NF-10-2-0062).

References

1. Hen, Y.W., Paramesran, R.: Single camera 3d human pose estimation: A review of current techniques. In: International Conference for Technical Postgraduates, pp. 1–8 (2009)
2. Poppe, R.: Vision-based human motion analysis: An overview. Computer Vision and Image Understanding 108, 4–18 (2007)
3. Sminchisescu, C.: 3d human motion analysis in monocular video: techniques and challenges. Computation Imaging and Vision 36, 185 (2008)
4. Balan, A., Black, M.: An adaptive appearance model approach for model-based articulated object tracking. In: CVPR, vol. 1, pp. 758–765 (2006)
5. Sigal, L., Isard, M., Haussecker, H., Black, M.: Loose-limbed people: Estimating 3d human pose and motion using non-parametric belief propagation. IJCV, 1–34 (2011)

6. Yang, Y., Ramanan, D.: Articulated pose estimation with flexible mixtures-of-parts. In: CVPR, pp. 1385–1392 (2011)

7. Perrone, J., Zanker, J., Zeil, J.: A closer look at the visual input to self-motion estimation. Motion Vision: Computational, Neural, and Ecological Constraints, 169–179 (2001)

8. Ryde, J., Waghmare, S., Corso, J., Fu, Y.: ISTARE quaterly report: Signal unit. Technical report, SUNY Buffalo (2011)

9. Andriluka, M., Roth, S., Schiele, B.: Monocular 3d pose estimation and tracking by detection. In: CVPR, pp. 623–630 (2010)

10. Hinton, G.E.: Products of experts. In: ICANN, pp. 1–6 (1999)

11. Soille, P.: Morphological Image Analysis: Principles and Applications. Springer (2003)

12. Mardia, K., Jupp, P.: Directional statistics. John Wiley & Sons Inc. (2000)

13. Anderson, F., Pandy, M.: Dynamic optimization of human walking. Journal of Biomechanical Engineering 123, 381 (2001)

14. Costa, L., Santo, I., Denysiuk, R., Fernandes, E.M.G.P.: Hybridization of a Genetic Algorithm with a Pattern Search Augmented Lagrangian Method. In: International Conference on Engineering Optimization (2010)

15. Goldberg, D.: Genetic algorithms in search, optimization, and machine learning. Addison-wesley (1989)

16. Schuldt, S.B.: A method of multipliers for mathematical programming problems with equality and inequality constraints. Journal of Optimization Theory and Applications 17, 155–161 (1975)

17. Powell, M.J.D.: An efficient method for finding the minimum of a function of several variables without calculating derivatives. The Computer Journal 7, 155–162 (1964)

18. Swann, W.H.: Report on the development of a new direct search method of optimization. Research Note (64)

19. Felzenszwalb, P., Girshick, R., McAllester, D., Ramanan, D.: Object detection with discriminatively trained part-based models. PAMI, 1627–1645 (2009)

20. Kumar, S., Agarwal, P., Corso, J., Krovi, V.: ISTARE proxy evaluation report: Human tracking. Technical report, SUNY Buffalo (2011)

21. Sigal, L., Black, M.: Humaneva: Synchronized video and motion capture dataset for evaluation of articulated human motion. International Journal of Computer Vision 87, 4–27 (2010)

Solving MRF Minimization by Mirror Descent

Duy V.N. Luong, Panos Parpas, Daniel Rueckert, and Berç Rustem

Department of Computing, Imperial College London, United Kingdom

Abstract. Markov Random Fields (MRF) minimization is a well-known problem in computer vision. We consider the augmented dual of the MRF minimization problem and develop a Mirror Descent algorithm based on weighted Entropy and Euclidean Projection. The augmented dual problem consists of maximizing a non-differentiable objective function subject to simplex and linear constraints. We analyze the convergence properties of the algorithm and sharpen its convergence rate. In addition, we also use the convergence analysis to identify an optimal stepsize strategy for weighted entropy projection and an adaptive stepsize strategy for weighted Euclidean projection. Experimental results on synthetic and vision problems demonstrate the effectiveness of our approach.

1 Introduction

MRF energy minimization is a central problem in many computer vision applications. State-of-the-art algorithms to solve the MRF problem can be classified in three methodological frameworks: graph cut [1], belief propagation [2] and LP relaxation. We concentrate on the LP relaxation model for MRF problem. The two common frameworks for the LP relaxation of MRF are based on tree-reweighted message passing [3] and dual decomposition [4]. Message-passing techniques exploit acyclic structures in the MRF models and are known to be efficient. However, the convergence properties of message-passing algorithm is not fully understood. In contrast, the dual decomposition approach is connected to the theory of convex optimization, thus the convergence analysis and suboptimality can be established. In the dual framework, the graphical model is decomposed into easy slave MRFs with favourable properties such as submodular graph, acyclic graph. These slaves can be solved efficiently via dynamic programming and their solutions are used to update the parameters of the master problem in a subgradient projection manner. Compared to other methods, the dual-based approach benefits from better convergence properties and has suboptimality guarantees. Recently, improvements to the Dual Decomposition Sub Gradient technique have been made, including Nesterov's smoothing [5], First Order Primal-Dual method [6], Improved Decomposition [7].

In this paper, we develop a projection algorithm to solve the dual problem of the LP relaxation using weighted Entropy and Euclidean distances. The method is based on Mirror Descent algorithm [8,9] and its generalization on "favourable geometry" domain [10]. We employ a dual decomposition technique as in Komodakis et al. [4] to obtain the dual framework with two types of problems.

G. Bebis et al. (Eds.): ISVC 2012, Part I, LNCS 7431, pp. 587–598, 2012.

The master problem solves a non-smooth objective function subject to linear constraints. The MRFs subproblems can be solved by dynamic programming independently. In the dual LP-based algorithm [4,7], subgradient projection is often used by the master to optimally distribute the data cost between the slaves. Main drawbacks of this approach are slow convergence rate and its sensitivity to the choice of stepsize. In order to address these drawbacks, we transform the domain of dual variables to the intersection of simplexes and linear constraints. The search is performed within the simplexes before proceeding with subgradient method. As a result, our method inherits faster convergence rate from the Mirror Descent algorithm with weighted entropy distance. For the second procedure we employ the weighted Euclidean projection with an adaptive stepsize that shows significant speed up in practice. Our method does not require more memory than any other dual-based methods. The sub problems and all variables are decoupled therefore parallelizing computation is fully supported. In the worst case, this method has an $O(\frac{1}{\epsilon^2})$ complexity whereas the method based on Nesterov's smoothing technique [5,11] provide a convergence rate of $O(\frac{1}{\epsilon})$. However, those methods run an inner loop to compute a good stepsize where each inner iterations require computations of sub MRF problems. It is important to stress that this theoretical comparison is only valid in the worst case. In practice, using good adaptive stepsize strategy for first order method significantly reduces the number of iterations.

The main contributions of this paper are:

- We reformulate the original dual problem and construct the ingredients required for the Mirror Descent algorithm, including weighted distance, weighted norm, dual norm and the local Lipschitz constants.
- We provide the solutions updates using Mirror Descent algorithm for our model.
- Through the convergence analysis, we show that sequential updates by performing Entropy projection before Euclidean projection is better than parallel updates. We also use the bounded optimality to identify the optimal stepsize for entropy projection and adaptive stepsize for euclidean projection.

2   Background

Discrete MRF minimization aims to solve a general graphical multi-labelling problem. Given a set of discrete labels L, the goal is to find a labelling configuration such that it returns the minimal energy on the MRF model specified by an undirected hypergraph $G = (V, E)$ where V and E denote the sets of nodes and edges respectively. Each node $a \in V$ must admit one label from L. By $\theta_{a,i}$, we denote the unary cost of assigning label $i \in L$ to node $a \in V$. The notation $\theta_{ab,ij}$ is used to denote the pairwise cost for edge $ab \in E$. The LP relaxation of the MRF problem is defined as follows:

$$\min_{x \in X} \sum_{a \in V} \sum_{i \in L} \theta_{a,i}.x_{a,i} + \sum_{ab \in E} \sum_{i \in L} \sum_{j \in L} \theta_{ab,ij}.x_{ab,ij} \tag{1}$$

where the constraint set X is known as the *local marginal polytope* [3]. Due to the special structure of the problem above, it turns out that the dual of (1) can be solved efficiently [4]. We write the LP problem compactly as:

$$E(\theta, x) := \min_{x \in X} \langle \theta, x \rangle \tag{2}$$

In the dual approach , the original graph G is decomposed into a collection of trees (*acyclic* graphs) T. Each tree $t \in T$ corresponds to a simpler MRF problem $E^t(\theta^t, x^t)$ that can be solved efficiently by the Max Product Belief Propagation algorithm. Without loss of generality, we assume each tree contains all nodes and every edge must appear only once in T. For example, in a 2D grid graph, one tree contains all horizontal edges and one contains all verticle edges. In this setting, no constraints apply to pairwise cost and the sum of unary costs across the trees must preserve the unary cost of the original graph, ie. $\sum_{t \in T} \theta^t_{a,i} = \theta_{a,i}$. The dual-based algorithm aims to distribute the right amount of unary costs for each tree in order to maximize the dual problem.

$$\max_{\{\theta^t\} \in \Theta} \sum_{t \in T} E^t(\theta^t, x^t) \quad \text{where} \quad \Theta = \left\{ \sum_{t \in T} \theta^t = \theta \right\} \tag{3}$$

It is well-known that the solution to problem (3) is the lower bound of the LP problem (2). The key property in dual-based algorithms is to maintain the feasibility set Θ.

Transformation of the Dual Domain: For computational reasons, most methods to solve (3) are based on the subgradient algorithm with Euclidean projection. One disadvantage of this approach is the slow convergent rate. The choice of stepsize significantly affects the algorithm and at every iteration, all unary costs are adjusted by the same amount. In order to address this issue, ie. adjusting the unary cost differently based on the cost itself, we transform the domain of the dual problem such that it still maintains the fesibile set Θ while accelerating the search procedure. Consider the following augmented dual problem:

$$\max_{\rho \in \Delta, \lambda \in \Lambda} F(\rho, \lambda) := \max_{\rho \in \Delta, \lambda \in \Lambda} \sum_{t \in T} E^t(\rho^t.\theta + \lambda^t, x^t) \tag{4}$$

where: $\quad \Delta = \left\{ \rho \,\middle|\, \sum_{t \in T} \rho^t = 1 \,,\, \rho \succeq 0 \right\} \quad ; \quad \Lambda = \left\{ \lambda \,\middle|\, \sum_{t \in T} \lambda^t = 0 \right\}$

It is easy to see that the sets Δ and Λ preserve Θ. The augmented model has the same optimal objective function value as the original dual problem. Notice that if we choose a constant $\rho \in \Delta$, then our model is equivalent to Komodakis et al. [4]. The objective function $F(\rho, \lambda)$ is linear in both variables; in addition, ρ and λ are completely decoupled.

3   Mirror Descent (MD)

Mirror Descent algorithm [8,10] is a generalization of the proximal algorithm [12] with a nonlinear distance function [9] and an optimal stepsize. In order to utilize the Mirror Descent algorithm with the special structure of our augmented dual model, we need to define the subgradient, the weighted distances and weighted norm which favour the problem's geometry. Instead of solving the augmented dual problem (4) directly, we generate a sequence of updates:

$$\begin{bmatrix} \rho^{k+1} \\ \lambda^{k+1} \end{bmatrix} = \arg\max_{\rho \in \Delta, \lambda \in \Lambda} \left\{ \begin{array}{l} \langle F'(\rho^k), \rho \rangle - \frac{1}{\tau} D_\Delta(\rho, \rho^k) \\ + \langle F'(\lambda^k), \lambda \rangle - \frac{1}{\eta} D_\Lambda(\lambda, \lambda^k) \end{array} \right\} \tag{5}$$

Since the function F is linear in both variables, ρ and λ are decoupled, the subgradients with respect to each variable are also disjoint. The weighted distances D_Δ, D_Λ and stepsizes τ, η are defined independently to exploit the geometry of each set. To simplify our notation, we define an index set to cover all unary terms: $I = \{(a, i) | \forall a \in V, \forall i \in L\}$. The domains Δ and Λ are built by taking the direct product of the disjoint subsets:

$$\Delta :=_\otimes \Delta_i \; ; \; \Lambda :=_\otimes \Lambda_i \; , \; \forall i \in I$$

Let $T(i)$ be the collection of trees that cover the same unary term i, then each subset reads:

$$\Delta_i = \left\{ \sum_{t \in T(i)} \rho_i^t = 1 \, , \, \rho_i^t \geq 0 \right\} \; ; \; \Lambda_i = \left\{ \sum_{t \in T(i)} \lambda_i^t = 0. \right\}$$

Subgradient: The following lemma shows how the subgradient is estimated in our algorithm.

Lemma 1. *Let $F'(\rho, \lambda)$ be defined as follows: $F'(\rho, \lambda) = [\theta.\bar{x}; \bar{x}]$, where $\bar{x} \in \arg\min_{x \in X} \langle \rho.\theta + \lambda, x \rangle$. Then $F'(\rho, \lambda) \in \partial F(\rho, \lambda)$, where $\partial F(\rho, \lambda)$ denotes the subgradient of $F(\rho, \lambda)$ at the point (ρ, λ).*

Proof. The point $\bar{x}$ is suboptimal for $\min_{x \in X} \langle \rho'.\theta + \lambda', x \rangle$, therefore:

$$F(\rho', \lambda') \leq \langle \rho'.\theta + \lambda', \bar{x} \rangle = \langle \rho.\theta + \lambda, \bar{x} \rangle + \langle \theta.\bar{x}, \rho' - \rho \rangle + \langle \bar{x}, \lambda' - \lambda \rangle$$
$$F(\rho', \lambda') \leq F(\rho, \lambda) + \langle \theta.\bar{x}, \rho' - \rho \rangle + \langle \bar{x}, \lambda' - \lambda \rangle$$

as required by subgradient inequality.   □

Distance Function: MD generates a projection based on nonlinear distance function. Let D_C^i denotes a Bregman distance function defined on a single closed convex set C_i:

$$D_C^i(u_i, v_i) = \psi_C^i(u_i) - \psi_C^i(v_i) - \langle \nabla \psi_C^i(v_i), u_i - v_i \rangle$$

where $u_i, v_i \in C_i$ and $\psi_C^i : C_i \to \mathbb{R}$ is a *1-strongly* convex *distance-generating-function (d.g.f)*. The weighted distance function D_C defined on the domain $C :=_\otimes C_i$ is thus given by:

$$D_C(u, v) = \sum_{i \in I} \alpha^i D_C^i(u_i, v_i) = \sum_{i \in I} \alpha^i \left[\psi_C^i(u_i) - \psi_C^i(v_i) - \langle \nabla \psi_C^i(v_i), u_i - v_i \rangle \right]$$
$$:= \psi_C(u) - \psi_C(v) - \langle \nabla \psi_C(v), u - v \rangle$$

where $\alpha_C^i > 0$ is the weighted parameter. The weighted d.g.f defined on C is:

$$\psi_C(u) = \sum_{i \in I} \alpha^i \psi_C^i(u_i) \tag{6}$$

Norm: Another requirement for MD is its *compatible* norm, ie. weighted d.g.f $\psi_C : C \to \mathbb{R}$ is *1-strongly* convex w.r.t the weighted norm $\|.\|_C$ [10]:

$$\|u\|_C = \sqrt{\sum_{i \in I} \alpha^i \|u_i\|_{C_i}^2} \tag{7}$$

In the formulation above, $\|.\|_{C_i}$ is a local norm that is defined based on the geometry of a subset C_i.

Dual Norm and the Local Lipschitz Constant:
From the definition of Dual Norm [13], we can derive the dual norm of (7):

$$\|u\|_{C*} = \sqrt{\sum_{i \in I} \|u_i\|_{C_i*}^2 / \alpha^i}$$

Let $\mathcal{L}_{C_i} := \sup_{u_i \in C_i} \|F_u'\|_{C_i*}$ then the *local* Lipschitz constant is given by:

$$\mathcal{L}_C = \sup_{u \in C} \|F_u'\|_{C*} = \sqrt{\sum_{i \in I} \mathcal{L}_{C_i}^2 / \alpha^i} \tag{8}$$

Note that in our notation, we refer the Lipschitz constant as *local* since it depends on the specific choice of subgradient.

Weighted Entropy Distance. With the general definitions of distance and norm, we define the *weighted entropy* distance D_Δ over the domain Δ using *entropy* d.g.f ψ_Δ^i and l_1-norm $\|.\|_1$ on individual set Δ_i:

$$\psi_\Delta^i(\rho_i) = \sum_{t \in T(i)} \rho_i^t \ln \rho_i^t \; ; \; \|\rho\|_\Delta = \sqrt{\sum_{i \in I} \alpha_\Delta^i \|\rho_i\|_1^2} \tag{9}$$

Lemma 2. *Let $\psi_\Delta : \Delta \to \mathbb{R}$ be the weighted d.g.f (6) defined with summand ψ_Δ^i. Then ψ_Δ is 1-strongly convex w.r.t the norm $\|.\|_\Delta$*

Proof. $\langle \nabla \psi_\Delta(u) - \nabla \psi_\Delta(v), u - v \rangle = \sum_{i \in I} \alpha_\Delta^i \langle \nabla \psi_\Delta^i(u_i) - \nabla \psi_\Delta^i(v_i), u_i - v_i \rangle$

$$\geq \sum_{i \in I} \alpha_\Delta^i \|u_i - v_i\|_1^2 = \|u - v\|_\Delta^2 \tag{10}$$

The inequality in (10) comes from the well known *1-strongly convex* property of entropy function ψ_Δ^i over the simplex Δ_i w.r.t l_1-norm [9]. □

Weighted Entropy Distance. Weighted Euclidean distance D_Λ is defined by summand ψ_Λ^i and l_2-norm on subset Λ_i as follow:

$$\psi_\Lambda^i(\lambda_i) = \frac{1}{2}\lambda_i^\top \lambda_i \; ; \; \|\lambda\|_\Lambda = \sqrt{\sum_{i\in I} \alpha_\Lambda^i \|\lambda_i\|_2^2} \tag{11}$$

Lemma 3. *Let* $\psi_\Lambda : \Lambda \to \mathbb{R}$ *be the weighted d.g.f* (6) *defined with summand* ψ_Λ^i. *Then* ψ_Λ *is 1-strongly convex w.r.t the norm* $\|.\|_\Lambda$.

Proof. The proof is similar to Lemma 2. □

Solution Updates: Using the weighted distance D_Δ and D_Λ above, we obtain the solutions to proximal sequence (5):

$$\rho_i^{k+1(t)} = \frac{\rho_i^{k(t)} \exp\left(F'(\rho^k)_i^t.\tau/\alpha_\Delta^i\right)}{\sum_{t\in T(i)} \rho_i^{k(t)} \exp\left(F'(\rho^k)_i^t.\tau/\alpha_\Delta^i\right)}$$
$$\lambda_i^{k+1(t)} = \frac{\eta}{\alpha_\Lambda^i}\left(F'(\lambda^k)_i^t - \frac{\sum_{t\in T} F'(\lambda^k)_i^t}{T_i}\right) \tag{12}$$

where T_i denotes the number of trees that cover the unary term i. It is straightforward to see that variable updates only happen at the nodes which are assigned different labels across the trees. In addition, the stepsize update for each unary term is affected by the weighting factor associated with that term. Through the convergence analysis below, we derive an optimal stepsize for entropy projection and adaptive stepsize for euclidean projection.

4   Convergence Analysis

The MD iterations (5) solve for ρ and λ independently. Since the two variables are disjoint, MD can either update them simultenously or sequentially. By examining the convergence analysis, we justify that updating sequentially provides a faster convergence rate, ie. MD updates ρ first, until there is no improvement in the dual, then it switchs to update λ. In addition, we define the optimal stepsize for entropy projection and an adaptive stepsize for euclidean projection based on the bounded sub-optimality.

4.1   Convergence Analysis

Lemma 4. *The proximal update* (5) *provides better sub-optimality in sequential manner than in parallel manner. It has the following worst case optimality:*

$$F^* - \max_{k=1..K}\{F_k\} \leq \frac{\sqrt{2}(\mathcal{L}_\Delta\sqrt{\Omega_\Delta} + \mathcal{L}_\Lambda\sqrt{\Omega_\Lambda})}{\sqrt{K}} \tag{13}$$

Proof. (Sketch) Assume we have a sequence of updates $\{\rho^k\}_{k=1}^{k_1}, \{\lambda^k\}_{k=k_1+1}^{k_1+k_2}$, follows the proof of Proposition 1.1 (i) as in [10] with ingredients of MD algorithm that we developed in Section 3, we can obtain the following inequality:

$$\langle F'(\rho_k), \rho^* - \rho^k \rangle \leq \frac{1}{\tau} D_\Delta(\rho^*, \rho^k) - \frac{1}{\tau} D_\Delta(\rho^*, \rho^{k+1}) + \frac{\tau \|F'(\rho^k)\|_{\Delta *}^2}{2} \quad (14)$$

$$\langle F'(\lambda_k), \lambda^* - \lambda^k \rangle \leq \frac{1}{\eta} D_\Lambda(\lambda^*, \lambda^k) - \frac{1}{\eta} D_\Lambda(\lambda^*, \lambda^{k+1}) + \frac{\eta \|F'(\lambda^k)\|_{\Lambda *}^2}{2} \quad (15)$$

Let $K = k_1 + k_2$, $\hat{F} = \max_k \{F_k\}$. Summing up (14) and (15) over K iterations:

$$K(F^* - \hat{F}) \leq \sum_{k=1}^{K}(F^* - F_k) \leq \sum_{k=1}^{k_1}\langle F'(\rho_k), \rho^* - \rho^k \rangle + \sum_{k=k_1+1}^{K} \langle F'(\lambda_k), \lambda^* - \lambda^k \rangle$$

$$\leq \frac{D_\Delta(\rho^*, \rho^1)}{\tau} + \frac{D_\Lambda(\lambda^*, \lambda^{k_1+1})}{\eta} + \frac{k_1 \tau \mathcal{L}_\Delta^2 + k_2 \eta \mathcal{L}_\Lambda^2}{2}$$

where $\mathcal{L}_\Delta$ and $\mathcal{L}_\Lambda$ are the local Lipschitz constants. Let $\Omega_\Delta = \max\limits_{\rho \in \Delta} D_\Delta$ and $\Omega_\Lambda = \max\limits_{\lambda \in \Lambda} D_\Lambda$, then:

$$F^* - \hat{F} \leq \frac{\Omega_\Delta}{K\tau} + \frac{\Omega_\Lambda}{K\eta} + \frac{1}{2}\left(\frac{k_1}{K}\tau \mathcal{L}_\Delta^2 + \frac{k_2}{K}\eta \mathcal{L}_\Lambda^2\right) \quad (16)$$

Inequality (16) gives the bounded sub-optimality when updates are done in sequential manner. Let B denotes the RHS of (16). If parallel updates are used, then $k_1 = k_2 = K$, and we have:

$$B \leq \frac{\Omega_\Delta}{K\tau} + \frac{\Omega_\Lambda}{K\eta} + \frac{1}{2}\left(\tau \mathcal{L}_\Delta^2 + \eta \mathcal{L}_\Lambda^2\right) \quad (17)$$

From inequality (17), we can justify that sequential updates provide better sub-optimal approximation than parrallel updates. Minimizing the RHS of (17) w.r.t τ and η gives:

$$\tau = \frac{\sqrt{2\Omega_\Delta}}{\mathcal{L}_\Delta\sqrt{K}} \quad ; \quad \eta = \frac{\sqrt{2\Omega_\Lambda}}{\mathcal{L}_\Lambda\sqrt{K}} \quad (18)$$

Hence, the rate of convergence is bounded by:

$$F^* - \hat{F} \leq \frac{\sqrt{2}(\mathcal{L}_\Delta\sqrt{\Omega_\Delta} + \mathcal{L}_\Lambda\sqrt{\Omega_\Lambda})}{\sqrt{K}}$$

Theorem 1. *The sequential updates generated by the MD algorithm provides the following bound on sub-optimality:*

$$F^* - \max_{k=1..K} \{F_k\} \leq \frac{\sqrt{2}\left(\sum_{i \in I} |\theta_i|\sqrt{\ln(T_i)} + |\lambda_i^*|T_i\right)}{\sqrt{K}} \quad (19)$$

Proof. We want to minimize the RHS of (13). The parameters associate with two disjoint domains can be computed independently. Consider minimizing an arbitrary term:

$$\Omega \mathcal{L}^2 = \left[\sum_{i \in I} \alpha^i \Omega^i\right]\left[\sum_{i \in I} \alpha_i^{-1} \mathcal{L}_i^2\right] \tag{20}$$

Optimising (20) w.r.t α, we obtain:

$$\alpha^i = \frac{\mathcal{L}_i}{\sqrt{\Omega^i}\left[\sum_{i \in I} \mathcal{L}_i \sqrt{\Omega^i}\right]}$$

Therefore, $\Omega = 1$ and $\mathcal{L} = \sum_{i \in I} \mathcal{L}_i \sqrt{\Omega_i}$.

The Lipschitz Constant over the Domain Δ

Each subset Δ_i is equipped with $\|.\|_1$, therefore $\mathcal{L}_i^\Delta = \sup_{\rho_i \in \Delta_i} \|F_i'(\rho)\|_\infty = |\theta_i|$ In addition, the maximum distance Ω_Δ^i over the simplex Δ_i is defined in Proposition 5.1 (c) [9]: $\Omega_\Delta^i = \ln(T_i)$. Hence, we have: $\mathcal{L}_\Delta = \sum_{i \in I} |\theta_i| \sqrt{\ln(T_i)}$

The Lipschitz Constant over the Domain Λ

Similarly, we can compute:

$$\mathcal{L}_i^\Lambda = \sup_{\lambda_i \in \Lambda_i} \|F_i'(\lambda)\|_2 = \sqrt{T_i} \ ; \ \Omega_\lambda^i = T_i(\lambda_i^*)^2 \ ; \ \mathcal{L}_\lambda = \sum_{i \in I} |\lambda_i^*| T_i$$

Note that, the amount λ_i^* and its maximum distance Ω_λ^i can only estimated due to the "unbounded" nature of the set Λ_i. Finally, we obtain the bound (19):

$$F^* - \hat{F}_{k=1..K} \leq \frac{\sqrt{2}\left(\sum_{i \in I} |\theta_i| \sqrt{\ln(T_i)} + |\lambda_i^*| T_i\right)}{\sqrt{K}}$$

Remarks. From Lemma 4, we have justified that sequential updates is better than parallel updates. Now, let us consider two other type of updates: using weighted entropy projection only and using weighted Euclidean projection only. Clearly, using weighted entropy only will get trapped into a local maxima because the set Δ does not cover the original feasible set Θ. However, in applications where the simplexes fully cover the original feasible set, we can obtain very fast convergence. If we use only weighted Euclidean projection to search within the same space defined by Δ, then in the worst-case, the sub-optimality is defined by:

$$\frac{\sqrt{2}\left(\sum_{i \in I} |\theta_i| T_i + |\lambda_i^*| T_i\right)}{\sqrt{K}}$$

It is easy to see that this bound is much larger than our optimal bound in the RHS of (19) as the size of the set I is very large.

4.2   Discussion

Switching Criteria: An intuitive idea is to derive switching criteria based on dual gap. When the MD sequence based on entropy projection finds a sub-optimal distribution in its domain, it will not improve the dual further. One can define a switching point when there is no improvement in the dual objective or dual gap. However, subgradient method often fluctuates the dual objective, thus, the dual gap appears to have zic-zagging behaviour. Therefore it is not efficient to detect switching point based on the dual gap. On the other hand, an important feature of the Dual Decomposition model is that as the method converges, the number of non-agreement nodes is decreasing. This observation works better in general since it does not fluctuate as much as the dual objective. We define a threshold $\sigma \in \{1, .., 20\}$ depends on applications for switching to Euclidean projection when the number non-agreement nodes does not decrease after σ iterations.

Implementation: The proximal updates are done in sequence where we solve for ρ until the switching criteria is met, then we solve for λ. With the ingredients developed sofar, the sequence (12) reduces to:

$$\rho^t = \frac{\theta_i^t}{\sum_{t \in T} \theta_i^t} \; ; \; \omega^t = \varepsilon . \operatorname{sign}(\theta_i^t . \bar{x}_i^t) \sqrt{2 \ln(T_i)/k} \; ; \; \rho^t = \frac{\rho^t \exp(\omega^t)}{\sum_{t \in T(i)} \rho^t \exp(\omega^t)}$$

$$\theta_i^t = \rho^t \left(\sum_{t \in T} \theta_i^t \right) \tag{21}$$

$$\theta_i^t = \theta_i^t + \sqrt{2 . \frac{\Omega_i}{k}} \left(\bar{x}_i^t - \frac{\sum_{t \in T} \bar{x}_i^t}{T_i} \right) \tag{22}$$

where $\varepsilon \in (0, 2)$ is a speed up parameter. The entropy projection updates the master's parameter by (21). Equation (22) is used for Euclidean projection. Since we can compute ρ based on the current value of unary terms, the memory required is not more than any other type of dual decomposition methods.

Adaptive Stepsize: The stepsize for entropy projection is optimal and can be computed analytically since we know the maximum distance of the simplex. However, we do not know the maximum distance Ω_i on the unbounded set Λ_i, therefore we estimate it by:

$$\sqrt{\frac{\Omega_i}{k}} \approx \frac{|\hat{E} - \hat{F}|}{T_i . L_k} \tag{23}$$

where $\hat{E}$ is the best primal solution after k iterations. At iteration k, there is L_k number of non-agreement nodes that need to be adjusted to reduce the dual gap $|\hat{E} - \hat{F}|$. The difference between primal and dual is distributed evenly for L_k nodes. In addition, for each node, this amount is dispensed evenly amongst the trees that cover the node.

5   Experiments

In order to demonstrate the effectiveness of our method, we present experimental results with synthetic data and segmentation problem with the UGM Matlab package [6]. In addition, we also examine our method with the Tsukuba

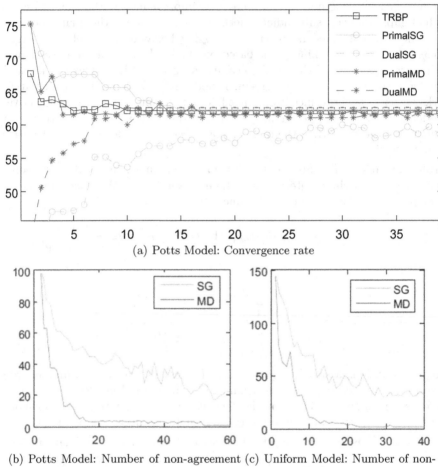

(a) Potts Model: Convergence rate

(b) Potts Model: Number of non-agreement (c) Uniform Model: Number of non-
Nodes   agreement Nodes

Fig. 1. Synthetic data

Stereo problem in MRF-Bechmark package [14]. In all experiments, we apply three methods: Tree-reweighted variants (TRBP in UGM and TRW-S in
MRF-benchmarks), Mirror Descent and Sub Gradient with adaptive stepsize
$\alpha = \frac{|\hat{E} - \hat{F}|}{\|F'_k\|^2}$ as suggested in [4].

Synthetic Data: For our synthetic experiments, we used a grid graph of size
20×20 and 5 labels. For the Potts model, $\theta_{a,i}$ was drawn from $\mathcal{U}(-1, +1)$, while
$\theta_{ab,ij} = \omega_{ab} * \mathbb{I}(i = j)$ and $\omega_{ab} = \mathcal{N}(0,1)$. For the Uniform model, we withdraw
all data from $\mathcal{U}(0,1)$, for the edge weight, we also use $\omega_{ab} = \mathcal{N}(0,1)$. For these
small tests, we set the switching threshold to 5.

Figures 1(a) and 1(b) shows the convergence of primal-dual gap and number of labels to fix for the Potts model. The switch between methods occur between iterations 20 and 25. All methods converge eventually, however our method

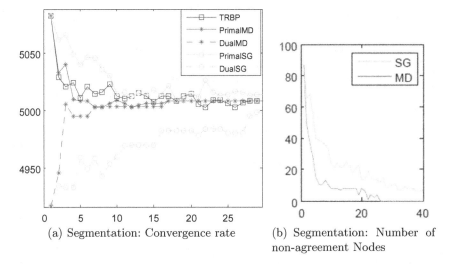

(a) Segmentation: Convergence rate

(b) Segmentation: Number of non-agreement Nodes

Fig. 2. Segmentation data

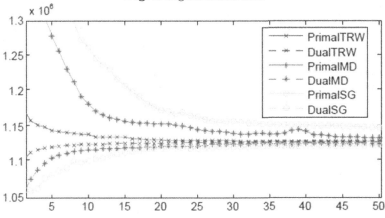

Fig. 3. Stereo: Convergence

outperforms the sub-gradient method significantly and obtains the optimal solution slightly before TRW. In the Uniform model (Figure 1(c)), the switch even not happen, MD can compute optimal labelling by entropy projection only.

Segmentation Probem: The segmentation problem is to recover a coloured X from its noisy image in the UGM package. Figures 2(a) and 2(b) show how the methods perform. Note the switching step happens between 15 and 20 iterations. After the switch to the Euclidean projection, with our adaptive stepsize, MD can recover the optimal solution at around iteration 25.

The Stereo Problem: Figure 3 shows the convergence rate for Tsukuba problem with three methods. We can see that TRW still converges fastest, with the MD method comes second. Both TRW and MD generate similar dual objective sequence after 30 iterations.

6   Conclusion

An efficient algorithm to solve the dual MRF minimization problem is presented. The method is based on Mirror Descent algorithm with weighted distance projections, weighted norms and local Lipschitz constants. After a careful analysis of the algorithm, we are able to sharpen the theoretical convergence rate as well as to improve the performance of the algorithm in practice. Mirror Descent can be applied efficiently on any bounded set, a direction of future research is to establish the relationship between the dual gap and feasible sets, and address the possibility of performing entropy projection on the unbounded set Θ.

References

1. Boykov, Y., Veksler, O., Zabih, R.: Fast approximate energy minimization via graph cuts. IEEE Trans. Pattern Anal. Mach. Intell. 23, 1222–1239 (2001)
2. Yedidia, J.S., Freeman, W.T., Weiss, Y.: Understanding belief propagation and its generalizations. In: Exploring Artificial Intelligence in the New Millennium, pp. 239–269 (2003)
3. Wainwright, M.J., Jaakkola, T.S., Willsky, A.S.: Map estimation via agreement on (hyper)trees: Message-passing and linear-programming approaches. IEEE Trans. on Information Theory 51, 3697–3717 (2005)
4. Komodakis, N., Paragios, N., Tziritas, G.: Mrf energy minimization and beyond via dual decomposition. IEEE Trans. Pattern Anal. Mach. Intell. 33, 531–552 (2011)
5. Savchynskyy, B., Schmidt, S., Kappes, J., Schnorr, C.: A study of nesterov's scheme for lagrangian decomposition and map labeling. In: Computer Vision and Pattern Recognition, pp. 1817–1823 (2011)
6. Schmidt, M.: Ugm: Matlab code for undirected graphical models (2011)
7. Jancsary, J., Matz, G.: Convergent decomposition solvers for tree-reweighted free energies. Journal of Machine Learning Research
8. Ben-tal, A., Margalit, T., Nemirovski, A.: The ordered subsets mirror descent optimization method with applications to tomography. SIAM Journal on Optimization 12 (2001)
9. Beck, A., Teboulle, M.: Mirror descent and nonlinear projected subgradient methods for convex optimization. Operations Research Letters (2003)
10. Juditsky, A., Nemirovski, A.: First order methods for nonsmooth convex large-scale optimization, i: General purpose methods. In: Optimization for Machine Learning. MIT Press (2010)
11. Jojic, V., Gould, S., Koller, D.: Accelerated dual decomposition for map inference. In: International Conference of Machine Learning, pp. 503–510 (2010)
12. Censor, Y., Zenios, S.A.: Proximal minimization algorithm with d-functions. Journal of Optimization Theory and Applications 73, 451–464 (1992)
13. Boyd, S., Vandenberghe, L.: Convex Optimization. Cambridge University Press (2004)
14. Szeliski, R., Zabih, R., Scharstein, D., Veksler, O., Kolmogorov, V., Agarwala, A., Tappen, M., Rother, C.: A comparative study of energy minimization methods for mrfs. IEEE Trans. Pattern Anal. Mach. Intell., 1068–1080 (2008)

Similarity Registration for Shapes Based on Signed Distance Functions

Sasan Mahmoodi[1], Muayed S. Al-Huseiny[2], and Mark S. Nixon[1]

[1] School of Electronic and Computer Science, University of Southampton, UK
[2] Computer and Software Engineering Department, University of Mustansiriyah, Iraq

Abstract. A fast algorithm for similarity registration for shapes with various topologies is put forward in this paper. Fourier transform and Geometric moments are explored here to calculate the rotation, scaling and translation parameters to register two shapes by minimizing a dissimilarity measure introduced in the literature. Shapes are represented by signed distance functions. In comparison with the algorithms in the literature, the algorithm proposed here demonstrates superior performance for the registration of two shapes with various topologies as well as two shapes, each containing various and different numbers of shape components. The registration process using this algorithm is robust in comparison with the shape registration algorithms in the literature and is as fast as a couple of FFTs.

1 Introduction

Shape registration may be regarded as the result of a point-wise transformation between a reference and an observed shape [2]-[3]. Registration algorithms established on matching contour points (contour-based) are cited in the literature due to their fast convergence (e.g. see [5], [6], and [7]). The very fact though that such algorithms rely on contour points only (require point correspondence) makes these methods vulnerable to topological changes in shapes. More recent work tends towards using SDF (SDF-based) (e.g. see [1], [4], [8] and [9]). These SDF-based algorithms usually minimize the distance between the SDFs iteratively for instance by using a gradient descent algorithm. The SDF-based shape registration methods (for example the seminal work in [1]) are in general capable of dealing with shapes that have various Euler characteristic numbers (various topologies). However, as the complexity of shapes increases, the possibility of convergence to local minima becomes more likely. These methods are somewhat slow and sometimes lack stability due to their iterative nature. Also, the implementation is slightly difficult to manage because there is no unique method for a universal stopping criterion that is applicable for every case (see [4] for more details). The algorithm proposed in this paper however is fast, reliable, robust to local minima problem and can perform registration successfully between shapes with various topologies. The paper is organized as follows: The registration problem is stated in section 2. The proposed algorithm is presented in section 3. Important considerations noted for implementation are discussed in section 4. The algorithm is evaluated in section 5, and finally the paper concludes in section 6.

G. Bebis et al. (Eds.): ISVC 2012, Part I, LNCS 7431, pp. 599–609, 2012.
© Springer-Verlag Berlin Heidelberg 2012

2   The Statement of the Problem

The signed distance function (SDF) of a shape p is defined as:

$$\phi_p(x,y) = \begin{cases} D_E((x,y),B), & (x,y) \in I_p, \\ -D_E((x,y),B), & (x,y) \in \Omega - I_p, \end{cases} \tag{1}$$

where Ω is the bounded domain, D_E stands for the minimum Euclidean distance between the perimeter B of the shape p and any point in domain Ω, and I_p is the subset of Ω representing the interior of the shape [1]. Registration between two shapes aims to retrieve transform parameters s, θ, T_x and T_y (scaling, rotation, and translations along x and y axes respectively) minimizing a dissimilarity measure between ϕ_p and ϕ_q introduced in [1] and given in (2),

$$E = \iint_\Omega \left| \phi_p(x,y) - \frac{1}{s}\phi_q\left(sR_\theta\left(x+T_x, y+T_y\right)\right) \right|^2 dxdy \tag{2}$$

such that,

$$\left(\hat{\theta}, \hat{s}, \hat{T}_x, \hat{T}_y\right) = \underset{\theta, s\,T_x, T_y}{\arg \min}\, E \tag{3}$$

where Ω, $\hat{\theta}, \hat{s}, \hat{T}_x, \hat{T}_y$ and R_θ are image domain, the estimated angle, scale, translation parameters, and a conventional rotation (transform) matrix respectively,

$$R_\theta = \begin{bmatrix} \cos\theta & -\sin\theta \\ \sin\theta & \cos\theta \end{bmatrix}.$$

3   Registration Method

The minimization of dissimilarity measure (2) with respect to the desired parameters can be directly and iteratively implemented as demonstrated in [1]. Such an implementation is slow, unreliable (may fall into local minima), and difficult to tune [4]-[11]. The algorithm presented here on the other hand, suggests linear methods to estimate the registration parameters minimizing the dissimilarity measure in (2).

3.1 Rotation

For simplicity, let the objective function in (2) be a function of only θ:

$$E_\theta = \iint_\Omega \left| \phi_p(x,y) - \phi_q\left(R_\theta(x,y)\right) \right|^2 dxdy \tag{4}$$

Rotation in Cartesian coordinates is equivalent to displacement of the angular component in polar coordinates [10].

Shapes p and q are initially centered at the origin of the coordinate system. Centralized shapes are then mapped to polar coordinates i.e., $\hat{\phi}_p(\rho,\omega)$ and $\hat{\phi}_q(\rho,\omega)$ such that $x = \rho\cos\omega$ and $y = \rho\sin\omega$. In theorem 1, we prove that the rotation angle minimizing term (5) minimizes term (4):

$$E_\theta = \int_0^{+\infty}\int_0^{2\pi}\left|\hat{\phi}_p(\rho,\omega) - \hat{\phi}_q((\rho,\omega+\theta))\right|^2 d\rho\, d\omega, \tag{5}$$

Theorem 1: *The rotation angle minimizing term (5) is the minimizer of term (4) where* $\Omega = R^2$.

The proof is presented in the appendix. We use the result obtained from theorem 1 to propose an algorithm to estimate the desired rotation angle θ. To this end, let us denote $\bar{\phi}_p$ a normalized instance of $\hat{\phi}_p$:

$$\bar{\phi}_p(\rho,\omega) = \frac{\hat{\phi}_p(\rho,\omega)}{\sqrt{\int_\rho\int_\omega\left|\hat{\phi}_p(\rho,\omega)\right|^2 d\rho\, d\omega}}. \tag{6}$$

Also, let β be the scale factor (β is a function of the desirable rotation angle (θ) between $\bar{\phi}_p$ and $\hat{\phi}_q$, i.e., $\beta(\theta) = \langle\hat{\phi}_q,\bar{\phi}_p\rangle = \int_\rho\int_\omega\left(\hat{\phi}_q(\rho,\omega)\bar{\phi}_p(\rho,\omega+\theta)\right)d\rho\, d\omega$.

The desirable rotation angle is estimated by minimizing the dissimilarity term E_θ between $\hat{\phi}_p$ and $\hat{\phi}_q$ in (5):

$$E_\theta = \int_\rho\int_\omega\left|\hat{\phi}_q - \bar{\phi}_p\right|^2 d\rho\, d\omega, \tag{7}$$

$$= \int_\rho\int_\omega\left(\left|\hat{\phi}_q\right|^2 - 2\hat{\phi}_q\bar{\phi}_p + \left|\bar{\phi}_p\right|^2\right)d\rho\, d\omega.$$

From (6), $\int_\rho\int_\omega\left|\bar{\phi}_p\right|^2 d\rho\, d\omega = 1$; also, as defined above, $\langle\hat{\phi}_q,\bar{\phi}_p\rangle = \beta$, therefore:

$$E_\theta = \int_\rho\int_\omega\left|\hat{\phi}_q\right|^2 d\rho\, d\omega - 2\beta(\theta). \tag{8}$$

Since the first two terms in (8) are independent of θ, the minimization of E_θ is equivalent to the maximization of β, i.e., the optimal rotation $\hat{\theta}$ is estimated by maximizing β.

The Fourier transform is employed here to compute the local maxima of β. Let the Fourier transform of $\hat{\phi}_q$ and $\bar{\phi}_p$ be $\hat{\psi}_q(\xi_1,\xi_2)$ and $\bar{\psi}_p(\xi_1,\xi_2)$ respectively, such that,

$$\hat{\psi}_q(\xi_1,\xi_2) = \int_\rho\int_\omega\hat{\phi}_q(\rho,\omega)e^{-i(\rho\xi_1+\omega\xi_2)2\pi}d\rho\, d\omega. \tag{9}$$

$$\bar{\psi}_p(\xi_1,\xi_2) = \int_\rho\int_\omega\bar{\phi}_p(\rho,\omega)e^{-i(\rho\xi_1+\omega\xi_2)2\pi}d\rho\, d\omega, \tag{10}$$

According to Parseval's theorem, expression (11) holds,

$$\beta(\theta) = \int_\rho \int_\omega \left(\hat{\phi}_q(\rho,\omega)\,\bar{\phi}_p(\rho,\omega+\theta)\right)d\rho\,d\omega$$

$$= \int_{\xi_1}\int_{\xi_2}\left(\hat{\psi}_q(\xi_1,\xi_2)\,\bar{\Psi}_p{}^*(\xi_1,\xi_2)e^{2\pi i\xi_2\theta}\right)d\xi_1\,d\xi_2, \tag{11}$$

where $(*)$ denotes the complex conjugate.

Hence, from (11), $\hat{\theta}$ is estimated as in (12),

$$\hat{\theta} = \underset{\theta}{\arg\max}\,\beta = \underset{\theta}{\arg\max}\,\int_{\xi_1}\int_{\xi_2}\left(\hat{\psi}_q\,\bar{\Psi}_p{}^*\,e^{i\theta}\right)d\xi_1\,d\xi_2. \tag{12}$$

3.2 Scale

Dissimilarity measure (2) is minimized with respect to s to compute the relative scale between the two given shapes. This measure is expressed in (13) in terms of s for the sake of simplicity:

$$E_s = \iint\left|\phi_p(x,y) - \frac{1}{s}\phi_q(s(x,y))\right|^2 dx\,dy, \tag{13}$$

where the relation between two shapes' SDFs which have different scales is known to be:

$$s\hat{\phi}_p(x,y) = \hat{\phi}_q(sx,sy), \tag{14}$$

where $\hat{\phi}_p$ and $\hat{\phi}_q$ are the centralized versions of ϕ_p and ϕ_q. In theorem 2, we prove that the scaling parameter s minimizing the following term is a minimizer of term (13):

$$E_s' = \left|\sum_{m=0}^{M}\sum_{n=0}^{N}\left(M_{mn}^p - \frac{M_{mn}^q}{s^{m+n+3}}\right)\right|^2 \tag{15}$$

where M_{mn}^p and M_{mn}^q are respectively the $(m+n)^{\text{th}}$ order geometrical moments of $\hat{\phi}_p(x,y)$ and $\hat{\phi}_q(x,y)$ defined as:

$$M_{mn}^p = \iint_\Lambda x^m y^n \hat{\phi}_p(x,y)dx\,dy$$

$$M_{mn}^q = \iint_\Lambda x^m y^n \hat{\phi}_q(x,y)dx\,dy$$

Theorem 2: *The scaling parameter minimizing term (13) is also the minimizer of term (15).*

The proof is presented in the appendix. This distance term given in (15) is not linear with respect to s. By using a change of variable, (15) is linearized with respect to $\hat{s} = \log s$,

$$\hat{E}_s = \left| \sum_{m=0}^{M} \sum_{n=0}^{N} \log\left(\frac{M_{m,n}^q}{M_{m,n}^p} \right) - \hat{s}\,(m+n+3) \right|^2, \tag{16}$$

Therefore, s minimizing E'_s, is computed as:

$$s = \exp\left(\frac{\sum_m \sum_n \log\left(\frac{M_{mn}^q}{M_{mn}^p} \right)}{\sum_m \sum_n (m+n+3)} \right). \tag{17}$$

3.3 Translation

The scale and rotation information computed previously in sections 3.1 and 3.2 are used to fix the scaling and rotation discrepancies between shapes. For translation parameters only, term (2) thus reduces to Eq.(18),

$$E_{T_x,T_y} = \iint \left| \phi_p(x,y) - \phi_q\big((x+T_x, y+T_y)\big) \right|^2 dx\,dy, \tag{18}$$

A similar approach to that employed in 3.1 is used here to estimate translation parameters. Let $\bar{\phi}_p$ denote a normalized version of ϕ_p, that is:

$$\bar{\phi}_p(x,y) = \frac{\phi_p(x,y)}{\sqrt{\int_{x,y} \left| \phi_p(x,y) \right|^2 dx\,dy}}. \tag{19}$$

By using a similar argument as the one employed in section 3.1, the translation parameters are calculated as:

$$\left[\hat{T}_x \ \hat{T}_y \right] = \operatorname*{argmax}_{T_x, T_y} \int_{\lambda_x} \int_{\lambda_y} \left(\psi_q(\lambda_x, \lambda_y)\, \bar{\psi}_p^*(\lambda_x, \lambda_y)\, e^{2\pi i (T_x \lambda_x + T_y \lambda_y)} \right) d\lambda_x\, d\lambda_y, \tag{20}$$

where $\hat{T}_x$, $\hat{T}_y$ $\psi_q(\lambda_x, \lambda_y)$, $\bar{\psi}_p(\lambda_x, \lambda_y)$, λ_x, λ_y and $*$ represent the estimated optimal translation parameters, 2D Fourier transform of ϕ_q and $\bar{\phi}_p$, spatial frequencies and complex conjugate respectively.

4   Implementation Issues

W notice that the theorems proved in this paper, indicate that the transformation parameters can be calculated by using a linear method. It is therefore important to note that this linearity is not an assumption and therefore it is applicable to general cases. The employment of continuous Fourier transform in the computation of the rotation and translation is not numerically tractable. Fast Fourier transform (FFT) is employed instead in this paper. Accordingly, the definition of SDFs is modified to accommodate the periodicity property associated with FFT: Let Ω be the shape domain. This domain is partitioned by the shape perimeter into two regions, the shape interior I and the background, and let $\phi : \Omega \rightarrow \Re^+$ be a Lipschitz function as defined in (21):

$$\phi(x, y) = \begin{cases} D_E((x, y), B), & (x, y) \in I, \\ 0, & (x, y) \in \Omega - I, \end{cases} \tag{21}$$

where D_E, as mentioned earlier, stands for the minimum Euclidean distance between shape perimeter B and any point inside the shape.

　　Registration parameters are computed in the following order: rotation, scaling and finally translation. Observed shapes are fixed accordingly after computing each parameter. In the case of scale it is important to remove the variance in shapes due to differences in translation (by centralizing the shapes) and rotation since these variances affect the computed moments. A Matlab® (version 7) implementation code of the algorithm proposed in this paper is available: at http://users.ecs.soton.ac.uk/sm3/SDFShapeRegistration.zip. The practical limitation of this approach is that it may fall into local minima, because dissimilarity measure (2) is also associated with the local minima. The last but not the least issue is that the order in which the registration parameters are calculated does not change the final result.

5　　Results

The proposed algorithm is evaluated by using a set of problems to address issues common to shapes. In the subsequent examples the moments up to the tenth order (up to $m = n = 5$) are used to compute the scale parameter s presented in section 3.2. Indeed, the higher the number of moments is, the more accurately s is estimated and the more expensive the algorithm becomes.

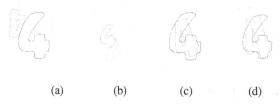

(a)　　　　(b)　　　　(c)　　　　(d)

Fig. 1. Registration of shapes with different topologies (size=300 X 300). (a) initial shapes, (b) registration by using the contour-based method in [6], (c) registration by using the SDF-based method in [1] (d) registration by using the approach proposed here.

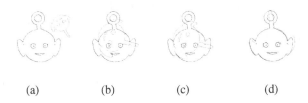

(a)   (b)   (c)   (d)

Fig. 2. Registration of shapes with different topologies (size = 300 X 300). (a) initial shapes, (b) registration by using the contour-based method in [6], (c) registration by using the SDF-based method in [1] (d) registration by using the approach proposed here.

For the sake of comparison, the results of the proposed technique are compared to those utilized by two well-established shape registration algorithms. The first uses contours for representing shapes, (e.g. see [6]) referred here as the contour-based method. The other algorithm used here for comparison is based on SDFs (see [1] for more details) and referred here as the SDF-based method. For better visualization, the boundaries of the shapes rather than the SDFs are used in all figures in this paper. Contours of the reference shapes are presented in red, and those of the observed shapes are shown in green. The first experiment investigates the impact of topological difference on the registration outcomes. In Figure (1-a), the reference shape is an open number 4 with Euler number unity. The observed shape is a closed number 4 with Euler number zero. The registration by using the contour-based method in [6] shown in Figure (1-b) is not satisfactory; the two shapes are completely misaligned here. A similar example is shown in Figure (2-a), the shapes of the two characters have different topologies. In this example, we refer to the results of applying the SDF-based registration algorithm in [1] to this problem. Figure (2-c) shows a typical case of the local minima issue usually attributed to the iterative procedure of the work presented in [1]. Figures (1-d) and (2-d) on the other hand demonstrate that in both cases the shapes are aligned optimally by using the approach proposed here. This robustness can be due to the fact that the algorithm proposed here employs regions (SDF) rather than boundaries to find the optimal values for registration.

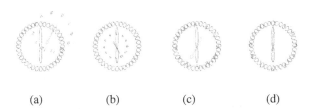

(a)   (b)   (c)   (d)

Fig. 3. Registration of clocks with different number of components (size = 300 X 300), (a) an initial state, and (b) registration of (a) by using contour-based technique in [6], (c) registration of (a) by using the SDF-based algorithm in [1], and (d) registration of (a) by using the algorithm proposed here

The next experiment concerns the registration of two complex shapes containing multiple simple shapes (components). The contour-based method in [6] requires point correspondence of the contours. Since the number of components in the two shapes in question (see Figure (3-a)) is different for each shape, the parts with no corresponding counterparts remain untouched. For the sake of comparison, these shapes are treated here as a single entity and registered directly by using the method proposed by the authors in [6]. Figure (3-b) depicts the meaningless alignment by using the contour-based algorithm caused primarily by the lack of contour points' correspondence. The algorithm in [1] on the other hand due to the use of SDFs is capable of handling complex shapes. However, the increased complexity increases the tendency towards stopping at local minima (see figure (3-c)). The results shown by the example in Figure (4) are also consistent with the previous analysis. In either case the algorithm proposed here possesses the capacity to register the shapes quickly and accurately as depicted in Figures (3-d), and (4-d). Our algorithm proposed here is as fast as a couple of FFT operations; however the iterative SDF based algorithm may take a long time to converge, depending on the complexity and size of the shapes.

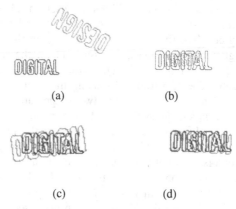

(a)   (b)

(c)   (d)

Fig. 4. Registration of words with various number of letters (size = 300 X 300), (a) original shapes (b) registration by using contour-based algorithm in [6], (c) registration by using SDF-based algorithm in [1]. (d) registration by using algorithm presented here.

6   Conclusion

This paper presents a fast reliable shape registration algorithm. It employs implicit representation of shapes by using signed distance functions (SDFs). The desirable parameters are calculated by minimizing a dissimilarity measure between two shapes. To achieve that, linear orthogonal transformations are employed to minimize the dissimilarity measures. This technique is based on spectral phase correlation and geometric moments to compute registration parameters individually. A modified signed distance function definition is adopted here to satisfy the requirements of the optimization strategy used in this context. The proposed approach is tested satisfactorily on problems such as complex shapes and shapes with various topologies. These shapes prove difficult to register by using contour based methods. This work is also examined in comparison with a SDF based iterative registration algorithm in the literature. The evaluation data demonstrate the reliability, stability and speed of convergence of the proposed registration algorithm. The registration technique presented in this paper is as fast as a couple of FFTs. For the future work, this algorithm can be developed to account for 3D shape registration problems

7   Appendix

Proof of Theorem 1:
Dissimilarity measure (4) can be written in polar coordinate system, i.e.:

$$E_\theta = \iint_{R^2} |\phi_p(x,y) - \phi_q(R_\theta(x,y))|^2 dxdy = \int_0^{+\infty}\int_0^{2\pi} |\phi_p(\rho,\omega) - \phi_q(R_\theta(\rho,\omega))|^2 \rho\,d\rho\,d\omega \tag{A-1}$$

where ρ and ω are polar coordinates so that $x = \rho\cos\omega$, and $y = \rho\sin\omega$.

In a polar coordinate system, term (A-1) can be written as:

$$E_\theta = \lim_{L\to+\infty} \int_0^L\int_0^{2\pi} |\phi_p(\rho,\omega) - \phi_q(\rho,\theta+\omega)|^2 \rho\,d\rho\,d\omega < \lim_{L\to+\infty} L\int_0^L\int_0^{2\pi} |\phi_p(\rho,\omega) - \phi_q(\rho,\theta+\omega)|^2 d\rho\,d\omega \tag{A-2}$$

It is easy to see from (A-1) and (A-2) that a parameter $\hat{\theta}$ minimizing

$$\int_0^{+\infty}\int_0^{2\pi} |\phi_p(\rho,\omega) - \phi_q(\rho,\theta+\omega)|^2 d\rho\,d\omega \quad \text{is} \quad \text{a} \quad \text{minimizer} \quad \text{of} \quad \text{the} \quad \text{term}$$

$$\iint_{R^2} |\phi_p(x,y) - \phi_q(R_\theta(x,y))|^2 dxdy \qquad \blacksquare$$

Proof of Theorem 2:
Before proving theorem 2, we need to visit theorem 3:

Theorem 3: *Let a geometrical moment with orders m and n of signed distance function (SDF)* $\phi_q(x,y):\Omega \to R$ *be* M_{mn}^q. *The geometrical moment with orders m and n of the scaled SDF with scaling parameter s is* $\dfrac{M_{mn}^q}{s^{m+n+3}}$.

Proof:

$$M_{mn}^q = \iint_\Omega x^m y^n \phi_q(x,y)dxdy \tag{A-3}$$

The mth and n^{th} order moment of the scaled SDF $\dfrac{1}{s}\phi_q(sx,sy)$ is therefore calculated as:

$$M_{mn}^{'q} = \iint_\Omega x^m y^n \left(\frac{1}{s}\phi_q(sx,sy)\right)dxdy \tag{A-4}$$

By changing the variables $X = sx$ and $Y = sy$, equation (A-4) is rewritten as:

$$M_{mn}'^q = \iint_\Omega \left(\frac{X}{s}\right)^m \left(\frac{Y}{s}\right)^n \left(\frac{1}{s}\phi_q(X,Y)\right)\frac{dXdY}{s^2} = \frac{1}{s^{m+n+3}}\iint_\Omega X^m Y^n \phi_q(X,Y)dXdY = \frac{M_{mn}^q}{s^{m+n+3}} \qquad (A-5)$$

Proof of theorem 2: The signed distance functions $\phi_p(x,y)$ and $\phi_q(x,y)$ can be approximated in terms of their geometrical moments, i.e.:

$$\phi_i(x,y) \approx \sum_{m=0}^{M}\sum_{n=0}^{N} M_{mn}^i x^m y^n \qquad (A-6)$$

where $M_{mn}^i = \iint_\Omega x^m y^n \phi_i(x,y)dxdy$ and i can be p or q. M and N are the total number of geometrical moments around the axes x and y. If equations (14) are substituted in equation (13), by using the result of theorem 3, one can obtain:

$$E_s \approx \iint_\Omega \left|\sum_{m=0}^{M}\sum_{n=0}^{N} M_{mn}^p x^m y^n - \sum_{m=0}^{M}\sum_{n=0}^{N}\frac{M_{mn}^q}{s^{m+n+3}}x^m y^n\right|^2 dxdy = \iint_\Omega \left|\sum_{m=0}^{M}\sum_{n=0}^{N}\left(M_{mn}^p - \frac{M_{mn}^q}{s^{m+n+3}}\right)x^m y^n\right|^2 dxdy \qquad (A-7)$$

By letting a_{mn} denote $\left(M_{mn}^p - \frac{M_{mn}^q}{s^{m+n+3}}\right)$, equation (A-7) can be written as:

$$E_s = \iint_\Omega \left|\sum_{m=0}^{M}\sum_{n=0}^{N} a_{mn}x^m y^n\right|^2 dxdy = \iint_\Omega \left(\sum_{m=0}^{M}\sum_{n=0}^{N} a_{mn}^2 x^{2m} y^{2n} + 2\sum_{m=0}^{M}\sum_{n=0}^{N}\sum_{\substack{m'=0\\m'\neq m}}^{M}\sum_{\substack{n'=0\\n'\neq n}}^{N} a_{mn}a_{m'n'}x^{m+m'}y^{n+n'}\right)dxdy \qquad (A-8)$$

Without loss of generality, for $\Omega = [0,L]\times[0,H]$, E_s in (A-8) can be calculated as:

$$E_s = \left(\sum_{m=0}^{M}\sum_{n=0}^{N}\left(a_{mn}^2 \frac{L^{2m+1}}{2m+1}\frac{H^{2n+1}}{2n+1}\right) + 2\sum_{m=0}^{M}\sum_{n=0}^{N}\sum_{\substack{m'=0\\m'\neq m}}^{M}\sum_{\substack{n'=0\\n'\neq n}}^{N}\left(a_{mn}a_{m'n'}\frac{L^{m+m'+1}}{m+m'+1}\frac{H^{n+n'+1}}{n+n'+1}\right)\right) \qquad (A-9)$$

It is easy to conclude from (A-9) that

$$E_s < L^{2M+1}H^{2N+1}\left(\sum_{m=0}^{M}\sum_{n=0}^{N}\left(a_{mn}^2\right) + 2\sum_{m=0}^{M}\sum_{n=0}^{N}\sum_{\substack{m'=0\\m'\neq m}}^{M}\sum_{\substack{n'=0\\n'\neq n}}^{N}\left(a_{mn}a_{m'n'}\right)\right) = L^{2M+1}H^{2N+1}\left|\sum_{m=0}^{M}\sum_{n=0}^{N} a_{mn}\right|^2 \qquad (A-10)$$

By recalling that the term $a_{mn} = \left(M_{mn}^p - \frac{M_{mn}^q}{s^{m+n+3}}\right)$ is the only term which is a function of the scaling parameter s and from inequality (A-10), it is straightforward to see that the scaling parameter s minimizing $\left|\sum_{m=0}^{M}\sum_{n=0}^{N}\left(M_{mn}^p - \frac{M_{mn}^q}{s^{m+n+3}}\right)\right|^2$ is a minimizer of E_s given in equation (13).

References

1. Paragios, N., Rousson, M., Ramesh, V.: Non-rigid registration using distance functions. Computer Vision and Image Understanding 89(2-3), 142–165 (2003)
2. Brown, L.G.: A survey of image registration techniques. ACM Comput. Surv. 24(4), 325–376 (1992)
3. Zitova, B.: Image registration methods: a survey. Image and Vision Computing 21(11), 977–1000 (2003)
4. Cremers, D., Osher, S., Soatto, S.: Kernel Density Estimation and Intrinsic Alignment for Shape Priors in Level Set Segmentation. International Journal of Computer Vision 69(3), 335–351 (2006)
5. Marques, J.S., Abrantes, A.J.: Shape alignment - Optimal initial point and pose estimation. Pattern Recognition Letters 18(1), 49–53 (1997)
6. Markovsky, I., Mahmoodi, S.: Least-Squares Contour Alignment. IEEE Signal Processing Letters 16(1), 41–44 (2009)
7. Hui, L., Manjunath, B.S., Mitra, S.K.: A contour-based approach to multisensor image registration. IEEE Transactions on Image Processing 4(3), 320–334 (1995)
8. Vemuri, B.C.: Image registration via level-set motion: applications to atlas-based segmentation. Medical Image Analysis 7(1), 1–20 (2003)
9. El Munim, H.A., Farag, A.A.: Shape Representation and Registration using Vector Distance Functions. In: IEEE Conference in Pattern Recognition (2007)
10. Casasent, D., Psaltis, D.: Position, rotation, and scale invariant optical correlation. Applied Optics 15(7), 1795–1799 (1976)
11. Al-Huseiny, M.S., Mahmoodi, S., Nixon, M.S.: Robust Rigid Shape Registration Method Using a Level Set Formulation. In: Bebis, G., Boyle, R., Parvin, B., Koracin, D., Chung, R., Hammound, R., Hussain, M., Kar-Han, T., Crawfis, R., Thalmann, D., Kao, D., Avila, L. (eds.) ISVC 2010. LNCS, vol. 6454, pp. 252–261. Springer, Heidelberg (2010)

Protrusion Fields for 3D Model Search and Retrieval Based on Range Image Queries

Konstantinos Moustakas[1,2], G. Stavropoulos[2], and Dimitrios Tzovaras[2]

[1] Electrical and Computer Engineering Department, University of Patras, Greece
[2] Informatics and Telematics Institute,
Centre for Research and Technology Hellas, Greece

Abstract. This paper presents a novel framework for 3D object search and retrieval based on a query-by-range-image approach. Initially, salient features are extracted for both the query range image and the 3D target model that is followed by the estimation of the protrusion field generated by the extracted salient points of the 3D objects. Then, based on the concept that for a 3D object and a corresponding query range image, there should be a virtual camera with such intrinsic and extrinsic parameters that would generate an optimum range image, in terms of minimizing an error function that takes into account the protrusion field of the objects, when compared to other parameter sets or other target 3D models, matching is performed via estimating dissimilarity within the protrusion field. Experimental results illustrate the efficiency of the proposed approach even in the presence of noise or occlusion.

1 Introduction

3D model recognition and matching is a very challenging research area that has been extensively addressed during the last decades. It has numerous application areas, including CAD, autonomous navigation, robotics, etc. Especially the problem of full 3D objects search and retrieval has been successfully addressed by many researchers in the past, while excellent, extensive surveys can be found in [1], [2].

1.1 Related Work

Focusing on the problem of recognizing a 3D object when only a part of its shape is available as query few approaches have been presented in the past. Reeb Graphs are topological and skeletal structures that are used as a search key that represents the features of a 3D shape. In [3] Reeb graphs that are obtained by using different quotient functions are obtained and highlight how their choice determines the final matching result. Other commonly used methods for 3D matching, that also support partial matching use *Local features* as described in [4], [5]. Moreover, in [6] the light field descriptor is presented that is based on the concept that if two objects correspond, then they should also correspond from every viewpoint. However, these approaches do not deal with partial matching.

G. Bebis et al. (Eds.): ISVC 2012, Part I, LNCS 7431, pp. 610–619, 2012.
© Springer-Verlag Berlin Heidelberg 2012

Germann et al. [4] initially precalculate a number of range images from different points of view. Shum et al. [7] map the surface curvature of 3D objects to the unit sphere with the use of a spherical coordinate system. By searching over a spherical rotation space, a distance between two curvature distributions is computed and used as a measure for the similarity of two objects. Range images [8], soft shape descriptors [9] and salient features [10], [11] are also used to provide compact representations of the 3D models, while in [12] point pattern matching and partial matching through structural descriptors is addressed respectively.

1.2   Motivation

The aforementioned approaches have a few drawbacks. In many cases, they use a priori information for registering the partial view with the complete 3D model, while they do not utilize the saliency information of the 3D objects. In the proposed method, protrusion fields are generated based on the saliency map of 3D models. The protrusion fields provide a stochastic saliency-based representation of the 3D models that is seen experimentally to efficiently compensate for errors generated by noise and occlusion in the query range image that are realistic application scenarios. The prosed scheme is experimentally seen to be robust to noise and to outperform current state-of-the-art approaches.

The rest of the paper is organized as follows. Section 2 describes the salient feature extraction and the protrusion field estimation procedure. The hierarchical approach for performing range image to 3D model matching is described in Section 3. Finally, the experimental results are presented in Section 4, while the conclusions are drawn in Section 5.

2   The Protrusion Field

As will be discussed in a following chapter, the identification of similar areas is computationally very expensive if matching is performed by comparing the query range image with another range image extracted from the 3D model. Therefore, a subset of features for the range image and the 3D model should eventually be used that should be easy to handle and representative for each object. In the present paper, salient features are used, that lie in general in the most protruding areas of a 3D surface, as described in Section 2.2.

2.1   3D Model Preprocessing

The 3D models of most databases are in general in various scales. In order to be able to easily compare the similarities between a query range image and a set of models, the models should be normalized to a common scale. As mentioned earlier, the spherical coordinate system is used, so the best choice would be to normalize each model to the *unit sphere* before extracting the *salient points* of the 3D models. It should be noticed that the database models are normalized to a common scale so as to ease the implementation of the matching algorithm, even if the normalization is not a crucial process.

2.2   Salient Feature Extraction

The developed method for salient feature extraction that correspond to sharp protruding areas of the object's surface is based on Hoffman and Singh's theory of salience [13]. A brief description of the method follows.

Initially, the dual graph $G = (V, E)$ of the given triangulated surface is generated, where V and E are the dual vertices and edges. A dual vertex is the center of mass of a triangle and a dual edge links two adjacent triangles. The degree of protrusion for each dual vertex results from the following equation:

$$p(\mathbf{u}) = \sum_{i=1}^{N} g(\mathbf{u}, \mathbf{v}_i) \cdot area(\mathbf{v}_i) \tag{1}$$

where N is the number of dual vertices in the entire surface, $p(\mathbf{u})$ is the protrusion degree for the dual vertex $\mathbf{u}$, $g(\mathbf{u}, \mathbf{v}_i)$ is the geodesic distance of $\mathbf{u}$ from dual vertex $\mathbf{v}_i$ and $area(\mathbf{v}_i)$ is the area of the triangle $\mathbf{v}_i$.

Using simple gradient based methods (i.e. steepest descent) all local maxima of the protrusion map $p(u)$ are obtained. Geodesic windows are then applied and only the global maxima inside the window are considered as salient. A geodesic window, GW, centered at the dual vertex u is defined as follows:

$$GW_{\mathbf{u}} = \{\mathbf{v} | \forall \mathbf{v} \in V, g(\mathbf{u}, \mathbf{v}) < \varepsilon\} \tag{2}$$

where ϵ defines the window size.

Consider now the generalization $GGW_{\mathbf{u}}$ of the geodesic window $GW_{\mathbf{u}}$ that refers to the union of all geodesic windows defined on the dual mesh that include $\mathbf{u}$. The generalized geodesic window is defined as follows:

$$GGW_{\mathbf{u}} = \bigcup GW_{\mathbf{a}} | \mathbf{u} \in GW_{\mathbf{a}}, \quad \forall \mathbf{a} \in V \tag{3}$$

Assuming the set of salient features S that is a subset of the set of dual vertices V, then a dual vertex $\mathbf{s}$ is characterized as salient

$$\mathbf{s} \in S \subset V \tag{4}$$

if and only if the following condition holds:

$$\exists GGW_{\mathbf{s}} \quad so\ that$$

$$p(\mathbf{s}) = \max(p(\mathbf{u})), \quad \forall \mathbf{u} \in GGW_{\mathbf{s}} \tag{5}$$

2.3   Generation of the Protrusion Field

Based on the extracted salient features, the proposed approach introduces the notion of the protrusion field that is a mapping of the 3D Euclidean space $\Re^3$ into positive real numbers $\Re^+$ and based on the definition of the salient field as follows.

Consider the salient field $S_{\mathbf{u}}$ as a mapping of the 3D Euclidean space $\Re^3$ into positive real numbers in the interval $[0, 1]$.

$$S_{\mathbf{u}} : \Re^3 \to [0, 1] \tag{6}$$

$$S_{\mathbf{u}}(\mathbf{x}) = e^{-\frac{1}{2}(\mathbf{x}-\mathbf{u})^T C^{-1}(\mathbf{x}-\mathbf{u})}, \quad \mathbf{x} \in \Re^3, \mathbf{u} \in V \tag{7}$$

where C is the covariance matrix. It is evident that the salient field $S_{\mathbf{u}}$ is maximized for $\mathbf{x} = \mathbf{u}$, while it vanishes for large distances of $\mathbf{x}$ and $\mathbf{u}$.

A simple choice for the covariance matrix C is described in equation (8)

$$C = p(\mathbf{u}) \begin{bmatrix} \sigma_x & 0 & 0 \\ 0 & \sigma_y & 0 \\ 0 & 0 & \sigma_z \end{bmatrix} \tag{8}$$

that could be reduced to $C = p(\mathbf{u})\mathbf{I}$, where $\mathbf{I}$ is the 3×3 identity matrix, for isotropic salient fields.

However, since it has been observed in previous works that variations in the position of the detected salient features in the presence of noise and occlusions are mainly seen to happen along the surface of the dual mesh, in the context of the proposed framework the covariance matrix is defined so as to provide a non-isotropic salient field that vanishes faster in the normal direction of the local surface around $\mathbf{u}$.

In particular the covariance matrix is defined as

$$C = p(\mathbf{u})\mathbf{R} \begin{bmatrix} \sigma_a & 0 & 0 \\ 0 & \sigma_b & 0 \\ 0 & 0 & \sigma_n \end{bmatrix} \tag{9}$$

where σ_n is the standard deviation along the normal direction of the local surface around $\mathbf{u}$ and σ_a, σ_b the standard deviations along the other orthogonal directions a, b of the local coordinate system, while $\mathbf{R}$ refers to a rotation matrix transforming the local coordinate system (a, b, n) into the global coordinate system (x, y, z). Moreover, in the context of the proposed framework the standard deviations are set as $\sigma_a = \sigma_b = \frac{\sigma_n}{2}$, thus generating a broader salient field along coordinate vectors a and b. It should be also emphasized that since $\sigma_a = \sigma_b$ the choice of a specific set of coordinate vectors a and b is irrelevant for the estimation of the salient field $S_{\mathbf{u}}$.

Then, based on the definition of the salient field in equations (6-7) the protrusion field F is a mapping of the 3D Euclidean space $\Re^3$ into positive real numbers $\Re^+$ and is defined as follows:

$$F : \Re^3 \to \Re^+ \tag{10}$$

$$F(\mathbf{x}) = \sum_{\mathbf{u} \in S} S_{\mathbf{u}}(\mathbf{x}) \tag{11}$$

Moreover, in order to avoid the artificial escalation of the protrusion field in the areas, where several salient features are present, the normalized protrusion field $\tilde{F}(\mathbf{x})$ is defined as:

$$\tilde{F} : \Re^3 \rightarrow [0, 1], \quad \tilde{F}(\mathbf{x}) = \min(1, F(\mathbf{x})) \tag{12}$$

2.4   Range Image

In order to extract salient features from the range image, a 3D surface should initially be formed. The surface is created using only a subset of the points of the image, so as to reduce the redundancy and size of the triangulated surface to be generated.

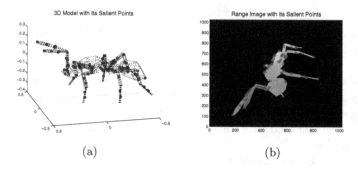

(a)   (b)

Fig. 1. Salient features of an Ant model extracted using its a) entire 3D model, b) 2D projected range image

Features on the range image are selected, as the ones with maximum minimal eigenvalue in a predefined window. After the 3D surface is formed, the salient features are extracted in the same way they are extracted for the 3D models. Figure 1 illustrates the salient features extracted from the 3D model of an Ant and from its 2D projected range image.

3   Matching

The concept of the proposed matching scheme that is based on hierarchical search in the feature space using only the salient points is briefly described in Section 3.1.

3.1   Matching Scheme

The basic element of the approach is a virtual camera, assumed to lie in the space of the examined 3D model. The proposed method searches in the parameter space for the set of camera parameters that capture a surface as similar as possible (identical in the ideal case) to the surface of the query range image. The above is encapsulated in the following hypothesis:

Hypothesis: *Assuming that the 3D model M and the partial surface that is described through the range image I, do correspond, then:*

$$\exists C_{P_i,P_e} \quad so \quad that \quad C_{P_i,P_e}(M) = I \tag{13}$$

where C_{P_i,P_e} is a range sensor (camera) with specific intrinsic (P_i) and extrinsic (P_e) parameters.

Consider the query range image I of an object M. If the object $\hat{M}$ (identical to M) is captured using a virtual camera $\hat{C}$ that captures a range image $\hat{I}$ using the camera parameters $\hat{P}_i, \hat{P}_e$ then $\hat{I} = \hat{C}_{\hat{P}_i,\hat{P}_e}(\hat{M})$. Under the assumption that the objects M and $\hat{M}$ are identical and if the correct camera parameters $\hat{P}_i, \hat{P}_e$ are estimated (i.e. $\hat{P}_i = P_i, \hat{P}_e = P_e$) then trivially $\hat{I} = I$.

For the non-ideal case of non identical objects ($\hat{M} \approx M$), in the presence of noise or occlusions ($I \approx C_{P_i,P_e}(M)$), or with an approximate only estimation of the camera parameters ($\hat{P}_i \approx P_i, \hat{P}_e \approx P_e$), the images $\hat{I}$ and I cannot be identical.

Utilizing the above hypothesis, the problem of identifying the correspondence between a range image and a 3D model, is reduced in finding the correct camera parameters, P_i and P_e, that minimize the error function:

$$\{P_i, P_e\} = arg_{min}\{\mathcal{E}(P_i, P_e)\} \tag{14}$$

where:

$$\mathcal{E}(P_i, P_e) = f(C_{P_i,P_e}(\hat{M}), I) \tag{15}$$

Finally, in the proposed framework, assuming a query range image, the algorithm searches for the best match in the parameter space that consist of all possible positions and orientations of the virtual camera. Since this is a computationally very expensive procedure an hierarchical approach is utilized to search for the best match.

The parameter space is defined by the position and "*look at*" of the virtual camera. The position of the camera is described using a spherical coordinate system with the following parameters; radius (ρ), longitude (ϕ), latitude (θ). In the proposed framework the camera typically looks at the origin of the coordinate system. Parameters roll (ω_α) yaw and pitch ($\omega_\beta, \omega_\gamma$) are used to refine where the camera looks at.

Matching is performed in a hierarchical manner by selecting a set of camera parameters and calculating the value of an error function (equation 15) defined by the protrusion field. The hierarchical matching algorithm proceeds by searching for the minimum error set of parameters (q_l) at the "l" level of the hierarchy, in the neighborhood of the minimum error set of parameters (q_{l-1}) of the upper level of the hierarchy.

3.2 Matching Using Protrusion Fields

If the virtual camera described in the hypothesis of Section 3.1 exists and the query range image corresponds to the 3D model, then if a set of salient points is extracted from the query range image and another one from the surface that the virtual camera captures, then these two sets of points should also have corresponding subsets of features. The same should also hold for their corresponding protrusion fields that is expected to result in a more resilient representation that could demonstrate robustness to noise, occlusions that normally appear in realistic situations.

A set of salient points (S^I) is extracted for the query range image (as described in Section 2.4), and another one (S^M) for the 3D model. The latter is transformed for each set of camera parameters (P_i, P_e), in order to be comparable with the set of salient points (S^I), thus producing the set of salient points $S^M(P_i, P_e)$. These sets of points are used to calculate the protrusion field and subsequently the similarity between the query range image and the 3D model.

More specifically the distance function $\mathcal{E}_S$ used to calculate the difference between the protrusion field of the target 3D model and the salient features of the query, stems from equation (16):

$$\mathcal{E}_S(P_i, P_e) = \frac{1}{N_s} \sum_{k=1}^{N_s} \tilde{F}(\mathbf{w}_k) \tag{16}$$

where N_s is the number of the salient points of the range image and $\tilde{F}(\mathbf{w}_k)$ is the evaluation of the protrusion field for a specific salient point $\mathbf{w}_k$. The value of N_s has been experimentally selected to be $N_s = 30$. As seen over the performed experiments the further increase of this parameter does not lead to more efficient retrieval performance. The function of equation (16) actually computes the normalized effect of the protrusion field over the salient features of the range image.

Finally, the set of camera parameters that minimizes the error function $\mathcal{E}_S$ is considered as the best match:

$$\{P_i, P_e\} = arg_{min}\{\mathcal{E}_S(P_i, P_e)\} \tag{17}$$

4   Experimental Results

The proposed method was tested on the 3D model database of the *Watertight model Track of Shape Retrieval Contest '07* (SHREC) [14]. The SHREC database consists of 400 models organized in 20 categories. From the database, 400 range images where created from different views (one for each model) using random angle parameters, and entered as query in the matching algorithm. Each range image was compared with all the 3D models.

To obtain comparable results the approach of Germann et. al. [4] is used since it is one of the very few approaches for partial matching using as query partial object views, although utilizing a very different method and making assumptions that are not made in the proposed approach like known object scale. Moreover, it should be emphasized that comparison with approaches like [5] that require fine detail in the query mesh would not be fair for the latter since the query range images do not exhibit these properties and the proposed scheme clearly outperforms them.

The evaluation is performed by computing the ranking during retrieval and using precision-recall diagrams, where precision is defined as the ratio of the relevant retrieved elements against the total number of the retrieved elements, and recall is the ratio of the relevant retrieved elements against the total relevant elements in the database.

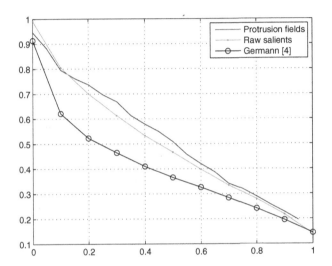

Fig. 2. Precision vs. Recall curves for the proposed approach compared to the approach using "raw salients" [10] and [4]. The proposed scheme clearly outperforms the approach in [4], while it also outperforms the "raw salients" approach [10] for most of the "Recall" range.

Figure 2 illustrates the Precision vs. Recall (P-R) curves for the proposed approach compared to "raw salients" [10] approach and [4], for the SHREC database. It can be clearly seen that the protrusion field demonstrates high performance for high recall values, while it also maintains its accuracy and outperforms the "raw salients" [10] approach for higher recall values. Moreover, it clearly outperforms the approach described in [4].

Figure 3 demonstrates the robustness of the proposed framework in the presence of 20% occlusion and additive gaussian noise (28dB PSNR) in the query range image. The performance of the proposed method remains high, compared to the approach using the "raw" salients, while the performance of the approach described in [4] drops significantly. The robustness to noise and occlusion of the proposed scheme is justified by the fact that it models saliency using a protrusion field. Therefore, even if using the raw salients [10] outstanding robustness

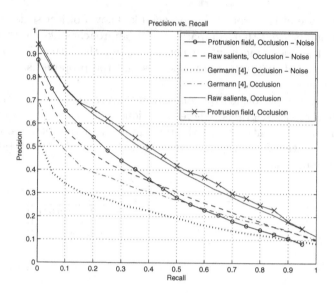

Fig. 3. Precision vs. Recall curves for the proposed approach compared to the "raw salients" [10] approach and [4] for queries exhibiting occlusion and additive Gaussian noise

to noise and occlusion can be demonstrated, the proposed scheme provides even superior results either due to high noise values or in areas where the occlusion boundaries are close to a salient feature of the underlying object.

The proposed approach is divided into two separate processing phases; the preprocessing of the 3-D model databases and the online processing of the users query. Concerning the preprocessing phase the average time needed to process a 3D model is 2s. Regarding the online matching procedure, the average time to perform pairwise matching is 10 ms.

5   Conclusions

In this paper a novel method for 3D object search and retrieval using range images as queries has been presented. The proposed approaches introduces a description of 3D objects using protrusion fields that provides a compact description of the underlying geometry. Matching is performed via an hierarchical view-based approach so as to reduce the computational complexity of matching pipeline. The approach has been tested in benchmark datasets and is seen to outperform relevant state-of-the-art approaches, while it is also seen to be robust to noise and occlusion. Besides, the efficiency and robustness of the proposed scheme, it can be further advanced through more sophisticated matching techniques so as to resolve queries in sublinear time. Moreover, the protrusion field descriptor could be potentially extended to articulated or deformable models utilizing the manifold surface of the underlying geometries that forms our major future research direction in the field of 3D object retrieval through range image queries.

Acknowledgements. This work was supported by the EU funded I-SEARCH STREP (FP7-248296).

References

1. Tangelder, J.W.H., Veltkamp, R.C.: A survey of content based 3D shape retrieval methods. Multim. Tools and Applications 39(3), 441–471 (2008)
2. Bustos, B., Keim, D.A., Saupe, D., Schreck, T., Vranic, D.V.: Feature-based similarity search in 3D object databases. ACM Computing Surveys 37(4), 345–387 (2005)
3. Biasotti, S., Marini, S., Mortara, M., Patane, G., Spagnuolo, M., Falcidieno, B.: 3D Shape Matching Through Topological Structures. In: Nyström, I., Sanniti di Baja, G., Svensson, S. (eds.) DGCI 2003. LNCS, vol. 2886, pp. 194–203. Springer, Heidelberg (2003)
4. Germann, M., Breirenstein, M.D., Park, I.K., Pfister, H.: Automatic Pose Estimation for Range Images on the GPU. In: Proc. 3D Digital Imaging and Modeling, 3DIM 2007, pp. 81–90 (2007)
5. Gal, R., Cohen-Or, D.: Salient geometric features for partial shape matching and similarity. ACM Transactions on Graphics 25(1), 130–150 (2006)
6. Chen, D.Y., Tian, X.P., Shen, Y.T., Ouhyoung, M.: On visual similarity based 3D model retrieval. Computer Graphics Forum 22(3) (2003)
7. Shum, H.-Y., Hebert, M., Ikeuchi, K.: On 3D Shape Similarity. In: Proc. IEEE Computer Vision and Pattern Recognition, pp. 526–531 (1996)
8. Stavropoulos, G., Moustakas, K., Tzovaras, D., Strintzis, M.G.: A Novel Approach for Range Image to 3D Model Partial Matching. In: Eurographics Workshop on 3D Object Retrieval, Crete, Greece (April 2008)
9. Darlagiannis, V., Moustakas, K., Tzovaras, D.: On Geometric and Soft Shape Content-Based Search. In: IEEE International Conference on Image Processing, ICIP 2010, Hong Kong, pp. 3157–3160 (September 2010)
10. Stavropoulos, G., Moschonas, P., Moustakas, K., Tzovaras, D., Strintzis, M.G.: 3-D Model Search and Retrieval From Range Images Using Salient Features. Proc. IEEE Transactions on Multimedia 12(7), 692–704 (2010)
11. Atmosukarto, I., Shapiro, L.G.: A Salient-Point Signature for 3D Object Retrieval. In: Proc. ACM Multimedia Information Retrieval, pp. 208–215 (2008)
12. Caetano, T.S., Caelli, T., Schuurmans, D., Barone, D.A.C.: Graphical Models and Point Pattern Matching. In: IEEE Transactions on Pattern Analysis and Machine Intelligence, vol. 28 (2006)
13. Moustakas, K., Tzovaras, D., Strintzis, M.G.: SQ-Map: Efficient Layered Collision Detection and Haptic Rendering. IEEE Transactions on Visualization and Computer Graphics 13(1), 80–93 (2007)
14. Giorgi, D., Biasotti, S., Paraboschi, L.: Shape retrieval contest 2007: Watertight models track. In: Remco C. Veltkamp, Frank B. ter Haar: SHREC 2007 3D Shape Retrieval Contest. Technical Report UU-CS-2007-015 (2007)

Object Recognition for Service Robots through Verbal Interaction about Multiple Attribute Information

Hisato Fukuda[1], Satoshi Mori[1], Yoshinori Kobayashi[1,2], and Yoshinori Kuno[1]

[1] Department of Information and Computer Science, Saitama University
255 Shimo-okubo, Sakura-ku, Saitama 338-8570, Japan
{fukuda,tree3mki,yosinori,kuno}@cv.ics.saitama-u.ac.jp
[2] Japan Science and Technology Agency (JST), PRESTO, 4-1-8 Honcho,
Kawaguchi, Saitama 332-0012, Japan

Abstract. In order to be effective, it is essential for service robots to be able to recognize objects in complex environments. However, it is a difficult problem for them to recognize objects autonomously without any mistakes in a real-world environment. Thus, in response to this challenge we conceived of an object recognition system that would utilize information about target objects acquired from the user through simple interaction. In this paper, we propose image processing techniques to consider the shape composition and the material composition of objects in an interactive object recognition framework, and introduce a robot using this interactive object recognition system. Experimental results confirmed that the robot could indeed recognize objects by utilizing multiple attribute information obtained through interaction with the user.

1 Introduction

As the numbers of elderly and handicapped persons in Japan continues to rise, the potential of service robots to offer assistance has increasingly attracted attention. In the process of developing a helper robot able to bring objects requested by users, we needed reliable object recognition technology, which such robotic systems depend upon in order to recognize objects correctly.

Object recognition has emerged as a major theme in computer vision. Recently, much progress has been made due to data representations utilizing invariant semi-local features such as SIFT [1] and statistical machine learning [2]. However, at present the technology is still at the stage where a recognition rate (in a category level) in excess of 70% is considered "good." Although rates will no doubt increase in the future, developing a system able to recognize objects without fail in various conditions remains a challenge. To address this problem, we are currently working on an interactive recognition system [3]. In this system, the robot asks the user to verbally provide information about the object it cannot detect.

Since Winograd's pioneering work [4], a great deal of research has been conducted on systems able to comprehend a scenario or tasks through interaction

G. Bebis et al. (Eds.): ISVC 2012, Part I, LNCS 7431, pp. 620–631, 2012.

with the user[5,6]. However, in most of these studies object recognition in the field of computer vision has not been the primary area of concern. Moreover, such studies have primarily dealt with objects that can be described sufficiently with simple word combinations, such as 'blue box' or 'red ball.' Roy et al. [7] proposed a system that can learn concepts from the user's verbal descriptions of various objects while the user indicates the object in question by pointing. For example, while indicating and describing a 'blue square' the user is able to teach the robot the meanings of 'blue' and 'square.' Unfortunately, this system too only dealt with simple objects. The application of these systems to the real world, filled with complex objects, is therefore more problematic.

The purpose of our research, in contrast, is to develop an interactive vision system that can recognize real daily objects through natural interaction. Such daily objects are seldom as simple as archetypal 'blue boxes' used in conventional studies. For example, snack packages often employ various colors. However, in our daily lives we appear to be able to indicate such complex objects through simple interaction. We first need to solve this issue, that is, the relationship between human descriptions and actual objects. Then, we need to develop computer vision techniques to detect objects described as such.

We examined how a human enables another to understand and retrieve a desired object when the second person is not supplied with the name of the object in question [3]. Our investigation revealed that humans primarily employ two methods to do this. One is to provide the second person with specific attributes of the object, such as its color, shape, and material. The second method is to mention the relation of the object to other objects spatially. In our previous work [3], we considered interaction about color and revealed that humans usually only mention one color when describing objects, even multicolored objects. Consequently, we developed a system that can recognize multicolored objects through the same simple interaction that humans use on a daily basis. We also investigated interaction about positional relationships among objects [3,8]. In this paper, we extend this system further by considering the shape composition and the material composition of objects in addition to color and spatial relationships. The use of such multiple attributes stands to significantly improve the capabilities of the system. We have implemented the system on a robot and performed experiments to confirm its usefulness.

2   Interaction about Shape

2.1   Shape Descriptions by Humans

In our human-human interaction experiment, the most frequent descriptor employed was color (about 50%), which was why we decided to pursue an investigation of color first [3]. The second most frequent descriptor was shape (about 25%). We consider this descriptor in this paper. However, shape descriptions by humans were far less straightforward. Thus, we performed additional experiments to investigate how human describe shapes [9]. We have obtained two findings from the experiments. We have found that humans mention three kinds of shapes: shapes of whole objects (3D, 2D, 1D), 2D shapes seen from a particular direction (from either front, upper, or side), and shapes of components. We have also found that the same expressions are used for describing geometrically different shapes depending on the situation. A typical expression example that we noticed is 'Marui' in Japanese. 'Maru' is the Japanese for circle and 'Marui' is its adjective form. 'Marui' is also used for describing a sphere. 'Marui' may be used for 2D circles, 3D spheres and sometimes various curved objects in Japanese. Since 'Marui' was often used in the experiments, we performed a further experiment to know the degree of roundness of various round objects. We adopted Thurstone's method of paired comparisons to make the scale of roundness. We generated 10 objects by computer graphics. We prepared 45 cards, each showing one of the possible 45 pairs from these 10 objects. We used 57 participants. All of them were Japanese students at our university and Japanese was their native language. We showed them these 45 cards and asked to indicate which they think more round for each card. Fig.1 shows the roundness scale computed from the result. In Japanese, 'Marui' is used to indicate both 2D circles and 3D spheres. Fig.1 shows that 3D objects are generally felt rounder than 2D objects. And, if most of objects in the scene are polygons and polyhedrons, objects with bit curved parts may be called round.

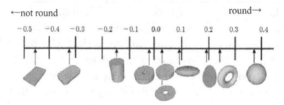

Fig. 1. Roundness scale

2.2   Object Detection Using Shape

Among the three kinds of shape descriptions, we consider the whole shape and the shape from a particular direction in this paper, because they are often used (more than 70 % cases in the experiment[9]). We use a Kinect Sensor [10] to obtain 3D scene information. Fig.2 shows an example of 3D scene data.

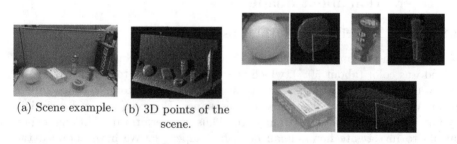

(a) Scene example.   (b) 3D points of the scene.

Fig. 2. Example of acquired 3D data   **Fig. 3.** Examples of 3D points of objects

The system analyzes the 3D volume data of objects to recognize whole shapes of objects described by such as "sphere" and "box". It first fixes a 3D coordinate system on each segmented volume data. It detects the plane on which the object lies and takes the normal direction of the plane as the direction of Y axis. Then, it finds the most fitted plane parallel to the Y axis by applying the plane model fitting method to 3D point data. It determines the normal direction of this plane as the direction of the X axis. It takes the origin as the centroid of the points. Fig.3 shows some examples.

The system categorizes objects into '3D objects', '2D objects', and '1D objects' depending on the distribution of point data along X, Y, and Z axes. Objects are classified as 3D objects if their point data are distributed along three axes. 2D and 1D objects are determined in similar ways. 2D objects correspond to 'plane objects', and 1D objects to 'line objects' of human description. We have prepared several processes to detect basic primitives such as sphere, cylinder, and box, by using the model fitting method based on RANSAC algorithm.

We have also prepared the processes to recognize the shapes seen from a particular direction by examining the point data projected onto either XY-plane, YZ-plane or ZX-plane.

Humans usually do not mention what shape description way they use. 'Marui' may mean that the whole object shape is sphere or that the object is round seen from a particular direction such as a cylinder. Moreover it may describe a circle as the 2D whole shape in Japanese. However, the experimental result shown in the previous subsection may indicate that the 'Marui' object should be the sphere in a scene with a sphere (ball) and a cylinder (can). The finding from the experiment can be schematically shown as in Fig.4. Spheres may be felt roundest as all 2D shapes viewed from three directions (from X, Y, and Z axes) are circle. When comparing the left and middle figures in Fig.4, the middle one may be considered rounder as the circular area viewed from above is much larger than the rest two non-circular shape views.

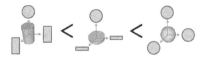
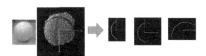

Fig. 4. Relation between the roundness order felt by humans and the shapes seen from the three directions

Fig. 5. Point data projected onto three orthogonal planes

We have devised the following method to compute the roundness measure. First we project the point data onto XY-plane, YZ-plane, and ZX-plane, and calculate the roundness R_p by Eqn.1 for each projection data.

$$R_p = \frac{\pi L_p^2}{4S_p} \tag{1}$$

where L_p is the maximum length of the region, and S_p is the area of the region. Then we make a summation of the three direction data with the weight of the area size.

$$w_p = \frac{S_p}{S_1 + S_2 + S_3} \tag{2}$$

$$degree\_of\_roundness = \sum_{p=1}^{3} w_p R_p \tag{3}$$

where w_p is the weight for each direction (Eqn.2). We use this value as the measure of roundness(Eqn.3). Of course, we cannot obtain complete three views only from a single observation. However, the method can give a rough degree of roundness. Fig.5 shows an example of the process.

Fig.6 shows the results for a scene with five objects. In this scene, the system detects a sphere, box, and two cylinders by the primitive detection. It also computes the roundness for these objects. The roundness is largest for the ball (sphere) and the lowest for the tissue box. The packing tape is considered rounder than the potato chips cylindrical packages.

Fig. 6. Shape recognition results

3   Interaction about Material

3.1   Material Descriptions by Humans

While in the human-human interaction experiment described in [3] color and shape were employed frequently, we did not observe many descriptions of an object's material. However, even if humans do not explicitly describe objects in terms of their material frequently, it is nevertheless important to use material information for object recognition. This is because object material often has a deep relationship with the names of objects. For example, a drinking glass is called a "glass" because of its material; likewise a "paper cup" describes a cup made of paper. Therefore, we decided to include material information in our interactive object recognition system.

3.2   Object Detection Using Material

It is possible to obtain information about the material of an object through observing reflection patterns with a calibrated light source and camera system. However, such a method may have some drawbacks: the target objects have to be composed of one material, and additional devices are required. Thus, we selected and subsequently developed a method that estimates object material from a single color image. This method estimates material likelihood by using observable features such as color, texture, shape, and reflection that may work effectively for material recognition. After extracting these features from an image, we convert the image into bag-of-visual-words [11] using the extracted features. We use Latent Dirichlet Allocation (LDA)[12], a probabilistic machine learning technique used for document clustering, to model the distribution of the words. These techniques were used in many previous object recognition methods and material recognition methods[13]. The method may work on objects composed of multiple materials because its estimation is based on semi-local features. Fig.7 shows the flow of the estimation of material likelihood, while Table 1 shows the visual words used for recognition. The specifics of each observable feature used in the material recognition process are as follows:

Color Feature. For the color feature we simply use the 27 dimensional vectors, arranging the RGB values in a 3×3 pixel window extracted from an RGB image.

Texture Feature. We use Scale-Invariant Feature Transform (SIFT) [1], which is often used in object recognition, as the texture feature. However, we do not use the key-point detector part of SIFT. We adopt SIFT for a feature descriptor. Keypoints are densely sampled on a reticular pattern, and then the feature vector is calculated using SIFT for each keypoint.

Shape Feature. We use Histograms of Oriented Gradients (HOG) [14] for the shape feature. HOG is a feature descriptor that describes the feature for a given local area based on the distribution of gradient directions of luminance in the area. It can therefore represent rough shapes, and is often used for human detection and so on [14].

Reflection Feature. In order to describe reflection, we also use HOG on specific areas. When highlights occur on an object's surface, we can observe large luminance changes in a particular direction in an image of the object. Thus, we divide an image into square windows, and in each window apply HOG on the rectangular region in the particular direction where the variance of luminance reaches its maximum.

We quantize the features into visual words and construct the classifier. Recognizing object material from RGB images is a difficult problem. However, in the interactive object recognition framework, we do not necessarily need to recognize the material of each object in a given scene. Since the user specifies the material of his/her target object, the problem that the system should solve is to arrange the objects in the scene in the order of likeliness of the specified material. Therefore, we can prepare a classifier for each material class that is optimized for the material.

Table 1. Visual words used in the object recognition process

Feature	Dim.	Feature num/image	Cluster num
Color	27	4315	150
Texture	128	1598	200
Shape	120	5504	100
Reflection	72	571	100

It may not always be good to use all features described above to construct classifiers. For example, in the case of fabric, since it may take various colors, employing color features cannot improve recognition; rather it might hinder recognition. Therefore, we choose the optimal combination of features for each material in the training stage, and estimate respective material likelihood by the classifier in the recognition stage.

We quantize the features into visual words and use the LDA method to select the optimal feature combination and learn per-class distributions for recognition. We apply the LDA method to the training images with a given number of topics k, images $d \in (d_1, \ldots, d_N)$ and topics $z \in (z_1, \ldots, z_k)$, and obtain the probability of each topic over each image $P(z|d)$, the probability of each material property over each topic $P(likely_i|z)$ using labeled training images.

In the step of estimating the likelihood, the probability of desired material d', $P(likely_i|d')$ is given by Eqn.4. Then the system detects the object that has the largest likelihood of that material as the first candidate of target object.

$$P(likely_i|d') = \sum_{z \in Z} P(likely_i|z)P(z|d') \qquad (4)$$

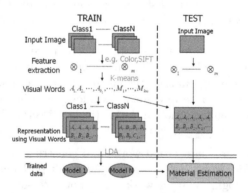

Fig. 7. Flow of estimating material likelihood

We performed an experiment to examine the optimal combination of features for each material. In this experiment, we used the Flickr Material Database[13]. This database consists of 100 natural images for each of 10 material classes

(Fabric, Foliage, Glass, Leather, Metal, Paper, Plastic, Stone, Water, and Wood), and mask images for natural images. We chose 50 images for each material in the training stage. In the LDA training step, we varied the number of topics k from 50 to 250 with step size 50 and picked the best one. In the recognition stage, we used the rest 50 images for each of 10 material classes that were not used for training. We prepared sets of 10 images by choosing an image for each material. We tested all possible combinations of features for 50 sets for each material to choose the most likable object made of the specified material.

Table 2 shows the experimental result where the optimal combination for each material is indicated by the red rectangle. Certain combinations of features depending on the material class yield better recognition rates. This shows the effectiveness of using the optimal feature combination. We cannot directly compare our recognition performance with others because our problem is easier than their classification problem. However, the values themselves are generally larger than those by Liu et al. [13], on average 45%. Although their rates about foliage and water are high (about 80 %), the rates about metal and fabric, which may often be target object materials, are low (about 20%).

It is still a difficult problem to classify the material. If we stick to the classification approach, the robot may often fail to detect the object made of the material specified by the user. However, our approach is to choose the object most probably made of the material specified by the user. This will increase the possibility to recognize the correct object. In addition, this approach allows us to improve the performance by constructing the classifier for each material by selecting the optimal combination of features. Experimental results have proved our approach promising.

4   Robot System

We have implemented our object recognition system into a robot system. We utilize a Robovie-Rver.3 [15] as a robot platform and a Kinect Sensor[10] to capture the sets of color and depth images, while employing Julius [16] for speech recognition. Fig.8 shows the system configuration. In this system, the robot is able to point at an object with its finger by using the 3D position of the object and the positional relationship between the Kinect Sensor and the robot.

We have added the interaction capabilities about shape and material described in the previous sections to our previous system [3], [8]. The system is designed to first attempt to recognize the object by autonomous object recognition. If this fails, the system then shifts to the interactive mode. Since interactions about color and positional relationships are most often used, the interactive object recognition starts either with the color or spatial relation module developed in the previous system.

As mentioned in the introduction section, humans often describe multicolored objects in terms of only one color, usually that of the background or that occupying the largest area of the object. Based on this finding, we developed a vision program to detect multicolored objects, where the color of the background or

Table 2. Recognition rate (%) for every possible combination of features

	Fabric	Foliage	Glass	Leather	Metal	Paper	Plastic	Stone	Water	Wood
Color	16	62	14	38	14	26	16	38	40	46
Texture	42	26	18	32	12	24	36	26	18	40
Shape	16	58	8	22	28	30	64	34	64	58
Reflection	22	24	28	50	4	24	56	28	50	42
Color+Texture	20	52	22	38	18	28	40	42	44	46
Color+Shape	42	76	16	32	34	36	32	38	46	56
Color+Reflection	10	62	6	34	8	38	32	26	34	48
Texture+Shape	30	54	16	20	20	34	64	32	48	46
Texture+Reflection	44	32	38	34	10	28	50	32	16	32
Shape+Reflection	36	54	10	18	24	14	60	38	52	52
Color+Texture+Shape	30	66	16	38	24	32	52	42	52	58
Color+Texture+Reflection	28	54	22	38	12	40	36	44	44	56
Color+Shape+Reflection	38	72	14	30	28	52	36	38	46	58
Color+Texture+Reflection	42	52	18	22	20	34	54	38	36	48
All set of features	42	76	16	26	24	36	46	38	60	52

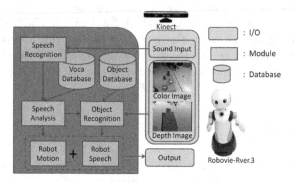

Fig. 8. System configuration

largest area is that specified by the user [3]. The background color is determined as the color of the region whose convex hull is the largest among neighboring regions in the color segmentation result. In a situation with objects of various colors present (number of colors ≥ 3 in the current implementation), the system asks the user the color of the target object.

In the interaction about the spatial relationships, the system asks the user the positional relationship between the target object and the objects that the system has autonomously recognized. If the number of recognized objects is large (≥ 3 in the current implementation), the system says that it knows that such-and-such objects are present.

If both color variation and the number of recognized objects exceeds their respective thresholds, the system takes the color-based interaction first in the current implementation. It asks the user about the shape or the material of the target object when the variation of colors is one, the number of known objects is zero, or if it has already asked about the color and spatial relationships of the target object. If the system has not asked about either attribute, the system takes the shape-based interaction in the current implementation.

In any case, if the robot can determine the target object by using the information given, it asks the user for confirmation by pointing at the recognized object. If it is unable to choose the target object following additional questions, the robot randomly chooses an object which has not yet been chosen, and confirms whether this is the target object both by speech and by pointing at it. By repeating these processes among the remaining possible target objects, the robot can finally recognize the target object.

Fig.9 depicts the interaction process utilized to detect the target object. In this scene, the user asks the robot, "Bring me the flashlight" (Fig.9(a)). Since the robot is not familiar with this object, it asks the user about its color, to which the user answers, "It is red." The robot finds three red objects in the setting. It randomly chooses the toy from among these, and points at it to confirm whether or not this is the target object. The user answers in the negative, telling the robot, "No, it is not what I want." The robot proceeds by asking the user about the material of the target object (Fig.9(b)). The user answers, "It is made of plastic." Using this information from the user, the robot estimates the plastic likelihood of the two remaining candidates, and finds that the likelihood is higher for the object located to its left. It then confirms this selection by pointing at it, and learns that this is indeed the target object.

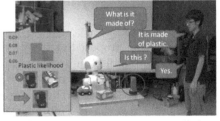

(a)Object detection using color   (b)Object detection using material

Fig. 9. Robot in operation

5 Experiment

We performed operation experiments using the robot. We used 26 participants, all of whom were undergraduate students in the Faculty of Liberal Arts at our university. We placed 30 objects on a table. In each session, the participant was asked to choose seven objects among them and to put them on the table in front of the robot. S/he was also asked to determine one of the seven objects as the target object. Then, the robot started to recognize the target object. In this experiment, we set up the system to recognize autonomously eight objects among 30 objects and asked the participants not to select them as their target objects. Fig.10 shows an experimental scene.

The robot found the target objects in all 26 sessions through some interactions. Fig.11 shows the distribution of the number of interactions (the number of questions that the robot system asked until detecting the correct target). The average number of interactions was 2.12. If the robot randomly chooses an object and makes confirmation through pointing at it, the expected number of interactions is 3.85. Our experimental result is statistically smaller than this value (p < 0.05). Since any interaction can reduce at least one candidate and may sometimes find the target, this reduction itself is apparent. It is left for our future work to develop the most efficient interaction strategy. However, as shown in Fig.11, the current system required one or two interactions in almost cases and can be said to work efficiently.

Fig. 10. Example of experiment scene

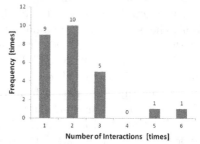

Fig. 11. Number of interactions necessary for recognition

6   Conclusion

In this paper, we proposed an interactive object recognition system using multiple attribute information (color, shape, and material), and explained how we implemented an integrated object recognition system on a robot system. The robot follows a process whereby if autonomous object recognition is not successful, it requests from the user information about the target object's attributes and spatial relationships. It then proceeds to detect the target object based on the information provided in response. The robot performs smooth object detection by changing its questions based on the variation of colors in the scene and the number of objects autonomously recognized. In the experiment using our robot system, we could confirm the effectiveness of our system.

An ongoing challenge is presented by the nature of human verbal descriptions, which may not always indicate the same physical concepts; rather their meanings may change depending on the situation. Hence, we intend to further study the relationships among human descriptions, various situations, and physical and geometrical concepts in order to more effectively realize robot systems that can smoothly interact with humans.

Acknowledgments. This work was supported by KAKENHI (23300065).

References

1. Lowe, D.G.: Object recognition from local scale–invariant features. In: Proc. ICCV 1999, pp. 1150–1157 (1999)
2. Ponce, J., Hebert, M., Schmid, C., Zisserman, A. (eds.): Toward Category-Level Object Recognition. LNCS, vol. 4170. Springer, Heidelberg (2006)
3. Kuno, Y., Sakata, K., Kobayashi, Y.: Object recognition in service robot: Conducting verbal interaction on color and spatial relationship. In: Proc. IEEE 12th ICCV Workshops, pp. 2025–2031 (2009)
4. Winograd, T.: Understanding Natural Language. Academic Press (1972)
5. McGuire, P., Fritsch, J., Steil, J.J., Roothling, F., Fink, G.A., Wachsmuth, S., Sagerer, G., Ritter, H.: Multi-modal human machine communication for instruction robot grasping tasks. In: Proc. IEEE/RSJ International Conference on Intelligent Robots and Systems, IROS 2002, pp. 1082–1089 (2002)
6. Kawaji, T., Okada, K., Inaba, M., Inoue, H.: Human robot interaction through integrating visual auditory information with relaxation method. In: Proc. Int. Conf. Multisensor Fusion on Integration for Intelligent Systems, pp. 323–328 (2003)
7. Roy, D., Scheile, B., Pentland, A.: Learning audio-visual associations using mutual information. In: Proc. ICCV 1999.Workshop on Integrating Speech and Image Understanding, pp. 147–163 (1999)
8. Cao, L., Kobayashi, Y., Kuno, Y.: Object spatial recognition for service robots: Where is the Fronts? In: Proc. International Conference on Mechatronics and Automation, ICMA 2011, pp. 875–880 (2011)
9. Mori, S., Kobayashi, Y., Kuno, Y.: Understanding the Meaning of Shape Description for Interactive Object Recognition. In: Huang, D.-S., Gan, Y., Bevilacqua, V., Figueroa, J.C. (eds.) ICIC 2011. LNCS, vol. 6838, pp. 350–356. Springer, Heidelberg (2011)
10. Kinect for Windows, `http://kinectforwindows.org/`
11. Csurka, G., Bray, C., Dance, C., Fan, L.: Visual categorization with bags of keypoints. In: Proc. ECCV Workshops on Statistical Learning in Computer Vision, pp. 1–22 (2004)
12. Blei, D.M., Ng, A.Y., Jordan, M.I.: Latent Dirichlet Allocation. Journal of Machine Learning Research 3, 993–1022 (2003)
13. Liu, C., Sharan, L., Adelson, E.H., Rosenholtz, R.: Exploring features in a bayesian framework for material recognition. In: CVPR 2010, pp. 239–246 (2010)
14. Dalal, N., Triggs, B.: Histograms of oriented gradients for human detection. In: CVPR 2005, pp. 886–893 (2005)
15. Intelligent Robotics and Communication Laboratories, `http://www.vstone.co.jp/products/robovie_r3/index-en.html`
16. Open-Source Large Vocabulary CSR Engine Julius, `http://julius.sourceforge.jp/index-en.html`

TCAS: A Multiclass Object Detector for Robot and Computer Vision Applications

Rodrigo Verschae and Javier Ruiz-del-Solar

Advanced Mining Technology Center, Universidad de Chile, Chile

Abstract. Building efficient object detection systems is an important goal of computer and robot vision. If several object types are to be detected, the most simple solution is to run several object-specific classifiers independently of each other (in parallel). This solution is computationally expensive if several object classes are to be detected. In this paper, *TCAS*, a new classifier structure designed to be used on multiclass object detection problems is introduced as an alternative solution. TCAS offers an efficient solution and reduces the aggregated false detection rate. TCAS extends cascade classifiers (introduced by Viola & Jones) to the multiclass case and corresponds to a nested coarse-to-fine tree of multiclass nested boosted cascades. Results for three different object detection problems are presented: face and hand detection, robot detection, and multiview face detection. In the experiments, the obtained TCAS have classification times about 2-times shorter than the ones obtained using parallel cascades, and have the same or lower number of false positives (for the same detection rate).

1 Introduction

The development of robust and real-time object detection systems is an important goal of the robot-vision and computer-vision communities. Applications of multiclass detection systems include human-robot interaction (e.g. face and hand detection), and autonomous drivings system (e.g. car and sign detection), among many others.

For detecting all instances of an object-class, the sliding window approach [1] requires to perform an exhaustive search over image patches at different scales and positions. The classification of all possible windows (patches) may require a high computational power. To achieve an efficient detection, less processing time should be spent on non-object windows than on object windows. This can be achieved using cascade classifiers [1] (Fig. 1 (a)). Starting from the seminal work of Viola & Jones [1] on cascade boosted classifiers for object detection, several improvements have been proposed, with the introduction of nested cascades [2] being one of the most important ones.

In order to detected multiple classes of objects, the most simple solution is to use several cascades in a a concurrent manner, with each cascade trained for each particular class independently of each other. The main problem with this approach is that the processing time and the false positive rate (*fpr*) aggregates with the number of classes. As a solution to these problems we propose a multiclass classifier with a tree structure. The proposed detector corresponds to a *multiclass* nested tree of multiclass nested cascades that compared to the use of parallel cascades presents a considerable gain in processing time. Results in 3 different detection problems are presented later.

G. Bebis et al. (Eds.): ISVC 2012, Part I, LNCS 7431, pp. 632–641, 2012.

TCAS learns an implicit partition of the target space, and performs an efficient coarse-to-fine (CTF) search in the object space, both, within each node of the tree, and while traversing it. On the contrary a cascade classifier learns a implicit partition of the target space that only allows to discriminate between the object class and non-object class. A *TCAS* is efficient because (1) it performs a CTF search, and (2) the feature evaluations and classifiers evaluations are shared between subsets of classes , reducing the computational complexity. The proposed *TCAS* structure (See Fig 1 (c) for an example) has similarities and differences with existing classifiers: Width-First-Search (WFS) trees [4], Cluster-Boosted-Trees (CBT) [5], and Alternating Decision Trees (ADT) [6].

A CBT has a similar structure to the a *TCAS*, but it is designed only for multiview detection problems. The main differences of CBT with *TCAS* are: (1) during training of a CBT, a partition of the object class is learnt, (2) the depth of CBT is the same for all branches, and (3) each node of the CBT contains a single "layer", while in a *TCAS* it contains a multiclass cascade. These differences reflect in that the *TCAS* performs a CTF search in the object space within each node of the tree, and the depth of each branch of the tree is variable.

In [4] a Width-First-Search (WFS) tree is proposed. The WFS tree structure is similar to the *TCAS*, but the main differences are that: (1) the WFS-tree is not nested, and (2) the WFS tree does not use CTF multiclass cascades. The main similarities are the use of Vectorboost and *WFS* search. The term WFS refers to the fact that at each node of the tree, any number of siblings can be followed while traversing the tree.

ADTs [6] are decision trees where each node does not only gives an output, but also a confidence, which is accumulated over all nodes that are visited. The main similarity between ADT and *TCAS* is that in an ADT, a leaf's output depends on all nodes of the branch the leafs belongs to (i.e. is "nested"), but it may also depend on nodes of other branches. This requires, in the case of the ADT, to maintain "global" weights during training, while in *TCAS* the weights are "local" to the node being trained.

In the following, background in presented in Section 2, followed by the proposed *TCAS* classifier (Section 3), an evaluation (Section 4), to finally conclude in Section 5.

2   Background

In [1] Viola and Jones proposed using cascade classifiers (Fig. 1 (a)) for the efficient detection of faces. Besides using rectangular features that can be efficiently evaluated, in the proposed cascade classifier the cascade's layers have an increasing computational complexity. Thus, because that most analized windows of any image do not contain faces, it can quickly reject windows that do not resemble faces. In [2] a nested cascade which reuses the confidence information from one layer in the next one was proposed. This nested structure is possible thanks of the additivity of the boosting model.

Multiclass Boosted Classifiers. In [7] nested cascades were extended to the multiclass case, with each layer corresponding to a multiclass classifier and having a vectorised [4] form (see Fig. 1 (b)): $H(x) = \sum_{t=0}^{T} h_t(f_t(x))$. The classifier $H(x)$ is trained using Vectorboost [4], which is a multiclass extension of Adaboost that assigns to each training example x_i an objective region in a vector space. An objective region is defined as the intersection of a set of half spaces, with each half-space defined by a vector, a, and a set of half-spaces defined by a parameter set $\mathbf{R}$. Under this setting, a sample x, belonging to class Y_q and represented by a parameter set $\mathbf{R}_q$, is classified correctly iff:

$$\forall a \in \mathbf{R}_q, \langle a, H(x) \rangle \geq 0, \tag{1}$$

thus a class Y_q is assigned to a new sample x, if all the inequalities of R_q are fulfilled.

The weak classifiers $h_t(f_t(x))$ are designed after the *domain-partitioning weak hypotheses* paradigm [8]: each feature domain $\mathbb{F}$ is partitioned into disjoint blocks $F_1, \ldots, F_J$, and a weak classifier $h(f(x), m)$ has a constant output for each partition block of its associated feature f. The weak classifier's components $\{h(f(x), m)\}_m$ can have different relations with each other (see [7] for details). In the present work we consider 2 cases: *independent* components and *coupled* components. For efficiency, weak classifiers are stored in look-up-tables (LUTs), which can be evaluated in constant time.

Multiclass Coarse-to-Fine Nested Cascades. A coarse-to-fine (CTF) nested multiclass cascade [9] is used as a *TCAS* classifier's building block (see Section 3). A CTF nested multiclass cascade performs a search in the object target space, and this allows to reduce the processing time and to increase the accuracy of the cascade classifier. To perform the CTF search in the object target space, an *active mask* that has a binary output and that represents a subsets of active components at the current layer of the multiclass cascade is used. At layer k of the multiclass cascade, the *active mask* $\mathbf{A}_k(\cdot)$, is defined component-wise by: $\mathbf{A}_k(x, m) = u\left(H_k(x, m)\right) \prod_{i=0}^{k-1} \mathbf{A}_i(x, m)$, with $u(x)$ the unit step function. Note that if $\mathbf{A}_k(x, m)$ is 0, then $\mathbf{A}_j(x, m)$ is 0 for all $j \geq k$. Using the *active mask*, the output of a layer, k, of the cascade is:

$$H_k(x) = \left[H_{k-1}(x) + \sum_{t=1}^{T_k} h_{k,t}(f_{k,t}(x)) \right] \odot \mathbf{A}_{k-1}(x) \tag{2}$$

with, $H_0(x) = 0$, $\mathbf{A}_0(x) = 1$, and with $\odot$ the point-wise product between two vectors. Eq.2 can be interpreted as verifying the condition of the input belonging to a particular class (Eq.1) at each layer of the cascade in a per component manner, and only for the subset of hypotheses that was already verified at the previous layers of the cascade.

Only non-zero components of $\mathbf{A}_{k-1}(x)$ need to be evaluated in Eq. 2 (at layer k). These non-zero components represent a subset of classes with positive output at the current layer (and potentially a positive output in the cascade). In this way, as a sample moves through the cascade, the output goes from a coarse output in the target space, to a finner one. Also, as in a standard cascade classifier, there is a CTF search on the non-object space: at each layer a window being analysed can be discarded.

3   *TCAS*: Nested Coarse-To-Fine Tree of Nested Cascades

The CTF multiclass cascade just described has the problem that if a subset of the classes to be detected is very "different" to another subset, it is not possible to train an efficient classifier, as too many weak classifier may need to be added to the cascade's layers.

The reason is that at each level of the system there are two competing goals: to discriminate between the background (the non-objects) and the objects, and also to discriminate among object classes. At the first levels of the tree, to discriminate between the objects and the non-objects is not difficult, but at later levels of the cascade, the non-objects to be discarded by the cascade resemble more the objects, thus features and weak classifiers that are more "specialized" to each class are needed. More precisely, this means that it becomes difficult to find features that can be used to discriminate between all objects classes from "all" background(s). The *TCAS* allows to partition the target space in a CTF way, in particular in later tree levels, but it also allows to share feature evaluations (whenever possible). Thus a trade-off between accuracy and efficiency is achieved.

A *TCAS* classifier, $\mathbb{T}$, corresponds to a directed tree, with each node having a variable number of siblings. A node $\mathbb{N}$ of $\mathbb{T}$ consists of: (a) a multiclass nested cascade classifier $H_{\mathbb{N}}^C$ (as in Eq. 2), (b) a mask $\mathbf{A}_{\mathbb{N}} \in \{0,1\}^M$, (c) a "pointer", $p_{\mathbb{N}}$, to its direct ancestor, and (d) $n_{\mathbb{N}}$ siblings, $\{\mathbb{N}_s\}_{s=\{1,...n_{\mathbb{N}}\}}$. Every node, $\mathbb{N}$, has a nested structure (depends on the output of its ancestors): $H_{\mathbb{N}}(x) = H_{p_{\mathbb{N}}}(x) \odot \mathbf{A}_{\mathbb{N}} + H_{\mathbb{N}}^C(x)$, with $H_{p_{\mathbb{N}}}(x)$ the output of the ancestor of $\mathbb{N}$, $p_{\mathbb{N}}$. Note that if $p_{\mathbb{N}}$ is the root of the tree then $H_{p_{\mathbb{N}}}(\cdot) = 0$. Only non-zero components of $\mathbf{A}_{\mathbb{N}}$ need to be evaluated in $H_{\mathbb{N}}^C(x)$, i.e. the CTF evaluation in the object target space is also used here, which further helps to achieve an efficient evaluation. The mask $\mathbf{A}_{\mathbb{N}}$ indicates which classifier's components are active at the current node and should be evaluated.

During training the following restriction is made (recall that $\mathbf{A} \in \{0,1\}^M$):

$$\mathbf{A}_{\mathbb{N}} = \sum_{s=Siblings(\mathbb{N})} \mathbf{A}_s \qquad (3)$$

This means that the (binary) masks of any two siblings of node $\mathbb{N}$, $\mathbb{N}_i$ and $\mathbb{N}_j$, with $i \neq j$, holds that $\mathbf{A}_{\mathbb{N}_i} \odot \mathbf{A}_{\mathbb{N}_j} = 0$. This restriction implies an efficient recursive implementation, and simplifies the training process, because nodes at different branches of the tree do not depend on each other. Thanks to the structure of the tree, the output of a *TCAS*, $\mathbb{T}$, can be defined as the sum of the output all of its leafs $\mathbb{N}_{(1)}, \ldots, \mathbb{N}_{(n_{leafs})}$:

$$H_{\mathbb{T}}(x) = \sum_{j=1,...,n_{leafs}} H_{\mathbb{N}_{(j)}}(x) \qquad (4)$$

Eq. 4 can be implemented efficiently (see Alg.1) thanks to Eq. 3.

Tree Classifier Training. The training of a *TCAS* (see Alg.2) is a top-down procedure, that recursively adds nodes to the tree starting from the root. Nodes are added by taking into account the nested multiclass cascade built by traversing the nodes' ancestors (line 3, using algorithm FLATTENING), and the training examples that have a target region defined by the active mask of the current node (lines 4 - 6). A node is trained and added (line 6, using algorithm TRAINNODE [9]) using the corresponding training examples and cascade. Later the components of a node are grouped (line 8), and the new siblings are initialised and added to the stack of nodes to be trained (lines 9 - 11). The training of the tree stops when the stack is empty (line 3). One reason for stopping adding nodes to a branch of the tree is that branch of the tree already fulfils the required false positive rate (*fpr*) (line 7). This means that the depth of the tree's branches is variable, which allows to adjust the complexity of each branch of the tree to the corresponding class.

The FLATTENING algorithm constructs a cascade associated to a tree's branch. Starting from a leaf node, it goes backward, up to the tree's root, taking only the components that are active at the leaf node. In the algorithm GROUPCOMPONENTS, the active components of a node are grouped and later each of these groups is assigned to a sibling. This procedure must fulfill two conditions: (1) the groups must have an empty intersection (Eq. (3)), and (2) the union of the groups must be equal to the input set. The 2nd condition indicates that if the input mask has only one active component, it must return the empty set (stop adding nodes to that branch) because the node does not have any sibling. In the present work the grouping is not done automatically, but it is preset.

4 Experimental Results

We present experimental results on three different detection problems: hand and face detection, multiview face detection, and robot detection. Each experiment seeks to analyse a different aspect of *TCAS*. In the first experiment (Face and Hand detection) we consider a multiclass detection problem and compare the use of several cascades versus the use of a *TCAS*. In the second experiment (multiview face detection) we analyse how *TCAS* scales up when increasing the number of classes. In the third experiment (robot detection) results on an multiview and multiclass setting showing the performance of *TCAS* when handling very different classes (humanoid and AIBO robots) are presented.

As in [9], in all following experiments, rectangular features are used in the first two levels of *TCAS* and modified LBP features ares used in later ones. Other parameter values are: *fpr* per layer $= 0.35$, *tpr* per layer per class $= 0.9995$, and target *fpr* $= 10^{-6}$.

Experiment 1: Face and Hand Detection. We evaluate *TCAS* in a multiclass problem: hand and face detection. More specifically, three classes under frontal views are considered: faces, left fists and right fists ("frontal" indicates low in-plane rotations, up to $\pm \sim 15°$). Given that there is no standard datasets for this problem, we only present ROC curves for the detection of frontal faces in a the BioID database (1,521 images and 1,521 faces, DB that does not contains any hand), which gives and idea how well a specific 1-class detector (face detector) compares to a 3-classes detector (hands and faces) in detecting a specific class (faces in this case). The classifiers were trained using 4,900 examples per class (left hand fist, faces and right hand fist). In all trained classifier, all used parameters and (initial) training sets were exactly the same.

Table 1 presents the built detectors and their corresponding processing times. Note that it was not possible to obtain two-classes nor three-classes multiclass cascades [9], because the training procedure did not converge (even after adding a large number of weak classifiers during boosting, the *fpr* did not decrease and the detection rate did not increase, thus the required accuracy was not achieved). On the contrary the *TCAS* classifiers were built for the two-classes and three-classes problems as well as one-class cascade classifiers. In Fig.2 (a) the obtained detection rates of a *TCAS* (trained to detect all three classes but used to detect only faces), compared to the use of a one-class cascade trained specifically for face are presented. It can be observed that the accuracy of the *TCAS* is slightly higher for all operation points, being up to 1 percentage point higher for 5 FP. Moreover, in Table 1 can be observed that *TCAS*, compared to the use of parallel cascades, has a gain in processing time close to 1.6 times, i.e., in this problem *TCAS* is both faster and has a better accuracy than parallel cascades.

Experiment 2: Number of Classes and Multiview Face Detection. We analyse the use of *TCAS* when handling an increasing number of classes. For this analysis we consider the multiview face detection problem with in-plane-rotations (RIP) in the range $[0, 360]$. This range was divided on non-overlapping view-ranges of $18°$, and each range is defined an object class. The base range $[-45, 45)$ is divided in to the following 5 ranges: $\{[-45, -27] + i * 18\}_{i=0,...,4}$. To cover the full range $[0, 360)$, classifiers are applied on rotated versions (0, 90, 180, and 270°) of the input or trained on classes that are defined by rotated versions of the 5 ranges. Different classifiers are trained to detected subsets of these classes: e.g in one case, twenty 1-class cascades are used in parallel, while in another case four 5-classes *TCAS* are used in parallel, and both systems are compared in a 20-classes problem. The CMU Rotated set [10], which consists of 50 images and 223 faces with in-plane rotations in [0,360], is used as test set.

We consider 4 cases, each one corresponding to a fixed number of classes ($n = 1, 5, 10, 20$) when training the classifiers. For the case $n = 1$, five 1-class cascades are trained, and they are applied 4 times (4 rotated images), which gives 20 cascades in total. Also, four 5-classes *TCAS* (applied to the image rotated in 0, 90, 180 and 270°), two 10-classes *TCAS* (applied to the image rotated in 0 and 180°), and one 20-classes *TCAS* (covering the whole range) are evaluated.

The training set for each of these ranges was built using the ISL WebImage dataset[1], with 800 labelled faces that were rotated randomly (uniform distribution) in the corresponding range. Also random translations and scale changes were applied, generating 1600 training samples per range. Per class, 3,200 negative examples were collected, i.e. when building a classifier for 5 objects classes, $16,000$ ($5 * 3,200$) negative examples and $8,000$ ($5 * 1,600$) positive examples are used. In the case of the 20-classes *TCAS* where 3,000 negative examples per class were used, because of memory restrictions. The maximum number of weak classifiers per node was set to $T_{max} = 50$; this triggers stopping the training of a node and starting the training the next tree's level).

The obtained processing times are presented in Table 2. The detection results are omitted for space reasons, but it can be stated that: the 5-classes *TCAS* and the 10-classes *TCAS* have a better performance than 20 parallel cascades, and a slightly worst performance is obtained by the 20-classes *TCAS*.

In terms of processing time, using four 5-classes *TCAS* is approximately 1.7 times faster than running 20 parallel cascades, using two 10-classes *TCAS* is 1.8 times faster than running parallel cascades, and the 20-classes *TCAS* is 1.2 times faster than running parallel cascades. The lower performance of the 20-classes *TCAS* may be due to the use of a slightly smaller training set, or it could be also due to a structural issue. Fig. 3 (a) shows a multiview detection example using the 10-classes TCAS classifier.

The training time grows quadratically with the number of classes, but if the size of training set is taken into account, it grows only linearly ($R^2 = 0.98$). This shows that the training procedure can scale up with the number of classes, and the main variable

---

[1] Available at
http://www.ecse.rpi.edu/$\sim$cvrl/database/database.html

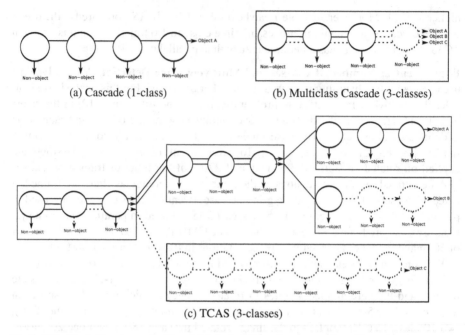

Fig. 1. Classifier's Structures. In (b) and (c) dashed lines show examples of branches that are not evaluated when performing a classification of a particular input.

Algorithm 1. EVALTREE($\mathbb{T}, \mathbf{A}_0, x$)

1: $\mathbf{O}(x) \leftarrow \mathbf{0}; Push\,(ST, [\mathbb{T}, \mathbf{A}_0 \odot A_{\mathbb{T}}])$ // Initialize output and stack
2: **while** $([\mathbb{N}, \mathbf{A}] \leftarrow Pop(ST))$ **do**
3:     $\mathbf{O}(x) \leftarrow \mathbf{O}(x) + \mathbf{A} \odot \mathbf{H}_{\mathbb{N}}^{C}$
4:     $\mathbf{A}_t \leftarrow \mathbf{A} \odot Binary(\mathbf{H}(x))$
5:     **if** $\mathbf{A}_t \neq 0$ **then**
6:         **for** each $\mathbb{N}_s \in Siblings(\mathbb{N})$ **do**
7:             **if** $\mathbf{A}_t \odot A_{\mathbb{N}_s} \neq 0$ **then**
8:                 $Push\,(ST, [\mathbb{N}_s, \mathbf{A}_t \odot Mask(\mathbb{N}_s)])$
9:             **end if**
10:        **end for**
11:    **end if**
12: **end while**
Output: $\mathbf{O}(x)$ // Only non-zero components of $\mathbf{A}$ need to be evaluated in $\mathbf{H}_t(x)$ at line 5.
13: $Binary(\mathbf{V})$ returns a vector which is, per component m: $(Binary(\mathbf{V}))_m = u\,((\mathbf{V})_m)$

Table 1. Processing times of frontal left and right hand, and frontal faces detectors

Classifier	Classes	Time [sec]
Three 1-class cascades	Face & Hands	1.70
1-class + 2-class *TCAS*	Face & (Left & Right Hand)	1.27
1-class + 2-class *TCAS*	Left Hand & (Right Hand & Face)	1.39
3-class *TCAS*	Face & Hands	**1.01**

Algorithm 2. TRAINTREE(S, F_C, D, PV, NV)

1: $\mathbb{T} \leftarrow \{\emptyset\}$; $Push(ST, [\mathbb{T}, \mathbf{1}])$; $S = \{(x_i, a_i)\}_{i=1,\ldots,n}$: Set of positive examples
2: **while** ($[\mathbb{N}, \mathbf{A}] \leftarrow Pop(ST)$) **do**
3:      $H_t \leftarrow$ FLATTENING($\mathbb{N}$)
4:      $S_t \leftarrow filterExamples(\mathbf{A}, S)$
5:      $PV_t, NV_t \leftarrow filterExamples(\mathbf{A}, PV), filterExamples(\mathbf{A}, NV)$
6:      $H_{\mathbb{N}} \leftarrow$ TRAINNODE($H_t, \mathbf{A}, S_t, F_C, PV_t, NV_t$)
7:      **if** $fpr\,(H_t + H_{\mathbb{N}} \odot \mathbf{A}) \geq F_C$ **then**
8:          $\mathbb{G} \leftarrow GroupComponents(\mathbf{A})$
9:          **for** each $\mathbf{g} \in \mathbb{G}$ **do**
10:             $Push(ST, [newSibling(\mathbb{N}), \mathbf{g}])$
11:         **end for**
12:     **end if**
13: **end while**
Output: $\mathbb{T}$
14: $newSibling(\mathbb{N})$ creates an empty node and adds an edge (new sibling) to $\mathbb{N}$.
15: F_C: global false positive rate; PV / NV: Positive/Negative Validation Set
16: $S_t \leftarrow filterExamples(\mathbf{A}, S)$: returns $S_t = \{(x_i, a_i) \in S \mid (a_i \odot \mathbf{A}) \neq \mathbf{0}\}$

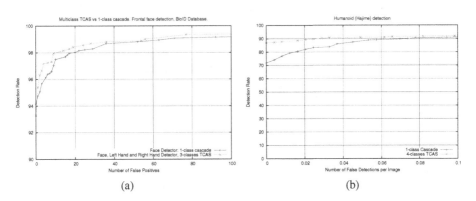

Fig. 2. ROC curves: (a) Hand and face detection: 3-classes *TCAS* (left hand, right hand and faces) vs a 1-class cascade (faces) for frontal face detection, and (b) Humanoid and Aibo Detection

defining the training time is the number of training examples. The number of weak classifiers at the first layer of the *TCAS* grows logarithmically with the number of classes, showing that feature evaluations are efficiently shared, reducing the computational complexity.

Experiment 3: Robot Detection. Finally, the proposed classifier is evaluated in a multiview and in a multiclass robot detection problem. SONY AIBO ERS7 robots are detected under 3 different views: frontal, lateral and back. The multi-class detection problem includes the multiview detection of AIBO robots plus a single-view of HA-JIME HR18 humanoid robots. Altogether, four classes are defined (3 associated with AIBOs and one with HAJIME robots) and we built 4 one-class cascade detectors for each of the 4 classes (frontal, lateral, and back AIBOs, and Humanoids), a 3-classes

Table 2. *TCAS* and number of classes. Multiview RIP detection. Each view covers a 18° range. Avg. running time for images of 320x240 pixels. * value summed up over the 5 cascades.

Classifier	Training [min]	Running [sec]	Weak classifiers: Total	Weak classifiers: First layer
1x 20-classes	1862	8.05	10037	23
2x 10-classes	461	5.76	6712	20
4x 5-classes	176	6.10	3913	17
4x5x 1-class	75*	9.64	3251*	78*

Table 3. Robot detection: 3 views of Aibos robots (frontal, lateral and back) and Humanoids. Processing times on the HumanoidDB (640x480, 244 images) and AiboDB (208x160, 724 images).

Classifier	Target views	Database	Processing Time [sec]
3-class TCAS	Frontal, Lateral and Back AIBOs	Aibo DB	0.079
3 Parallel Cascades	Frontal, Lateral and Back AIBOs	Aibo DB	0.126
4-class TCAS	Humanoids	Humanoid DB	0.464
1-class Cascade	Humanoids	Humanoid DB	0.411
4-class TCAS	AIBOs & Humanoids	Humanoid DB	0.870
4 Parallel Cascades	AIBOs & Humanoids	Humanoid DB	1.767

(a) Face detection with view estimation   (b) Humanoid Detection

Fig. 3. Robot and face detection examples. View estimation for in-plane rotations in $[-90, 90]$.

TCAS is used to detect the AIBOs under the 3 views, and a 4-classes TCAS classifier that solves the defined multi-class robot detection problem. The classifiers are evaluated on 2 databases: AIBO DB and Humanoid DB (see Tab. 3 for details).

Table 3 (first two rows) presents the processing times for the multi-view AIBO detection problem. The obtained *TCAS* is 1.6 times faster than using the 3 one-class cascades in parallel. Fig. 3 (b) shows a detection example of using the 4-classes *TCAS* in the detection of HAJIME humanoids. Fig. 2 (b) presents ROC curves when the 4-class *TCAS* and a one-class cascade classifier are used for the detection of HAJIME robots only. Table 3 (last four rows) presents the processing time of the 4-classes *TCAS* used to detect all four classes, and the use of four one-class cascades. The 4-classes *TCAS* used to detect HAJIME humanoids has a clearly better performance than the cascade trained to detect just that class, with a gain of up to 17 percentual points for 0 false positives. In addition, the processing speed of both classifiers is very similar, with the *TCAS* being slightly slower when compared to the one-class cascade for humanoid detection. More importantly, when the 4-classes *TCAS* is used to detect Humanoids, and AIBOs (with frontal, lateral and back views), its processing speed is two times faster than the one required by four one-class cascades performing the same task.

5   Conclusions

In the present paper, a methodology for building multiclass object detectors was proposed. The proposed *TCAS* classifier corresponds to a multiclass classifier with a nested tree structure that makes use of coarse-to-fine (CTF) multiclass nested cascades and variants of multiclass weak classifiers in combination with a CTF tree structure.

TCAS was used to build multiclass and multiview objects detectors for three different problems. The obtained *TCAS* are efficient, being 1.5 to 2 times faster than the use of parallel cascades in all problems, and having the same or a lower number of false positives (for the same detection rate). Regarding the training time, the *TCAS* can be built efficiently, with training times of only hours for problems with a few classes (3-5) and mid-size training sets (5,000 samples per class) on modern workstation computers.

Future research directions include: (1) automatically grouping the classes instead of using predefined groups, (2) evaluating the system using other features, and on problems (e.g car and pedestrian), and (3) testing using larger number of classes.

References

1. Viola, P., Jones, M.: Rapid object detection using a boosted cascade of simple features. In: Proc. of the IEEE Conf. on Computer Vision and Pattern Recognition, pp. 511–518 (2001)
2. Wu, B., Ai, H., Huang, C., Lao, S.: Fast rotation invariant multi-view face detection based on real adaboost. In: 6th Int. Conf. on Face and Gesture Recognition, pp. 79–84 (2004)
3. Torralba, A., Murphy, K.P., Freeman, W.T.: Sharing visual features for multiclass and multi-view object detection. IEEE Transactions on PAMI 29, 854–869 (2007)
4. Huang, C., Ai, H., Li, Y., Lao, S.: High-performance rotation invariant multiview face detection. IEEE Trans. on Pattern Analysis and Machine Intell. 29, 671–686 (2007)
5. Wu, B., Nevatia, R.: Cluster boosted tree classifier for multi-view, multi-pose object detection. In: Proc. 11th IEEE Int. Conf. on Computer Vision, ICCV 2007 (2007)
6. Freund, Y., Mason, L.: The alternating decision tree learning algorithm. In: Proc. 16th International Conf. on Machine Learning, pp. 124–133 (1999)
7. Verschae, R., Ruiz-del-Solar, J.: Multiclass Adaboost and Coupled Classifiers for Object Detection. In: Ruiz-Shulcloper, J., Kropatsch, W.G. (eds.) CIARP 2008. LNCS, vol. 5197, pp. 560–567. Springer, Heidelberg (2008)
8. Schapire, R., Singer, Y.: Improved boosting using confidence-rated predictions. Machine Learning 37, 297–336 (1999)
9. Verschae, R., Ruiz-del-Solar, J.: Coarse-to-fine multiclass nested cascades for object detection. In: Int. Conference on Pattern Recognition, pp. 344–347 (2010)
10. Rowley, H.A., Baluja, S., Kanade, T.: Neural network-based detection. IEEE Transactions on Pattern Analysis and Machine Intelligence 20, 23–28 (1998)

Augmented Multitouch Interaction upon a 2-DOF Rotating Disk

Xenophon Zabulis, Panagiotis Koutlemanis, and Dimitris Grammenos

Institute of Computer Science, Foundation for Research and Technology - Hellas,
Herakleion, Crete, Greece

Abstract. A visual user interface providing augmented, multitouch interaction upon a non-instrumented disk that can dynamically rotate in two axes is proposed. While the user manipulates the disk, the system uses a projector to visualize a display upon it. A depth camera is used to estimate the pose of the surface and multiple simultaneous fingertip contacts upon it. The estimates are transformed into meaningful user input, availing both fingertip contact and disk pose information. Calibration and real-time implementation issues are studied and evaluated through extensive experimentation. We show that the outcome meets accuracy and usability requirements for employing the approach in human computer interaction.

1 Introduction

The emerging trend of smart environments entails the need for direct interaction with non-instrumented physical surfaces. In this context, often referred as "surface computing" [1], systems combine the projection of a user interface on a surface (e.g., tabletop, wall), with visual sensing of finger contacts with the surface to provide multi-touch interaction. The use of non-instrumented surfaces simplifies the application and maintenance of such systems. Recent availability of consumer depth cameras has reinforced the interaction capabilities upon virtually any surface, as the shape of interaction surfaces and the location of user hands can be accurately found in 3D.

We explore the issues arising when enabling augmentation and interaction upon a non-instrumented dynamically moving surfaces and, in particular, upon a disk surface mounted on two concentric gimbals, providing two degrees of freedom (DOF). Besides the provision of augmented multitouch interaction, such an achievement can serve two distinct functions. On one hand, the rotation mechanism can be used as a means for easily and intuitively browsing and interacting with alternative, dynamically changing, projection views. On the other, the high flexibility and extensive range of projection poses supported by the system can be used in order to dynamically personalize the physical properties of an interactive projection surface to the ergonomic preferences and needs of users.

The system employs a circular planar surface, or a disk, mounted on two gimbals (see Fig. 1). The outer gimbal rotates about the vertical axis at an angle θ (yaw). The inner gimbal is horizontal, dependent on the outer gimbal,

G. Bebis et al. (Eds.): ISVC 2012, Part I, LNCS 7431, pp. 642–653, 2012.

and rotates along with it; the disk is mounted at joints $q_{1,2}$, and has an elevation (pitch) angle of ϕ. These axes intersect at the center c of the disk. A projector above the disk fully covers it with its projection. A depth camera overlooks this scene acquiring its depth map D. The two gimbals can be freely rotated by the user. Due to the finite resolution of the camera and the projector, the operation of the system is limited at very oblique angles as, then, the disk corresponds to very few pixels in the camera and the projector.

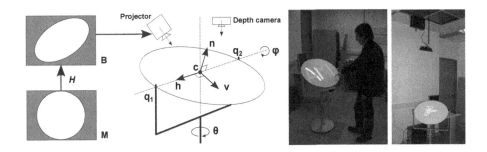

Fig. 1. System overview. *Left, middle:* illustration of system geometry and dataflow. The virtual display M is mapped upon the disk devoid of projective distortions by transforming M, according to the pose of the disk, prior copying it to the projector buffer B. Fingertip contacts are mapped back in M as multitouch events. *Right:* two photographs of a user testing a "finger-paint" pilot application, where drawings remain on the disk regardless of its posture, and an image of the system setup.

The system creates the following user interface metaphor. Let M be a virtual display buffer that renders a circular display.[1] The projector renders M upon the disk devoid of projective distortions, as if M was a multitouch display upon the disk's surface. This is achieved by continuously estimating the disk's pose and updating the projection appropriately. The pitch and yaw estimates can be used as additional user interfaces, by associating their values to (two) application variables.

Thus, a central system component is the real-time estimation of the disk's pose from depth map D. At each camera frame, depth map D is updated yielding a new pose estimate. To support brisk and accurate interactivity upon the surface, it is essential that this operation is performed in real-time and robustly.

By distorting M according to this pose, prior its copy into the projector's pixel buffer B, M appears undistorted on the disk. Intuitively, rotating the disk while M is displaying a static image would create the illusion of the image being "painted" upon the disk. In the projection, the row axis of M is aligned with the disk's intrinsic horizontal rotation axis h. Correspondingly, the column axis v lies on the disk surface and is perpendicular to h. When user hands are in contact with the surface, touch events are generated and attributed with the 3D

---

[1] As displays are rectangular, the display region outside the circle is inactive.

contact coordinates of this contact. These coordinates are transformed into M's, 2D, reference frame, implementing a multitouch display upon the disk.

The remainder of this paper is organized as follows. In Sec. 2 related work is reviewed. Disk pose estimation is described in Sec. 3. A way to spatially calibrate of the above setup is proposed in Sec. 4. The display and contact estimation modules are described in Sec. 5. In Sec. 6, experiments which evaluate the accuracy, performance and usability of the approach are presented. In Sec. 7, conclusions and directions for future work are provided.

2   Related Work

To date, several touch-based interactive surfaces exist both in the form of research prototypes [2–7], but also as commercially available products [8–10]. Such systems use a static planar surface as an interaction surface.

In early approaches towards augmented interactive surfaces [11, 12], the dynamic component concerned steering a projector to display upon the surface of choice. The display was adapted using prior modeling of the surfaces and limited hand interaction was based on the input of a color camera. More recent approaches have used a more detailed 3D model of the whole scene, availing the ability to project at virtually any geometry of surfaces [13], but used a stylus instrumented with an infra-red beacon to enable single user interaction. The recent growth of depth cameras enabled the dynamic modeling, augmentation and touch interaction upon arbitrary surfaces [14]. Still, a training time is required in order for the system to model the interaction surfaces which, additionally, should remain static during interaction. The proposed approach constantly estimates the interaction surface and, thus, does not require an adaptation time.

Another aspect of dynamically moving interaction surfaces is that the location or pose of the surface itself can avail valuable information to the user interface. In [15], a coarse estimate of the inclination of a handheld surface (a piece of cardboard) provides input to an interactive game. In [16, 17], a similar surface is used to explore a maps. As the above approaches use a conventional camera, they offer limited (or none) touch interaction. To estimate surface pose, they rely on visual markers, thus being sensitive to marker occlusions by user hands and illumination artifacts. The proposed approach overcomes such limitations using depth information and, furthermore, is able to support multitouch interaction.

3   Disk Pose Estimation

Disk pose estimation, is used both in the real-time operation as well as the calibration of the proposed system. It is comprised of the two processes described below. The first estimates the plane that the disk lies upon, despite outliers arising from user hands and noisy pixels in D. To meet real-time requirements, the method is parallelized in the GPU. The second disambiguates disk yaw when the disk is approximately parallel to the ground plane.

3.1 Parallel and Robust Plane Estimation

To estimate disk pose the plane $\mathcal{E}$ is robustly fit to 3D points originating from depth map D. Only points originating from an elliptical region of interest (ROI) within depth map D, are considered. This ROI is large enough to image the entire disk and is predicted from camera calibration and disk geometry (see Sec. 4.1). Plane $\mathcal{E}$ has equation $n \cdot (x - c) = 0$, where n is the normal of the plane. The spherical coordinates of n, ϕ and θ, indicate disk pose. A rotation of π about the horizontal axis is considered to bring the disk to the same posture and, thus, $\phi \in [0, \pi/2)$ and $\theta \in [0, 2\pi)$, where $\phi = 0$ corresponds to a posture parallel to the ground plane.

Significant amounts of outlier points are included in the data. Some occur due to sensor noise. Others occur as within the ROI more surfaces besides the disk are imaged, such as the user's hands or body. A robust plane estimation is obtained using RANSAC [18]; a threshold of $1\,cm$ distance to the plane is set to characterize a point as an inlier.

Conventionally, RANSAC iterates by selecting a random triplet of points and evaluating the number of inliers for the plane they define, until it finds a good fit or a maximum number of iterations is reached. The method is parallelized in the GPU by performing all trials in parallel threads and selecting the triplet with the most inliers. Finally, using least squares, a plane is fit to all the inlier points to this plane which constitutes the final result. By convention, the normal of this plane is set to be in the direction of gravity.

A singularity, known also as the "gimbal lock" problem, is met when the $\phi = 0$. Then, any value of θ produces the same $n = [0\,0\,1]^T$ and, in this case, the value of θ is determined using the method in Sec. 3.2.

3.2 Horizontal Axis Estimation

To disambiguate its pose when ϕ is approximately 0, the horizontal axis of the disk h is found by estimating the line through the 3D locations of the inner gimbal joints, q_1 and q_2. These joints are detected in D and their 3D locations extracted by the corresponding 3D depth values.

Candidate points for this detection are sought in the periphery of the ellipse (or circle) at which the disk appears in D. As above this ellipse is predicted for the particular pose from camera calibration and disk geometry (see Sec. 4.1). In addition, the 3D coordinates of candidate points are required to approximately validate the current plane equation $\mathcal{E}$. In D, candidate points are grouped into blobs by a Connected Component Labeling process. Each blob is represented by its centroid, which comprises a candidate point.

Spurious candidates can arise from user hands occurring at the periphery of the disk. The pair of centroids selected as q_1 and q_2 is the one at which the two candidates are diametrically opposed across the disk center c. By considering the 3D coordinates of the candidate pair of centroids, the pair that defines the line segment with the least distance from c is selected.

As this line does not specify the direction of the horizontal axis, this is determined to be the same with that of the previous frame, assuming that the framerate of the system is frequent enough for the user to perform a rotation of π about the vertical axis during a single frame. The assumption is reasonable as the framerate of the system operation is $60\,Hz$ (see Sec. 6.

4   Calibration

System calibration includes the estimation of center c and the spatial modeling of the projection, both in the depth camera's coordinate frame. An additional color camera is used during the second part of this calibration (see Sec. 4.2) providing image I; we conveniently use the *Kinect* depth camera that already incorporates this additional sensor. A calibration of the depth and color cameras is assumed, based on [19].

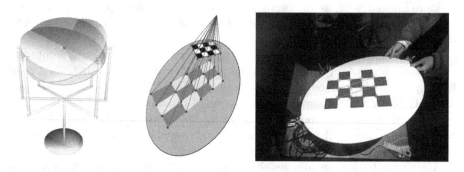

Fig. 2. Calibration geometry. *Left:* The intersection of all plane estimates during calibration yields an estimate of center c. *Middle:* By projecting a calibration pattern and finding the 3D coordinates of its reference points, the projector is calibrated. *Right:* The projected pattern detected in image I.

4.1   Disk Center

Center c is estimated from the depth maps acquired, while the disk is freely rotated about both its axes.

At each frame, the method in Sec. 3.1 estimates the plane approximating the disk. Angles ϕ and θ are discretized, at a step of $1°$, yielding $k = 360 \times 90$ potential planes. A $k \times 4$ matrix, is employed as a lookup table (LUT) and each time a plane is estimated its 4 parameters are copied into the corresponding LUT entry. Thus, very similar planes are considered once.

The result is obtained as the intersection of all, let m, estimated planes and computed as the point that minimizes the sum of its squared distances from all estimated planes. As c should ideally validate the equations of all planes we seek x so that $\|Ax - B\| = 0$ is minimized, where A is a $m \times 3$ matrix containing in each row the first 3 parameters α_i, β_i, γ_i of the estimated planes, B is a $m \times 1$ matrix containing values $-\delta_i$ in each row, and $i \in [1, m]$. A least-squares solution is then found through the SVD decomposition of A. Typically, less than a minute is required to fill a sufficient proportion of the LUT ($\approx 20\%$) for an accurate estimate of c.

4.2 Projector Calibration

The projector operation is modeled by a 3×4 perspective projection matrix P that predicts the pixel coordinates in B that will illuminate a given homogeneous point in 3D space. Given the high quality optics of projectors and the usage of only its central portion of the display, the lens distortion of the projector is assumed to be negligible. During the calibration process both color I and depth D images of the sensor are used. Using their calibration, a registration of these two images is obtained and, thus, 3D coordinates of the surfaces imaged in the color camera can be obtained.

Matrix P is estimated as follows. The image of a calibration target, a checkerboard, is constantly projected upon the disk. This image appears distorted upon the disk, according its posture. As in Sec. 4.1, the disk is freely rotated in both angles, while the system is acquiring frames. For each frame, the projected checkerboard upon the disk surface is detected in I and the image coordinates of its corners identified using a checkerboard detector [20]. The 3D coordinates of these points are then found at the same pixel coordinates of, registered, image D.

Using then these 2D-3D correspondences the method in [21, p.181] estimates matrix P. Inclusion of lens distortion in this optimization is left for future work.

5 Real-Time Application

The real-time system is comprised of three modules that operate at the camera's framerate. Using the module in Sec. 3.1 a pose estimate of the disk is availed for each camera frame. Using this estimate, the transformation that undistorts display M upon the disk is computed. At the same time, a second module estimates finger contacts with the disk and produces multitouch user interface events in M's, 2D, coordinate frame.

5.1 Display

This module computes the distortion that M must undergo before projection so that (i) it appears undistorted upon the disk and (ii) each of its pixels illuminates the same physical region upon the disk, regardless of its posture. The distortion that an image undergoes when projected upon the disk is modeled through a 3×3 homography matrix, let H.

The homography can be defined if we determine its required physical limits on the disk surface and predict their coordinates in B (see Fig. 3). These limits are predicted as the corners of a hypothetical square that lies upon $\mathcal{E}$, encloses the disk, and is aligned with its intrinsic axes, v, h which are computed from the estimates of ϕ and θ.

Once the 3D coordinates of these hypothetical points are determined, their coordinates in B are predicted through P. By associating these points with the points in B that define the limits of the displayed content H is calculated. Image M is warped using H and the result is copied into B.

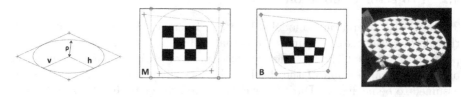

Fig. 3. Display geometry (left to right). (a) The disk enclosed in a hypothetical square. (b) Establishing point correspondences for homography computation: M is shown annotated with (a) the 4 points that correspond to the corners of the hypothetical square and (b) the same corners projected upon the projector buffer. (c) Warping of M according to H makes M appear undistorted upon the disk and aligned with its intrinsic axes. (d) M is set to contain a checkerboard image and visualize axes h (red) and v (green); when warped and projected on the disk checkers appear undistorted and axes accurately aligned with the disk intrinsic axes.

5.2   Touch Detection

This process detects which physical points of the disk that are in contact with the fingertips of the user, as well as, the coordinates of these points in M.

The points of fingertip contact upon the disk are found as follows. Given the current plane equation $\mathcal{E}$, c, and ρ, the ellipse upon which the disk projects in D is predicted. The pixels of D within this ellipse are transformed into 3D points. For each of these points, its distance to $\mathcal{E}$ is computed. As in [22], we also use two thresholds to detect contact. The first (d_{max}) indicates if a pixel is closer to the camera than the disk surface. However, only this constraint will include points belonging to the user's arm as well. The second threshold (d_{min}) eliminates points that are overly far from the surface to be considered part of object in contact.

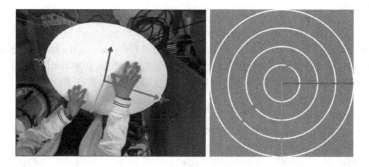

Fig. 4. Touch detection. *Left:* a user touches the surface with all his fingers and the projections of the detected contact points on I are superimposed with red. For reference, estimates of h, c, and n are also superimposed in red, green, and blue respectively. *Right:* contact blobs in the M's reference frame; column (red) and row (green) axes correspond to h and c.

The 3D points found in contact with the surface are then mapped using H^{-1} into an image, let T, that has the same size as M (see Fig. 4). In T, a pixel is 1 if it considered to signify contact at the corresponding point of the surface and 0 otherwise. Typically, a blob of pixels corresponds to the contact of each finger. A blob tracker [23] is employed upon images T to track individual fingers with ids that are consistent across frames.

6 Experiments

The experiments focused in validating the suitability of the proposed system as a multitouch interactive surface. In the first set of experiments, the accuracy of disk pose estimation and projection upon the disk were measured. In the second, the accuracy of contact detection for single and multiple fingertips was measured. Finally, the effectiveness of the system in supporting interactive applications was qualitatively assessed by implementing a pilot application and performing a formative usability evaluation.

The experiments were performed on a conventional PC equipped with $nVidia$ $GeForce\ GTX\ 260$ $1.2\ GHz$ GPU. The system is adequately fast to keep up with the image acquisition and image projection rates, which occur at $60\ Hz$. In fact, in offline experiments the system's operation rate is ≈ 120. In our setup, the disk has a radius of $\rho = 30\ cm$ and is placed $150\ cm$ from the ground. A Microsoft Kinect sensor (640×480 pixel resolution) and a projector (1280×800 pixels) overlook the disk from a height of $149\ cm$ and $102\ cm$, respectively.

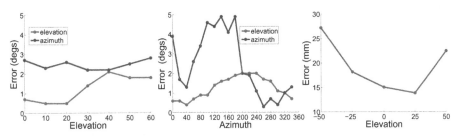

Fig. 5. Accuracy experiments. *Left, middle:* Elevation and azimuth error in degrees per elevation (left) and per azimuth (right). *Right:* Distance error (mm) per elevation.

6.1 Ground Truth Experiments

In this experiment, the accuracy of disk pose estimation was evaluated. The disk was placed at known poses and the error was measured as the difference between the ground truth values and the estimation of the elevation and azimuth angles ϕ and θ, respectively. The disk's elevation range from $0°$ to $60°$ was discretized in 7 steps of $10°$ each. The full azimuth range ($360°$) was discretized in 18 steps of $20°$. Ground truth was measured using a digital inclinometer, temporarily mounted on the disk. No occlusions of the disk occurred in the experiment. The results are presented in Fig. 5(left, middle).

The accuracy of projection, based on the disk's pose estimation was assessed as follows. A checkerboard pattern was displayed in M and correspondingly projected upon the disk. The aspect ratio of checkers was confirmed to be 1 upon the disk, for the above range of elevations as above, meaning that M's appearance on the disk is devoid of projective distortions for that range of elevations. We observed that though the aspect ratio was preserved up to angles as steep as 70°. However, images projected in more oblique angles than $\approx 50°$ were poor in quality and, thus, limited operation of the system up to that obliqueness.

6.2 Interaction Experiments

To assess the system's usability as a touch display two interaction experiments were performed. In the first, the accuracy of touch detection was measured and, in the second, tracking of multiple fingers in simultaneous contact was evaluated. In essence, the evaluation concerned the accuracy of registration between the projected display and contact localization estimates. The two experiments were performed by 5 users each that were naive to the experimental hypotheses.

In the first experiment, users had to touch a sequence of dots, of .75 cm radius, that were appearing upon the disk (see Fig. 6, left). The users were instructed to touch the dots at their centers. By comparing the distance of the touch event in M with the projected location of the dot, the error was measured. The sequence consisted of 70 dots, laid out on 7 concentric circles, centered at the disk's center. The entire surface of the disk was used. The dots were presented to the users one at a time and in random order. The experiment was repeated for 5 elevation angles, in the range of $[-50°, 50°]$, in steps of 25°. Fig. 5(right) presents the average error in millimeters, for each elevation angle.

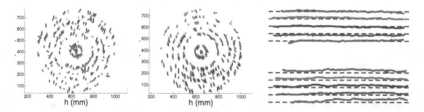

Fig. 6. Interaction experiments. *Left:* Two plots of the projected dots and the estimated contact locations for $\theta = 30°$ and $\theta = -40°$. *Right:* Two plot of the projected lines (straight dashed lines) and traced contact trajectories for the same posture. Both examples are shown in M's reference frame.

In the second experiment, the users were presented with 5 vertical, parallel line segments, which were instructed to trace with their fingertips (see Fig. 6, right). The stripes appeared at a distance of 3 cm from each other. The experiment was repeated for each of the same 5 elevation angles of the previous experiment (see above). The trajectories were recorded and using the stripes as ground truth, the Multiple Object Tracking Accuracy (MOTA) metric [24] was calculated to evaluate the accuracy of multiple simultaneous contacts to be 1 in the whole range of system operation, that is when $\phi \in [0°, 50°]$.

6.3 Pilot Application

To test the system in a realistic setting, through a set of representative user tasks, an interactive application was developed. The application supports the exploration of ancient artifacts in 360°. The data of the application include datasets where ancient artifacts have been placed on a turntable and photographed from the side, in 360 steps of 1°. For each step, the direction of illumination was mechanically modulated to follow an arc trajectory above the artifact in 20 steps. Using images corresponding to the same set of illuminations, a sparse reconstruction of the artifact was obtained using [25] and regions of interest were defined upon this point, corresponding to "hotspots" on the surface of the artifact.

In the application, a photograph of the artifact is initially projected on the surface of the metal disk, as can be seen from the angle that the metal disk is rotated. By rotating the disk around the vertical axis (θ), the user can see 360 different views of the artifact, as if the actual object was placed behind the disks surface, thus creating a 3D visualization effect. By tilting the disk surface, the user can access alternative lighting settings, revealing different details of the artifact. When the user touches the metal surface, hotspot areas of the current view are presented. Upon touching a hotspot, related information is presented. Additionally, using two fingers, the user can zoom in the image.

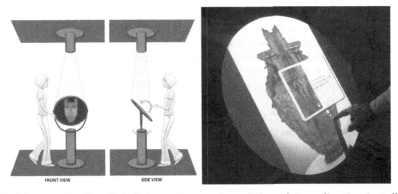

Fig. 7. Pilot application. *Left:* Schematic overview of the pilot application installation *Right:* Working prototype of the pilot application

7   Conclusion and Future Work

A visual approach that creates an interactive multitouch surfaces upon a 2-DOF rotating surface and its implementation have been presented and evaluated.

The results from experiments show that disk pose is estimated very accurately despite the presence of sensor noise and user hand interaction. Correspondingly, an accurate metaphor of a display upon it is created. Furthermore, user interface input originating from angles ϕ and θ can be reliably used for fine operations. For example, in the pilot application modulation of viewpoint and illumination occurred smoothly and accurately when users moved the disk.

The accuracy of multiple contact detection and localization is sufficiently accurate for conventional multitouch screen interaction. Indeed, this accuracy decreases at oblique angles of the disk, and originates mainly from the reduction in area that the disk undergoes in the sensor's (depth) image. At such angles the projector's limitations are also reached, as less pixels can be used to form an image upon the disk surface. Improvements could use steerable projectors (i.e. [12]) and sensors to compensate for this obliqueness, by rotating in accordance to the elevation (ϕ) of the disk.

A limitation of the approach is met in multitouch and multi-hand interaction and in particular when using more than one hands upon the surface. In such cases, user digits that may be in contact with the disk are occluded from the sensor and missed. Though this topic could be addressed with additional sensors a topic of future work is to investigate whether a more retentive tracking of user hand position would suffice user requirements.

Another topic of future work is a more robust way to disambiguate the azimuth (θ) of the disk when it is approximately parallel to the ground plane. The contour of a disk with less symmetries (i.e. an ovaloid, or an irregular shape) could provide a more global and, thus, more robust orientation cue than the localization of disk joints.

References

1. Rowell, L.: Scratching the surface. NetWorker 10, 26–32 (2006)
2. Rekimoto, J.: Smartskin: an infrastructure for freehand manipulation on interactive surfaces. In: CHI, pp. 113–120 (2002)
3. Streitz, N., Tandler, P., Muller-Tomfelde, C., Konomi, S.: Roomware: Towards the next generation of human-computer interaction based on an integrated design of real and virtual worlds (2001)
4. Wilson, A.: Playanywhere: a compact interactive tabletop projection-vision system. In: UIST, pp. 83–92 (2005)
5. Han, J.: Low-cost multi-touch sensing through frustrated total internal reflection. In: UIST, pp. 115–118 (2005)
6. Gross, T., Fetter, M., Liebsch, S.: The cuetable: cooperative and competitive multitouch interaction on a tabletop. In: CHI, pp. 3465–3470 (2008)
7. Gaver, W., Bowers, J., Boucher, A., Gellerson, H., Pennington, S., Schmidt, A., Steed, A., Villars, N., Walker, B.: The drift table: designing for ludic engagement. In: CHI, pp. 885–900 (2004)
8. Microsoft (Microsoft surface), http://www.surface.com
9. Dietz, P., Leigh, D.: Diamondtouch: a multi-user touch technology. In: UIST, pp. 219–226 (2001)
10. SMART: Smart table (2008), http://www.smarttech.com/
11. Pinhanez, C.: Using a steerable projector and a camera to transform surfaces into interactive displays. In: CHI, pp. 369–370 (2001)

12. Kjeldsen, R., Pinhanez, C., Pingali, G., Hartman, J., Levas, T., Podlaseck, M.: Interacting with steerable projected displays. In: FG (2002)
13. Jones, B., Sodhi, R., Campbell, R., Garnett, G., Bailey, B.: Build your world and play in it: Interacting with surface particles on complex objects. In: ISMAR, pp. 165–174 (2010)
14. Harrison, C., Benko, H., Wilson, A.: Omnitouch: wearable multitouch interaction everywhere. In: UIST, pp. 441–450 (2011)
15. Song, P., Winkler, S., Tedjokusumo, J.: A tangible game interface using projector-camera systems. In: HCI, pp. 956–965 (2007)
16. Grammenos, D., Michel, D., Zabulis, X., Argyros, A.: Paperview: augmenting physical surfaces with location-aware digital information. In: TEI, pp. 57–60 (2011)
17. Reitmayr, G., Eade, E., Drummond, T.: Localisation and interaction for augmented maps. In: ISMAR, pp. 120–129 (2005)
18. Fischler, M., Bolles, R.: Random sample consensus: A paradigm for model fitting with applications to image analysis and automated cartography. Communications of the ACM 24, 381–395 (1981)
19. Zhang, C., Zhang, Z.: Calibration between depth and color sensors for commodity depth cameras. In: ICME, pp. 1–6 (2011)
20. Vezhnevets, V., Velizhev, A., Chetverikov, N., Yakubenko, A.: GML C++ camera calibration toolbox (2011),
http://graphics.cs.msu.ru/en/science/research/calibration/cpp
21. Hartley, R., Zisserman, A.: Multiple View Geometry in Computer Vision, 2nd edn. Cambridge University Press (2004)
22. Wilson, A.: Using a depth camera as a touch sensor. In: ACM Int. Conf. on Interactive Tabletops and Surfaces, pp. 69–72 (2010)
23. Argyros, A.A., Lourakis, M.I.A.: Real-Time Tracking of Multiple Skin-Colored Objects with a Possibly Moving Camera. In: Pajdla, T., Matas, J(G.) (eds.) ECCV 2004. LNCS, vol. 3023, pp. 368–379. Springer, Heidelberg (2004)
24. Bernardin, K., Stiefelhagen, R.: Evaluating multiple object tracking performance: the clear mot metrics. Journal of Image and Video Processing, 1–10 (2008)
25. Snavely, N., Seitz, S., Szeliski, R.: Photo tourism: exploring photo collections in 3D. In: SIGGRAPH, pp. 835–846 (2006)

On Making Projector Both a Display Device and a 3D Sensor

Jingwen Dai and Ronald Chung

Department of Mechanical and Automation Engineering
The Chinese University of Hong Kong, Hong Kong
{jwdai,rchung}@mae.cuhk.edu.hk

Abstract. We describe a system of embedding codes into projection display for structured light based sensing, with the purpose of letting projector serve as both a display device and a 3D sensor. The challenge is to make the codes imperceptible to human eyes so as not to disrupt the content of the original projection. There is the temporal resolution limit of human vision that one can exploit, by having a higher than necessary frame rate in the projection and stealing some of the frames for code projection. Yet there is still the conflict between imperceptibility of the embedded codes and the robustness of code retrieval that has to be addressed. We introduce noise-tolerant schemes to both the coding and decoding stages. At the coding end, specifically designed primitive shapes and large Hamming distance are employed to enhance tolerance toward noise. At the decoding end, pre-trained primitive shape detectors are used to detect and identify the embedded codes – a task difficult to achieve by segmentation that is used in regular structured light methods, for the weakly embedded information is generally interfered by substantial noise. Extensive experiments including evaluations of both code imperceptibility and decoding accuracy show that the proposed system is effective, even with the prerequisite of incurring minimum disturbance to the original projection.

1 Introduction

The improving performance and declining price of digital video projectors make it possible to use them prevalently. Being able to generate arbitrarily large display is a feature of projectors that makes them exceedingly attractive, especially in applications that demand portability.

On the other hand, the adoption of structured light illumination has been proven to be an effective and accurate means for 3D information perception [1]. Recently, the availability of pico projectors with average dimensions of $4 \times 2 \times 1$ inches has widely extended the application domain of structured light system. There are already pocket DCs, DVs and cellular phones in the consumable market that have both projector and camera built-in, making it possible to implement structured light system in hand-held consumer electronic products.

For these reasons projector-camera (ProCam) system has been actively researched in the last few years. Many research groups apply projectors in unconventional ways to develop new and innovative information displays that go beyond simple screen presentation [2].

G. Bebis et al. (Eds.): ISVC 2012, Part I, LNCS 7431, pp. 654–664, 2012.

Some researchers designed structured light system in the non-visible spectrum [3]. That way the media for regular projection and structure light sensing can be made separate. However, if structured light and regular projection can be achieved through the same projector, additional hardware demand could be reduced and device cost and size could be diminished. This leads to the concept of Imperceptible Structured Light (ISL). ISL makes use of a projection frame rate that is beyond the perceptibility of human vision, so that some of the projected frames can be "stolen" for structured light use without the user perceiving it. Specifically, it modulates the projected display either spatially or temporally to embed code patterns into the projection for structured light sensing. Due to the limitation of human visual perception, such embedded code patterns can be made largely or entirely unnoticeable to the user (the degree of unnoticeability depends upon how fast is the projection frame rate and how wide is the intensity contrast between the intended projection and the embedded code), but cameras synchronized to the modulation are able to reconstruct the embedded codes for structured light sensing.

There is however challenge in embedding codes into user-specified arbitrary projection. While the codes should be made as undetectable as possible to the user, they have to be decodable to the camera for the purpose of structured light sensing. On top of the dilemma, there is the inevitable fact that the displayed signals are generally corrupted by substantial noise that arises from the nonlinearity of the projector, the sensing resolution and other limits of the camera, and the variation of the ambient illumination. The objective of this work is to deal with the dilemma.

This article describes a novel method of embedding imperceptible structured codes into arbitrarily intended projection. Through precise projector-camera synchronization, structured codes consisting of three primitive shapes are embedded into the projection, in a way that is imperceptible to viewers but extractable from the "difference" of successive images captured by a camera. To make the decoding process more robust against noise, we do not extract the codes by region segmentation in the image domain. Instead we employ specially trained classifiers to detect and identify the codes. To enhance the error tolerance further, specially designed primitive shapes and large Hamming distance are adopted in the spatial coding. Even with some bits of the codewords missed or wrongly coded, the correct correspondence could still be derived correctly.

The remainder of this paper is structured as follows. In Section 2, related works on imperceptible structured light sensing are briefly reviewed. The principle of embedding imperceptible codes along with robust coding and a noise-tolerant decoding mechanism are described in Section 3. In Section 4, system setup and experimental results are shown. Conclusion and possible future work are offered in Section 5.

2   Related Works

A proof of concept for embedding invisible structured light patterns into DLP (Digital Light Processing) projections first appeared in the "Office of the Future" project [4]. In this work, binary codes are embedded by projecting temporally alternating code images and their complements. Provided that the frequency of projection reaches the *flicker fusion threshold* ($\geq 75Hz$), the pattern and the inverse pattern are visually integrated over time in human perception, and the illumination has the appearance of a flat field ("white" light) to humans. However, the demonstration required significant modification effort on the projection hardware and firmware, including removal of the color wheel and reprogramming of the controller. The resulting images were also in greyscale only. The implementation of such a setting was impossible without mastering and full access to the projection hardware.

Cotting et. al. introduced a coding scheme [5] that synchronizes a camera to a specific time slot of a DLP micro-mirror flipping sequence in which imperceptible binary patterns are embedded. However, not all mirror states are available for all possible intensities, and the additional hardware, DVI repeater with tapped vertical sync signal, is not an off-the-shelf instrument.

However, with the development of digital projection technology, some so-called 3D compatible DLP projectors with fresh rate of $120Hz$ or higher emerged recently. This makes it possible to implement imperceptible structured light without any hardware modification or extra assisting hardware. Many researcher began to study how to determine the embedded intensity properly to guarantee code imperceptibility.

In [6], subjective evaluation results and their statistical analysis on the visual perceptibility of embedded codes in different ways were reported. The factors affecting code visibility are also concluded. Park et al. [7] presented a technology for adaptively adjusting the intensity of the embedded code with the goal of minimizing its visibility. It was regionally adapted depending on the spatial variation of neighboring pixels and their color distribution in the YIQ color space. The final code intensity was then weighted by the estimated local spatial variation. Since two manually defined parameters adjusted the overall strength of the integrated code, the system was not able to automatically calculate an optimized intensity. Grundhofer et al. [8] proposed a method considering the capabilities and limitations of human visual perception for embedding codes. It estimated the Just Noticeable Differences (JND) based on the human contrast sensitivity function and adapted the code intensity on the fly through regional properties of the projected image and code, such as luminance and spatial frequencies. The shortcoming of this method was that some parameters need be pre-measured using some optical devices (e.g. photometer), which were not accessible to nonprofessional users.

To the best of our knowledge, up to now, few works focus on the decoding method in imperceptible code embedding configuration, especially when huge external noise could exist.

3   Methodology

3.1   Principle of Embedding Imperceptible Codes

The fundamental principle behind imperceptible structured code embedding is the temporal integration achieved by projecting each image twice at high frequency: a first image containing actual code information (e.g., by adding or subtracting a certain amount (Δ) to or from the pixels of the original image, depending upon the code) and a second image that compensates for the distortion in the first image. The vital aspects of ISL sensing are code embedding and projector-camera synchronization.

Since projection is generally in color, it is possible to embed color code through three different channels theoretically. However, to enhance code robustness toward noise, we use binary code and embed it into all three color channels simultaneously. Let B, O, I and I' be the binary code image, the original image, the projected image, and the complementary image (that is also projected) respectively. Then the projected image and complementary image could be formulated as

$$I_i(x, y) = O_i(x, y) + P(x, y), \tag{1}$$

$$I'_i(x, y) = O_i(x, y) - P(x, y), \tag{2}$$

$$P(x, y) = \begin{cases} \Delta, & when \quad B(x, y) = 1; \\ 0, & when \quad B(x, y) = 0. \end{cases} \tag{3}$$

where $i = \{R, G, B\}$ indicates red, green and blue channels, Δ is the embedded intensity.

To avoid intensity saturation at lower and higher intensity levels when adding or subtracting Δ, the original image needs to have the intensity range in each color channel compressed to between Δ to $255 - \Delta$. Since the embedded intensity required in the coding is small enough, the visual degradation due to contrast reduction is generally negligible.

The degree of imperceptibility thus depends upon the embedded intensity. A larger intensity ensures that the code be more tolerant toward noise and more readable in the image of the projection, whilst a smaller intensity makes the embedded codes more invisible. In our design, code imperceptibility has higher priority, and thus embedded intensity is set to a very small value.

In order to achieve imperceptible structured light projection, the frequency of projection must exceed the flicker fusion threshold, which is $75Hz$ for most of the people. The embedded codes could be internally and simply extracted from the "subtraction image"[1] between consecutively captured images as

$$S(x, y) = \max_i [C_i(x, y) - C'_i(x, y)], \quad i = \{R, G, B\}. \tag{4}$$

In principle, the subtraction image should be a binary image that has intensity values between 2Δ and 0. However, the subtraction image in reality is generally disturbed by rather substantial external noise. Since the embedded intensity is always small, the subtraction image has low signal-to-noise ratio. It is generally nontrivial to retrieve the embedded codes. In the rest of this section, we describe how robust coding and noise-tolerant decoding approaches can help tackle the issue.

3.2 Design of Embedded Pattern

Considering the constraints of imperceptible code embedding, we employ the spatial multiplex scheme to design our pattern. Due to the choice of using binary code for robust code embedding, the symbols cannot be coded with different colors. We thus use an alphabet set comprising three different geometrical primitives: cross, sandglass, and rhombus, as shown in Fig. 1. There are three advantages of this configuration. First, all the shapes own a natural center point, which simplifies the shape identification process in the decoding stage. Then, there are sufficient variations between different

---

[1] All the subtraction images in this article are scaled to [0, 255] for illustration purpose.

658   J. Dai and R. Chung

shapes; even with large disturbance from noise on the shapes, the decoding method could distinguish them. Moreover, the directional information carried by the cross shape could rectify the observation window during the step of neighborhood detection without enforcing any other constraint.

Fig. 1. The primitive shapes: cross, sandglass and rhombus

In the decoding stage, the centroid of each detected primitive is considered as the feature point position, and the 9-bit codeword associated to each feature point is composed of the elements in the 3×3 window centered on it. In traditional structured light methods, the uniqueness of the codeword is usually assured by M-arrays (perfect maps), which are random arrays of dimensions $r \times v$ in which a sub-matrix of dimensions $n \times m$ appears only once in the whole pattern [9]. The M-arrays give a total of $rv = 2^{nm} - 1$ unique sub-matrices in the pattern and a window property of $n \times m$. However, the Hamming distance between the codewords is 1, which is generally too small for our code embedding scenario in which the codeword retrieval error could be large due to noise. In our system, we generate a matrix of dimensions 27×29 using the method proposed by Albitar [10], in which 95.97% of the codewords have a Hamming distance higher than 3 and the average Hamming distance is $\bar{H} = 6.0084$, so that even some bits in the codeword are missed or incorrectly coded, the codeword is still distinguishable. On the basis of this matrix, the binary code image composed of the primitive shapes appears like the one illustrated in Fig. 2, in which the size of each primitive shape is a collection of 11×11 pixels while the interval between each shape is 11 pixels. The total number of feature points is 783.

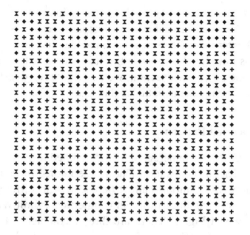

Fig. 2. The embedded binary code image

3.3 Primitive Shape Identification and Decoding

In the decoding stage, the existence of intense noise (due to influence from projector projection, camera sensing, ambient illumination and object surface reflection) makes it impossible to segment the primitive shape by the integrated use of region segmentation and edge or contour detection as in ordinary structured light methods. Here, we regard the primitive shapes as objects to "identify" and "detect" rather than "segment".

Compared with other object identification or recognition methods, the machine learning approach proposed by P. Viola [11] has been shown to be capable of processing images rapidly with high detection rates for visual object detection. The approach is adopted here for training a detector that identifies the three primitive shapes. Below we use cross shape as an example to describe the procedure of detector training.

The performance of any training-based detector has a great deal to do with the availability of training samples. Unlike generic objects like human face, body, or vehicle, which have a large number of samples in a great many of public databases, we have to collect the specific training samples ourselves in the required configuration. 500 color images with different content were collected from Internet, and 40 cross shapes were embedded in those images at different positions to generate 500 pairs of projected images and complementary images. By projecting them to a locally smooth textureless surface with orientation variations, 500 subtraction images could be derived from image capture exercises. The sub-images containing cross shapes were then segmented by manual labeling, which were considered as positive training samples. The background images with holes filled by random noise were divided into small patches to generate negative training samples. The training sample preparation process is shown in Fig. 3.

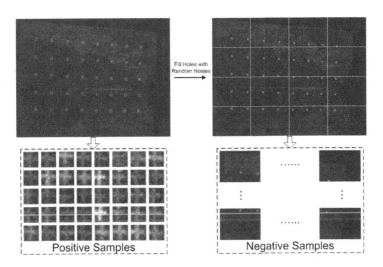

Fig. 3. Training sample preparation

To obtain the optimal performance, the positive samples were resized to 20×20, and the extended haar-like features and Gentle Adaboost algorithm were employed, following the suggestion in [12]. Eventually, from over 7000 positive samples and 3000 negative samples, a 16-stage cascade classifier for cross detection was trained. Following the same procedure, the detectors for sandglass and rhombus shapes were derived as well.

By using the pre-trained primitive shape detectors, the centroid of each primitive, i.e., the position of each feature point, can be determined. Once a feature point is extracted from the image, its codeword can be produced from the associated 3×3 intensity window centered on the feature point. Its corresponding point on the projector's display panel is known *a priori*. This way 3D position on the object surface can be determined via triangulation. The above is the 3D sensing step we use in the system.

4   Experiments

To assess the feasibility of the proposed method for embedding imperceptible codes in regular projection, we conducted experiments on both imperceptibility evaluation and accuracy evaluation.

The projector-camera system we used consisted of a DLP projector (Mitsubishi EX240U projector) of 1024×768 resolution and $120Hz$ refresh rate, and a camera (Adimec OPAL-1000 CCD camera with Myutron FV1520 f15mm lens) of 1024×1024 resolution and $123fps$ frame rate, both being off-the-shelf equipments. The focal length of the camera was fixed ar $15mm$, while that of the projector was in the range of $25 - 31mm$. The ILS was configured for a working distance (the distance from the camera to the mean position of the working area) of about $800mm$.

4.1   Imperceptibility Evaluation

Embedded code imperceptibility and user satisfaction are of the first priority in the system design. We conducted a subjective evaluation based on a questionnaire. Ten persons were invited to participate in this experiment, of which six were male and four were female, and seven wearing glasses. 500 images were collected from Google Image randomly, in which our proposed pattern was embedded with different intensities. The viewers were seated in front of a white planar screen at a distance of about $2m$, and asked to comment on the images projected to the screen. The questions asked were simplified from the questionnaire in [6], focusing on the feeling of flickering, the recognition of image deterioration, and the overall satisfaction for projection quality. The score for each question was divided into 10 levels.

The average scores of the subjective evaluation are illustrated in Fig. 4. When the embedded intensity is small, i.e., $\Delta = 5, 10$, the viewer could rarely notice the embedded codes and were satisfied with the projection quality. With the increase of the embedded intensity, the viewers' sense of flickering and image degradation became stronger. When $\Delta = 25$, almost every viewer was not satisfied with the projection quality.

In practice, because it was difficult to retrieve weakly embedded codes with the standard commercial cameras, we choose $\Delta = 10$ in our configuration, striking a compromise between user satisfaction and code imperceptibility.

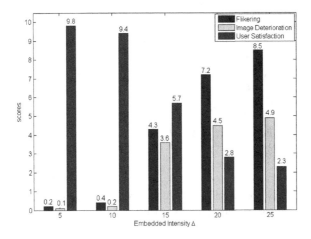

Fig. 4. Subjective evaluation result on code imperceptibility

4.2 Accuracy Evaluation

After code imperceptibility evaluation, the experiments on code retrieval accuracy were carried out. To assess accuracy, experimental data with ground-truth were required. Three different primitives and the spatially coded pattern image were embedded into the 500 images used for imperceptibility evaluation respectively, with intensity $\Delta = 10$. Then the projected and complementary images were projected successively to a smooth surface, while the camera conducted synchronized capture. The surface was adjusted to different positions and orientations with respect to the camera to involve sufficient shape distortion in the test data. Then the subtraction images embracing embedded codes information were derived for accuracy evaluation. The ground-truth was obtained by manual labeling in the image data captured under binary pattern illumination.

Experimental results in some subtraction images are presented in Fig. 5(a). The four sub-figures display the cross (top-left), sandglass (top-right), rhombus (bottom-left) shapes, and the spatially coded pattern (bottom-right) respectively. For qualitative evaluation, the detected features are indicated by rectangles, and in bottom-right sub-figure, the cross,sandglass and rhombus shapes are separately marked by red, green and blue rectangles. The average feature point detection errors along the x-axis and y-axis (as shown in Fig. 5 (b)) are formulated as $\epsilon_X = \frac{1}{N}\sum_{i=1}^{N}|X_d - X_g|_i$, $\epsilon_Y = \frac{1}{N}\sum_{i=1}^{N}|Y_d - Y_g|_i$, where N is the total number of embedded shapes, (X_d, Y_d) and (X_g, Y_g) are the detected position coordinates and ground-truth respectively. More detailed quantitative testing results are listed in Table 1. Through the proposed method, 91.23% of the embedded feature points could have their correspondences found correctly. By analyzing the missed and false detection cases, we find that the mistakes were mainly caused by large noise that occludes the embedded codes, implying that external noise has the greatest influence on the decoding process.

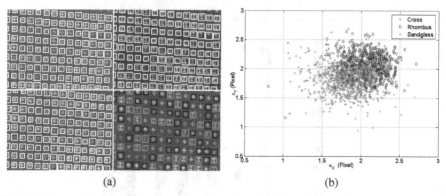

(a)   (b)

Fig. 5. (a) Some qualitative experiment results on (embedded) code detection accuracy. (b) The average detection error upon the three primitive shapes.

Table 1. The quantitative experiment results on (embedded) code detection accuracy

	Hits(%)	Missed(%)	False(%)	$[\epsilon_X, \epsilon_Y]$(pixel)	Corr. Acc.(%)
Cross	86.21	11.63	2.16	[1.931, 1.927]	—
Rhombus	85.83	12.57	1.60	[2.056, 2.051]	—
Sandglass	87.49	11.64	0.87	[1.816, 1.821]	—
Whole Pattern	86.33	11.06	2.61	[2.013, 2.043]	91.23

4.3   3D Reconstruction Accuracy Evaluation

To evaluate the accuracy of the proposed method in the 3D reconstruction task, we conducted an experiment to compare the performance of our method with that of a classical structured light method using visible patterns. As shown in Fig. 6-(a1)(b1)(c1) and Fig. 6-(a2)(b2)(c2), three objects (sphere, cone and cylinder) with known dimensions were illuminated by visible binary pattern image (the same as Fig. 2) and code embedded normal projection respectively.

In the classical structured light scenario, some feature points were extracted by segmentation and shape identification using the method proposed in [10]; whilst in our code embedded normal projection scenario, the feature points were detected and classified through the pre-trained primitive shape detectors. The depth value of each feature point was calculated through triangulation using the intrinsic and extrinsic parameters of the projector and camera. Then on the basis of point clouds calculated through our method, surfaces were rendered as illustrated in Fig. 6-(a3)(b3)(c3). Since the dimensions of the objects are known, we could conduct quantitative accuracy assessment. The residual mean error E_μ and standard deviation E_σ of the calculated 3D points with respect to the ground-truth were listed in Table 2. It is evident that our method has almost the same performance as that of the classical structured light method in 3D reconstruction. By the reason that the textures on the cylindrical object obstruct code retrieval, the reconstruction error on the particular object is greater than those of the other two objects. It is worth pointing out that in our method the decoding process was conducted in the subtraction image, which would alleviate the texture's influence to a certain extent.

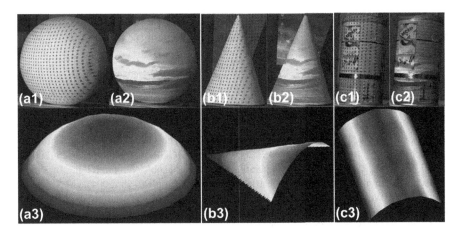

Fig. 6. Some results of 3D reconstruction

Table 2. 3D reconstruction accuracies on a variety of shapes

Object	General SL [10]		Our Method	
	$E_\mu(mm)$	$E_\sigma(mm)$	$E_\mu(mm)$	$E_\sigma(mm)$
Sphere	1.502	0.576	1.410	0.587
Cylinder	2.054	0.824	1.939	0.762
Cone	1.383	0.557	1.391	0.564

5   Conclusion and Future Work

We have described a novel system of embedding imperceptible structured codes into user-define arbitrary projection, that strikes the balance between imperceptibility and detectability of the codes. Through precise projector-camera synchronization, structured codes consisting of three primitive shapes are embedded into the regular projection, in a way that is imperceptible to the user but extractable by a camera (via the difference of successive images). Disturbance from various external sources makes it difficult to retrieve the codes by the region segmentation approaches adopted in general structured light systems. Instead of segmenting the codes, specially trained classifiers are employed to detect and identify them. To increase the robustness of code extraction, large Hamming distance are adopted in spatial coding. Even if some bits are missed or wrongly decoded, the correct correspondence between the projection panel and the image plane could still be arrived at correctly for structured light sensing. Extensive experimentation shows that the method is a promising one.

In the current system, the image capture interval is $10ms$. In sensing object that moves fast, the substantial displacement between successive images will result in blur or destruction of the embedded codes in the difference image. Some compensation methods need be in place to deal with the problem. In addition, the embedded code could be denser for more precise 3D sensing. New coding scheme capable of generating denser patterns should be used. The proposed method enables a regular projector to serve the dual role of a display device as well as a 3D sensor. That provides a platform for more natural user interface schemes. Our future work will lie on these directions.

References

1. Salvi, J., Fernandez, S., Pribanic, T., Llado, X.: A state of the art in structured light patterns for surface profilometry. Pattern Recognition 43, 2666–2680 (2010)
2. Bimber, O., Iwai, D., Wetzstein, G., Grundhöfer, A.: The visual computing of projector-camera systems. In: ACM SIGGRAPH 2008 Classes. SIGGRAPH 2008, pp. 1–25 (2008)
3. Fofi, D., Sliwa, T., Voisin, Y.: A comparative survey on invisible structured light. In: Proc. of Machine Vision Applications in Industrial Inspection XII, pp. 90–98 (2004)
4. Raskar, R., Welch, G., Cutts, M., Lake, A., Stesin, L., Fuchs, H.: The office of the future: A unified approach to image-based modeling and spatially immersive displays. In: Proc. of SIGGRAPH 1998, pp. 179–188 (1998)
5. Cotting, D., Naef, M., Cross, M., Fuchs, H.: Embedding imperceptible patterns into projected images for simultaneous acquisition and display. In: Proc. of IEEE and ACM ISMAR, pp. 100–109 (2004)
6. Park, H., Seo, B.-K., Park, J.-I.: Subjective evaluation on visual perceptibility of embedding complementary patterns for nonintrusive projection-based augmented reality. IEEE Trans. Circuits Syst. Video Technol. 20(5), 687–696 (2010)
7. Park, H., et al.: Content adaptive embedding of complementary patterns for nonintrusive direct-projected augmented reality. In: HCI International, pp. 132–141 (2007)
8. Grundhofer, A., Seeger, M., Hantsch, F., Bimber, O.: Dynamic adaptation of projected imperceptible codes. In: Proc. of IEEE and ACM ISMAR, pp. 1–10 (2007)
9. Etzion, T.: Constructions for perfect maps and pseudorandom arrays. IEEE Transactions on Information Theory 34, 1308–1316 (1988)
10. Graebling, P.: Robust structured light coding for 3d reconstruction. In: Proc. of IEEE ICCV, pp. 1–6 (2007)
11. Viola, P., Jones, M.: Rapid object detection using a boosted cascade of simple features. In: Proc. of IEEE CVPR, pp. 511–518 (2001)
12. Lienhart, R., Kuranov, A., Pisarevsky, V.: Empirical analysis of detection cascades of boosted classifiers for rapid object detection. In: Proc. of DAGM PRS, pp. 297–304 (2003)

Moving Object Detection via Robust Low Rank Matrix Decomposition with IRLS Scheme

Charles Guyon, Thierry Bouwmans, and El-Hadi Zahzah

Laboratoire MIA (Mathematiques, Image et Applications)
University of La Rochelle
thierry.bouwmans@univ-lr.fr

Abstract. Moving object detection is a key step in video surveillance system. Recently, Robust Principal Components Analysis (RPCA) shows a nice framework to separate moving objects from the background when the camera is fixed. The background sequence is then modeled by a low rank subspace that can gradually change over time, while the moving objects constitute the correlated sparse outliers. In this paper, we propose to use a low-rank matrix factorization with IRLS (Iteratively Reweighted Least Squares) scheme for RPCA decomposition and to address in the minimization process the spatial connexity of the pixels. Experimental results on different datasets show the pertinence of the proposed method.

1 Introduction

The detection of moving objects is the basic low-level operations in video analysis. The basic operation consists of separating the moving objects called "foreground" from the static information called "background" [1][2][3]. Recent reseach on robust PCA shows qualitative visual results with the background variations appromatively lying in a low dimension subspace, and the sparse part being the moving objects. First, Candes et al. [4] proposed a convex optimization to address the robust PCA problem. The observation matrix is assumed represented as $A = L + S$ where L is a low-rank matrix and S must be sparse matrix with a small fraction of nonzero entries. This research seeks to solve for L with the following optimization problem:

$$\min_{L,S} \ ||L||_* + \lambda||S||_1 \quad \text{subj} \quad A = L + S \tag{1}$$

where $||.||_*$ and $||.||_1$ are the nuclear norm (which is the L_1 norm of singular value) and l_1 norm, respectively, and $\lambda > 0$ is an arbitrary balanced parameter. Under these minimal assumptions, this approach called Principal Component Pursuit (PCP) solution perfectly recovers the low-rank and the sparse matrices. Candes et al. [4] showed results on face images and background modeling that demonstrated encouraging performance. However, PCP is limited to the low-rank component being exactly low-rank and the sparse component being exactly sparse but the observations in real applications are often corrupted by noise

G. Bebis et al. (Eds.): ISVC 2012, Part I, LNCS 7431, pp. 665–674, 2012.

affecting every entry of the data matrix. Therefore, Zhou et al. [5] proposed a stable PCP (SPCP) that guarantee stable and accurate recovery in the presence of entry-wise noise. However, PCP and SPCP present the following limitation: When few columns of the data matrix are generated by mechanims different from the rest of the columns, the existence of these outlying columns tends to destroy the low-rank structure of the data matrix.

In this paper, we propose a novel algorithm for moving object detection based on a robust matrix factorization. For a data matrix A, we assume that it is approximatively low-rank, and a small part of this matrix is corrupted by the outliers. The aim of the proposed method is to alleviate the limitation of PCP and SPCP by addressing the spatial connexity of the pixel to obtain a robust estimation of the true low-rank and the sparse structure of the matrices L and S. Our contributions can be summarized as follows: 1) Addition of spatial constraint to minimization process, 2) IRLS alternating scheme for weighted the 2-parameters $||.||_{\alpha,\beta}$ for matrix low rank decomposition. The rest of this paper is organized as follows. In Section 2 and 3, we present the proposed method based on a robust low-rank matrix factorization which allows us to detect moving objects in dynamic backgrounds. Then, in Section 4, we present comparison and evaluation versus the state-of-the-art methods. Finally, the conclusion is provided in Section 5.

2   1-D Case: L_1 Minimization with Spatial Constraint

In most applications, video surveillance data is assumed to be compose of background and moving objects components. We improved the decomposition (1) by adding a third component which models the noise. We believe the regression task as a crucial part of the proposed decomposition algorithm and illustrate the main keys through a simple 1-D example shown in Fig. 1. Consider the following minimization problem, where A is a dictionary matrix (row order) and b is a row vector,

$$\underset{x}{\text{argmin}} \ ||Ax - b||_\alpha + \mu||\nabla_s \phi(E)||_1, \quad E = |b - Ax| \tag{2}$$

By this, we impose the error E should be a connexe shape, through the TV (Total Variation) of the error must be small, where ∇_s is the spatial gradient. Note that the problem is convexe for $\alpha > 1$ and the standard IRLS (Iteratively reweighted least squares) scheme is given by,

$$\underset{x}{\text{argmin}} \ ||Ax - b||_\alpha \quad \left| \begin{array}{l} D^{(i)} = \text{diag}((\varepsilon + |b - Ax^{(i)}|)^{\alpha-2}) \\ x^{(i+1)} = (A^t D^{(i)} A)^{-1} A^t D^{(i)} b \end{array} \right. \tag{3}$$

As note in [6] and [7], it was proven that a suitable IRLS method is convergent for $1 < \alpha < 3$. Some modifications allow to get more numericaly stability and let us to choose freely $\alpha \in [1, \infty[$ with an adapted λ_{opt}.

$$\begin{array}{l} r^{(i)} = b - Ax^{(i)} \\ D = \text{diag}((\epsilon + |r^{(i)}|)^{\alpha-2}) \\ y^{(i)} = (A'DA)^{-1} A'Dr^{(i)} \\ x^{(i+1)} = x^{(i)} + (1 + \lambda_{opt})y^{(i)} \end{array} \tag{4}$$

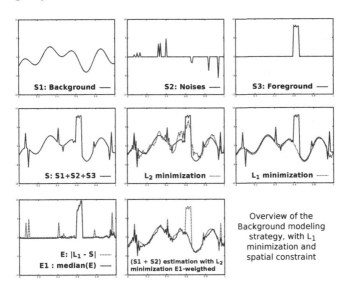

Fig. 1. Schematic example for the 1-D case: We illustrate different fitting strategies with random dictionary basis (cosine function, wavelets, ...) and a composite signal S

The algorithm is twice iterative, where we try to get an optimal x and an optimal λ at each step.

$$
\begin{aligned}
c^{(i)} &= Ay^{(i)} \\
d^{(i)} &= b - A(x^{(i)} + y^{(i)}) \\
\lambda_{opt} &= \operatorname*{argmin}_{\lambda} \ ||c^{(i)}\lambda - d^{(i)}||_{\alpha}
\end{aligned}
$$

$$
\begin{aligned}
\lambda^{(0)} &= \Lambda(\alpha) \\
s^{(k)} &= d - c\lambda^{(k)} \\
E &= \operatorname{diag}((\epsilon + |s^{(k)}|)^{\alpha-2}) \\
\lambda^{(k+1)} &= \lambda^{(k)} + (1 + \Lambda(\alpha))\frac{c^t E s^{(k)}}{c^t E c}
\end{aligned}
\tag{5}
$$

With a fixed λ_{opt}, we should choose $\lambda_{opt} = \Lambda(\alpha)$ to allow $1 < \alpha < \infty$ and distinguish three cases:

$$
\Lambda(\alpha) = \begin{cases}
5/6 & \text{if } \alpha \leq 1 \\
-\frac{2}{3}\alpha + \frac{3}{2} & \text{if } 1 < \alpha < 1 + \frac{3}{4} \\
\frac{1}{\alpha-1} - 1 & \text{if } \alpha \geq 1 + \frac{3}{4}
\end{cases}
\tag{6}
$$

Observe for case $\alpha > 2$, convergence is achieved when $0 < 1 + \lambda < \frac{2}{\alpha-1}$. More generally, we can solve matrix regression problem with two parameters norm (α, β) and a weighted matrix (W).

$$
\min_{X} \ ||AX - B||_{\alpha,\beta} \quad \text{where} \quad ||M_{ij}||_{\alpha,\beta} = \left(\sum_{i=1}^{n}\left(\sum_{j=1}^{m} W_{ij} |M_{ij}|^{\beta}\right)^{\frac{\alpha}{\beta}}\right)^{\frac{1}{\alpha}}
\tag{7}
$$

This is solved by the following algorithm:

$$
\begin{aligned}
&\text{Until X is stable, repeat on each } k\text{-columns} \\
&\left|
\begin{aligned}
R &\leftarrow B - AX \\
S &\leftarrow \varepsilon + |R| \\
D &\leftarrow \operatorname{diag}(S_{ik}^{\beta-2} \circ (\textstyle\sum_{j}(S_{ij}^{\beta} \circ W_{ij}))^{\frac{\alpha}{\beta}-1} \circ W_{ik}) \\
X_{ik} &\leftarrow X_{ik} + (1 + \Lambda(\max(\alpha,\beta)))(A^t D A)^{-1} A^t D R_{ik}
\end{aligned}
\right.
\end{aligned}
\tag{8}
$$

3 Moving Object Detection via Robust Low-Rank Matrix Factorization

Let us denote the training video sequences $A = \{I_1, ... I_m\}, A \in \mathbb{R}^{n \times m}$ where I_j is a vectorized frame at time j and m is the number of training frames. Let each pixel (x,y) be characterized by its intensity in the gray scale. The decomposition involves the following model:

$$A = L + S = BC + S \tag{9}$$

where B is a low-rank matrix corresponding to the background model plus noise, and C allows us to reconstruct L by linear combination. S is a matrix which corresponds to the moving objects. The model involves the error reconstruction determined by the following constraints:

$$\min_{B \in \mathbb{R}^{n \times p}, C \in \mathbb{R}^{p \times m}} ||(A - BC) \circ W||_{\alpha, \beta} + \mu ||BC||_* \tag{10}$$

where $||.||_*$ denote the nuclear norm. The decomposition is split into two parts. firstly, we track 1-Rank decomposition since the first eigen-vector is strongly dominant in video surveillance.

$$
\begin{array}{l|l}
R_1 = A - B_1 C_1 & (1)\ \min_{B_1, C_1} ||R_1||_{1,1} \\
R = A - B_1 C_1 - B_r C_r & (2)\ \min_{B_r, C_r} ||R \circ \phi(R_1)||_{2, 1 \to 0}
\end{array} \tag{11}
$$

We use $||.||_{2,1 \to 0}$ instead of usual $||.||_{1,1}$ because it forces spatial homogeneous fitting. $\beta = (1 \to 0)$ means the β parameter decreases during iteration. First, we search a solution of the convex problem $||.||_{2,1}$, then use the solution as an initial guess for non-convex problem $||.||_{2,(1-\epsilon)}$. Finally, we find a local minimum of $||.||_{2,0}$ and hope that is near of the global minimum. In the case where $\alpha = \beta = 2$, the decomposition is usually solved by a SVD (Singular Value Decomposition). Thus, our SVD algorithm can be seen as an iterative regression. The proposed scheme determines alternatively the optimal coefficients, it means searching C for B fixed and searching B for C fixed.

$$
\begin{aligned}
C^{(k+1)} &= (A^t A)^{-1} A^t B^{(k)} \\
\bar{C}^{(k+1)} &= C^{(k+1)} \sqrt{C^{t(k+1)} C^{(k+1)}}^{-1} \\
B^{(k+1)} &= (A^t A)^{-1} A^t \bar{C}^{(k+1)}
\end{aligned} \tag{12}
$$

Additionnaly this alternating regression framework allows to associate a weigthed matrix W which is entrywise multiplied to the error ($\circ$ denotes the Hadamard product).

$$\min_{B,C} ||(A - BC) \circ W||_{\alpha, \beta} \tag{13}$$

We define a function ϕ that have two goals, smooth the error (like spatial median filtering) and transform the error for obtain a suitable mask for regression.

$$W = \exp(-\sigma \phi(|A - BC|)) \tag{14}$$

By including local penalty as a constraint in a RPCA decomposition, this explicitly increases local coherence of the error (therefore foreground/moving object).

4 Experimental Results

We have compared the proposed approach with recent RPCA approaches: LBD [8], LRR [9], SADAL [10] and GRASTA [11] algorithms. The experiments were conducted qualitatively and quantitatively on the Wallflower[1] dataset [12], I2R[2] dataset [13] and CDW[3] dataset [14]. The algorithms are implemented in Matlab.

4.1 Wallflower Dataset

This dataset provided by Toyama et al. [12] consists of seven video sequences, with each sequence presenting one of the difficulties a practical task is likely to encounter. The images are 160×120 pixels. For each sequence, the ground truth is provided for one image when the algorithm has to show its robustness to a specific change in the scene. Thus, the performance is evaluated against hand-segmented ground truth. The Fig. 2 and Fig. 3 show the qualitative results. For the quantitative evaluation, we used metrics based on the true negative (TN), true positive (TP), false negative (FN), false positive (FP) detections. Then, we computed the detection rate, the precision and the F-measure [15]. A good performance is then obtained when the F-measure is closed to 1. Table 1 shows the results obtained for each algorithms. The F-measure value of "Moved Object" sequence can't be computed due to the absence of true positives in its ground-truth. We have highlighted when one algorithm outperforms the other ones. As these encouraging results are obtained by using one ground-truth image, we have evaluated the proposed method on a dataset with more ground-truth images in the following sub-section.

Table 1. F-measure in percentage for LBD [8], LRR [9], SADAL [10], GRASTA [11], and our method (direct one-to-one correspondence with Fig. 3)

Dataset	Sequence	Frame	LBD	LRR	SADAL	GRASTA	Our method
	Bootstrap	00299	**70.18**	69.40	67.02	55.37	58.78
	Foreground Aperture	00489	60.69	50.10	71.55	75.14	**75.19**
	Light Switch	01865	57.74	36.76	**69.33**	28.35	58.49
Wallflower	Moved Objects	00985	-	-	-	-	-
	Time of Day	01850	71.43	54.41	**80.84**	79.80	80.43
	Waving Trees	00247	62.65	50.74	81.67	84.16	**84.70**
	Camouflage	00251	70.58	70.49	76.00	70.34	**82.27**

4.2 I2R Dataset

This dataset provided by [13] consists of nine video sequences, which each sequence presenting dynamic backgrounds or illumination changes. The size of the images is 176×144 pixels. For each sequence, the ground truth is provided for 20 images. Among this dataset, we have chosen to show results on six sequences due the limitation of pages (see Fig. 4 and Fig 5). Tab. 2 shows the corresponding qualitative results.

---

[1] http://research.microsoft.com/en-us/um/people/jckrumm/wallflower/testimages.htm
[2] http://perception.i2r.a-star.edu.sg/bk_model/bk_index.html
[3] http://www.changedetection.net/

Fig. 2. Wallflower dataset. From left to right (split in two columns): The original image, the background model computed by our method and their difference with enhanced variance.

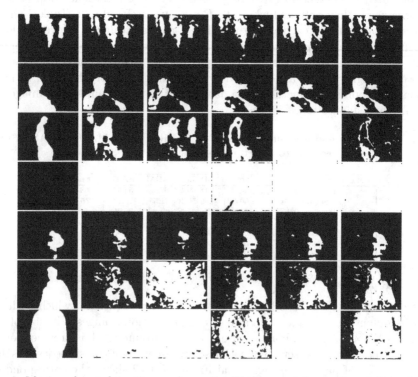

Fig. 3. Moving object detection masks on the Wallflower dataset. From left to right: Ground Truth, LBD, LRR, SADAL, GRASTA, our method. For *Moved Object* and *Camouflage* sequences, two fails occurs due of breaking *a priori*: Area of background must be bigger than foreground and inherently of evaluation procedure (*i.e.* F-measure), ground truth must not be void.

Table 2. F-measure in percentage for LBD [8], LRR [9], SADAL [10], GRASTA [11], and our method (direct one-to-one correspondence with Fig. 5)

Dataset	Sequence	Frame	LBD	LRR	SADAL	GRASTA	Our method
I2R	Campus	01650	59.93	59.99	68.14	61.84	**75.07**
	Curtain	22772	**91.08**	88.34	91.01	89.88	90.73
	Escalator	02424	65.28	63.41	59.81	**68.31**	63.68
	Hall	02926	73.58	69.75	77.57	78.25	**79.26**
	Shopping Mall	01862	80.15	77.88	**82.80**	79.54	81.94
	Water Surface	01499	90.37	83.57	**92.01**	90.97	91.02

Fig. 4. I2R dataset. From left to right (split in two columns): The original image, the background model and their difference.

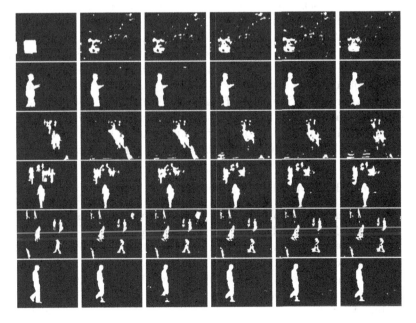

Fig. 5. Moving object detection masks on the I2R dataset. From left to right: Ground Truth, LBD, LRR, SADAL, GRASTA, our method.

4.3   Change Detection Workshop Dataset

This dataset provided by [13] consists of 31 video sequences. We have chosen to show results of ten videos, corresponding of the subsets *baseline* and *dynamic backgrounds* categories. The size of the images is variable. For each sequence, the ground truth is provided for each frame. Among this dataset, the results (see Fig. 5 and Fig. 6). Tab. 3 shows the corresponding qualitative results.

Table 3. F-measure results of LBD [8], LRR [9], SADAL [10], GRASTA [11], and our method apply on partial CDW dataset (direct one-to-one correspondence with Fig. 6)

Dataset	Sequence	Frame	LBD	LRR	SADAL	GRASTA	Our method
Baseline	PETS2006	00500	56.50	55.25	74.09	69.20	**78.12**
	highway	00800	87.48	83.99	93.82	92.29	**94.44**
	office	01886	76.01	58.42	75.67	78.36	**80.71**
	pedestrians	00471	**82.74**	79.02	81.79	81.24	81.54
Dynamic Back.	boats	02000	84.01	82.29	83.36	**84.02**	83.39
	canoe	00966	78.82	75.47	58.69	**79.73**	68.19
	fall	01492	62.22	59.46	78.49	65.27	**88.39**
	fountain01	00718	35.85	32.82	38.14	37.25	**54.46**
	fountain02	00740	85.21	83.93	87.24	**87.25**	86.23
	overpass	02529	72.01	72.55	**78.83**	78.65	70.33

4.4   Results Interpretation

The proposed approach outperforms the other algorithms for three sequences for Wallower dataset, five sequences for CDW dataset and obtain similar performance with SADAL for I2R dataset. Globally, our method provide better result for ten sequences, followed by SADAL and for the rest results are still acceptable. For this comparison, we used default parameters setting provided by the authors for all the algorithms and computed low rank decomposition on time window of 200 consecutive frames. Evaluation of time computing is difficult because thoses algorithms are implemented in non-optimized matlab codes, but we can give a line of thought (in timing of algorithm by timing of SVD): LBD 50, LRR 10, SADAL 70, GRASTA 240, and our method 110 (in Time/(SVD Time)).

4.5   Note on the Evaluation Procedure

Only grayscale images are computed, otherwise green channel is selected. Some high resolution video are subsampled, reasonably for memory allocating since for example, a 200 frames video of 640×480 cost 492 Mb. Additionally, a 5 × 5 median filter is postprocessed in order to suppress peak noise but this does not offer much advantage over a noise sensitive algorithm. For each sequence, the thresholding value is automatically choose for maximize the F-measure. That is an optimistic evaluation and may differ from real case, but algorithms are compared on the same manner and are not favored by a particular threshold selection strategy.

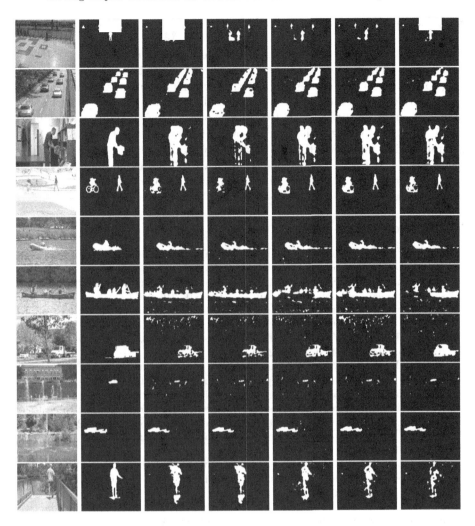

Fig. 6. Moving object detection masks on the CDW dataset. From left to right: Original, Ground Truth, LBD, LRR, SADAL, GRASTA, our method.

Fig. 7. Background modeling of selected sequences. By row: GRASTA and our method. Edges of shape are less fuzzy on the second row due to the add of the local constraint.

5 Conclusion

In this paper, we have presented a moving object detection method based on robust matrix factorization. This IRLS alternating scheme is conceptually simple, easy to implement and efficient for matrix low rank decomposition.

Furthermore, experiments on video surveillance datasets show that this approach is more robust than RPCA in the presence of dynamic backgrounds and illumination changes. Further research consists in developping an incremental version to update the model at every frame and to achieve real-time requirements. IRLS seems an good alternative to Lagrangian-based approach and is generally well suited to integrate an add-on process of acceleration of convergence.

References

1. Piccardi, M.: Background subtraction techniques: a review. In: IEEE International Conference on Systems, Man and Cybernetics
2. Bouwmans, T., Baf, F.E., Vachon, B.: Background modeling using mixture of gaussians for foreground detection - a survey. RPCS 1, 219–237 (2008)
3. Bouwmans, T.: Recent advanced statistical background modeling for foreground detection: A systematic survey. RPCS 4, 147–176 (2011)
4. Candes, E., Li, X., Ma, Y., Wright, J.: Robust principal component analysis? International Journal of ACM 58 (2011)
5. Wright, J., Peng, Y., Ma, Y., Ganesh, A., Rao, S.: Robust principal component analysis: Exact recovery of corrupted low-rank matrices by convex optimization. In: Neural Information Processing Systems, NIPS 2009 (2009)
6. Daubechies, I., Devore, R., Fornasier, M., Gntrk, C.S.: Iteratively reweighted least squares minimization for sparse recovery. Comm. Pure Appl. Math. (2008)
7. Osborne, M.R.: Finite algorithms in optimization and data analysis. John Wiley & Sons (1985)
8. Tang, G., Nehorai, A.: Robust principal component analysis based on low-rank and block-sparse matrix decomposition. In: Annual Conference on Information Sciences and Systems, CISS 2011 (2011)
9. Lin, Z., Liu, R., Su, Z.: Linearized alternating direction method with adaptive penalty for low-rank representation. In: NIPS 2011 (2011)
10. Ma, S.: Algorithms for sparse and low-rank optimization: Convergence, complexity and applications. Thesis (2011)
11. He, J., Balzano, L., Szlam, A.: Incremental gradient on the grassmannian for online foreground and background separation in subsampled video. In: International Conference on Computer Vision and Pattern Recognition (CVPR) (June 2012)
12. Toyama, K., Krumm, J., Brumitt, B., Meyers, B.: Wallflower: Principles and practice of background maintenance. In: ICCV 1999, pp. 255–261 (1999)
13. Li, L., Huang, W., Gu, I., Tian, Q.: Statistical modeling of complex backgrounds for foreground object detection. IEEE Transaction on Image Processing, 1459–1472 (2004)
14. Jodoin, P., Porikli, F., Konrad, J., Ishwar, P.: Change detection benchmark web site. In: IEEE Workshop on Change Detection (June 2012)
15. Maddalena, L., Petrosino, A.: A fuzzy spatial coherence-based approach to background foreground separation for moving object detection. Neural Computing and Applications, 1–8 (2010)

Comprehensible and Interactive Visualizations of GIS Data in Augmented Reality

Stefanie Zollmann[1], Gerhard Schall[1], Sebastian Junghanns[2], and Gerhard Reitmayr[1]

[1] Graz University of Technology
[2] GRINTEC Gmbh

Abstract. Most civil engineering tasks require accessing, surveying and modifying geospatial data in the field and referencing this virtual, geospatial information to the real world situation. Augmented Reality (AR) can be a useful tool to create, edit and update geospatial data representing real world artifacts by interacting with the 3D graphical representation of the geospatial data augmented in the user's view.

One of the main challenges of interactive AR visualizations of data from professional geographic information systems (GIS) is the establishment of a close linkage of comprehensible AR visualization and the geographic database that allows interactive modifications. In this paper, we address this challenge by introducing a flexible data management between GIS databases and AR visualizations that maintain data consistency between both data levels and consequently enables an interactive data roundtrip. The integration of our approach into a mobile AR platform enables us to perform first evaluations with expert end-users from utility companies.

1 Introduction

Geographic information systems (GIS) support civil engineering companies in managing existing or future utility infrastructures. Locating existing assets during construction work (e.g. gas or water pipes), surveying and visualizing the planned construction in the context of existing structures are some of the tasks which benefit from GIS. Efficient utility location tools and computer-assisted management practices can largely reduce costs and therefore are worth to be continuously improved.

Some companies already employ mobile GIS systems for on-site inspection (e.g. ARCGIS for Android[1]). However, current visualization techniques implemented in these tools do not show the relation of GIS data to the real world context and still involve the tedious task of referencing assets correctly to the real world.

Using Augmented Reality (AR) as an interface to extend mobile GIS systems has the potential to provide significant advances for the field of civil engineering by supporting the visual integration of real worlds and existing assets. AR is an emerging user interface technology superimposing registered 3D graphics over the user's view of the real world in real-time [1]. The visualization of both, real and virtual geospatial information, at the same time in reference to each other has a big potential to avoid errors

---

[1] http://www.arcgis.com

G. Bebis et al. (Eds.): ISVC 2012, Part I, LNCS 7431, pp. 675–685, 2012.

Fig. 1. Interactive planning and surveying with mobile AR. Left: Users with setup. Middle left: Creating and surveying a new cable with the setup. Middle right: Manipulating the cable. Right: Comprehensible visualization showing an excavation along the cable.

and to decrease workload. The goal of this project is to enable - with AR technology - information access, interactive planning and surveying.

AR can simplify such tasks by presenting an integrated view of the geospatial models in 3D and providing immediate feedback to the user and intuitive ways for data capturing, data correction and surveying. Merging the view of the real world with the presentation of geospatial objects raise some major challenges in terms of visualization: such as information clutter, depth perception issues or wrong interpretations of information. While GIS have a well standardized symbology used for representing geographical features, it is not adapted for 3D AR visualizations. To address these issues, we propose new comprehensible AR visualization techniques for GIS data following these three goals:

- The visualized information should be easily interpretable.
- The spatial arrangement of real and virtual structures should be understandable.
- Interactive modifications should be consistent in the AR visualization and the GIS database.

Our work contributes to the field of AR, specifically in the domain of visualization of GIS data and digital surveying applications. The core contributions are thereby (1) a novel transcoding layer that ensures data consistency between the GIS database and the displayed content during interactive modifications; (2) novel visualization techniques that help to interpret the GIS data as well as support the spatial understanding of the augmented scene. In addition we present (3) interactive manipulation techniques that are applied to the displayed GIS data and automatically update the GIS database through our transcoding layer. Finally, (4) the integration of all the aforementioned features into an outdoor AR system that was evaluated with expert users.

2   Related Work

Since the introduction of stand-alone geographic information systems in the late 1970s, there were many advancements in research and development on the visualization of geographic data. Mobile GIS already extends GIS from the office to the field by combining mobile devices, wireless communication and positioning systems. Whereas mobile GIS enable on-site capturing, storing, manipulating, analyzing and displaying of geographical data, the user still has to build the reference between geospatial data and real-world, as well as data interpretations by himself. Several research groups worked on bridging this gap by using AR for visualizing geographic data on-site. The group of Roberts were among the first that propose an AR overlay of underground assets over a live video image on a mobile device [2]. Later on, Schall et al. built an AR platform for experimenting with the visualization of underground infrastructure in the Vidente project [3]. The potential of AR for the Architecture, Engineering, and Construction (ACE) industry was also identified by Shin et al. [4]. For the construction industries, some research groups showed potential applications. For instance, Hakkariainen et al. describe an AR system for the visualization of Building Information Models (BIM) [5] and Golparvar-Fard et al. proposed an approach for visualizing construction progress [6].

On the other site, Mobile AR has lately gained more interest as research field. While the Touring Machine as one of the first mobile AR systems combing tracking position and orientation with differential GPS and a magnetometer was quite bulky [7], research groups have been working on making AR systems more compact and mobile such as the Tinmith system by Piekarski et al. [8] or the Vesp'R system by Veas and Kruijff [9]. Further research moved AR applications towards low-end devices such as cell phones [10].

Besides registration techniques and hardware design, an important aspect for the presentation of complex data in AR are visualization methods that support the comprehension of the presented information. Kalkofen et al. addressed this issues by implementing comprehensible visualization techniques based on Focus and Context techniques, filtering and stylization [11]. To enable the comprehensible visualization of GIS data, Mendez et al. introduced a transcoding pipeline that allows the mapping of different stylizations to GIS data [12]. Since these methods only work unidirectional, they can only provide passive viewing functions. The main goal of our work is a bi-directional data management allowing interactive manipulations and comprehensible visualization of GIS data at the same time. For this purpose we maintain data consistency between the GIS data base and the visualized geometric representations. This approach opens new prospects for outdoor AR-GIS by enabling interactive surveying in the field with visual references to the real-world.

3   Approach

To enable interactive modifications of geospatial data in the AR view while keeping consistency with the GIS database, we introduce two different types of data levels in our architecture: the *GIS-database level* and the *comprehensible 3D geometry level*. The *GIS-database level* consists of a set *features*, each describing one real world object with a 2D geometry and a set of attributes. Attributes are stored as key-value pairs and provide a description of various properties of the feature, such as type, owner or status. The *comprehensible 3D geometry level* consists of a set of 3D geometric representations of real world objects, such as extruded circles, rectangles, polygons and arbitrary 3D models; visualizing pipes, excavations, walls or lamps respectively.

To support the consistency between both data levels, we add a new data layer that serves as transmission layer between both. We call the additional layer *transcoding layer*, since it supports the bi-directional conversion of data between the comprehensible 3D geometry level and GIS database level. Each feature of the GIS database is stored

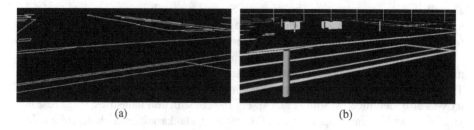

(a)   (b)

Fig. 2. (a) GIS data model, all data is represented by lines and points, which makes interpretations difficult. (b) Advanced geometries, GIS data is transcoded to a comprehensible representation showing cylindrical objects representing trees and cylindrical objects representing pipes. Color coding enables the user to interpret semantics.

as scene graph object with a set of attributes. Interactive modification in our system are conducted at this level and automatically propagated to the two other levels. Applying manipulations at the transcoding layer allows to manipulate feature data directly and avoids that manipulations are only applied to specific 3D geometries. For instance, the exclusive manipulation of a excavation representation of a pipe makes no sense without modifying the line feature representing the pipe. Furthermore, the transcoding layer has still access to the semantic information of a feature, which is important since interaction methods can depend on the type of object.

We introduce a bi-directional transcoding pipeline that creates the transcoding layer and the comprehensible 3D geometry level automatically from the geospatial data and updates the database with manipulations applied in the AR View. The pipeline is working as follows: (1) The conversion of GIS data into the transcoding layer and into specific comprehensible 3D geometries (Section 4). (2) Interaction techniques such as selection, manipulation and navigation allow the user to manipulate the various features (Section 5). The data connections between the 3 data layers guarantee data coherency while interacting with the data. To avoid administration overhead, modifications are recorded through tracing and only changed features will be written back to the GIS database.

4   From Geospatial Data to Comprehensible AR Visualization

Real-time comprehensible AR visualization and manipulation of geospatial objects requires a different data model than traditional geospatial data models. Abstract line and point features need to be processed to create 3D geometry representing more the actual shape than the surveyed line of points (compare Figure 2, (a) and (b)). Additional geometries have to be created automatically based on the features attributes to improve the comprehension of the presented information in the AR visualization. For instance, virtual excavations should help to understand the spatial arrangement of subsurface objects. All of these geometries need to be interactive and changeable, so that interactive manipulation allows updating the features. To support these operations we developed a bi-directional transcoding pipeline that realizes the conversion from GIS features to

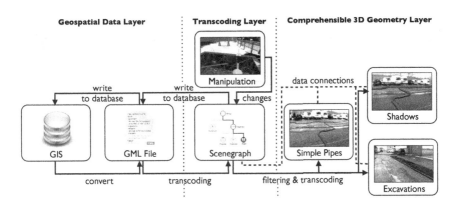

Fig. 3. Overview of the bi-directional transcoding pipeline. Data from the geospatial database is converted to a simple GML exchange format. The GML file is imported to the application and transcoded into the transcoding layer representation. Filtering and transcoding operations map the transcoding layer data to comprehensible 3D geometry. Data connections between the transcoding layer scene graph and the comprehensible geometry scene graph keep the visualization up-to-date. User interaction is applied directly to the transcoding layer.

comprehensible 3D data and back (Figure 3). A transcoding operation using a commercial tool extracts and translates geospatial data into a simpler common format for the mobile AR client, listing a set of features and their geometries (section 4.1). The AR system further filters the features for specific properties and applies transformations to generate 3D data from them (section 4.2). The 3D data structures are functionally derived from the geospatial data and stay up-to-date when it changes. Interactions operate directly on the feature data synchronizing the 3D visualization and features.

4.1 From Geospatial Data to the Transcoding Layer

For extracting the features from the geo-database, we use FME[2] which is an integrated collection of tools for spatial data transformation and data translation. FME represents a GIS utility that help users converting data between various data formats as well as process data geometry and attributes. The user interactively selects objects of interest in the back-end GIS, which is then exported to a GML3-based file format. A GML file represents a collection of features, where each feature describes one real world object. Geometric data being only available in 2D is converted to 3D representations by using a digital elevation model (DEM) and known laying depths of the subsurface objects. This step has to be done offline before starting the AR system since it requires external software and interactive selection of the export area. All following steps of the transcoding pipeline can be done during runtime but are not in realtime.

Finally, the GML file is converted into a scene graph format representing the data in the transcoding layer. For each feature, we create a scene graph object representing the semantic attributes and geometric properties of the feature. We support the main

---

[2] The Feature Manipulation Engine: http://www.safe.com

Fig. 4. Different visualizations of an electricity line feature. (a) A yellow rectangular extrusion as graphical representation. (b) A red cylindrical extrusion. (c) Showing an excavation along the electricity line. (d) Showing virtual shadows cast on the ground plane.

standard features of GML such as GMLLineStrings, GMLLinearRings, GMLPoint and GMLPolygon in the conversion step. In our current implementation, we use COIN3D[3] to implement the scene graph because it is easily extendable, but the approach can be easily adapted to be used with other scene graphs.

4.2 From Transcoding Layer Data to Comprehensible Geometries

The second step is the creation of comprehensible 3D geometries from the data of the transcoding layer. The final visualization of the geospatial data strongly depends on the application, application domain and the preferences of the user (e.g. color, geometry symbology or geometry complexity). For instance, a pipe could be represented in several ways, such as a normal pipe using an extruded circle (Figure 4(b)) or as an extruded rectangle to show an excavation around the pipe (Figure 4(c)). We call the conversion from the transcoding layer data representation to comprehensible geometries *geometry transcoding*. Different types of transcoding operations are called *transcoders* and each transcoder can be configured offline or during runtime to create different geometries from the same geospatial data. Each comprehensible 3D geometry is independent from other 3D representations of the corresponding feature but connected to the feature data in the transcoding layer (Figure 3).

The implementation of different visualization styles for different feature types is supported by a filtering-transcoding concept. The filtering step searches for a specific object type from attributes stored in the transcoding layer and the transcoding step transforms the data into specific geometric objects, which can later be displayed by the rendering system. The separation of the two steps allows for a very flexible system that can support many applications.

Filter Operations. The filtering step searches for specific tags in the semantic attributes of features in the transcoding layer and extracts the corresponding features. For instance, features can be filtered by a unique id, a class name, class alias, or type. The matching is implemented through regular expressions testing against the string values of the attributes. The features extracted by filtering can then be processed by an assigned transcoder. Filter rules and transcoding operations can be configured by the application

---

[3] http://www.coin3d.org

designer using a script or during runtime. The mapping of filters and transcoding operations has to be implemented in the application and allows to not only configure the visualization methods for specific data, but also a filtering of the presented information.

Transcoding Operation. Each transcoding operation depends on the type of transcoding and the transcoding parameters. The transcoding type assigns the underlying geometric operation for deriving the 3D geometry, for instance converting a line feature into a circular extrusion representing a pipe. The transcoding parameters configure visualization specifications such as color, width, radius, height, textures and 3D models of the objects. Multiple transcoders can be used to create different visualizations of the same data. The user selects the appropriate representation during runtime and the geometry created by the filtering and transcoding is finally rendered by the AR application. The transcoding is usually performed during start-up time of the application and takes several seconds for thousands of features.

Comprehensible Visualization Techniques. The filtering-transcoding concept allows us to create various geometric objects from the same semantic data. This is important since the comprehensible visualization of underground infrastructures on-site poses several challenges:

- Semantic interpretation: Geometric and appearance of visualized objects should fit to the requirements of users and application areas and is mostly achieved by using adequate colors and shapes or meaningful geometric models.
- Depth perception: A comprehensible arrangement of virtual and real objects in the augmentation is important to improve the comprehension of the visualized information and can be achieved by providing additional depth cues or avoiding clutter.

For AR visualization of underground infrastructure, depth perception is particularly challenging. Using simple overlays for visualizing underground objects via an AR X-Ray view can cause perceptual issues, such as the impression of underground objects floating over the ground. To avoid these problems, it is essential to either decide which parts of the physical scene should be kept and which parts should be replaced by virtual information [13] or which kind of additional virtual depth cues can be provided [14]. A comprehensible visualization should provide users with essential perceptual cues to understand the relationship of depth between hidden information and the physical scene.

The flexible data management allows us to address both challenges by using the filtering-transcoding pipeline to create easily interpretable geometric objects and additional depth cues. While we support the semantic interpretation by using different color codings (e.g red for electricity cables) and adequate geometric representations (e.g cylinders for trees), depth perception issues are addressed by creating cutaways (Figure 4(c)), reference shadows projecting the pipe outlines to the surface, or connection lines visualizing the connection between the object and the ground (Figure 4(d)).

The advantage of our method of separating the GIS data level and the comprehensible geometry level with the transcoding layer is that we can create as many different visualization objects as needed by an application. There is no additional effort for updating the various geometric objects, since modifications are automatically applied due to the connection with the transcoding layer. The data connections between the comprehensible 3D data level and the transcoding layer are implemented as field connections in COIN3D to ensure data consistency.

5 Interaction Techniques

The bi-directional transcoding pipeline allows an interactive data round trip of GIS data in an AR application and therefore the interactive manipulation of the geospatial data itself. The direct AR visualization allows to align and modify geospatial data with immediate visual feedback. In-field tasks that can benefit from this direct interaction include planning, inspection and surveying of new structures. Following taxonomy from Bowman et al. [15], we divided the interaction techniques in selection, manipulation and navigation tasks.

5.1 Selection

The selection of GIS features is the starting point for information access, manipulation and surveying. While selecting features consisting of one feature point is unambiguous, for features consisting of multiple feature points, the user may want to select all corresponding feature points or only a subset. For instance, for surveying it may be useful to select only a single vertex to manipulate it, while for planning tasks it may be useful to select and manipulate all corresponding vertices at once, e.g. to move a complete pipe.

Therefore, we provide two selection options: Object-based selection for selecting a complete object and feature point-based selection for selecting single feature points. For the object-based selection, we compute the object that contains the intersection point and add all corresponding feature points to the selection. For the feature-point based selection, we compute the closest feature point of the intersection object to the intersection point. After selecting feature points, information about these features can be accessed or can be manipulated.

5.2 Manipulation

The manipulation of data is important since as-built objects have to be surveyed, already documented objects may have to be updated, planned data should be adjusted after checks against the real-world situation, or has to be adjusted after being surveyed.

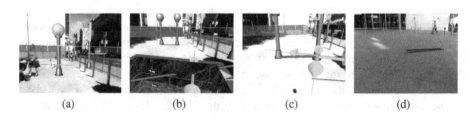

(a)   (b)   (c)   (d)

Fig. 5. Manipulations: (a) Constrained transformation of lamps. (b) Constrained transformation of cables. (c) Single-Point Manipulation. (d) Surveying of a physical cable.

Transformation. After selecting features, transformations can be applied by manipulating a reference geometry with the mouse pointer. A number of different manipulators in COIN3D support the easy integration of different manipulation methods. The resulting manipulation matrix is directly applied to the transcoding layer on feature vertex level. That means all selected feature vertices are transformed by the transformation matrix of the current manipulation operation. Due to the direct data connection between transcoding layer and comprehensible geometries, all corresponding visualizations are updated by the same transformation.

Surveying. The AR-Surveying allows to survey features directly on-site. To provide accurate measurements, we use a laser measurement device. Points measured by the device are mapped into the global coordinate system and used as input for surveying features points. Since the measurement only provides information about distances in a fixed direction, the device has to be calibrated in relation to the camera. That allows to compute a 3D point for every distance measurement. The measured 3D points are then visualized and used to survey point features and line features by measuring single or multiple points respectively (Figure 5(d)). Additionally, we integrated a surveying method based on pen input. In this case, the surveyed 3D point is calculated by re-projecting screen-coordinates onto a 3D-digital terrain model describing the surface of the surrounding environment.

5.3 Navigation

Navigation in AR differs to the navigation in virtual environments. Usually, users move physically in the real world (1:1 motion mapping) updating their point of view with their movement. However, in our application, the surveying of new features may be physically engaging, if a feature is exceptionally large. For instance, pipe features

Fig. 6. Augmented Multi-views

may be distributed over large areas, which makes it challenging to measure them directly by laser surveying from one location, since start and end points may be not visible from the same camera view. To overcome this problem, we provide a method called multi-views [16] that allows capturing camera views and their poses and switch between these stored views. The user can access these multi-views to view and edit objects from remote locations. Furthermore they can be stored as KML[4] files and viewed in external virtual globe viewers for documentation purposes.

6   Results

We integrated our approach into a mobile AR platform designed for high-accuracy outdoor usage to test the interactive data roundtrip and analyze the surveying accuracy. Additionally, we gained first feedback with the setup in workshops with expert users from civil engineering companies.

---

[4] Keyhole Markup Language.

Mobile AR Platform. The AR prototype hardware setup consist of a tablet PC (Motion J3400) with 1.6GHz Pentium CPU and sunlight viewable screen capable for real-world outdoor conditions. We equipped the tablet PC with different sensors such as a camera, a 3DoF orientation sensor, and a Novatel OEMV-2 L1/L2 Real-Time Kinematic (RTK) receiver for achieving a positional accuracy within the centimeter range.

The laser distance measurement device for surveying was mounted on the back of the camera and calibrated with respect to the camera. To calibrate the laser device, we captured several images showing the laser-dot and a standard checkerboard calibration target and determined the position of the dots in relation to the camera. A set of determined positions with known distance measurements allows to compute the ray of the laser in relationship to the camera. Finally, a 3D point can be computed from each distance measurement by mapping the distance back into a 3D point on the ray.

Field Tests. To test the system with real-world data in field tests, we used data from conventional GIS provided by two civil engineering companies. We performed first field-trials with the surveying system with 16 expert participants (12m/4f) from a civil engineering company. Users were asked to survey an existing pipe with the system and afterwards to complete a short questionnaire. Results of the test showed that users rated the suitability of the AR system for "As-built" surveying over average on a 7-point Likert scale (avg. 5.13, stdev 1.14) and compared to traditional surveying techniques as quite equivalent (avg. 4.43, stdev 1.03). The simplicity of surveying new objects was rated above average (avg. 5.44, stdev 0.96). And while the outdoor suitability of the current setup was rated low due to the prototypical character (avg. 3.28, stdev 1.20), the general usefulness of the AR application was rated high (avg. 5.94, stdev 1.19).

For assessing the surveying accuracy of the surveying application, we performed experiments measuring a known reference point. The reference point was surveyed with the AR setup from more than 20 different positions and directions. The results (Easting: avg. 0.02, stdev 0.19, Northing: avg. 0.18, stdev 0.14, Height: avg. -0.12, stdev 0.16) show that the accuracy is better than 30 centimeters (the minimum accuracy required by our end users from the civil engineering sector). The observed inaccuracies are caused by orientation sensor and laser calibration errors.

7   Conclusion and Outlook

In this paper, we demonstrated how existing workflows such as on-site planning, data capture and surveying of geo-spatial data can be improved through a new approach enabling interactive, on-site AR visualizations. Surveying tasks benefit by the immediate visualization of preview geometries and correction/surveying of the geospatial objects through showing the known and captured features in context. First field trials with expert users from the utility industry showed promising results. To achieve this level of functionality, several technical advances were necessary. For the visualization of geospatial data we implemented a data roundtrip which allows a comprehensible AR visualization and still being flexible for modifications. Furthermore, the integration of interaction tools for creating, editing and surveying features shows how planning and surveying structures can be simplified. Currently, we explore the commercial development of the prototype in a set of pilot projects with industrial partners. The aim is to adapt the prototype to industrial needs to realize a novel on-site field GIS providing a simpler, yet more appealing way to address specific productive industrial workflows.

Acknowledgements. This work was supported by the Austrian Research Promotion Agency (FFG) FIT-IT projects SMARTVidente (820922) and Construct (830035).

References

1. Azuma, R.: A survey of augmented reality. Presence: Teleoperators and Virtual Environments 6, 355–385 (1997)
2. Roberts, G.W., Evans, A., Dodson, A., Denby, B., Cooper, S., Hollands, R.: The use of augmented reality, GPS, and INS for subsurface data visualization. In: FIG XXII International Congress, pp. 1–12 (2002)
3. Schall, G.: Handheld Augmented Reality in Civil Engineering. In: Proc. ROSUS 2009, pp. 19–25 (2009)
4. Shin, D., Dunston, P.: Identification of application areas for Augmented Reality in industrial construction based on technology suitability. Automation in Construction 17, 882–894 (2008)
5. Hakkarainen, M., Woodward, C., Rainio, K.: Software architecture for mobile mixed reality and 4D BIM interaction. In: Proc. 25th CIB W78 Conference, pp. 1–8 (2009)
6. Golparvar-Fard, M., Pena-Mora, F., Savarese, S.: D4AR- A 4-Dimensional augmented reality model for automating construction progress data collection, processing and communication. Journal of Information Technology 14, 129–153 (2009)
7. Feiner, S., MacIntyre, B., Hollerer, T., Webster, A.: A touring machine: prototyping 3D mobile augmented reality systems for exploring the urban environment. In: ISWC 1997, pp. 74–81 (1997)
8. Piekarski, W., Thomas, B.H.: Tinmith-metro: New outdoor techniques for creating city models with an augmented reality wearable computer. In: ISWC 2001, pp. 31–38 (2001)
9. Veas, E., Kruijff, E.: Vesp'R: design and evaluation of a Handheld AR device. In: Proc. ISMAR 2008, pp. 43–52. IEEE Computer Society (2008)
10. Wagner, D., Schmalstieg, D.: First steps towards handheld augmented reality. In: ISWC 2003, pp. 127–135 (2003)
11. Kalkofen, D., Mendez, E., Schmalstieg, D.: Comprehensible visualization for Augmented Reality. IEEE Trans. Vis. Comput. Graphics 15, 193–204 (2009)
12. Mendez, E., Schall, G., Havemann, S., Fellner, D., Schmalstieg, D., Junghanns, S.: Generating semantic 3D models of underground infrastructure. IEEE Comput. Graph. Appl. 28, 48–57 (2008)
13. Zollmann, S., Kalkofen, D., Mendez, E., Reitmayr, G.: Image-based ghostings for single layer occlusions in augmented reality. In: Proc. ISMAR 2010, pp. 19–26 (2010)
14. Feiner, S., Seligmann, D.: Cutaways and Ghosting: Satisfying visibility constraints in dynamic 3D illustrations. The Visual Computer 8, 292–302 (1992)
15. Bowman, D.A., Kruijff, E., LaViola Jr., J.J., Poupyrev, I.: 3D User Interfaces: Theory and Practice, 1st edn. Addison-Wesley (2004)
16. Veas, E., Grasset, R., Kruijff, E., Schmalstieg, D.: Extended overview techniques for Outdoor Augmented Reality. IEEE Trans. Vis. Comput. Graphics 18, 565–572 (2012)

Sketch-Line Interactions
for 3D Image Visualization and Analysis

T. McInerney and Y.S. Shih

Dept. of Computer Science, Ryerson University, Toronto, ON, Canada, M5B 2K3

Abstract. This paper explores the effectiveness of an interaction model based on user-sketched line segments and curve segments, together known as sketch-lines. Directly sketching line segments on image slices, or curve segments on the surfaces of objects in volume rendered or surface rendered 3D data, is an effective means by which to quickly and simply transfer accurate information, such as position, object surface orientation, and surface region width, from the user to a visualization algorithm. This information can be used for fast, intuitive and precise object-relative image slice positioning and for precise, editable region of interest (ROI) delineation in a volume image. The results from two user studies are presented that analyze the efficiency, intuitiveness and precision of the sketch-line interaction model as well as quantitatively and qualitatively compare aspects of the model to other interaction techniques.

1 Introduction

Efficient and intuitive user interactions that aid in the visualization and analysis of 3D data can often be effectively realized using data surface relative input actions that mimic familiar constrained physical actions such as drawing, sliding, and painting [1]. Furthermore, several 3D view generation and region delineation tasks (for example, for surgical planning or for segmenting objects in noisy images) not only require efficiency and simplicity, but also precise control and accuracy. These additional requirements can complicate the design of a 3D interaction model as they often conflict with efficiency and simplicity.

This paper explores the effectiveness of an interaction model based on user-sketched line segments and curve segments, together known as sketch-lines. Directly sketching line segments on image slices, or curve segments on volume rendered or surface rendered object surfaces, is an effective means by which to quickly and simply transfer accurate information, such as position, object surface orientation, and surface region width, from the user to a visualization algorithm. In the context of volume image visualization and analysis, this information can then be used for fast, intuitive and precise object-relative image slice positioning or for precise, editable region of interest (ROI) delineation. These two primary user operations underlie many common volume image visualization tasks, including 3D image slice-based exploration and inspection, visualization via cutaway, surgical planning, and deformable model-based segmentation [2]. We combine the sketch-line interaction model with the use of an interpolating spline curve,

G. Bebis et al. (Eds.): ISVC 2012, Part I, LNCS 7431, pp. 686–697, 2012.

an interpolating subdivision surface and an extrusion process to support the implementation of the two primary user interactions. We demonstrate the use of sketch lines with several examples. We also present results from two user studies in order to gain insight into the efficiency, intuitiveness and precision of the sketch-line interaction model as well as compare it to other interaction techniques. A conclusion and future work section summarize the advantages and current drawbacks of the model and its implementation.

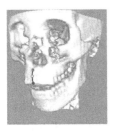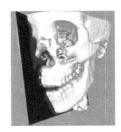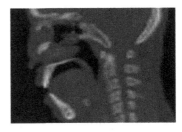

Fig. 1. Left: a sketch-line is drawn on the volume rendered skull surface and an image slice (middle) is instantly positioned and oriented such that it is approximately orthogonal to the surface. Right: the image slice is also displayed in a 2D window.

2   Related Work

Image slice views remain an integral part of the detailed inspection and analysis of volume images. Many volume image visualization packages provide widgets that support the translation and arbitrary rotation of a 3D image slice. The 3D image slice is typically rendered together with a volume rendering of anatomical structures and optionally as a 2D image slice rendered in a separate window. These 2D/3D view combinations are effective for volume image navigation and exploration [3]. It is often useful to orient the image slices orthogonally to a curved surface region of an anatomical structure in order to examine and measure these "natural" object cross sections. Image slice widgets typically provide controls to "push" the slice plane in a direction along its normal vector as well as controls to rotate it. However, direct links between the slice plane manipulation and the 3D object typically do not exist. Orienting and positioning are performed in "absolute" 3D world coordinates. Repeatedly positioning the slice back and forth along a structure in order to explore it therefore relies on a series of independent world space manipulations. Direct object surface relative positioning, on the other hand, is potentially more efficient and intuitive [1].

Selecting or enveloping an arbitrarily shaped volume of interest (VOI) in volumetric data sets in a simple, intuitive and precise way is a complex 3D interaction task. This task is further complicated if the user wishes to represent the VOI with a precisely and easily editable model. Common representations of a VOI include a triangle mesh, a set of voxels, or blended implicitly functions. These VOI representations can be too low level, often have limited editing support, and often cannot be precisely defined without considerable user effort. A selected VOI can be cutaway to remove unwanted data in order to visualize hidden objects. In addition, a geometric representation of the VOI can be used as input to a segmentation algorithm to constrain the algorithm when it is applied to noisy images, in an effort to prevent "leaks" into neighboring objects.

Most visualization systems support standard tools such as cropping boxes, controlled using widgets, that can be used to select a box-shaped VOI in real time. Fuchs [4] extended this idea to allow the generation of a non-convex polyhedral VOI. Many 3D region selection techniques use some sort of interaction metaphor. Sketching [5], tracing [6], painting [7], sculpting [8], are among the most common. In the sketching metaphor, the user draws lines or contours on the screen. These contours are then connected and "inflated" to form a 3D envelope. The sculpting metaphor simulates cutting tools with convex-shaped tool "tips" that are positioned and/or moved within the volume. Any voxels inside the tool tip are selected and sculpted away. Similar to sculpting, painting defines a surface region, such as a circle, as the "brush tip" and as the brush is moved along the data object surface, voxels inside the brush region are rendered transparent. In tracing the user draws a contour on the object surface to outline a region. Although many of these interaction metaphors support fast, approximate VOI specification, most do not support precise and efficient editing of the VOI, often due to the low level VOI representation.

3   Sketch Line Methodology and Examples

The sketch-line interaction model presented in this paper is an extension of the sketch-line model in [2] used for image segmentation. The basic idea is to maximize the amount of accurate information (position, orientation, width, surface normal etc.) from the user to the algorithm with the least amount of effort by using only simple strokes. Sketching lines is a simple and familiar action for users and the lines can be quickly and precisely drawn.

The sketch-line system was implemented using VTK 5.6.0. The sketch-line system interface is designed with two window views: the left window displays a volume rendering of the data and a 3D image slice plane that can be positioned, scaled and rotated using a sketch line or using standard widget controls, and the right window shows a 2D image slice view (Figure 1). The 2D window camera position and orientation are linked to the 3D slice plane orientation so that a consistent 2D window view "up" direction is maintained.

3.1   Sketching Line Segments on Image Slices

Sketch-lines are formed by clicking and holding the left mouse button (or finger/stylus), dragging the mouse to another location to stretch out the line segment, and finally releasing the button. Sketch-lines can be connected together with a spline to form a closed contour (Figure 2). To maximize the amount of user-provided information needed to create a contour envelope that accurately delineates an object cross-section, sketch-lines should ideally be drawn across

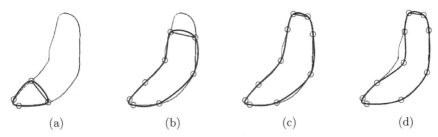

(a)   (b)   (c)   (d)

Fig. 2. (a)-(c) Sketch lines are drawn across the width of an object and connected to form a closed spline. Only a few lines are need to quickly and accurately delineate complex contours and the spline can be input to a segmentation algorithm. (d) The spline control points provide precise editing capability.

the object cross-section so that they are approximately orthogonal to its primary medial axis (Figure 2). As is clear in the figure, the resulting spline curve accurately delineates the object cross-section using very few control points. This sketching process takes less than a second per sketch line and has a short learning curve - after some practice it is often obvious where to place the lines and how many to use. As the user adds a sketch-line, the spline curve is updated and rendered. Most simple shaped contours require only 2 or 3 sketch lines (in Figure 2 6 are used). The straight line segment joining the front two control points in Figure 2 is called the *active edge*. When a new line is sketched, its endpoints are connected to the active edge endpoints and a new spline contour is created. The user can click on the spline contour in another location to change the active edge, allowing for the creation of more complex shapes. Finally, the spline contour can be intuitively edited, at any time, by selecting and dragging the control points (Figure 2d).

3.2   Sketch-Lines on Object Surfaces

A sketch- "line" is drawn on the surface of the volume rendered object using a similar mouse click and drag action as the image slice case. Because object surfaces are curved, a sketch-line is visually represented as an interpolating quadratic spline curve segment so that it conforms to the surface (Figure 1). The spline curve is constructed from 3 sampled surface points. Along with surface point samples, object surface normals are also sampled along the sketch-line path and are used to construct a 3D coordinate system for an image slice (Section 3.3).

Surface Sketch-Line Construction. To construct the spline curve segment, the first point picked on the object surface (e.g. Figure 3, leftmost sphere) is used as the start control point. As the user drags the mouse and moves the cursor along the surface, surface points and normals are sampled and an average surface normal is calculated. Similar to an image-slice sketch line, as the user drags the cursor over the surface the end control point (e.g. Figure 3, rightmost sphere) is

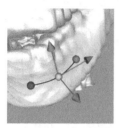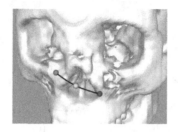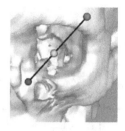

Fig. 3. Examples of sketch lines with the spline control points highlighted. The left image shows the basis vectors of a image slice coordinate system. The middle and right images show sketch lines automatically spanning concave regions.

continuously updated (and the spline continuously redisplayed). A 3rd middle control point (e.g. Figure 3, middle sphere) is continuously calculated as well using the sampled surface point currently furthest away (in a positive average surface normal direction) from the 3D line segment formed by subtracting the start and end control points (that is, furthest above the surface).

The average surface normal is calculated "on the fly" from surface normals sampled roughly along the surface path from the start control point to the end control point. The surface sampling process is as follows. The 2D screen window points corresponding to the 3D start and end control points are connected to form a line in screen space. Points are evenly sampled along this screen space line and are projected back onto the object surface and the normal samples are gathered at these locations. Similar to the image slice sketch lines, this sampling process allows users to stretch and drag the spline segment along the surface until they are satisfied with its position, length and orientation.

The sketch-line construction process was designed to create a spline curve that is always visible on the surface while also conforming to the surface shape. If the sketch-line is drawn over a concave region (i.e. an indentation such as the eye socket (Figure 3 middle) or a hole in the surface), the sketch-line will be constructed across the region rather than bend inwards. From experiments, 5-10 sampled surface normals are sufficient to form an accurate average surface normal. To increase accuracy and reduce susceptibility to noise, the number of sample points along the line can be increased, surface points within a small neighborhood of these sample points can be included, and/or a pre-smoothed image volume can be used.

3.3   Image Slice Positioning Using Surface Sketch Lines

To position and orient an image slice using a sketch line (Figure 1) such that it is roughly orthogonal to the object surface, an image slice coordinate system is constructed (Figure 3 left) using the average surface normal as the u axis (Figure 3 left, vector coming away from the surface), the normalized line segment formed from subtracting the spline start control point from the end control point as the initial "up" vector (v axis - Figure 3) left, vector pointing to the right) and the cross product of these two vectors as the slice plane normal (n axis - Figure 3 left, vector pointing up). The standard orthogonal image slice plane (sagittal, axial, coronal) closest to this coordinate system is determined and this slice is then translated and rotated to match the constructed coordinate system. The initial up vector of the image slice is then adjusted by rotating the slice around its normal vector until the u and v axes are parallel to the sides of the volume image bounding box. This image slice "spin angle" constraint makes the slice appear "upright" in the 3D window from the user's perspective rather than tilted and establishes an up direction in the 2D window view. In a final step, the image slice is translated with respect to the u and v axes so that it is centered within the volume image bounding box. The image slice may then be edited by the user with the standard widget controls.

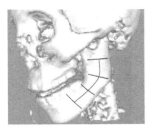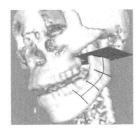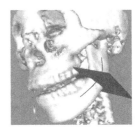

Fig. 4. Left: a series of sketch-lines (4, in this example) are drawn along the curving surface of the lower jaw and their midpoints are connected to form a smooth spline. Middle, Right: the user can slide the slice along the curve and it is constrained to remain approximately orthogonal to the jaw surface.

Image Slice Plane Sliding. If the user wishes to examine object cross-sections along a curving part of a target object surface, they would typically use widget controls to push and rotate the image slice in 3D world coordinates until the slice is visually orthogonal to the surface, and then repeat this process back and forth at different points along the target region. Using sketch-lines, this inspection of various cross-sections orthogonal to the target surface is simple and efficient. The user quickly draws a series of sketch-lines along the target region (4 sketch lines in Figure 4 left). The midpoint of each sketch-line is automatically connected and used to form an interpolating spline. The average surface normal calculated from each sketch-line is also smoothly interpolated using this interpolating spline. The user can now push the image slice as usual and the slice center point automatically follows the spline path while the slice normal is automatically smoothly interpolated between the sketch-lines. The image slice can now slide back and forth along the object surface and is automatically constrained to remain approximately orthogonal. This is an example of how sketched input typically contains more "structured" information that can be interpreted by the visualization algorithm.

3.4   VOI Selection Using Sketch Lines

Similar to the contour constructed from lines sketched on an image slice, multiple sketch-lines can be quickly and precisely drawn on the object surface along the length of a region of interest (Figure 5). The corresponding control points of the 3D spline curves can be connected to form a mesh of triangles. This control mesh is fed into an interpolating subdivision surface algorithm [2] which finely subdivides the triangles to form a smooth surface (Figure 5 top row). This subdivision surface delineates and/or envelopes the region and can be used to either cut the region away or it can be converted to a deformable surface model [2] and fitted to the region (segmenting it).

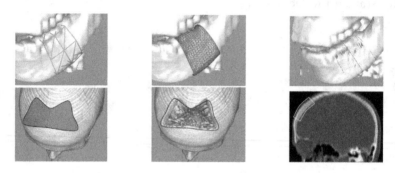

Fig. 5. Top row: 3 sketch lines are converted to profile curves and connected to form a sleeve control mesh. Middle: subdivided mesh and (right) the result of fitting the mesh to the jaw. Bottom row: a patch is created with 3 sketch lines, extruded inward and used to cut away the skull. The right figure shows a cross sectional view.

The sketch-line system currently supports 3 types of meshes: a closed mesh, an open cylindrical mesh known as a "sleeve", and an open surface mesh known as a "patch" (Figure 5). A patch can be used to delineate a curving region of the target object surface such as the top of the skull. The patch can be copied, the copy moved into the object surface, and the two patches connected together to form a closed "thin shell" mesh envelope. This type of envelope is useful for cutting away thin shell structures such as the skin or the skull. A sleeve is an open cylinder mesh that can be used to define a section of a curving cylindrical structure such as an artery. The sleeve can be fitted to the object section using the deformable surface model segmentation algorithm [2], segmenting it. The open cylinder sleeve mesh can also be capped at its ends to form a closed mesh which can then be used to cutaway sections such as a portion of the jaw. Finally, a closed profile curve can be constructed from the open sketched spline segment. The 3 control points of the sketched spline curve are copied and moved inward (i.e. in the direction of the negative average surface normal vector) a user-defined distance (using a slider) into the object. The copied spline curve is also reflected across the axis formed by the line joining the spline end control points. The copied spline curve segment is connected to the sketched spline to form the closed curve. A series of 3D sketch-lines can be converted to profile curves, connected and subdivided to form the smooth closed surface mesh.

The subdivision surface envelope or patch mesh can be flexibly edited in several ways, either in the 3D window or 2D window via the profile curves. The envelope control points can be repositioned and the envelope is updated in real time. Entire profile curves can be translated/rotated via the slice plane in which they are embedded. If the initially sketched envelope does not quite cover the ROI or if the depth of the envelope needs adjusting, the user can quickly adjust control points to correct it. Finally, sketch lines can be quickly undone to return the envelope to its previous state and then redrawn if desired. Further details on editing can be found in [9].

4   Evaluation

Two user studies were conducted to quantitatively measure the efficiency of the sketch-line interaction technique compared to several other common interaction models, as well as to qualitatively assess its effectiveness. The first user study was conducted to compare the sketch-line slice plane positioning and orienting technique with a standard widget push/rotate technique. A volume rendered skull was used as the target object for slice positioning as it contains easily visible surface features. In the push/rotate technique, the user selects any point on the slice plane and moves the mouse to "push" the slice in its normal vector direction. To rotate the slice for both push/rotate and for fine-tuning the result of the sketch line slice positioning, a rotation model using precisely controllable circular rotation handles was used [9].

A target slice plane is presented to the users, positioned and oriented approximately orthogonally to some part of the skull surface. For the push/rotate technique, a reference slice plane in a standard position is provided and the users must manipulate it to match the target plane. For the sketch-line technique the user sketches a line on the skull surface near the target slice in order to match it. The users are permitted to fine-tune the sketch-line result using the standard push/rotate technique. A total of 3 trials were performed, each consisting of 10 target slice test cases. Users were allowed to practice the two techniques before the trials. The presentation order of the target slices, as well as the technique used first, was randomized.

The basic hypothesis in this study was that the sketch-line technique would need fewer mouse clicks and less overall time to position and orient the slice plane with respect to the skull surface. The hypothesis was verified ($p < 0.03$) (see [9] for details of the study). As expected, the Push/Rotate technique was dominated by slice rotation input actions whereas the sketch-line technique was almost entirely dominated by the quick curve sketching action and seldom required fine tuning with the standard widget controls. A questionnaire asked the participants whether the sketch-line and push/rotate techniques were easy to control and easy to learn. The results (Table 1) were roughly equal with Push/Rotate scoring better on both questions. However, the participants were also asked to state their

favorite technique. The results were that 2/11 listed push/rotate as their favorite
while 9/11 favored sketch-line. The two who preferred push/rotate commented
that they found it difficult to know what to expect for the slice position and
orientation after sketching.

Table 1. Sketch-lines vs. Push/Rotate questionnaire results

5-Point Likert Scale	Push/Rotate	Sketch-Line
Easy to Learn	Avg. 4.64 (Std. 0.67)	Avg. 4.18 (Std. 0.75)
Easy to Control	Avg. 4.18 (Std. 0.87)	Avg. 3.91 (Std. 0.54)

4.1   Sketch-Lines versus Tracing and Painting

A second user study compared sketch-lines against two other common interaction metaphors: tracing and painting. Users were asked to delineate 2D contours
using the 3 techniques. We were interested not only in efficiency and simplicity but also in issues of control and precision. The time taken to delineate, the
editing time required, as well as other interaction model specific quantities were
measured. A 2D delineation task was chosen for several reasons. The most basic
reason was that equivalent systems of 3D tracing and painting that were readily
comparable to our VTK-based implementation were not available at the time
of the user study. We were also concerned that different systems would have
different interaction controls and/or differences in interaction model and volume
rendering efficiency and quality, especially compared to our somewhat restrictive
and slow VTK-based prototype system. We conjectured that a 2D contour delineation task, where all methods were implemented within our VTK framework,
would use highly similar input actions as their 3D counterparts and therefore
useful information could still be gleaned from such an experiment. Furthermore,
there is no depth control requirement and highly similar interface controls using
a mouse could be used for all 3 models. In addition, the 2D contour delineation
task is very simple to explain and to grasp by naive users. Finally, designing
target contours, measuring delineation time and precision is very simple in 2D.

In our tracing implementation the user used a mouse to move the cursor along
the target contour, tracing out a smooth spline curve with a user-definable control
point spacing. The user can, at any time, stop and select a spline control point and
edit the shape of the spline by repositioning the control point. The user may also
use a slider positioned in a window region to the immediate right of the target
contour to change the distance between control points as the tracing proceeds. In
the painting implementation, the uses moves a circle (a "paintbrush tip") around
the interior of the target contour and the circles are blended to form a smooth
closed contour. To edit the painted contour, the user holds down the right mouse
button and the circle becomes an "eraser". The user may also use a slider to change
the size of the brush tip. Tracing and Painting were chosen for comparison as they
are both common, are familiar to non-expert users, and are simple to learn.

Eleven people (undergrad. and grad. computer science students, with no overlap from user study 1) participated in the within-subjects study, 8 males and

3 females with an average age of 22 years, 35 hours of mouse usage per week and 8 hours of video games per week. The study took approximately one hour to complete. Each trial consisted of 10 contour matching tasks, with contours of different shapes and lengths and shape complexity (see [9] for details). The users were asked to delineate each target contour as quickly and as precisely as possible, initially without editing it. In the last two trials, the users could edit their result. Each user performed 3 trials and the delineation technique (sketch, trace, paint) as well as the presentation order of the contours was randomized. Each user was allowed to practice each of the 3 techniques. The first

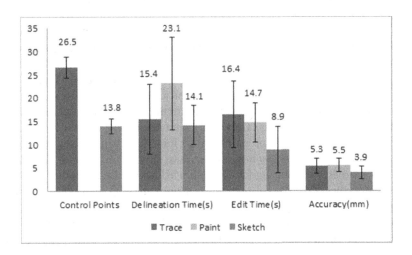

Fig. 6. Sketch-lines vs. Tracing and Painting

hypothesis we formed was that the sketch-line model would be more precise (without contour editing) than paint and trace due to the simple and precisely controllable input actions. Similarly, a second hypothesis was that sketching would be faster (with editing) than paint and trace and that the editing time would be less. A final secondary hypothesis was that sketching would require fewer control points to form the spline curve than tracing, assuming no post processing of the traced spline curve. The results are summarized in Figure 6. The first hypothesis was validated ($p < 0.005$ for sketch vs. trace and $p < 0.003$ for sketch vs. paint). The sketched contour was more accurate than the traced or painted contour. Sketch-lines were also faster (without editing) and the variation in contour delineation times smaller than either trace or paint but this result was only statistically significant for sketch vs. paint ($p < 0.02$). The second hypothesis was also validated. When editing time is included, both the average time required to delineate the contours using sketch-lines as well as the average editing time required were significantly less for sketch-lines than trace ($p < 0.003$) and paint ($p < 0.001$). The final hypothesis was also validated (see "Control Points" in Figure 6). Sketch-lines required significantly fewer control points to

Table 2. Averaged responses to a questionnaire. Standard deviation is in brackets

5-Point Likert Scale	Trace	Paint	Sketch
Easy to Learn	4.91 (0.30)	4.36 (0.92)	4.82 (0.42)
Easy to Control	4.09 (0.94)	3.18 (1.33)	4.27 (0.79)

accurately delineate the contours than tracing ($p < 3.9E - 07$). According to a questionnaire (Table 2), all participants scored Tracing and Sketching easy to control and learn. Painting scored slightly lower on easy to control although this result must be considered in the context of the precise delineation task. The participants were also asked to state their favorite technique. Since tracing and painting techniques are similar to simple pen tracing and painting on paper, we expected sketch-lines might not be chosen as a favorite. Nevertheless 4/11 users preferred sketching while 2/11 liked painting and 5/11 preferred tracing. Some comments were that painting is easy to use for a quick rough outline of a target contour. People who liked sketch commented that it was fast, precise, easy to edit, easy to learn, easy to visualize and convenient. Those who liked tracing commented that it was efficient and easy to focus resulting in fewer mistakes. Participants were also asked to name their least favorite technique. The results were that 7/11 disliked painting for the delineation task commenting that paint was hard to predict and control precisely without making mistakes. This result reinforces the idea that a painting style interaction is perhaps best used for fast exploration where precision is not as important. Only one user disliked sketching because it required the user to sketch precisely. For tracing, 3/11 disliked it and commented that it was hard to trace using the mouse and get an accurate result. Our experience also found that long tracing-like input actions where precision is important are tedious and require a higher level of user concentration.

5   Conclusion

The sketch-line interaction model supports simple, direct and precise input actions that nonetheless transfer a significant amount of user information to volume visualization algorithms. It complements and augments existing image slice widget interaction by allowing slices to be quickly positioned orthogonally to object surfaces. The object-relative sketch input actions provides a more seamless bridge between a volume rendered data views and (oblique) image slice views. By combining the interaction model with a spline curve, a subdivision surface and an extrusion process, the user is able to sketch editable geometric meshes that accurately delineate or surround regions of interest. Two user studies support the effectiveness and efficiency of the sketch-line model for tasks that require precise control and accuracy. The algorithm is not limited to volume images - sketch lines can be drawn on triangulated surface meshes.

There are many areas for improvement and extension. Firstly, it is somewhat difficult to sketch on narrow anatomical structures, such as small arteries.

An accurate volume image interpolation and zoom feature is needed. Also, in noisy medical images, the volume rendering of objects results in "fuzzy" surfaces that are also difficult to sketch on. More surface samples within a larger region around the cursor and a smoothed volume image may help with this issue. The VTK-based implementation is rather limiting and needs to be replaced with a more powerful volume rendering system. While envelopes can be created and manipulated in real time, both the VTK cutaway operation and the deformable model segmentation algorithm require GPU acceleration. Finally, local or global self-intersection prevention of the various envelopes is currently not implemented and relies on user editing.

References

1. Hinckley, K., Pausch, R., Goble, J., Kassell, N.: A survey of design issues in spatial input. In: 7th Annual ACM Symposium on User Interface Software and Technology (UIST 1994), pp. 213–222 (1994)
2. Aliroteh, M., McInerney, T.: SketchSurfaces: Sketch-Line Initialized Deformable Surfaces for Efficient and Controllable Interactive 3D Medical Image Segmentation. In: Bebis, G., Boyle, R., Parvin, B., Koracin, D., Paragios, N., Tanveer, S.-M., Ju, T., Liu, Z., Coquillart, S., Cruz-Neira, C., Müller, T., Malzbender, T. (eds.) ISVC 2007, Part I. LNCS, vol. 4841, pp. 542–553. Springer, Heidelberg (2007)
3. Tory, M., Moller, T., Atkins, M., Kirkpatrick, A.: Combining 2d and 3d views for orientation and relative position tasks. In: CHI 2004: Proceedings of the SIGCHI Conference on Human Factors in Computing Systems, pp. 73–80 (2004)
4. Fuchs, R., Welker, V., Hornegger, J.: Non-convex polyhedral volume of interest selection. Computerized Medical Imaging and Graphics 34, 105–113 (2010)
5. Pühringer, N.: Sketch-based modelling for volume visualization. Master's thesis, Vienna University of Technology (2009)
6. Bruyns, C., Senger, S.: Interactive cutting of 3d surface meshes. Computer & Graphics 25, 635–642 (2001)
7. Chen, H., Samavati, F., Sousa, M.: Gpu-based point radiation for interactive volume sculpting and segmentation. Visual Computer 24, 689–698 (2008)
8. Weiskopf, D., Engel, K., Ertl, T.: Interactive clipping techniques for texture-based volume visualization and volume shading. IEEE Transactions on Visualization and Computer Graphics 9, 298–312 (2003)
9. Shih, Y.: A sketch-line interaction model for image slice-based examination and region of interest delineation of 3d image data. Master's thesis, Dept. of Computer Science, Ryerson University, Toronto, ON, Canada (2012)

Fast Illustrative Visualization of Fiber Tracts

Jesús Díaz-García and Pere-Pau Vázquez

MOVING Group - Universitat Politècnica de Catalunya

Abstract. The visualization of human brain fibers is becoming a new challenge in the computer graphics field. Nowadays, with the aid of DTI and fiber tracking algorithms, complex geometric models consisting of massive sets of polygonal lines can be extracted. However, rendering such massive models often results in non-detailed, cluttered visualizations. In this paper we propose two methods (one object-space and another image-space) for the fast rendering of fiber tracts by including illustrative effects such as halos and ambient occlusion. We will show how our approaches provide extra visible cues that enhance the final result by removing clutter, thus revealing fibers' shapes and orientations. Moreover, the use of ambient-occlusion based techniques improves the perception of their absolute and relative positions in space.

1 Introduction

Diffusion MRI (Magnetic Resonance Imaging) is a method that produces in vivo images of biological tissue by using the internal local micro structure of water. Different acquisition techniques are used depending on the way water diffuses in the medium; the most common are Diffusion Weighted Imaging and Diffusion Tensor Imaging. Tractography is the procedure to demonstrate neural tracts by means of MRI techniques, computer-based image analysis and its later graphical representation. We focus our work in the last stage of tractography, implementing techniques to render polygonal lines extracted by fiber tracking algorithms.

Modern GPUs are able to render large models in real time. Unfortunately, due to the geometry of fiber tracts, when represented with lines, the result is a cluttered visualization where depth relations are not depicted. Shaded cylindrical representations have also traditionally been used for this purpose. However, this increases the rendering burden due to the amount of triangles added to the, initially simple, geometry, and thus, frame rates are reduced.

In this paper we present two algorithms (image-space and object-space, respectively) that enhance the visualization of fiber tracts by adding halos and ambient occlusion. This way, we improve the perception of fibers' relative positions and orientations. All the process is carried out in an optimized way, making use of GPUs. This yields fast algorithms, even for large window sizes and when the geometry covers a large portion of the viewport. Next Section reviews previous work. Section 3 describes the methods proposed for illustrative brain fiber rendering. Visual and performance results are discussed in Section 4, and Section 5 draws some conclusions and points to some lines for future resesearch.

G. Bebis et al. (Eds.): ISVC 2012, Part I, LNCS 7431, pp. 698–707, 2012.
© Springer-Verlag Berlin Heidelberg 2012

2   Previous Work

Visualization of neural tracts and brain fibers is not a very evolved field in computer graphics, although some effort has already been devoted to this area of study. Zhang *et al.* present a geometrical approach by means of streamtubes and streamsurfaces in order to distinguish between linear or planar anisotropy regions [1]. Another geometrically-based technique is introduced in [2], where the authors generate surfaces wrapping bundles of lines, achieving a more intuitive representation of white matter neural tracts. In [3], another illustrative method for the rendering of white matter fiber bundles is presented; it consists in the generation of silhouettes and contours that reduces the original rendered geometry to a set of hint lines wrapped by a colored and contoured region. Tarini *et al.* [4] use impostors for the visualization of molecular structures. In this case, renderings are also enhanced with ambient occlusion and edge cueing. In the field of illustrative rendering of streamlines, a method that creates colored depth-dependent halos around lines to group tight sets of fibers and deemphasize separated ones is due to Everts et al [5]. They pre-process all the streamlines in the model to duplicate their vertices; this way they can convert stream lines into eye-facing triangle strips by expanding these pairs of vertices in the vertex shader stage. After this expansion, all the rasterized fragments are given a color (black for the fiber and white for the halo) depending on whether or not its distance to the center of the fiber exceeds a certain threshold, after which the fragment is considered to be part of the halo. Furthermore, fragments representing the halo are displaced deep into the screen plane as they get far from the center of the fiber (this way fibers that are far from each other are de-emphasized by the halo, while fibers that lay very close to each other become grouped). This method, although improving over previous line methods, still requires data pre-processing, which doubles the amount of geometry to be processed by the GPU. Moreover, its frame-rate decays when geometry footprint is high.

3   Efficient Illustrative Rendering of Fiber Tracts

We have developed two different techniques that provide high quality renderings by performing basic Phong shading which helps to better understand fiber orientations, as well as adding ambient occlusion and halo calculations that generate perceptual cues on the absolute and relative positions of fibers. These techniques perform a multi-pass rendering algorithm that accomplishes two main tasks: i) Stream tube simulation, and ii) Final shading. The first task is shared by both algorithms. The differences come from the final shading stage.

3.1   First Rendering Pass: Stream Tubes Simulation

In order to obtain a high quality shading, we *simulate* stream tubes, instead of building them. This is achieved by creating view aligned triangle strips (VATS) along each of the lines of the model, and shading the resulting triangles in a per-fragment basis. Thus, we can simply obtain results that resemble stream tubes with a couple of triangles. This process is depicted in Figure 1. The geometry shader creates the new triangles by duplicating the vertices of the input line strips and moving them away on the plane perpendicular to the viewing direction (indicated as *widen direction*). Normals and texture coordinates of each vertex are also computed and exported so that the tube curvature can be later simulated. Then, the fragment shader computes per-fragment normals and writes both the normal and the fragment depth into a G-Buffer[6]. Everts *et al.* [5] also generate VATS, but in a preprocess. On the contrary, we perform the triangle expansion on the fly (in the geometry shader), which allows us to avoid geometry duplication and reduces the cost.

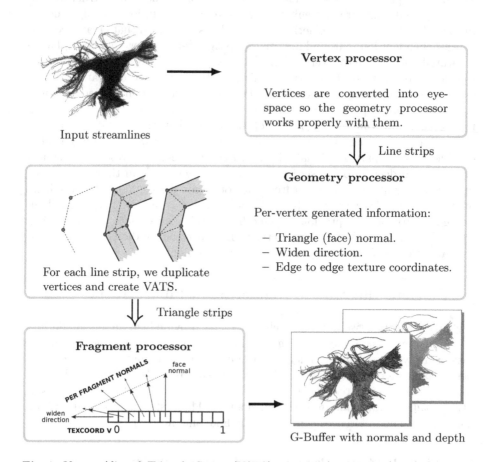

Fig. 1. Vertex Aligned Triangle Strips (VATS) creation (middle row) and G-Buffer generation (bottom row)

3.2   Screen Space Ambient Occlusion-Based Halos

Screen-space ambient occlusion (SSAO) estimation has been used previously for
the simulation of halos (e. g. Díaz *et al.* [7]). Inspired by this idea, we also
compute an ambient occlusion factor and use it for rapid halo generation. In [7],
the authors approximate the occlusion by using Summed Area Tables. However,
this is a costly process that requires either downloading the depth buffer to
GPU, or implementing a multiple-pass algorithm such as in Hensley *et al.* [8].
Since we do not require large sampling regions, the savings we may obtain using
Summed Area Tables are limited, and therefore a more classical approach yields
faster timings. Thus, we adapted the method by Filion and McNaughton [9]. As a
result, spatial features such as location and relation between grouped neighboring
fibers are more clearly depicted thanks to the soft shadows provided by the
occlusion of light, and soft halos around fibers rendered over distant ones also
help to understand the model's shape.

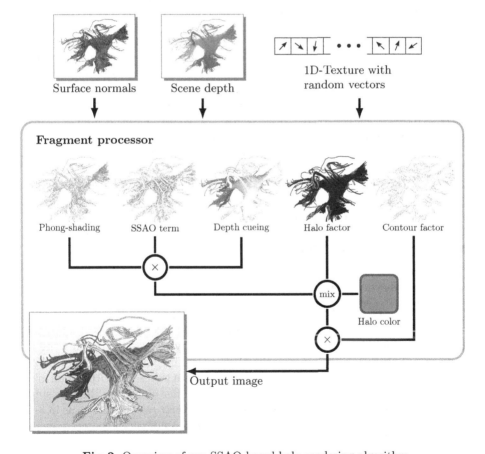

Fig. 2. Overview of our SSAO-based halo rendering algorithm

For a given pixel ambient occlusion (AO_f) is usually calculated by visiting all (or a subsampled set of) neighboring pixels, and comparing the difference in depth between the target pixel and its neighbors. We use these samples to also compute a halo factor H_f, whose calculations are exactly the same as in the case of AO_f but using a different function of the difference in depth. Our algorithm, as depicted in Figure 2, performs the following steps:

1. Render the geometry and store its normals and depth in a separate buffer.
2. Color, AO, halo and contour factors computation and final color composition.
   (a) Flip offset vectors pointing against the surface normal.
   (b) Reconstruct pixel eye-space position.
   (c) Add offset vectors to this position.
   (d) Remap these sample points back into screen space.
   (e) Obtain samples to compute ambient occlusion and halo factors.
   (f) Extra accesses to immediate neighbors to compute a contour factor C_f.
   (g) Compose final color $F_{col} = (P_{col}AO_fD_f(1-H_f)+H_{col}H_f)C_f$ by rendering a screen-filling quad, using Phong color (P_{col}), Ambient Occlusion factor (AO_f), distance factor (D_f), and halo color (H_c) and weight (H_f).

The occlusion factor (AO_f) is computed as $1 - \sum_1^n of(\Delta depth)/n$, where of is a typical decreasing curve (see [9]). The contour factor (C_f) is calculated like AO_f and H_f but just sampling the immediate neighboring pixels, and it is used to obtain thin and sharp contours. D_f is a factor that linearly darkens fragments as they lay deeper in the scene thus providing an extra depth cue. It is computed as $D_f = K((F+depth_{eye})/(F-N)-1/2)+1/2$, where N and F are the *ZNear* and *ZFar* distances, and K is a factor to reduce the darkening effect.

In our implementation we do not perform randomization of the offset vectors used to access the neighborhood of each fragment processed as in the original paper [9]. Moreover, our sampling is quite dense. As a consequence, no distracting visible patterns are noticeable due to the lack of random accesses. Therefore, we do not require blurring the resulting image to reduce noise. As shown later, this implementation is very efficient and may even outperform other geometry-based approaches even in large viewports with large geometry footprint.

3.3 Geometry-Aware Halos

Ambient occlusion-based halos might perform sub-optimally for simple models rendered in very large screen sizes. In such cases, a geometry-based approach may be superior. The main problem with previous systems is the depth-write in the fragment shader. This prevents the GPU to use its efficient Early-Z Culling system. We accelerate rendering by simulating stream tubes but with a multiple-pass algorithm that takes advantage of the depth buffer created in the first step.

This technique takes three rendering passes. The *first pass* is exactly as in 3.1. Figure 3 shows subsequent passes. The *second pass* renders the geometry again with an expanded fiber radius, in order to generate soft halos, whose intensity at each fragment is computed as a function of the difference in depth with the

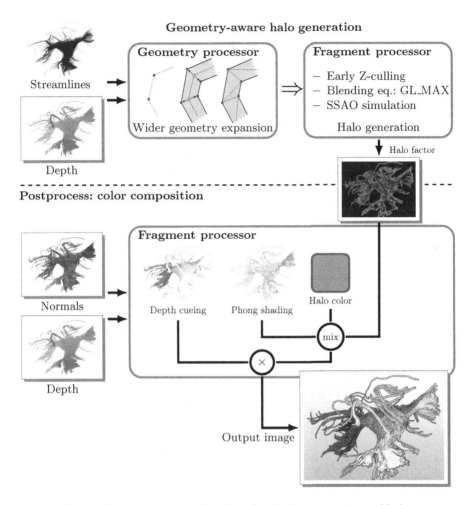

Fig. 3. Geometry-Aware algorithm for the fast generation of halos

previous depth map. This pass renders in a blending mode that only retains the brighter fragment processed at each pixel. An additive blending mode could seem a better choice in order to accumulate overlapping halo intensities, but the high density of fibers makes this approach over-accumulate halos. Finally, the *third rendering pass* is only used to compose the final image by blending the outputs of the previous passes. To sum up, the processes carried out are:

1. Render the geometry and store its normals and depth in a separate buffer.
2. Render the geometry with extended radius to generate the halo in a buffer.
   (a) Disable Z writing but use the previous depth buffer for Early-Z culling.
   (b) Activate a blending mode that only stores the brightest color.
   (c) The geometry shader generates wider VATS in most cases, but,

(d) It generates square billboards instead of VATS for those line segments whose direction is too similar to the viewing direction.

(e) The fragment shader sets an intensity for the halo H_f based on:

 i. A falloff factor that depends on the distance from the center of the VATS or billboard.

 ii. The difference between current fragment depth and scene depth.

3. Render a screen filling quad and compose the final image by blending color and halos: $F_{col} = (P_{col}(1 - H_f) + H_{col}H_f)D_f$.

In the previous formula, H_f is computed as a falloff factor that vanishes as if gets far from the fiber, multiplied by a function of the difference in depth between the halo and the original scene, represented by a curve that simulates ambient occlusion. D_f is the same darkening factor as in our previous method.

4 Results

In order to assess the validity of our approach, we compare our final rendering with other previous approaches in Figure 4. Stream tubes are better than line rendering but they have a prohibitive cost, since the geometry to render

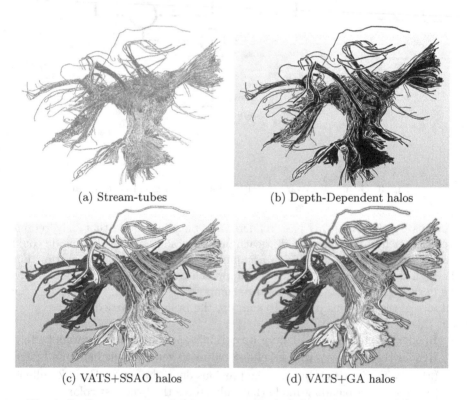

(a) Stream-tubes   (b) Depth-Dependent halos

(c) VATS+SSAO halos   (d) VATS+GA halos

Fig. 4. Quality comparison between our methods and previous state of the art

is considerably increased. Depth-dependent halos also improve over pure line rendering, but there is still some sensation of visual clutter due to big regions that end up being rendered with the same color. Moreover, it is still difficult to distinguish distant fibers from those located near the observer.

Our techniques improve the perception of both the orientation and relative positions of fibers while still maintaining high frame rates. Note how ambient occlusion provides fine details on the fibers' shapes (Figure 5-c). Distant fibers are darkened, and therefore its relative position is also easy to distinguish. The last method yields an ambient occlusion-like effect, but without the need of an intensive fragment processing algorithm. Darkest hollows are limited to be as dark as the darkest (the most intense) computed halo over them (due to the blending equation).

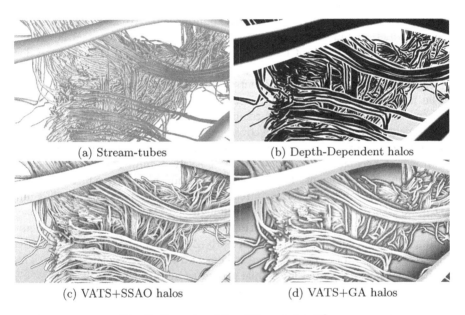

(a) Stream-tubes   (b) Depth-Dependent halos

(c) VATS+SSAO halos   (d) VATS+GA halos

Fig. 5. Zoom-in of the different algorithms

We have prepared several tests cases that include three different models (one of which is significantly complex) and two different scenarios: In the first one (*zoom out*) the whole geometry fits in screen, and its coverage of the viewport is relatively small, while in the second one (*zoom in*) the geometry covers most part of the viewport. The different models we tested have 901K, 15.6K, and 20.7K vertices, respectively. We ran the experiments on a i7 CPU running at 2.8GHz, 8GB RAM, and a GeForce GTX 470 GPU. We can see the timings in Table 1.

We may see in Table 1 that our screen-based approach (VATS+SSAO) never gets under 100 fps. The geometry-aware (VATS+GA) approach also reaches over 100 fps with ease, and only in the case of the larger window size (1280×1024) and large viewport coverage, the frame rate drops to 77 fps. Note how stream tubes

Table 1. Performance in fps for Model 1 (900K vertices) with different algorithms under different screen resolutions. Both new methods maintain high frame rates independently on the viewport resolution and coverage.

Zoom	out	in	out	in	out	in	out	in
Plain lines	846	847	846	841	845	823	840	667
Stream tubes	62	63	62	63	62	63	62	63
Depth-Dependent halos	603	183	603	123	566	78	470	48
VATS+SSAO halos	300	257	280	217	251	166	211	118
VATS+GA halos	207	171	203	139	197	106	183	77
Resolution	640x480		800x600		1024x768		1280x1024	

are clearly vertex limited, whilst our algorithms that generate streamtube-like geometry with a significantly lower vertex count are not. We have also compared with other methods in literature, such as *Depth Dependent Halos*, although the final results are very different in quality. In the case of Depth Dependent Halos, its performance drops when the viewport coverage is large, due to the effect of the fragment depth write, that prevents the GPU from using Early-Z rejection.

The performance of our screen-space algorithm (VATS+SSAO) linearly depends on the window size, since it requires several texture lookups (we use 32) per fragment. However, in all large coverage scenarios it outperforms geometry-based methods that rely on Z modification at fragment level (except for simple line drawing) because it takes advantage of Early-Z culling, and its computation is simple and efficient. Furthermore, it avoids blending, which is a time consuming operation because it requires fragment read before writing, thus limiting parallelism. On the other hand, although the Geometry Aware approach (VATS+GA) requires multiple rendering steps (geometry is rendered twice) and blending, it still benefits from the use of Early-Z, which helps to keep frame rates high enough, even in the case of large screens and large viewport coverage. This algorithm is specially suited for simple geometry models (see Table 2).

Table 2. Comparison of screen-space and geometry-based approaches for a low vertices model (Model 3: 20.7K vertices) under different screen resolutions. Note that in this case, the VATS+GA outperforms VATS+SSAO, no matter the size of the viewport.

Zoom	out	in	out	in	out	in	out	in
VATS+SSAO halos	1753	1016	1270	684	870	436	541	263
VATS+GA halos	2482	1119	1922	782	1336	509	867	326
Resolution	640x480		800x600		1024x768		1280x1024	

5   Conclusions and Future Work

We have designed two efficient algorithms for the illustrative visualization of fiber tracts. We focused on high quality renderings, that provide visual cues for the proper perception of fibers' absolute and relative positions and orientations. This is achieved by using halos. The algorithms we presented improve previous stream tubes visualizations both in quality and performance. We build View Aligned Triangle Strips (like [5]) in a more efficient way, by creating them in the in the geometry shader, thus avoiding the preprocess, and reducing by half the amount of vertices that are processed at vertex shader level. Our second step adds high quality halos with little cost. This is achieved in two different ways: the first approach computes halos using a screen-space technique based on ambient occlusion, and the second, computes halos with a multiple-pass geometry-based approach. Our approach provides a better perception on features like fiber orientation, location and density. We can also clearly perceive the relative positions of fibers. The screen-space method is most suitable for large line sets while the geometry-based method performs better when the number of lines is relatively small. In future we want to focus on applying these techniques to groups of fibers, such as when they are clustered, or labeled with uncertainty information.

Acknowledgments. This project has been supported by TIN2010-20590-C02-01 Project of the Spanish Government.

References

1. Zhang, S., Demiralp, C., Laidlaw, D.H.: Visualizing diffusion tensor mr images using streamtubes and streamsurfaces. IEEE Transactions on Visualization and Computer Graphics 9, 454–462 (2003)
2. Enders, F., Sauber, N., Merhof, D., Hastreiter, P., Nimsky, C., Stamminger, M.: Visualization of white matter tracts with wrapped streamlines. In: Visualization, VIS 2005, pp. 51–58. IEEE (2005)
3. Otten, R., Vilanova, A., van de Wetering, H.: Illustrative white matter fiber bundles. Comput. Graph. Forum 29, 1013–1022 (2010)
4. Tarini, M., Cignoni, P., Montani, C.: Ambient occlusion and edge cueing for enhancing real time molecular visualization. IEEE Trans. Vis. Comput. Graph. 12, 1237–1244 (2006)
5. Everts, M.H., Bekker, H., Roerdink, J.B.T.M., Isenberg, T.: Depth-dependent halos: Illustrative rendering of dense line data. IEEE Transactions on Visualization and Computer Graphics 15, 1299–1306 (2009)
6. Saito, T., Takahashi, T.: Comprehensible rendering of 3-d shapes. SIGGRAPH Comput. Graph. 24, 197–206 (1990)
7. Díaz, J., Yela, H., Vázquez, P.: Vicinity occlusion maps: Enhanced depth perception of volumetric models. In: Computer Graphics International 2008, pp. 56–63 (2008)
8. Hensley, J., Scheuermann, T., Coombe, G., Singh, M., Lastra, A.: Fast summed-area table generation and its applications. Computer Graphics Forum 24, 547–555 (2005)
9. Filion, D., McNaughton, R.: Effects & techniques. In: ACM SIGGRAPH 2008 Classes, SIGGRAPH 2008, pp. 133–164. ACM, New York (2008)
10. Díaz, J., Vázquez, P., Navazo, I., Duguet, F.: Real-time ambient occlusion and halos with summed area tables. Computers & Graphics 34(4), 337–350 (2010)
11. Mittring, M.: Finding next gen: Cryengine 2. In: ACM SIGGRAPH 2007: Courses, pp. 97–121. ACM, New York (2007)

Practical Volume Rendering in Mobile Devices

Marcos Balsa Rodríguez[1]
and Pere Pau Vázquez Alcocer[2]

[1] CRS4, Visual Computing Group, Italy
mbalsa@crs4.it
http://www.crs4.it/vic/
[2] UPC, MOVING Graphics group, Spain
ppau@lsi.upc.edu
http://moving.upc.edu/

Abstract. Volume rendering has been a relevant topic in scientific visualization for the last two decades. A decade ago the exploration of reasonably big volume datasets required costly workstations due to the high processing cost of this kind of visualization. In the last years, a high end PC or laptop was enough to be able to handle medium-sized datasets thanks specially to the fast evolution of GPU hardware. New embedded CPUs that sport powerful graphics chipsets make complex 3D applications feasible in such devices. However, besides the much marketed presentations and all its hype, no real empirical data is usually available that makes comparing absolute and relative capabilities possible. In this paper we analyze current graphics hardware in most high-end Android mobile devices and perform a practical comparison of a well-known GPU-intensive task: volume rendering. We analyze different aspects by implementing three different classical algorithms and show how the current state-of-the art mobile GPUs behave in volume rendering.

1 Introduction

Discrete 3D scalar fields are used in a variety of scientific areas like geophysics, meterology, fluid flow simulations and medicine; its exploration allows scientists to extract different types of relevant information. The most outstanding property of this kind of data is the availability of information in the whole volume, with each space portion having differentiate values. Mesh-based visualization [1] is not well suited to explore all this information since it is typically surface-oriented. This led the development of a new branch of scientific visualization to focus on volume rendering with the objective of enabling exploration of this enormous sources of data. The last two decades volume rendering has been a very active research topic generating many publications covering different issues like illumination, compression or massive model exploration.

One of the main difficulties of volume rendering is the amount of information to deal with. In the 1990s costly workstations were required to work with volume models using software rendering or special-purpose hardware. It was in the later 2000s that interactive volume visualization became possible in high-range

G. Bebis et al. (Eds.): ISVC 2012, Part I, LNCS 7431, pp. 708–718, 2012.

desktop and laptop computers by exploiting the texture functionality present in consumer graphics hardware [2].

Nowadays, laptops are being replaced by lighter and smaller embedded devices, like smartphones or tablets, for everyday's work. These devices have become powerful enough to run complex 3D applications previously only available to high end PCs and laptops. Current generation of mobile devices is able to run 3D games with quality comparable to the previous generation of console games.

Many vendors often praise the horsepower of the new CPUs and GPUs sported in mobile phones. However, it is difficult to predict their performance for graphics intense tasks, since it depends on a complex combination of computation power, memory, bandwidth, and several other factors. Therefore, we decided to evaluate the suitability of most modern devices for one well-known GPU-consuming scenario: volume rendering. The contributions introduced in this work are:

- An exhaustive analysis of the most recent mobile platforms and mobile devices currently available in the market.
- An in-depth analysis of the performance of three state-of-the-art volume rendering methods on a subset of the most relevant graphics hardware available in modern mobile devices.

2   Previous Work

2.1   Volume Rendering Algorithms

Volume rendering is a set of techniques used to display a 2D projection of a 3D discretely sampled dataset. These 3D datasets can come from different sources: fluid simulation, geological exploration, medical images or industry object scans. We will focus on medical models. The two most popular volume rendering techniques are RayCasting and Texture Slicing.

RayCasting works by tracing rays from the camera into the volume and solving the rendering integral along this rays. Volume ray casting was introduced by Levoy[3] two decades ago; Krüger[4] et al. presented one of the first GPU implementations one decade ago. Ray casting was done in software for many years due to the lack of hardware support, which was introduced with programmable shader functionality and 3D texture support. A modern GPU implementation of this technique relies on the fragment shader to perform the tracing of rays from eye view into the volume.

Together with ray casting, **texture slicing** is the second most popular technique for GPU based volume rendering. It is an *object-order* approach. The proxy geometry used to render the volume data are 2D slices or quads. The slices are projected onto the image plane and combined according to the composition scheme. Slices can be sorted front-to-back or back-to-front, although probably the most popular is back-to-front order and relying on hardware color blending. Since this technique only relies on standard 2D textures and texture blending, which are available in graphics hardware since many years, it is the most compatible and efficient technique. These techniques, as other *object-order*

algorithms, use simpler addressing arithmetics because of working in storage order and so have better performance without complex improvements. These slices
can be *object-aligned* or *view-aligned*. Object-aligned slices require having three
slice sets in GPU, one for each axis, since we need to render the slice set that
is most perpendicular to the view direction. There are graphic glitches when
switching from one slice set to another. *View-aligned slices* do not have these
problems; however, *view-aligned slices* require 3D texture support. The proxy
geometry must be calculated each frame depending on the view position; for
this purpose, a bounding box is intersected with planes perpendicular to the
view direction and regularly arranged.

2.2   Mobile Devices Graphics Hardware

Mobile devices, especially high-end models, have been typically accompanied
by graphics acceleration hardware, only 2D acceleration was supported at first
but current devices typically include 2D and 3D acceleration. As of today, it is
common in high-end mobile devices to have at least a resolution of 480 pixels
width and 800 pixels height, while while next generation will introduce resolutions around 1200x720 (Samsung Galaxy Nexus and most Android tablets) or
2048x1536 (iPad3). This resolution increase comes together with a great increase
in graphics hardware performance. This is possible thanks to the powerful CPU
and GPU that they all have built-in. All of the devices included in the comparison have support for OpenGL ES 2.0 enabling the use of shaders for graphics
programming. There are mainly five dominating architectures in the market (see
Table 1):

- Qualcomm. Qualcomm chipsets have been implemented in many devices
  from a wide range of manufacturer's, with HTC being its most dedicated
  customer. The SoC solutions provided by Qualcomm come with a graphics
  solution of its own called Adreno. There are three generations of Adreno
  GPU's: 200 (Nexus One), 205 (Htc Desire HD) and 220 (HTC Sensation);
  each generation easily doubles the graphic performance of its antecessor.
- Texas Instruments. TI is one of the most well known embedded device manufacturers and has been present in many Motorola devices and also in recent
  LG devices (LG Optimus 3D). TI has frequently used the Power SGX 535
  and Power SGX 540 GPUs from Imagination Technologies.
- Samsung. In the last years, has also developed a couple of ARM chipsets:
  the Hummingbird implemented in the Samsung Galaxy S and accompanied
  by a Power SGX 540 GPU, and the Exynos dual-core implemented in the
  Samsung Galaxy S2 with the Mali-400MP GPU from ARM.
- NVIDIA. Last year NVIDIA introduced its Tegra 2 platform which is an
  implementation of ARM's instruction set and accompanied by an Ultra-low
  voltage (ULV) GeForce graphic chipset also by NVIDIA. This was one of
  the very first dual-core solutions for mobile devices and is present in the
  majority of Android tablets sold the past 12 months.

– Apple. Initially integrated chipset solutions from other companies but with the iPhone 4 they started developing their own chipsets, like the A4 in iPhone 4. This ARM processor is complemented with a Power SGX 535 GPU.

Table 1. Comparison of most extended mobile GPU hardware

Model	MTris/sec	MPix/sec	3D textures	Manufacturer
Adreno 200	22	133	Yes	Qualcomm
Adreno 205	41	245	Yes	Qualcomm
Adreno 220	88	500	Yes	Qualcomm
Power SGX 535	14	500	No	Imagination Technologies
Power SGX 540	28	1K	No	Imagination Technologies
Power SGX 543	40-200	1K	No	Imagination Technologies
Power SGX 543MP	40-532	1K-16K	No	Imagination Technologies
Mali 400MP	30	300-1K	No	ARM
Tegra2	71	1.2K	No	NVidia

In Table 1 there is a comparison of the most advanced graphic chipsets used in current high-end mobile devices. There are two predominant manufacturers in this table: Qualcomm and its Adreno family of GPUs, and Imagination Technologies with its Power SGX GPUs. Last year, the Tegra2 chipset from NVIDIA has gained importance, specially for being included in almost all new Android tablets. The Mali-400MP GPU is ARM's proposal for their reference design. All the GPUs present in this table offer support for OpenGL ES 2, and so shader programming and state-of-the-art graphics. The only GPUs offering 3D texture support are the ones from Qualcomm: the Adreno family. The numbers in this table are mostly given by the manufacturers and should be taken only as peak values, since they are very conditioned by hardware configuration parameters such as clock frequency. For the sake of comparison, let's say that current generation consoles peak numbers are not too far from those provided by mobile hardware: the XBOX 360 has a peak of 500 million of triangles per second, while the PS3 peak is at 250 million of triangles per second. And for the mobile GPUs let's mention the Power SGX 540, that could reach 90 million of triangles per second by increasing clock's frequency to 400Mhz, and the NVIDIA Tegra 2 with a peak theoretical limit of 71 million of polygons per second. On the other hand, for the purposes of our empirical comparisons, the most relevant numbers are those related to fill rate (millions of pixels per second) because volume rendering applications use little geometry and depend mostly on fragment processing. Comparing typical mobile GPU fill rate of 1 billion of pixels per second with medium range Desktop GPU (like NVIDIA GTX 460) fill rate around 37 billion of pixels per second, shows that there is still an important gap to be filled. It must be taken into account that typical mobile devices have to render 384.000 pixels for a screen of 480x800, while a Full HD desktop monitor would require rendering 2,073.600 pixels, which is 5.4X the number of pixels in a mobile device. Anyway, including these 5.4X factor in the comparison still gives 6.8 times more fill rate to medium-range desktop GPUs.

3 Implementation Details

Our application implements two different rendering techniques for volume rendering: object-aligned slices and ray casting. The *object-aligned slices* technique is implemented both for *2D textures* and *3D textures* in order to make the application compatible with more hardware. Not all the techniques work on every device since 3D textures are not common in embedded graphics chipsets, but having these different implementations allows us to compare which one performs better on different hardware. We have also implemented a benchmark thought to be executed on many different devices with heterogeneous hardware configurations in order to compare performance.

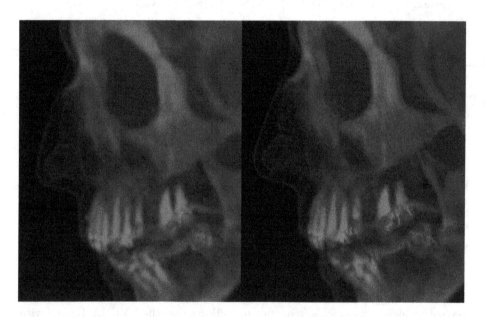

Fig. 1. These images compare a detail in two renditions of the CT head dataset at half viewport resolution (left) and full viewport resolution (right)

Since the frame rates are not high, as we will show later, we need to add some improvements in order to make the application interactive. Being the application so pixel intensive, we took to two different approaches for level of detail: reducing the number of slices and reducing the viewport resolution. When low viewport resolution is enabled, static renders are also done in half resolution; thanks to

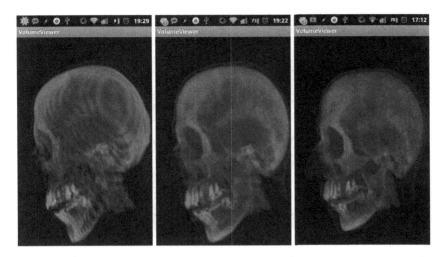

Fig. 2. These images show renditions at full viewport size of the CT head dataset (with dimensions 256x256x113) using 64, 128, and 256 slices, respectively

linear interpolation using half the resolution of the viewport still give good visual quality slightly alleviating the aliasing (see Figure 1).

On the other hand, reducing the number of slices without using lower resolution viewport produces more noticeable artifacts (see Figure 2). For this reason, we propose using the maximum number of slices to match dataset dimensions and simply enable low viewport resolution to improve interaction. We offer the user the option to define a lower slice count for rendering while there is user interaction for slowest devices, but in our tests it has only been needed for ray casting where the hardware still performs at low frame rates.

We have also taken care of the usability of our approach. For volume models, the definition of a transfer function is a very important step, since it determines which information is visible and how. We implemented a transfer function editor tailored to small screens with numerous visual feedbacks for selection, such as the selection highlighting in yellow, or the mini-zoom tool tailored to perform fine selection whilst avoiding the problem of finger occlusion (see Figure 3).

4   Results

We have performed a variety of performance tests with different configurations to be able to extract meaningful information about different hardware restrictions:

- **Sampling Resolution.** We have run benchmarks with different volume sampling frequencies (number of slices for slice-based renderers or number of samples per voxel for the ray cast renderer) to analyze the impact of voxel color composition on performance. We have run the tests with the Engine dataset with dimensions 256x256x256 because most of the devices

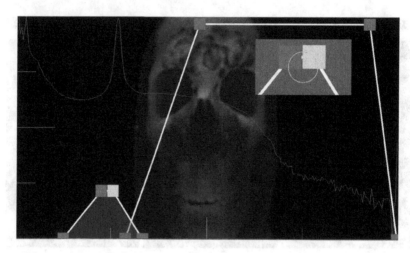

Fig. 3. Edition of the transfer function with the mini-zoom helper. This function defines two sub-functions associating transparency and color to the scalar values.

tested haven't been able to load larger datasets. Only for the 3D texture slice renderer and for the ray caster we have run the benchmark with the Sheep dataset with dimensions 352x352x256.

- **Viewport Resolution**. We have run benchmarks in full resolution (480x800) and half resolution (240x400) to analyze the effect of fill rate.

The benchmarks consist of a series of camera positions that are rendered consecutively. These camera positions start very close to the viewer with the volume covering the full viewport and gets farther and farther progressively while rotating exactly five times 360 degrees until the volume only covers about 1/8 of the screen. This way we get an averaged frame time from all the views of the volume covering the full viewport and only covering a small portion.

We have defined four different qualities based on sampling frequency. For slice-based renderers, quality 0 uses at most 64 slices per axis, doubling the slice count until quality 3 where 512 slices per axis is the limit (depending on the dataset). Quality is defined as the number of samples per voxel for the ray casting renderer, with 0 being 0.25 samples per voxel (taking into account only 1 of every 4 voxel) and 3 being 1 sample per voxel. It must be noted that for the 2D texture slice renderer we resample the textures in order to reduce the texture data size. This allows us to load bigger volumes although the rendering will be in lower resolutions (ie. the slices in a 512^3 dataset in quality 2 would be resampled to 256^3). The maximum effective volume size we have been able to load in most devices is the 256^3 *Engine* dataset for this renderer; it has the big disadvantage of requiring 3 slice sets, one for each axis, and so uses three times more GPU memory. For the 3D texture slice renderer and the ray cast renderer the limits are imposed by the drivers and the largest volume dataset that we have been able to load is the 352x352x256 *Sheep* dataset.

Table 2. Benchmark results of the implemented volume renderers on different mobile devices in full viewport resolution and half viewport resolution. Quality is the number of slices [64, 128, 256, 512] for slice-based renderers and the number of samples per voxel [0.25, 0.5, 0.75, 1] for the ray casting renderer. The values in the table are frames per second.

		high resolution				low resolution			
	Qual.	Galaxy S	Advent Vega	HTC desire	HTC desire Z	Galaxy S	Advent Vega	HTC desire	HTC desire Z
slices 2d	0	6.28	7.41	5.75	13.25	11.49	22.73	15.15	34.48
	1	3.24	3.65	3.08	7.35	6.85	12.66		22.47
	2	1.91	1.92	2.2	5.38	3.64	7.14	5.92	9.95
	3								
slices 3d	0			5.08	10.31			6.76	11.24
	1			2.65	5.15			4.37	6.62
	2			1.84	2.33			2.33	3.71
	3			1.81	1.63			2.43	2.94
ray cast	0			0.42	2.06			1.71	5.38
	1				1.54			0.95	4.02
	2				0.96			0.65	3.06
	3				0.77				1.96
GPU		Power SGX 540	Tegra 2	Adreno 200	Adreno 205	Power SGX 540	Tegra 2	Adreno 200	Adreno 205

Table 2 shows the frame rate values obtained from running the benchmark on different devices. The 256^3 Engine model has been used for all the benchmarks. All these devices have a screen resolution of 480x800 and the results are for low (half the viewport resolution) and for high resolution (full viewport resolution). From this table, we can infer that the *Samsung Galaxy S*, with a performance boost of $2X$, is less affected by reducing the resolution than the HTC desire, which gets a boost of $3X$. The *Adreno 205* GPU in the *HTC Desire Z* and the *Tegra 2* in the *Advent Vega* show the same $3X$ performance boost with half resolution. The *Advent Vega* has not achieved the results that one could expect from the *Tegra 2* chipset (while the *LG Optimus 2X* also including the *Tegra 2* performs significantly better, as will be seen later).

4.1 Device Performance Comparison

We have run our benchmarks on many devices to illustrate the results of the 2D texture slice renderer, which is the most compatible approach; this will give us a good overview of current mobile GPUs performance.

In Figure 4 we can see benchmark results for all the devices we have tested with the slice renderer based on 2D textures. There are four differentiated groups:

– The *Samsung Galaxy S2*, with a dual-core processor including the Mali-400MP GPU, achieved the best performance by almost doubling its nearest

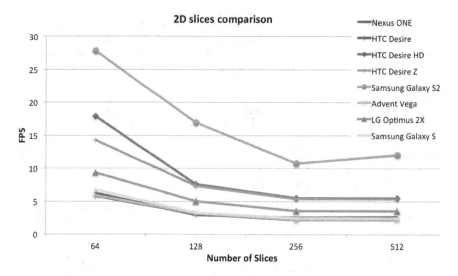

Fig. 4. Benchmark results of various devices for the 2D texture slice renderer. Lowest lines show Nexus One, Samsung Galaxy S with previous generation single core hardware and the Advent Vega, a low-cost Tegra 2 device.

competitor. This is one of the latest devices in the market at the time of writing. On the other hand, they have not support for 3D textures.

- The *HTC Desire HD/Z* are two recent devices from HTC; they implement the second generation of Adreno processors, the Adreno 205. This GPU family is the only one that we know to have 3D texture support. Its performance is in the second range quite below the Galaxy S2.

- The *LG Optimus 2X*, with a dual-core CPU based on NVIDIA Tegra 2 platform, is half way between the Qualcomm first and second generations of GPUs. The Tegra 2 platform was expected to perform much better, but seems that for our benchmark implementation this is not the case.

- The *Nexus ONE, HTC Desire, Samsung Galaxy S and the Advent Vega* are in the last group. The first two integrate the Adreno 200 from the first generation of Qualcomm chipsets, while the Samsung Galaxy S sports a powerful Power SGX 540 from Imagination. The Advent Vega seems to lack some driver optimizations since the same platform in the Optimus 2X has performed significantly better. Taking the Advent Vega apart, this last group if composed of some of the most extended single-core Android devices.

4.2   Comparing the Volume Rendering Implementations

We see how the slice-based renderers have much better performance than the ray casting; this is mainly because the work done in the fragment shader is much lighter and most of the work is done in the composition phase performed just after that using the typical hardware pipeline which is much more optimized.

For the 2D texture slice renderer we have used the 256^3 Engine dataset and so quality 2 and 3 have the same performance. The trilinear filtering used by 3D textures is one of the reasons of the performance gap between the 2D and 3D slice renderer, trading visual quality for performance. For the ray casting renderer the intensive use of fragment shaders takes the GPU to its limits.

Mobile GPUs are not yet as capable as their desktop and laptop counterparts due mainly to low graphic unit count and no dedicated graphic memory. Latest mobile GPUs typically have between 4 and 12 processing units or shaders, depending on them having unified shader architecture or implementing Vertex/Fragment shaders, and shared system memory with some fraction of this memory reserved for the GPU. On the other hand, latest mobile devices sporting these GPUs also include high resolution screens (ranging from 800x400 to 1280x720 and 5" to 10"). These two facts: low processing unit count with no dedicated memory and high resolution screens introduce a big bottleneck into the fragment shader phase. We have seen that typically using partial resolution is enough for giving good results in such small screens, specially while interacting, mainly because human eye is not able to exploit this high resolution at the typical usage distance (around 15-30 cm).

5   Conclusions and Future Work

We have achieved interactive frame rates on most high-end mobile devices currently available. We also implemented a transfer function editor that is specially designed for small screens. However, many aspects could be improved on both sides. The object-aligned slices approach produces disturbing graphic glitches when changing the point of view from one axis to another; also there are noticeable artifacts when looking at steep angles. For the 2D texture slice renderer, requiring three slice sets is unavoidable but by using the volume rendering schema presented by Krüger[5] the graphic glitches produces by slice set changes should be unnoticeable. Also the usage of multiple textures per slice and implementing trilinear filtering on the shader should improve visual quality. All these extensions will add a considerable CPU and GPU cost. For the 3D texture slice renderer the most noticeable problem is due to the object-aligned slices again, requiring high number of slices to produce a high quality view. In this case, implementing view-aligned slices would give much better results at the expenses of some performance loss due to the calculation of new slices each frame. The ray casting renderer we have implemented is very basic and can be greatly improved both for quality and performance. Using too few samples produces many artifacts, but this is unavoidable due to the low performance of this technique on current hardware. To improve performance, we could use empty-space skipping [6] to avoid sampling empty regions at the expenses of using another low resolution 3D texture and one extra texture access at each sample point. However, these are improvements we plan for next generation devices. At the moment of the implementation, the most advanced devices and were LG Optimus 2x and Samsung Galaxy SII. All the implemented techniques could be greatly improved by adding shadows. However, this would add quite many calculations on the fragment shader and could not be feasible.

Acknowledgments. This project has been supported by TIN2010-20590-C02-01 Project of the Spanish Government and People Programme (Marie Curie Actions) of the EU's 7th Framework Programme FP7/2007-2013/ under REA grant agreement n290227.

References

1. Akenine-Moller, T., Haines, E., Hoffman, N.: Real-Time Rendering, 3rd edn. A K Peters (2008)
2. Engel, K., Kraus, M., Ertl, T.: High-quality pre-integrated volume rendering using hardware-accelerated pixel shading. In: Proceedings of the ACM SIGGRAPH/EUROGRAPHICS Workshop on Graphics Hardware, HWWS 2001, pp. 9–16. ACM (2001)
3. Levoy, M.: Display of surfaces from volume data. IEEE Comput. Graph. Appl. 8, 29–37 (1988)
4. Krüger, J., Westermann, R.: Acceleration techniques for gpu-based volume rendering. In: Proceedings IEEE Visualization 2003 (2003)
5. Krüger, J.: A new sampling scheme for slice based volume rendering. In: Volume Graphics, pp. 1–4 (2010)
6. Li, W., Mueller, K., Kaufman, A.: Empty space skipping and occlusion clipping for texture-based volume rendering. In: Proc. IEEE Visualization 2003, pp. 317–324 (2003)

Real-Time Visualization of a Sparse Parametric Mixture Model for BTF Rendering

Nuno Silva[1,*], Luís Paulo Santos[2], and Donald Fussell[3]

[1] Centro de Computação Gráfica, Campus de Azurém, Guimarães, Portugal
[2] Dep. Informática, Universidade do Minho, Braga, Portugal
[3] The University of Texas at Austin, Texas, USA

Abstract. Bidirectional Texture Functions (BTF) allow high quality visualization of real world materials exhibiting complex appearance and details that can not be faithfully represented using simpler analytical or parametric representations. Accurate representations of such materials require huge amounts of data, hindering real time rendering. BTFs compress the raw original data, constituting a compromise between visual quality and rendering time. This paper presents an implementation of a state of the art BTF representation on the GPU, allowing interactive high fidelity visualization of complex geometric models textured with multiple BTFs. Scalability with respect to the geometric complexity, amount of lights and number of BTFs is also studied.

1 Introduction

Digital representations of complex materials can have a major role in footwear and textile industries, assisting designers and artists in virtual prototyping new products and providing end-users realistic visualizations of such products. For the designer, the usefulness of these tools is greatly dependent on both the representation quality and high fidelity interactive visualization rates; achieving both these requirements is a challenging task.

A material's appearance depends on the way radiant flux is scattered when it hits a surface, and varies, among others, according to incoming light and observation directions [1]. Parametric Bidirectional Reflectance Distribution Functions (BRDF) are often used to model a material's appearance, but they cannot simulate many complex lighting phenomena such as self-shadowing, self-occlusion, sub-surface scattering and inter-reflections. Instead, image based approaches, such as the Bidirectional Texture Function (BTF), are becoming ever more popular due to the realism they can provide, the improvement in acquisition systems quality and the increasing computational power of GPUs [1–3].

The BTF is a 6D function [4] that models a material's appearance at a given point on the surface by recording several images captured under different lighting

---

* Work partially funded by QREN project nbr. 13114 TOPIC Shoe and by National Funds through the FCT - Fundação para a Ciência e a Tecnologia (Portuguese Foundation for Science and Technology) within project PEst-OE/EEI/UI0752/2011.

G. Bebis et al. (Eds.): ISVC 2012, Part I, LNCS 7431, pp. 719–728, 2012.

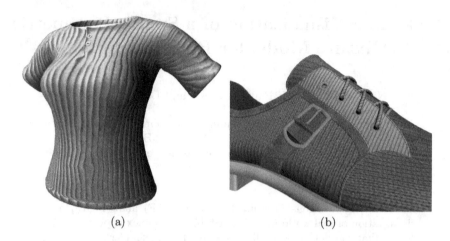

(a)   (b)

Fig. 1. High fidelity interactive rendering: a corduroy shirt (a), and a shoe with multiple BTFs (b)

and viewing directions. The images are stored in large tables, and rendering involves simple look-ups within these. A single BTF can take up several gigabytes of storage, too much to be of practical use in real-time rendering. To tackle this problem, the BTF data must be transformed into a compact and efficiently renderable representation, without compromising image fidelity [2, 3, 5–7].

Wu et al. [8] presented a novel general representation for BTFs, the Sparse Parametric Mixture Model (SPMM). They demonstrate their approach using a parallel ray tracer, which, although achieving high quality visualizations, is far from interactive. This paper presents a rasterization oriented visualizer for SPMMs that achieves interactive frame rates for complex models involving multiple BTFs, without compromising on visual quality. Rendering times are dependent on the number of geometric primitives, number of fragments mapped with an SPMM and number of light sources; scalability is studied with respect to these parameters. The proposed interactive visualizer is currently being used by project partners in the Portuguese footwear industry.

This paper is organized as follows: first, background on material appearance and their representations is presented, along with the SPMM and other related work. Then, our approach to achieve real-time rates is described, followed by a discussion of results. The paper terminates with conclusions and future work.

2   Appearance Modeling and Visualization

Most real world materials exhibit complex appearance that can be described at three levels [2, 9]. The macroscale is the large scale geometry of the object, traditionally modeled with explicit representations, such as polygon meshes. The microscale level relates to interactions of light with a point on the surface of the material, and can be represented using BRDFs. The mesoscale level is in between these two and comprises various subtle lighting effects such as self-shadowing, self-occlusion, sub-surface scattering and inter-reflections, which cannot be faithfully represented with the BRDF; instead, image based approaches are used, the most common method being texture mapping. Spatially-Varying BRDFs (SVBRDF) [10], which can be seen as a combination of texture mapping and BRDFs, account for materials that have different BRDFs throughout their surfaces; while addressing some of the issues raised at the mesoscale level, they cannot capture self-shadowing and self-occlusion. For that, BTFs, an image driven approach, are often used instead. A Bidirectional Subsurface Scattering Reflectance Distribution Function (BSSRDF) can model all these light phenomena, but is much too complex to be usable in interactive rendering pipelines, and current capturing systems only allow to measure subsets of this function [1, 2].

It must be noted that appearance can be reproduced by following a procedural approach, i.e., hand-tuning an algorithm and mathematical functions until the desired effect is achieved [11]. This is, however, a time consuming task that demands high level of expertise and might still fail to accurately simulate real world materials. Image based approaches are ever more popular because appearance is measured using cameras, demanding no particular expertise, thus being more adequate for project partners in the Portuguese footwear industry.

2.1   Bidirectional Texture Function

The BTF emerged as an alternative that represents both mesoscale and microscale levels. For each wavelength, it models the material appearance based on a point on the surface (x), and the incident and reflection directions (ω_i, ω_o).

BTFs model the appearance of a material from several images captured under different observation and lighting directions; they can be seen as a special class of the SVBRDF since surfaces are assumed to be planar [12]. A good quality BTF, such as the ones in the Bonn database [7], encodes 81×81 images for light and viewing directions, each consisting of 256^2 texels with three spectral values (RGB). This corresponds to roughly 1.2GB of raw data for a single BTF, not including High Dynamic Range (HDR) effects.

Achieving interactive visualization rates of objects with multiple BTFs requires compressing the raw data, while preserving as many of the relevant features of the BTF as possible. Compression must exploit the redundancy in the data in an efficient way, and allow fast decompression for real-time rendering. Refer to [2] and [3] for further details on BTF modeling.

2.2   The Sparse Parametric Mixture Model

In the SPMM representation proposed by Wu et al. [8], the captured data is analyzed and fit into a number of different parametric functions, each defined as a cosine-weighted rotated BRDF. Equation 1 describes such functions, where $f_j(k_j, \cdot)$ is one analytical BRDF model, with parameters k_j. R is a rotation that transforms a vector into the local coordinate system defined by local normal n_j.

$$\rho_j(\omega_i, \omega_o) = f_j(k_j, R(\omega_i), R(\omega_o))(n_j \cdot \omega_i). \tag{1}$$

The original data at a point x can be approximated by using a weighted linear combination of m such functions (see equation 2 and figure 2), each with weight α_j. Subtle appearance details that cannot be fit into the parametric functions are stored as a residual function, ϵ_x, which is obtained by subtracting the linear combination of parametric functions from the original BTF.

$$\mathrm{BTF}_x(\omega_i, \omega_o)(n_x \cdot \omega_i) = \sum_{j=1}^{m} \alpha_j \rho_j(\omega_i, \omega_o) + \epsilon_x(\omega_i, \omega_o). \tag{2}$$

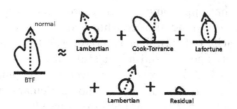

Fig. 2. Illustration of the SPMM representation. The BTF is approximated by a sum of analytical models and a residual part. Each analytical model has its own local frame. Adapted from Wu et al. [8].

Since fitting all the texels of the BTF is computationally expensive, spatial coherence of the original data is exploited through multilevel k-means clustering. The full fitting algorithm is applied to some selected representative texels, and the resulting parametric functions are used as a dictionary to accelerate the fitting of the other texels in the cluster. The residual function ϵ_x is also computed on a per-cluster basis. It is suggested that ϵ_x can be improved by storing a few additional basis error functions and respective coefficients obtained from Local Principal Component Analysis (LPCA) [5]; this allows marginal improvements of image quality at the cost of increased memory requirements. We support both options, allowing adaptation of image quality, and thus rendering times, to the available computing power on the graphics board (see sections 3.1 and 4).

By fitting BTF data into a sum of parametric functions, the SPMM provides a general representation that allows the volume of data to be significantly reduced, enables efficient rendering and intuitive editing of parameters. Wu et al. tested the SPMM with the Bonn database [7] of BTFs; our GPU implementation is based on those representations.

3   Implementation

Our interactive visualizer uses the OpenGL API to communicate with the GPU and GLSL to program the shading process. Efficient use of the GPU requires

SPMM data to be stored in the device texture memory, as 1D, 2D or 3D arrays [13, 14], allowing fast random accesses and high bandwidth. Thus, texture memory stores the cluster identifier for each texel, the parametric functions ρ_j, their corresponding parameters k_j and weights α_j, and the residual function ϵ_x.

The SPMM is not geared towards optimal GPU performance because it consists of a linear combination of different analytical BRDFs, and directly translating the CPU shader into a GPU shader results in a lot of branching and loop instructions (and a very large shader code base). Transforming the data structures in order to fit the GPU streaming programming model is also challenging.

First, OpenGL only allows to store basic data types into texture memory: 1, 2, or 4 byte words, fixed or floating point. The data format used is a crucial aspect in GPUs because it determines the amount of storage and bandwidth required; we minimize the number of bytes required for each texture. SPMMs generated from BTFs in the Bonn database [7] group texels into 32 clusters, thus the cluster identifier can be stored with a single byte. Other textures that use fixed point values store indexes or identifiers and so they are represented with 2 bytes. Textures that store the parametric function weights, their parameters and the residual function are in the floating point format, encoded in half-precision with 2 bytes. In our experiments this does not produce visible artifacts whilst greatly reducing memory requirements and improving render time.

The second issue is that the number of parametric functions m is not the same across all texels, and dynamically sending this information to the GPU would make real-time rendering hard to achieve. Calculating the maximum number of functions for all texels and letting all fragment shaders do roughly the same amount of work resulted in poor performance. Our implementation precomputes m for each texel and stores it in an additional texture. This resulted in higher frame rates, especially noticeable when the number of parametric functions varies greatly from one texel to another, i.e., when the BTF exhibits locally complex reflectance variations.

The third problem arises from the sparse discretization of the view/lighting directions during the BTF measurement. In order to render the BTF under novel viewing or lighting directions, interpolation of the closest view/lighting slots must be performed. The residual function must be interpolated in order to avoid the appearance of artifacts when these slots change. The parametric functions are not affected, since they are defined over the entire upper hemisphere. Those direction slots can be interpreted as 3D points and projected onto a set of points on the XY plane by ignoring the Z component. We then apply a Delaunay triangulation and store the resulting triangles in a texture. Interpolating now consists of a ray-triangle test using barycentric coordinates [15], resulting in the appropriate interpolation weights in case of a hit. This must be done separately for the view and lighting directions, for a total of nine interpolation weights.

Finally, 2D arrays are transformed into slices of 3D arrays to avoid exceeding the capabilities of the hardware. Figure 3 depicts the used data structures (some of them presented in the following section) and the data flow to render the full SPMM.

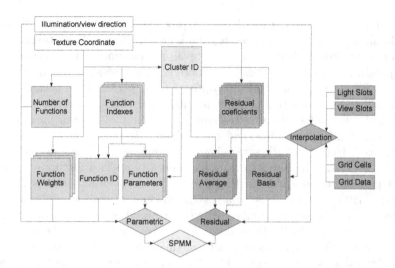

Fig. 3. The data flow in the fragment shader. The small rectangles, the squares and the stacked squares depict, respectively, 1D, 2D and 3D textures. The diamonds represent the combination of the inputs.

3.1   Optimizations

Profiling analysis indicated a bottleneck in the interpolation of the residual function. Since this is based on a ray-triangle test and the triangles are static and well distributed over the unit circle, a regular grid is used to quickly discard triangles, thus limiting intersections test to a (very small) subset of the triangles. A compact grid structure with minimal memory requirements [16] is built once in the CPU and uploaded to the GPU using 2 1D textures.

Equation 3 is used to perform the ray-triangle intersection test, where λ is the vector of barycentric coordinates, r is the ray vector, and T is a matrix formed by the Cartesian coordinates of the triangle vertices. This allows precomputation of the inverse matrix T^{-1} for each triangle, reducing the intersection test on the GPU to a vector subtraction and a matrix-vector multiplication. Additionally, to maximize data locality, the inverse matrix, the Cartesian coordinates of each vertex and its corresponding ID are packed together, for a total of 16 floats per triangle. With both these optimizations the computational cost of evaluating the residual function is roughly the same as evaluating the BRDFs, whereas before optimizations, the former was around 3 times longer than the later.

$$\begin{pmatrix} \lambda_1 \\ \lambda_2 \end{pmatrix} = T^{-1}(r - v_3);$$
$$\lambda_3 = 1 - \lambda_1 - \lambda_2. \tag{3}$$

To take full advantage of the bandwidth of the GPU, we exploit the SIMD patterns shader instructions, and strive to perform texel fetches of the RGBA channels of each texture. However, not all textures can have data packed in order to use all the available channels, and some demonstrated performance losses with this new arrangement of data. As we are aware this can change between GPU vendors and families of the same vendor, not much effort was put into fine tuning the data layout of textures.

Finally, at rendering time the number of LPCA components used to evaluate the residual function (see section 2.2) can be limited by a user defined parameter, allowing the program to adapt to the computing capabilities of the host machine. In our experiments, this can greatly increase the frame rate whilst having marginal impact on image quality.

4   Results

Experiments were conducted on a workstation equipped with an Intel 2.4GHz quad-core processor, 4GB RAM and with a Nvidia GeForce GTX 580 GPU, driver version 301.32. All tests were performed using the SPMM representation of the BTFs in the Bonn Database [7], which have a spatial resolution of 256×256 texels and an angular resolution of 81×81 directions. All the presented tests use the Wool SPMM (except when otherwise explicitly stated), using half precision floating point values (16 bits) which corresponds to 12.39MB of texture data; all the other SPMMs demonstrate similar results. The render target size is 512×512 pixels, all the LPCA coefficients in the residual function are evaluated for maximum visualization quality, and all the reported values are the mean of a 60 second run profiled using Nvidia Parallel Nsight 2.2.

4.1   Results Analysis

Figure 4(a) presents the frame rates achieved as a function of the percentage of pixels in the render target covered by a SPMM fragment. It is clear that our application is fill limited, since the most complex part of the code, the SPMM evaluation, is completely written in a GLSL fragment shader. Nevertheless, even with 80% of the pixels requiring an SPMM evaluation we achieve above 200 fps with our hardware configuration.

Figure 4(b) depicts performance variation with the number of visible SPMMs. The shoe model in figure 1(b) was initially mapped with the wool SPMM on five different materials; each new test replaced one of those materials with a different SPMM, until five were being used simultaneously, this way maintaining the same number of pixels covered by SPMMs in all tests. Results demonstrate that the number of SPMMs does not significantly affect frame rates, small differences being due to variations in the parameters that define each SPMM.

Since GPU hardware changes considerably with each new family, and associated compilers also differ accordingly, we used the GPU ShaderAnalyser tool from AMD to analyze and predict the fragment shader performance on various GPUs. We configured the analyzer to assume branch coherence of 90%, an

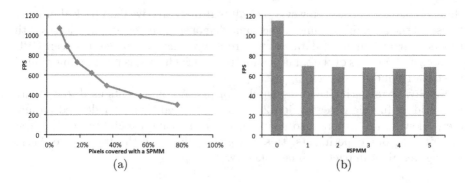

(a)   (b)

Fig. 4. (a) Performance variation with the amount of pixels covered by the Wool SPMM. (b) Performance variation with the number of SPMMs for a fixed number of pixels covered by SPMMs.

average loop count of 4 and maximum loop count of 8. The reported results for 4 diferent AMD Radeon GPU families are presented in table 1. These show that the shader is currently compute bound (see ALU:TEX and Bottleneck columns), and throughput increases sharply with the higher clock rates and core count available in newer GPU families. It is therefore to be expected that our visualizer performance will continue to scale across new generations of GPUs.

Table 1. GPU ShaderAnalyzer output for various AMD Radeon GPUs. Avg - Average number of cycles the shader is expected to take; ALU - The number of ALU instructions in the shader; TEX - The number of texture fetch instructions in the shader; CF - The number of control flow instructions in the shader; Throughput - Millions of pixels per second; CR - Clock rate in MHz; CC - Number of stream processors .

Name	Avg	ALU	TEX	CF	ALU:TEX	Bottleneck	Throughput	CR	CC
HD3870	263.06	826	105	213	**2.41**	ALU Ops	47	775	320
HD4890	105.22	840	109	213	**2.13**	ALU Ops	129	800	850
HD5870	55.94	845	109	212	**1.07**	ALU Ops	258	850	1600
HD6970	48.07	940	109	215	**1.17**	ALU Ops	293	880	1536

4.2   Scalability Analysis

Scalability of the visualizer with respect to the model and illumination complexity is of paramount importance on an industrial setting. Figure 5 depicts rendering times for various geometric complexity and directional light sources.

Rendering times increase linearly with the number of light sources. This is due to the entire evaluation of the fragment shader for each light, but as indicated in figure 3, not all data structures depend on the light direction; exploiting this fact in future implementations can improve scalability with the number of lights.

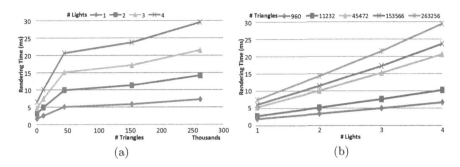

Fig. 5. Rendering times with increasing geometric complexity (a) and number of directional light sources (b)

The number of geometric primitives also has a significant impact on rendering time. Although the SPMM is entirely calculated in a fragment shader, thus independent from the vertex processing stage, our visualizer is not prepared to handle large amounts of vertex data. Since the final goal is to integrate it on a footwear CAD system developed by a project partner, this will feed the visualizer with only the visible geometric primitives. Utilization on other contexts is possible by applying appropriate culling techniques.

Nvidia Parallel Nsight reported, for all experiments, that the GPU is busy processing the workload 92% of the time of each frame, which reinforces evidence that the shader is compute bound and computing resources are being used near their peak performance.

5   Conclusion and Future Work

We presented a GPU visualizer that combines the compaction benefits of the original SPMM approach, a state of the art BTF representation format, with the performance benefits of more GPU friendly approaches, enabling high fidelity visualization at previously unreachable interactive rendering rates. It was shown that performance is fill rate limited, that the main bottleneck is the number of ALU operations in the shaders and the number of rendered SPMMs does not affect performance. By precomputing barycentric coordinate matrices and using acceleration structures we were able to further increase rendering rates, exploiting on average 92% of computational resources.

As future work we would like to improve scalability and expand the visualizer to allow real-time editing of SPMM parameters. We believe this can be of great use for digital designers and artists, assisting in rapid virtual prototyping of new products. Additionally, it would be interesting to support multiple GPUs and progressive rendering in order to adapt to the compute capabilities of the host machine. Also, mipmapping can boost performance and reduce aliasing artifacts.

References

1. Weyrich, T., Lawrence, J., Lensch, H., Rusinkiewicz, S., Zickler, T.: Principles of appearance acquisition and representation. In: ACM SIGGRAPH 2008 Classes, SIGGRAPH 2008, pp. 80:1–80:119. ACM, New York (2008)
2. Müller, G., Meseth, J., Sattler, M., Sarlette, R., Klein, R.: Acquisition, synthesis, and rendering of bidirectional texture functions. Computer Graphics Forum 24, 83–109 (2005)
3. Filip, J., Haindl, M.: Bidirectional texture function modeling: A state of the art survey. IEEE Transactions on Pattern Analysis and Machine Intelligence 31, 1921–1940 (2009)
4. Dana, K.J., van Ginneken, B., Nayar, S.K., Koenderink, J.J.: Reflectance and texture of real-world surfaces. ACM Trans. Graph. 18, 1–34 (1999)
5. Müller, G., Meseth, J., Klein, R.: Compression and real-time rendering of measured btfs using local pca. In: Ertl, T., Girod, B., Greiner, G., Niemann, H., Seidel, H.P., Steinbach, E., Westermann, R. (eds.) Vision, Modeling and Visualisation 2003, pp. 271–280. Akademische Verlagsgesellschaft Aka GmbH, Berlin (2003)
6. Ma, W.C., Chao, S.H., Chen, B.Y., Chang, C.F., Ouhyoung, M., Nishita, T.: An efficient representation of complex materials for real-time rendering. In: Proceedings of the ACM Symposium on Virtual Reality Software and Technology, VRST 2004, pp. 150–153. ACM, New York (2004)
7. Sattler, M., Sarlette, R., Klein, R.: Efficient and realistic visualization of cloth. In: Eurographics Symposium on Rendering 2003 (2003)
8. Wu, H., Dorsey, J., Rushmeier, H.: A sparse parametric mixture model for btf compression, editing and rendering. Computer Graphics Forum 30, 465–473 (2011)
9. Suykens, F., Berge, K.V., Lagae, A., Dutr, P.: Interactive rendering with bidirectional texture functions. Computer Graphics Forum 22, 463–472 (2003)
10. McAllister, D.K., Lastra, A., Heidrich, W.: Efficient rendering of spatial bidirectional reflectance distribution functions. In: Proceedings of the ACM SIGGRAPH/EUROGRAPHICS Conference on Graphics Hardware, HWWS 2002, pp. 79–88. Eurographics Association, Aire-la-Ville (2002)
11. Ebert, D.S., Musgrave, F.K., Peachey, D., Perlin, K., Worley, S.: Texturing and Modeling: A Procedural Approach, 3rd edn. Morgan Kaufmann Publishers Inc., San Francisco (2002)
12. Lawrence, J.: Acquisition and representation of material appearance for editing and rendering. PhD thesis, Princeton, NJ, USA, AAI3214568 (2006)
13. Fernando, R.: GPU Gems: Programming Techniques, Tips and Tricks for Real-Time Graphics. Pearson Higher Education (2004)
14. Pharr, M., Fernando, R.: GPU Gems 2: Programming Techniques for High-Performance Graphics and General-Purpose Computation (Gpu Gems). Addison-Wesley Professional (2005)
15. Pharr, M., Humphreys, G.: Physically Based Rendering: From Theory to Implementation, pp. 125–130. Morgan Kaufmann Publishers Inc., San Francisco (2004)
16. Lagae, A., Dutré, P.: Compact, fast and robust grids for ray tracing. In: Computer Graphics Forum (Proceedings of the 19th Eurographics Symposium on Rendering), vol. 27, pp. 1235–1244 (2008)

Author Index